SUSTAINABLE NANOREMEDIATION

Modern Technologies for a Clean Environment

SUSTAINABLE NANOREMEDIATION

Modern Technologies for a Clean Environment

Edited by:

Fernanda Maria Policarpo Tonelli, PhD

Arpita Roy, PhD

Khalid Rehman Hakeem, PhD

APPLE ACADEMIC PRESS

First edition published 2025

Apple Academic Press Inc.
1265 Goldenrod Circle, NE,
Palm Bay, FL 32905 USA

760 Laurentian Drive, Unit 19,
Burlington, ON L7N 0A4, CANADA

CRC Press
2385 NW Executive Center Drive,
Suite 320, Boca Raton FL 33431

4 Park Square, Milton Park,
Abingdon, Oxon, OX14 4RN UK

Library and Archives Canada Cataloguing in Publication

Title: Sustainable nanoremediation : modern technologies for a clean environment / edited by: Fernanda Maria Policarpo Tonelli, PhD, Arpita Roy, PhD, Khalid Rehman Hakeem, PhD.
Names: Tonelli, Fernanda Maria Policarpo, editor. | Roy, Arpita (Researcher in Environmental Science), editor | Hakeem, Khalid Rehman, editor
Description: First edition. | Includes bibliographical references and index.
Identifiers: Canadiana (print) 20240448537 | Canadiana (ebook) 20240448553 | ISBN 9781774916100 (hardcover) | ISBN 9781774916117 (softcover) | ISBN 9781003468950 (ebook)
Subjects: LCSH: Nanostructured materials—Environmental aspects. | LCSH: Nanotechnology—Environmental aspects. | LCSH: Environmental health—Technological innovations.
Classification: LCC TA418.9.N35 S87 2025 | DDC 620.1/15—dc23

Library of Congress Cataloging-in-Publication Data

CIP data on file with US Library of Congress

ISBN: 978-1-77491-610-0 (hbk)
ISBN: 978-1-77491-611-7 (pbk)
ISBN: 978-1-00346-895-0 (ebk)

About the Editors

Fernanda Maria Policarpo Tonelli, PhD
Assistant Professor, Federal University of São João del Rei, Divinópolis, Minas Gerais, Brazil

Fernanda Maria Policarpo Tonelli, PhD, is an Assistant Professor at Federal University of São João del Rei, Divinópolis, Minas Gerais, Brazil, where she teaches graduate and postgraduate students. She specializes in biotechnology/molecular biology and has been studying nanomaterials applications such as delivery and remediation as well as green synthesis protocols and allelopathic potential of secondary metabolites from plants. Her research projects have already resulted in 10 patent applications, three of which have already been analyzed and granted. She has published 12 books and authored 18 scientific articles and more than 50 book chapters with international publishers. She has presented and participated in many national and international conferences and has helped in organizing various scientific events. Dr. Tonelli has also dedicated herself to the promotion of science and technology through co-funding a nongovernmental organization. Dr. Tonelli is active in advocacy groups that support gender equality in science. Her efforts as a researcher have been recognized with various awards, including "For Women in Science Brazil" (awarded by L'Oréal Brasil, UNESCO-Brazil, and Brazilian Academy of Sciences) and "Under 30 Brazil" (Forbes) as well as various certificates of merit.

Arpita Roy, PhD
Assistant Professor, School of Engineering and Technology (Department of Biotechnology), Sharda University, Greater Noida, India

Arpita Roy, PhD, is working as an Assistant Professor in the Department of Biotechnology in the School of Engineering and Technology at Sharda University, Greater Noida, India. Her specialization is in plant biotechnology, nanobiotechnology, environmental biotechnology, and microbiology. She has been teaching graduate and postgraduate students of biotechnology. She has taught topics related to plant biotechnology, microbiology, bioprocess engineering, and solid waste management. She has edited two books and authored more than 25 scientific articles and 18 book chapters from international publishers. She has presented and participated in numerous state, national, and international conferences, seminars, workshops, and symposiums. She received a commendable research award for excellence in research during the years 2019 and 2020. She has served as an editorial board member and reviewer for several reputed international journals.

Khalid Rehman Hakeem, PhD
Professor, King Abdulaziz University, Jeddah, Saudi Arabia

Khalid Rehman Hakeem, PhD, is working as a Professor at King Abdulaziz University, Jeddah, Saudi Arabia. He is currently the Head (Plant Biotechnology Research Unit) at the Princess Dr. Najlaa Bint Saud Al Saud Center for Distinguished Research in Biotech, King Abdulaziz University, Jeddah, Saudi Arabia. Currently, he is also working as a visiting Professor at Daffodil International University, Dhaka, Bangladesh. He has initially worked as an Assistant Professor at the University of Kashmir, Srinagar, India. Later, he joined the Universiti Putra Malaysia, Selangor, Malaysia, and worked there as a Postdoctoral Fellow (2012) and Fellow Researcher (Associate Prof.) (2013 to 2016). Prof. Hakeem has more than 14 years of teaching and research experience in plant ecophysiology, biotechnology, molecular biology, medicinal plant research, plant-microbe-soil interactions as well as in environmental studies. He is the recipient of several fellowships at both national and international levels. He was recently elected as a fellow of the Royal Society of Biology, London. He has served as the visiting Scientist at Fatih Universiti, Istanbul, Turkey, as well as at Jinan University, Guangzhou, China. Currently, he is involved in several international research projects with different government organizations. To date, he has authored and edited more than 80 books with international publishers, including Springer Nature, Academic Press (Elsevier), CRC Press, etc. He has also to his credit more than 170 research publications in peer-reviewed international journals and 70 book chapters in edited volumes with international publishers. At present, he serves as an editorial board member and reviewer for several high-impact international scientific journals from Elsevier, Springer Nature, Taylor and Francis, Cambridge Press, and John Wiley. He is included in the advisory board of Cambridge Scholars Publishing, UK. He is also Fellow of the Plantae Group of the American Society of Plant Biologists, member of the World Academy of Science, member of the International Society for Development and Sustainability, Japan, and member of the Asian Federation of Biotechnology, Korea. He completed his PhD (Botany; specialization in Plant Ecophysiology and Molecular Biology) at Jamia Hamdard, New Delhi, India.

Contents

Contributors

Irfan Ahmad
Division of Fish Genetics and Biotechnology, Faculty of Fisheries, Sher-e-Kashmir University of Agricultural Sciences and Technology of Kashmir, Rangil, Ganderbal, Jammu and Kashmir, India

Sirajuddin Ahmed
Department of Environmental Sciences, Jamia Millia Islamia, New Delhi, India

Brijesh Ajjappala
Department of Plant Biotechnology, University of Agricultural Sciences, GKVK, Bangalore, Karnataka, India

Motasem Y.D. Alazaiza
Department of Civil and Environmental Engineering, A'Sharqiyah University, Ibra, Oman

Ahmed Albahnasawi
Department of Environmental Engineering, Gebze Technical University, Gebze-Kocaeli, Turkey

Salim Albukhaty
Department of Chemistry, College of Science, University of Misan, Maysan, Iraq
College of Medicine, University of Warith Al-Anbiyaa, Karbala, Iraq

Hassan Al-Karagoly
Department of Internal and Preventive Medicine, Veterinary Medicine College, University of Al-Qadisiyah, Al-Diwaniyah, Iraq

Ahmad Ali
Department of Life Sciences, University of Mumbai, Vidyanagari, Mumbai, India

Saima Andleeb
Division of Fish Genetics and Biotechnology, Faculty of Fisheries, Sher-e-Kashmir University of Agricultural Sciences and Technology of Kashmir, Rangil, Ganderbal, Jammu and Kashmir, India

Djaber Aouf
Laboratory of Dynamic Interactions and Reactivity of Systems, University of Kasdi Merbah, Ouargla, Algeria

Sumel Ashique
Department of Pharmaceutics, Bharat Institute of Technology (BIT), School of Pharmacy, Meerut, Uttar Pradesh, India

Sana Ashraf
College of Earth and Environmental Sciences, Quaid-e-Azam Campus, University of the Punjab, Lahore, Pakistan

Sobia Ashraf
Department of Environmental Sciences and Engineering, Government College University, Faisalabad, Pakistan

Humera Aziz
Department of Environmental Sciences and Engineering, Government College University, Faisalabad, Pakistan

Prairna Balyan
Department of Life Sciences, University of Mumbai, Vidyanagari, Mumbai, Maharashtra, India

Ruqia Bano
Department of Environmental Sciences, Jamia Millia Islamia, New Delhi, India

Eneyew Tilahun Bekele
Department of Applied Chemistry, School of Applied Natural Science, Adama Science and Technology University, Adama, Ethiopia

Linkon Bharali
Department of Chemistry, National Institute of Technology, Silchar, Assam, India

Hemavathi Brijesh
Department of Biotechnology, School of Applied Sciences, REVA University, Bangalore, Karnataka, India

Ishani Chakrabartty
Department of Applied Biology, School of Biological Sciences, University of Science and Technology Meghalaya (USTM), Ri-Bhoi, Baridua, Meghalaya, India

Ankita Chatterjee
Department of Biotechnology, School of Applied Sciences, REVA University, Bangalore, Karnataka, India

Siddhartha S. Dhar
Department of Chemistry, National Institute of Technology, Silchar, Cachar, Assam, India

Vinicius Marx Silva Delgado
Department of Biochemistry, UFSJ/CCO, Divinópolis, Brazil

Murat Eyvaz
Department of Environmental Engineering, Gebze Technical University, Gebze-Kocaeli, Turkey

Arshad Farid
Gomal Center of Biochemistry and Biotechnology, Gomal University, Dera Ismail Khan, Pakistan

Sana Farooq
Department of Environmental Sciences and Engineering, Government College University, Faisalabad, Pakistan

Fares Fenniche
Laboratory of Dynamic Interactions and Reactivity of Systems, University of Kasdi Merbah, Ouargla, Algeria

Ashish Garg
Department of P.G. Studies and Research in Chemistry and Pharmacy, Rani Durgavati University, Jabalpur, Madhya Pradesh, India

Ruchira Ghosh
Centre for Sustainable Technologies, Belfast School of Architecture and the Built Environment, Ulster University, Northern Ireland, United Kingdom

Suresh Ghotekar
Department of Chemistry, Smt. Devkiba Mohansinhji Chauhan College of Commerce and Science, University of Mumbai, Silvassa, Dadra and Nagar Haveli (UT), India

Yashodhara Goswami
Department of Applied Biology, School of Biological Sciences, University of Science and Technology Meghalaya (USTM), Ri-Bhoi, Baridua, Meghalaya, India

Nithin Gowda
School of Basic and Applied Sciences (SBAS), Dayananda Sagar University, Bengaluru, Karnataka, India

Ercan Gürbulak
Department of Environmental Engineering, Gebze Technical University, Gebze-Kocaeli, Turkey

Mehak Hafeez
Division of Fish Genetics and Biotechnology, Faculty of Fisheries, Sher-e-Kashmir University of Agricultural Sciences and Technology of Kashmir, Rangil, Ganderbal, Jammu and Kashmir, India

Fumeena Hashim
Department of Environmental Science, Jamia Millia Islamia, New Delhi, India

Atif Husain
Department of Chemistry, Integral University, Dasauli, Lucknow, India

Abid Hussain
Department of Soil and Environmental Sciences, Muhammad Nawaz Shareef University of Agriculture, Multan, Pakistan

Afzal Hussain
Department of Pharmaceutics, College of Pharmacy, King Saud University, Riyadh, Saudi Arabia

Shazia Iqbal
Department of Soil and Environmental Sciences, College of Agriculture, University of Sargodha, Pakistan

Jahangeer
Department of Environmental Sciences, Jamia Millia Islamia, New Delhi, India

Venkatesh S. Joshi
School of Basic and Applied Sciences (SBAS), Dayananda Sagar University, Bengaluru, Karnataka, India

Nisreen Jawad Kadhim
Department of Biomedicine, Faculty of Medicine, Warith Al-Anbiyaa University, Iraq

Juri Kalita
Department of Chemistry, National Institute of Technology, Silchar, Cachar, Assam, India

Sureshkumar Kempahanumakkagari
Department of Chemistry, B.M.S. Institute of Technology and Management, Doddaballapur Main Road, Avalahalli, Yelahanka, Bengaluru, Karnataka, India

Yasmina Khane
Université de Ghardaia, Ghardaia, Algeria
Laboratory of Applied Chemistry (LAC), University of Bouchaib Belhadj, Ain Temouchent, Algeria

Jaya Lakkakula
Amity Institute of Biotechnology, Amity University Maharashtra, Mumbai, Maharashtra, India
Centre for Computational Biology and Translational Research, Amity Institute of Biotechnology, Amity University Maharashtra, Mumbai, Maharashtra, India

Moline Severino Lemos
Department of Cell Biology – ICB/UFMG – Belo Horizonte, Brazil

Vitória De Oliveira Lourenço
Department of Biochemistry, UFSJ/CCO, Divinópolis, Brazil

Ramya Manjunath
Department of Biotechnology, School of Applied Sciences, REVA University, Bangalore, Karnataka, India

Shilpa Borehalli Mayegowda
[2]Department of Psychology, Christ University, Kengeri Campus Mysore Road Kumbalgodu, Kanmanike, Bengaluru, Karnataka, India

Neeraj Mishra
Department of Pharmaceutics, Amity Institute of Pharmacy, Amity University, Gwalior, Madhya Pradesh, India

N. G. Manjula
School of Basic and Applied Sciences (SBAS), Dayananda Sagar University, Bengaluru, Karnataka, India

Ghulam Murtaza
Institute of Soil and Environmental Sciences, University of Agriculture Faisalabad, Pakistan

H. C. Ananda Murthy
Department of Applied Chemistry, School of Applied Natural Science, Adama Science and Technology University, Adama, Ethiopia

Malik Nasibullah
Department of Chemistry, Integral University, Dasauli, Lucknow, India

Jessica Pandohee
Telethon Kids Institute, Nedlands, WA, Australia

T. Ramakrishnappa
Department of Chemistry, B.M.S. Institute of Technology and Management, Doddaballapur Main Road, Avalahalli, Yelahanka, Bengaluru, Karnataka, India

C. R. Ravikumar
Department of Applied Chemistry, School of Applied Natural Science, Adama Science and Technology University, Adama, Ethiopia

Umair Riaz
Department of Soil and Environmental Sciences, Muhammad Nawaz Shareef University of Agriculture, Multan, Pakistan

Mohd. Rihan
Department of Pharmacology, National Institute of Pharmaceutical Education and Research (NIPER), S.A.S. Nagar, Punjab, India

Ahmed Ali Ali Romeh
Plant Production Department, Faculty of Technology and Development, Zagazig University, Zagazig, Egypt; Department of Applied Chemistry, School of Applied Natural Science, Adama Science and Technology University, Adama, Ethiopia

Arpita Roy
Department of Biotechnology, School of Engineering & Technology, Sharda University, Greater Noida, India

Madhura Roy
Centre for Translational and Clinical Research, School of Chemical and Life Sciences, Jamia Hamdard University, India

Vedha Sahithya
Amity Institute of Biotechnology, Amity University Maharashtra, Mumbai, Maharashtra, India

Satyajit Singh Saini
Department of Environmental Sciences, Jamia Millia Islamia, New Delhi, India

Neha Sami
Department of Environmental Sciences, Jamia Millia Islamia, New Delhi, India

Shafiya
Department of Environmental Science, Jamia Millia Islamia, New Delhi, India

Mehreen Shah
Department of environmental science and engineering, Jamia Millia Islamia, New Delhi, India

Zoya Shaikh
Department of Life Sciences, University of Mumbai, Vidyanagari, Mumbai, India

Harish Byrapura Siddappa
Department of Biotechnology, University of Horticultural Sciences, Bangalore, Karnataka, India

Christopher Santos Silva
Department of Biochemistry – UFSJ/CCO - Divinópolis, Brazil

Yilkal Dessie Sintayehu
Department of Applied Chemistry, School of Applied Natural Science, Adama Science and Technology University, Adama, Ethiopia

Ishita Sharan Srivastava
Department of Environmental Science, Jawaharlal Nehru University, New Delhi, India

Vani Tiwari
Amity Institute of Biotechnology, Amity University Maharashtra, Mumbai, Maharashtra, India

Ali Athafah Tomah
Department of Plant Protection, College of Agriculture, University of Misan, Maysan, Iraq
State Key Laboratory of Rice Biology, Institute of Biotechnology, Zhejiang University, Hangzhou, China

Fernanda Maria Policarpo Tonelli
Department of Biochemistry, UFSJ/CCO, Divinópolis, Brazil
Campus Centro Oeste Dona Lindu, Federal University of São João del Rei (UFSJ), Divinópolis, Minas Gerais (MG), Brazil

Flávia Cristina Policarpo Tonelli
Estácio de Sá University, Divinópolis, Brazil

Tanveer Ul Haq
Department of Soil and Environmental Sciences, Muhammad Nawaz Shareef University of Agriculture, Multan, Pakistan

Nilesh Wagh
Amity Institute of Biotechnology, Amity University Maharashtra, Mumbai, Maharashtra, India

Abbreviations

AA	2-aminoanthracene
AFM	atomic force microscopy
Ag	silver
AgNP	silver nanoparticles
AM	arbuscular mycorrhizal
AMD	acid mine drainage
AOPs	advanced oxidation processes
ASC	ascorbic acid
BCP	bromocresol purple
BIT	Bharat Institute of Technology
BNC	bacterial nanocellulose
BNs	bimetallic nanomaterials
BOD	biochemical oxygen demand
BT	bayberry tannin
BTB	bromothymol blue
CFCs	chlorofluorocarbons
CIC	Carbo-Iron® colloids
CMC	carboxymethyl cellulose
CNCs	cellulose nanocrystals
CNFs	cellulose nanofibers
CNTs	carbon nanotubes
COD	chemical oxygen demand
COPD	chronic obstructive pulmonary disease
CR	Congo red
CR	corallene red
CuO	cupric oxide
CV	crystal violet
CVD	chemical vapor deposition
DAF	dissolved air floatation
DDE	dichlorodiphenyldichloroethylene
DDT	dichlorodiphenyltrichloroethane
DLS	dynamic light scattering
DOC	dissolved organic carbon
EC	emerging contaminant
EDAX	energy dispersion analysis of X-ray
EDC	endocrine disruptive chemical
EDX	energy dispersive X-ray analysis

EFB	empty fruit bunch
EGCG	epigallocatechin gallate
EN	engineered nanoparticles
EPA	Environmental Protection Agency
EY	eosin Y
FDA	Food and Drug Administration
FESEM	field emission scanning electron microscope
FE-TEM	field emission transmission electron microscope
FTIR	Fourier transform infrared spectroscopy
GA	glutaraldehyde
GC	gas chromatograph
GSH	glutathione
GT	green tea
HCH	hexachlorocyclohexane
HHMs	hazardous metals
HRAP	high-rate algae ponds
HRTEM	high-resolution transmission electron microscopy
LS	light scattering
LSPR	localized surface plasmon resonance
MB	methylene blue
MCB	monochlorobenzene
MG	malachite green
MNs	metal nanomaterials
MO	methyl orange
MONPs	metal oxide nanoparticles
MONs	metal oxide nanomaterials
MSNs	mesoporous silica nanomaterials
MWCNTs	multiwalled carbon nanotubes
NC	nanocellulose
ndl-PCB	non-dioxin-like polychlorinated biphenyl
NF	nanofibers
NHAP	nanohydroxyapatite
NMPR	nanomycorrhizo phytoremediation
NMs	nanomaterials
NOx	oxides of nitrogen
NPs	nanoparticles
NRP	nonribosomal peptide
NT	nanotubes
NW	nanowires
NZVI	nano-zero-valent iron
ODZ	ornidazole
OT	oolong tea
PAA	polyacrylic acid

PAH	polyallylamine
PAH	polycyclic aromatic hydrocarbon
PAHs	polycyclic aromatic hydrocarbons
PAN	peroxyacetyl nitrate
PB	Prussian blue
PBNP	Prussian blue nanoparticle
PCB	polychlorinated biphenyls
PDDA	poly(diallyldimethylammonium chloride
PEG	polyethylene glycol
PEI	polyethyleneimine
PFC	perfluorochemical
PGPR	plant growth-promoting rhizobial
P-L	photoluminescence
PM	particulate matter
PMM	porous matrix membrane
POPs	persistent organic pollutants
PPCP	pharmaceutical and personal care product
PVA	polyvinylalcohol
PVP	polyvinyl pyrrolidone
RB	rhodamine B
RD	*Rosa damascena*
RhB	rhodamine B
Rn	radon
ROS	reactive oxygen species
SCB	solochrome black
SDS	sodium dodecyl sulfonate
SEM	scanning electron microscopy
SMT	selenocysteine methyl transferase
SOD	superoxide dismutase
SPION	synthesized superparamagnetic iron oxide nanoparticles
SS	suspended solids
TCE	trichloroethylene
TEM	transmission electron microscopy
TiO_2	titanium dioxide
TP	tissue paper
TPHs	total petroleum hydrocarbons
TV	*Thymus vulgaris*
TY	titan yellow
UV	ultraviolet light
VOCs	volatile organic compounds
WHO	World Health Organization
XPS	X-ray photoelectron microscopy
XRD	X-ray diffraction

YP	yam
ZnO	zinc oxide
Z-P	zeta-potential
ZVI	zero-valent iron

Foreword

The development of scientific knowledge is supposed to offer elegant and efficient solutions to problems that negatively affect living beings and ecosystems. Environmental pollution is one of these problems. Harmful contaminants from diverse chemicals in nature have been accumulating in the ecosystems as with a negative consequence of development that ignores sustainability principles worldwide.

The field of nanotechnology provides interesting tools to remediate water, soil, and air pollution, restoring polluted spots. In this sense, green nanomaterials deserve to be highlighted, as they can be produced by applying green synthesis protocols using bio-based raw materials and avoiding the generation of toxic waste and the consequence of toxic reagents. They consist of low-cost eco-friendly tools to fight environmental pollution caused by inorganic, organic, and/or biologic contaminants.

The book, *Sustainable Nanoremediation: Modern Technologies for a Clean Environment,* includes 23 chapters written by scientists from diverse countries and is divided into three parts to offer readers a multidisciplinary and comprehensive addressing on the theme.

The first part of the book, Nanoremediation: Conjugated Strategies and Nanoparticles' Potential, contains six chapters. The first chapter, Environmental Pollution: Core Contaminants and Threats to Human's Life, reviews important environmental pollutants and the threats they represent to the life quality and survival of human beings. Chapter 2, Nanophytoremediation of Environmental Pollution, reviews strategies using plants and nanomaterials to remediate environmental contamination. Chapter 3, Nanoremediation: Prospects and Amendments, addresses modifications and amendments to enhance nanoremediation's efficiency. Chapter 4, Nanoremediation Associated with Microorganisms, discusses and elaborates on microbial-based nanoremediation. Chapter 5, Nanoparticles as Tools to Remediate Environmental Pollution, offers different examples of nanoparticles as nanoremediators and their potential to be explored in this field. Chapter 6, Nanoremediation of the Environment: Nanoparticles' Characteristics, Advantages and Applications, addresses the advantageous characteristics of nanoparticles and their main applications to remediate different environs.

The second part, Green-Based Nanoremediation, presents eight chapters and discusses topics related to green synthesis of nanosctructures. Chapter 7, A Brief History of the Green Synthesis of Nanomaterials, reviews important events of the history of nanostructures' green synthesis, highlighting the sustainability-related aspects. Chapter 8, Green Methods to Synthesize Nanomaterials to Remediate Polluted Environs, addresses bio-derived methods (based in bacteria, algae, fungus, and plants) to generate green nanomaterial, with a section also dedicated to metal-based nanomaterials. Chapter 9, Green-Synthesized Nanomaterials' Potential for Environmental Pollution Remediation, explores the potential of green nanomaterials as nanoremediators. Chapter 10, Principles of Eco-Friendly Synthesis of Silver Nanoparticles and Their Promising Antibacterial Properties, explores silver nanoparticles by focusing on aspects related to green synthesis and the capacity of the nanomaterials to fight biological contaminants such as bacteria. Chapter 11, Green Synthesis of Nanomaterials from Algae Materials to Remediate Environmental Pollution, presents an experimental approach using algae material, green silver, zinc, iron, and copper nanoparticles; the nanoparticles were characterized and were assayed to different applications including photocatalytic activity. Chapter 12, Algae-Based Nanomaterials for Contaminated Environs Nanoremediation, focuses on exploring general aspects of green nanomaterials (especially silver, gold, and iron nanoparticles) production from algae material to nanoremediate pollution. Chapter 13, Plant-Based Nanomaterials for Contaminated Environs Nanoremediation, reviews the current state of knowledge on plant-based nanoparticles, presenting the potential for environmental cleanup. Chapter 14, Plant-Derived Nanocellulosic Material: A Promising Technology Application in Environmental Bioremediation, focuses specifically on nanocellulosic material structure, properties, isolation, and adsorbent classes, highlighting their applications in environmental remediation.

The third part, Nanoremediation Based on Green Nanoparticles, presents nine chapters. From these, seven chapters address specific nanoparticles as nanoremediators. Chapter 15, Green Iron Nanoparticles for Nanoremediation, has been drafted by Indian authors. Chapter 16, Green Silver Nanoparticles for Nanoremediation, is written by Brazilian authors. Chapters 17, Green Palladium Nanoparticles for Nanoremediation, Chapter 18, Green Copper Nanoparticles for Nanoremediation, and Chapter 19, Green Prussian Blue Nanoparticles for Nanoremediation, are written by authors from

India, Chapter 20 Green Titanium Oxide Nanoparticles for Nanoremediation has been contributed by authors from Ethiopia, and Chapter 21, The Novel Technique and Capacity of Green-Gold Nanoparticles in Bioremediation, is written by Indian authors. The last two chapters discuss biocompatibility and practical difficulties in using green nanomaterials. Chapter 22, Green Nanomaterials: Biocompatible Materials in Nanoscale?" is written by Brazilian authors, and Chapter 23, Practical Difficulties in the Application of Green Nanoparticles to Environmental Pollution Remediation, has been produced by authors from India.

The most important aspects regarding green nanomaterials to remediate environmental pollutants in an efficient, low-cost, and eco-friendly manner are covered in this volume. The book presents the potential to provide the global scientific community the opportunity to access an up-to-date material that offers a comprehensive coverage of most of what addresses this theme. The editors are to be highly complimented for their sincere effort in bringing this volume together.

—Prof. Munir Ozturk
Department of Botany, Ege University, Izmir, Turkey

Preface

Sustainability principles have been ignored worldwide during the rapid industrialization/urbanization that the world has experienced. Consequently, a large array of negative consequences has had an adverse impact on the living beings and ecosystems as a whole.

It is crucial to develop and apply efficient strategies to deal with harmful menaces that threaten the survival on Earth.

Strategies to manage environmental pollution while preserving the ecosystem are necessary as the harmful consequences of pollution caused by human actions without taking into consideration the sustainability aspects threaten the survival of life on Earth. The efficient remediation of inorganic, organic, and biologic contaminants needs to be performed.

In this sense, the field of nanotechnology possesses a large potential to contribute. Nanomaterials present interesting characteristics such as good reactivity and the possibility to undergo functionalization protocols to suffer chemical modification, aiming enhancement of specific abilities like promoting nanoremediation.

In fact, green-synthetized nanomaterials, more specifically, are an elegant solution. This innovative approach presents numerous advantages when compared with conventional physicochemical methods by avoiding the use of harmful reagents and reducing the toxic waste production. Microbes and plants, for example, can be used to generate eco-friendly nanomaterials capable of remediating environmental contamination.

Sustainable Nanoremediation: Modern Technologies for a Clean Environment is a book dedicated to discuss the low-cost and eco-friendly green-synthetized nanomaterials as advantageous technology to promote efficient nanoremediation of environmental pollution, restoring polluted areas. It is an attempt to offer to researchers and scientists as well as undergraduate students and researchers a multidisciplinary and comprehensive knowledge on green nanoremediation from a biotechnological point-of-view.

The 23 chapters address themes such as main contaminants associated with environmental pollution and their threats to humans; strategies of remediation that link nanoremediation with microbes, plants, or amendments; nanoparticles as tools to nanoremediation and their advantageous characteristics to perform this role; the brief history of nanomaterials' green synthesis;

the potential of these nanomaterials to nanoremediate; green methods to synthesize nanoremediators; algae-based and plant-based nanomaterials capable of remediating environmental pollution; different types of green nanoparticles capable of dealing with environmental pollutants, such as iron, silver, palladium, cooper, Prussian blue, titanium, and gold nanoparticles; biocompatibility of green nanomaterials; and practical difficulties in the application of green nanoparticles to nanoremediate.

We are extremely grateful to authors who have contributed chapters in this project and to AAP for their generous cooperation and for publishing this book.

CHAPTER 1

Environmental Pollution: Core Contaminants and Threats to Human Life

HUMERA AZIZ[1], UMAIR RIAZ2, SOBIA ASHRAF[1], SANA ASHRAF[3], SANA FAROOQ[1], SHAZIA IQBAL[4], ABID HUSSAIN[2], TANVEER UL HAQ[2], and GHULAM MURTAZA[5]

[1]*Department of Environmental Sciences and Engineering, Government College University, Faisalabad, Pakistan*

[2]*Department of Soil and Environmental Sciences, Muhammad Nawaz Shareef University of Agriculture, Multan, Pakistan*

[3]*College of Earth and Environmental Sciences, Quaid-e-Azam Campus, University of the Punjab, Lahore, Pakistan*

[4]*Department of Soil and Environmental Sciences, College of Agriculture, University of Sargodha, Pakistan*

[5]*Institute of Soil and Environmental Sciences, University of Agriculture Faisalabad, Pakistan*

ABSTRACT

Environmental pollution is not a new term; it is the biggest threat to mankind and causes diseases or mortality worldwide. In recent years, the widespread use of chemicals in production and life has led to an increase of chemical pollutants in the environment, such as pesticides, polycyclic aromatic hydrocarbons (PAHs), phthalates, and heavy metal ions, which may pose a serious threat to human health and have become a global concern. The

Sustainable Nanoremediation: Modern Technologies for a Clean Environment.
Fernanda Maria Policarpo Tonelli, Arpita Roy, and Khalid Rehman Hakeem (Eds.)

human activities like urbanization, industrialization, mining, and exploration activities also have great impact on environmental pollution. All types of pollution directly or indirectly affect human health. Although during industrialization, soil has been used more and more as a sink for pollutants despite being a nonrenewable reserve. Some pollutants in soil are present naturally but others are introduced by anthropogenic activities. By inspecting the source of pollution in soil, it is controlled as early as feasible from its origin (point or nonpoint source) to minimize the harm of that pollutant to the soil. In this chapter, we have summarized environmental pollution, its main types, and its impact on human life. The purpose of this chapter is to provide an overview of the ecotoxicology of major classes of environmental contaminants, with respect to sources, environmental chemistry, and toxicity. This chapter highlights the concept of environmental pollution. The categories of pollutants have also been explained. The widely recognized types of pollution (namely, land/soil, air, and water) and their respective toxic and socioeconomic impacts have been discussed in detail in this chapter.

1.1 ENVIRONMENTAL POLLUTION BACKGROUND AND GENERAL CONCEPTS

Environmental pollution is world's biggest reason for disease or premature mortality (Landrigan et al., 2019). Universally, the expression "environmental pollution" had been known a long time before or its effects upon humankind have extraordinarily been felt in general, and its current circumstance, whether purposely or in any case, starting from the primary gathering and gatherings lived for some significant time-frame in one specific area of the Earth (Easterbrook, 1995; Botkin, 1995). Archeologists have found a few remaining parts of instruments, shell hills, dress, and rubble load of such individuals (Keller, 1988). Because there was huge accessibility of area for people and human community to dwell in, pollution was not viewed as a difficult issue throughout their time on earth. Despite the fact that many, though not all, actions contributed to pollution, people also weren't negatively impacted either by threat and pollution they created at the time (Miller, 1997, 2004; Ajibade et al., 2021).

Commonly, environmental pollution is worse in underdeveloped countries than developed countries. This may be because of factors like poverty, shoddy laws, and a lack of awareness of pollution sources. People probably encounter pollutants on such a regular basis despite noticing this, or maybe our quickly lifestyles have made us resistant to all of it (Muralikrishna and Manickam, 2017). As improbable as it might appear, ignorance about pollution causes mankind to engage in activities that produce harmful pollutants in amounts and forms that the environment is no more capable of combating and thus leading to utter structural distortion. For example, inappropriate electronics collection or handling, wood fires, unloading of domestic and agricultural squanders into waterways, utilization of synthetics to capture marine life, and deforestation, each leads to water, soil, and air pollution. Similarly, as human populace grow, so do anthropogenic activities and their corresponding effects on the environment. Their effects are on people as well as on other amphibian and earthly creatures containing microorganisms. That is because their variety and excessive quantity will generally keep up with their biogeochemical capability essential for supporting the environment (Ukaogo et al., 2020).

According to history, in antiquated time, the two urban communities and significant towns where earliest major civilizations originated in antiquity became terrible places, polluted by trash or human excrement. History has it that, in ancient times, both cities and major towns where major civilization emerged from were baneful areas, soiled by debris and the wastes from human. This situation became worse from the start of 17th century until the turn of the 19th century by processing of coal into coke, which was utilized for iron manufacturing. All across the late-19th century, this threat gradually spread into waterways. Within most cases, part solid waste accumulation in populated metropolitan regions was the issue throughout the 19th century. But because of human populace's fast industrialization and

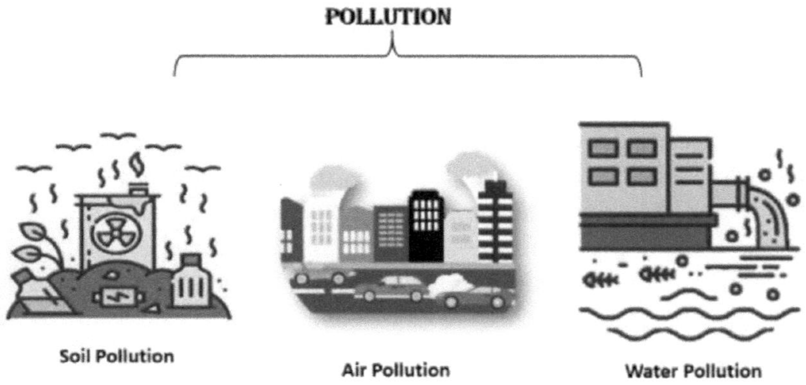

FIGURE 1.1 Main types of pollution.

exponential progress, most particularly in metropolitan urban communities, contamination has turned into a widespread issue and is posing significant danger to human life (Alina, 2018; Sulaymon et al., 2020). According to research by Landrigan et al. (2017), pollution causes nine million more fatalities annually or three times as many as are caused by influenza, TB, and AIDS combined. This accounts approximately 16% from all deaths globally. Almost a fatality out of every four was caused by environmental pollution (Landrigan et al., 2019). Furthermore, Suk et al. (2016) reported the kids were incredibly sensitive to pollution.

1.2 CATEGORIES OF ENVIRONMENTAL POLLUTANTS

1.2.1 POLLUTANTS

Understanding exactly what pollutants are and the ways that can actually influence our ecosystem are crucial. When they are not correctly described, such term "pollutant" or "contaminant" can give rise to certain misunderstandings. Although each relates with undesirable components, most people reciprocally have used the words "pollution" and "contamination." Mondal (2018) argues that it is relatively convenient to refer to materials that generate pollution as "pollutants" and others which induce contamination as "contaminants." Like an illustration, any waste product inside a water reservoir could also be referred to as a contamination or pollutant. Additionally, various cleaning methods used in remediation were comparable to those used in contamination, however, related in such a manner where pollution was caused by the injection of dangerous chemicals (Mondal, 2018). Owing of its more severe effects opposed to contamination, pollution was associated with much more pessimism. Although those activities were linked or identical, there are still significant variances.

Every element either chemically or physically present or released into the atmosphere, soil, or water in a proportion which threats individuals—their pleasure, and their legitimate working in any part of the climate—and the environment's ability to function properly is considered a pollution. Any pollutant is a material that is prevalent inside an environment at high levels that is above the usual threshold level for such region as well as the creatures within the environment. These pollutants can be chemical, biological, or radioactive.

1.2.2 CATEGORIES OF POLLUTANTS

Pollutants are typically be divided into two categories. Despite the fact that there are occasionally more classifications based on how things seem in existence, if they are thoroughly examined, they will fall into either of these two groups. Pollutants, both natural and artificial, exist. Alternatively, you may refer to them as quantitative and qualitative contaminants, correspondingly.

1.2.2.1 NATURAL POLLUTANTS

They include chemicals and compounds which are present in nature and yet have acquired the status of pollutants, if their quantity and concentrations inside a particular environmental area increases as a consequence of human senseless deeds or behaviors. Although these are natural, it doesn't indicate that humans are not contributing in some way; man is merely not conscious of it. As a fact, whenever a gas like CO_2 (produced through vehicles or industry) is found in the environment in amounts more than the recommended level, it has unimaginable impacts on living things like flora, fauna, and people.

1.2.2.2 ARTIFICIAL (MAN-MADE) POLLUTANTS

These are all elements or chemicals which don't exist in nature natively but are instead put there by the actions of individuals. These could be categorized as primary or secondary pollutants based on the form they take upon being released into nature. Pollutants may also be divided into biodegradable and nondegradable pollutants dependent on how they naturally decompose in relation to ecosystems.

The following is a short explanation of each one:

1.2.2.2.1 Primary Pollutants

All of such pollutants are already produced through their different sources immediately through their ecosystem, because typically they continue to exist within this state. They may be released either naturally or as a result of human activity. Nitric oxide, sulfur oxide, dust, gases, soot, smoke, and hydrocarbon are all examples of this subcategory. Primary pollutants generated naturally include fumes and ash from volcanic eruptions. Although carbon monoxide (CO) gas through automobiles or generators is another major pollution emitted by human activity. Such releasing of primary pollutants into the atmosphere is extremely damaging. It can result in major environmental and health issues like acid rain, global warming, and respiratory ailments. Automobiles or factories are the main producers of primary pollutants, which are frequent counterparts to secondary pollutants.

1.2.2.2.2 Secondary Pollutants

Instead of being absorbed by the environment directly, these pollutants are produced by primary pollutants. It could happen either through a response with some airborne elements or through the chemical reaction among two or more main contaminants. Several pollutants, such as ozone, aldehydes, ketones, sulfur oxide, and nitrogen dioxide, are examples.

1.2.2.2.3 POINT SOURCE POLLUTION

The term "point source" refers to a single, distinct place, or set of installations that generate pollution. Motor vehicle engine, industry, trench, tunnels, chimneys, and conduit are a few examples. Monitoring and regulating the emission of pollutants are simple when it comes to such type of pollution. Also, in order to restrict the flow from any single point and even to inspect and maintain how effluent levels meet with the established norms, regulatory authorities like the United States Environmental Protection Agency (EPA) have created guidelines for such discharge of certain chemicals and substances.

1.2.2.2.4 NONPOINT SOURCE POLLUTION

Nonpoint sources are described as broad and extensive because they lack a clear site for concentrated unregulated dispersion of pollutants or chemicals into ecosystem. The majority of these contaminants are dispersed further into airways by wind, but others are carried toward waterways by rainwater. Others appear out of nowhere, which include oil spilling from broken-down cars on the road, pesticides sprayed on the grass and fields, animal and pet excrement, and even disturbed topsoil via digging. Regulation of nonpoint source emissions is challenging since pollution release cannot be detected at the source instead at the final destination. For contaminated places, samples are taken or analyzed through any emission sources. Since it would be impossible in this situation, the EPA addressed their regulations at towns rather than any particular people or companies. States were obliged for cleaning their wastewater till it satisfies the federal regulations for allowed levels of contaminants present in drinking water. The categories of pollutants are discussed in Figure 1.2.

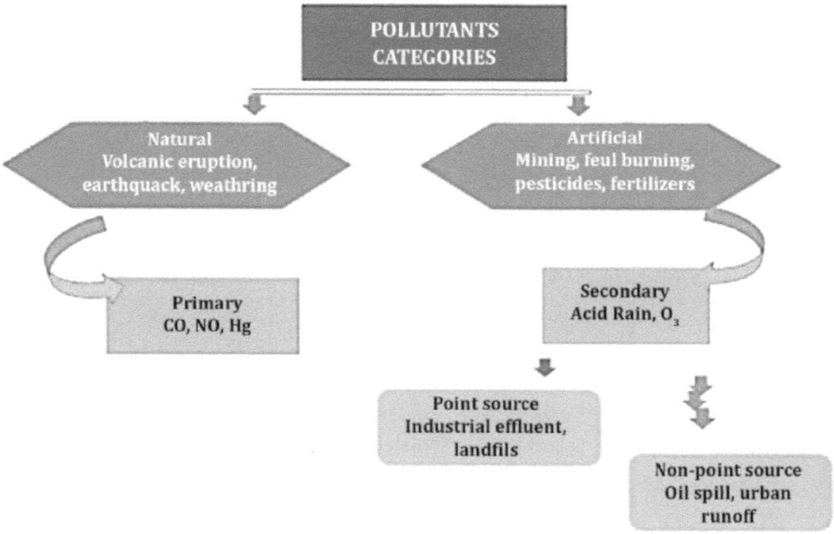

FIGURE 1.2 Pollutants categories.

1.3 TYPES OF ENVIRONMENTAL POLLUTION

Several main types of pollution and the specific pollutants pertinent to all of these are discussed below.

1.3.1 AIR POLLUTION

Natural and man-made aerial compounds which are put into the atmosphere at ratios large enough to cause discernible impact on people, animals, plants, or construction materials are called air pollutants. Major air pollutants contain carbon monoxide, sulphur dioxide, fluorinated gases (CFCs), and nitrous oxide emitted through industries and vehicles. Additionally, some compounds that are typically found in particular levels of the atmosphere, such as ozone in the stratosphere, become pollutants once they enter the troposphere (Havugimana et al., 2017).

1.3.2 WATER POLLUTION

The major danger to the world as a whole or, more crucially, to human life arises if water is contaminated (Inyinbor et al., 2018). Water pollution may come from a variety of sources, but the majority of them are caused by a collection of natural and inorganic chemicals, like O_2 wastes, microbes, micronutrients, suspended particles, ions, heavy metals, insecticides, volatile compounds (VOCs), or persistent organic pollutants (POPs) (Ajibade1et al., 2021).

1.3.3 SOIL POLLUTION

The primary contributors to soil contamination include agriculture's usage pesticides, increased industrialization, poor maintenance, and ineffective sewage treatment, which all present difficulties and a larger need for cleanup (Chibuike and Obiora, 2014). If pollutants are discharged or seep beneath, then soil pollution occurs. Heavy metals, fertilizers, and pesticides, especially chlorinated solvents, are some of those most important soil pollutants (Bishop and Flack, 1997).

1.3.4 RADIOACTIVE POLLUTION

Radioactive pollution is a result of 21st-century atomic science operations, including nuclear energy production and the development, production, and use of nuclear weapons.

1.3.5 NOISE POLLUTION

Any disruptive or undesired sound that obstructs or endangers people, animals, or the environment is referred to as noise pollution (Jain, 2015). Noise pollution includes noise from traffic, aircraft, machinery, and elevated radar.

1.3.6 THERMAL POLLUTION

Thermal pollution would be any variation in natural temperature, which might include cold water releases entering waterways below massive ponds and increased temperatures brought on by industrial cooling processes. Temperature changes in natural reservoirs brought on by anthropogenic activities, for example, when water is employed as a cooling in a power station, are referred to as thermal pollution (Verones et al., 2010).

1.4 SOIL POLLUTION

Soil is an all-inclusive sink beginning from living life forms and materials of inorganic substances, which bears the most weight of natural contamination (Ajibade et al., 2019b). According to Spellman (2017), soil is the top layer of eroded minerals and organic materials that can sustain life because it mainly consists of oxygen, water, and minerals. For the circulation of cyclical nutrients, soil is essential. Agricultural land makes up only 11.3% of the world's 13,073,109 hectares area; grassland land 24.6%; forest 34.1%; and the other 31% is made up of households, cities, factories, and highways. Because possible harmful compounds migrate through the food supply chain into aquifers, that were then used and eaten by humans, soil contamination offers a significant risk to human health together with other deteriorating processes like eroding, floods, and urbanization (Akinbile et al., 2016a, 2016b; Adewumi and Ajibade, 2015). According to Okrent (1999), soil pollution involves the accumulation of harmful substances such as pesticides, ions, radioactive isotopes, and disease-causing substances that have a negative impact on crop, man, and animal life. Graveyards, dumps, sewage treatment, wet and dry boreholes, underwater storage vessels, or pipes are a few examples of sources for pollutants in soil and sublayers.

Additionally, mining activities, drilling excavations, irrigation water trenches, and channels can all lead to soil polluting underneath the groundwater (Spellman, 2017). Soil contamination seems to be more widespread inside the world's developed countries than it is in less-advanced societies (Alloway, 2001). Chemicals pollute the soil via nutrient cycling, and these contaminates also affect the atmosphere. Sulfur oxides, nitric oxides (NO, NO_2, NH_3), hydrogen sulfide, hydrocarbons, carbon oxide, or ozone are a few of those contaminants. These impacts of soil upon air pollution were typically ignored while discussing air quality, though, according to the insignificant changes that occur inside the soil. Heavy metal deposition in blast furnaces and diesel exhaust from vehicles and manufacturing machines emissions in urban areas are two common sources of air particle soil pollution (Spellman, 2017).

1.4.1 SOURCES OF SOIL POLLUTION

There are two main categories of soil pollution—that which is caused by agricultural sources and that which is caused by nonagricultural sources, such like garbage, industrial discharges, and some other urban activities. Such factors collectively have an impact on the soil profile, subsurface soil, and top soil (Ashraf et al., 2013).

1.4.1.1 AGRICULTURAL SOURCES

Various agricultural methods, including the use of various rural practices like utilization of composts, pesticides, insect sprays, and herbicides, are the source of this kind of pollution (Aziz et al., 2021). Mass amounts of animal feces frequently contaminate the soil. When they are reintroduced back into the soil, some of these crop remnants, and even animal manures, are highly helpful. Contamination could be brought on by careless storage and disposal. A study was carried out in a region of West Java, which is significant for agroforestry or tourists; tea plantations were the subject of this study. The goal of this study was to determine how air pollution from cars affected the soil quality. These findings of the soil assessment revealed that lead concentration in the soil grew along major roads in the crops. The amount of lead pollution within soil—the majority of it caused by the burning of fuel—depended upon how far the area was away from the major highway. Its distance from the major road had no effect on the amount of cadmium within soils, though. This suggests that the cadmium amount found in the soil might not have come via air pollution but rather from the use of phosphate fertilizer at high rates in certain places.

1.4.1.2 NONAGRICULTURAL SOURCES

This kind typically results from urbanization brought by a fast population explosion that in turn raises the amount of waste produced per person. Such wastes are carelessly thrown upon soil that eventually make their way into it (Oluwatuyi et al., 2020). As a consequence of its prolong persistence or deposition, they are polluting sources because they concentrate hazardous substances. Inorganic hazardous chemicals are a few of the most significant soil contaminants (Ashraf et al., 2013). According to a survey by Kurnia et al. (2000), boron, cadmium, and lead quantities in the soil have been found to be extremely high. Both the production of rice and grower income fell in such locations as a result of declining soil fertility. The usual rice production had dropped by around 80% after 20 years of pollution. Their initial production of 4–6 metric tons of rice has decreased to 1 metric tons. Nevertheless, relative to nonpolluted soil, total amount of heavy metals found in the soil has risen by approximately 18–98%. High quantities of lead, cadmium, copper, chromium, and boron were identified in plant cells, roots, or rice grains, according to a greenhouse study employing damaged soil; however, majority of the contaminants had collected within rhizome.

1.5 INORGANIC POLLUTANTS AND THEIR TOXIC EFFECTS ON HUMANS

Elements or substances which exist in nature within parent rock and have manmade origins are included in this category of pollutants. Following is a description of the main inorganic pollutants, including asbestos, radionuclides, and trace elements.

1.5.1 TRACE ELEMENTS

A class of ubiquitous elements known as "trace elements" are those that are hazardous for life but typically present at extremely low levels throughout the environment. Iron (Fe), copper (Cu), zinc (Zn), manganese (Mn), nickel (Ni), boron (B), selenium (Se), or molybdenum (Mo) were several examples of metals that are necessary micronutrients for soil microbes, crops, and animals (Viets, 1962), but lead (Pb), cadmium (Cd), and mercury (Hg) have no recognized metabolic activity (Eisler, 2006; Flora et al., 2012).

World Health Organization (WHO, 1996) makes a clear distinction among trace elements which are:

a. Essential: chromium, copper, molybdenum, selenium, and zinc
b. Possibly essential: manganese, nickel, and vanadium
c. Hazardous elements: aluminum, arsenic, cadmium, lead, lithium, mercury, and tin.

Usually, soil and crops contain trace amounts of these elements. This same net amount of trace elements inside a particular region relies upon that natural origin substance, its locality, land mechanisms like weathering, leaching, or eroding (these helps mobilize and disperse trace elements inside the ecosystem), as well as human impacts like fertilizers, sludge, sewerage, industrial discharges, solid wastes, urban runoff, and also spray drift (Huang et al., 2015).

Trace elements are released through the atmosphere as a result of human activity including mineral extraction, casting, ore handling, oil works, and also other type of manufacture of steel, sewerage treatment, fertilizer manufacture or fossil fuel consumption during generating electricity. It happens as a result of direct exposure into soil including industrial effluent and air emissions (Cho et al., 2019; Hotra et al., 2020). Waste material that is dumped illegally and without sufficient management endangers the soil ecosystem in addition to air pollution (Bailon et al., 2018; Hotra et al., 2020).

Owing to its severe toxicity in small concentrations for people or other creatures as well as their widespread dispersion, the trace metals arsenic, cadmium, chromium, copper, mercury, lead, manganese, nickel, and zinc are of particular concern (Li et al., 2018; Rahman and Singh, 2019). Table 1.1 provides a thorough overview of these trace elements, together with their artificial and native origins or applications. For comprehending pollutant trends, analyzing periodic global cycling, or making judgments regarding soil pollution treatment, it is crucial to know all source for trace elements in the ecosystem (Cloquet et al., 2006).

TABLE 1.1 Important Trace Metals That Threaten Human's Health.

Heavy metals	Source	Effects on human health	References
Cadmium	Mining, electroplating, fertilizers, vehicles exhaust	Lung diseases, lung cancer, never disorder, bone defects	(Unaegbu et al., 2016; Engwa et al., 2015; Mudgal et al., 2010)
Chromium	Tanning, dying, mining, plating	Mental retardation in children, liver and kidney damage	(Shekhawat et al., 2015; Ghani, 2011)
Arsenic	Mining of arsenic, fungicide, timber industry, forest fires	Internal cancer, bronchitis, skin lesions	(Huy et al., 2014)
Lead	Paint, smoking, textile, battery manufacturing, insecticides	Anemia, hearing deficit, brain and nervous system damage	(Wani et al., 2015; Martin and Griswold, 2009)

1.5.2 RADIONUCLIDES

These pollutants which release radiation are induced when active atoms decompose; radionuclides are present as serious threat for both ecosystem and living things. Radioisotope exposure in soil results through human actions like past radium use for anticancer therapy and uranium refining, usage of nuclear phosphorus and cobalt-ores, nuclear weapons experiments, even nuclear incidents, as well as natural phenomena like parental weathering of rocks and volcanism. Plants absorb radioactive particles from the soil, making them accessible for more transfer inside the food web.

Three different types of radionuclides are found in soil.

a. **Primordial radionuclides:** They are still present as from Earth's origin, their half-life for primordial radionuclides, are often in the tens or thousands of years range.

b. **Cosmogenic radionuclides:** Its persistent production results from a cosmic radiation bombarding upon stable isotopes, especially inside the air. They had shorter half-life then primordial.

c. **Anthropogenic activities:** Radioactive incidents just like Chernobyl disaster or atmosphere nuclear of armament tests deposits are examples of manmade actions. Radionuclide soil pollution might even result from inappropriate nuclear waste storage. Strontium could displace calcium, build up with in bones, or lead toward neuronal loss within skeletal or synovial systems (Dutov and Yermolayev, 2013; Litvinov et al., 2008).

1.5.3 ASBESTOS

The name "asbestos," refers to a broad variety of natural sources that are moist crystal silicate fibers that are part of a sinuous or amphibole mineral families that create rocks. When they produce loosened fibers out of natively existing asbestos to be used in industry, it is mined. Most typical varieties of economically accessible asbestos are:

- **Chrysolite:** White, hydrated magnesium silicate mineral known as chrysolite is more widely in use and is of less dangerous type.
- **Amosite:** It is mild toxic and in color it is brown or grey like iron-magnesium silicate.

- **Crocidolite:** It is the worst toxic and is mineral blue in color like sodium-iron silicate.

Asbestos was already utilized in various industrial products, including lagging, emergency fire doors, pipework or casing, strengthening of concrete and construction industry (fiber), rubber products, including thermal, acoustics, and even chemically insulators (Frank and Joshi, 2014). Most asbestos fibers already have ability for being toxic for people, but the impacts vary depending just on asbestos fiber's usage, quality, position, or exposure. According to estimates, 0.001% of asbestos in dry soil may surpass workplace exposure guidelines which is 0.1 fibers/mL of air. Mesothelioma is a disorder (cancer) that develops inside the respiratory system, belly, and cardiac as a result of asbestos exposure (Molinari and Stevenson, 2020). Cancers of the larynx or the ovaries also are linked with asbestos exposure.

1.6 ORGANIC POLLUTANTS AND THEIR TOXIC EFFECTS ON HUMANS

1.6.1 POLYCYCLIC AROMATIC HYDROCARBONS

The sole elements present in PAHs are carbon and hydrogen. These molecules are arranged in two or perhaps more annulated phenyl ring which are connected across similar surfaces. The terminating hydrogen ions of the PAH core can be replaced by a variety of side groups, because of their comparable chemical components and unique compounds—referred to as homologous—which display identical qualities. The mixture of homologous compounds produced is dependent on the variables present even throughout process like the combustion temperature. According to the National Research Council (US) Committee on Pyrene and Selected Analogues from 1983, thousands of PAHs could exist naturally and could be emitted during a variety of previous and current human activities, including the extraction and burning of hydrocarbons (Kim et al., 2013). Additionally, the preparation of several food products, for example, roasted veggies, toasted bread, meat, or barbecued foods, produces PAHs (Singh et al., 2016).

Numerous PAHs are harmful for reproductive organs, and are mutagenic and cancerous (especially those bigger homologous). Benzo(a)pyrene, which has teratogenic, mutagenesis, and cancerous properties, are among the most dangerous. Due to their strong lipid solubility or ability to cross the crosses the placenta, PAHs could be taken through maternal epidermis, intestines, and lungs with an impact on the neuronal function of the fetus (Lawal, 2017).

1.6.2 VOLATILE ORGANIC COMPOUNDS

This term "volatile organic compounds" (VOCs) refers to the group of research disciplines that are mainly found as extremely flammable liquids or possibly gases above ambient temperature (Leo et al., 1971). Several BTEX chemicals (benzene, toluene, ethyl benzene, and xylene) were acyclic aromatics that are frequently found within soils and thus are readily biodegradable. Numerous VOCs were significant pollutants in industrial areas and therefore are frequently utilized in organic solvents within industries, for instance, in rust remover, laundry service, publishing, or painting processes. VOCs also cause a variety of harmful reactions, a few of which are teratogenic, mutagenic, or cancerous. Among major symptoms linked to severe BTEX contamination include hematologic consequences, acute myeloid leukaemia, or genetic defects, which are confirmed by better information (Galbraith et al., 2010).

1.6.3 PHENOLS, CHLOROBENZENES, AND CHLOROPHENOLS

The chlorine compounds from benzene and phenol are phenols, chlorobenzenes, and chlorophenols. A benzene ring contains a single or even more hydrogens that have been replaced with chloride ions producing chlorobenzenes that are cyclical aromatic compounds. Each of these substances pose a danger for environments as possible sources for dioxins that can have hazardous effects on people and other living things (Khatymov et al., 2003).

Phenolic chemicals are readily taken up through the epidermal, pulmonary, and digestive systems, but after entering the organism, these undergo biochemical breakdown within liver, stomach, and nephron, which leads to the production of new active molecules. These have high neurotoxic potency. Chlorophenols were regarded as having the possibility of causing human cancer. Both people and animals have been implicated with endocrine disruption. Overall consequences from acute or chronic contact to such a group of toxins for people are not well studied, despite very limited atmospheric exposure to these contaminants (Anadarko and Pennea, 2020).

1.6.4 POLYCHLORINATED BIPHENYLS

In addition to being extensively spread pollutants, nondioxin-like polychlorinated biphenyls (ndl-PCBs) too are extremely long-lasting in the ecosystem and can build up throughout human food chain. These quickly enter the bloodstream through the digestive system, and are transported to and built up in hepatic or abdominal tissue. PCBs could build up within embryo or baby because they enter through the placenta and thus are ingested in breast milk. PCBs was utilized in a variety of commercial processes due to their great chemical durability and electric conductivity, notably liquids for capacitors and transformers, hydraulic oils, retardants, even wetting agents (Faroon and Ruiz, 2016).

1.6.5 PESTICIDES

According to the FAO and WHO (2016), as well as Rodriguez Eugenio, McLaughlin, and Pennock (2018), "pesticide" refers as an active substance and combination of chemicals used only for eliminating, eradicating, or treating the insect that poses a threat to people or interferes in industrial activities like manufacturing, preparation, preservation, shipping, or advertising of food, timber and furniture, textiles, or other agricultural goods. Within healthcare field, pesticides are often employed as just a biocontrol for diseases like malaria (WHO and FAO, 2019). Another biggest issue is the existence of warehouses full of outdated pesticides throughout many regions. Such dumps still generate severe soil contamination due to spills and leaks.

Because pesticides are widely used, interact to many toxins, are challenging to develop incidental links among, or receive little investment, there are very few observational studies relating chronic conditions to environment exposed to pesticides. Nevertheless, research data suggests a connection among acute poisoning and chronic illnesses like cancer, asthma, allergies, and developmental problems (Aziz et al., 2021).

1.7 AIR POLLUTION

Air pollution has been documented as a pronounced public health and environmental hazard globally (Landrigan, 2017). It is the existence of such elements in the atmosphere that have detrimental impacts on human wellbeing and other living entities as well as adversely affects the Earth's climate and materials. It is a significant environmental and health dilemma, especially in the developing nations. The outdoor air pollution may be instigated by various gaseous pollutants such as oxides of carbon, nitrogen, sulfur, as well as particulates (Sheehan et al., 2016). It is well established in the literature that acute and chronic exposure to polluted air results in increased rates of mortality and disease incidence among human population (Liu et al., 2019).

The WHO has estimated about 4.2 million casualties each year due to breathing in ambient air pollution (WHO, 2018). The most casualties are happening in the low- and middle-income states and among them China and India are with the most casualties internationally (Liu and Liu, 2011). The high-income states of America, Europe, Eastern Mediterranean, and Western Pacific are also suffering from air pollution with 0.38 million casualties every year. The WHO has urged that it is the need of hour to restrict the emissions of air pollutants and enhance air quality globally (WHO, 2021). In this regard, different air pollution abatement techniques and policies are accessible to decrease air pollution (Fensterstock et al., 2012).

1.7.1 TYPES OF AIR POLLUTANTS

The WHO has documented the six foremost air pollutants, viz., carbon monoxide (CO), lead (Pb), nitrogen oxides (NO_x), ozone (O_3), particulate matter (PM), and sulfur dioxide (SO_2). These air pollutants can have devastating effects on all environmental constituents including air, soil, and water, posing serious hazards to living organisms and are categorized into (i) primary and (ii) secondary air pollutants. The primary air pollutants are generally released by the processes like ash from a volcanic eruption, CO gas from automobile exhausts, and SO_2 gas discharged from processing units. The secondary air pollutants are not directly discharged; actually, they are produced in the atmosphere when primary pollutants react with each other. The ground level O_3 is a well-known explanation of a secondary air pollutant. Few other pollutants are considered as both primary and as well as secondary owing to their direct release and generation by some other primary pollutants (Wilson and Suh, 1997).

1.7.1.1 PRIMARY AIR POLLUTANTS

Carbon dioxide (CO_2) is a natural constituent of the air and is being increased in earth's atmosphere as a potent greenhouse gas (Vaidyanathan, 2014). It is indispensable for all plant life and also discharged by the living being's respiratory system. Sulfur dioxide is released from volcanoes and from several industrial units. Fossil fuels such as petroleum and coal mostly have sulfur elements and release SO_2 on burning. Additional oxidation of SO_2, usually in the presence of a catalyst like nitrogen dioxide (NO_2) results in sulfuric acid (H_2SO_4) formation and consequently the acid rain. This is among the main reasons for concern regarding environmental effect of the usage of these fuels as an energy source. Nitrogen oxides (NO_x) particularly NO_2 is one of the main air pollutants and released as a result of combustion at elevated temperature. It is also instigated during unusual weather conditions like thunderstorms which produce electric discharge that can be observed in the form of a brown haze directly above the cities. The CO is a toxic gas. It is produced as a result of incineration of fuels like coal, natural gas, or wood. The vehicular gaseous emissions are the main source of CO discharge in atmosphere. It leads toward smog formation in the air that causes many lung diseases (Vallero, 2007).

Methane (CH_4) is the most potent greenhouse gas that causes global warming and climate change. Particulate matter or fine particles are small solid or liquid particles that remain suspended in the air. Some particulates are present naturally that originate from dust hurricanes, forest fires, living vegetation, sea sprays, and volcanoes. Human acts like fossil fuel burning in different industrial units, power plants, and vehicles also produce aerosols in high concentrations. High level of minute particles in the air is associated with health hazards like heart disorders, changed lung functions, and even lung cancer. Persistent free radicals with minute particles results in cardio-pulmonary disorder (Vallero, 2007).

The chlorofluorocarbons (CFCs) are damaging to the ozone sheet and are discharged from air conditioners, aerosol sprays, and refrigerators. After discharge into air from their sources, CFCs rise above in the stratosphere and react with other gaseous constituents and hence destroy the ozone protective layer. Consequently, damaging ultraviolet rays reach the troposphere and cause human health problems like eyes disorders, skin cancer, and can also destroy the soil. Ammonia (NH_3) is mostly released by agricultural waste and also a component for the formulation of several pharmaceuticals. Ammonia in the atmosphere combines with nitrogen and sulfur oxides (NOx and SOx) resulting in the production of secondary air pollutants (Int Panis et al., 2008). Smells from garbage, sewage, and industrial procedures also deteriorate the air quality. Radioactive air pollutants released by nuclear reactors, explosions, and natural processes like radioactive decay of radon (Rn) are significant air pollutants (Vallero, 2007).

1.7.1.2 SECONDARY AIR POLLUTANTS

The secondary air pollutants contain the particulates produced by gaseous primary pollutants and constituents present in the photochemical smog.

Photochemical and chemical reactions are involved in the formation of smog, a type of air pollution. The mixture of smoke and SO_2 generates classical smog when coal is burned in high quantity in any area. Photochemical smog is a blend of pollutants that forms when NOX and VOCs react with sunlight, generating a haze that is common over cities. The ground level O_3 that is not good is also produced from NOx and VOCs. Peroxyacetyl nitrates (PANs) also form from the NOx and VOCs. The persistent organic pollutants (POPs) are toxic substances composed of organic complexes and are usually resilient to environmental decomposition by chemical, biological, and photolytic methods. Hence, they get accumulated in animal and human tissues and also biomagnified in all food chains, thus pose hazardous effects to human and environmental health (Vallero, 2007).

1.7.2 SOURCES OF AIR POLLUTION

Sources of air pollution can be considered as natural and anthropogenic.

1.7.2.1 NATURAL SOURCES

The most important natural causes of air pollution include the following:

- Dust particles from natural sources generally vast land areas with slight or no vegetation cover (Goldstein et al., 2009).
- Methane is emitted during food digestion by animals and are also being released from agricultural activities (Vallero, 2007).
- Radon (Rn) is a colorless, nonaromatic, and radioactive gas which is naturally released by radioactive decay occurring in the Earth's crust. It is also released during the decay process of radium (Rd). It can be stored in buildings, particularly in confined ranges like basements. It causes dangerous health impacts and is the second main source of lung cancer after smoking (Goldstein et al., 2009).
- Smoke and CO gas, both are released from wildfires. During wildfires, the smoke released from the unrestrained biomass burning can make up nearly 75% of all air pollution (Fischetti, 2014).
- In some areas, vegetations such as black gum, oak, poplar, and willow trees emit high concentrations of VOCs during hot days which produce the ground level O3. These VOCs combine with

primary man-made contaminants such as NOx and SO2 to yield a seasonal haze (Fischetti, 2014).
- Volcanic activity release sulfur, chlorine, and ash particulates in the atmosphere (Goldstein et al., 2009).

1.7.2.2 ANTHROPOGENIC SOURCES

The anthropogenic sources are typically linked with the fuel burning but there are also the sources other than combustion. The following are the anthropogenic causes of air pollution:

- The stationary sources comprise the smoke stacks of fossil fuel-based power stations like coal industry, manufacturing units, waste incinerators, furnaces, and other fossil fuels-based plants. In the developing and poor states, the chief cause of air pollution is combustion of biomass, comprising of crop waste, dung, and wood (WHO, 2002).
- Mobile sources consist of motor vehicles and the aircraft (Bruce et al., 2000).
- Controlled burning practices for agriculture and forest management are also main anthropogenic sources of air pollution. The controlled burning is a practice occasionally employed for forest management, farmland and grassland reclamation, and for reduction in greenhouse gases. Fire is a natural process that occurs in both forest and grassland ecosystems and controlled fire can be a management practice for foresters and farmers. Controlled burning enhances the growth of interested forest trees, hence renewing the forest (WHO, 2002).
- The aerosol sprays, hair sprays, paints, varnish, and other solvents can be significant sources of fumes (Diep, 2018).
- Waste deposition in landfills produce CH4 gas, which is very combustible and may form explosive mixtures in the air. The CH4 gas can displace the oxygen in a confined environment and consequently suffocation will result if the oxygen level becomes less than 19.5% (WHO, 2014).
- Military objects like nuclear weapons, poisonous gases, and rocketry also contribute to air pollution (WHO, 2014).
- Fertilized farmlands may be a key source of NOx emissions to cause air pollution in nearby areas (Diep, 2018).

1.7.3 HEALTH PERSPECTIVES OF AIR POLLUTANTS

The poor air quality due to air pollutants has extensive effects on human health but primarily disturbs the human's cardiovascular and respiratory systems. The most usual causes of air pollution comprise of particulates, O_3, NO_2, and SO_2. The children below five years age in developing nations are the most susceptible population for severe health impacts and even death due to indoor and outdoor air pollution (Pratit et al., 2020). The WHO has projected that approximately seven million premature deaths annually are caused by air pollution globally. It is also documented that air pollution is linked with increased mortality rate from coronary stroke (Andersen et al., 2012).

The research has also established the increased chance of developing asthma in human beings from air pollution (Gehring et al., 2010). Contact with air pollution causes lung cancer in nonsmokers. The people living in excessive traffic experience high mortality rate due to cardiovascular diseases, lung cancer, and overall nonaccidental deaths that can be linked with air pollution (Chen et al., 2008). Health index of people living in the zones with comparatively low concentration of air pollutants is low, as huge number of people breathes in these pollutants (Di et al., 2017). Scientific reports demonstrate that breathing in polluted air also disturbs the central nervous system of human beings (Bos et al., 2014).

Another most dangerous air pollutant is particulate matter (PM) which gets deposited into the respiratory system and can diffuse into human lungs and blood circulation leading to devastating health impacts (Maragkidou, 2018). Scientific literature extensively suggests that contact with PM in air can instigate acute health consequences including asthma, respiratory and cardiovascular problems, oxidative stress, chronic obstructive pulmonary diseases (COPDs), and even lung cancer in humans (Hime et al., 2018). The individual's response to air pollutants depends on nature of the pollutant that person is exposed to, exposure time, and health status of individual (Vallero, 2007). Different air pollutants and their health effects are given in Table 1.2.

1.8 WATER POLLUTION

Human survival on Earth mainly depends on three basic natural resources—air, water, and soil. Among these, water is the most crucial commodity to mankind, forming the basic medium for origin of life. Its importance is

TABLE 1.2 Different Air Pollutants and Their Health Effects.

Air pollutant	Human health effects	References
Particulate matter (PM)	The short and long-time contact with PM results in nasopharyngitis cardiovascular ailments and infant mortality	Kelishadi and Poursafa, 2010; Zhang et al., 2019
Ozone (O_3)	O3 adversely affects the skin and tear ducts	WHO, 2019; Thiele et al., 1997
Carbon monoxide (CO)	CO hazards include headache, weakness, nausea, hypoxia, ischemia and cardiovascular ailments	Emberson et al., 2018
Nitrogen dioxide (NO_2)	NO2 induces respiratory disorders, coughing, wheezing, and bronchospasm when inhaled at high concentration	Chen et al., 2007
Sulfur dioxide (SO_2)	The main health issues linked with SO2 include the respiratory problems, skin redness, eye damage, and cardiovascular malfunctioning	Chen et al., 2007
Lead (Pb)	Pb via inhalation accumulates in the blood, tissues, organs, and cardiovascular and reproductive systems. It is neurotoxicant and induces learning disabilities, memory loss, hyperactivity, and mental obstruction in children	Farhat et al., 2013
Polycyclic aromatic hydrocarbons (PAHs)	PAHs are considered as hazardous, mutagenic, and carcinogenic and hence cause lung cancer	Abdel-Shafy and Mansour, 2016
Volatile organic compounds (VOCs)	Short-time contact with VOCs results in irritation of eyes, nose, throat, and mucosal membranes, although long-time contact results in toxic reactions	Ebersviller et al., 2012
Dioxins	Short-time exposure to dioxin causes pigmentation on the skin, while long time contact may result in developmental problems, impairment of nervous, and reproductive systems and cancer	WHO, 2019

increasing due to rapid industrialization, urbanization, and uncontrolled population growth. Globally, available water resources vary greatly geographically and about 80% world population is facing water security threat (Jayaswal et al., 2018).

Water is essential for life and its exposure to pollution has become worldwide problem which has threatened human health and infected a large number of people, sometimes in the form of a pandemic (Baldwin et al., 2020). In living organisms, water plays multiple roles such as solvent, metabolite, living environment, temperature buffer, and lubricants (Hanslmeier, 2011).

Pollution is the introduction of contaminants into the environment originating from various types of commercial, industrial, and agricultural activities by human. Water is considered polluted when excessive amounts of any substance or condition changes the basic characteristics of water in such a way rendering it useless for its intended purpose. Water pollution may pose serious threats to our environment and its living components and potential impacts of various types of pollutants may vary depending upon their source of origin and type. For instance, various organic pollutants, heavy metals, dyes, etc., are known carcinogens while hormones, personal care products, pharmaceuticals, and cosmetics are known as endocrine disruptive chemicals (Adeogun et al., 2016). These pollutants enter into the surface or groundwater by various anthropogenic activities, either directly or indirectly, and pose various hazards to the environment (Shakouri and Yazdi, 2014).

The following are the sources of water pollution:

Most commonly, water gets contaminated through two main sources, i.e., natural (wind and water erosion, floods, and storms) and anthropogenic sources (Karri et al., 2021). Water pollution originates mainly from: (i) point, (ii) nonpoint, and (iii) transboundary sources. Various human activities causing pollution of water bodies are as follows (Owa, 2013):

i. Animal waste
ii. Building visionaries and lavatories over running water bodies
iii. Combustion
iv. Deforestation
v. Eroded sediments
vi. Failing septic systems
vii. Heavy metals
viii. High population density
ix. House hold chemicals
x. Industrial waste dumped into water reservoirs
xi. Littering
xii. Processing plants for minerals such as production of coal
xiii. Mining
xiv. Oil spillage
xv. Pesticides, herbicides, and fertilizers
xvi. Aquifers contamination by drilling activities
xvii. Radioisotopes
xviii. Sewage leakages
xix. Toxic waste disposal at sea
xx. Waste deposition into water bodies by floods during monsoon season

Several factors cause pollution of our precious water resources, but the most prominent and major ones are as follows:

1. Urbanization
2. Industrialization
3. Population growth
4. Agricultural activities
5. Poor wastewater management system

1.8.1 CLASSIFICATION OF WATER POLLUTANTS

Generally, water pollutants are found in multiple forms such as liquid, solid, and semisolid and are categorized into physical, chemical, and biological (Verma and Ratan, 2020); but on the basis of their compositions, they are classified into (1) biodegradable, (2) nonbiodegradable, and (3) recalcitrant (Thongam and Chaturvedi, 2021).

Water pollutants may be classified as conventional (ions and heavy metals) and emerging (steroids, surfactants, hormones, pharmaceutical, and personal care products) water pollutants (Ahamad et al., 2020), and are broadly categorized into organic and inorganic pollutants (Borah et al., 2020).

Types of water pollutants:

1. Organic pollutants
2. Inorganic pollutants
3. Nutrients
4. Heavy metals and halogens
5. Dyes and pigments
6. Suspended solids and sediments
7. Emerging contaminants (ECs)
8. Endocrine disruptive chemicals (EDCs)
9. Microbial contaminants
10. Microplastics
11. Accidental oil leakage
12. Radioactive pollution
13. Thermal pollution

1.8.2 ORGANIC WATER POLLUTANTS

Organic pollutants such as dyes, pesticides, plasticizers, fertilizers, pharmaceuticals, carbohydrates, proteins, oil and grease, etc., are very wide in variety including a group of chemicals called endocrine disruptive chemicals that are also classified as emerging contaminants pose immense range of toxicity to aquatic organisms, plants, and humans (Jung et al., 2015). Organic pollutants are broadly categorized as: (i) oxygen demanding wastes, and (ii) synthetic organic pollutants. Oxygen demanding wastes are biodegradable, comprising of organic compounds in the form of colloidal, dissolved, and suspended particles which are most commonly discharged from effluents and wastewaters from domestic outlets, municipal sewage, butcheries, breweries, distilleries, tanneries, food, and beverage industries as well as paper and canning industries. These pollutants create anoxic conditions in the ecosystem by consuming all the oxygen present in the water bodies, while synthetic organic pollutants include synthetic insecticides, detergents, pesticides, fibers, plastics, food additives, paints, solvents, pharmaceuticals, and VOCs. They stay in water bodies for longer periods of time. Among organic pollutants posing immense risk to living organisms are cosmetics, dyes, and petroleum products causing chemical burns, nausea, convulsions, and even cancer.

1.8.3 INORGANIC WATER POLLUTANTS

Inorganic pollutants are naturally found in our environment but sometimes their concentration becomes significantly high or gets imbalanced in the open environment owing to various human activities. These could be any compound of trace elements or minerals, inorganic salts, metals, and part of metal-organic frameworks. These pollutants in water bodies come from various human activities like industry, mining, agriculture, etc., and these inorganic pollutants along with their organic forms have devastating negative health impacts on flora and fauna of earth's atmosphere, geosphere, and hydrosphere. In humans, they exposure results in severe health problems such as abnormal growth, diabetes, high risk of cancer, obesity, etc. Release of inorganic chemicals from chemical industry and wastes from automobile exhaust are among the major causes of water resources contamination. Besides, petrochemical companies also release inorganic contaminants into air which after getting settled down lead to contamination of water resources (Borah et al., 2020).

1.8.4 NUTRIENTS

Nitrate is the most vital element for plant growth, mostly applied in the form of nitrogenous fertilizers. It finds its way into water channels through different human actions such as utilization of nitrogenous manure, domestic and industrial wastewater, seepage from septic systems, disposal of urban drainage, sewage ejection, reject dumps, polluted ground and sanitation, dung and urine of farming animals, landfills, and soluble compounds of nitrogen contained by geological materials (Taneja et al., 2019). High nitrate content in water causes health issues like stomach cancer, gastrointestinal diseases, and birth imperfections in adults and methemoglobinemia in newborns, leading to mortality by asphyxiation (Wolfe and Patz, 2002; Zhai et al., 2017). Phosphate too is considered an essential nutrient for the growth of plants, animals as well as human beings. Phosphate is available in the form of (i) organic phosphorous, (ii) orthophosphate, and (iii) polyphosphates. Phosphate enters into water bodies through both point (sewage wastewater and industrial effluents) and nonpoint (natural disintegration of minerals and rocks, discharge of animal excreta, sedimentation, and erosion) sources (Singh, 2016). Excess phosphates may boost algal bloom development and growth of aquatic plants, resulting in eutrophication of aquatic ecosystems (Handa, 1990; Rao and Prasad, 1997).

1.8.5 HEAVY METALS AND HALOGENS

Heavy metals are metals or metalloids, having high atomic weight and specific density such as lead (Pb), iron (Fe), zinc (Zn), manganese (Mn), arsenic (As), chromium (Cr), cobalt (Co), aluminium (Al), cadmium (Cd), nickel (Ni), mercury (Hg), etc. These are toxic when present even at very low concentrations and are generally introduced into the water bodies geogenically by chemical weathering of rocks, decomposition of animal and plant material, and atmospheric deposition (dry or wet) and anthropogenically by mining, disposal of industrial effluents, fertilizer use in agriculture, roadside and automobiles, etc. (Duruibe et al., 2007; Eshagberi, 2012). Heavy metals and many other inorganic pollutants like inorganic salts, trace elements, mineral acids, sulfates, cyanides, metals and their complexation with organic compounds, acid mine drainage pollute surface, and groundwater sources. All these inorganic pollutants are nonbiodegradable and considered great menace for public health as well as aquatic fauna and flora (wasewar et al., 2020).

Halides (chloride, fluoride, bromide, and iodide) are very soluble ions and are found in all forms of water reservoirs, originating naturally from weathering of hard rocks and anthropogenically by the application of pesticides and fertilizers as well as industrial runoff (Tiffany et al., 1969).

1.8.6 DYES AND PIGMENTS

Dyes and pigments are widely used as colorants in textile, food, paper, paints, plastics, cosmetics, and pharmaceuticals, producing vast quantity of undesirable-colored effluents polluting natural water channels. More than one million tons of dyes used in these industries not only cause aesthetic pollution, deteriorate the quality of the environment, deplete dissolved oxygen content of water bodies, and reduce soil productivity but also creates serious health problems for humans, plants, and animals (Gurses et al., 2021; Karri et al., 2021).

1.8.7 SUSPENDED SOLIDS (SS) AND SEDIMENTS

Amount of organic and inorganic particulate matter (<62 mm) detained by a waterway such as lakes, streams, ponds, and rivers due to instability is termed as suspended solids (Waters, 1995). Suspended solids negatively impact the aquatic ecosystems particularly survival of benthic flora by obstructing the incoming sunlight necessary for photosynthetic activities. Harmful effects of SS depend upon their quantity, geochemistry, particle size, and contact time. Finer SS particles adversely affect the fish gills leading to asphyxiation (Bilotta and Brazier, 2008).

1.8.8 EMERGING CONTAMINANTS (ECS)

Emerging contaminants (ECs) are a newly discovered group of chemicals which are widely distributed in the surface and groundwaters. In general, ECs are divided into three categories: (i) all those compounds which are currently introduced into our environment such as industrial additives, (ii) compounds that have been existing in the environment for past many years but their presence have been detected in the recent past due to their increasing significance and interest such as pharmaceuticals, and (iii) compounds that have been known since a long ago but their potential negative impacts on man and environment were realized presently such as hormones (Barbosa Junior et al., 2016; Abdulrazaq et al., 2020; Ramirez-Malule et al., 2020). Although, ECs are not newly created chemicals but include different groups of contaminants, for example, pharmaceuticals, nanomaterials, personal care products, fragrances, pesticides, flame retardants, surfactants, and hormones. Most of these compounds are not effectively removed from wastewater treatment plants and disposal of their residual concentrations results in pollution of above and underground water resources (Ramirez-Malule et al., 2020). Some ECs are known due to their toxic effects and endocrine disrupting activities such as:

1. Pharmaceutical and personal care products (PPCP)
2. Hormones and steroids
3. Perfluorochemicals (PFCs)
4. Nanomaterials
5. Flame retardants and surfactants
6. Pesticides and synthetic fertilizers
7. Industrial compounds (dyes, heavy metals)
8. Endocrine disruptive chemicals (EDCs)

Water polluted by endocrine-disruptive chemicals (EDCs) also results in harmful effects on environment and ecosystem health. Widespread occurrence of EDCs in surface and groundwaters as well as wastewater has been reported since the last two decades (Geissen et al., 2015; Luo et al., 2018). These emerging contaminants adversely affect the animal and human health by directly disrupting their endocrine system which block the production of natural hormones responsible for the normal functioning of their body (Vieira et al., 2020). Chronic exposure to EDCs creates health problems in human life and wildlife. For example, in humans EDCs cause hormonal imbalance that results in abnormal homeostatic and development, irregular endocrine behavior, impairment of reproductive and thyroid functioning, as well as amplifies the cancer hazard (Jung et al., 2015), whereas in wildlife produces reproductive disorders, interferes with hatchability, and reduces vitellogenin (Kasonga et al., 2021).

1.8.9 MICROBIAL CONTAMINANTS

Pathogens enter into water bodies most commonly through untreated sewage discharge, human and animal excreta, and industrial effluents resulting in water-borne diseases due to various types of viruses and bacteria such as Shigella, Cryptosporidium, Salmonella, Campylobacter, etc., while coliforms, coliphage, Escherichia coli, Clostridium perfringens, and fecal coliforms/streptococci/enterococci are the most common water pathogens found under fecal contamination. The most common diseases caused by pathogenic water pollutants in human beings are hepatitis, dysentery, cholera, typhoid, and polio. Exposed water is easily get contaminated by saprobic microbes when comes in contact with air, soil, and sediments (Inyinbor Adejumoke et al., 2018). Microbial contamination of water systems can be tested mainly through analysis of Escherichia coli, coliforms, and other heterotrophic bacteria (Madhav et al., 2020).

1.8.10 MICROPLASTICS

Microplastics are minute plastic particles having size less than 5 mm and are broadly distributed throughout the world in fresh and marine water resources. These are generally divided into primary and secondary microplastics, on the basis of how they are produced. Primary microplastics are directly released into the environment by domestic and industrial effluents, sewage, and spills, whereas indirectly by an overspill. Microplastic particles affect human health through dermal contact, orally, and by inhalation (Rochman et al., 2013; Li et al., 2020).

1.8.11 ACCIDENTAL OIL SPILLAGE OR LEAKAGE

Oil spills are also considered one of the major sources of water pollution resulting in huge environmental damage, loss of aquatic biodiversity, economic destruction, and disturbance of ecological framework of water bodies (Singh et al., 2020). Accidental leakage of oil from oil tankers during operation or shipping refinement causes adverse impacts on the aquatic environment by causing chemical toxicity and physical/ecological changes. During oil spill, huge quantity of oil pollutes the sea as it is not dissolved and causes unreliable harm to local marine wildlife. Toxic impacts may be acute or chronic depending upon the amount and physicochemical properties of oil spilled, size of the water body, sensitivity of the living biota, ambient conditions of the environment, time of exposure, and toxicity of the contaminant (Blumer, 1969). Due to low density than water, oil floats on the surface of water, over time it spreads out as a thin layer, obstructing the dissolution of oxygen from air into water, therefore ensuing in decline of dissolved oxygen content. It also reduces the sunlight infiltration through layers of surface water, thus resulting in reduced photosynthesis carried out by aquatic plants. Oil leakage also affects aquatic food chains leading to poisoning of aquatic life and ecosystem biodiversity and is great threat for endangering coastal plants and water birds due to oil shells (Nelson-Smith, 1971).

1.8.12 RADIOACTIVE POLLUTANTS

Radionuclides (uranium, thorium, plutonium) are naturally present in the environment in microscopic quantities but anthropogenic activities has spiked their level by releasing effluents from nuclear power and fuel reprocessing plants, uranium mining, nuclear weapons and accidents, defensive and research activities, clinical and nuclear medicine laboratories, geological repositories, and medical (treatment and diagnosis), and industrial uses of radio isotopes (Amer et al., 1974). Radioactive substances are also released from unused liquid materials from radiotherapy, contaminated glassware, laboratory research, packages or absorbent paper, urinary discharge or excreta of patients during different treatments using radionuclides. Contamination of water bodies with radioactive waste materials poses great threat to environment and human health. Harmful effects of these pollutants include nail and hair loss, blood cells deformation, subcutaneous bleeding, and different metabolic functions (Karri et al., 2021).

1.8.13 HEAT OR THERMAL POLLUTION

Water is used as a coolant in thermal or nuclear power plants and industries which ultimately release hot water into water reservoirs is known as thermal pollution, causing adverse impacts on aquatic life and hydrogeochemical processes. Increasing temperature of water bodies reduces the amount of dissolved oxygen, which further get amplified in the presence of organic matter due to enhanced bacterial proliferation which ultimately adversely affect the aquatic existence (Davidson and Bradshaw, 1967; Clark, 1969).

1.8.14 WATER POLLUTION AND HUMAN HEALTH

Water pollution is the most common cause of health problems in human beings. According to UNESCO (2003), globally about 2.3 billion people suffer from water-related diseases causing 60% of infant mortality and more than 2.2 million people die annually due to drinking of contaminated water and inadequate sanitation (Ullah et al., 2009; WHO and UNISEF, 2000). Each year, millions of deaths are caused by water-related diseases having different harmful effects due to varying nature and method of transmission (Naushad, 2019). Approximately 14,000 deaths occur per day due to the contamination of potable water by untreated sewage discharge. Broadly, these are categorized into: (1) water-borne diseases, (2) water-based diseases, and (3) water-related vector diseases.

1.8.14.1 WATER-BORNE DISEASES

Water contamination by biological/chemical waste materials originating from human or animal excreta cause different types of diseases such as diarrhea, typhoid, polio, hepatitis (A, E), cholera, and meningitis. Chronic diseases especially cancer is mainly caused by different kinds of industrial chemicals, fertilizers, pesticides, etc.

1.8.14.2 WATER-BASED DISEASES

Aquatic organisms, for example, flukes, helminths, worms, etc., residing in polluted or unpolluted water and as parasites of animal's cause water-based diseases like bilharzia, guinea worm, and paragonimiasis.

1.8.14.3 WATER-RELATED VECTOR DISEASES

Vectors/insects breeding near water sources either polluted or unpolluted, transmit infections and cause diseases like dengue fever, malaria, filariasis, sleeping sickness, etc., mainly due to changing climatic conditions, migration, and generation of new breeding places.

1.8.15 PREVENTION OF WATER POLLUTION

Water pollution can be prevented by taking following actions or measures (Owa, 2013; Wakchaure et al., 2015):

1. Mass education people regarding sources and hazards of water pollution should be organized.
2. Proper and safe disposal of domestic and industrial sewage ensuring no leakage.
3. For pulling out water from any water reservoir, clean buckets and ropes should be used.
4. Ban on animal's entrance into water bodies and cloth washing in ponds.

5. Utilization of nontoxic cleaning agents.
6. During or after flooding, water (river/pond/well) should be treated prior to its use.
7. Always wash vehicles far away from storm water drains.
8. Avoid throwing trash and toxic chemicals into sewer drains.
9. Inspection of septic systems every 3–5 years.
10. Less use of pesticides and fertilizers that can run off into water channels.
11. Cleansing of oil spills with kitty litter.
12. Avoid washing brushes used for paint in the sinks.
13. Never throw or dump carcasses near water reservoirs.

1.9 CONCLUSION

Environment refers to the overall status of somebody's natural neighborhood and serves as the fundamental habitat from all living organisms. It has become more difficult for man to live comfortably as a result of the environment becoming degraded through various activities that were primarily produced by man himself through time. This same idea of environmental pollution, a generally acknowledged categories of pollution notably, soil, air, water and their specific effect on health, categories, and types of pollution or its health impacts have all been clearly discussed in this chapter. Soil has been used more and more as a sink of pollutants despite being a nonrenewable reserve. Some pollutants in soil are present naturally but others are introduced by anthropogenic activities. By inspecting the source of pollution in soil, it is controlled as early as feasible from its origin (point or nonpoint source) to minimize the harm of that pollutant to the soil. Air pollution is contamination of the environment (either indoor or outdoor) by any physical, chemical, or biological agent that modifies the natural characteristics of the atmosphere. Industrial exhausts, vehicle emissions, household combustion devices, and forest fires are among some important sources of air pollution. Most commonly occurring water pollutants include various types of organic and inorganic pollutants, heavy metals, pathogens, nutrients, heat, suspended solids, nanomaterials, microplastics, and many other recently emerging contaminants comprising of wide variety of compounds. All these types of water pollutants are the root cause of infections and diseases in humans, animals, and plants. Recent reports on environmental impacts of pollutants are unraveling more evidence on the relationship between pollution and various fatal human health conditions. Out of these studies, the ratio of reports proving human ill effects due to exposure through air pollutants is increasing at an alarming rate. There is a need to raise the awareness regarding the pollutants damaging effects on human health and all hands must be decked to prevent such anthropogenic actions resulting in contamination of various compartments of environment in order to facilitate remediation measures.

KEYWORDS

- **pollutants**
- **sources**
- **ecotoxicity**
- **exposure**
- **human health**
- **soil pollution**
- **water pollution**
- **air pollution**

REFERENCES

Abdel-Shafy, H. I.; Mansour, M. S. A Review on Polycyclic Aromatic Hydrocarbons: Source, Environmental Impact, Effect on Human Health and Remediation. *Egypt. J. Pet.* **2016,** *25* (1), 107–123.

Abdulrazaq, Y.; Abdulsalam, A.; Rotimi, A. L.; Abdulbasit, A. A.; Clifford, O.; Abdulsalam, O. A.; Racheal, O. N.; Joy, A. A.; Victor, F. O.; Johannes, Z. M.; Bilal, M. Classification, Potential Routes and Risk of Emerging Pollutants/Contaminant. In *Emerging Contaminants*; Intechopen, 2020.

Adeogun, A. O.; Ibor, O. R.; Adeduntan, S. D.; Arukwe, A. Intersex and Alterations in Reproductive Development of a Cichlid, Tilapia Guineensis, from a Municipal Domestic Water Supply Lake (Eleyele) in Southwestern Nigeria. *Sci. Total Environ.* **2016,** *541,* 372–382.

Ahamad, A.; Madhav, S.; Singh, A. K.; Kumar, A.; Singh, P. Types of Water Pollutants: Conventional and Emerging. In *Sensors in Water Pollutants Monitoring: Role of Material*; Springer: Singapore, 2020; pp 21–41.

Ajibade, F. O.; Adelodun, B.; Lasisi, K. H.; Fadare, O. O.; Ajibade, T. F.; Nwogwu, N. A.; Wang, A. Environmental Pollution and their Socioeconomic Impacts. In *Microbe Mediated Remediation of Environmental Contaminants*; 2021; pp 321–354.

Ajibade, F. O.; Adelodun, B.; Ajibade, T. F.; Lasisi, K. H.; Abiola, C.; Adewumi, J. R. The Threatening Effects of Open Dumping on Soil at Waste Disposal Sites of Akure City, Nigeria. *Int. J. Environ. Waste Manage.* **2019b,** *27* (2), 127–146.

Akinbile, C. O.; Ajibade, F. O.; Ofuafo, O. Soil Quality Analysis for Dumpsite Environment in a University Community in Nigeria. *FUTA J. Eng. Eng. Technol.* **2016a,** *10,* 68–73.

Akinbile, C. O.; Erazua, A. E.; Babalola, T. E.; Ajibade, F. O. Environmental Implications of Animal Wastes Pollution on Agricultural Soil and Water Quality. *Soil. Water Res.* **2016b,** *11,* 172180.

Alina, B. Pollution Facts and Types of Pollution [Online]. https://www.livescience.com/22728-pollution-facts.html. 2018.

Alloway, B. J. Soil Pollution and Land Contamination. In *Pollution: Causes, Effects and Control*, 4th ed.; Alloway, B. J., Arthur, D., Ayres, J., Chester, R., Clift, R., Crathorne, B., Eds.; the Royal Society of Chemistry: Cambridge, 2001; pp 352–377.

Amer, N. M.; Graven, R. M.; Budnitz, R. J.; Mack, D. A. Instrumentation for Environmental Monitoring in Biological Systems. **1974.**

Anderko, L.; Pennea, E. Exposures to Per-and Polyfluoroalkyl Substances (PFAS): Potential Risks to Reproductive and Children's Health. *Curr. Prob. Pediatr. Adolesc. Health Care* **2020,** *50* (2), 100760.

Andersen, Z. J.; Kristiansen, L. C.; Andersen, K. K.; Olsen, T. S.; Hvidberg, M.; Jensen, S. S.; Raaschou-Nielsen, O. Stroke and Long-Term Exposure to Outdoor Air Pollution from Nitrogen Dioxide: A Cohort Study. *Stroke* **2012,** *43* (2), 320–325.

Ashraf, M. A.; Rehman, M.; Alias, Y.; Yusoff, I. Removal of Cd(II) onto Raphanus Sativus Peels Biomass: Equilibrium, Kinetics, and Thermodynamics. *Desalin. Water Treat.* **2013,** *51,* 4402–4412.

Aziz, H.; Wang, X.; Murtaza, G.; Ashar, A.; Hussain, S.; Abid, M.; Murtaza, B.; Saleem, M. H.; Fiaz, S.; Ali, S. Evaluation of Compost and Biochar to Mitigate Chlorpyrifos Pollution in Soil and their Effect on Soil Enzyme Dynamics. *Sustainability* **2021,** *13* (17), 9695.

Aziz, H.; Murtaza, G.; Saleem, M. H.; Ali, S.; Rizwan, M.; Riaz, U.; Niaz, A.; Abualreesh, M. H.; Alatawi, A. Alleviation of Chlorpyrifos Toxicity in Maize (Zea Mays L.) by Reducing its Uptake and Oxidative Stress in Response to Soil-Applied Compost and Biochar Amendments. *Plants* **2021,** *10* (10), 2170.

Bailon, M. X.; David, A.; Park, Y.; Kim, E.; Hong, Y. Total Mercury, Methyl Mercury, and Heavy Metal Concentrations in Hyeongsan River and its Tributaries in Pohang City, South Korea. *Environ. Monit. Assess.* **2018,** 190.

Baldwin, W. S.; Bain, L. J.; Di Giulio, R.; Kullman, S.; Rice, C. D.; Ringwood A. H.; Den Hurk P. V. 20th Pollutant Responses in Marine Organisms (PRIMO 20): Global Issues and Fundamental Mechanisms Caused by Pollutant Stress in Marine and Freshwater Organisms. *Aquat. Toxicol.* **2020,** 227, 105620.

Barbosa Junior, F.; Campiglia, A.; Rocha, B.; Cyr, D. Contaminants of Emerging Concern: From the Detection to their Effects on Human Health. *Biomed. Res. Int.* **2016.**

Bilotta, G. S.; Brazier, R. E. Understanding the Influence of Suspended Solids on Water Quality and Aquatic Biota. Water Res. **2008,** *42* (12), 2849–61.

Bishop, G. P.; Flack, J. L. E. Land Pollution. In *Sensor Systems for Environmental Monitoring*; Springer, 1997; pp 1–57.

Blumer, M. Oil Pollution of the Ocean. In *Oil on the Sea*; Springer: Boston, MA, 1969; pp 5–13.

Borah, P.; Kumar, M.; Devi, P. Types of Inorganic Pollutants: Metals/Metalloids, Acids, and Organic Forms. In *Inorganic Pollutants in Water*; Elsevier, 2020; pp 17–31.

Bos, I.; De Boever, P.; Int Panis, L.; Meeusen, R. Physical Activity, Air Pollution and the Brain. *Sports Med.* **2014,** *44* (11), 1505–1518.

Botkin, D. B. *Environmental Science: Earth as a Living Planet*. John Wiley & Sons: New York, 1995.

Bruce, N.; Perez-Padilla, R.; Albalak, R. Indoor Air Pollution in Developing Countries: A Major Environmental and Public Health Challenge. *Bull. World Health Org.* **2000,** *78* (9), 1078–1092.

Chen, T. M.; Kuschner, W. G.; Gokhale, J.; Shofer, S. Outdoor Air Pollution: Nitrogen Dioxide, Sulfur Dioxide, and Carbon Monoxide Health Effects. *Am J Med Sci.* **2007,** *333* (4), 249–256.

Chibuike, G. U.; Obiora, S. C. Heavy Metal Polluted Soils: Effect on Plants and Bioremediation Methods. *Appl. Environ. Soil Sci.* **2014,** 1–12.

Cho, I.G.; Park, M. K.; Cho, H. K.; Jeon, J. W.; Lee, S. E.; Choi, S. D. Characteristics of Metal Contamination in Paddy Soils from Three Industrial Cities in South Korea. *Environ. Geochem. Health* **2019,** *41* (5): 1895–1907.

Clark, J. R. Thermal Pollution and Aquatic Life. *Sci. Am.* **1969,** *220* (3), 18–27.

Cloquet, C.; Carignan, J.; Libourel, G.; Sterckeman, T.; Perdrix, E. Tracing Source Pollution in Soils Using Cadmium and Lead Isotopes. *Environ. Sci. Technol.* **2006,** *40* (8): 2525–2530.

Davidson, B.; Bradshaw, R. W. Thermal Pollution of Water Systems. *Env. Sci. Technol.* **1967,** *1* (8), 618–630.

Development Report. 2003. United Nations Educational, Scientific and Cultural

Di, Q.; Wang, Y.; Zanobetti, A.; Wang, Y.; Koutrakis, P.; Choirat, C.; Dominici, F.; Schwartz, J. D. Air Pollution and Mortality in the Medicare Population. *N. Engl. J. Med.* **2017,** *376* (26), 2513–2522.

Diep, F. California's Farms are an Even Larger Source of Air Pollution than We Thought. 2018 [Online]. https://psmag.com/environment/californias-farms-are-an-evenlarger-source-of-air pollution-than-we-thought.

Duruibe, J. O.; Ogwuegbu, M. O. C.; Egwurugwu, J. N. Heavy Metal Pollution and Human Biotoxic Effects. *Int. J. Phy. Sci.* **2007,** *2* (5), 112–118.

Dutov, O. I.; Yermolayev, M. M. Radioecological Aspects of the Use of Soils, Contaminated With Radionuclides. *Bull. Agra. Sci.* **2013** (2), 51–54.

Easterbrook, G. *A Moment on the Earth: The Coming Age of Environmental Optimism*; Viking: Bergenfield, NJ. 1995.

Ebersviller, S.; Lichtveld, K.; Sexton, K. G.; Zavala, J.; Lin, Y. H.; Jaspers, I.; Jeffries, H. E. Gaseous Vocs Rapidly Modify Particulate Matter and Its Biological Effects–Part 1: Simple Vocs and Model PM. *Atmos Chem. Phys. Discuss.* **2012,** *12* (24), 12277–12292.

Eisler, R. *Mercury Hazards to Living Organisms*; CRC/Taylor & Francis: Boca Raton, FL. 2006. p 312.

Emberson, L. D.; Pleijel, H.; Ainsworth, E. A.; Van Den Berg, M.; Ren, W.; Osborne, S.; Mills, G.; Pandey, D.; Dentener, F.; Büker, P.; Ewert, F. Ozone Effects on Crops and Consideration in Crop Models. *Eur. J. Agron.* **2018,** *100,* 19–34.

Engwa, A. G.; Ihekwoaba, C. J.; Ilo, U. S.; Unaegbu, M.; Ayuk, L. E.; Osuji, A. G. Determination of Some Soft Drink Constituents and Contamination by Some Heavy Metals in Nigeria. *Toxic. Rep.* **2015,** *2,* 384–390

Eshagberi, G. O. Toxic Effects of Heavy Metals on Crop Plants. *Multi. J. Empir. Res.* **2012,** 10.

FAO & WHO. Manual on Development and Use of FAO and WHO Specifications for Pesticides. Food and Agriculture Organization of the United Nations [Online] 2016. http://www.fao.org/agriculture/crops/thematic-sitemap/theme/pests/jmps/manual/en/

Farhat, A.; Mohammadzadeh, A.; Balali-Mood, M.; Aghajanpoor-Pasha, M.; Ravanshad, Y. Correlation of Blood Lead Level in Mothers and Exclusively Breastfed Infants: A Study on Infants Aged Less Than Six Months. *Asia Pac. J. Med. Toxicol.* **2013,** *2* (4), 150–152.

Faroon, O.; Ruiz, P. Polychlorinated Biphenyls: New Evidence from the Last Decade. *Toxico. Indust. Health* **2016,** *32* (11), 1825–1847.

Fensterstock, J. C.; Kurtzweg, J. A; Ozolins, G. Reduction of Air Pollution Potential through Environmental Planning. *J. Air Pollut. Control Assoc.* **1971,** *21* (7), 395–399.

Fischetti, M. Trees that Pollute. *Sci. Am.* **2014,** *310* (6), 14.

Flora, G.; Gupta, D.; Tiwari, A. Toxicity of Lead: A Review with Recent Updates. *Interdiscip. Toxicol.* **2012,** *5* (2), 47–58.

Frank, A. L.; Joshi, T. K. The Global Spread of Asbestos. *Ann. Glob. Health* **2014,** *80* (4), 257–262.

Galbraith, D.; Gross, S. A.; Paustenbach, D. Benzene and Human Health: A Historical Review and Appraisal of Associations with Various Diseases. *Crit. Rev. Toxicol.* **2010,** *40* (Suppl 2), 1–46.

Gehring, U.; Wijga, A. H.; Brauer, M.; Fischer, P.; De Jongste, J. C.; Kerkhof, M.; Oldenwening, M.; Smit, H. A.; Brunekreef, B. Traffic-Related Air Pollution and the Development of Asthma and Allergies During the First 8 Years of Life. *Am. J. Respir. Crit. Care Med.* **2010,** *181* (6), 596–603.

Geissen, V.; Mol, H.; Klumpp, E.; Umlauf, G.; Nadal, M.; Van Der Ploeg, M.; Van De Zee, S. E.; Ritsema, C. J. Emerging Pollutants in the Environment: A Challenge for Water Resource Management. *Int. Soil Water Conserv. Res.* **2015,** *3* (1), 57–65.

Ghani, A. Effect of Chromium Toxicity on Growth, Chlorophyll and Some Mineral Nutrients of Brassica Juncea L. Egyptian Acadmic. *J. Biol. Sci.* **2011,** *2* (1), 9–15.

Goldberg, M. A Systematic Review of the Relation between Long-Term Exposure to Ambient Air Pollution and Chronic Diseases. *Rev. Env. Health* **2008,** *23* (4), 243–298.

Goldstein, A. H.; Koven, C. D.; Heald, C. L.; Fung, I. Y. Biogenic Carbon and Anthropogenic Pollutants Combine to Form a Cooling Haze Over the Southeastern United States. *Proc. Natl Acad. Sci.* **2009,** *106* (22), 8835–8840.

Gurses, A.; Guneş, K.; Şahin, E. Removal of Dyes and Pigments from Industrial Effluents. In *Green Chemistry and Water Remediation: Research and Applications*; Elsevier, 2021, pp 135–187.

Handa, B. K. Contamination of Groundwaters by Phosphates. *Bhu-Jal News.* **1990,** *5,* 24–36.

Hanslmeier, A. Life and Water. In *Water in the Universe*; Springer: Dordrecht, 2011, pp 25–36.

Havugimana, E. R. N. E. S. T. E.; Bhople, B. S.; Kumar, A. N. I. L.; Byiringiro, E. M. M. A. N. U. E. L.; Mugabo, J. P.; Kumar, A. Soil Pollution–Major Sources and Types of Soil Pollutants. *Environ. Sci. Eng.* **2017,** *11,* 53–86.

Health Organization and United Nations Children's Fund; 2000. Available At: http://

Hime, N. J.; Marks, G. B.; Cowie, C. T. A Comparison of the Health Effects of Ambient Particulate Matter Air Pollution from Five Emission Sources. *Int. J. Environ. Res. Public Health* **2018,** *15* (6), 1206.

Hołtra, A.; Zamorska-Wojdyła, D. The Pollution Indices of Trace Elements in Soils and Plants Close to the Copper and Zinc Smelting Works in Poland's Lower Silesia. *Environ. Sci. Poll. Res.* 2020. *27* (14), 16086–16099.

Huang, Y.; Li, T.; Wu, C.; He, Z.; Japenga, J.; Deng, M.; Yang, X. An Integrated Approach to Assess Heavy Metal Source Apportionment in Peri-Urban Agricultural Soils. *J. Hazard. Materi.* **2015,** 299, 540–549.

Huy, T. B.;Tuyet-Hanh, T. T.; Johnston, R.; Nguyen-Viet, H. Assessing Health Risk Due to Exposure to Arsenic in Drinking Water in Hanam Province, Vietnam. *Inter. J. Environ. Res. Public Health.* **2014,** *11,* 7575–7591.

Int Panis, L.; Provost, E. B.; Cox, B.; Louwies, T.; Laeremans, M.; Standaert, A.; Dons, E.; Holmstock, L.; Nawrot, T.; De Boever, P. Short-Term Air Pollution Exposure Decreases Lung Function: A Repeated Measures Study in Healthy Adults. *Environ. Health* **2017,** *16* (1), 1–7.

Inyinbor, A. A.; Adebesin, B. O.; Oluyori, A. P.; Adelani-Akande, T. A.; Dada, A. O.; Oreofe Toyin, A. Water Pollution: Effects, Prevention, and Climatic Impact. In *Water Challenges of an Urbanizing World*; Vol. 33, 2018; pp 33–47.

Jain, R. *Environmental Impact of Mining and Mineral Processing: Management, Monitoring, and Auditing Strategies*; Butterworth-Heinemann, 2015.

Jayaswal, K.; Sahu, V.; Gurjar, B. R. Water Pollution, Human Health and Remediation. In *Water Remediation*; Springer: Singapore, 2018, pp 11–27.

Jung, C.; Son, A.; Her, N.; Zoh, K. D.; Cho, J.; Yoon, Y. Removal of Endocrine Disrupting Compounds, Pharmaceuticals, and Personal Care Products in Water Using Carbon Nanotubes: A Review. *J. Ind. Eng. Chem.* **2015,** *27,* 1–11.

Karri, R. R.; Ravindran, G.; Dehghani, M. H. Wastewater—Sources, Toxicity, and their Consequences to Human Health. In *Soft Computing Techniques in Solid Waste and Wastewater Management*; Elsevier, 2021, pp 3–33.

Kasonga, T. K.; Coetzee, M. A.; Kamika, I.; Ngole-Jeme, V. M.; Momba, M. N. B. Endocrine-Disruptive Chemicals as Contaminants of Emerging Concern in Wastewater and Surface Water: A Review. *J Environ. Manage.* **2021,** *277,* 111485.

Kelishadi, R.; Poursafa, P. Air Pollution and Non-Respiratory Health Hazards for Children. *Arch Med Sci.* **2010,** *6* (4), 483–495.

Keller, E. A. *Environmental Geology*; Merrill: Columbus, OH, 1988.

Khatymov, R. V.; Muftakhov, M. V.; Mazunov, V. A. Phenol, Chlorobenzene and Chlorophenol Isomers: Resonant States and Dissociative Electron Attachment. *Rapid Commun. Mass Spectrom.* **2003,** *17* (20), 2327–2336.

Kurnia, U. S.; Sutono, N.; Markus, A.; Sulaeman, A. M.; Kurniawansyah, S. H. *Pengkajian Baku Mutu Tanah Pada Lahan Pertaniah. Laporan Akhir Kerjasama Penelitian Bapedal–Puslitbangtanak*, 2000.

Landrigan, P. J. Air Pollution and Health. *Lancet Public Health* **2017,** *2* (1), E4–E5.

Landrigan, P. J.; Fuller, R.; Acosta, N. J. R.; Adeyi, O.; Arnold, R.; Basu, N. N. The Lancet Commission on Pollution and Health. *Lancet* **2017,** *391* (10119), 462–512.

Landrigan, P. J.; Fuller, R.; Fisher, S.; Suk, W. A.; Sly, P.; Chiles, T. C. Pollution and Children's Health. Sci. *Total Environ.* **2019,** *650* (Part 2), 2389–2394.

Lawal, A.T. Polycyclic Aromatic Hydrocarbons. A Review. *Cogent Environ. Sci.* **2017,** *3* (1)

Leo, A.; Hansch, C.; Elkins, D. Partition Coefficients and their Uses. *Chem. Rev.* **1971,** *71* (6): 525–616.

Li, C.; Busquets, R.; Campos, L. C. Assessment of Microplastics in Freshwater Systems: A Review. *Sci. Total Environ.* **2020,** 707, 135578.
Li, W., Li, Z.; Jennings, A. Regulatory Performance Dataset Constructed from U.S. Soil Jurisdictions Based on the Top 100 Concerned Pollutants. *Data Brief* **2018,** *21,* 36–49.
Litvinov, I. V.; Pavlova, O. Y.; Yuzmukhamedov, D. N.; Lavrishchev, A. V. The Migration Capacity of Stable Strontium in Sod-Podzolic Soils of the Russian Northwest. *Eur. Soil Sci.* **2008** (5), 568–575.
Liu, C.; Chen, R.; Sera, F.; Vicedo-Cabrera, A. M.; Guo, Y.; Tong, S.; Coelho, M. S.; Saldiva, P. H.; Lavigne, E.; Matus, P.; Valdes Ortega, N. Ambient Particulate Air Pollution and Daily Mortality in 652 Cities. *N. Engl. J. Med.* **2019,** *381* (8), 705–715.
Liu, G.; Liu, H. Using Insulation in China's Buildings: Potential for Significant Energy Savings and Carbon Emission Reductions. *Low. Carbon Econ.* **2011,** *2* (4), 220–223.
Luo, J.; Zhang, Q.; Cao, J.; Fang, F.; Feng, Q. Importance of Monitor and Control on New-Emerging Pollutants in Conventional Wastewater Treatment Plants. *J. Geosc. Environ. Prot.* **2018,** *6* (01), 55.
Madhav, S.; Ahamad, A.; Singh, A. K.; Kushawaha, J.; Chauhan, J. S., Sharma, S.; Singh, P. Water Pollutants: Sources and Impact on the Environment and Human Health. In *Sensors in Water Pollutants Monitoring: Role of Material*; 2020, pp 43–62.
Maragkidou, A. Exposure to Coarse Particles and Floor Dust Biological and Chemical Contamination Inside Jordanian Indoor Environments. Ph.D. Thesis, University of Helsinki, Helsinki, 2018, http: //hdl.handle.net/10138/241380.
Martin, S.; Griswold, W. Human Health Effects of Heavy Metals. *Environ. Sci. Technol. Briefs Citizens.* **2009,** *15,* 1–6.
Measures. *Environ. Prob. Plant.* **2016,** *3,* 44–56.
Miller, G. T. Environmental Science: Working With the Earth, Fifth Ed.1997.
Molinari, M.; Stevenson, J. Malignant Mesothelioma Cancer: Stages, Prognosis, Treatment. 2020.
Mondal, P. Pollutants: Definition and Classification of Pollutants [Online] 2018. http://www.yourarticlelibrary.com/essay/pollutants-definition-and-classification-of-pollutants/28277.
Mudgal, V.; Madaan, N.; Mudgal, A.; Singh, R. B.; Mishra, S. Effect of Toxic Metals on Human Health. *Open Nutraceut. J.* **2010,** *3,* 94–99.
Muralikrishna, I. V.; Manickam, V. *Analytical Methods for Monitoring Environmental Pollution. Environmental Management*; Butterworth Heinemann: Elsevier 2017, pp 495–570.
National Research Council (US) Committee on Pyrene and Selected Analogues. *Lists of Polycyclic Aromatic Hydrocarbons*; National Academies Press (US), 1983.
Naushad, M. *A New Generation Material Graphene: Applications in Water Technology*; Springer International Publishing: Cham, 2019; pp 21–42.
Nelson-Smith, A. The Problem of Oil Pollution of the Sea. *Adv. Mar. Biol.* **1971,** *8,* 215–306.
Okrent, D. On Intergenerational Equity and Its Clash with Intragenerational Equity and on the Need for Policies to Guide the Regulation of Disposal of Wastes and Other Activities Posing Very Long-Time Risks. *Risk Anal.* **1999,** *19,* 877–901.
Oluwatuyi, O. E.; Ajibade, F. O.; Ajibade, T. F.; Adelodun, B.; Olowoselu, A. S.; Adewumi, J. R. Total Concentration, Contamination Status and Distribution of Elements in a Nigerian State Dumpsites Soil. *Environ. Sustain. Indic.* **2020,** 100021.
Organization (UNESCO) and Berghahn Books.

Owa, F. D. Water Pollution: Sources, Effects, Control and Management. *Mediterr. J. Soc. Sci.* **2013,** *4* (8), 65–65.

Pratiti, R.; Vadala, D.; Kalynych, Z.; Sud, P. Health Effects of Household Air Pollution Related to Biomass Cook Stoves in Resource Limited Countries and Its Mitigation by Improved Cookstoves. *Environ. Res.* **2020,** *186,* 109574.

Rahman, Z.; Singh, V. The Relative Impact of Toxic Heavy Metals (THMs) (Arsenic (As), Cadmium (Cd), Chromium (Cr)(VI), Mercury (Hg), and Lead (Pb)) on the Total Environment: An Overview. *Environ. Monit. Assess.* **2019,** *191,* 419.

Ramirez-Malule, H.; Quinones-Murillo, D. H.; Manotas-Duque, D. Emerging Contaminants as Global Environmental Hazards. A Bibliometric Analysis. *Emerg. Contam.* **2020,** *6,* 179–193.

Rao, N. S.; Prasad, P. R. Phosphate Pollution in the Groundwater of Lower Vamsadhara River Basin, India. *Environ. Geol.* **1997,** *31* (1), 117–122.

Rochman, C. M.; Hoh, E.; Hentschel, B. T.; Kaye, S. Long-Term Field Measurement of Sorption of Organic Contaminants to Five Types of Plastic Pellets: Implications for Plastic Marine Debris. *Environ. Sci. Technol.* **2013,** *47* (3), 1646–1654.

Rodríguez Eugenio, N.; Mclaughlin, M. J.; Pennock, D. *Soil Pollution: A Hidden Reality*. Food and Agriculture Organization of the United Nations: Rome, Italy. 2018, p 156.

Shakouri, B.; Yazdi, S. K. Environment and Water Pollution. *Adv. Environ. Biol.* **2014,** 1328–1333.

Sheehan, M. C.; Lam, J.; Navas-Acien, A.; Chang, H. H. Ambient Air Pollution Epidemiology Systematic Review and Meta-Analysis: A Review of Reporting and Methods Practice. *Environ. Int.* **2016,** *92,* 647–656.

Shekhawat, K.; Chatterjee, S.; Joshi, B. Chromium Toxicity and Its Health Hazards. *Int. J. Adv. Res.* **2015,** *7* (3), 167–172.

Singh, A. L. *Nitrate and Phosphate Contamination in Water and Possible Remedial.*

Singh, H.; Bhardwaj, N.; Arya, S. K.; Khatri, M. Environmental Impacts of Oil Spills and their Remediation by Magnetic Nanomaterials. *Environ. Nanotechnol. Monit. Manag.* **2020,** *14,* 100305.

Singh, L.; Varshney, J. G.; Agarwal, T. Polycyclic Aromatic Hydrocarbons' Formation and Occurrence in Processed Food. *Food Chem.* **2016,** *199,* 768–781.

Spellman, F. R. Soil Pollution. In *The Science of Environmental Pollution*, 3rd ed.; Spellman, F. R., Ed.; NW: CRC Press, Taylor & Francis Group: Boca Raton, FL, London, New York, 2017; pp 335–361.

Suk, W. A.; Ahanchian, H.; Asante, K. A.; Carpenter, D. O.; Diaz-Barriga, F.; Ha, E. H. Environmental Pollution: An Underrecognized Threat to Children's Health, Especially in Low- and Middle-Income Countries. *Environ. Health Perspect.* **2016,** *124* (3), A41-A45.

Sulaymon, I. D.; Mei, X.; Yang, S.; Chen, S.; Zhang, Y.; Hopke, P. K. PM 2.5 in Abuja, Nigeria: Chemical Characterization, Source Apportionment, Temporal Variations, Transport Pathways and the Health Risks Assessment. *Atmos. Res.* **2020,** *237,* 104833.

Taneja, P.; Labhasetwar, P.; Nagarnaik, P. Nitrate in Drinking Water and Vegetables: Intake and Risk Assessment in Rural and Urban Areas of Nagpur and Bhandara Districts of India. *Environ. Sci. Pollut. Res.* **2019,** *26* (3), 2026–2037.

Thiele, J. J.; Traber, M. G.; Tsang, K.; Cross, C. E.; Packer, L. In Vivo Exposure to Ozone Depletes Vitamins C and E and Induces Lipid Peroxidation in Epidermal Layers of Murine Skin. *Free Radic. Biol. Med.* **1997,** *23* (3), 385–391.

Thongam, D. D.; Chaturvedi, H. Nanomaterials for Climate Change and Water Pollution Mitigation. In *Water Conservation in the Era of Global Climate Change*; Elsevier, 2021; pp 277–314.

Tiffany, M. A.; Winchester, J. W.; Loucks, R. H. Natural and Pollution Sources of Iodine, Bromine, and Chlorine in the Great Lakes. *J. Water Pollut. Control Fed.* **1969,** 1319–1329.

Ukaogo, P. O.; Ewuzie, U.; Onwuka, C. V. Environmental Pollution: Causes, Effects, and the Remedies. In *Microorganisms for Sustainable Environment and Health*; Elsevier, **2020;** pp 419–429.

Ullah, R.; Malik, R. N.; Qadir, A. Assessment of Groundwater Contamination in an Industrial City, Sialkot, Pakistan. *Afr. J. Environ. Sci. Technol.* **2009,** *3* (12).

Unaegbu, M.; Engwa, G. A.; Abaa, Q. D.; Aliozo, S. O.; Ayuk, E. L.; Osuji, G. A. Heavy Metal, Nutrient and Antioxidant Status of Selected Fruit Samples Sold in Enugu, Nigeria. *Int. J. Food Contam.* **2016,** *3* (7), 1–8.

UNESCO. *Water for People Water for Life*. The United Nations World Water.

Vaidyanathan, G. The Worst Climate Pollution is Carbon Dioxide [Online], 2014. https://www.scientificamerican.com/article/the-worst-climate pollution is carbon dioxide/.

Vallero, D. *Fundamentals of Air Pollution*, 4th ed.; Academic Press, 2007; p 968.

Verma, P.; Ratan, J. K. Assessment of the Negative Effects of Various Inorganic Water Pollutants on the Biosphere—An Overview. In *Inorganic Pollutants in Water;* 2020, pp 73–96.

Verones, F.; Hanafiah, M. M.; Pfister, S.; Huijbregts, M. A.; Pelletier, G. J.; Koehler, A. Characterization Factors for Thermal Pollution in Freshwater Aquatic Environments. Environ. *Sci. Technol.* **2010,** *44* (24), 9364–9369.

Vieira, W. T.; de Farias, M. B.; Spaolonzi, M. P.; da Silva, M. G. C.; Vieira, M. G. A. Removal of Endocrine Disruptors in Waters by Adsorption, Membrane Filtration and Biodegradation. A Review. *Environ. Chem. Lett.* **2020,** *18* (4), 1113–1143.

Wadsworth, Belmont, CA. Miller, G. T. *Environmental Science*, 10th ed.; 2004.

Wakchaure, R.; Ganguly, S.; Praveen, P. K. Role of Water in Livestock. *Rec. Adv. Acad. Sc. J.* **2015,** *1,* 53–56.

Wani, A. L.; Ara, A.; Usmani, J. A. Lead Toxicity: A Review. *Interdis. Toxic.* **2015,** *8* (2):55–64.

Wasewar, K. L.; Singh, S.; Kansal, S. K. Process Intensification of Treatment of Inorganic Water Pollutants. In *Inorganic Pollutants in Water*; Elsevier, 2020; pp 245–271.

Waters, T. F. Sediment in Streams. Sources, Biological Effects, and Control. *Am. Fish. Soc. Monogr.* **1995**.

Webster.Com. Definition from Webster Dictionary. 2010, 08-13 Retrieved 2010-08-26.

WHO & FAO. Global Situation of Pesticide Management in Agriculture and Public Health. *Report of a 2018 WHO–FAO Survey*. World Health Organization and Food and Agriculture Organization of the United Nations: Geneva, Rome. 2019, p 73.

WHO (World Health Organization). Dioxins and their Effects on Human Health [Online]. https://www.who.int/news-room/fact-sheets/detail/dioxins-and-their-effects-onhuman-health (accessed Oct 5, 2019).

WHO (World Health Organization). World Health Organization Releases New Global Air Pollution Data [Online], 2018. Geneva, Switzerland. http://www.ccacoalition.org/en/n ews/world-health-organization-releases-new-global-air-pollution-data

WHO. *Trace Elements in Human Nutrition and Health*; Geneva, World Health Organization, 1996 p 343.

WHO, UNISEF. *Global Water Supply and Sanitation Assessment 2000 Report*. USA: World.

WHO. 7 Million Premature Deaths Annually Linked to Air Pollution [Online]. 2014. https://www.who.int/mediacentre/news/releases/2014/air-pollution/en/.

WHO. Health Risks of Ozone from Long-Range Transboundary Air Pollution [Online]. 2019. http://www.euro.who.int/data/assets/pdf_file/0005/78647/e91843.pdf

WHO. Reducing Risks, Promoting Healthy Life[Online]. 2002. https://www.who.int/whr/2002/en/.

WHO: WHO Global Air Quality Guidelines. Particulate Matter (PM2.5 and PM10), Ozone, Nitrogen Dioxide, Sulfur Dioxide and Carbon Monoxide [Online], 2021. ISBN 978-92-4-003422–8 (Electronic Version), https://apps.who.int/iris/handle/10665/345329

Wilson, W. E.; Suh, H. H. Fine Particles and Coarse Particles: Concentration Relationships Relevant to Epidemiologic Studies. *J. Air Waste Manag. Assoc.* **1997,** *47* (12), 1238–1249.

Wolfe, A. H.; Patz, J. A. Reactive Nitrogen and Human Health: Acute and Long-Term Implications. *Ambio: J. Human Environ.* **2002,** *31* (2), 120–125.

www.who.int/water_sanitation_health/monitoring/jmp2000.pdf

Zhai, Y.; Zhao, X.; Teng, Y.; Li, X.; Zhang, J.; Wu, J.; Zuo, R. Groundwater Nitrate Pollution and Human Health Risk Assessment by Using HHRA Model in an Agricultural Area, NE China. Ecotoxicol. *Eenviron. Saf.* **2017,** *137,* 130–142.

Zhang, L.; Yang, Y.; Li, Y.; Qian, Z. M.; Xiao, W.; Wang, X.; Rolling, C. A.; Liu, E.; Xiao, J.; Zeng, W.; Liu, T. Short-Term and Long-Term Effects of PM2. 5 on Acute Nasopharyngitis in 10 Communities of Guangdong, China. *Sci. Total Environ.* **2019,** *688,* 136–142.

CHAPTER 2

Nano-Phytoremediation of Environmental Pollution

AHMED ALI ALI ROMEH

Plant Production Department, Faculty of Technology and Development, Zagazig University, Zagazig, Egypt

ABSTRACT

Numerous contaminants contribute to environmental pollution resulting from global industrialization or human activities in the global environment and ecosystem, for instance, persistent organic compounds and heavy metals. These harm the ecosystem, wildlife, and human health around the world. As a result, rehabilitation of polluted environments is a principle global challenge that must be addressed as soon as possible. To deal with this problem, effective remediation procedures are needed, including nano-phytoremediation. Phytoremediation, supported by nanomaterials, is a unique technology to clean the ecosystem in an economical and long-term manner. Therefore, it's no surprise that nanotechnology-assisted phytoremediation is attracting academics' attention as they research ways to purify contaminated land, water, and air. This is a new field with a promising technique for sustainable development. By eliminating pollutants directly from the ecosystem, encouraging plant development, and boosting phyto-availability of pollutants, nanomaterials can make phytoremediation systems more successful. The utilization of hyperaccumulating plants together with appropriate nanomaterials as adsorbent materials may be a favorable strategy for removing hazardous metal (HMs) from polluted soil. Thanks to its high specific surface area and great photoelectric properties, nanomaterials participate a key role in the removal of heavy metals via several mechanisms. Organic

Sustainable Nanoremediation: Modern Technologies for a Clean Environment.
Fernanda Maria Policarpo Tonelli, Arpita Roy, and Khalid Rehman Hakeem (Eds.)

and inorganic nanoparticles (NPs) are the two chemical classifications of NPs used in nano-phytoremediation to remediate environments contaminated by heavy metals. NPs such as TiO_2-NPs, nZVI (nano zerovalent iron), SiO_2-NPs, Ag NPs, and carbon nanotubes (CNTs) have little or no effect on higher plants at low concentrations, indicating the possibility of phytoremediation of these chemicals in the ecosystem. Nano-mycorrhiza-assisted phytoremediation (NMPR) technology has enhanced the phytoremediation effectiveness of bioenergy plants growing in heavy metal-contaminated soil in a synergistic manner. Combining microbiological, nanotechnological, and transgenic plants has the potential to pave the way for the elimination of metals from contaminated locations. Nanomaterial-facilitated phytoremediation of organic contaminants by eliminating pollutants directly, encouraging plant growth, and enhancing the phyto-availability.

2.1 INTRODUCTION

Pollutants cause ecosystem pollution in numerous ways resulting from global industrialization or human activities within the global environment and ecosystem like heavy metals and different metalloids, agrochemical residues, organic compounds that are persistent, especially polychlorinated biphenyl (PCBs), phenols, aromatic polycyclic hydrocarbons, phthalates, dioxins (polychlorinated dibenzodioxins and dibenzofurans), organochlorine pesticides, 2, 4, 6-trinitrotoluene (TNT), radioactive wastes, other biological wastes, and hazardous by-products of industrial (Midhat et al., 2019; Ghazaryan et al., 2020; Anerao et al., 2022). These chemical pollutants are internationally significant because they have the potential to slowly transfer environmental persistence, bioaccumulation, and biomagnification efficacy across the food chain. These harm the ecosystem, wildlife, and health of humans around the world (Anerao et al., 2022). As a result, rehabilitation of polluted environments could be a principle global challenge that has to be addressed as soon as possible. Conventional approaches, like physico-chemical and microorganism-based processes, are expensive and less effective at removing harmful compounds from the ecosystem (Zekic et al., 2018).

To deal with this problem, effective remediation procedures are needed, including nano-phytoremediation, plants, and related microorganisms from rhizosphere macrosystems. Furthermore, genetically modified plants are utilized in phytoremediation to eliminate pollution from various ecosystems like soil, water, and air through many physical and chemical processes such as accumulation, adsorption, transport, and chemical reactions, particularly enzymatic reactions, photocatalysis, and filtration (Das et al., 2022). Furthermore, high-cost technologies for eliminating heavy metals, such as carbon adsorption, air stripping, and biosorption, are being used, but the importance of plants cannot be overlooked because they help to remove heavy metal contamination without harming the environment (Singh et al., 2022). Consequently, phytoremediation is a crucial environmental technology in reducing pollutants, especially with nanomaterials, which is called nano-phytoremediation. Nanomaterials are cost-efficient and ecologically beneficial technology that have a lesser hazardous impact on the environment and better microbial activity on specific wastes and toxic materials (Das et al., 2022). The new technology of using phytoremediation supported by nanomaterials is a unique technology to clean the ecosystem in an economical and long-term manner. The fundamental principle of this technology is that phytoremediation with nanotechnology is an environmentally friendly and efficient two-step process. To begin, the nanomaterials used for adsorbing pollutants, particularly lipophilic pollutants, and break them down into more hydrophilic metabolites, enhancing their phyto-availability and promoting plant growth. Plant roots acquire the degraded products, which are eventually translocated by plant leaves (Mustafa et al., 2022). Nano-phyto-remediation is a newly emerging technique for pollution remediation employing biosynthesized nanoparticles (NPs) and is gaining popularity in the world of nanotechnology. For this, phytoremediation strategies using plants with nanotechnology to remove dangerous organic substances and harmful metals from polluted sites to clean the environment and understanding approaches to enhance the efficacy of this process are extremely important. Future research should specialize in boosting the rate of nano-enabled phytoremediation processes. To increase effectiveness even further, nano-phytoremediation can combine agronomic management, chemical/biological additive treatment, inoculation with microorganisms within the rhizosphere area, and genetic engineering (Hou et al., 2019).

2.2 NANO-PHYTOREMEDIATION OF HEAVY METALS

2.2.1 INTEGRATED NANOMATERIALS-PHYTOREMEDIATION TECHNIQUES FOR THE REMOVAL OF DANGEROUS METAL

The ecosystem and the health of the biota are both at risk because of metals building up in diverse environmental compartments. Arable soil dangerous metals pollution is becoming more of a worldwide concern, posing a major risk to food safety, human health, the food chain, and the ecosystem (da

Conceição Gomes et al., 2016; Zamani et al., 2020). Over 10 million sites covering more than 20 million ha of land worldwide are classified as soil polluted, with >50% of them contaminated with hazardous metals (HMs) at a cost of more than US $10 billion per year (He et al., 2015; Kumar et al., 2019). Over the years, a variety of soil cleanup procedures have been used, being classified as physical, chemical, and biological approaches. It is widely acknowledged that traditional metal-contaminate soil treatment has significant limits (da Conceição Gomes et al., 2016). New and improved solutions are needed to clean up mineral-contaminated soils, as well as bioremediation techniques like phytoremediation and bioremediation that are considered the best because they are natural, ecological strategies that do not have a negative impact on the environment (Ullah et al., 2015). The effectiveness of the remediation procedure is determined by the types of metal contamination, amounts of contamination, nanomaterials employed, and plant species involved. Green remediation, also known as phytoremediation, is a new method of extracting, removing, filtering, volatilizing, degrading, or stabilizing xenobiotics waste from mines, contaminated soils, sediments, and resources for water-using plants (Moameri and Abbasi Khalaki, 2019). As a result, the environment can use phytoremediation to lower high levels of different contaminants, including organic chemicals and metals (da Conceição Gomes et al., 2016). Pollutant removal methods in phytoremediation are complicated, and they are not limited to the direct metabolism of contaminants by plants (Halder and Ghosh, 2014; Zand and Tabrizi, 2021). Indirect mechanisms like the metabolism of pollutants by microorganisms associated to the plant within the rhizosphere and/or changes in the contaminated environment caused by plants are also involved in phytoremediation (Zand and Tabrizi, 2021). Although there are more than 400 plants capable of phytoremediation of dangerous heavy metals from the ecosystem, it frequently takes a long time to complete this task (Zhu et al., 2019). Attempts are still ongoing to increase the effectiveness of phytoremediation and minimize plant growth inhibition and oxidative stress, including agronomic management, chemical additive treatment, inoculation of rhizospheric bacteria, genetically modified plants, and the use of nanotechnology to increase phytoremediation capabilities, which has received more attention in this regard (Alia et al., 2017; da Conceição Gomes et al., 2016; Gerhardt et al., 2017; Shirani Bidabadi, 2020). A successful method for creating better phytoremediators is to introduce unique features for the uptake and accumulation of contaminants into high biomass plants (Buhari et al., 2016). A more practical way to remove heavy metals is by combining phyto-nanotechnology. The employment of nanotechnology alongside phytoremediation is a revolutionary strategy for developing and implementation of cleaning methods and innovations that will lead to better, less-expensive, and faster site cleanups. Unsurprisingly, nanotechnology-assisted phytoremediation is attracting academics' attention as they try to figure out how to purify polluted air, water, and soil (Song et al., 2019). Utilizing both nanotechnology and phytoremediation has given heavy metals remediation new dimensions. Technologies for nano-phytoremediation were developed, with one based on sophisticated nanotechnology and the other on the well-established subject of phytoremediation. This is a new field with a promising technique for sustainable improvement based on better phytoremediation earlier techniques (Bhati and Rai 2018). The use of hyperaccumulating plants together with appropriate nanomaterials as adsorbent materials may be a promising strategy for removing HMs from polluted soil (Rajput et al., 2022).

Plant metabolism is influenced by NPs because they provide micronutrients, altering genes, as well as interfere with a variety of oxidative processes in plants, which can lead to an oxidative burst, as illustrated in Figure 2.1 (Jiamjitrpanich et al., 2013). Studies on accumulation and translocation were done to find out how the NP affected plant growth/development, such as shoots and roots, as well as the uptake of dangerous metals. NPS can be used to treat various polluted media by altering pollutants synthetically or as a super adsorbent (Liu and Lal, 2015). The size and movement of NPs within the tissue of plants, through intercellular spaces (apoplast), or through plasmodesmata region in the cell (symplast) depends on the physico-chemical and biological ambient factors (Srivastav et al., 2018). Due to their large size and large surface area-to-unit mass ratio, short intra-particle diffusion distance, magnetic properties, surface modifiability, biocompatibility, tunable surface chemistry, increased adsorption sites, reusability, increased porosity, higher gas permeability, ease of separation, enhanced catalytic activity, greater dispersion degree, and comparatively low manufacturing cost, NMs or NPs have got a lot of attention (Ali and Khan, 2018). Since the nanomaterial has a large specific surface area and excellent photoelectric characteristics, it can be used to remove heavy metals physical adsorption through intermolecular forces (such as van der Waals force, electrostatic attraction, and chemical adsorption (chemisorption) efficiency) through electron transfer or electron pair sharing to form chemical bonds or surface coordination compounds (Ding et al., 2021). Functional groups which include hydroxyl, amino, and carboxyl groups are frequently found on the surface of nanomaterials that can chemically adsorb heavy metal ions. CNTs, for example, primarily fix heavy metal ions on their surfaces via intermolecular forces such as van der Waals force and electrostatic attraction (Ding et al., 2021). Heavy metals are removed by adsorption and redox reaction, the most studied being NZVI, which is considered a good reducing agent (Ding et al., 2021). Furthermore, co-precipitation by nanomaterials especially nZVI can develop heavy metal removal (Liu et al., 2015; Ding et al., 2021). Phytoremediation, which supports nanotechnology in environmental protection, has focused primarily on the development of new materials for the restoration of heavy metals in soil, like nanomaterials, nanometal compounds, carboxymethyl cellulose (CMC) stable nanometal compounds, and composite nanomaterials (Chen et al., 2021). Studies have found that the presence of NPs, which include TiO_2-NPs, nZVI (nano zero-valent iron), SiO_2-NPs, Ag NPs, and carbon nanotubes (CNTs) at low concentrations has little or no effect on higher plants, indicating the possibility of phytoremediation of these chemicals within the environment (Rastogi et al., 2017). Multiple techniques like combination of nano with the degradation of pollutants by plants (nano-phytodegradation), combination of nano with the stabilization process by plants (nano-phytostabilization), and combination of nano with the accumulation process by plants (nano-phytoaccumulation) occurs simultaneously within plant (Nwadinigwe and Ugwu, 2018; Ghazaryan et al., 2019). Organic NPs and inorganic NPs are the two chemical classifications for NPs. Carbon NPs (fullerenes) are among the former, whereas magnetic NPs, metallic element NPs (such as gold and silver), and semiconductor NPs are among the latter (e.g., titanium dioxide and zinc oxide) (Rastogi et al. 2017).

2.2.2 APPLICATIONS OF NANOMATERIAL-ASSISTED PHYTOREMEDIATION FOR REMOVAL OF DANGEROUS METALS

2.2.2.1 INORGANIC NANOMATERIALS

NMs have a dose-dependent toxicity; at high concentrations; they hinder growth and have several other deleterious effects, whereas at lower concentrations, they can increase the phyto-availability of pollutants, remove pollutants directly, as well as increase the efficiency of phytoremediation systems (Hou et al., 2019). Table 2.1 provides some illustrations of phytoremediation with nanomaterial assistance. Applying nanomaterials could improve phytoavailability of soil contaminated with heavy metals, especially Cd, CHR, Pb, Ni, and Zn (Tripathi, et al., 2015; Khan and Bano, 2016; Liang, et al., 2017; Vítková et al., 2018; Mustafa et al., 2022).

2.2.2.1.1 TiO_2-NPs

The presence of titanium dioxide NPs (TiO_2-NPs) contributes to microbial processes considered important in plant productivity and biogeochemical strategies (Simonin et al., 2016). Nano-TiO_2 boosted the dry weight of plant, photosynthetic rate, chlorophyll production, and the activity of several enzymes such as rubisco carboxylase activity with no notable harmful effects (Zhang et al., 2005; Doshi et al., 2008). In *Glycine max*, TiO_2 NPs spiking in soil at 100–300 mg/kg resulted in Cd accumulation in shoots of 1.9–2.6 folds and roots of 2.5 and 3.3 folds (Singh and Lee, 2016). Also, Goutam et al. (2018) showed that nano-TiO_2 with leaf extract of *Jatropha curcas* L. caused photocatalytic degradation for chemical oxygen demand (COD) and chromium (Cr). Cibane (2021) suggested that seed treated with nano-TiO_2 before sowing could improve metal uptake selectivity in phytoremediation. By increasing photosynthetic rate and plant growth in soybeans, TiO_2-NPs promoted Cd accumulation from polluted soils and reduced Cd toxicity (Singh and Lee, 2016).

2.2.2.1.2 nZVI-NPs

Because of its outstanding technical applications in detoxifying contaminated soil and groundwater, nanoscale Fe^0 is one of the most-researched nanomaterials for facilitating phytoremediation (Song et al., 2019). The usage of nZVI as a practical and promising technique for heavy metals-polluted soil remediation has been proposed. According to Fajardo et al. (2019), nZVI has 5% success in removing Pb and Cd from contaminated soil. Cucumber, wheat, and tomato plants have all been shown to benefit from Fe NPs' favorable effects on reducing the plant toxicity of HMs (Konate et al., 2017; Rahmatizadeh et al., 2019). Plants treated with low levels of nZVI (100–500 mg/kg) increased Pb accumulation in *Lolium perenne* to 1175.40 g per pot at one hundred mg/kg and improved Pb uptake in ragweed, *Kochia scoparia*, up to 857.18 g per pot at 500 mg/kg (Huang et al., 2018; Zand and Tabrizi, 2021). The rationale for this is often because of the improvement of the growth characteristics of the plant at lower levels of applied nZVI, whereas greater rates of nZVI caused nanotoxicity (Brasili et al., 2020). Also, *Sorghum bicolor* L. has exhibited uptake of Cd, Pb, and Zn from acidic soil using arbuscular mycorrhizal (AM) fungus, *A. mellea*, and nZVI (Cheng et al., 2021). *Brassica napus* with organic modification and nZVI greatly increased the effectiveness and practical diversity of the soil microbial community, according to Lacalle et al.

(2018). Increased activity of soil enzyme and reduced toxicity of microbial were also observed in HMs-contaminated soil that was remediated with 1% nZVI (Kim et al., 2014). Wheat plants' uptake of iron oxide NMs reduced the hazardous of Cd (II) and Cr (VI) in the crops (Lopez-Luna et al., 2016). Although the utilization of Fe^0 substantially improved growth of plant and inspired plant defense mechanisms in reaction to HMs toxicity by reducing malondialdehyde content, H2O2 generation, and electrolyte leakage while increasing reduced relative growth rate, chlorophyll fluorescence, and total antioxidative capacity (the phytoremediation capabilities of Salvia masterpieces were negatively affected using Fe^0 because of its reduced antioxidant capacity (Shirani Bidabadi, 2020). The Cd accumulation within the roots, stems, and leaves of ramie seedlings was raised by 16%, 50%, and 31%, 73%, respectively, through the supplement starch-stabilized nZVI (100, 500, and 1000 mg/kg) (Gong et al., 2017b). According to Vtkova et al. (2018), adding Fe^0 particles had a favorable impact on the stability of arsenic in sunflower rhizosphere.

2.2.2.1.3 *SiO_2-NPs*

Nano-SiO_2 in soil polluted by four heavy metals, Cd, Pb, Cu, and Zn, was shown by Wang et al. (2016) to enhance rice growth, improve yield, and reduce heavy metal accumulation in rice, particularly cadmium in grain. Also, Nano-SiO_2 increased resistance of seed coat and nutrition accessibility in maize, *Zea mays*, according to Suriyaprabha et al. (2012). On the other hand, the highest levels of Pb accumulation by *Secale montanum* roots were found in pots supplemented with nano silica 500 (Moameri and Abbasi Khalaki, 2019).

2.2.2.1.4 *Ag NPs*

Hyperaccumulator plants with suitable NPs, which includes Ag NPs, might be a possible potential strategy for contaminated soil repair. The accumulation and uptake of heavy metals like Ni, Pb, Na, Zn, and Cu can effectively minimize by Ag NPs when maximize gibberellin, abscisic, and PGPR in the leaves of *zea mays* (Khan and Bano, 2016). Through seedlings of *Brassica juncea*, silver NPs establish an invitro state (Shekhawat and Arya, 2009). The electrochemical reduction potential of silver NPs is strong, and they have good catalytic characteristics (Chaudhary et al., 2021).

2.2.2.2 ORGANIC NANOMATERIALS

2.2.2.2.1 Multi-Walled Carbon Nanotubes (MWCNTs)

The interactions among NMs and plants are anticipated to affect heavy metal uptake in plants in addition to plant responses to metal-induced toxicity. The augmentation of metal accumulation and reduction of metal toxicity caused by CNTs will help with phytoremediation of metal-contaminated locations. Several factors affect the HMs removal efficacy which includes pH, temperature, contact time, and CNT dose (Rajput et al., 2022). Multi-walled carbon nanotubes (MWCNTs) at the concentration of 500 mg kg^{-1} improved the buildup and uptake of Cd in ramie seedlings, *Boehmeria nivea*, and reduced Cd-induced toxicity. They did this by reducing oxidative stress, increasing specific antioxidant content, activating antioxidant enzymes, and stimulating plant growth. These findings showed that using MWCNTs at the right concentrations could boost phytoremediation efficacy in dangerous metal contaminated river sediment restoration (Gong et al., 2019). MWCNTs at 10 mg L^{-1} increased the buildup of Cd and Pb in *Cannabis sativa* L, according to Oloumi et al. (2018). Cd may be translocated through plant tissues by MWCNT transporters and translocation channels, according to the theory (Gogos et al., 2016). Metal accumulation in plants is primarily determined by two factors: metal concentration and plant biomass (Gong et al., 2019). The utilization of nano-graphene oxide (GO-NP) in the removal of dangerous heavy metals from soil to improve As uptake in wheat plants has been explored. By converting AsO_4^{3-} to AsO_3^{3-} and thus reducing oxidative stress in seedlings of wheat, this process can improve the effected biomass of plant, root branches exposed to toxins, disrupting the urea cycle, and fat metabolism, in comparison to heavy metal exposure without remediation alone (Hu et al., 2014). According to Cheng et al. (2014), surface-modified nanoscale carbon black (MCB) might be used to remediate Cu and Zn-polluted soils. Shoot biomass of ryegrass and roots rose as the quantity of surface-MCB added increased. Cu and Zn buildup in ryegrass shoots in soil was dramatically reduced as the amount of MCB added increased. NPs of MCB significantly boosted *Suaeda salsa* plant growth while lowering Cd and Ni uptake (Cheng et al., 2019). One of the best alternatives is to use multiwall CNTs since they have a larger metal-ion sorption capacity (3–4 times) than powder and granular activated carbon (Yano et al., 2018).

2.2.2.2.2 Nano-Hydroxyapatite (NHAP)

The use of nano-hydroxyapatite (NHAP) in the soil lowered the bioavailability of lead. In addition, NHAP gained the fresh weight of shoots and roots, as well as promoting ryegrass growth (Ding et al., 2017). After 1.5 months, the supplement of 0.2% (w/w) nano-hydroxyapatite considerably increased the uptake of Pb in the ryegrass shoots (Liang et al., 2017).

2.2.3 INTEGRATED NANO-MYCORRHIZO-PHYTOREMEDIATION (NMPR) STRATEGY

Nano-mycorrhizo assisted phytoremediation (NMPR) technology has enhanced the phytoremediation effectiveness of bioenergy plants growing in heavy metal contaminated soil during a synergistic manner (Khan et al., 2016; Khan, 2020; Khan, 2021). Nanomaterials research that enhances the beneficial effects of arbuscular mycorrhizal fungi (AMF) and associated plant growth-promoting rhizobial (PGPR) microbiota by increasing productivity and enhancing soil fertility by regulating the movement of nutrients from rhizospheres to plants via external mycelial hyphae and offering protection against hazardous pollutants like heavy metals in polluted areas. This combination realizes a lot of potential in sustainable agriculture (Khan, 2020). Understanding the interactions between rhizosphere microorganisms and plant roots under heavy metal limitations is critical, as is using this information to co-inoculate plants like Vetiver grass for nano-phytoremediation optimization. Rhizobacteria, including PGPR, and AMF, can influence HM-bioavailability by releasing microbial acids, exuding a range of nanomaterials into the rhizosphere, modifying physico-chemical properties of soil, and influencing redox potential shifts. Processes of remediation with plants and NPs in the presence of PGPR stimulate the plant's secretion of NPs, which play an effective role in transforming soil pollutants from polluted sites (Srivastav et al., 2018). Plant growth and metal accumulation were enhanced when AMF/PGPR was added to HM-polluted soils, and proteome analysis of plant leaves revealed molecular processes driving plant adaptation to HMs difficulties, such as increased production of photosynthesis, Calvin cycle enzymes, antioxidant enzymes, and proteins involved in GHS metabolism (Guarino et al., 2014). Many authors have backed up mycorrhiza-assisted phytoremediation with evidence. The uptake of Ni and Zn by Glycine max and Lens culinaris inoculated with AMF, *Glomus mosseae* under glass house conditions increased over a 12-week period (Jamal et al., 2002). At soil contaminated with low level of Pb or Zn, *Glomus mosseae* and *G. intraradices* as AM fungi boosted Pb and Zn uptake by *Chrysopogon zizanioides*, but decreased Pb and Zn uptake at higher concentrations, according to Wong et al. (2007). It has been reported that aquatic macrophytes, Vetiver grass, and their root-associated microbes, AMF, play an effective role in protecting against pollutants during environmental stresses and pollutants, via the bonds between cysteine and heavy metals, phytochelators, and the formation of HM-complexes (Khan, 2021). AMF adapted to HMs not only improves Vetiver grass growth, resulting in more biomass for bioenergy generation, but also uptakes and stabilizes HMs, as seen in NMPR. Certain free-living rhizosphere bacteria that encourage plant development in heavy metal-contaminated soils improve phytoremediation and reduce metal toxicity (Khan et al., 2016). Strains of *Pseudomonas fluorescence* achieved a significant role in bioremediation of metal and use (Singh et al, 2013). Rhizobacteria that promote growth have developed several methods to help them survive in metal-stressed environments. These include mobilizing or converting metals into an inert state to allow heavy metal ions to be taken up (Khan et al., 2016). Metal ions are either totally inhibited by microbial populations due to their various metabolic activities, or organisms can acquire tolerance mechanisms to greater metal concentrations (Khan et al., 2016). The PGPR-achieved higher root area, length, peroxidase, and catalase activity of plants treated with municipal wastewater by the Ag-nano particles. The application of Ag-nano particles changed the levels of ABA (34%), IAA (55%), and GA (82%), enhanced proline synthesis (70%) and caused oxidative stress, enhancing the bioremediation ability of PGPR for Pb, Cd, and Ni (Khan et al., 2016). Figure 2.1 shows the role of NMPR for improving the removal of heavy metals-contaminated soils.

2.2.4 USING NANOREMEDIATION, BIOTECHNOLOGICAL, AND TRANSGENIC APPROACHES TO IMPROVE PHYTOREMEDIATION EFFICIENCY OF HEAVY METALS

Combining microbiological and nanotechnological techniques with phytoremediation could pave the way for metal removal from contaminated locations by extracting, degrading, modifying, sequestering, and stabilizing a wide range of soil contaminants (Srivastav et al., 2018; Yadav et al., 2018). Heavy-metal stress can be responded to microorganisms by transport

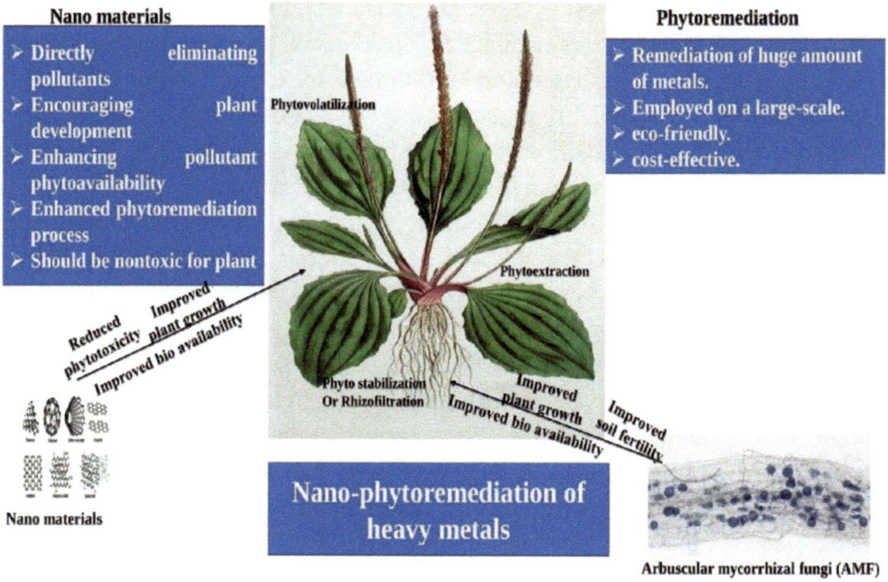

FIGURE 2.1 Nano-phytoremediation strategy for heavy metals.

across the cell membrane, trapping in extracellular capsules, precipitation, biosorption to cell walls, complexation, and redox reactions in certain bacteria. Furthermore, microorganisms can also protect themselves against heavy metal stress by employing a variety of defense mechanisms, including compartmentalization, exclusion, complex formation, and the creation of binding proteins (metallothionein's and phytochelatins) (Mejare and Bulow, 2001; Malik, 2004). Nanomaterials (NMs) can increase metal accumulation by increasing the permeability of cell wall, co-translocation of heavy metals and NMs, and modulation of transporter gene expression (Kumar et al., 2019). Co-uptake between NMs, plants, and pollutants results in the elimination of hazardous materials from soil. When plants taken up NPs, damage to the cell membrane and cell walls can occur, resulting in a loss of membrane integrity, stress, and cell function, allowing contaminants from to enter the plant body (Gong et al., 2017). The ability of plants to take in and accumulate heavy metals is strongly influenced by its genetic potential (Ovecka and Takac, 2014; Kumar et al., 2019). Plants have improved heavy metal accumulation to hyperaccumulation, uptake capacity from root-to-shoot, sequestration mechanisms, and detoxification as a result of gene transfer, making them more suitable for/phytomining. (Duan et al., 2011; Kumar et al., 2019).

Arabidopsis thaliana, Nicotiana spp, Oryza sativa, Brassica juncea, and *Populus* species have made the most progress in the production of genetically engineered plants with better phytoremediation capabilities (Kumar et al., 2019). Based on the availability of genetic sequences, Arabidopsis, tobacco, Indian mustard, and sugar beet improved their ability to survive heavy metal stress by increasing biomass accumulation (Kumar et al., 2019). Transgenic lines with overexpression of the YCF1 gene in *Arabidopsis thaliana* were able to improve transgenic plant absorption and accumulation efficiency of heavy metals (Lee et al., 2010). Transgenic *Brassica juncea* has the gene for selenocysteine methyl transferase (SMT) expression, which causes it to become a Se hyperaccumulator, but *A. bisulcatus* has more Se uptake and tolerance than wild kinds (Huysen et al., 2004). In transgenic plants, overexpression of *E. coli* GSH1and GSH2 genes in *Brassica juncea* and *Populus canescens* led to glutathione and phytochelatines increase and improved Cd and Zn tolerance, respectively (Zhu et al., 1999).

2.3 NANO-PHYTOREMEDIATION OF ORGANIC POLLUTANTS

Now, phytoremediation technology in cooperation with nanotechnology is considered a viable alternative to traditional remediation methods. Furthermore, this technology aims to create a more efficient, faster, and environmentally benign processing process than individual methods (Kumar et al., 2021; Mustafa et al., 2022). Organic pollutants such as polycyclic aromatic hydrocarbons, chlorinated hydrocarbons, PCBs, organochlorine insecticides, phenols, and their derivatives respond well to nanomaterial-assisted phytoremediation, and site contaminants are removed more quickly. To achieve this, many techniques were introduced to this technology, such as nanotechnology with phytoremediation and bioremediation. Nanotechnology improves phytoremediation efficiency, and NPs can be utilized to clean up soils and water that have been contaminated with organic contaminants (Srivastav et al., 2018). For removing TNT (2,4,6-TNT)-contaminated soil, phytoremediation combined with nanoremediation (nano-phytoremediation) is more effective than nanoremediation or phytoremediation (Srivastav et al., 2018).

The use of nanomaterials has increased the efficiency of phytoremediation of soil contaminated with the organic compounds, like trichloroethylene, endosulfan, and TNT (Ma and Wang, 2010; Jiamjitrpanich et al., 2012; Pillai and Kottekottil, 2016). Table 2.1 provides some illustrations of phytoremediation with nanomaterial assistance.

TABLE 2.1 Nano-Phytoremediation Used for Removal of Contaminants.

Plant species	NPs used	Pollutants	Removal capacity	References
		Organic pollutants		
Plantago major	nZVIs	Chlorpyrifos and the metabolite 3, 5, 6-trichloro-2-pyridinol (TCP)	≈100%	Romeh (2021)
Plantago major	nZVIs on activated charcoal (Ach)	Chlorfenapyr	86% in water 71.22% in soil	Romeh and Saber (2020)
Plantago major	Nano brassica-Ag on activated charcoal (Ach)	Chlorfenapyr	79.70% in water 73.10% in soil	Romeh and Saber (2020)
Plantago major	(AgNPs)	Fipronil and degradation products	63.48–95.45% in water, 30.99–68.8% in soil	Romeh (2018)
Alpinia calcarata	nZVIs	Endosulfan	≈100 (%)	Pillai and Kottekottil (2016)
Ocimum sanctum	nZVI	Endosulfan	65.08–86.16%	Pillai and Kottekottil (2016)
Panicum maximum	nZVI	Trinitrotoluene	85.7–100% through 120 days	Jiamjitrpanich et al. (2012)
Populus deltoides	Fullerene nanoparticles	Trichloroethylene	26% and 82% with 2 and 15 mg/L	Ma and Wang (2010)
		Heavy metals		
Lolium perenne	Nanohydroxyapatite	Pb	Pb content in the root and shoot dropped by 2.86–21.1% and 13.19–20.3%	Ding et al. (2017)
Glycine max	TiO2	Cd	100–300 mg/kg, with 128.5–507.6 μg	Singh and Lee (2016)
Lolium perenne	nZVI	Pb	With 100 mg/kg nZVI, the deposition of Pb in the root and shoot reached 1175.4 lg per pot	Huang et al. (2018)
Helianthus, annuus, and lolium perenne	nZVI	As, Cd, Pb, and Zn	50–60% less of the concentrations were found in the roots and shoots	Vıtkova et al. (2018)

2.3.1 MECHANISMS OF NANOMATERIALS FOR PHYTOREMEDIATION

Nanomaterial-facilitated phytoremediation of organic pollutants by directly acting on the pollutants and plants or through interactions between the pollutants and plants indirectly influences the effectiveness of the ultimate remediation process (Song et al., 2019). Nanomaterials can play a role in the phytoremediation system by eliminating pollutants directly, encouraging plant growth, and enhancing the phyto-availability of pollutants in this combination (Song et al., 2019; Ding et al., 2021).

2.3.1.1 NANOMATERIALS FOR DIRECT REMOVAL OF ORGANIC POLLUTANTS

The use of reactive NPs for the transformation and detoxification of polluted environments has been explored using nanoremediation approaches. These NPs are particularly reactive and catalytic due to their high surface-to-volume ratio (Wu et al., 2020). Nanomaterials, with their large surface area, multiple active sites, and high adsorption capacity, are ideal for soil remediation (Bhat et al., 2021). By reducing the energy needed for their breakdown due to special surface properties, nanoscale process regulation facilitates the adsorption and breakdown of pollutants through catalyze reactions, leading to advanced remediation methods that integrate nanotechnology with phytoremediation and bioremediation. The NPs can adsorb and accelerate pollutant degradation through a variety of mechanisms including redox reactions, surface processes, adsorption, ion exchange, surface complexation, and electrostatic contact (Medina-Pérez et al., 2019). The nano-bioremediation (NBRs) main idea is depicted as the degradation of organic wastes using nano-catalysts as a medium for them to reach deep within pollutants, allowing the entire process to be carried out safely without affecting the environment (Chauhan et al., 2020). Particularly some hydrophobic organic pollutants can be effectively adsorbent using CNTs through π-π bonding, lipophilic interaction, and electrostatic attraction. On the other hand, heavy metals exhibit complex formation, electrostatic attraction, physical adsorption, and surface precipitation (Zhang, et al., 2017; Kang et al., 2018; Song et al., 2018). Chlorinated organic pollutants, such as organochlorine insecticides, can be reductively dechlorinated using nZVI as an electron donor through redox processes (Song et al., 2019). In phytoremediation systems, nanomaterials can directly remove a portion of the contaminants, which lowers the phytotoxicity brought on by the stress of high concentrations of pollutants such as heavy metals and persistent organic compounds (Chai et al., 2013; Liang et al., 2017).

2.3.1.2 NANOMATERIALS FOR PLANT GROWTH

Nanomaterials as nZVI-NPs, Ag-NPs, ZnO-NPs, and CNTs can also stimulate plant development in the phytoremediation system by increasing photosynthetic rate by enhancing water and nutrient absorption, regulating the soil microbial community, and increasing the tolerance to pollutants and reducing abiotic stress (such as excessive salinity and drought) (Singh and Lee, 2016; Ding et al., 2017; Praveen et al., 2018; Ding et al., 2021). TiO_2-NPs can enter chloroplasts and quicken electron transport and light adaption (Singh and Lee, 2016). Surprisingly, NMs may reduce toxicity via altering biological processes such as antioxidation, detoxification, or biodegradation, as well as removing contaminants, by antagonizing the toxicity of coexisting contaminants through adsorption/reductive–oxidation interactions (Deng et al., 2017). According to Khodakovskaya et al. (2013), CNTs can activate a plant's reproductive system, which will promote tomato development. Other methods, such as employing transgenic plants, plant growth regulators, and plant growth promoting rhizobacteria (PGPR), are employed in phytoremediation strategy to increase plant biomass and growth rate (Ma et al., 2016; Aderholt et al., 2017; Yadu et al., 2018).

2.3.1.3 NANOMATERIALS FOR PHYTO-AVAILABILITY OF ORGANIC POLLUTANTS

Furthermore, when contaminants penetrate cells, NPs can act as a transporter, improving bioavailability (Su et al., 2013). Pollutants' chemical speciation and distribution in soil have a large influence on how bioavailable they are to plants. Additionally, phyto-availability of contaminants is also influenced by soil physicochemical qualities, physicochemical properties of organic pollutants, and plant physiological traits (Ren et al., 2018; Romeh, 2022). There are several important strategies in promoting phytoremediation such as use of agricultural management, solubility-enhancing compounds, rhizosphere bacteria, genetic engineering, and nanomaterials; however, they are just a few of the proposed techniques to increase the effectiveness of phytoremediation (Romeh and Hendawi, 2017; Jacobs et al., 2018; Romeh, 2022).

When contaminants reach the cell, nanomaterials can act as a transporter, enhancing the bioavailability. Additionally, the bioavailability of pollutants may be decreased by adsorption of contaminants onto nanomaterials outside of living things (Su et al., 2013; Glomstad et al., 2016). Numerous studies are being conducted on nanomaterials to increase the phyto-availability of contaminants (Romeh, 2018; Romeh and Saber 2020; Romeh, 2022). Plants could ingest CNTs, Fe_3O_4-NPs, TiO_2-NPs, silica-NPs, and quantum dots (Wang et al., 2016).

Nano-encapsulated enzymes offer a higher potential for treating persistent insecticides (organochlorines) and long-chain hydrocarbons that are difficult to breakdown using microbial or plant-based methods (Chauhan et al., 2020). The utilization of nanomaterials to soil for the remediation strategy can be done in a variety of ways. Injection under pressure, recirculation, pressure pulse technology, and hydraulics are examples of these methods (Bardos et al., 2014). In nano-phytoremediation investigations, a variety of plant species and NP variations are utilized. The phytoremediation procedure is frequently limited by hydrophobic pollutants' (log K_{ow} > 3.0) low phyto-availability, restricting the removal of soil polluted with these substances (Jing et al., 2018). Nanomaterials, solubility enhancing chemicals, surfactants, and PGPB cloud reduce the log K_{ow} range between 0.5 and 3.0 and thus hydrophobic pollutants can translocate from the soil to the plant shoots (Romeh, 2020; Romeh and Saber 2020; Romeh, 2022).

2.3.2 APPLICATIONS OF NANOMATERIAL-ASSISTED PHYTOREMEDIATION FOR REMOVAL OF ORGANIC POLLUTANTS

There are many examples of the use of nanomaterials to enhance plant availability, which are explained below.

2.3.2.1 NZVI-NP

The most often utilized remediation method is iron NPs (Fe^o-NPs). Because iron NPs have the potential to be employed in phytoremediation of a variety of pollutants and are a highly reactive reducing agent as an electron donor; they are used as a powerful reductant for contaminants that need a reduction process to be degraded (Sun et al., 2006; Rajput et al., 2022). NZVI can be employed as an electron donor to slow down the degradation of pollutants or to stabilize them (Ding et al., 2021). *Plantago major*, when treated with the proper Ficus–FeNPs (F–Fe^0), Ipomoea–Ag (Ip–Ag^0), and Brassica–AgNPs (Br–Ag^0), supported on activated charcoal was able to remove 93.7%, 91.30%, and 92.92% of chlorfenapyr, respectively (Romeh and Saber, 2020). The nZVI supported on seed pod fibers of sycamore tree (nZVIΘSTSPF) exhibited a higher removal capability for malachite green oxalate (92.59 mg/g), methyl violet 2B (92.59 mg/g), and methylene blue hydrate (140.80mg/g) compared to the raw biosorbent (43.67 mg/g, 25.32 mg/g and 126.60 mg/g), respectively (Parlayıcı, and Pehlivan, 2019). Through hydrogenolysis and progressive dehalogenation, nano-phytoremediation with *Alpinia calcarata* and nZVI removed 82% of endosulfan in just seven days (Pillai and Kottekottil, 2016). Pesticides (chlorfenapyr, chlorpyrifos, atrazine, and lindane), pentachlorophenol, trichloroethylene (TCE), and pyrene are examples of organic pollutants could be rapidly degraded by nZVI, magnetite NPs (nFe_3O_4), and bimetallic NP integrated with phytoremediation (Singh et al., 2012; Reddy et al., 2013; Romeh and Saber, 2020; Romeh, 2021). According to Pillai and Kottekottil (2016), *Cymbopogon citratus* and nZVIs can be employed to remediate endosulfan pollution with an efficiency rate of 86.16 %. The aquatic local life is negatively impacted by oily sewage. With improved findings of 88.34%, iron NPs are effective at removing total petroleum hydrocarbons (TPHs) from water; thus, treatments based on nanotechnology can deliver high-performance treated water with fewer pollutants and hazardous compounds (Murgueitio et al., 2018). Phytoremediation efficiency of trinitrotoluene's, TNT-contaminated soil could be significantly increased using nZVI particles. The TNT-nZVI ratio of 1/10 was found to have the optimum removal performance after 60 days at 100 mg/kg TNT (Jiamjitrpanich et al., 2012).

2.3.2.2 SIO2-NP

PAHs tend to collect in the soil due to their hydrophobic and lipophilic properties, disrupting not only normal soil functions but also damaging the soil ecosystem (Qin et al., 2019). Because of their limited bioavailability in soil, PAHs are inadequately attenuated during phytoremediation.

Combination of surfactants with SiO_2-NP for plant improves PAH aqueous solubility, resulting in increased bioavailability (Lu et al. 2019). Surfactants can aid in the desorption of PAHs from soils, converting them to the aqueous phase and so enhancing their bioavailability. Triton X-100 in combination with SiO_2-NP protects plants, *Erigeron annuus*, from the stress of polycyclic aromatic hydrocarbons (PAHs) and may improve phytoremediation of PAH-contaminated soils (Zuo et al., 2020). Because nano-SiO_2 absorbed the PAHs, increased PAH breakdown rates in plants occurred, resulting in lower biotoxicity. (Włóka et al., 2019). Nano-SiO_2 improved tomato (*Lycopersicum esculentum*) seed germination and aided *Secale montanum's* phytoremediation capabilities in Pb-contaminated soils (Moameri and Abbasi Khalaki, 2019).

2.3.2.3 TiO2-NP

In soil, nano-TiO_2 degrades organic pollutants (diuron, p,p′-DDT, pyrene, phenanthrene) and reduces pyrene and phenanthrene half-lives (Li et al., 2016). Transformation NPs coated with titanium dioxide (TiO_2) can detoxify deoxynivalenol, DON-contaminated cereal grains (Wu et al., 2020).

2.3.2.4 AG^0-NP

Broad leaf plant, *Plantago major*, as phytoremediation and the combination of Ficus ZVI (F–Fe^0), ipomoea–silver (Ip–Ag^0), and brassica–silver NPs (Br–Ag^0) as green nanotechnology have played a key role in the cleanup of chlorfenapyr-polluted water and soil (Romeh and Saber, 2020). *Plantago major* and Ag^0-NP are also important in the cleanup of fipronil-contaminated water and flooded soil, with *P. major* playing a key role in the uptake of the polar break product, fipronil-amide (Romeh, 2018). Ag^0-NP stimulated the production of ABA and GA phytohormones in plants allowing them to endure stress and increase nutrient and water intake to boost growth (Khan and Bano, 2016).

2.4 CONCLUSION

Metals and persistent organic pollutants accumulation in various environmental compartments is becoming more of a major worldwide concern, posing a major threat to food safety, human health, the food chain, and the ecosystem. Phytoremediation combined with nanotechnology appears to be a viable alternative to traditional remediation methods. By eliminating pollutants directly from the environment, encouraging plant development, and boosting phyto-availability of pollutants, nanomaterials can improve the effectiveness of phytoremediation systems. Because of their large specific surface area and excellent photoelectric properties, nanomaterials are critical in the removal of heavy metals via a variety of methods. By extracting, degrading, altering, sequestering, and stabilizing a variety of soil pollutants, combining microbiological, nanotechnological, and transgenic plants could pave the way for the elimination of metals from contaminated area. The two chemical categories of NPs employed in nano-phytoremediation to clean up settings contaminated with heavy metals are organic and inorganic NPs. Pollutants removal from contaminated sites might be made possible by combining NMPR technology.

KEYWORDS

- **organic nanoparticles**
- **inorganic nanoparticles**
- **phytoremediation**
- **heavy metals**
- **organic pollutants**

REFERENCES

Aderholt, M.; Vogelien, D. L.; Koether, M.; Greipsson, S. Phytoextraction of Contaminated Urban Soils by Panicum Virgatum L. Enhanced with Application of a Plant Growth Regulator (BAP) and Citric Acid. *Chemosphere* **2017,** *175,* 85–96.

Ali, H.; Khan, E. Trophic Transfer, Bioaccumulation, and Biomagnification of Nonessential Hazardous Heavy Metals and Metalloids in Food Chains/Webs-Concepts and Implications for Wildlife and Human Health. *Human Ecol. Risk Assess.: Int. J.* **2018,** *8,* 1–24.

Ali, I.; Peng, C.; Naz, I.; Khan, Z. M.; Sultan, M.; Islam, T.; Abbasi, I. A. Phytogenic Magnetic Nanoparticles for Wastewater Treatment: A Review. *RSC Adv.* **2017,** *7* (64), 40158–40178.

Anerao, P.; Kaware, R.; Kumar Khedikar, A.; Kumar, M.; Sin Gh, L. Phytoremediation of Persistent Organic Pollutants: Concept Challenges and Perspectives. In *Phytoremediation Technology for the Removal of Heavy Metals and Other Contaminants from Soil and Water*; Elsevier, 2022; pp 375–404

Bardos, P.; Bone, B.; Daly, P.; Elliott, D.; Jones, S.; Lowry, G.; Merly, C. A Risk/Benefit Appraisal for the Application of Nanoscale Zero Valent Iron (NZVI) for the Remediation of Contaminated Sites. *WP9 Nanorem* **2014**.

Bertrand, N.; Desgranges, C.; Gauvain, D.; Monceau, D.; Poquillon, D. Low Temperature Oxidation of Pure; Iron: Growth Kinetics and Scale Morphologies. In *Materials Science Forum*, Vol 461; Trans Tech Publications Ltd, 2004, pp 591–598.

Bhat, R. A.; Tonelli, F. M. P.; Dar, G. H.; Hakeem, K. R., Eds. *Phytoremediation: Biotechnological Strategies for Promoting Invigorating Environs*; Elsevier, 2021.

Bhati, M.; Rai, R. Nano-Phytoremediation Application for Water Contamination. In *Phytoremediation*; Springer: Cham, 2018; pp 441–452.

Brasili, E.; Bavasso, I.; Petruccelli, V.; Vilardi, G.; Valletta, A.; Dal Bosco, C.; Di Palma, L. Remediation of Hexavalent Chromium Contaminated Water Through Zero-Valent Iron Nanoparticles and Effects on Tomato Plant Growth Performance. *Sci. Rep.* **2020,** *10* (1), 1–11.

Buhari, M.; Sulaiman, B.; Vyas, N. L.; Badaru, S.; Harisu, U. Role of Biotechnology in Phytoremediation. *J. Bioremed. Biodegrad.* **2016,** *7* (2).

Chai, M.; Shi, F.; Li, R.; Liu, L.; Liu, Y.; Liu, F. Interactive Effects of Cadmium and Carbon Nanotubes on the Growth and Metal Accumulation in a Halophyte Spartina Alterniflora (Poaceae). *Plant Growth Regul.* **2013,** *71* (2), 171–179.

Chaudhary, K.; Khan, S.; Saraswat, P. K. Newer Approaches in Phytoremediation: An Overview. In *Research Anthology on Synthesis, Characterization, and Applications of Nanomaterials*; 2021; pp 1785–1808.

Chauhan, R.; Yadav, H. O.; Sehrawat, N. Nanobioremediation: A New and a Versatile Tool for Sustainable Environmental Cleanup-Overview. *J. Mater. Environ. Sci.* **2020,** *11* (4), 564–573.

Chen, S.; Chen, Y.; Feng, T.; Ma, H.; Liu, X.; Liu, Y. In *Application of Nanomaterials in Repairing Heavy Metal Pollution Soil, IOP Conference Series: Earth And Environmental Science;* IOP Publishing, 2021; *621* (1), 012123.

Cheng, J. M.; Liu, Y. Z.; Wang, H. W. Effects of Surface-Modified Nano-Scale Carbon Black on Cu and Zn Fractionations in Contaminated Soil. *Int. J. Phytoremediation* **2014,** *16* (1), 86–94.

Cheng, J.; Sun, Z.; Yu, Y.; Li, X.; Li, T. Effects of Modified Carbon Black Nanoparticles on Plant-Microbe Remediation of Petroleum and Heavy Metal Co-Contaminated Soils. *Int. J. Phytoremediation.* **2019,** *21* (7), 634–642.

Cheng, P.; Zhang, S.; Wang, Q.; Feng, X.; Zhang, S.; Sun, Y.; Wang, F. Contribution of Nano-Zero-Valent Iron and Arbuscular Mycorrhizal Fungi to Phytoremediation of Heavy Metal-Contaminated Soil. *Nanomaterials* **2021,** *11* (5), 1264.

Cibane, N. S. The Efficiency of Phytoremediation Using Panicum Maximum and TiO_2 Nanoparticles, Doctoral Dissertation, 2021.

da Conceição Gomes, M. A.; Hauser-Davis, R. A.; De Souza, A. N.;Vitória, A. P. Metal Phytoremediation: General Strategies, Genetically Modified Plants and Applications in Metal Nanoparticle Contamination. *Ecotoxicol. Environ. Saf.* **2016,** *134,* 133–147.

Das, S.; Mondal, U. S.; Paul, S. Nanophytoremediation Technology: A Better Approach for Environmental Remediation of Toxic Metals and Dyes from Water. In *Phytoremediation Technology for the Removal of Heavy Metals and Other Contaminants from Soil and Water*; Elsevier, 2022; pp 459–481.

Deng, R.; Lin, D.; Zhu, L.; Majumdar, S.; White, J. C.; Gardea-Torresdey, J. L.; Xing, B. Nanoparticle Interactions with Co-Existing Contaminants: Joint Toxicity, Bioaccumulation, and Risk. *Nanotoxicology* **2017,** *11* (5), 591–612.

Ding, L.; Li, J.; Liu, W.; Zuo, Q.; Liang, S. X. Influence of Nano-Hydroxyapatite on the Metal Bioavailability, Plant Metal Accumulation and Root Exudates of Ryegrass for Phytoremediation in Lead-Polluted Soil. *Int. J. Environ. Res. Public Health.* **2017,** *14* (5), 532.

Ding, N.; Han, M.; He, Y.; Wang, X.; Pan, Y.; Lin, H.; Guo, Y. In *Advances in Application of Nanomaterials in Remediation of Heavy Metal Contaminated Soil*, E3S Web of Conferences; EDP Sciences, 2021, *261*.

Doshi, R.; Braida, W.; Christodoulatos, C.; Wazne, M.; O'Connor, G. Nano-Aluminum: Transport Through Sand Columns and Environmental Effects on Plants and Soil Communities. *Environ. Res.* **2008,** *106* (3), 296–303.

Duan, G. L.; Hu, Y.; Liu, W. J.; Kneer, R.; Zhao, F. J.; Zhu, Y. G. Evidence for a Role of Phytochelatins in Regulating Arsenic Accumulation in Rice Grain. *Environ. Exp. Bot.* **2011,** *71,* 416–421.

Fajardo, C.; Costa, G.; Nande, M.; Martín, C.; Martín, M.; Sánchez-Fortún, S. Heavy Metals Immobilization Capability of Two Iron-Based Nanoparticles (NZVI and Fe3O4): Soil and Freshwater Bioassays to Assess Ecotoxicological Impact. *Sci. Total Environ.* **2019,** *656,* 421–432.

Gerhardt, K. E; Gerwing, P. D.; Greenberg, B. M. Opinion: Taking Phytoremediation from Proven Technology to Accepted Practice. *Plant Sci.* **2017,** *256,* 170–185.

Ghazaryan, K.; Movsesyan, H.; Gevorgyan, A.; Minkina, T.; Sushkova, S.; Rajput, V.; Mandzhieva, S. Comparative Hydrochemical Assessment of Groundwater Quality from Different Aquifers for Irrigation Purposes Using IWQI: A Case-Study from Masis Province in Armenia. *Groundw. Sustain. Dev.* **2020,** *11,* 100459.

Ghazaryan, K. A.; Movsesyan, H. S.; Minkina, T. M.; Sushkova, S. N.; Rajput, V. D. The Identification of Phytoextraction Potential of Melilotus Officinalis and Amaranthus Retroflexes Growing on Copper- and Molybdenum-Polluted Soils. *Environ. Geochem. Health* **2019,** *43,* 1327–1335.

Glomstad, B.; Altin, D.; Sørensen, L.; Liu, J.; Jenssen, B. M.; Booth, A. M. Carbon Nanotube Properties Influence Adsorption of Phenanthrene and Subsequent Bioavailability and Toxicity to *Pseudokirchneriella Subcapitata*. *Enviro. Sci. Technol.* **2016,** *50* (5), 2660–2668.

Gogos, A.; Moll, J.; Klingenfuss, F.; Van Der Heijden, M.; Irin, F.; Green, M. J.; Bucheli, T. D. Vertical Transport and Plant Uptake of Nanoparticles in a Soil Mesocosm Experiment. *J. Nanobiotechnol.* **2016,** *14* (1), 1–11.

Gong, X.; Huang, D.; Liu, Y.; Peng, Z.; Zeng, G.; Xu, P.; Cheng, M.; Wang, R.; Wan, J. Remediation of Contaminated Soils by Biotechnology with Nanomaterials: Bio-Behavior, Applications, and Perspectives. *Crit. Rev. Biotechnol.* **2017a,** *38,* 455–468.

Gong, X.; Huang, D.; Liu, Y.; Zeng, G.; Wang, R.; Wan, J.; Xue, W. Stabilized Nanoscale Zerovalent Iron Mediated Cadmium Accumulation and Oxidative Damage of *Boehmeria nivea* (L.) Gaudich Cultivated in Cadmium Contaminated Sediments. *Enviro. Sci. Technol.* **2017b,** *51* (19), 11308–11316.

Gong, X.; Huang, D.; Liu, Y.; Zeng, G.; Wang, R.; Xu, P.; Chen, S. Roles of Multiwall Carbon Nanotubes in Phytoremediation: Cadmium Uptake and Oxidative Burst in *Boehmeria nivea* (L.) Gaudich. *Environ. Sci.: Nano.* **2019,** *6* (3), 851–862.

Goutam, S. P.; Saxena, G.; Singh, V.; Yadav, A. K.; Bharagava, R. N.; Thapa, K. B. Green Synthesis of TiO_2 Nanoparticles Using Leaf Extract of *Jatropha curcas* L. for Photocatalytic Degradation of Tannery Wastewater. *Chem. Eng. J.* **2018,** *336,* 386–396.

Guarino, C.; Conte, B.; Spada, V.; Arena, S.; Sciarrillo, R.; Scaloni, A. Proteomic Analysis of Eucalyptus Leaves Unveils Putative Mechanisms Involved in the Plant Response to a Real Condition of Soil Contamination by Multiple Heavy Metals in the Presence or Absence of Mycorrhizal/Rhizobacterial Additives. *Enviro. Sci. Technol.* **2014,** *48* (19), 11487–11496.

Halder, S.; Ghosh, S. Wetland Macrophytes in Purification of Water. *Int. J. Environ. Sci.* **2014,** *5* (2), 432–437.

He, F.; Zhao, D. Manipulating the Size and Dispersibility of Zerovalent Iron Nanoparticles by Use of Carboxymethyl Cellulose Stabilizers. *Environ. Sci. Technol.* **2007**, *41,* 6216–6221.

Hou, J.; Lin, D.; White, J. C.; Gardea-Torresdey, J. L.; Xing, B. Joint Nanotoxicology Assessment Provides a New Strategy for Developing Nanoenabled Bioremediation Technologies. **2019,** 7927–7929.

Hu, X.; Kang, J.; Lu, K. Graphene Oxide Amplifies the Phytotoxicity of Arsenic in Wheat. *Sci Rep.* **2014,** *4,* 6122.

Huang, D.; Qin, X.; Peng, Z.; Liu, Y.; Gong, X.; Zeng, G.; Hu, Z. Nanoscale Zero-Valent Iron Assisted Phytoremediation of Pb in Sediment: Impacts on Metal Accumulation and Antioxidative System of Lolium Perenne. *Ecotoxicol. Environ. Saf.* **2018,** *153,* 229–237.

Huysen, T.; Terry, N.; Pilon-Smits, E. A. H. Exploring the Selenium Phytoremediation Potential of Transgenic Indian Mustard Overexpressing ATP Sulfurylase or Cystathionine-γ-Synthase. *Int. J. Phytorem.* **2004,** *6,* 111–118.

Jacobs, A.; De Brabandere, L.; Drouet, T.; Sterckeman, T.; Noret, N. Phytoextraction of Cd and Zn with Noccaea Caerulescens for Urban Soil Remediation: Influence of Nitrogen Fertilization and Planting Density. *Ecol. Eng.* **2018,** *116,* 178–187.

Jamal, A.; Ayub, N.; Usman, M.; Khan, A. G. Arbuscular Mycorrhizal Fungi Enhance Zinc and Nickel Uptake from Contaminated Soil by Soybean and Lentil. *Int. J. Phytorem.* **2002,** *4* (3), 205–221.

Jiamjitrpanich, W.; Parkpian, P.; Polprasert, C.; Kosanlavit, R. Trinitrotoluene and Its Metabolites in Shoots and Roots of Panicum Maximum in Nano-Phytoremediation. *Int. J. Environ. Sci. Dev.* **2013,** *4* (1), 7.

Jiamjitrpanich, W.; Parkpian, P.; Polprasert, C.; Kosanlavit, R. In *Enhanced Phytoremediation Efficiency of TNT-Contaminated Soil by Nanoscale Zero Valent Iron,* 2nd International Conference On Environment And Industrial Innovation IPCBEE; 2012, *35,* pp 82–86.

Jing, R.; Fusi, S.; Kjellerup, B. V. Remediation of Polychlorinated Biphenyls (PCBs) in Contaminated Soils and Sediment: State of Knowledge and Perspectives. *Front. Environ. Sci.* **2018,** *6,* 79.

Kang, J.; Duan, X.; Wang, C.; Sun, H.; Tan, X.; Tade, M. O.; Wang, S. Nitrogen-Doped Bamboo-Like Carbon Nanotubes with Ni Encapsulation for Persulfate Activation to Remove Emerging Contaminants with Excellent Catalytic Stability. *Chem. Eng. J.* **2018,** *332,* 398–408.

Khan, A. G. Promises, and Potential of in Situ Nano-Phytoremediation Strategy to Mycorrhizo-Remediate Heavy Metal Contaminated Soils Using Non-Food Bioenergy Crops (*Vetiver Zizinoides* and *Cannabis sativa*). *Int. J. Phytoremediation.* **2020,** *22* (9), 900–915.

Khan, A. G. Nano-Phytoremediation of Heavy Metals Contaminated Wastewater Ecosystems and Wetlands by Constructed Wetlands Planted with Waterlogging-Tolerant Mycorrhizal Fungi and Vetiver Grass. *Environ. Sci. Proceed.* **2021,** *6* (1), 25.

Khan, N.; Bano, A. Role of Plant Growth Promoting Rhizobacteria and Ag-Nano Particle in the Bioremediation of Heavy Metals and Maize Growth Under Municipal Wastewater Irrigation. *Int. J. Phytoremediation.* **2016,** *18* (3), 211–221.

Khan, N.; Bano, A. Modulation of Phytoremediation and Plant Growth by the Treatment with PGPR, Ag Nanoparticle and Untreated Municipal Wastewater. *Int. J. Phytoremed.* **2016,** *18* (12), 1258–1269.

Khodakovskaya, M. V.; Kim, B. S.; Kim, J. N.; Alimohammadi, M.; Dervishi, E.; Mustafa, T.; Cernigla, C. E. Carbon Nanotubes as Plant Growth Regulators: Effects on Tomato Growth, Reproductive System, and Soil Microbial Community. *Small* **2013,** *9* (1), 115–123.

Kim, J. H.; Lee, Y.; Kim, E. J.; Gu, S.; Sohn, E. J.; Seo, Y. S.; Chang, Y. S. Exposure of Iron Nanoparticles to Arabidopsis Thaliana Enhances Root Elongation by Triggering Cell Wall Loosening. *Enviro. Sci. Technol.* **2014,** *48* (6), 3477–3485.

Konate, A.; He, X.; Rui, Y. K.; Zhang, Z. Y. In *Magnetite (Fe_3O_4) Nanoparticles Alleviate Growth Inhibition and Oxidative Stress Caused by Heavy Metals in Young Seedlings of Cucumber (Cucumis Sativus L), ITM Web of Conferences*; EDP Sciences. 2017, Vol. 12; p 03034.

Kumar, P.; Kumar, A.; Kumar, R. Phytoremediation and Nanoremediation. In *New Frontiers of Nanomaterials in Environmental Science*; Kumar, R., Kumar, R., Kaur, G., Eds.; Springer: Singapore, 2021; pp 281–297.

Kumar, S.; Prasad, S.; Yadav, K. K.; Shrivastava, M.; Gupta, N.; Nagar, S.; Malav, L. C. Hazardous Heavy Metals Contamination of Vegetables and Food Chain: Role of Sustainable Remediation Approaches—A Review. *Environ. Res.* **2019,** *179,* 108792.

Lacalle, R. G.; Gómez-Sagasti, M. T.; Artetxe, U.; Garbisu, C.; Becerril, J. M. Brassica Napus has a Key Role in the Recovery of the Health of Soils Contaminated with Metals and Diesel by Rhizoremediation. *Sci.Total Environ.* **2018,** *618,* 347–356.

Lee, K.; Bae, D. W. Kim, S. H. Han, H. J. Liu, X.; Park, H. C.; Lim, C. O.; Lee, S. Y. Chung, W. S. Comparative Proteomic Analysis of the Short-Term Responses of Rice Roots and Leaves to Cadmium. *J. Plant Physiol.* **2010,** *167,* 161–168.

Li, Q.; Chen, X.; Zhuang, J.; Chen, X. Decontaminating Soil Organic Pollutants with Manufactured Nanoparticles. *Environ. Sci. Pollut. Res.* **2016,** *23* (12), 11533–11548.

Liang, C.; Xiao, H.; Hu, Z.; Zhang, X.; Hu, J. Uptake, Transportation, and Accumulation of C60 Fullerene and Heavy Metal Ions (Cd, Cu, and Pb) in Rice Plants Grown in an Agricultural Soil. *Environ. Pollut.* **2018,** *235,* 330–338.

Liang, J.; Yang, Z.; Tang, L.; Zeng, G.; Yu, M.; Li, X.; Luo, Y. Changes in Heavy Metal Mobility and Availability from Contaminated Wetland Soil Remediated with Combined Biochar-Compost. *Chemosphere* **2017,** *181,* 281–288.

Liu, M.; Wang, Y.; Chen, L.; Zhang, Y.; Lin, Z. Mg $(OH)_2$ Supported Nanoscale Zero Valent Iron Enhancing the Removal of Pb (II) from Aqueous Solution. *ACS Appl. Mater. Interfaces*. **2015,** *7* (15), 7961–7969.

Liu, R.; Lal, R. Potentials of Engineered Nanoparticles as Fertilizers for Increasing Agronomic Productions. *Sci. Total Environ.* **2015,** *514,* 131–139.

Lopez-Luna, J.; Silva-Silva, M. J.; Martinez-Vargas, S. Magnetite Nanoparticle (NP) Uptake by Wheat Plantsand Its Effect on Cadmium and Chromium Toxicological Behavior. *Sci. Total Environ.* **2016,** *565,* 941–950.

Lu, H.; Wang, W.; Li, F.; Zhu, L. Mixed-Surfactant-Enhanced Phytoremediation of PAHs in Soil: Bioavailability of PAHs and Responses of Microbial Community Structure. *Sci. Total Environ.* **2019,** *653,* 658–666.

Ma, X.; Wang, C. Fullerene Nanoparticles Affect the Fate and Uptake of Trichloroethylene in Phytoremediation Systems. *Environ. Eng. Sci.* **2010,** *27* (11), 989–992.

Ma, Y.; Rajkumar, M.; Zhang, C.; Freitas, H. Inoculation of Brassica Oxyrrhina with Plant Growth Promoting Bacteria for the Improvement of Heavy Metal Phytoremediation under Drought Conditions. *J. Hazard. Mater.* **2016,** *320,* 36–44.

Malik, A. Metal Bioremediation through Growing Cells. *Environ. Int.* **2004,** *30,* 261–278.

Matos, M. P.; Correia, A. A. S.; Rasteiro, M. G. Application of Carbon Nanotubes to Immobilize Heavy Metals in Contaminated Soils. *J. Nanopart. Res.* **2017,** *19* (4), 1–11.
Medina-Pérez, G.; Fernández-Luqueño, F.; Vazquez-Nuñez, E.; López-Valdez, F.; Prieto-Mendez, J.; Madariaga-Navarrete, A.; Miranda-Arámbula, M. Remediating Polluted Soils Using Nanotechnologies: Environmental Benefits and Risks. *Polish J. Environ. Stud.* **2019,** *28* (3).
Mejáre, M.; Bülow, L. Metal-Binding Proteins and Peptides in Bioremediation and Phytoremediation of Heavy Metals. *Trends Biotechnol.* **2001,** *19,* 67–73.
Midhat, L.; Ouazzani, N.; Hejjaj, A.; Ouhammou, A.; Mandi, L. Accumulation of Heavy Metals in Metallophytes from Three Mining Sites (Southern Centre Morocco) and Evaluation of their Phytoremediation Potential. *Ecotoxicol. Environ. Saf.* **2019,** *169,* 150–160.
Moameri, M.; Abbasi Khalaki, M.; Capability of Secale Montanum Trusted for Phytoremediation of Lead and Cadmium in Soils Amended with Nano-Silica and Municipal Solid Waste Compost. *Environ. Sci. Pollut. Res.* **2019,** *26* (24), 24315–24322.
Moameri, M.; Abbasi Khalaki, M. Capability of Secale Montanum Trusted for Phytoremediation of Lead and Cadmium in Soils Amended with Nano-Silica and Municipal Solid Waste Compost. *Environ. Sci. Pollut. Res.* **2019**, *26* (24), 24315–24322.
Mustafa, K.; Kanwal, J.; Farrukh, S.; Mussaddiq, S.; Saddiq, N.; Younas, M. Nano-Phytoremediation Technology in Environmental Remediation. In *Phytoremediation Technology for the Removal of Heavy Metals and Other Contaminants from Soil and Water*; Elsevier, 2022; pp 433–459.
Murgueitio, E.; Cumbal, L.; Abril, M.; Izquierdo, A.; Debut, A.; Tinoco, O. Green Synthesis of Iron Nanoparticles: Application on the Removal of Petroleum Oil from Contaminated Water and Soils. *J. Nanotechnol.* **2018,** *2018.*
Mustafa, K.; Kanwal, J.; Farrukh, S.; Mussaddiq, S.; Saddiq, N.; Younas, M. Nano-Phytoremediation Technology in Environmental Remediation. In *Phytoremediation Technology for the Removal of Heavy Metals and Other Contaminants from Soil and Water*; Elsevier, 2022; pp 433–459.
Nwadinigwe, A. O.; Ugwu, E. C. Overview of Nano-Phytoremediation Applications. In *Phytoremediation*; Springer: Cham, 2018; pp 377–382.
Oloumi, H.; Mousavi, E. A.; Nejad, R. M. Multi-Wall Carbon Nanotubes Effects on Plant Seedlings Growth and Cadmium/Lead Uptake In Vitro. *Russ. J. Plant Physiol.* **2018,** *65* (2), 260–268.
Ovecka, M.; Takac, T. Managing Heavy Metal Toxicity Stress in Plants: Biological and Biotechnological Tools. *Biotechnol. Adv.* **2014,** *32,* 73–86.
Parlayıcı, Ş.; Pehlivan, E. Fast Decolorization of Cationic Dyes by Nanoscale Zero Valent Iron Immobilized in Sycamore Tree Seed Pod Fibers: Kinetics and Modelling Study. *Int. J. Phytoremediation* **2019,** *21* (11), 1130–1144.
Pillai, H. P.; Kottekottil, J. Nano-Phytotechnological Remediation of Endosulfan Using Zero Valent Iron Nanoparticles. *J. Environ. Prot.* **2016,** *7* (5), 734–744.
Praveen, A.; Khan, E.; Ngiimei, D. S.; Perwez, M.; Sardar, M.; Gupta, M. Iron Oxide Nanoparticles as Nano-Adsorbents: A Possible Way to Reduce Arsenic Phytotoxicity in Indian Mustard Plant (*Brassica juncea* L.). *J. Plant Growth Regul.* **2018,** *37* (2), 612–624.
Qin, Z.; Zhao, Z.; Adam, A.; Li, Y.; Chen, D.; Mela, S. M.; Li, H.The Dissipation and Risk Alleviation Mechanism of PAHs and Nitrogen in Constructed Wetlands: The Role of Submerged Macrophytes and their Biofilms-Leaves. *Environ. Int.* **2019,** *131,* 104940.

Rahmatizadeh, R.; Arvin, S. M. J.; Jamei, R.; Mozaffari, H.; Reza Nejhad, F. Response of Tomato Plants to Interaction Effects of Magnetic (Fe3O4) Nanoparticles and Cadmium Stress. *J. Plant Interact.* **2019,** *14* (1), 474–481.

Rajput, V. D.; Minkina, T.; Upadhyay, S. K.; Kumari, A.; Ranjan, A.; Mandzhieva, S.; Verma, K. K. Nanotechnology in the Restoration of Polluted Soil. *Nanomaterials* **2022,** *12* (5), 769.

Rastogi, A.; Zivcak, M.; Sytar, O.; Kalaji, H. M.; He, X.; Mbarki, S.; Brestic, M. Impact of Metal and Metal Oxide Nanoparticles on Plant: A Critical Review. *Front. Chem.* **2017,** *5,* 78.

Reddy, A.; Madhavi, V.; Reddy, K. G.; Madhavi, G. Remediation of Chlorpyrifos-Contaminated Soils by Laboratory-Synthesized Zero-Valent Nano Iron Particles: Effect of PH and Aluminium Salts. *J. Chem.* **2013**.

Ren, X.; Zeng, G.; Tang, L.; Wang, J.; Wan, J.; Liu, Y.; Deng, R. Sorption, Transport and Biodegradation–An Insight Into Bioavailability of Persistent Organic Pollutants in Soil. *Sci. Total Environ.* **2018,** *610,* 1154–1163.

Romeh, A. A. Synergistic Use of Plantago Major and Effective Microorganisms, EM1 To Clean Up the Soil Polluted with Imidacloprid Under Laboratory and Field Condition. *Int. J. Phytoremediation* **2020,** *22* (14), 1515–1523.

Romeh, A. A. Synergistic Effect of Ficus-Zero Valent Iron Supported on Adsorbents and Plantago Major for Chlorpyrifos Phytoremediation from Water. *Int. J. Phytoremediation* **2021,** *23* (2), 151–161.

Romeh, A. A. A. Integrated Application of Green Nanotechnology, Bioremediation, and Solubility Enhancing Chemicals for Improving Phytoremediation Efficiency: A Case Study in Egypt. In *Pesticides Bioremediation*; Springer: Cham, 2022; pp 455–478.

Romeh, A. A. A. Green Silver Nanoparticles for Enhancing the Phytoremediation of Soil and Water Contaminated by Fipronil and Degradation Products. *Water Air Soil Pollut.* **2018,** *229* (5), 1–13.

Romeh, A. A.; Hendawi, M. Y. Biochemical Interactions Between Glycine Max L. Silicon Dioxide (SiO_2) and Plant Growth-Promoting Bacteria (PGPR) for Improving Phytoremediation of Soil Contaminated with Fenamiphos and Its Degradation Products. *Pestic. Biochem. Phys.* **2017,** *142,* 32–43.

Romeh, A. A.; Saber, R. A. I. Green Nano-Phytoremediation and Solubility Improving Agents for the Remediation of Chlorfenapyr Contaminated Soil and Water. *J. Environ. Manag.* **2020,** *260,* 110104.

Said, I.; Salman, S. A. E. R.; Samy, Y.; Awad, S. A.; Melegy, A.; Hursthouse, A. S. Environmental Factors Controlling Potentially Toxic Element Behaviour in Urban Soils, El Tebbin, Egypt. *Environ. Monit. Assess.* **2019,** *191* (5), 1–15.

Shekhawat, G. S.; Arya, V. Biological Synthesis of Ag Nanoparticles through in Vitro Cultures of *Brassica juncea C. zern.* In *Advanced Materials Research* Vol 67; Trans Tech Publications Ltd., 2009; pp 295–299.

Shirani Bidabadi, S. The Role of Fe-nano Particles in Scarlet Sage Responses to Heavy Metals Stress. *Int. J. Phytoremediation.* **2020,** *22* (12), 1259–1268.

Simonin, M.; Richaume, A.; Guyonnet, J. P.; Dubost, A.; Martins, J. M.; Pommier, T. Titanium Dioxide Nanoparticles Strongly Impact Soil Microbial Function by Affecting Archaeal Nitrifiers. *Sci. Rep.* **2016,** *6* (1), 1–10.

Singh, D. V.; Bhat, R. A.; Dervash, M. A.; Qadri, H.; Mehmood, M. A.; Dar, G. H.; Rashid, N. Wonders of Nanotechnology for Remediation of Polluted Aquatic Environs. In *Fresh Water Pollution Dynamics and Remediation*; Springer: Singapore, 2020; pp 319–339).

Singh, J.; Lee, B. K.; Influence of Nano-TiO_2 Particles on the Bioaccumulation of Cd in Soybean Plants (*Glycine Max*): A Possible Mechanism for the Removal of Cd from the Contaminated Soil. *J. Environ. Manag.* **2016,** *170*, 88–96.

Singh, R.; Misra, V.; Mudiam, M. K. R.; Chauhan, L. K. S.; Singh, R. P. Degradation of γ-HCH Spiked Soil Using Stabilized Pd/Fe^0 Bimetallic Nanoparticles: Pathways, Kinetics, and Effect of Reaction Conditions. *J. Hazard. Mater.* **2012,** *237,* 355–364.

Singh, S.; Kumar, V.; Dhanjal, D. S.; Parihar, P.; Ramamurthy, P. C.; Singh, J. Phytoremediation of Heavy Metals, Metalloids, and Radionuclides: Prospects and Challenges. In *Phytoremediation Technology for the Removal of Heavy Metals and Other Contaminants from Soil And Water*; 2022; pp 253–276.

Singh, Y.; Ramteke, P. W.; Shukla, P. K. Isolation and Characterization of Heavy Metal Resistant *Pseudomonas spp.* and their Plant Growth Promoting Activities. *Adv. Appl. Sci. Res.* **2013,** *4,* 269–272.

Song, B.; Xu, P.; Chen, M.; Tang, W.; Zeng, G.; Gong, J.; Ye, S. Using Nanomaterials to Facilitate the Phytoremediation of Contaminated Soil. *Critic. Rev. Environ. Sci. Technol.* **2019,** *49* (9), 791–824.

Song, B.; Xu, P.; Zeng, G.; Gong, J.; Zhang, P.; Feng, H.; Ren, X. Carbon Nanotube-Based Environmental Technologies: the Adopted Properties, Primary Mechanisms, and Challenges. *Rev. Environ. Sci. Bio Technol.* **2018,** *17* (3), 571–590.

Srivastav, A.; Yadav, K. K.; Yadav, S.; Gupta, N.; Singh, J. K.; Katiyar, R.; Kumar, V. Nano-Phytoremediation of Pollutants from Contaminated Soil Environment: Current Scenario and Prospects. In *Phytoremediation*; Springer: Cham, 2018, pp 383–401.

Su, Y.; Yan, X.; Pu, Y.; Xiao, F.; Wang, D.; Yang, M. Risks of Single-Walled Carbon Nanotubes Acting as Contaminants-Carriers: Potential Release of Phenanthrene in Japanese Medaka (Oryzias latipes). *Enviro. Sci. Technol.* **2013,** *47* (9), 4704–4710.

Sun, Y. P.; Li, X. Q.; Cao, J.; Zhang, W. X.; Wang, H. P. Characterization of Zero-Valent Iron Nanoparticles. *Adv. Colloid Interface Sci.* **2006,** *120,* 47–56.

Suriyaprabha, R.; Karunakaran, G.; Yuvakkumar, R.; Rajendran, V.; Kannan, N. Silica Nanoparticles for Increased Silica Availability in Maize (*Zea Mays.* L) Seeds under Hydroponic Conditions. *Curr. Nanosci.* **2012,** *8* (6), 902–908.

Tripathi, D. K.; Singh, V. P.; Prasad, S. M.; Chauhan, D. K.; Dubey, N. K. Silicon Nanoparticles (Sinp) Alleviate Chromium (VI) Phytotoxicity in *Pisum sativum* (L.) Seedlings. *Plant Physiol. Biochem.* **2015,** *96,* 189–198.

Ullah, A.; Heng, S.; Munis, M. F. H.; Fahad, S.; Yang, X. Phytoremediation of Heavy Metals Assisted by Plant Growth Promoting (PGP) Bacteria: A Review. *Environ. Exp. Bot.* **2015,** *117,* 28–40.

Vítková, M.; Puschenreiter, M.; Komárek, M. Effect of Nano Zero-Valent Iron Application on As, Cd, Pb, and Zn Availability in the Rhizosphere of Metal (loid) Contaminated Soils. *Chemosphere* **2018,** *200,* 217–226.

Wang, S.; Wang, F.; Gao, S.; Wang, X. Heavy Metal Accumulation in Different Rice Cultivars as Influenced by Foliar Application of Nano-Silicon. *Water Air Soil Pollut.* **2016,** *227* (7), 1–13.

Wang, X.; Liu, Y.; Zhang, H.; Shen, X.; Cai, F.; Zhang, M.; Tao, S. The Impact of Carbon Nanotubes on Bioaccumulation and Translocation of Phenanthrene, 3-CH 3-Phenanthrene and 9-NO 2-Phenanthrene in Maize (*Zea mays*) Seedlings. *Environ. Sci.: Nano* **2016,** *3* (4), 818–829.

Wong, C. C.; Wu, S. C.; Kuek, C.; Khan, A. G.; Wong, M. H.The Role of Mycorrhizae Associated with Vetiver Grown in Pb-/Zn-Contaminated Soils: Greenhouse Study. *Restor. Ecol.* **2007,** *15* (1), 60–67.

Wu, S.; Wang, F.; Li, Q.; Wang, J.; Zhou, Y.; Duan, N.; Wang, Z. Photocatalysis and Degradation Products Identification of Deoxynivalenol in Wheat Using Up Conversion Nanoparticles@ TiO_2 Composite. *Food Chem.* **2020,** *323,* 126823.

Yadav, K. K.; Gupta, N.; Kumar, A.; Reece, L. M.; Singh, N.; Rezania, S.; Khan, S. A. Mechanistic Understanding and Holistic Approach of Phytoremediation: A Review on Application and Prospects. *Ecol. Eng.* **2018,** *120,* 276–298.

Yadu, B.; Chandrakar, V.; Korram, J.; Satnami, M. L.; Kumar, M.; Keshavkant, S. Markers in Fluoride Stressed *Cajanus cajan* L. *J. Hazard. Mater.* **2018,** *353,* 44–52.

Yano, H.; Omura, H.; Honma, Y.; Okumura, H.; Sano, H.; Nakatsubo, F. Designing Cellulose Nanofiber Surface for High Density Polyethylene Reinforcement. *Cellulose* **2018,** *25* (6), 3351–3362.

Zamani, N.; Mehrpour, O.; Hassanian-Moghaddam, H.; Jalali, M.; Amirabadizadeh, A.; Samie, S.; Sabeti, S.; Kolahi, A. A, A Preliminary Report on the Largest Ongoing Outbreak of Lead Toxicity in Iran. *Sci. Rep.* **2020,** *10,* 11797.

Zand, A. D.; Tabrizi, A.M. Effect of Zero-Valent Iron Nanoparticles on the Phytoextraction Ability of Kochia Scoparia and Its Response in Pb Contaminated Soil. *Environ. Eng. Res.* **2021,** *26* (4): 200227.

Zhang, J.; Gong, J. L.; Zeng, G. M.; Yang, H. C.; Zhang, P. Carbon Nanotube Amendment for Treating Dichlorodiphenyltrichloroethane and Hexachlorocyclohexane Remaining in Dong-Ting Lake Sediment—An Implication For In-Situ Remediation. *Sci. Total Environ.* **2017,** *579,* 283–291.

Zheng, L.; Hong, F.; Lu, S.; Liu, C. Effect of Nano-TiO_2 on Strength of Naturally Aged Seeds and Growth of Spinach. *Biol. Trace Elem. Res.* **2005,** *104* (1), 83–91.

Zhu, Y.; Xu, F.; Liu, Q.; Chen, M.; Liu, X.; Wang, Y.; Zhang, L. Nanomaterials, and Plants: Positive Effects, Toxicity and the Remediation of Metal and Metalloid Pollution in Soil. *Sci. Total Environ.* **2019,** *662,* 414–421.

Zhu, Y. L.; Pilon-Smits, E. A.; Jouanin, L.; Terry, N. Overexpression of Glutathione Synthetase in Indian Mustard Enhances Cadmium Accumulation and Tolerance. *Plant Physiol.* **1999,** *119,* 73–80.

Zuo, R.; Liu, H.; Xi, Y.; Gu, Y.; Ren, D.; Yuan, X.; Huang, Y. Nano-SiO_2 Combined with A Surfactant Enhanced Phenanthrene Phytoremediation by Erigeron Annuus (L.) Pers. *Environ. Sci. Pollut. Res.* **2020,** *27* (16), 20538–20544.

Zekić, E.; Vuković, Z.; Halkijević, I. Application of Nanotechnology in Wastewater Treatment. *Građevinar* **2018,** *70* (4), 315–323.

CHAPTER 3

Nanoremediation: Prospects and Amendments

RUQIA BANO#, SATYAJIT SINGH SAINI#, JAHANGEER, NEHA SAMI, and SIRAJUDDIN AHMED

Department of Environmental Sciences, Jamia Millia Islamia, New Delhi, India

#Both authors have contributed equally.

ABSTRACT

Eco-friendly management of pollutants present in different environmental arenas including soil, water, and air is the need of the hour. Although the bioremediation technique has emerged as a knight in shining armor, it is ineffective in remediating higher concentrations of pollutants which include heavy metals (HMs) and salts because they pose toxicity to microbes. With the advent of nanotechnology and fast-paced development in studies of nanomaterials, their innovative role in providing effective remediation remedies to a broad spectrum of environmental challenges is increasingly gaining importance. Microbes integrated with nanomaterials improve the HMs reduction which makes nanomaterials more effective comparable to individual applications of them. Nanoremediation relies on the characteristic properties of nanosized materials, that is, the high surface area along with high reactivity, making it applicable for a wide range of pollutants that are hazardous to the environment, including HMs, hydrocarbons, and organo-halogenated compounds. It can be used under different environmental conditions and it offers easy and fast remediation. Despite being advantageous, it has been observed that the complete removal of hazardous contaminants is still uncertain. Additionally, the persistence of nanomaterials in the environment, its probable risks regarding the generation of toxic end products, and the issues related to human safety are the least studied, which might lead to unexpected

Sustainable Nanoremediation: Modern Technologies for a Clean Environment.
Fernanda Maria Policarpo Tonelli, Arpita Roy, and Khalid Rehman Hakeem (Eds.)

consequences. A few dimensions that can be further focused on are the green synthesis of nanomaterials, assessment of sustainability, nanosensing, use of carbon nanotubes (CNTs), zero-valent iron, etc. Further, the nanomaterials can be supplemented or substituted with other metallic/nonmetallic oxide nanoparticles (NPs) for dealing with a high concentration of pollutants. Thus, nanoremediation needs modifications and amendments to enhance its efficacy and acceptance among environmentalists and scientists as a better alternative.

3.1 INTRODUCTION

Eco-friendly management of pollutants present in different environmental arenas including soil, water, and air is the need of the hour. Although the bioremediation technique has emerged as a knight in shining armor, it is ineffective in remediating higher concentrations of pollutants which include heavy metals (HMs), and salts because they pose toxicity to microbes. With the advent of nanotechnology and fast-paced development in studies of nanomaterials, their innovative role in providing effective remediation remedies to a broad spectrum of environmental challenges is increasingly gaining importance. Microbes integrated with nanomaterials improve the HMs reduction which makes nanomaterials more effective and comparable to individual applications of them. Nanoremediation relies on the characteristic properties of nanosized materials, that is, the high surface area along with high reactivity, making it applicable to a wide range of pollutants that are hazardous to the environment, including HMs, hydrocarbons, and organohalogenated compounds. It can be used under different environmental conditions and it offers easy and fast remediation.

Despite being advantageous, it has been observed that the complete remediation of hazardous contaminants is still questionable and its probable risks regarding toxicity, etc. are being least studied that might lead to unexpected consequences, which is why certain amendments are necessary for nanoremediation. A few dimensions that can be further focused upon are the green synthesis of nanomaterials, nanosensing, current practices employing carbon nanotubes (CNTs), zero-valent iron, etc. can be supplemented or substituted with graphene oxide (GO) nanoparticles (NPs) for arsenic contamination, photocatalysis, assessment of sustainability, etc. can be added along while doing nanoremediation for pollution free environment (Table 3.1). Thus, nanoremediation needs modifications and amendments to enhance its efficacy and acceptance among environmentalists and scientists as a better alternative. The present chapter will explain the benefits, risks, and amendments, of nanoremediation.

TABLE 3.1 Terminology Associated with Nanoremediation.

Term	Description	References
Nanotechnology	Nanoscale leveled technology in which components such as materials or systems are produced when the matter is limited to the nanoscale length for stimulating the properties that are unique to the material.	Kreyling et al., 2010
Nanomanufacturing	Manufacturing at nanoscale and manifested *via* top-down or bottom-up methodologies.	Bhushan, 2017
Nanoscale	Scale of range 1–100 nm.	Nasrollahzadeh et al., 2019
Nanomaterial	Nanomaterial necessarily has at least one dimension (1D) in the nanoscale range of 1–100 nm.	Trotta et al., 2019
Nanoobject	Discrete pieces of a particular material with one to three external dimensions ranging in level of nanoscale.	Jeevanandam et al., 2018
NP	An object having all of its dimensions in the nanoscale range.	Trotta et al., 2019
Aspect ratio	The ratio of the length to the width of the major axis and minor axis, respectively.	Murphy et al., 2002
Nanosphere	An NP having aspect ratio 1.	Murphy et al., 2002
Nanorod	This term is used in the case of different lengths of shortest and longest axes. Nanorods have a width ranging from 1 to 100 nm with aspect ratio always greater than 1.	Jeevanandam et al., 2018
Nanofiber	When two dimensions have a length in the nanoscale range and the third dimension is significantly large.	Trotta et al., 2019
Nanowire	These are analogous to nanorods, however, having a higher aspect ratio.	Murphy et al., 2002
Nanotube	Hollow nanofibers are referred to as nanotubes.	Trotta et al., 2019
Nanostructured material	Materials having molecules, structural elements, crystallites, or clusters with dimensions range-bound within 1–100 nm.	Moriarty et al., 2001
Engineered nanomaterials	Deliberately developed materials having one or more dimensions of 100 nm or lesser.	Kreyling et al., 2010
Nanocomposite	This includes multicomponent material with many different phase domains, of which at least one of the phases has the least 1D of nanometers.	Chen et al., 2008

3.2 METHODS FOR NANOPARTICLE SYNTHESIS

The particles are scaled at the nanometer level and at least 1D is to be kept at 100 nm or less. NPs these days are produced by various methods which include chemical, biological, and physical. Environmental-friendly NPs are biologically synthesized out of various organisms including fungi, plants, and bacteria (Yadav et al., 2017), which are utilized in diverse applications in many sectors. However, plants happen to be more appropriate in comparison to fungi or bacteria for the synthesis of NPs, as they require less time for incubation for the reduction of metal ions. The mechanism behind the synthesis can be categorized as (Fig. 3.1):

a. Outside to inside also known as top down, it turns the bigger bulk material into smaller particles.
b. Bottom-to-top or bottom–up, this methodology involves small particles building bigger bulk material.

Precisely the synthesis of NPs involves inert gas condensation, hydrothermal synthesis, coprecipitation, ion sputtering scattering, sol–gel, microemulsion, pulse laser ablation, microwave, sonochemical, biological synthesis, template synthesis, and spark discharge.

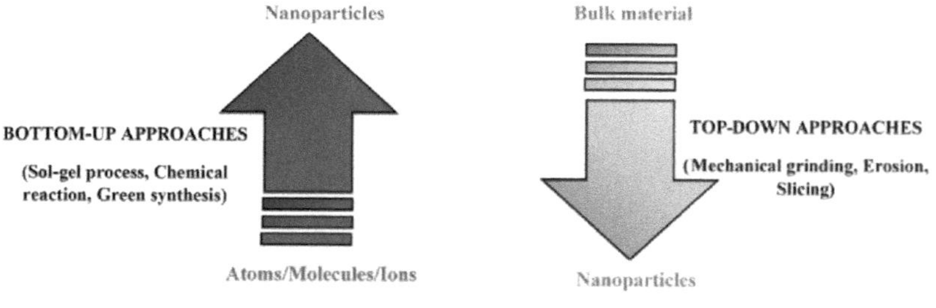

FIGURE 3.1 Approaches for synthesis of NPs.

As per top–down approach, bulk materials get broken to generate nanostructured particles and it includes electrospinning, mechanical milling, sputtering, lithography, and laser ablation. Some of them have been discussed in the following subsections.

3.2.1 ELECTROSPINNING

For nanostructured particle synthesis, electrospinning is among the simplest of top-down methods. It can be used for generation of nanofibers from wide range of materials, usually the polymers. Coaxial electrospinning was a major breakthrough in the domain of electrospinning. In this, the spinneret is comprised of two coaxial capillaries having two viscous liquids, or one viscous liquid as shell and another nonviscous liquid facilitating as core, which can easily be used to form core–shell type of nanoarchitectures in an electric field. On a large scale, core–shell ultrathin fibers can be obtained by coaxial electrospinning in an effective manner. The length of such ultrathin nanomaterial is extendable to a few centimeters linearly. For manufacturing of core–shell hollow polymer, organic, hybrid, and inorganic materials, this method is used.

3.2.2 MECHANICAL MILLING

It is a very efficient methodology with respect to the cost of production of NPs. For developing different phase blends and in producing nanocomposites, mechanical milling is very effective. Oxide and carbide-strengthened alloys of aluminum, aluminum/magnesium/nickel/copper-based nanoalloys, wear-resistant spray coatings, and many other nanocomposite materials are developed by this method. Ball-milled carbon nanomaterials are seen as a novel category of nanomaterial, efficient in performing environmental site remediation, energy conversion, and energy storage demands. Overall, mechanical milling is a prime method for commercial use.

3.2.3 SPUTTERING

High-energy plasma particles or gas particles with high energy are bombarded on solid surfaces. Sputtering is efficient methodology for production of thin films made up of nanomaterials (Ayyub et al., 2001). When gaseous ions with high energy are bombard on the target surface, they cause physical ejection of small clusters of atoms depending upon the energy of incident gas-ion (Son et al., 2017; Wender et al., 2013). This particular technique is considered to be interesting because the composition of both sputtered nanomaterial and the target material remains the same but with few impurities and also it is cost effective than electron–beam lithography, comparably (Nie et al., 2009).

3.2.4 LITHOGRAPHY

Nanoarchitectures are primarily developed by a focused beam of electrons or light. Lithography is divided into two main types (Pimpin et al., 2012):

1. Masked lithography
2. Maskless lithography

In masked nanolithography, a specific template or mask is used to transfer nanopatterns over a large surface area. Masked lithography has methodologies such as photolithography (Szabó et al., 2013), nanoimprint lithography (Kuo et al., 2003) and soft lithography (Yin et al., 2000). Maskless lithography has methodologies like focused ion–beam lithography (Matsumoto, et al., 2020), electron–beam lithography, and scanning probe lithography (Xu and Chen, 2020). In maskless lithography, without the mask, arbitrary nanopattern writing is done. Implantation of ions with a focused ion beam with a wet chemical etching is used to achieve 3D freeform micronanofabrication. (Garg et al., 2020).

As per bottom–up approaches, small particles build bigger bulk material. Bottom–up methodology includes the reverse micelle method, sol–gel method, chemical vapor deposition (CVD), and soft and hard templating methods.

3.2.5 THE SOL–GEL METHOD

For the production of nanomaterials this methodology uses wet–chemical technique. High-quality metal oxide-based nanomaterials are produced by the sol–gel method. The liquid precursor is transformed to a sol, and the sol is finally converted into a network structure which is known as gel. Hence, the name sol–gel got derived (Danks et al., 2016). Conventional precursors are metal alkoxides for the production of nanomaterials by this methodology. The synthesis takes several steps. Firstly, in water or with help of alcohol the hydrolysis of metal oxide takes place for sol formation. The next step includes condensation, resulting in porous structure because of increase in the solvent viscosity. During the polycondensation process, oxo (M–O–M) or hydroxo (M–OH–M) bridges get formed, which results in metal–oxo- or metal–hydroxo-polymer formation (Tseng et al., 2010). Polycondensation continues with changes to the properties, porosity, and structure during the aging process. Porosity continuously decreases and the distance among colloidal particles increases during aging. Then water and organic solvents are removed from the gel in the drying process. Lastly, to achieve NPs, calcination is performed (Parashar et al., 2020). Aging time, hydrolysis rate, precursor nature, and the molar ratio of H_2O to precursor affect the final product obtained (Coelho-Escobar et al. 2014). To generate composites and complex nanostructures, this sol–gel method has proven to be cost effective and advantageous because the homogeneous output and the processing temperature is low (Parashar et al., 2020).

3.2.6 REVERSE MICELLE METHODS

When the desired shape and size of nanomaterial is to be obtained, this method becomes useful. Emulsion of oil in water brings out normal category micelles, having hydrophobic tails aimed toward a core having oil droplets trapped. However, in the other case of emulsion of water in oil, reverse micelles would be formed having hydrophilic heads pointing toward water containing core (Malik et al., 2012). The core facilitates as a nanoreactor in the synthesis of NPs. For producing nanomaterials, it behaves like the water pool. The size of synthesized NP is controlled by changing the water-to-surfactant proper ratio. Decreased water concentration results in small water drops, ultimately resulting in NPs formation which are small (Soleimani Zohr Shiri et al., 2019). Thus, this method provide facility to precisely control the size of NP. NPs developed through them are amazingly superfine and monodispersed in nature, NPs can be obtained by the reverse micelle method (Nguyen, 2013).

3.2.7 CHEMICAL VAPOR DEPOSITION (CVD)

The chemical vapor deposition method plays a vital role in the development of carbon-based nanomaterials. In the CVD, the chemical reaction of vapor phase precursors led to the formation of a thin film over the surface of the substrate (Jones and Hitchman, 2009). High chemical purity, adequate volatility, good stability during evaporation, nonhazardous nature, a long shelf-life, and low cost are characteristics of precursors. Also, it should not lead to residual impurities after decomposition. For example, during the generation of CNTs, a substrate is put in the oven and exposed to high temperature. Subsequently, a precursor generally carbon-containing gas such as hydrocarbons is slowly introduced to the system. The decomposition of the gas releases carbon atoms at high temperatures, which recombine and form CNTs, on the substrate (Shah and Tali, 2016). The type of catalyst determines the type and morphology of nanomaterial generated. In the CVD based preparation of graphene, Cu catalyst provides monolayer graphene whereas nickel (Ni) and cobalt (Co) catalysts provide multilayer graphene

(Ago, 2015). CVD is the perfect method for producing high-quality nanomaterials (Machac et al., 2020), and it is respectfully known for the generation of 2D nanomaterials.

3.3 ROLES OF NANOMATERIALS FOR REMEDIATION OF POLLUTION

The process of environmental remediation utilizes a number of methods for the removal of pollutants from the environment. Formerly, a bulk of nanomaterials capable of environmental remediation applications have been designed and employed. The unique specific physiochemical, surface, and optical electronic characteristics of NPs compared to other macroscale materials enable them to tackle issues that seemed formerly difficult to resolve with conventional methods. NPs, for instance, have been employed to remediate various sites polluted by hazardous wastes such as chlorinated solvents, and for the remediation of oil spills (Table 3.2).

3.3.1 METAL-BASED NANOMATERIALS

Metals and metal oxides such as gold, silver, titanium oxide, and iron oxide are the eminently surveyed NPs for environmental remediation (Khin et al., 2012). AgNPs with antibacterial properties are extensively recommended in the application of wastewater treatment (Bosetti et al., 2002; Chou et al., 2005). Gold NPs assist in treating drinking water contaminated with inorganic mercury. TiO_2 NPs are being widely employed for the oxidative and reductive conversion of impurities in the air and water media, attributed to their properties such as semi-conductivity, nontoxicity, photocatalysis, and energy conversion (Adesina, 2004; Mahendra et al., 2008). Zinc oxide (ZnO), is used as a strong adsorbent for the removal of HMs (Wang et al., 2010) showing a higher removal rate of copper ions and a good capacity for absorbing lead (Khin et al., 2012).

3.3.2 SILICA-BASED NANOMATERIALS

Mesoporous silica nanomaterials (MSNs) because of their remarkable attributes such as excessive surface area, adjustable pore size, considerable pore volume, and adjustable pore size have been found to have greater application in the field of environmental remediation(Tsai et al., 2016). MSNs perform as excellent adsorbents and have been functionalized with thiol and amino functional groups and used in the deduction of toxic metals and organic dyes from wastewater (Son et al., 2008). Amine-xerogels of silica-based NPs have been applied for the remediation of pollutants such as hydrogen sulfide (H_2S) and carbon dioxide (CO_2) from natural gas (Huang et al., 2003).

3.3.3 CARBON-BASED NANOMATERIALS

Carbon nanomaterials include grapheme, fullerenes, and carbon nanotubes (CNTs). Fullerenes and their composites are reported to be used in water disinfection because of their antibacterial properties. In addition, fullerenes can be potentially utilized in membrane technology because of their ease of functionalization, their strength, electron affinity, etc. (Baby et al., 2019). CNTs, either single-walled CNTs (SWCNTs) or multi-walled CNTs (MWCNTs), are the chiefly explored carbon nanomaterials being used in the remediation of environmental pollution. They have an excellent adsorption capacity and are extensively used as selective and powerful adsorbents for the sterilization of adulterated air and drinking water (Tyagi et al., 2017; Ajayan and Zhou, 2001). Graphene with a hexagonal structure of carbon atoms is said to have enormous strength for the adsorption of HM ions from water.

3.4 NANOREMEDIATION OF SOIL, WATER, AND AIR MEDIA

3.4.1 NANOREMEDIATION OF SOIL

Soil contamination is among the big environmental challenges posing a threat to the production of agriculture and the security of food (Bakshi and Abhilash, 2020). Furthermore, the accumulation of toxic substances such as HMs, persistent organic pollutants (POPs), and herbicides augments the issue. The biomagnification of these pollutants, in turn, toxify the food chain in the ecosystem. Thus, remediating the soil of such pollutants is the dire need of the hour. One of the most favored approaches for soil remediation is the adsorption or immobilization of the contaminants. On-site immobilization provides a feasible, economical, and environmental-friendly choice for the decontamination of soil pollutants (Bhadouria et al., 2020). Nanomaterials such as carbon-based nanotubes, oxides of metals such as titanium dioxide, and various nanocomposites are approved to be the outstanding adsorbents of various

TABLE 3.2 Nanoparticles and their Application in Environmental Remediation (Modified from Thangavelu et al., 2022).

Nanomaterial	Application	References
	Metal-based nanomaterials	
Silver NPs (AgNPs)	Wastewater treatment due to antibacterial properties	Bosetti et al., 2002;
		Chou et al., 2005;
		Jin et al., 2012
Iron NPs	Remediation of groundwater by reducing metal ions effectively	Savage and Diallo, 2005
Gold NPs	Eliminates inorganic mercury from drinking water	Pradeep et al., 2009
Zinc oxide (ZnO)	Adsorption of HMs such as lead and copper	Wang et al., 2010
		Khin et al., 2012
Titanium dioxide (TiO2)	Air and wastewater purification	Adesina, 2004
TiO2-based nanomaterials Removal of SO2 from air		Chen et al., 2013
	Silica-based nanomaterials	
Mesoporous silica	The adsorbent of air pollutants	Tsai et al.,2016
Nanomaterials (MSN)		
SiO2	Removes mercury from aqueous solution	Thangavelu et al., 2022
Amine-modified xerogels	Removal of carbon dioxide and H2S from natural gas	Huang et al., 2003
	Carbon-based nanomaterials	
Fullerenes (C60)	Utilized for killing microbes via a photocatalytic process	Brunet et al., 2009
MWCNTs	Removes HM ions	Yang et al., 2009
		Ntim et al., 2012
		Tang et al., 2012
		Luo et al., 2013
Graphene	Adsorption of metal contaminants from wastewater	Tabish et al., 2018
		Zhang et al., 2018

soil impurities. Ferric oxide NPs, for example, have exceptional potential of adsorbing HMs such as cadmium (Cd) and arsenic (As) from the soils (Gaur et al., 2014). Magnetic nanomaterials offer a long-term usage for the treatment of pollution via adsorption and catalysis. (Ajmal et al., 2018). Sodium carboxymethyl cellulose-stabilized nanozero-valent iron (NZVI) has been reported to withdraw 80% of the chromium [Cr (VI)] bound in the soil(Wang et al., 2014). Nanomaterials employ a reductive mechanism for the removal of redox-sensitive pollutants. Adding NZVI to pyrene-adultered soil remediates the soil of pyrene by the mechanism of reduction through the generation of redox conditions. Also, the NPs are utilized in the advanced oxidation process for the decontamination of organic soil contaminants (Usman et al., 2020). Green NPs have a large opportunity in the application of localized treatment of the contaminated sites. In addition to the HM removal, ferrous NPs are effective as eco fertilizers synchronizing the nutrient uptake of plants while decreasing the contamination of groundwater (Prasad et al., 2010).

3.4.2 NANOREMEDIATION OF WATER

The wastewater discharge from various factories and industries and its influx into the water bodies is hazardous to both, the health of humans and to the environmental health. In this regard, the primary pollutants include organic compounds, HMs, and oils among others making the water unfit for consumption (Lu et al., 2016). Nanomaterials as adsorbent systems effectively withdraw contaminants from the water. Metal oxide NPs such as TiO_2 and ZnO are employed in water purification owing to their higher reactivity, adsorbent, and photolytic properties (Aragaw et al., 2021). Iron NPs are effective in treating dyes due to their stability in the suspension medium. Electrospun polyether sulfone nanofibers containing vanadium NPs are employed for the extraction of methylene blue dye from water. However, adsorption is not a complete solution for dye removal. One of the most explored realistic destructive methods for dyes is photocatalysis. Metal oxide NPs as semiconductors work as excellent photodecomposers capably degrading the dye contaminants. Nanomaterials and their composites function as supreme filtration membranes in permeable aggregating units (Bansal et al., 2015). The effectiveness of nanofiltration is enhanced by using membranes as support for other materials. For example, an increment in salt rejection rate was observed using carboxylated MWCNTs in polyamide membranes (Al-Hobaib et al., 2017).

3.4.3 NANOREMEDIATION OF AIR

Air pollution is among the notable environmental issues faced worldwide impacting global climate change and public health. Nanotechnology has a role to play in the control of air pollution in terms of remediation, pollution prevention, detection, and sensing. Nanocatalysts work by accelerating the conversion of hazardous vapors from automobiles and industries into nontoxic gases. Surface modification of silica nanomaterials offers an efficient strategy for fabricating novel catalysts and adsorbents. Mesoporous silica surface modified with amine groups has been reported to trap H_2S and CO_2 from natural gas (Huang et al., 2003). Filters are beneficial for air pollution control which being porous allows the gas to pass through while trapping the particles inside. Nanofilters because of nanoscale pores are more competent than traditional ones in removing a variety of microbes and organic contaminants. GO membrane on account of its enormous surface area and continuous pore structure has been applied for capturing particulate matter (PM 2.5) from the air (Jung et al., 2018; Zuo et al., 2019). Nanosensing is yet another application of nanotechnology for air pollution remediation which can identify a detectable physical change at nanoscale dimensions. Modified MWCNTs or SWCNTs detect H_2S and SO_2 (Zhang et al., 2012), CO and NH_3 (Dong et al., 2013), NO_2 and NH_3 (Kim et al., 2016), NO_2 (Park et al., 2019), VOCs (Amade et al., 2014), NO_x, and CO_2 (Su et al., 2009).

3.5 BENEFITS OF NANOREMEDIATION

It utilizes NPs which have certain characteristics and properties, which make this process beneficial. Let us look at a few such qualities.

3.5.1 EXTRAORDINARY THERMAL AND ELECTRICAL CONDUCTIVITY

Compared to bulk counterparts, extraordinary electrical and thermal conductivity can be exhibited at the nanoscale level. For example, graphene obtained from graphite (Krishnan et al., 2019).

3.5.2 CATALYST SUPPORT

2D sheets of various nanomaterials aid the right dispersion of NPs of the active catalyst, by which, the performance of the catalyst substantially enhanced (Zhu et al. 2020). Catalysts when atomically dispersed on nanomaterial 2D sheets act like performance boosters (Liu et al. 2020).

3.5.3 QUANTUM EFFECTS

Such effects are enhanced at the level of the nanoscale. However, the optimum size for these effects to appear firm depends on the nature of the material of semiconductor (Geoffrion and Guisbiers, 2020).

3.5.4 SURFACE AREA

Generally, the surface area is considerably high in nanomaterials when compared with bulk counterparts (Tomar et al., 2020).

3.5.5 ANTIMICROBIAL ACTIVITY

Some nanomaterials have a great capacity to deal with pathogen-related diseases as they possess antifungal, antiviral, and antibacterial properties (Makvandi et al., 2020; Castro et al., 2017).

3.5.6 MAGNETISM

At the nanoscale, elements display changed magnetic behavior. A nonmagnetic element might become a magnetic element at the nanoscale size (Roduner, 2006).

3.5.7 AMAZING MECHANICAL PROPERTIES

Nanomaterial, generally, exhibits great mechanical properties which outshine their macroscopic other counterparts (Wu et al., 2020).

Because of such reasons, nanoremediation becomes very helpful in many ways such as enhanced extent and speed of contaminant clearance, source term treatment capability, compatibility with other treatments, extended range of treatable contaminants, extended range of environmental conditions, and most importantly, it offers in situ treatment.

3.6 RISKS AND RECOMMENDATIONS

3.6.1 RISKS

Nanoremediation, no doubt, is an emerging economical and facile onsite remediation technique but it lacks certification as a complete sustainable approach. There is a lack of awareness about the behavior of NPs in the environment. The uncertainty associated with the conveyance and fate of nanomaterials is posing a hindrance in the assessment of risks associated with their use. Risks are associated with their use due to uncertainties about their persistence, dispersal, bioaccumulation, ecotoxicity, and reversibility. Furthermore, there is a lack of legislative and regulatory guidelines for the use of nanomaterials, at both the national and global levels. (Corsi et al., 2018, Patil et al., 2016). Nanoscale materials unexpectedly spread to large distances. This easy dispersal is attributed to their smaller size and higher mobility causing ecotoxicity. Nanozero-valent iron (NZVI), upon application into the medium rapidly aggregate forming clusters and losing effectiveness as NP. Besides self-aggregation, the nanoscale particles may agglomerate onto the suspended solids and sediments leading to their bioaccumulation in the soil. This in turn ensures their entry into the food chain and leads to their biomagnification in the organisms at higher trophic levels. In the case of plants, a higher concentration of NZVI has been found to reduce the rate of transpiration and affect the translocation of food to the shoots causing stunted growth and finally the death of the plant. Besides the health of the ecosystem, NPs have implications for human health as well. Exposure of humans to nanomaterials is found to be linked with genotoxicity, fat peroxidation, oxidative stress, and pulmonary diseases (Sharma et al., 2007).

3.6.2 RECOMMENDATIONS

As discussed in the section above, the process of nanoremediation besides being rapid and competent has various risks and challenges associated therein. The issues related with the process can be managed through ensuring better and effective solutions and recommendations.

3.6.2.1 ECOTOXICOLOGICAL TESTING AND PREDICTIVE SAFETY ASSESSMENT TOOLS

A comprehensive ecologically safe prognostic estimation approach should be executed which shall direct the following essential features:

a. Estimating the fate of NMs in the media which are to be decontaminated.
b. Considering the nature of the contaminants and the characteristics of the contaminated media.

c. Identifying all possible targets of NMs toxicity and providing an ecotoxicity evaluation.

3.6.2.2 GREENER NANOSOLUTIONS FOR REMEDIATION

Green nanotechnology refers to the use of green nanoproducts for the environmental sustainability. Utilization of green synthesized nanomaterials produced from plant parts decreases the release of noxious derivatives into the surrounding environment thus curtailing the ecotoxicity. Likewise, utilizing bio-NPs obtained from microorganisms can provide an efficient approach for the treatment of acid mine water.

3.6.2.3 ECOLOGICAL SAFETY AND INDUSTRIAL COMPETITIVENESS

A strong collaboration is absolutely needed in the discipline of environmental remediation between the research and the industrial sector. Researchers and industries must together play a role in the choice of approaches permitting the expansion of the material and the methodologies developed.

3.7 AMENDMENTS IN NANOREMEDIATION

3.7.1 NANOBIOREMEDIATION

Nanobioremediation is an economical process of employing plants and microbes in combination with several NPs for the disintegration of various environmental contaminants, which eventually improves the health of soil besides combatting pollution. This technology is attracting major attention as a facile technique for the sustainable cleanup of the environment. Most of the techniques available for environmental remediation work on the principle of excavation of the contaminated substance and then its ex situ treatment using traditional methods. Such methods are highly energy demanding and thus are uneconomical. Among various recently developed approaches to remediation are physical and chemical remediation, incineration, and bioremediation. Bioremediation, among others, presents an eco-friendly and cost-effective approach to the reduction of environmental pollution (Singh and Walker, 2006). Bioremediation has been studied in detail in the past and with time attempts were made to improve the efficiency of the process using various chemical additives or through augmentation via biotechnology, (Gerhardt et al., 2017), nonetheless, the emerging nanotechnology has further put in a newer aspect to the process (Gong et al., 2018). Nanobioremediation employs nanocatalysts which get penetrated deeply into the contaminants and treat them safely with the help of microbes. The major concept of nanobioremediation is defined by the size of the NPs and can produce better results compared to the microparticles by penetrating deep into the contaminated sites and the cleanup process is much accelerated compared to the normal bioremediation technology. The technique of nanobioremediation is a two-stage process involving both the abiotic and biotic processes (Usman et al., 2020). The first phase involves the integration of NPs into the system and the exposure of contaminants to various physicochemical processes such as adsorption and photocatalysis (Abebe et al., 2018). The second phase involves biotic processes such as bioaccumulation, biostimulation, and biotransformation (Desiante et al., 2021; Filote et al., 2021). These second-phase processes are crucial for the eradication of pollutants from the contaminated system. Although the novel approach of nanobioremediation has gained massive attention but a handful of available literature will not allow the framing of a categorical decision.

3.7.2 NANOBIOREMEDIATION OF HEAVY METALS

HMs are categorized as the major inorganic impurities manifesting considerable toxicity to biological life even at the lowest possible concentrations (Alissa and Ferns, 2011; Jaishankar et al., 2014). HMs primarily affect the plants and the lower soil organisms by generating reactive oxygen species (ROS), which result in damage to biomolecules such as nucleic acids (Jaishankar et al., 2014), damage to crops and vegetation, their nutritional quality, and their associated ecological aspects. Site stabilization is one of the principal techniques for the removal of HMs wherein the HMs are immobilized site specifically decreasing their transportation and availability in the soil, thereby preventing their leaching across different possible sites (Suman et al., 2018). The use of nanotechnology along with biotechnology, has been an attractive avenue for the removal of HMs (Rajput et al., 2021). Bioremediation of the Pb and Cd contaminated soil brought about by Gram-negative bacteria, *Halmonas* sp., was successfully enhanced using polyvinylpyrrolidone (PVP) coated FeO NPs. When compared to the separate treatment by bacteria or by NPs alone, this approach was notably found to remove almost 100% of Pb after 24 h and Cd after 48 h of treatment (Alabresm et al., 2018). A recovery rate of 83.0–89.5% for Cu^{2+}, 99.4–100% for Pb^{2+}, and 92.6–97.5% for Ni^{2+} was observed when the contaminated site was treated with a biosorbent of magnetic Fe_3O_4 NPs integrated with *Staphylococcus aureus*, with a surface encapsulated with phthalic acid. This work revealed that both the NP core and the functional groups on the microbial surface had a remarkable impact on the elimination of the contaminants (Mahmoud et al., 2016).

A threefold approach is now gaining importance as the HM pollutants can be selectively employed by microbes to prepare biogenic NPs (resource recovery), thereby eliminating them from the environment (remediation) and generating utility for the waste (effective waste utilization). A study on the biorecovery of Pd using *Enterococcus faecalis* with Pd NPs reported that the resulting bionanocatalyst showed good catalytic efficiency and was potentially useful for the treatment of industrial effluents (Cao et al., 2020).

3.7.3 NANOBIOREMEDIATION OF PERSISTENT ORGANIC POLLUTANTS

The POPs are potentially harmful to both the environment and to human health. Many of them tend to bioaccumulate in the adipose tissue and are potential carcinogens. Thus, the elimination of such pollutants is challenging and mandatory. A bacterial strain (NM05 of *Sphingomonas*) when applied along with Pd/Fe0 NPs (CMC-Pd/nFe0) to hexachlorocyclohexane (HCH) contaminated site synergistically enhanced the degradation of HCH by about 1.7–2.1 fold compared to the individual application of both (Xie et al., 2011).

The degradation of PAHs on the marine sediments was notably boosted by applying perovskite ($LaFeO_3$) NPs and biochar from water caltrop (*Trapa natans*) shells. The study reported an increase in the microbial diversity of sediment and the major observed phylum was *Proteobacteria* initially, however after the process, *Hyphomonas* was predominantly observed. In the case of soil remediation, the pollution load can also be removed by enriching the microbial community via application of NPs. Microbial colonization and biomass have been reported to increase after application of Si NPs thereby improving soil health (Srivastava et al., 2021; Gajić et al., 2018). However, prolonged exposure to nanomaterials can affect the nutrient and organic matter content of the soil.

3.7.4 NANOMYCOREMEDIATION

Nanomycoremediation is the application of fungi in the elimination of toxins from the contaminated site. It is highly effective in HM contamination. Nanomycoremediation can become a savior in biological and chemical warfare. Research are going on in this direction. Toxic material becomes a threat to the survival of all species and the ecosystem as a whole gets damaged. The role of fungi has been taken to the fore through its application at the nanoscale. NP is a complete unit in the context of its transport and the general properties they exhibit.

Fungi develop as a mycelial mesh that enables them to bear tough conditions including agitation and flow pressure. Filamentous fungi grow on cheap and easily available substrates. They are capable of producing a wide range of commercially utilized metabolites. For metal NP fabrication, various fungal strains are used as efficient resources such as *Penicillium*, *Fusarium*, *Verticillium*, and *Aspergillus*. Different species of fungi are employed for the synthesis of metal NPs in both ways, extracellularly and intracellularly.

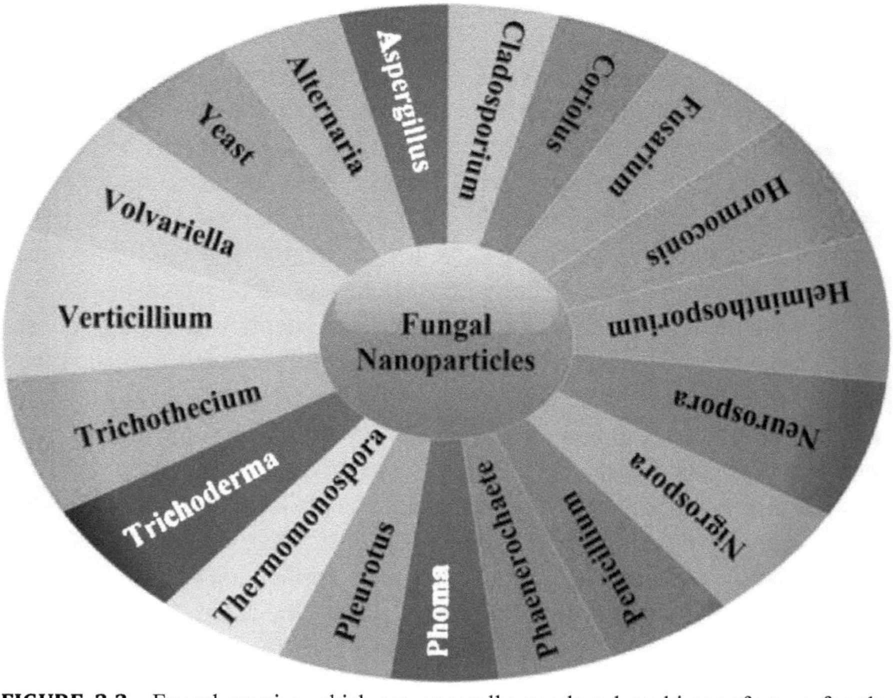

FIGURE 3.2 Fungal species which are generally employed as bionanofactory for the synthesis of NPs.

Nanomycoremediation can be employed for toxic weapons such as TNT and nerve gases such as VX and sarin. Naturally present microbes act in coherence with the fungi to break down the pollutants, to H_2O and CO_2. Along the shoreline of San Francisco, a cargo ship spilled 58,000 gallons of fuel in 2007 (Incardona et al., 2008). Human hair woven mats were used as sponges which soaked up the oil that was spilled. Further layering with straw and oyster mushrooms was done: oil was broken down by the mushrooms and within a few weeks the ensuing soil was found to be fairly clean which was good to be employed for landscaping purposes. Toxic components such as aromatic contaminants from petroleum and chlorinated persistent pesticides are decomposed effectively by wood-degrading fungi (Rhodes et al., 2014). Likewise, mycofiltration is another method, which employs mycelia as a filter which is used to treat water in the soil, by removing contaminants.

3.7.5 NANOPHYTOREMEDIATION

Phytoremediation is a cost-effective "green" technique used for the remediation of polluted environmental media contaminated with pesticides, crude oil, explosives, metals, and other pollutants. Several plants act as hyper-accumulators absorbing large amounts of pollutants at a site based on the nature of the pollutant present and other conditions of the site. Pollutants can be removed by the plants or converted into harmless products by way of bioaccumulation, extraction, degradation, or immobilization. Integrated usage of nanotechnology and phytoremediation can be observed in the process of nanophytoremediation. It has been previously reported that plants tend to produce NPs under appropriate conditions. Nanophytoremediation is an economically feasible process, and its proper utilization may offer an environmentally-friendly solution for hazardous site remediation. Literature reports that the application of NPs has an affirmative effect on plants. Application of nano-TiO_2 and nano-SiO_2 to soybean (*Glycine max*) increased the activity of nitrate reductases thereby increasing the propagation of the plant. Nanoderivatives such as carbon dots have been found to promote growth in mung bean by inducing physiological responses in the plant.

The identification and the transfer of hyper-accumulator exclusive genes to the other species of transgenic plants can improve the capacity of phytoremediation. Plants have an inbuilt capability to reduce metals through their distinct metabolic pathways (Handy et al., 2008). *Brassica juncea* seedlings are being utilized for the in vitro production of Ag NPs (Shekhawat and Arya,

2009). Gold NPs are reported to be synthesized from *Terminalia catappa* leaf extract in an aqueous medium (Ankamwar et al., 2010). Despite the fact that the research on synthesis of NPs from plants is in a nascent stage, more qualitative work is needed to realize the physiological, biochemical, and molecular mechanistic process relative to NPs.

Various measured steps have been taken through the different facets of phytoremediation to control the menace of pollution but there is always scope for further improvement which can be applied in this scenario. Nanophytoremediation is an emerging and encouraging technology that has gathered a wider acceptance. A number of plant families have been employed in the biological synthesis of NPs. Thus, conclusively, the technique of nanophytoremediation can be a harmonious biological clean-up choice that maintains the environmental sustainability.

3.7.6 ADVANCED POLYMERIC NANOCOMPOSITES

Nanotechnology in recent times has come ahead to change our lives and the way we live. Atoms are being manipulated to create particles of our utility. With advanced properties of electrical and thermal conductivity, tensile strength, scratch and impact resistance, fire resistance, and thermal stability, the advanced nanocomposite materials are put to meet the rising demand for high strength and lightweight automotive parts, the need for sustainable packing materials and conservation of flavor in the drinks, food and beverages industries and defense aspirations such as fire retardation, electromagnetic shielding, and ballistic protection.

Advanced Polymer nanocomposites are hot products for industry as they serve the needs efficiently and are hence well known in the nanotechnology field. Inorganic nanomaterials and polymeric matrices are being researched upon via light magnetic behaviors, toughness, electrical/thermal conductivity, stiffness, and mechanical strength. The inorganic quantum dots/NPs, 2D materials (such as graphene-based nanosheets), and nanorods/nanotubes can be affixed in the polymer-matrix by physical blending or chemical synthesis for enhanced performance. Thus, the fabricating technology for the homogeneous dispersion of the fillers in the polymer matrix has been crucial. It can be titled as "breakthrough" technology in this domain which can be utilized in optoelectronics and sensing, green energy, coating, and biomedical materials. It opens the horizon of special interest including but not limited to material selection, morphology, chemical synthesis, structural design, characterization, and applications of advanced polymer nanocomposites. The burgeoning field of technology will reach its zenith with the advancement in research and utility improvement by such advanced polymeric nanocomposites.

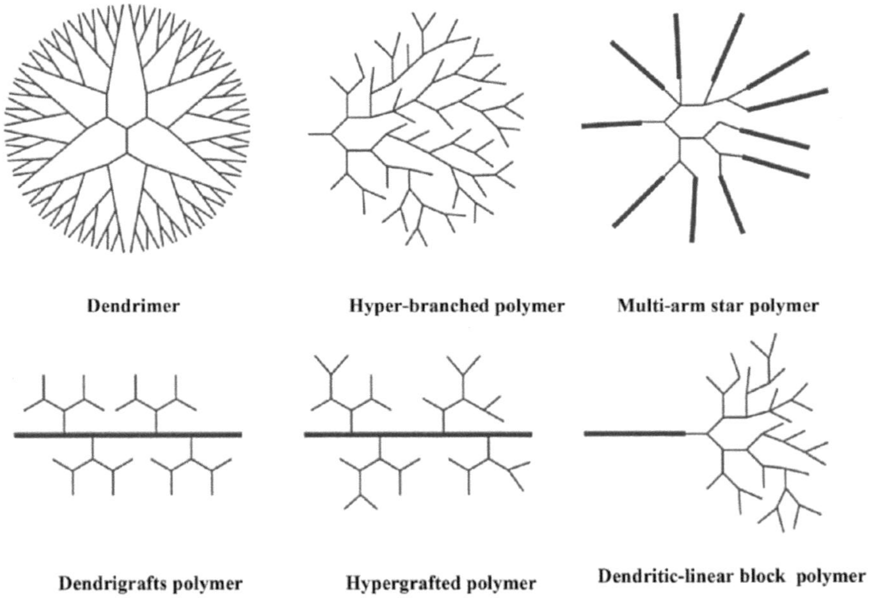

FIGURE 3.3 Examples of advanced polymeric nanocomposites.

Nanofillers and/or NPs are dispersed in polymeric matrices such as resins, dendrimers, cellulose, etc. in polymer nanocomposites to improve their thermophysical, mechanical, and physicochemical properties. Various techniques can be used to produce polymer nanocomposites. The most used methods are melt-mixing, mixing, in situ polymerization, selective laser sintering techniques, and electrospinning. Evolving technologies for polymer nanocomposite development counts on selective laser sintering and microwave-assisted techniques, are offering solutions to aggregation challenges and other defects such as morphological defects. Such techniques aim to produce durable, efficient, and cost-effective polymer nanocomposites with minimal defects and uniform dispersion. The polymer nanocomposites can be utilized as adsorbents and filtering membranes to remove chemical contaminants from aqueous media efficiently.

3.7.6.1 CLAY BASED NANOCOMPOSITES

As the name suggests, clay nanocomposites are materials of nanoscale that are produced from minerals found in clay. Because of their unique properties and high surface area, they are having a high variety of applications. These are obtained very cost effectively. Because of their abundance in nature, biological application, biocompatibility, and simplicity of construction, they are being taken seriously to put to use (Theng, 1974). They have nutrition potential also, as since early times they have been utilized for the treatment and protection of mankind (Chung et al., 2011). Clay NPs are used in many modern products such as clothing, cosmetics, sunscreen, bone implantation, electronics, food, tires, medicine, sports equipment, polymer composites, controlled drug delivery systems, protective coatings (such as antibacterial, anticorrosion, or anti-molding), and for synthesis of the materials (Floody et al., 2009). The clay-based nanocomposites act as adsorbents while putting on the task of contaminated water treatment. Plastic nanocomposites are developed for producing unique devices for the next generation of biological applications, which include cancer treatment, drug delivery, antimicrobial agents, etc. (Suresh et al., 2010). Bone tissue engineering and drug delivery have been done by nanoclay composites.

3.7.7 NANOBIOCHAR SUPPORTED REMEDIATION

Biochar is material rich in carbon, produced from plant or animal biomass, under limited oxygen conditions (Wang et al., 2017). Biochar has gained much attention as a multifunctional adsorbent having diverse applications in the environment and agriculture, owing to its higher surface area and a variety of surface functional groups (Inyang et al., 2016; Rodrigueza et al., 2016; Yao et al., 2016). Enhancing the abilities of biochar for broad-scale applications, biochar is engineered using a variety of methods to improve the physicochemical and biological properties of raw biomass (Li et al., 2014). In particular, embodying functional NPs in the biochar surface is a growing research field. Using biochar as a substrate for nanocomposite production has several advantages over others including the abundance and low cost of feedstock (Lehmann and Joseph, 2015; Sohi, 2012), low energy requirements being produced at lower temperature conditions (Meyer et al., 2011), and the generation of biofuels and syngas from the thermal and chemical treatment of biomass (Lehmann and Joseph, 2015). Thus four integrated goals can be achieved by synthesizing biochar-based nanocomposites including pollutant removal, management of waste, sequestration of carbon, and production of energy (Sohi, 2012; Tan et al., 2015).

Based on various kinds of nanomaterials used for the functionalization of biochar, there are three general categories of biochar-based nanocomposites which include nanometal oxide/hydroxide-biochar composites, magnetic biochar composites, and functional NPs-coated biochar. The nanomaterials employed in these nanocomposites are nanometal oxide/hydroxide, magnetic iron oxide, and functional NPs (e.g., chitosan, graphene, GO, CNTs, ZnS nanocrystals, layered double hydroxides, nanoscale zero-valent iron), respectively (Table 3.3).

3.7.7.1 APPLICATION FOR ENVIRONMENTAL REMEDIATION

3.7.7.1.1 Wastewater Decontamination

Nanocomposites based on biochar have been applied for the removal of HMs, organic contaminants, and inorganic pollutants from the wastewater. As compared to the raw biochar, the amount of oxygen-containing functional groups such as carboxylate-COOH, hydroxyl-OH, and carboxyl –C=O, may show an increment on the surface of the biochar after the incorporation of nanomaterials (Song et al., 2014; Wang et al., 2015a). The enrichment of functional groups improves the interactions between biochar and contaminants by forming surface complexes, hydrogen bonding, electrostatic attraction, and ion-exchange, enhancing the adsorption capacity. Reports show that the adsorption of Pb(II) onto chitosan-modified biochars and CNTs-coated biochars is primarily through the interactions with the amine functional groups of chitosan (Zhou et al., 2013) and the complexation with oxygen-containing functional groups of the CNTs respectively (Inyang et al., 2015). So, generally the functional groups and the NPs on the surface of biochar both offer the adsorption sites for the HM ion in the aqueous solution.

3.7.7.1.2 Soil Restoration

The interesting characteristics of biochar-based nanocomposites can be advantageous for application in soil restoration, such as allowing lesser aggregation of NPs, adsorbing metal ions in the biochar or NPs, eliminating or disintegrating contaminants, preventing the bioavailability of derivatives, and improving soil fertility (Wu et al., 2016). An interesting application of biochar nanocomposites for soil restoration is that the pollutants removed can be applied as fertilizers (Li et al., 2016; Yao et al., 2013). For instance, removing $PO_4^{3?}$ from the polluted water body and applying it in the field as fertilizer has been done using alginate and chitosan-based materials. The

TABLE 3.3 Recent Research on Synthesis, Characteristics, and Application of Nanobiochar Composites (Modified from Tan et al., 2016)

Feedstock	Pyrolytic conditions (temperature, residence time)	Technique used	Characteristics	Contaminants	Effect on removal capacity	References
Corn	300,450,600 (3 h)	Pyrolysis of $MgCl_2$-pretreated biomass	Increased numerous Mg NPs	Phosphate	More effective than pristine biochar	Fang et al., 2014
Sewage sludge	500 (1 h)	One-step pyrolysis with citric acid-$ZnCl_2$ mixed fabricating pore agents	More micropores and mesopores, and a higher surface area	Benzene derivatives	Effectively sorb many kinds of benzene derivatives in aqueous solution	Kong et al., 2014
Rice husk	450–1050 (30 min)	Amended with iron oxide using dissolved ferric nitrate	Increased surface area	Arsenic	Nearly two times higher than those reported for iron oxide-amended sand	(Cope et al., 2014)
Cotton weed	600 (1 h)	Pyrolysis of $FeCl_3$-treated biomass	Magnetic biochar containing colloidal nanosized Fe_2O_3 particles in porous char structures	As(V)	Comparable to or that of many commercial adsorbents	Zhang et al., 2013a
Orange peel	250,400,700 (6 h)	Chemical coprecipitation of Fe^{3+}/Fe^{2+} on orange peel powder	Nanosize magnetic particles and amorphous biochar	HOCs and phosphate	Higher sorption capacity than nonmagnetic biochars	Chen et al., 2011
Oakwood and oak bark	400,450 (?)	Chemical coprecipitation of iron oxides onto the char	Magnetic biochar with increased surface area	Pb(II) and Cd(II)	Better compared to other biochars reported earlier	Mohan et al., 2014a
Cotton weed	600 (1h)	The biomass was pretreated with graphene/pyrene derivative	Graphene coated on biochar and the thermal stability of the biochar improved	Methylene blue	More than 20 times higher than pristine biochar	Zhang et al., 2012b

TABLE 3.3 *(Continued)*

Feedstock	Pyrolytic conditions (temperature, residence time)	Technique used	Characteristics	Contaminants	Effect on removal capacity	References
Wheat straw	600 (1 h)	Slow pyrolysis of graphene-pretreated biomass	Large surface area, more functional groups, and increased thermal stability	Phenanthrene	Nearly 2 times higher than pristine biochar	Tang et al., 2015
Bamboo	300,450,600 (?)	Chitosan was used as a dispersing and soldering reagent to attach fine ZVI particles onto bamboo biochar surfaces	Loaded with large amounts of the chitosan and ZVI	Pb(II),Cr(IV), cAs(V), phosphate and methylene blue	Excellent ability to remove various contaminants	Zhou et al., 2014b

magnetic biochar composite was found to act as an effective catalyst for the removal of high-ring PAHs as opposed to low-ring PAHs, because of the affinity of high-ring PAHs to the biochar surface. This selectivity of biochar-based nanocomposite for structures in the PAHs group offers possibilities for other prospective applications. Despite the satisfying results the role of biochar-based nanomaterials is not well understood in the degradation of organic contaminants in the soil.

3.7.8 NITROCELLULOSE-BASED MATERIALS FOR ENVIRONMENTAL REMEDIATION

To date, the most extensively studied category for the synthesis of green NPs are the plant extracts. The use of natural products such as cellulose has facilitated the way for overcoming the challenges concerned with the biodegradability of material, its renewability, cost, and energy. Nanoscale cellulose has gained margin over conventional cellulose fibers owing to its higher surface area and aspect ratio, Young's modulus (Suopajärvi et al., 2014) Water purification using nanocellulose is recognized as an excellent choice for energy intensive and costly technologies for wastewater treatment, due to its high adsorption affinity toward organic pollutants. Unlike its micrometric counterparts the relatively smaller size and a larger surface area of nanometric cellulose limits its internal diffusion and quantum size effects.

Recently, several researchers have demonstrated the use of cellulose-based metal oxide nanostructures for photocatalytic treatment of wastewater in the form of thin film, membrane, fiber materials under visible and UV light irradiation. The rate of photocatalytic degradation of organic pollutants has been improved using nanocellulose–metal oxide composites as compared to individual materials. Also, an electrospun cellulose acetate nanofibrous membrane with high permeability and porosity was reported to be a competent support for photocatalysts. This synergistically enhanced the catalytic performance of the individual materials. Cellulose nanofibers are being applied as construction materials in membranes for the treatment of wastewater, gas separation, HM removal, and removal of dyes through adsorption, ultrafiltration, and ion exchange.

Nanocellulose although appears as a promising material for environmental restoration, it is still in the developing stage. So far very limited literature is available and there are still many challenges affecting the application of this nanomaterial. The challenges lie in the scalability, cost, and environmental impact, attributed to the selection of routes for the production of nanocellulose. Further studies should aim to execute improved treatment approaches to effectively extract nanocellulose and its functionality as an environmental remediation agent in terms of its performance and economic viability.

3.7.9 NANOCRYSTALLINE ZEOLITES

Crystalline microporous solids of nanosized crystals with physicochemical characteristics quite similar to those of micron-sized crystals. This decrease in crystal size brings advantage and increases the applications of the zeolites. They have distinguished internal and external surface reactivity. Nanocrystalline zeolites, such as ZSM-5, Y, and silicalite are synthesized and highly characterized by powder electron microscopy, nitrogen adsorption isotherms, X-ray diffraction, and dynamic light scattering. Extensive structural information about external and internal surface sites are provided by spectroscopic characterization of the nanocrystalline zeolites by solid-state NMR and FTIR. To form hollow and larger zeolite structures encapsulating organic or metal species, the nanocrystalline zeolites were also used as building blocks. Functionalization of surface silanol groups was done to customize the surface properties of nanocrystalline zeolites and hollow zeolite structures. For the photoreduction of Cr(VI) to Cr(III) and the selective catalytic reduction of NOx in aqueous solution, the nanocrystalline zeolites would be applied. Aluminosilicates of nanoscale are of very big use, as they have acid sites, presence of framework cations, which can be explored for even further applications.

KEYWORDS

- **contamination**
- **nanoremediation**
- **sustainability**
- **pollution**
- **amendments**
- **integrated approach**

REFERENCES

Abebe, B.; Murthy, H. A.; Amare, E. Summary on Adsorption and Photocatalysis for Pollutant Remediation: Mini Review. *J. Encapsulation Adsorpt. Sci.* **2018,** *8* (4), 225–255.

Adesina, A. A. Industrial Exploitation of Photocatalysis: Progress, Perspectives and Prospects. *Catal. Surv. Asia* **2004,** *8* (4), 265–273.

Ago, H. CVD Growth of High-Quality Single-Layer Graphene. In *Frontiers of Graphene and Carbon Nanotubes*; Springer: Tokyo, 2015; pp 3–20.

Ajayan, P. M.; Zhou, O. Z. Applications of Carbon Nanotubes. In *Carbon Nanotubes*; 2001; pp 391–425.

Ajmal, Z.; Muhmood, A.; Usman, M.; Kizito, S.; Lu, J.; Dong, R.; Wu, S. Phosphate Removal from Aqueous Solution Using Iron Oxides: Adsorption, Desorption and Regeneration Characteristics. *J. Colloid Interface Sci.* **2018,** *528,* 145–155.

Alabresm, A.; Chen, Y. P.; Decho, A. W.; Lead, J. A Novel Method for the Synergistic Remediation of Oil-Water Mixtures Using Nanoparticles and Oil-Degrading Bacteria. *Sci. Total Environ.* **2018,** *630,* 1292–1297.

Alissa, E. M.; Ferns, G. A. Heavy Metal Poisoning and Cardiovascular Disease. *J. Toxicol.* **2011,** *2011.*

Amade, R.; Hussain, S.; Ocaña, I. R.; Bertran, E. Growth and Functionalization of Carbon Nanotubes on Quartz Filter for Environmental Applications. *J. Environ. Eng. Ecol. Sci.* **2014,** *3* (2), 1–7.

Ankamwar, B. Biosynthesis of Gold Nanoparticles (Green-Gold) Using Leaf Extract of Terminalia Catappa. *E-J. Chem.* **2010,** *7* (4), 1334–1339.

Aragaw, T. A.; Bogale, F. M.; Aragaw, B. A. Iron-Based Nanoparticles in Wastewater Treatment: A Review on Synthesis Methods, Applications, and Removal Mechanisms. *J. Saudi Chem. Soc.* **2021,** *25* (8), 101280.

Baby, R.; Saifullah, B.; Hussein, M. Z. Carbon Nanomaterials for the Treatment of Heavy Metal-Contaminated Water and Environmental Remediation. *Nanosc. Res. Lett.* **2019,** *14* (1), 1–17.

Bakshi, M.; Abhilash, P. C. Nanotechnology for Soil Remediation: Revitalizing the Tarnished Resource. In *Nano-Materials as Photocatalysts for Degradation of Environmental Pollutants*; Elsevier, 2020; pp 345–370.

Bansal, P.; Chaudhary, G. R.; Mehta, S. K. Comparative Study of Catalytic Activity of ZrO2 Nanoparticles for Sonocatalytic and Photocatalytic Degradation of Cationic and Anionic Dyes. *Chem. Eng. J.* **2015,** *280,* 475–485.

Bhadouria, R.; Mishra, D.; Singh, V. K.; Singh, P.; Srivastava, P.; Tripathi, S.; Singh, R.; Nanocatalyst Types and their Potential Impacts. In *Agroecosystems: An Overview. Nano-Materials as Photocatalysts for Degradation of Environmental Pollutants*; 2020; pp 323–344.

Bhushan, B. Introduction to Nanotechnology. In *Springer Handbook of Nanotechnology*; Springer: Berlin, Heidelberg, 2017; pp 1–19.

Bosetti, M.; Massè, A.; Tobin, E.; Cannas, M. Silver Coated Materials for External Fixation Devices: In Vitro Biocompatibility and Genotoxicity. *Biomaterials* **2002,** *23* (3), 887–892.

Bosetti, M.; Massè, A.; Tobin, E.; Cannas, M. Silver Coated Materials for External Fixation Devices: In Vitro Biocompatibility and Genotoxicity. *Biomaterials* **2002,** *23* (3), 887–892.

Brunet, L.; Lyon, D. Y.; Hotze, E. M.; Alvarez, P. J.; Wiesner, M. R. Comparative Photoactivity and Antibacterial Properties of C60 Fullerenes and Titanium Dioxide Nanoparticles. *Environ. Sci. Technol.* **2009,** *43* (12), 4355–4360.

Bussamara, R.; Eberhardt, D.; Feil, A. F.; Migowski, P.; Wender, H. de Moraes, D. P.; Machado, G.; Papaléo, R. M.; Teixeira, S R.; Dupont, J. *Chem. Commun.* **2013,** *49,* 1273–1275.

Cao, X.; Alabresm, A.; Chen, Y. P.; Decho, A. W.; Lead, J. Improved Metal Remediation Using a Combined Bacterial and Nanoscience Approach. *Sci. Total Environ.* **2020,** *704,* 135378.

Castro, E.; Garcia, A. H.; Zavala, G.; Echegoyen, L. Fullerenes in Biology and Medicine. *J. Mater. Chem. B* **2017,** *5* (32), 6523–6535.

Chauhan, R.; Yadav, H. O.; Sehrawat, N. Nanobioremediation: A New and a Versatile Tool for Sustainable Environmental Clean Up-Overview. *J. Mater. Environ. Sci*, **2020,** *11* (4), 564–573.

Chen, B.; Chen, Z.; Lv, S. A Novel Magnetic Biochar Efficiently Sorbs Organic Pollutants and Phosphate. *Biores. Technol.* **2011,** *102* (2), 716–723.

Chen, B.; Evans, J. R.; Greenwell, H. C.; Boulet, P.; Coveney, P. V.; Bowden, A. A.; Whiting, A. A Critical Appraisal of Polymer–Clay Nanocomposites. *Chem. Soc. Rev.* **2008,** 37 (3), 568–594.

Chen, C.; Tsyusko, O. V.; Mcnear, D. H. Jr.; Judy, J.; Lewis, R. W.; Unrine, J. M. Effects of Biosolids from a Wastewater Treatment Plant Receiving Manufactured Nanomaterials on Medicago Truncatula and Associated Soil Microbial Communities at Low Nanomaterial Concentrations. *Sci. Total Environ.* **2017,** *609,* 799–806.

Chen, X.; Cen, C.; Tang, Z.; Zeng, W.; Chen, D.; Fang, P.; Chen, Z. The Key Role of PH Value in the Synthesis of Titanate Nanotubes-Loaded Manganese Oxides as a Superior Catalyst for the Selective Catalytic Reduction of NO with NH3. *J. Nanomater.* **2013,** 2013.

Chou, K. S.; Lu, Y. C.; Lee, H. H. Effect of Alkaline Ion on the Mechanism and Kinetics of Chemical Reduction of Silver. *Mater. Chem. Phys.* **2005,** *94* (2–3), 429–433.

Chou, K. S.; Lu, Y. C.; Lee, H. H. Effect of Alkaline Ion on the Mechanism and Kinetics of Chemical Reduction of Silver. *Mater. Chem. Phys.* **2005,** *94* (2–3), 429–433.

Chung, H. E.; Kim, I. S.; Baek, M.; Yu, J.; Choi, S. J. Long-Term Cytotoxicity Potential of Anionic Nanoclays in Human Cells. *Toxicol. Environ. Health Sci.* **2011,** *3* (2): 129–133.

Cope, C. O.; Webster, D. S.; Sabatini, D. A. Arsenate Adsorption onto Iron Oxide Amended Rice Husk Char. *Sci. Total Environ.* **2014,** *488,* 554–561.

Corsi, I.; Winther-Nielsen, M.; Sethi, R.; Punta, C.; Della Torre, C.; Libralato, G.; Lofrano, G.; Sabatini, L.; Aiello, M.; Fiordi, L.; Cinuzzi, F. Ecofriendly Nanotechnologies and Nanomaterials for Environmental Applications: Key Issue and Consensus Recommendations for Sustainable and Ecosafe Nanoremediation. *Ecotoxicol. Environ. Saf.* **2018,** *154,* 237–244.

Corsi, I.; Winther-Nielsen, M.; Sethi, R.; Punta, C.; Della Torre, C.; Libralato, G.; Lofrano, G.; Sabatini, L.; Aiello, M.; Fiordi, L.; Cinuzzi, F. Ecofriendly Nanotechnologies and Nanomaterials for Environmental Applications: Key Issue and Consensus Recommendations for Sustainable and Ecosafe Nanoremediation. *Ecotoxicol. Environ. Saf.* **2018,** *154,* 237–244.

Danks, A. E.; Hall, S. R.; Schnepp, Z. J. M. H. the Evolution of 'Sol–Gel'chemistry as a Technique for Materials Synthesis. *Mater. Horiz.* **2016,** *3* (2), 91–112.

de Coelho Escobar, C.; dos Santos, J. H. Z. Effect of the Sol–Gel Route on the Textural Characteristics of Silica Imprinted with Rhodamine B. *J. Sep. Sci.* **2014,** *37* (7), 868–875.

Desiante, W. L.; Minas, N. S.; Fenner, K. Micropollutant Biotransformation and Bioaccumulation in Natural Stream Biofilms. *Water Res.* **2021,** *193,* 116846.

Dong, K. Y.; Choi, J.; Lee, Y. D.; Kang, B. H.; Yu, Y. Y.; Choi, H. H.; Ju, B. K. Detection of A CO and NH 3 Gas Mixture Using Carboxylic Acid-Functionalized Single-Walled Carbon Nanotubes. *Nanosc. Res. Lett.* **2013,** *8* (1), 1–6.

Du, P.; Song, L.; Xiong, J.; Li, N.; Xi, Z.; Wang, L.; Jin, D.; Guo, S.; Yuan, Y. Coaxial Electrospun TiO2/ZnO Core–Sheath Nanofibers Film: Novel Structure for Photoanode of Dye-sensitized Solar Cells. *Electrochim. Acta* **2012,** *78,* 392–397.

Enamala, M. K.; Sruthi, P. D.; Sarkar, S.; Chavali, M.; Vasavi, I.; Kuppam, C. Nanobioremediation: A Novel and Sustainable Biological Advancement for Ecological Cleanup. In *Nanotechnology in Biology and Medicine*; CRC Press, 2019; pp 245–257.

Fang, C.; Zhang, T.; Li, P.; Jiang, R. F.; Wang, Y. C. Application of Magnesium Modified Corn Biochar for Phosphorus Removal and Recovery from Swine Wastewater. *Int. J. Environ. Res. Public Health* **2014,** *11* (9), 9217–9237.

Filote, C.; Roşca, M.; Hlihor, R. M.; Cozma, P.; Simion, I. M.; Apostol, M.; Gavrilescu, M. Sustainable Application of Biosorption and Bioaccumulation of Persistent Pollutants in Wastewater Treatment: Current Practice. *Processes* **2021,** *9* (10), 1696.

Floody, M. C.; Theng, B. K. G.; Reyes, P.; Mora, M. L. Natural Nanoclays: Applications and Future Trends: A Chilean Perspective. *Clay Miner.* **2009,** *44,* 161–176.

Gajić, G.; Djurdjević, L.; Kostić, O.; Jarić, S.; Mitrović, M.; Pavlović, P. Ecological Potential of Plants for Phytoremediation and Ecorestoration of Fly Ash Deposits and Mine Wastes. *Front. Environ. Sci.* **2018,** *6,* 124.

Ganie, A. S.; Bano, S.; Khan, N.; Sultana, S.; Rehman, Z.; Rahman, M. M.; Sabir, S.; Coulon, F.; Khan, M. Z. Nanoremediation Technologies for Sustainable Remediation of Contaminated Environments: Recent Advances and Challenges. *Chemosphere* **2021,** *275,* 130065.

Garg, V.; Mote, R. G.; Fu, J. Facile Fabrication of Functional 3D Micro-Nano Architectures with Focused Ion Beam Implantation and Selective Chemical Etching. *Appl. Surf. Sci.* **2020,** *526,* 146644.

Gaur, N.; Flora, G.; Yadav, M.; Tiwari, A. A Review with Recent Advancements on Bioremediation-Based Abolition of Heavy Metals. *Environ. Sci.: Proc. Impacts* **2014,** *16* (2), 180–193.

Geoffrion, L. D.; Guisbiers, G. Quantum Confinement: Size on the Grill!. *J. Phys. Chem. Solids* **2020,** *140,* 109320.

Gerhardt, K. E.; Gerwing, P. D.; Greenberg, B. M. Opinion: Taking Phytoremediation from Proven Technology to Accepted Practice. *Plant Sci.* **2017,** *256,* 170–185.

Gong, J. L.; Wang, B.; Zeng, G. M.; Yang, C. P.; Niu, C. G.; Niu, Q. Y.; Zhou, W. J.; Liang, Y. Removal of Cationic Dyes from Aqueous Solution Using Magnetic Multi-Wall Carbon Nanotube Nanocomposite as Adsorbent. *J. Hazard. Mater.* **2009,** *164* (2–3), 1517–1522.

Gong, X.; Huang, D.; Liu, Y.; Peng, Z.; Zeng, G.; Xu, P.; Cheng, M.; Wang, R.; Wan, J. Remediation of Contaminated Soils by Biotechnology with Nanomaterials: Bio-behavior, Applications, and Perspectives. *Crit. Rev. Biotechnol.* **2018,** *38* (3), 455–468.

Handy, R. D.; Owen, R.; Valsami-Jones, E. The Ecotoxicology of Nanoparticles and Nanomaterials: Current Status, Knowledge Gaps, Challenges, and Future Needs. *Ecotoxicology* **2008,** *17* (5), 315–325.

Huang, H. Y.; Yang, R. T.; Chinn, D.; Munson, C. L. Amine-Grafted MCM-48 and Silica Xerogel as Superior Sorbents for Acidic Gas Removal from Natural Gas. *Indus. Eng. Chem. Res.* **2003,** *42* (12), 2427–2433.

Huang, H. Y.; Yang, R. T.; Chinn, D.; Munson, C. L. Amine-Grafted MCM-48 and Silica Xerogel as Superior Sorbents for Acidic Gas Removal from Natural Gas. *Indus. Eng. Chem. Res.* **2003,** *42* (12), 2427–2433.

Inyang, M.; Gao, B.; Zimmerman, A.; Zhou, Y.; Cao, X. Sorption and Cosorption of Lead and Sulfapyridine on Carbon Nanotube-Modified Biochars. *Enviro. Sci. Pollut. Res.* **2015,** *22* (3), 1868–1876.

&Inyang, M. I.; Gao, B.; Yao, Y.; Xue, Y.; Zimmerman, A.; Mosa, A.; Pullammanappallil, P.; Ok, Y. S.; Cao, X. A Review of Biochar as a Low-Cost Adsorbent for Aqueous Heavy Metal Removal. *Crit. Rev. Environ. Sci. Technol.* **2016,** *46* (4), 406–433.

Jaishankar, M.; Tseten, T.; Anbalagan, N.; Mathew, B. B.; Beeregowda, K. N. Toxicity, Mechanism and Health Effects of Some Heavy Metals. *Interdiscip. Toxicol.* **2014,** *7* (2), 60.

Jeevanandam, J.; Barhoum, A.; Chan, Y. S.; Dufresne, A.; Danquah, M. K. Review on Nanoparticles and Nanostructured Materials: History, Sources, Toxicity and Regulations. *Beilstein J. Nanotechnol.* **2018,** *9* (1), 1050–1074.

Jin, G.; Prabhakaran, M. P.; Nadappuram, B. P.; Singh, G.; Kai, D.; Ramakrishna, S. Electrospun Poly (L-Lactic Acid)-co-Poly (ϵ-Caprolactone) Nanofibres Containing Silver Nanoparticles for Skin-Tissue Engineering. *J. Biomater. Sci. Polym. Ed.* **2012,** *23* (18), 2337–2352.

Jones, A. C.; Hitchman, M. L. Overview of Chemical Vapour Deposition. Chemical Vapour Deposition: Precursors. *Proc. Appl.* **2009,** *1,* 1–36.

Jung, W.; Lee, J. S.; Han, S.; Ko, S. H.; Kim, T.; Kim, Y. H. An Efficient Reduced Graphene-Oxide Filter for PM 2.5 Removal. *J. Mater. Chem. A* **2018,** *6* (35), 16975–16982.

Karn, B.; Kuiken, T.; Otto, M. Nanotechnology and In Situ Remediation: A Review of the Benefits and Potential Risks. *Environ. Health Perspect.* **2009,** *117* (12), 1813–183.

Khin, M. M.; Nair, A. S.; Babu, V. J.; Murugan, R.; Ramakrishna, S. A Review on Nanomaterials for Environmental Remediation. *Energy Environ. Sci.* **2012,** *5* (8), 8075–8109.

Khin, M. M.; Nair, A. S.; Babu, V. J.; Murugan, R.; Ramakrishna, S. A Review on Nanomaterials for Environmental Remediation. *Energy Environ. Sci.* **2012,** *5* (8), 8075–8109.

Kim, J.; Choi, S. W.; Lee, J. H.; Chung, Y.; Byun, Y. T. Gas Sensing Properties of Defect-Induced Single-Walled Carbon Nanotubes. Sens. *Actuators B Chem.* **2016,** *228,* 688–692.

Km, A. H. A. A. S.; Shaik, M. R. Al-Suhybani MS. *Appl. Water Sci.* **2017,** *7,* 4341–4350.

Kong, L.; Xiong, Y.; Sun, L.; Tian, S.; Xu, X.; Zhao, C.; Luo, R.; Yang, X.; Shih, K.; Liu, H. Sorption Performance and Mechanism of a Sludge-Derived Char as Porous Carbon-Based Hybrid Adsorbent for Benzene Derivatives in Aqueous Solution. *J. Hazard. Mater.* **2014,** *274,* 205–211.

Kreyling, W. G.; Semmler-Behnke, M.; Chaudhry, Q. A Complementary Definition of Nanomaterial. *Nano Today* **2010,** *5* (3), 165–168.

Kuo, C. W.; Shiu, J. Y.; Cho, Y. H.; Chen, P. E. I. L. I. N. Fabrication of Large-Area Periodic Nanopillar Arrays for Nanoimprint Lithography Using Polymer Colloid Masks. *Adv. Mater.* **2003,** *15* (13), 1065–1068.

Lehmann, J.; Joseph, S. Biochar for Environmental Management: An Introduction. In *Biochar for Environmental Management*; Routledge, 2015; pp 1–13.

Li, Q.; Mahendra, S.; Lyon, D. Y.; Brunet, L.; Liga, M. V.; Li, D.; Alvarez, P. J. Antimicrobial Nanomaterials for Water Disinfection and Microbial Control: Potential Applications and Implications. *Water Res.* **2008,** *42* (18), 4591–4602.

Li, R.; Wang, J. J.; Zhou, B.; Awasthi, M. K.; Ali, A.; Zhang, Z.; Lahori, A. H.; Mahar, A. Recovery of Phosphate from Aqueous Solution by Magnesium Oxide Decorated Magnetic

Biochar and Its Potential as Phosphate-Based Fertilizer Substitute. *Bioresour. Technol.* **2016,** *215,* 209–214.

Li, W.; Zheng, P.; Guo, J.; Ji, J.; Zhang, M.; Zhang, Z.; Zhan, E.; Abbas, G. Characteristics of Self-Alkalization in High-Rate Denitrifying Automatic Circulation (DAC) Reactor Fed with Methanol and Sodium Acetate. *Bioresour. Technol.* **2014,** *154,* 44–50.

Liu, X.; Xu, M.; Wan, L.; Zhu, H.; Yao, K.; Linguerri, R.; Chambaud, G.; Han, Y.; Meng, C. Superior Catalytic Performance of Atomically Dispersed Palladium on Graphene in CO Oxidation. *Acs Catalysis* **2020,** *10* (5), 3084–3093.

Lu, H.; Wang, J.; Stoller, M.; Wang, T.; Bao, Y.; Hao, H. An Overview of Nanomaterials for Water and Wastewater Treatment. *Adv. Mater. Sci. Eng.* **2016,** *2016.*

Luo, C.; Tian, Z.; Yang, B.; Zhang, L.; Yan, S. Manganese Dioxide/Iron Oxide/Acid Oxidized Multi-Walled Carbon Nanotube Magnetic Nanocomposite for Enhanced Hexavalent Chromium Removal. *Chem. Eng. J.* **2013,** *234,* 256–265.

Machac, P.; Cichon, S.; Lapcak, L.; Fekete, L. Graphene Prepared by Chemical Vapour Deposition Process. *Graphene Technol.* **2020,** *5* (1), 9–17.

Mahmoud, M. E.; Abdou, A. E.; Mohamed, S. M.; Osman, M. M. Engineered Staphylococcus Aureus via Immobilization on Magnetic Fe3O4-Phthalate Nanoparticles for Biosorption of Divalent Ions from Aqueous Solutions. *J. Environ. Chem. Eng.* **2016,** *4* (4), 3810–3824.

Makvandi, P. C. Y. Wang, EN Zare, A. Borzacchiello, L. n. Niu, FR Tay. *Adv. Funct. Mater.* **2020,** *30,* 1910021.

Malik, M. A.; Wani, M. Y.; Hashim, M. A. Microemulsion Method: A Novel Route to Synthesize Organic and Inorganic Nanomaterials: 1st Nano Update. *Arab. J. Chem.* **2012,** *5* (4), 397–417.

Matsumoto, R.; Adachi, S.; Sadki, E. H. S.; Yamamoto, S.; Tanaka, H.; Takeya, H.; Takano, Y. Maskless Patterning of Gallium-Irradiated Superconducting Silicon Using Focused Ion Beam. *ACS Appl. Electron. Mater.* **2020,** *2* (3), 677–682.

Meyer, S.; Glaser, B.; Quicker, P. Technical, Economical, and Climate-Related Aspects of Biochar Production Technologies: A Literature Review. *Environ. Sci. Technol.* **2011,** *45* (22), 9473–9483.

Mohan, D.; Kumar, H.; Sarswat, A.; Alexandre-Franco, M.; Pittman, C. U. Jr. Cadmium and Lead Remediation Using Magnetic Oak Wood and Oak Bark Fast Pyrolysis Bio-Chars. *Chem. Eng. J.* **2014,** *236,* 513–528.

Moriarty, P. Nanostructured Materials. *Rep. Prog. Phys.* **2001,** *64* (3), 297.

Murphy, C. J.; Jana, N. R. Controlling the Aspect Ratio of Inorganic Nanorods and Nanowires. *Adv. Mater.* **2002,** *14* (1), 80–82.

Nasrollahzadeh, M.; Sajadi, S. M.; Sajjadi, M.; Issaabadi, Z. An Introduction to Nanotechnology. In *Interface Science and Technology*, Vol. 28 ; Elsevier, 2019; pp 1–27.

Nguyen, T. D. From Formation Mechanisms to Synthetic Methods toward Shape-Controlled Oxide Nanoparticles. *Nanoscale* **2013,** *5* (20), 9455–9482.

Nie, M.; Sun, K.; Meng, D. D. Formation of Metal Nanoparticles by Short-Distance Sputter Deposition in a Reactive Ion Etching Chamber. *J. Appl. Phys.* **2009,** *106* (5), 054314.

Ntim, S. A.; Mitra, S. Adsorption of Arsenic on Multiwall Carbon Nanotube–Zirconia Nanohybrid for Potential Drinking Water Purification. *J. Colloid Interface Sci.* **2012,** *375* (1), 154–159.

Parashar, M.; Shukla, V. K.; Singh, R. Metal Oxides Nanoparticles via Sol–Gel Method: A Review on Synthesis, Characterization and Applications. *J. Mater. Sci.: Mater. Electron.* **2020,** *31* (5), 3729–3749.

Park, S.; Byoun, Y.; Kang, H.; Song, Y. J.; Choi, S. W. ZnO Nanocluster-Functionalized Single-Walled Carbon Nanotubes Synthesized by Microwave Irradiation for Highly Sensitive NO2 Detection at Room Temperature. *ACS Omega* **2019,** *4* (6), 10677–10686.

Patil, S. S.; Shedbalkar, U. U.; Truskewycz, A.; Chopade, B. A.; Ball, A. S. Nanoparticles for Environmental Clean-Up: A Review of Potential Risks and Emerging Solutions. *Environ. Technol. Innovat.* **2016,** *5,* 10–21.

Peng, F.; Su, Y.; Zhong, Y.; Fan, C.; Lee, S. T.; He, Y. Silicon Nanomaterials Platform for Bioimaging, Biosensing, and Cancer Therapy. *Acc. Chem. Res.* **2014,** *47* (2), 612–623.

Pillai, H. P.; Kottekottil, J. Nano-Phytotechnological Remediation of Endosulfan Using Zero Valent Iron Nanoparticles. *J. Environ. Prot.* **2016,** *7* (5), 734.

Pimpin, A.; Srituravanich, W. Review on Micro- and Nanolithography Techniques and their Applications. *Eng. J.* **2012,** *16* (1), 37–56.

Pivato, R. V.; Rossi, F.; Ferro, M.; Castiglione, F.; Trotta, F.; Mele, A. B-Cyclodextrin Nanosponge Hydrogels as Drug Delivery Nanoarchitectonics for Multistep Drug Release Kinetics. *ACS Appl. Polym. Mater.* **2021,** *3* (12), 6562–6571.

Pradeep, T. Noble Metal Nanoparticles for Water Purification: A Critical Review. *Thin Solid Films* **2009,** *517* (24), 6441–6478.

Prado-Audelo, D.; García Kerdan, I.; Escutia-Guadarrama, L.; Reyna-González, J. M.; Magaña, J. J.; Leyva-Gómez, G. Nanoremediation: Nanomaterials and Nanotechnologies for Environmental Cleanup. *Front. Environ. Sci.* **2021,** *645.*

Prasad, R.; Jain, V. K.; Varma, A. Role of Nanomaterials in Symbiotic Fungus Growth Enhancement. *Curr. Sci.* **2010,** *99* (9), 1189–1191.

Qi, G.; Wang, Y.; Estevez, L.; Duan, X.; Anako, N.; Park, A. H. A.; Li, W.; Jones, C. W.; Giannelis, E. P.; *Energy Environ. Sci.* **2011,** *4,* 444. RSC; (b) Huang, H. Y.; Yang, R. T.; Chinn, D.; Munson. C. L. *Ind. Eng. Chem. Res.* **2003,** *42,* 2427.

Rajput, V. D.; Minkina, T.; Kimber, R. L.; Singh, V. K.; Shende, S.; Behal, A.; Sushkova, S.; Mandzhieva, S.; Lloyd, J. R. Insights into the Biosynthesis of Nanoparticles by the Genus Shewanella. *Appl. Environ. Microbiol.* **2021,** *87* (22), E01390–E01321.

Rajput, V. D.; Minkina, T.; Upadhyay, S. K.; Kumari, A.; Ranjan, A.; Mandzhieva, S.; Sushkova, S.; Singh, R. K.; Verma, K. K. Nanotechnology in the Restoration of Polluted Soil. *Nanomaterials* **2022,** *12* (5), 769.

Rhodes, L.; Harwood, T.; Smith, K.; Argyle, P.; Munday, R. Production of Ciguatoxin and Maitotoxin by Strains of *Gambierdiscus australes, G. Pacificus* and *G. polynesiensis* (Dinophyceae) Isolated from Rarotonga, Cook Islands. *Harmful Algae* **2014,** *39,* 185–190.

Rodrigueza, O.; Peralta-Hernandeza, J. M.; Goonetillekeb, A.; Bandalac, E. R. Treatment Technologies for Emerging Contaminants in Water: A Review, 2016.

Rodriguez-Narvaez, O. M.; Peralta-Hernandez, J. M.; Goonetilleke, A.; Bandala, E. R. Biochar-Supported Nanomaterials for Environmental Applications. *J. Indus. Eng. Chem.* **2019,** *78,* 21–33.

Roduner, E. Size Matters: Why Nanomaterials are Different. *Chem. Soc. Rev.* **2006,** *35* (7), 583–592.

Roy, A.; Sharma, A.; Yadav, S.; Jule, L. T.; Krishnaraj, R. Nanomaterials for Remediation of Environmental Pollutants. *Bioinorg. Chem. Appl.* **2021,** *2021.*

Savage, N.; Diallo, M. S. Nanomaterials and Water Purification: Opportunities and Challenges. *J. Nanopart. Res.* **2005,** *7* (4), 331–342.

Shah, K. A.; Tali, B. A. Materials Science in Semiconductor Processing Synthesis of Carbon Nanotubes by Catalytic Chemical Vapour Deposition: A Review on Carbon Sources, Catalysts and Substrates. *Mater. Sci. Semicond. Process* **2016,** *41,* 67–82.

Shak, K. P. Y.; Pang, Y. L.; Mah, S. K. Nanocellulose: Recent Advances and Its Prospects in Environmental Remediation. *Beilstein J. Nanotechnol.* **2018,** *9* (1), 2479–2498.

Sharma, C. S.; Sarkar, S.; Periyakaruppan, A.; Barr, J.; Wise, K.; Thomas, R.; Wilson, B. L.; Ramesh, G. T. Single-Walled Carbon Nanotubes Induces Oxidative Stress in Rat Lung Epithelial Cells. *J. Nanosci. Nanotechnol.* **2007,** *7* (7), 2466–2472.

Sharma, J. Nanoremediation. *Int. J. Life Sci. Technol.***2019,** *12* (1), 1–6.

Shekhawat, G. S.; Arya, V. Biological Synthesis of Ag Nanoparticles through In Vitro Cultures of Brassica Juncea C. Zern. In *Advanced Materials Research,* Vol. 67 ; Trans Tech Publications Ltd, 2009; pp 295–299.

Singh, B. K.; Walker, A. Microbial Degradation of Organophosphorus Compounds. *FEMS Microbiol. Rev.* **2006,** *30* (3), 428–471.

Sohi, S. P. Carbon Storage with Benefits. *Science* **2012,** *338* (6110), 1034–1035.

Soleimani Zohr Shiri, M.; Henderson, W.; Mucalo, M. R. A review of the lesser-studied microemulsion-based synthesis methodologies used for preparing nanoparticle systems of the noble metals, Os, Re, Ir and Rh. *Materials* **2019,** *12* (12), 1896.

Son, H. H.; Seo, G. H.; Jeong, U.; Kim, S. J. Capillary Wicking Effect of a Cr-Sputtered Superhydrophilic Surface on Enhancement of Pool Boiling Critical Heat Flux. *Int. J. Heat Mass Transf.* **2017,** *113,* 115–128.

Son, W. J.; Choi, J. S.; Ahn, W. S. Adsorptive Removal of Carbon Dioxide Using Polyethyleneimine-Loaded Mesoporous Silica Materials. *Microporous Mesoporous Mater.* **2008,** *113* (1–3), 31–40.

Song, W.; So, S. K.; Cao, L. *Appl. Phys. A: Mater. Sci. Process,* 2001.

Song, Z.; Lian, F.; Yu, Z.; Zhu, L.; Xing, B.; Qiu, W. Synthesis and Characterization of a Novel Mnox-Loaded Biochar and Its Adsorption Properties for Cu2+ in Aqueous Solution. *Chem. Eng. J.* **2014,** *242,* 36–42.

Srivastava, S.; Shukla, A.; Rajput, V. D.; Kumar, K.; Minkina, T.; Mandzhieva, S.; Shmaraeva, A.; Suprasanna, P. Arsenic Remediation through Sustainable Phytoremediation Approaches. *Minerals* **2021,** *11* (9), 936.

Su, F.; Lu, C.; Cnen, W.; Bai, H.; Hwang, J. F. Capture of CO2 from Flue Gas via Multiwalled Carbon Nanotubes. *Sci. Total Environ.* **2009,** *407* (8), 3017–3023.

Suman, J.; Uhlik, O.; Viktorova, J.; Macek, T. Phytoextraction of Heavy Metals: A Promising Tool for Clean-Up of Polluted Environment? *Front. Plant Sci.* **2018,** *9,* 1476.

Suopajärvi, T.; Koivuranta, E.; Liimatainen, H.; Niinimäki, J. Flocculation of Municipal Wastewaters with Anionic Nanocelluloses: Influence of Nanocellulose Characteristics on Floc Morphology and Strength. *J. Environ. Chem. Eng.* **2014,** *2* (4), 2005–2012.

Suresh, R.; Borkar, S. N.; Sawant, V. A.; Shende, V. S.; Dimble, S. K. Nanoclay Drug Delivery System. *Int. J. Pharm. Sci. Nanotechnol.* **2010,** *3,* 901–905.

Szabó, Z.; Volk, J.; Fülöp, E.; Deák, A.; Bársony, I. Regular ZnO Nanopillar Arrays by Nanosphere Photolithography. *Photonics Nanostruct.-Fundam. Appl.* **2013,** *11* (1), 1–7.

Tabish, T. A.; Memon, F. A.; Gomez, D. E.; Horsell, D. W.; Zhang, S. A Facile Synthesis of Porous Graphene for Efficient Water and Wastewater Treatment. *Sci. Rep.* **2018,** *8* (1), 1–14.

Tai, X. H.; Chook, S. W.; Lai, C. W.; Lee, K. M.; Yang, T. C. K.; Chong, S.; Juan, J. C. Effective Photoreduction of Graphene Oxide for Photodegradation of Volatile Organic Compounds. *RSC Adv.* **2019,** *9* (31), 18076–18086.

Tan, X.; Liu, Y.; Zeng, G.; Wang, X.; Hu, X.; Gu, Y.; Yang, Z. Application of Biochar for the Removal of Pollutants from Aqueous Solutions. *Chemosphere* **2015,** *125,* 70–85.

Tan, X. F.; Liu, Y. G.; Gu, Y. L.; Xu, Y.; Zeng, G. M.; Hu, X. J.; Liu, S. B.; Wang, X.; Liu, S. M.; Li, J. Biochar-Based Nano-Composites for the Decontamination of Wastewater: A Review. *Bioresour. Technol.* **2016,** *212,* 318–333.

Tang, J.; Lv, H.; Gong, Y.; Huang, Y. Preparation and Characterization of a Novel Graphene/Biochar Composite for Aqueous Phenanthrene and Mercury Removal. *Bioresour. Technol.* **2015,** *196,* 355–363.

Tang, W. W.; Zeng, G. M.; Gong, J. L.; Liu, Y.; Wang, X. Y.; Liu, Y. Y.; Liu, Z. F.; Chen, L.; Zhang, X. R.; Tu, D. Z. Simultaneous Adsorption of Atrazine and Cu (II) from Wastewater by Magnetic Multi-Walled Carbon Nanotube. *Chem. Eng. J.* **2012,** *211,* 470–478.

Thangavelu, L.; Veeraragavan, G. R.; Mallineni, S. K.; Devaraj, E.; Parameswari, R. P.; Syed, N. H.; Dua, K.; Chellappan, D. K.; Balusamy, S. R.; Bhawal, U. K. Role of Nanoparticles in Environmental Remediation: An Insight into Heavy Metal Pollution from Dentistry. *Bioinorg. Chem. Appl.* **2022,** *2022.*

Thangavelu, L.; Veeraragavan, G. R.; Mallineni, S. K.; Devaraj, E.; Parameswari, R. P.; Syed, N. H.; Dua, K.; Chellappan, D. K.; Balusamy, S. R.; Bhawal, U. K. Role of Nanoparticles in Environmental Remediation: An Insight into Heavy Metal Pollution from Dentistry. *Bioinorg. Chem. Appl.* **2022,** *2022.*

Theng, B. K. G. *The Chemistry of Clay-Organic Reactions*; 1974.

Tomar, R.; Abdala, A. A.; Chaudhary, R. G.; Singh, N. B. Photocatalytic Degradation of Dyes by Nanomaterials. *Mater. Today: Proc.* **2020,** *29,* 967–973.

Tsai, C. H.; Chang, W. C.; Saikia, D.; Wu, C. E.; Kao, H. M. Functionalization of Cubic Mesoporous Silica SBA-16 With Carboxylic Acid via One-Pot Synthesis Route for Effective Removal of Cationic Dyes. *J. Hazard. Mater.* **2016,** *309,* 236–248.

Tsai, C. H.; Chang, W. C.; Saikia, D.; Wu, C. E.; Kao, H. M.; Functionalization of Cubic Mesoporous Silica SBA-16 with Carboxylic Acid via One-Pot Synthesis Route for Effective Removal of Cationic Dyes. *J. Hazard. Mater.* **2016,** *309,* 236–248.

Tseng, T. K.; Lin, Y. S.; Chen, Y. J.; Chu, H. A Review of Photocatalysts Prepared by Sol–Gel Method for Vocs Removal. *Int. J. Mol. Sci.* **2010,** *11* (6), 2336–2361.

Tyagi, I.; Gupta, V. K.; Sadegh, H.; Ghoshekandi, R. S.; Makhlouf, A. H. Nanoparticles as Adsorbent; a Positive Approach for Removal of Noxious Metal Ions: A Review. *Sci. Technol. Devel.* **2017,** *34* (3), 195–214.

Usman, M.; Farooq, M.; Wakeel, A.; Nawaz, A.; Cheema, S. A.; ur Rehman, H.; Ashraf, I.; Sanaullah, M. Nanotechnology in Agriculture: Current Status, Challenges and Future Opportunities. *Sci. Total Environ.* **2020,** *721,* 137778.

Usman, M.; Farooq, M.; Wakeel, A.; Nawaz, A.; Cheema, S. A.; Ur Rehman, H.; Ashraf, I.; Sanaullah, M. Nanotechnology in Agriculture: Current Status, Challenges and Future Opportunities. *Sci. Total Environ.* **2020,** *721,* 137778.

Wang, B.; Gao, B.; Fang, J. Recent Advances in Engineered Biochar Productions and Applications. *Crit. Rev. Environ. Sci. Technol.* **2017,** *47* (22), 2158–2207.

Wang, H.; Gao, B.; Wang, S.; Fang, J.; Xue, Y.; Yang, K. Removal of Pb (II), Cu (II), and Cd (II) from Aqueous Solutions by Biochar Derived from KMnO4 Treated Hickory Wood. *Bioresour. Technol.* **2015,** *197,* 356–362.

Wang, X.; Cai, W.; Lin, Y.; Wang, G.; Liang, C. Mass Production of Micro/Nanostructured Porous ZnO Plates and their Strong Structurally Enhanced and Selective Adsorption Performance for Environmental Remediation. *J. Mater. Chem.* **2010,** *20* (39), 8582–8590.

Wang, X.; Cai, W.; Lin, Y.; Wang, G.; Liang, C. Mass Production of Micro/Nanostructured Porous ZnO Plates and their Strong Structurally Enhanced and Selective Adsorption Performance for Environmental Remediation. *J. Mater. Chem.* **2010,** *20* (39), 8582–8590.

Wang, Y.; Fang, Z.; Liang, B.; Tsang, E. P. Remediation of Hexavalent Chromium Contaminated Soil by Stabilized Nanoscale Zero-Valent Iron Prepared from Steel Pickling Waste Liquor. *Chem. Eng. J.* **2014,** *247,* 283–290.

Wu, J.; Yi, Y.; Li, Y.; Fang, Z.; Tsang, E. P. Excellently Reactive Ni/Fe Bimetallic Catalyst Supported by Biochar for the Remediation of Decabromodiphenyl Contaminated Soil: Reactivity, Mechanism, Pathways and Reducing Secondary Risks. *J. Hazard. Mater.* **2016,** *320,* 341–349.

Wu, Q.; Miao, W. S.; Gao, H. J.; Hui, D. Mechanical Properties of Nanomaterials: A Review. *Nanotechnol. Rev.* **2020,** *9* (1), 259–273.

Wu, Q.; Wongwiriyapan, W.; Park, J. H.; Park, S.; Jung, S. J.; Jeong, T.; Lee, S.; Lee, Y. H.; Song, Y. J. In Situ Chemical Vapor Deposition of Graphene and Hexagonal Boron Nitride Heterostructures. *Curr. Appl. Phys.* **2016,** *16* (9), 1175–1191.

Xie, Y.; He, Y.; Irwin, P. L.; Jin, T.; Shi, X. Antibacterial Activity and Mechanism of Action of Zinc Oxide Nanoparticles against Campylobacter Jejuni. *Appl. Environ. Microbiol.* **2011,** *77* (7), 2325–2331.

Xu, K.; Chen, J. High-Resolution Scanning Probe Lithography Technology: A Review. *Appl. Nanosci.* **2020,** *10* (4), 1013–1022.

Yadav, K. K.; Singh, J. K.; Gupta, N.; Kumar, V. J. J. M. E. S. A Review of Nanobioremediation Technologies for Environmental Cleanup: A Novel Biological Approach. *J. Mater. Environ. Sci.* **2017,** *8* (2), 740–757.

Yang, S.; Li, J.; Shao, D.; Hu, J.; Wang, X. Adsorption of Ni (II) on Oxidized Multi-Walled Carbon Nanotubes: Effect of Contact Time, Ph, Foreign Ions and PAA. *J. Hazard. Mater.* **2009,** *166* (1), 109–116.

Yao, C.; Pan, Y.; Lu, H.; Wu, P.; Meng, Y.; Cao, X.; Xue, S. Utilization of Recovered Nitrogen from Hydrothermal Carbonization Process by Arthrospira Platensis. *Bioresour. Technol.* **2016,** *212,* 26–34.

Yao, Y.; Gao, B.; Chen, J.; Yang, L. Engineered Biochar Reclaiming Phosphate from Aqueous Solutions: Mechanisms and Potential Application as a Slow-Release Fertilizer. *Environ. Sci. Technol.* **2013,** *47* (15), 8700–8708.

Yi, S.; Dai, F.; Zhao, C.; Si, Y. A Reverse Micelle Strategy for Fabricating Magnetic Lipase-Immobilized Nanoparticles with Robust Enzymatic Activity. *Sci. Rep.* **2017,** *7* (1), 1–9.

Yin, Y.; Gates, B.; Xia, Y. A Soft Lithography Approach to the Fabrication of Nanostructures of Single Crystalline Silicon with Well-Defined Dimensions and Shapes. *Adv. Mater.* **2000,** *12* (19), 1426–1430.

Zhang, C. Z.; Chen, B.; Bai, Y.; Xie, J. A New Functionalized Reduced Graphene Oxide Adsorbent for Removing Heavy Metal Ions in Water via Coordination and Ion Exchange. *Sep. Sci. Technol.* **2018,** *53* (18), 2896–2905.

Zhang, M.; Gao, B.; Varnoosfaderani, S.; Hebard, A.; Yao, Y.; Inyang, M. Preparation and Characterization of a Novel Magnetic Biochar for Arsenic Removal. *Bioresour. Technol.* **2013,** *130,* 457–462.

Zhang, M.; Gao, B.; Yao, Y.; Xue, Y.; Inyang, M.; Synthesis, Characterization, and Environmental Implications of Graphene-Coated Biochar. *Sci. Total Environ.* **2012b,** *435,* 567–572.

Zhang, X.; Yang, B.; Dai, Z.; Luo, C. The Gas Response of Hydroxyl Modified Swcnts and Carboxyl Modified Swcnts to H2S and SO2. Prz. Elektrotech. **2012,** *88* (7B), 311–314.

Zhou, Y.; Gao, B.; Zimmerman, A. R.; Chen, H.; Zhang, M.; Cao, X. Biochar-Supported Zerovalent Iron for Removal of Various Contaminants from Aqueous Solutions. *Bioresour. Technol.* **2014,** *152,* 538–542.

Zhou, Y.; Gao, B.; Zimmerman, A. R.; Fang, J.; Sun, Y.; Cao, X. Sorption of Heavy Metals on Chitosan-Modified Biochars and Its Biological Effects. *Chem. Eng. J.* **2013,** *231,* 512–518.

Zhu, W.; Guo, Y.; Ma, B.; Yang, X.; Li, Y.; Li, P.; Zhou, Y.; Shuai, M. Fabrication of Highly Dispersed Pd Nanoparticles Supported on Reduced Graphene Oxide for Solid Phase Catalytic Hydrogenation of 1, 4-bis (Phenylethynyl) Benzene. *Int. J. Hydrog. Energy* **2020,** *45* (15), 8385–8395.

Zhuang, S.; Lee, E. S.; Lei, L.; Nunna, B. B.; Kuang, L.; Zhang, W. Synthesis of Nitrogen-Doped Graphene Catalyst by High-Energy Wet Ball Milling for Electrochemical Systems. *Int. J. Energy Res.* **2016,** *40* (15), 2136–2149.

Znaidi, L. Sol–Gel-Deposited ZnO Thin Films: A Review. *Mater. Sci. Eng. B* **2010,** *174* (1–3), 18–30.

Zou, W.; Gu, B.; Sun, S.; Wang, S.; Li, X.; Zhao, H.; Yang, P. Preparation of a Graphene Oxide Membrane for Air Purification. *Mater. Res. Exp.* **2019,** *6* (10), 105624.

CHAPTER 4

Nanoremediation Associated with Microorganisms

JESSICA PANDOHEE[1], RUCHIRA GHOSH[2], PRAIRNA BALYAN[3], ZOYA SHAIKH[3], and AHMAD ALI[3]

[1]*Telethon Kids Institute, Nedlands, WA, Australia*

[2]*Centre for Sustainable Technologies, Belfast School of Architecture and the Built Environment, Ulster University, Northern Ireland, United Kingdom*

[3]*Department of Life Sciences, University of Mumbai, Vidyanagari, Mumbai, India*

ABSTRACT

Environmental pollution has a detrimental effect on human health. Short- and long-term exposure to harmful pollutants can lead to serious ailments and incurable diseases. Consequently, it is necessary to manage the environmental contaminants to mitigate or even eliminate the deleterious effects to living beings. Nanotechnology field can contribute to offering innovative tools to deal with these substances in an efficient and sustainable manner. This chapter deals with various aspects of nanotechnology in bioremediation as a biogenic synthesis of nanomaterials, the innovative technological approach of nanomaterials in abatement of pollution, application of microbial-based nanoremediation of xenobiotics, pesticide residues, and metal toxicity, and prospects of nanobioremediation.

Sustainable Nanoremediation: Modern Technologies for a Clean Environment.
Fernanda Maria Policarpo Tonelli, Arpita Roy, and Khalid Rehman Hakeem (Eds.)

4.1 INTRODUCTION

Environmental pollution has a detrimental effect on human health. Short-term and long-term exposure to harmful pollutants can lead to serious ailments and incurable diseases. The severity of diseases depends on the nature, concentration, and exposure duration of each pollutant or mixture of pollutants. For instance, heavy metal pollution poses a very serious threat to mankind as continuous exposure to the pollutants can magnify their impact on the food chain, through a process called biomagnification (Domingo and Nadal, 2019). In an aerial ecosystem, air pollutants such as particulate matter, nitrogen oxide, sulfur dioxide, volatile organic compounds (VOCs), dioxins, and polycyclic aromatic hydrocarbons are notorious for their small diameter and capacity to penetrate the human respiratory system via inhalation. Air pollutants are known to lead to severe respiratory and cardiovascular diseases such as chronic obstructive pulmonary disease, asthma, bronchiolitis, lung cancer, cardiovascular diseases, and central nervous system dysfunction (Manisalidis et al., 2020).

Moreover, pollution of the drinking water system can lead to serious and sometimes irreversible health impacts, including chronic diseases (e.g., cancer and cardiovascular disease), adverse reproductive consequences, and impacts on children's health (e.g., neurodevelopment). Endocrine-disrupting chemicals (EDCs), in particular, are a global problem for both environmental and human health (Yilmaz et al., 2020). EDCs can cause developmental and fertility problems both in men and women by interfering with hypothalamo–pituitary–thyroid and adrenal axes, disrupting endocrine functioning, and hormone-specific cancers. Well known carriers of EDCs are pesticides, fungicides, industrial chemicals, plasticizers, nonylphenols, metals, pharmaceutical agents and phytoestrogens.

Unfortunately, it is difficult to deny that one of the main causes of environmental pollution is human activity. Factors such as extensive migration, increased urban complexity and mismanagement of water resources are only a few examples of human activities aggravating environmental pollution (Filho et al., 2019; Levallois and Villanueva, 2019; P. Li and Wu, 2019).

Moreover, the global consumption of fossil fuels, biomass, metals, and minerals is anticipated to increase 2-fold and the recycling sector is likely to triple in size from 2017 to 2060 (OECD, 2018). Simultaneously the waste generation across the world is expected to reach 3.40 billion ton by 2050 (The World Bank, 2018). Altogether, this will result in a larger proportion of the human population being exposed to pollutants and developing serious diseases and mutations (Domingo and Nadal, 2019; G. Li et al., 2018; Sillmann et al., 2021; Zamora-Ledezma et al., 2021; Zeng et al., 2019). To avoid reaching a point of no return, it is imperative that new technologies are developed to reduce environmental pollution.

Remediation consists of the decontamination of polluted systems. It can be performed through several means including phytoremediation, bioremediation, or more targeted methods such as nanotechnology. The latter is believed to be more sustainable, noninvasive, eco-friendly, and cost efficient. The use of covalent organic framework-based adsorption method is one of the most popular methods of environmental remediation (N. Zhang et al., 2020). This approach has been an efficient technique to recover nuclear pollutants such as uranium through sorbents made of porous inorganic materials (e.g., clay minerals, metal oxides, and mesoporous silica), organic polymers (e.g., resins, cellulose, and chitosan), and other carbon materials (e.g., mesoporous carbon, carbon nanotubes, and graphene oxides) (He and Li, 2020). Although some conventional methods such as photocatalysis are still in use (Djellabi et al., 2021), there has been an increased interest in advanced technologies including sulfate radical-based oxidation (Ushani et al., 2020) and zirconium-based metal–organic frameworks photocatalysis (X. Zhang et al., 2021).

Bioremediation is the use of living organisms (mostly microorganisms) to remove pollutants from soil, water, and air. The appeal of bioremediation is that it is less intrusive and more sustainable. For instance, the removal of organic matter and heavy metal using magnetostatic bacteria as adsorbents is a novel technology promising enormous advantages in smart city development toward sustainable living (Wang et al., 2020). Other bioremediating agents such as biogenic manganese oxides have also been used to remove pollutants through their oxidizing and catalyzing properties (Zhou and Fu, 2020). Thus, the use of biogenic nanomaterials (bioinspired nanoscale adsorbents and catalysts) is gaining wide acceptance due to their high chemical reactivity, cost and energy effectiveness, high surface area, and other mechanical properties derived from plants, bacteria, fungi, and algae (Gautam et al., 2019). It is important to note that the choice of the microorganism type and species is crucial for the development of bioremediation technologies. Microorganisms are not always remediating agents and, in some cases, can act as pathogens and lead to diseases. For example, most Gram-negative bacteria release endotoxins which are very harmful as environmental pollutants (Rasuli et al., 2022).

The use of nanotechnology in bioremediation is an advanced and effective innovation promising little or no side effects and assuring a sustainable solution to environmental pollution (Bhavya et al., 2021). Biologically synthesized nanoparticles are eco-friendly and effective in mitigating the environmental pollutants, through their photo and enzymatic catalysis, electrostatic interaction, and surface-active site interaction properties. Recently, the biogenic metal nanoparticles (Bio–Me–NPs) induced with microbes have been in use to deal with water pollutants such as heavy metals. Thus, with a huge opportunity to explore for the upcoming generation, this chapter deals with various aspects of nanotechnology in bioremediation as a biogenic synthesis of nanomaterials, the innovative technological approach of nanomaterials in abatement of pollution, application of microbial-based nanoremediation of xenobiotics, pesticide residues, and metal toxicity, and prospects of nanobioremediation.

4.2 BIOGENIC SYNTHESIS OF NANOMATERIALS FOR REMEDIATION

Biogenic synthesis of nanoparticles is a bottom–up strategy that employs simple unicellular to large multicellular biological entities. The process is driven through microbial-mediated synthesis. Microbial culture filtrates are utilized as a reducing agent to produce nanoparticles. Microbes such as bacteria, fungi, yeast, and algae are metal tolerant and flourish in extreme environmental circumstances and use these intrinsic properties to tolerate, accumulate, and transform metal into metal ions. However, nanoparticles are typically generated in the following manner: metal ions are originally trapped on the surface of microbial cells or within them, and these metal ions are then reduced to nanoparticles in the presence of enzymes (Patil and Chandrasekaran, 2020; Prairna et al., 2020). In general, microbes have two unique effects on mineral formation: First, they can change the solution's composition to make it more saturated; and second, they can influence mineral formation by producing organic polymers, which have the ability to influence the nucleation (Benzerara et al., 2011).

Several factors influence the use of biological approaches for the synthesis and engineering of nanoparticles. The most crucial variable is the shape of the metal nanoparticle that will be manufactured. The organisms' resistance to a small number of metals limits their choice of species (Patra and Baek, 2014). The following are some of the microbial resources (algae, fungi, bacteria, viruses, and yeast) that have been used to produce most of the commonly studied metal and metal salt nanoparticles: Copper, silver, gold, cadmium, platinum, palladium, cadmium sulfide, titanium dioxide, and zinc oxide as shown in Figure 4.1.

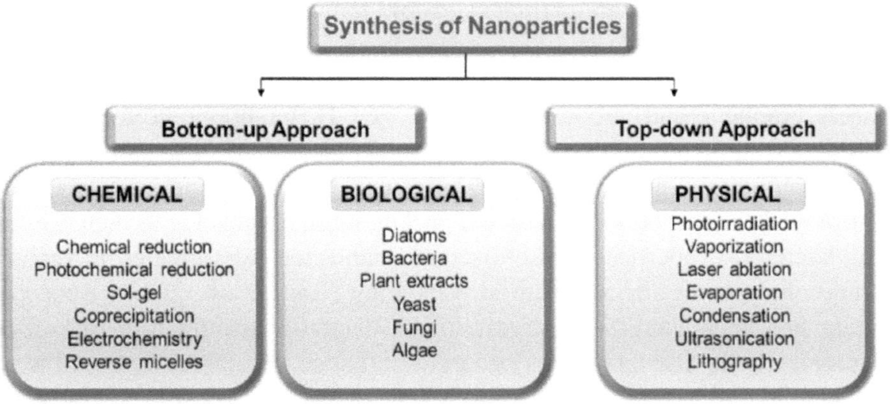

FIGURE 4.1 Approaches to nanoparticle synthesis.

4.2.1 NANOMATERIAL FORMED BACTERIA

Bacteria are thought to be a possible biofactory to produce various nanoparticles such as silver, palladium, platinum, magnetite, titanium, and cadmium sulfide (Table 4.1). Since most metal ions are poisonous to bacteria, bioreduction of ions or the creation of water-insoluble complexes is a bacterial defensive mechanism for overcoming such toxicity. *Bacillus subtilis* was used to create the first bacterial gold nanoparticles (Beveridge and Murray, 1980). Since then, several approaches to synthesized bionanoparticles have been developed. Mullen and his colleagues (1989) investigated the capability of *B. subtilis*, *Bacillus cereus*, *Pseudomonas aeruginosa*, and *Escherichia coli* to remove Cd^{2+}, Ag^{+}, La^{3+}, and Cu^{2+} from the solution. They reported that the bacterial cells have the ability to bind to significant amounts of metallic cations. Additionally, some bacteria, such as magnetostatic bacteria, can manufacture inorganic substances like intracellular magnetite nanoparticles (Lovely et al., 1989).

The extracellular synthesis of silver nanoparticles was reported by Kalishwaralal et al. (2008) via bioreduction of Ag^{+} ions using *Bacillus licheniformis*. Wei et al. (2012) used solar irradiation of *Bacillus amyloliquefaciens* cell-free extracts as well as silver nitrate to produce crystalline silver nitrate nanoparticles ($AgNO_3$). Green biosynthesis of silver nanoparticles (20–40 nm) was achieved using *B. cereus* isolated from *Garcinia xanthochymus*. Antibacterial property against pathogenic microorganisms such as *E. coli, Klebsiella pneumoniae*, *P. aeruginosa*, and *S. typhi* were also reported (Sunkar and Nachiyar, 2012). One research team demonstrated the green production of silver nanoparticles at room temperature using *Lactobacillus casei* (Korbekandi, 2012).

Carbon nanomaterials, for example, carbon nanoparticles, carbon nanotubes, graphene, and carbon dots, are discovered to be superior to conventional food packaging materials. Carbon nanomaterials are safe alternatives for the environment and are simple to manufacture. Carbon nanomaterials can be employed as food preservatives and packaging materials due to their natural antibacterial properties. An antibacterial nanosheet was created to replace traditional plastic for active food packaging (Raul et al., 2022). Recently, there has been increased interest in the bacterial manufacture of carbon nanomaterials, also known as bacterial cellulose. Excellent hydrophilicity, biodegradability, renewability, and mechanical characteristics characterize these newly developed natural biopolymers (Hamimed et al., 2022). Bacterial cellulose is an intriguing polymer having a three-dimensional (3D) structure generated by pure cellulose nanofibers manufactured and structured as a coherent network by some bacteria, for example, *Azotobacter*, *Acetobacter*, *Gluconacetobacter*, *Salmonella, Sarcina ventriculi*, and *Pseudomonas*. Scientific research into the use of this material in a variety of applications, including biomedical and electrical ones, has been inspired by its exceptional mechanical properties and biocompatibility. However, numerous investigations on the creation of bacterial cellulose nanocomposites for a variety of applications have been stimulated by the capacity to include nanosized components into the coherent 3D network structure given by BC (Torres et al., 2019).

By employing 3D nanostructured biomass as a precursor, a molybdenum carbide (Mo_2C)-based hydrogen evolution reaction electrocatalyst was created, according to Wu et al. (2016), using ultrafine Mo_2C nanoparticles enmeshed in 3D N-doped carbon nanofiber networks synthesized from bacterial cellulose. After being pyrolyzed, carbon nanofibers generated from bacterial cellulose are employed as high-capacity anode materials in lithium–ion batteries because of their large surface area, mechanical flexibility, structural stability, and open porosity structure (Illa et al., 2018). One-step pyrolysis of inexpensive bacterial cellulose with the aid of urea was reported to produce N-doped carbon nanofibers (Huang et al., 2019).

Carbon nanotubes with linked porosity can be fabricated using *Magnetospirillum magneticum* (AMB-1), a biotemplate. Due to its high porosity and strong pore connectivity, the resulting porous scaffold displayed remarkable mechanical stability and a huge surface area (Dubey et al., 2011). Moreover, there is evidence pointing to the role of nanotube structures in mediating intracellular chemical exchange between nearby cells of the same species and those of different species (Baidya et al., 2018).

The antimicrobial capabilities of lead sulfide nanopowders demonstrated their efficacy as an antibacterial agent by inhibiting the growth of pathogenic bacterial and fungal strains, making them appropriate for nanodrug development as well as medical and pharmaceutical applications (Suganya and Balu, 2017). *E. coli* can produce intracellular cadmium sulfide nanocrystals when incubated with cadmium chloride and sodium sulfide (Sweeney et al., 2004). A biomimetic poison nanosponge that serves as a toxin decoy in vivo was demonstrated by Hu et al. (2013). They demonstrated that the red blood cell membranes that surround a polymeric nanoparticle core in the nanosponge absorb toxins that damage membranes and direct them away from their intended cellular targets. Under ultraviolet (UV) light, titanium dioxide exhibits a potent oxidative action and is used as a photocatalytic disinfectant in the biomedical area. Titanium dioxide-based photocatalysts have the potential to be employed even in interior environments to inactivate bacteria and viruses since they are potent antibacterial and antiviral agents in ambient circumstances (Prakash et al., 2022).

TABLE 4.1 Nanoparticle Synthesis Using Bacteria.

Bacteria	NPs	Size (nm)	Morphology	References
B. cereus	Ag	20–40	Spherical	Sunkar and Nachiyar (2012)
B. subtilis	Ag	5–60	Spherical and triangular	Saifuddin et al. (2008)
Desulfobacteraceae	ZnS	2–5	Spherical	Labrenz et al. (2000)
E. coli	CdS	2–5	Wurtzite crystal	Sweeney et al. (2004)
E. coli	Ag	8–9	Spherical	Mahanty et al. (2013)
Lactobacillus strains	Ti	40–60	Spherical	Prasad et al. (2007)
Magnetospirillum magnetotacticum	Magnetite	-	Cluster	Philipse and D. Maas (2002)
Rhodopseudomonas palustris	CdS	8.01 ± 0.25	Crystalline	Bai et al. (2009)

4.2.2 NANOMATERIAL FORMED FUNGI

Fungi have several benefits over other microorganisms. They ease of isolation from soil or plants, as opposed to uncommon bacteria and actinomycetes, which need specialized enrichment techniques for isolation (Xiaowen et al., 2019) Fungi also secrete a lot of metabolites and extracellular enzymes, which help metal ions become reduced into nanoparticles and they further simple to scale up because they are growing quickly (Shah et al., 2015) Moreover, fungi have a broad growth range in terms of pH, temperature, and sodium chloride, making it easier to alter the culture conditions to create homogenous nanoparticles (Guilger-Casagrande, 2019).

Due to their ability to tolerate, absorb, and accumulate metals, fungi have gained increased attention in recent years for the production of nanoparticles (Moghaddam et al., 2015). *Fusarium oxysporum* uses nitrate reductase and anthraquinones for the production of silver nanoparticles (Gholami-Shabani et al., 2014). Zinc oxide nanoparticles are created using the culture filtrate of the fungus *Alternaria tenuissima*, which was isolated from *Erythrophleum fordii* (Abdelhakim et al., 2020). According to Clarance et al. (2020), the *Fusarium solani*, an endophytic fungus, isolated from the *Chonemorpha fragrans* is employed in the production of gold nanoparticles. Bagur et al. (2019) identified *Exserohilum rostrata*, an endophyte from the plant *Ocimum tenuiflorum*, and this strain was employed to produce spherical silver nanoparticles. In a study, carbon nanotubes were attached to fungal hyphae using a biological assembly process to create a sphere The fungal hyphae and carbon nanotube composite is a flexible adsorbent for water pollution treatment (Zhu et al., 2018).

Furthermore, it has been shown that the use of titanium dioxide and cesium dioxide nanofiber mixtures in biomedical applications since they exhibit intriguing antifungal characteristics (Firmino et al., 2020). The manganese-oxidizing fungus was immobilized onto nanofiber surfaces to aid in the clean-up of heavy metal-contaminated water. *Coniothyrium* sp. catalyzed the conformal deposition of manganese oxide across its hyphae and nanofibers after fungal attachment to manganese chloride-seeded nanofibers, whereas *Coprinellus* sp. catalyzed manganese oxide solely along its hyphae. Hybrid fungi–nanofibers removed different heavy metals from water (Park et al., 2020). *Mucor indicus*, a member of the zygomycetes family of filamentous fungi, is one of the most well-known fungal strains for producing chitin and its deacetylated by-product chitosan. To increase the rate at which fungal chitin is produced, chitin nanofibers were isolated from *M. indicus* using a purification technique and mechanical treatment (Salehinik et al., 2020). Some of the examples of nanoparticles formed using fungi are summarized in Table 4.2.

TABLE 4.2 Nanoparticle Synthesis Using Fungi.

Fungi	NPs	Size (nm)	Morphology	References
Aspergillus fumigatus	ZnO	1.2–6.8	Spherical and hexagonal	Raliya and Tarafdar (2013)
Aspergillus niger	Au	10–20	Triangular and hexagonal	Xie et al. (2007)
Aspergillus tubingensis	Ca3P2O8	28.2	Spherical	Tarafdar et al. (2012)
Aureobasidium pullulans	Au	29	Spherical	Zhang et al. (2011)
F. solani	Au	40–45	-	Clarance et al. (2020)
Penicillium chrysogenum	MgO	5–12	-	El-Sayyad et al. (2018)
Rhizopus oryzae	MgO	20.38	Crystalline	Hassan et al. (2021)
Trichoderma harzianum	Ag	21.49	Cubic crystal	Konappa et al. (2021)

4.2.3 NANOMATERIAL FORMED YEAST

In yeast, nanoparticles can be produced intracellularly or extracellularly. The capacity of yeast to take up significant amounts of harmful metals from its environment makes it possible to carry out the synthesis, which results in the formation of nanoparticles through various processes, such as bioprecipitation, detoxification, chelation, biosorption, and extracellular sequestration. Numerous yeast species, including *Schizosaccharomyces pombe*, *Candida glabrata*, *Pichia jadinii*, *Torulopsis* sp., *and MKY3*, are employed to synthesize nanoparticles as shown in Table 4.3. It was reported that extracellular production of hexagonal-shaped silver nanoparticles during the log phase of growth was possible using the silver-tolerant yeast strain MKY3 (Kowshik et al., 2003). Additionally, using yeast cells from *Magnusiomyces ingens* LH-F1, gold nanoparticles were biochemically synthesized in a study (Zhang et al., 2016). Several species of yeast such as *P. jadinii, S. pombe, C. glabrata*, and *Y. lipolytica* are also involved in the synthesis of Cd, Ag, and Au nanoparticles as shown in Table 4.3.

In a study, zirconium phosphate was produced using yeast and used as a biotemplate to create an air electrode with exceptional electrocatalytic activity for oxygen decrease (Tian et al., 2010). Therefore, yeasts were used as biobased templates in the manufacture of zinc phosphate nanopowders. Similarly, chloroauric acid concentrations varied throughout the incubation of *Y. lipolytica* cells, and cell-related gold nanoparticles and nanoplates were produced (Pimprikar et al., 2009). Yan et al. (2008) demonstrated the synthesis of zinc phosphate powders having butterfly-like microstructures that ranged in size from 80–200 nm in length and 10–80 nm in width. Cadmium sulfide and lead sulfide nanoparticles have been effectively made using yeast as a synthesis agent. When exposed to aqueous Pb^{2+} ions, Kowshik et al. (2002) demonstrated that the *Torilopsis* species may generate nanoscale PbS (intracellularly), and CdS Quantum dots were seen to form in *S. pombe* yeast cells.

TABLE 4.3 Nanoparticle Synthesis Using Yeast.

Yeast	NPs	Size (nm)	Morphology	References
Candida albicans	Ag	2.0–7.3	Spherical crystalline	Ananthi et al. (2018)
M. ingens	Se	70–90	Hexagonal crystalline	Lian et al. (2019)
Meyerozyma guilliermondii	Ag	2.5–30	Cubic crystalline	Alamri et al. (2018)
Phaffia rhodozyma	Au	2.22 ± 0.7	Spherical	Rónavári et al. (2018)
Pichia kudriavzevii	ZnO	10–61	-	Moghaddam et al. (2017)
Pichia pastoris	Se, Ag	70–180	Spherical	Elahian et al. (2017)
Saccharomyces cerevisiae	Au	9.99 ± 1.63	Spherical	Shi et al. (2017)
Yarrowia lipolytica	Au	104	-	Pimprikar et al., 2009

4.2.4 NANOMATERIAL FORMED ALGAE

Microalgal nanomaterials offer a wide range of uses in the fields of reaction catalysis and medical treatment. The hydrophilic surface groups such as sulfate, carboxyl, and hydroxyl that are present in algae-mediated nanomaterials allow a special range of applications (Rahman et al., 2020). According to Xie et al. (2007), a chloroauric acid aqueous solution was treated at room temperature with an extract of the single-celled *Chlorella vulgaris* to create single-crystalline gold nanoplates. Hassan et al. (2021) reported the green production of silver nanoparticles using freshwater *C. vulgaris* algae. The perfectly manufactured silver nanoparticles were only accessible by the ratio 3:1 of *C. vulgaris*: silver nitrate, with scanning electron microscope imaging revealing a size of 16.99 ± 0.3 nm. The algae were first dried and pulverized to create the 100-ppm aqueous extract, then added with silver nitrate in various ratios. Regarding the pathogens *P. aeruginosa*, *K. pneumonia*, and *E. coli*, these silver nanoparticles demonstrated enormous promise. Other than these, different algae including *Caulerpa peltate*, *Bifurcaria bifurcate*, *Sargassum myriocystum*, *Sargassum muticum*, *Hypnea Valencia*, among others have been employed to synthesize nanoparticles (Al Nadhari et al., 2021).

Spirulina platensis has successfully biosynthesized crystallized silver nanoparticles in an aqueous setting. Live *S. platensis* biomass was added to an aqueous solution of silver ions to induce the synthesis of single-nucleotide polymorphism (SNPs). These NPs displayed an absorption peak in the UV-visible spectrum at 430 nm that was associated with SNP plasmon resonance. Transmission electron micrographs of nanoparticles in an aqueous solution demonstrated the generation of silver nanoparticles (12 nm) by *S. platensis* (Mahdieh et al., 2012). Some of the examples of nanoparticles formed using algae are summarized in Table 4.4.

TABLE 4.4 Nanoparticle Synthesis Using Algae.

Algae	NPs	Size (nm)	Morphology	References
Amphiroa rigida	Ag	20–30	Spherical	Gopu et al. (2021)
Anabaena cylindrica	ZnO	40–60	Spherical	Bhattacharya et al. (2020)
Botryococcus braunii	Ag	40–90	Cubical, spherical, and truncated triangular shaped	Arya et al. (2019)
C. vulgaris	Ag	16.99 ± 0.3	Crystalline	Hassan et al. (2021)
C. vulgaris	Ag	40–90	Cubical, spherical, and truncated triangular shaped	Mahajan et al. (2019)
Gelidiella acerosa	Au	5.81–117.59	Spherical	Kumar et al. (2019)
Sargassum wightii	Ag	18.45–41.59	Spherical	Deepak et al. (2018)
Ulva armoricana	Ag	12.5	Crystalline	Massironi et al. (2019)

4.3 APPLICATION OF MICROBIAL-BASED NANOREMEDIATION

4.3.1 *MICROBIAL NANOREMEDIATION OF HEAVY METALS*

The potential of microbes in environmental clean-up is validated by microbe-mediated nano bioremediation of heavy metals. As they have significant harmful effects on biota even at the lowest concentrations, heavy metals make up a significant portion of inorganic pollutants (Alissa and Ferns, 2011). Reactive oxygen species, which are produced because of heavy metal exposure, damage macromolecules such as proteins and nucleic acids. This damage is the primary mechanism by which heavy metals impact plants and lower soil creatures (Jaishankar et al., 2014). Crops, vegetation, their nutritional value, and the ecological elements related to them are all impacted by the presence of heavy metals in the soil. Depending on the crop species, soil physicochemical properties, and type, heavy metals have different effects on different types of crops (Chaplygin et al., 2021). Site stabilization, which immobilizes heavy metals at a particular site to reduce their mobility and availability in the soil and prevents them from leaking across sites, is one of the important approaches for the eradication of heavy metals (Suman et al., 2018). The elimination of heavy metals has drawn a lot of attention to the utilization of various nanoparticles, including biogenic. Biological organisms are used in the creation of biological nanoparticles (Rajput et al., 2022).

Morganella psychrotolerans produce well-known biogenic nanoparticles (Sharma et al., 2019). *Halomonas* sp., a Gram-negative bacterium, has successfully improved bioremediation of soil contaminated with lead and cadmium by using nanoparticles of iron (II) oxide coated with polyvinyl-pyrrolidone. In comparison to removal by bacteria or nanoparticles only, this method considerably reduced lead and cadmium after 24 and 48 hours, respectively (Rajput et al., 2022; Alabresm et al., 2018). For the removal of copper, nickel, and lead, a biosorbent composed of magnetic iron (III) oxide nanoparticles treated with *S. aureus* and having a surface coated with phthalic acid was utilized. The adsorptive removal of 795, 1355, and 985 μmol g^{-1} for copper, lead, and nickel, respectively, was accomplished. Cu^{2+} recovery rates ranged from 83.0 to 89.5%, Pb^{2+} recovery rates ranged from 99.4 to 100%, while Ni^{2+} recovery rates ranged from 92.6 to 7.5%. The specialized core of the nanoparticle and the functional groups present on the microbial surface both appeared to be important in the removal of heavy metals, according to a comparison study using dried *S. aureus* (Mahmoud et al., 2016). As a result, this research demonstrated that the functional groups on the microbial surface as well as the nanoparticle core had a significant influence on the removal of pollutants (Rajput et al., 2022).

According to a recent study on the removal of copper, cadmium, chromium, and lead using heavy metal-resistant bacteria such as *B. cereus*, and *Lysinibacillus macroides*, zinc oxide nanoparticles at 5 mg $L-^{1}$ remove the chromium by 60%, the copper by 70%, and the lead by 85%, as opposed to *B. cereus* (80 and 60%) and *L. macroides* (55 and 50%) at neutral pH, respectively (Baragaño et al., 2020). However, at lower pH, the heavy metals precipitate as hydroxides, and then hydrogen ions compete for binding with adsorbents (Xie et al., 2011). At neutral pH, the surface of zinc oxide nanoparticles exhibits negative charges that facilitate electrostatic interactions with metal cations. It has also been noted that the *B. cereus* strain $XMCr^{6}-$ reduces the Cr^{6+} through an enzyme-mediated mechanism. It was discovered that the reduced Cr^{3+} formed coordination bonds with the functional group on the surface of the bacterial cell wall to have a binding affinity to cells. On the cell surface, chromium oxide nanoparticles were discovered as a by-product (Dong et al., 2013; Laslo et al., 2022; Rajput et al., 2022; Bhatkalkar et al., 2018).

Investigations were also conducted on the utilization of probiotic bacteria (*L. casei and Limosilactobacillus fermentum*) to absorb cadmium from water in conjunction with Se^{5+} and selenium nanoparticles. In this work, the increased cadmium absorption by *L. casei* with Se^{4+} ions (65%) as compared to selenium nanoparticles (55.90%) was found, and it was linked to the higher solubility of Se^{5+} as compared to selenium nanoparticles. Comparing L. fermentum with *L. casei*, it was shown that *L. fermentum* had a substantially better efficiency of cadmium absorption (50.87%) than *L. casei* (43.78%). When combined with selenium nanoparticles, *L. casei* does not significantly alter its cadmium adsorption %. However, in the presence of *L. casei,* cadmium absorption was marginally raised from 5.49 to 16.54 with higher selenium nanoparticles ratio percentages (Dong et al., 2013; Rajput et al., 2022).

As heavy metal pollutants can be utilized by specific bacteria to create biogenic nanoparticles (resource recovery), their removal from the environment (remediation), and the creation of value from the waste, this strategy is currently gaining favor (effective waste utilization). According to a study, lead nanoparticles were biorecovered utilizing *Enterococcus faecalis*, and both intracellular and extracellular lead was produced (membrane-bound). Transmission electron microscopy revealed a lead nanoparticles size range of 10 nm, however environmental factors including temperature, pH, and biomass affected the lead nanoparticles size. The produced lead nanoparticles have tremendous potential as a bionanocatalyst because they have remarkable catalytic efficiency (6.3 mg lead nanoparticles totally reduced 5.0 mol Cr^{6+} in 12 h), and the application could be helpful in treating industrial effluents (Cao et al., 2020; Rajput et al., 2022).

Exopolysaccharides from *Lysinibacillus sphaericus* were used in their study as a more effective reducing, stabilizing, and capping agent since they have several binding sites for different metal ions. The magnetic oxide nanoparticles with exopolysaccharides functionalization showed an improved ability to absorb Cr (VI). Algae, on the contrary, have shown useful in nanobioremediation. *Chlorococcum* sp. role in mediating the synthesis of iron nanoparticles revealed a notable reduction of 4 $mgL-^{1}$ Cr (VI) to Cr (III) of about 92% (Subramaniyam et al., 2015). Iron nanoparticles that were produced through photosynthesis and mediated by biomolecules from this algal cell exhibited high reactivity, higher stability, and effective toxicity reduction in the environment. *C. vulgaris* was added to ultrafine bi-metallic (TiO_2/Ag) chitosan nanofiber mats as a functionalized agent, which clarified the significance of algae in the photoremoval approach of Cr (VI) under UV irradiation (L. Wang et al., 2017). *C. vulgaris* was reported to have released organic compounds (chlorophylls, carboxylate acids), which improved the photocatalytic reduction of Cr (VI) on TiO^2/Ag chitosan nanofiber mats, confirming the synergistic relationship between algae and TiO^2/Ag hybrid nanomaterial (Torimiro et al., 2021).

The ability of *Rhodosporidium diobovatum* ability to produce lead sulfide nanoparticles supports the idea that fungus could easily convert Pb (II) ions contaminants found in the environment into less harmful and useful forms. The ability to biosynthesize chitosan nanoparticles from *Cunninghamella elegans* employing a bioactive polymer (chitosan) is demonstrated by research findings (Alsharari et al., 2018). According to a study, generated chitosan nanoparticles outperformed bulk chitosan in biosorbing and bioremediating Pb (II) and Cu (II) ions in polluted environments. The ability of *P. aeruginosa* to increase cadmium bioreduction, which in turn sped up the manufacture of cadmium sulfide nanoparticles, was demonstrated by the effective removal of cadmium from cadmium-contaminated water (Raj et al., 2016).

Under aerobic and anaerobic circumstances, the conversion of elemental mercury to insoluble mercuric selenide via biogenic nanoselenium yielded bioremediation values of 39.1–48.6 and 45.8–57.1%, respectively. It has been documented that industrial effluents with nickel electroplating can be cleaned up using nanobiotechnology. The introduction of *Microbacterium* sp. in this investigation enabled the removal of nickel compounds from the effluent, which produced and recovered nickel oxide nanoparticles (Sathyavathi et al., 2014). The ability of fungi to biologically reduce harmful metals, particularly nickel in pollutants, was discovered by a study on the fungus *Hypocrea lixii.* This led to the development of the biosynthesis of nickel oxide nanoparticles from the trash for further use (Salvadori et al., 2015; Torimiro et al., 2021)

4.3.2 AGRICULTURAL FARMLANDS AND RESTORATION OF POLLUTED SOIL AND ITS PRODUCE

In agricultural fields, pesticides are chemicals that are used to both kill and control weeds and pests. The overuse of pesticides poses harm to humans as well as to the environment, lowering biodiversity, decreasing insect pollinator populations, endangering endangered species, and damaging bird habitats (affecting the nervous system, exhibiting carcinogenic effects, mimicking the hormones, and leading to death in severe cases). The most dangerous pesticides for both the environment and human health include dichlorodiphenyltrichloroethane (DDT), dichlorodiphenyldichloroethylene (DDE), parathion, malathion, chlordane, atrazine, and others (Gaur et al., 2022).

Scientists are creating materials with dimensions in the nanoscale range for this purpose by employing the nanotechnology method. As a result, due to the very specialized nature of this type of approach, this technology has recently attracted researchers from all over the world for the detection, degradation, and removal of dangerous pesticides. Nanomaterials may detect extremely specific pesticides because of their high surface-to-volume ratios, great target specificity, and small nanoscale sizes. As a result, these characteristics are crucial to the breakdown of pesticides (Tharmavaram et al., 2017).

Various physical and chemical techniques have been utilized to create nanoparticles, but these traditional synthesis methods require toxic substances that are bad for the environment and other living things. To solve this issue, several biotic sources, including plants, bacteria, yeast, and fungi, have been used to create an efficient and environmentally acceptable method of producing nanoparticles. Similar to this, microorganisms have typically been utilized to produce iron nanoparticles. In a study, the thermophilic bacteria *Thermoanaerobacter* were used to make greigite and maghemite nanoparticles. Thus, it may be argued that green nanobioremediation offers a potentially effective substitute for traditional remediation techniques (Gaur et al., 2022).

It is thought that the optimal soil for a plant's growth and optimum health is one that is rich in essential nutrients and micronutrients. Polyaromatic hydrocarbons, polychlorinated biphenyls, volatile organic compounds (VOCs), heavy metals, agrochemicals (pesticides, fungicides, and fertilizers), as well as occasionally too many nutrients, have all been introduced into the soil as a result of human-made activities. Urbanization and industrialization have also increased the amount of solid waste, different types of chemicals, and solvents in the environment as well as in agricultural soil (Rajput et al., 2022).

By breaking down toxic molecules with the help of microbes, nanobioremediation can improve soil quality and cut down on pollution. The procedure may be able to eliminate, retain, or lessen the amount of pollutants present by breaking down contaminants in the soil (Chandra et al., 2017). The effectiveness of bioremediation has previously been investigated and improved utilizing biotechnology or chemical additives (Gerhardt et al., 2017), but nanotechnology further strengthened the process with a novel feature (Gong et al., 2018). Due to the specific physicochemical characteristics of the nanoparticles, which also serve as catalysts and aid in lowering the activation energy needed to break down the compounds, the target pollutants are adsorbed, degraded, or changed (Mehndiratta et al., 2013).

Analyzing a study on the removal and bioremediation of soil contaminated with cadmium, it was found that using nanohydroxyapatite and *B. subtilis* together efficiently eliminated the cadmium (Liu et al., 2018) encouraging the expansion of bacterial diversity and the rhizosphere community in the soil that has been cleaned up. It has been documented that *Halomonas* sp. isolated from oil-polluted soil interacted with polyvinylpyrrolidone-coated iron oxide nanoparticles (Alabresm et al., 2018) in an assessment of a biofunctionalized method. There may be a way to bioremediate both lead and cadmium in the soil, according to a recent study (Zhu et al., 2020). *E. coli* and metallic nanoparticles were used in conjunction to remove the heavy metals according to this microbe-mediated technique. The bioremediation of mercury-contaminated soil using selenium nanoparticles was found to be successful, and the selenium nanoparticles were created with the presence of *Citrobacter freudii* (X. Wang et al., 2017).

The development of nanoparticles may illuminate the environmentally favorable and sustainable road to hasten the clearance of harmful chemicals from polluted soils. Numerous initiatives have been taken to boost the effectiveness of phytoremediation, including the use of rhizobacteria, genetic engineering, and chemical additions. The incorporation of nanotechnology with bioremediation in this setting has opened new avenues for rethinking remediation techniques. Since the nanoscale process regulation enhances the adsorption and degradation of contaminants, advanced remediation systems combine nanotechnological and biological remediation strategies. Due to their special surface qualities, nanoparticles can absorb/adsorb a wide range of pollutants and accelerate reactions by reducing the energy needed to break them down. As a result, this remediation procedure prevents pollutants from building up and spreading from one medium to another (Rajput et al., 2022).

4.3.3 PHARMACEUTICAL PRODUCTS AND XENOBIOTICS

It is extremely concerning when antibiotics and antiseptics are frequently released into wastewater. This comes from both the home and industrial sectors of society. In addition to polluting the environment, it has accelerated the spread of bacteria that are resistant to antibiotics in wastewater. However, these pharmaceutical chemicals can be eliminated using a bionanoremediation technique (Torimiro et al., 2021).

1,2,4-triazole can be successfully removed from wastewater by a bacterial consortium led by *Pseudomonas* and *Bacillus* spp., which are responsible for the manufacture of manganese oxide (Wu et al., 2017). The elimination of several refractory contaminants from biotreated chemical industrial effluent using biogenic manganese oxide nanoparticles was shown to be possible. Similar to picric acid, 2,4,6-trinitrophenol is a helpful component in the creation of antiseptics but also poses a concern to the environment as a pollutant in aqueous solution. A study showed how *P. aeruginosa*-mediated iron (III) oxide nanoparticles were gradually applied as a fragment of multiwalled carbon nanotubes for the synthesis of nanocomposite mostly employed for picric acid bionanoremediation (Yousefi et al., 2020).

However, the significance of fungi (*Trametes versicolor*) as a significant biofunctionalized agent for bimetallic (Pd/Fe) nanoparticles for the removal of triclosan in liquid effluents was established (Bokare et al., 2010). Triclosan micro-accumulation, which has been implicated as a risk factor for cancer, has frequently found its use in antibacterial and antiseptic agents. It was discovered that the *T. versicolor*-derived laccase enzyme had a crucial part to play in the two-step redox method, which involved successive oxidation of 2-phenoxyphenol and anaerobic dichlorination. Similar to this, yeast (*Candida* sp.) was described as biofunctionalizing magnesium oxide nanoparticles (Adikesavan and Nilanjana, 2016). It was discovered that this myconano approach sped up the process of treating and degrading cefdinir in an aqueous media. An alga called *Turbinaria conoides*, which is useful as an antimicrofouling agent, facilitated the production of both gold and silver nanoparticles (Vijayan et al., 2014). The effective removal of hydrogen peroxide, a common medicinal component but an environmental contaminant (Momeni and Nabipour, 2015). This was accomplished by the chemical being reduced electrocatalytically in industrial waste effluents with the help of palladium nanoparticles made from *Sargassum bovinum*.

Chemicals that an organism is exposed to that are extrinsic to the regular metabolism of that organism are known as xenobiotics. The heavy metals that have been building up for years in the river's sediments are the most dangerous of the various xenobiotics poisoning the river (Joutey et al., 2013). In a recent study, alumina was used in conjunction with greenly produced silver nanoparticles assisted by *B. cereus* for an efficient bioremediation procedure. The elimination of 98.13% and 98.76% of chromium and lead, respectively, which were emitted as waste effluents from pharmaceutical businesses, was validated by the nanoadsorbent approach mediated by bacterial cells. Cr (VI) elimination strategy using UV rays (Torimiro et al., 2021; L. Wang et al., 2017).

4.3.4 WASTEWATER AND POLLUTED BODIES OF WATER

Water is necessary for life to exist on Earth, it is equally important to clean it of pollution. For the sustainable development of companies and the environment, these industrial effluents must be treated in a precise and economical manner. Diverse electrochemical, sophisticated oxidation and valorization techniques have been used to lessen the toxicity of wastewater effluents and to make their use more sustainable (Shukla, 2020). The advancement of nanotechnology and nanoscience has made it possible to treat water pollution in efficient and innovative ways. The smaller size, high surface area to volume ratio, and higher chemical characteristics of the nanotechnological routes make them more effective than their conventional counterparts (Baruah et al., 2019). A path toward the environmentally friendly remediation of contaminants has been created by the synthesis of green nanomaterials from microbes and extracts of other creatures. Iron nanoparticles are green nanoparticles that are employed in remediation because of their nontoxic makeup, magnetic susceptibility, and redox potential when reacting with water (Shukla, 2020).

Nanomaterials attached to membranes can remove wastewater efficiently as well. Nanomaterials present novel roles for pollution degradation and enhance membrane permeability, stink resistance, and mechanical and thermal robustness. Additionally, nanocatalysts are crucial for accelerating degrading events (Corsi et al., 2018). Metal–organic frameworks are used for the removal of heavy metals from wastewater in addition to membranes and nanocatalysts. The coordination of organic ligands with metal ion precursors produces these metal–organic frameworks. The coordination of functional groups with metals rather than the organic ligand can increase the efficiency of metal–organic frameworks which is attributed to the metals' reduced steric barrier (Deshpande et al., 2020).

Nanomaterials are useful for the treatment of wastewater because of their tiny size. Their unique chemical, physical, and biological characteristics make them more useful in a variety of applications. For the removal of effluent from wastewater, many nanomaterials have been employed, including carbon-based (nanocomposites or nanotubes), metal- and oxide-based nanomaterials. The techniques used in wastewater treatment include photocatalytic degradation, adsorption, filtration using nanoparticles, and observation of various pollutants and contaminants (Palit and Hussain, 2020).

The utilization of microorganisms concurrently with the biofabrication of nanoparticles increases the sustainability and environmental friendliness of nanotechnology. The usage of chemicals and self-agglomeration in aqueous solutions are two potential drawbacks of chemically generated nanoparticles. Therefore, a potential remedy could be the green manufacturing of nanoparticles using bacterial, fungal, and plant extracts. They produce metallic nanoparticles and serve as reductive agents for the metal complex salt. Due to co-precipitation or by coating the outer face of the nanoparticles with proteinaceous and bioactive materials, these nanoparticles achieve enhanced firmness in an aqueous environment.

A. tubingensis (STSP 25) biofabricated iron oxide nanoparticles were produced from the rhizosphere of *Avicennia officinalis* in the Sundarbans of India. With a five-cycle regeneration capacity, the synthesized nanoparticles were able to remove more than 90% of the heavy metals Pb (II), Ni

(II), Cu (II), and Zn (II) from wastewater. In endothermic processes, the metal ions were chemically adsorbed onto the surface of the nanoparticles (Mahanty et al., 2020). Exopolysaccharides from *C. vulgaris* were employed in a different investigation to co-precipitate with iron oxide nanoparticles. It was discovered by Fourier-transform infrared spectroscopy analysis that exopolysaccharides functional groups had successfully modified nanoparticles. Additionally, it was shown that the nanocomposite was able to remove 85% of NH_4^+ and 91% of PO_4^{3-} (Govarthanan et al., 2020).

Using microbes to create nanoparticles has proven to be both economical and environmentally benign. *Escherichia* sp. was used to create copper nanoparticles because it is copper resistant. It has been demonstrated that the biogenic nanoparticles decompose azo dye and textile wastewater. Reactive black-5, Congo red, direct blue-1, and malachite green were reduced by 83.61, 97.07, 88.42, and 90.55%, respectively, at a dosage of 25 mgL^1, but they were reduced by 76.84, 83.90, 62.32, and 31.08% at a concentration of 100 mgL^{-1}. The industrial effluent was also treated, and treated samples showed a decrease in suspended particles as well as chloride and phosphate ions (Shukla, 2020).

The biological reduction of 80 mgL^{-1} Cr (VI) to soluble Cr (III) by the immobilized cells was the cause of an 84% reduction, which further suggests a full reduction of Cr (VI) after 24 hours of interaction. It has also been shown that *Shewanella oneidensis* immobilized on calcium alginate beads stabilized carbon nanotubes is a good biofunctionalized mechanism for the conversion of hazardous Cr (VI) to Cr (III) in wastewater (Yan et al., 2013). When compared to microorganisms and calcium alginate beads alone, this configuration was said to have demonstrated four times higher reduction rates, demonstrating the inclusion of the three ingredients for efficient biological reduction processes in environmental bioremediation and wastewater treatment. Magnetic oxide nanoparticles were produced with the help of a different bacterium (*L. sphaericus*) with a goal of eradicating hexavalent chromium in the environment (Kumar et al., 2019).

The amount of dye that build up in the waterways prevents light from reaching the surface, which hinders photosynthesis in plants. Aquatic creatures and plants perish as a result of the water's reduced oxygen concentration. The importance of dye has long been recognized in a variety of industries, including textile and cosmetics. However, it is primarily released into the environment as liquid effluent, which has been shown to be harmful to living things (Asaduzzaman et al., 2016). It has been shown that textile industry dyes can undergo microbe-mediated nano bioremediation. According to a study, *Chromobacterium violaceum* and silver nanoparticles together have a biosorption action that can be used to clean up washing water used to process cotton clothing (Durán et al., 2010).

The technique ensured the efficient removal of both organic chemicals and dyes used in the manufacture of garments. Additionally, this therapy demonstrated its effectiveness for the removal of used silver nanoparticles and the recovery of microorganisms, posing less harm to the environment. Silver nanoparticles made from *Bacillus pumilis* have also been used to remove Congo red, a color used on cotton products, from wastewater (Modi et al., 2015). This technique was shown to be effective for getting rid of Congo red, which is less resistant to light and washing while recovering the most silver material from the effluents to minimize environmental harm. In a study, silver nanoparticles made from the green algal species *Caulerpa serrulata* were successfully used to catalyze the production of Congo red dye (Aboelfetoh et al., 2017).

Even though methyl orange dye is sensitive to acids and is therefore rarely used in textiles, it is nonetheless useful as a wool dye and is therefore a type of wastewater pollutant. *Micrococcus lylae*, *Micrococcus aloeverae, and Cellulosimicrobium* sp. were evaluated using a nanobioremediation method for their ability to produce titanium oxide nanoparticles (Fulekar et al., 2018). In a reactor, active methyl orange dye degradation under the effect of UV radiation was accomplished. These rhizospheric roots-associated bacteria demonstrated the viability and efficacy of natural sources for nanoparticle biosynthesis, as well as the significance of a photocatalytic technique for a 99.9% photocatalytic degradation of the methyl orange dye (Torimiro et al., 2021).

A similar discovery was made about the production of silver nanoparticles by the algal *Hypnea musciformis* and its utility in actively breaking down methyl orange solution when illuminated by visible light (Ganapathy Selvam and Sivakumar, 2015). Using palladium nanoparticles made from *Klebsiella oxytoca*, an attempt was made to bioreduce azo dyes, significant synthetic colorants that have been widely utilized in textile, printing, and paper manufacture (P.-T. Wang et al., 2018). The liquid effluents' synthetic organic colorants were properly bioreduced and recovered. *Chlamydomonas reinhardtii* and synthetic polysulfone nanofibrous web were reported to work well together to remove reactive colors from wastewater (San Keskin et al., 2015).

It has been demonstrated that *S. cerevisiae's* production of palladium nanoparticles allows for the adequate photocatalytic breakdown of direct blue dye in liquid effluents (Sriramulu and Sumathi, 2018). Silver nanoparticles made from the microalgae *Chlorella pyrenoidosa* (Aziz et al., 2015) and *Caulerpa racemosa* (Edison et al., 2016) were reported to have produced significant results, reducing the level of the pollutants in their controlled experiments, for the photocatalytic degradation of methylene blue and the treatment of effluent containing hazardous dye. For bionanoremediation of other toxic chemicals through microbes, the ability of *Pseudomonas putida* to biosynthesize manganese oxide nanoparticles to effectively remove organic micropollutants was recognized (Furgal et al., 2015). For the elimination of bisphenol A, another study established the combined impact of silica nanoparticles biofunctionalized with an enzyme (laccase) obtained from *Thielavia* sp. (Gasser et al., 2012). The extensively used industrial chemical bisphenol A, which is used to make the polymers and resins needed for the storage of food and beverages, has started to irritate the environment (R. Wang et al., 2017).

C. vulgaris was used in a catalytic investigation that was focused on removing nitrate from liquid effluents. Palladium nanoparticles are produced biochemically by algae, and their biogenic synthesis combined with their immobilization on electrospun nanofiber mats increases the complex's catalytic activity for nitrate removal from liquid effluents (Eroglu et al., 2013). Industrial effluents from the production of nitro compounds as solvents or chemical intermediates are fairly large in volume. For the biogenic generation of gold nanoparticles that were used for the catalytic reduction of nitro compounds in wastewater, nanobioremediation mechanism evaluated *T. conoides* and *Sargassum tenerrimum* (Ramakrishna et al., 2016).

4.4 CONCLUSION

The paradigm shift in urban resource flow is certain to bring in a complex network in its consumption pattern. This would result in severe environmental pollution at both micro- and macrolevels, thus, remediation processes using microbes show a promising future with less to no side effects. Besides, using the conventional physio-chemical remediating process can alter the natural property of the water, or soil system which can destroy the natural habitat of many floras and faunas and microcosm can also be disrupted, and all of these hold a vital role in maintaining the food chain balance. The nanotechnology imbibed bioremediation is not much popular among city planners. Hence, the chapter has brought out the pros of microbial powerhouse and their infusion in nanotechnology to remediate environmental pollution. With the least side effects and future feasibility, and being parallel to the sustainable development goals, this method can be widely used to deal with environmental pollution. The chapter has focused on current methods of enhancing nanobioremediation focusing on the innovative remediation technologies. Thus, microbes if used in effective manner can do wonders, even if they are small in size but have immense capacity to mitigate pollution without disturbing the natural property of water, soil, or air ecosystems. The remediation measures could range from remedying acid mine drainage to drinking water treatment, to reducing particulate matters, to mitigating heavy metal and nuclear pollution. Further, the chapter presents the application of nanoremediation in reducing heavy metal toxicity which has a severe detrimental effect on the entire food chain if the toxicity gets magnified, which is common in the case of heavy metal pollution. Besides, the pollution from pesticide or xenobiotic introduction would not only alter the human body's functioning but also has the potential to impact in tangible way to sustainable living. Hence, the nanoremediation method presented here in this chapter is a key pathway for urban planners and utility service providers to understand the extent of pollution, the trend, and the rate of increase or decrease of pollution and plan the urban utilities accordingly not only for this generation but also for coming generations. Therefore, with a huge opportunity lying with nanotechnology in pollution abatement, it is vital to adopt this widely, explore more, and research further toward smart city development and a sustainable future for all.

KEYWORDS

- **microbial-based nanoremediation**
- **environmental pollution**
- **bioremediation**
- **biogenic synthesis of nanomaterials**
- **nanoremediation**

REFERENCES

Aboelfetoh, E. F.; El-Shenody, R. A.; Ghobara, M. M. Eco-friendly Synthesis of Silver Nanoparticles Using Green Algae (*Caulerpa serrulata*): Reaction Optimization, Catalytic and Antibacterial Activities. *Environ. Monit. Assess.* **2017,** *189* (7), 1–15.

Adikesavan, S.; Nilanjana, D. Degradation of Cefdinir by C Andida Sp. SMN 04 and MgO Nanoparticles—An Integrated (Nano-Bio) Approach. *Environ. Prog. Sustain. Energy* **2016,** *35* (3), 706–14.

Alabresm, A.; Chen, Y. P.; Decho, A. W.; Lead, J. A Novel Method for the Synergistic Remediation of Oil-water Mixtures Using Nanoparticles and Oil-degrading Bacteria. *Sci. Total Environ.* **2018,** *630,* 1292–1297.

Alissa, E. M.; Ferns, G. A. Heavy Metal Poisoning and Cardiovascular Disease. *J. Toxicol.* **2011,** *2011.*

Alsharari, S. F.; Tayel, A. A.; Moussa, S. H. Soil Emendation with Nano-Fungal Chitosan for Heavy Metals Biosorption. *Int. J. Biol. Macromol.* **2018,** *118,* 2265–2268.

Asaduzzaman, M. M.; Hossain, F.; Li, X.; Quan, . A Study on the Effects of Pre-treatment in Dyeing Properties of Cotton Fabric and Impact on the Environment. *J. Textile Sci. Eng.* **2016,** *6* (2).

Aziz, N.; Faraz, M.; Pandey, R.; Shakir, M.; Fatma, T.; Varma, A.; Barman, I.; Prasad, R. Facile Algae-derived Route to Biogenic Silver Nanoparticles: Synthesis, Antibacterial, and Photocatalytic Properties. *Langmuir* **2015,** *31* (42), 11605–11612.

Baragaño, D.; Rodríguez-Valdés, E.; Peláez, A. I.; Boente, C.; Matanzas, N.; García, N.; Gallego, J. L. R. Environmental Forensic Study and Remediation Feasibility in an Abandoned Industrial Site. *Proceedings* **2020,** *2,* 1503.

Baruah, A.; Chaudhary, V.; Malik, R.; Tomer, V. K. Nanotechnology based Solutions for Wastewater Treatment. In *Nanotechnology in Water and Wastewater Treatment*; Elsevier, 2019; pp 337–368.

Bhavya, G.; Belorkar, S. A.; Mythili, R.; Geetha, N.; Shetty, H. S.; Udikeri, S. S.; Jogaiah, S. Remediation of Emerging Environmental Pollutants: A Review Based on Advances in the Uses of Eco-friendly Biofabricated Nanomaterials. *Chemosphere* **2021,** *275.* https://doi.org/10.1016/j.chemosphere.2021.129975

Bhatkalkar, S. G.; Kumar, D.; Ali, A.; Sachar, S. Surface Dynamics Associated with Zinc Oxide Nanoparticles and Biomolecules in Presence of Surfactants. *J. Mol. Liquids* **2018,** *268,* 1–10.

Bokare, V.; Murugesan, K.; Kim, Y.-M.; Jeon, J.-R.; Kim, E.-J.; Chang, Y. Seok. Degradation of Triclosan by An Integrated Nano-bio Redox Process. *Bioresour. Technol.* **2010,** *101* (16), 6354–6360.

Cao, X.; Alabresm, A.; Chen, Y. P.; Decho, A. W.; Lead, J. Improved Metal Remediation Using a Combined Bacterial and Nanoscience Approach. *Sci. Total Environ.* **2020,** *704,* 135378.

Chandra, R.; Dubey, N. K.; Kumar, V. *Phytoremediation of Environmental Pollutants*; CRC Press, 2017.

Chaplygin, V. A.; Minkina, T. M.; Mandzhieva, S. S.; Nazarenko, O. G.; Zimulina, I. V.; Bauer, T. V.; Litvinov, Y. A.; Rajput, V. In *Heavy Metals in Agricultural Crops of Rostov Region Through the Example of Soft Wheat (Triticum aestivum)*, IOP Conference Series: Earth and Environmental Science; 2021.

Corsi, I.; Winther-Nielsen, M.; Sethi, R.; Punta, C.; Torre, C. D.; Libralato, G.; Lofrano, G.; Sabatini, L.; Aiello, M.; Fiordi, L. Ecofriendly Nanotechnologies and Nanomaterials For Environmental Applications: Key Issue and Consensus Recommendations for Sustainable and Ecosafe Nanoremediation. *Ecotoxicol. Environ. Saf.* **2018,** *154,* 237–244.

Djellabi, R.; Giannantonio, R.; Falletta, E.; Bianchi, C. L. SWOT Analysis of Photocatalytic Materials Towards Large Scale Environmental Remediation. *Curr. Opin. Chem. Eng.* **2021,** *33*. https://doi.org/10.1016/j.coche.2021.100696

Domingo, J. L.; Nadal, M. Human Exposure to Per- and Polyfluoroalkyl Substances (PFAS) Through Drinking Water: A Review of the Recent Scientific Literature. *Environ. Res.* **2019,** *177*. https://doi.org/10.1016/j.envres.2019.108648

Deshpande, B. D.; Agrawal, P, S.; Yenkie, M. K. N.; Dhoble, S. J. Prospective of Nanotechnology in Degradation of Waste Water: A New Challenges. *Nanostruct. Nanoobj.* **2020,** *22,* 100442.

Dong, G.; Wang, Y.; Gong, L.; Wang, M.; Wang, H.; He, N.; Zheng, Y.; Li, Q. Formation of Soluble Cr (III) End-products and Nanoparticles during Cr (VI) Reduction by *Bacillus cereus* Strain XMCr-6. *Biochem. Eng. J.* **2013,** *70,* 166–172.

Durán, N.; Marcato, P. D.; Alves, O. L.; Da Silva, Joao P. S. G.; De Souza, I. H.; Rodrigues, F. A.; Esposito, E. Ecosystem Protection by Effluent Bioremediation: Silver Nanoparticles Impregnation in a Textile Fabrics Process. *J. Nanopart. Res.* **2010,** *12* (1), 285–292.

Edison, T. N. J. I.; Atchudan, R.; Kamal, C.; Lee, Y. R. *Caulerpa racemosa*: A Marine Green Alga for Eco-friendly Synthesis of Silver Nanoparticles and Its Catalytic Degradation of Methylene Blue. *Bioprocess Biosyst. Eng.* **2016,** *39* (9), 1401–1408.

Eroglu, E.; Chen, X.; Bradshaw, M.; Agarwal, V.; Zou, J.; Stewart, S. G.; Duan, X.; Lamb, R. N.; Smith, S. M.; Raston, C. L. Biogenic Production of Palladium Nanocrystals Using Microalgae and their Immobilization on Chitosan Nanofibers for Catalytic Applications. *RSC Adv.* **2013,** *3* (4), 1009–1012.

Filho, W. L.; Balogun, A. L.; Olayide, O. E.; Azeiteiro, U. M.; Ayal, D. Y.; Muñoz, P. D. C.; Nagy, G. J.; Bynoe, P.; Oguge, O.; Yannick Toamukum, N.; Saroar, M.; Li, C. Assessing the Impacts of Climate Change in Cities and their Adaptive Capacity: Towards Transformative Approaches to Climate Change Adaptation and Poverty Reduction in Urban Areas in a Set of Developing Countries. *Sci. Total Environ.* **2019,** *692*. https://doi.org/10.1016/j.scitotenv.2019.07.227

Fulekar, J.; Dutta, D. P.; Pathak, B.; Fulekar, M. H. Novel Microbial and Root Mediated Green Synthesis of TiO2 Nanoparticles and Its Application in Wastewater Remediation. *J. Chem. Technol. Biotechnol.* **2018,** *93* (3), 736–743.

Furgal, K. M.; Meyer, R. L.; Bester, K. Removing Selected Steroid Hormones, Biocides and Pharmaceuticals from Water by Means of Biogenic Manganese Oxide Nanoparticles In Situ at ppb Levels. *Chemosphere* **2015,** *136,* 321–326.

Ganapathy Selvam, G.; Sivakumar, K. Phycosynthesis of Silver Nanoparticles and Photocatalytic Degradation of Methyl Orange Dye Using Silver (Ag) Nanoparticles Synthesized from *Hypnea musciformis* (Wulfen) JV Lamouroux. *Appl. Nanosci.* **2015,** *5* (5), 617–622.

Gasser, C. A.; Hommes, G.; Corvini, P. F. X. Removal of Phenolic Pollutants from Municipal Wastewater through Immobilized Laccase Enzymes. *Environ. Eng. Manag. J.* **2012,** *11* (3), S123.

Gautam, P. K.; Singh, A.; Misra, K.; Sahoo, A. K.; Samanta, S. K. Synthesis and Applications of Biogenic Nanomaterials in Drinking and Wastewater Treatment. *J. Environm. Manag.* **2019,** *231*. https://doi.org/10.1016/j.jenvman.2018.10.104

Gaur, N.; Diwan, B.; Choudhary, R. Bioremediation of Organic Pesticides Using Nanomaterials. In *Nano-Bioremediation: Fundamentals and Applications*; Elsevier, 2022; 517–540.

Gerhardt, K. E.; Gerwing, P. D.; Greenberg, B. M. Opinion: Taking Phytoremediation from Proven Technology to Accepted Practice. *Plant Sci.* **2017,** *256,* 170–185.

Gong, X.; Huang, D.; Liu, Y.; Peng, Z.; Zeng, G.; Xu, P.; Cheng, M.; Wang, R.; Wan, J. Remediation of Contaminated Soils by Biotechnology with Nanomaterials: Bio-behavior, Applications, and Perspectives. *Crit. Rev. Biotechnol.* **2018,** *38* (3), 455–468.

Govarthanan, M.; Jeon, C.-H.; Jeon, Y.-H.; Kwon, J.-H.; Bae, H.; Kim, W. Non-toxic Nano Approach for Wastewater Treatment using *Chlorella vulgaris* Exopolysaccharides Immobilized in Iron-magnetic Nanoparticles. *Int. J. Biol. Macromol.* **2020,** *162,* 1241–1249.

He, X.; Li, P. Surface Water Pollution in the Middle Chinese Loess Plateau with Special Focus on Hexavalent Chromium (Cr6+): Occurrence, Sources and Health Risks. *Exp. Health* **2020,** *12* (3). https://doi.org/10.1007/s12403-020-00344-x

Jaishankar, M.; Tseten, T.; Anbalagan, N.; Mathew, B. B. Corresponding and Krishnamurthy N. Beeregowda, 2014.

Joutey, N. T.; Bahafid, W.; Sayel, H.; El Ghachtouli, N. Biodegradation: Involved Microorganisms and Genetically Engineered Microorganisms. *Biodegrad. Life Sci.* **2013,** *1,* 289–320.

Kumar, H.; Sinha, S. K.; Goud, V. V.; Das, S. Removal of Cr (VI) by Magnetic Iron Oxide Nanoparticles Synthesized from Extracellular Polymeric Substances of Chromium Resistant Acid-tolerant Bacterium *Lysinibacillus sphaericus* RTA-01. *J. Environ. Health Sci. Eng.* **2019,** *17* (2), 1001–1016.

Laslo, V.; Pinzaru, S. C.; Zaguła, G.; Kluz, M.; Vicas, S. I.; Cavalu, S. Synergic Effect of Selenium Nanoparticles and Lactic Acid Bacteria in Reduction Cadmium Toxicity. *J. Mol. Struc.* **2022,** *1247,* 131325.

Levallois, P.; Villanueva, C. M. Drinking Water Quality and Human Health: An Editorial. *Int. J. Environ. Res. Public Health 16* (4), **2019**. https://doi.org/10.3390/ijerph16040631

Li, G.; Sun, G. X.; Ren, Y.; Luo, X. S.; Zhu, Y. G. Urban Soil and Human Health: A Review. *Eur. J. Soil Sci. 69* (1), **2018**. https://doi.org/10.1111/ejss.12518

Li, P.; Wu, J. Drinking Water Quality and Public Health. *Exp. Health* **2019,** *11* (2). https://doi.org/10.1007/s12403-019-00299-8

Liu, W.; Zuo, Q.; Zhao, C.; Wang, S.; Shi, Y.; Liang, S.; Zhao, C.; Shen, S. Effects of *Bacillus subtilis* and Nanohydroxyapatite on the Metal Accumulation and Microbial Diversity of Rapeseed (Brassica campestris L.) for the Remediation of Cadmium-Contaminated Soil. *Environ. Sci. Pollut. Res.* **2018,** *25* (25), 25217–25226.

Mahanty, S.; Chatterjee, S.; Ghosh, S.; Tudu, P.; Gaine, T.; Bakshi, M.; Das, S.; Das, P.; Bhattacharyya, S.; Bandyopadhyay, S. Synergistic Approach Towards the Sustainable Management of Heavy Metals in Wastewater Using Mycosynthesized Iron Oxide Nanoparticles: Biofabrication, Adsorptive Dynamics and Chemometric Modeling Study. *J. Water Proc. Eng.* **2020,** *37,* 101426.

Mahmoud, M. E.; Abdou, A. E. H.; Mohamed, S. M. S.; Osman, M. M. Engineered Staphylococcus Aureus via Immobilization on Magnetic Fe3O4-phthalate Nanoparticles for Biosorption of Divalent Ions from Aqueous Solutions. *J. Environ. Chem. Eng.* **2016,** *4* (4), 3810–3824.

Manisalidis, I.; Stavropoulou, E.; Stavropoulos, A.; Bezirtzoglou, E. Environmental and Health Impacts of Air Pollution: A Review. *Front. Public Health* **2020,** *8*. https://doi.org/10.3389/fpubh.2020.00014

Mehndiratta, P.; Jain, A.; Srivastava, S.; Gupta, N. Environmental Pollution and Nanotechnology. *Environ. Pollut.* **2013,** *2* (2), 49.

Modi, S.; Pathak, B.; Fulekar, M. H. Microbial Synthesized Silver Nanoparticles for Decolorization and Biodegradation of Azo Dye Compound. *J. Environ. Nanotechnol.* **2015,** *4* (2), 37–46.

Momeni, S.; Nabipour, I. A Simple Green Synthesis of Palladium Nanoparticles with Sargassum Alga and their Electrocatalytic Activities Towards Hydrogen Peroxide. *Appl. Biochem. Biotechnol.* **2015,** *176* (7), 1937–1949.

OECD. *Perspectives on Global Development 2019 Rethinking Development Strategies* [Online] 2018. https://doi.org/10.1787/persp_glob_dev-2019-en

Palit, S.; Hussain, C. M. Functionalization of Nanomaterials for Industrial Applications: Recent and Future Perspectives. In *Handbook of Functionalized Nanomaterials for Industrial Applications*; Elsevier, 2020; pp 3–14.

Raj, R.; Dalei, K.; Chakraborty, J.; Das, S. Extracellular Polymeric Substances of a Marine Bacterium Mediated Synthesis of CdS Nanoparticles for Removal of Cadmium from Aqueous Solution. *J. Colloid Interface Sci.* **2016,** *462,* 166–175.

Rajput, V. D.; Minkina, T.; Upadhyay, S. K.; Kumari, A.; Ranjan, A.; Mandzhieva, S.; Sushkova, S.; Singh, R. K.; Verma, K. K. Nanotechnology in the Restoration of Polluted Soil. *Nanomaterials* **2022,** *12* (5), 769.

Ramakrishna, M.; Babu, D. R.; Gengan, R. M.; Chandra, S.; Rao, G. N. Green Synthesis of Gold Nanoparticles using Marine Algae and Evaluation of their Catalytic Activity. *J. Nanostruc. Chem.* **2016,** *6* (1), 1–13.

Rasuli, L.; Dehghani, M. H.; Aghaei, M.; Mahvi, A. H.; Mubarak, N. M.; Karri, R. R. Occurrence and Fate of Bacterial Endotoxins in the Environment (Air, Water, Wastewater) and Remediation Technologies: An Overview. *Chemosphere* **2022,** *303*, 135089. https://doi.org/10.1016/J.CHEMOSPHERE.2022.135089

Salvadori, M. R.; Ando, R. A.; Nascimento, C. A. O.; Correa, B. Extra and Intracellular Synthesis of Nickel Oxide Nanoparticles Mediated by Dead Fungal Biomass. *PLoS One* **2015,** *10* (6), e0129799.

Keskin, S.; Oya, N.; Celebioglu, A.; Uyar, T.; Tekinay, T. Microalgae Immobilized by Nanofibrous Web for Removal of Reactive Dyes from Wastewater. *Indus. Eng. Chem. Res.* **2015,** *54* (21), 5802–5809.

Sathyavathi, S.; Manjula, A.; Rajendhran, J.; Gunasekaran, P. Extracellular Synthesis and Characterization of Nickel Oxide Nanoparticles from Microbacterium sp. MRS-1 towards Bioremediation of Nickel Electroplating Industrial Effluent. *Bioresour. Technol.* **2014,** *165,* 270–273.

Sillmann, J.; Aunan, K.; Emberson, L.; Büker, P.; Van Oort, B.; O'Neill, C.; Otero, N.; Pandey, D.; Brisebois, A. Combined Impacts of Climate and Air Pollution on Human Health and Agricultural Productivity. *Environ. Res. Lett.* **2021.** *16* (9). https://doi.org/10.1088/1748-9326/ac1df8

Sharma, D.; Kanchi, S.; Bisetty, K. Biogenic Synthesis of Nanoparticles: A Review. *Arab. J. Chem.* **2019,** *12* (8), 3576–3600.

Shukla, P. Microbial Nanotechnology for Bioremediation of Industrial Wastewater. *Front. Microbiol.* **2020,** *11,* 590631.

Sriramulu, M.; Sumathi, S. Biosynthesis of Palladium Nanoparticles Using *Saccharomyces cerevisiae* Extract and Its Photocatalytic Degradation Behaviour. *Adv. Nat. Sci. Nanosci. Nanotechnol.* **2018,** *9* (2), 025018.

Subramaniyam, V.; Subashchandrabose, S. R.; Thavamani, P.; Megharaj, M.; Chen, Z.; Naidu, R. Chlorococcum sp. MM11—A Novel Phyco-Nanofactory for the Synthesis of Iron Nanoparticles. *J. Appl. Phyc.* **2015,** *27* (5), 1861–1869.

Suman, J.; Uhlik, O.; Viktorova, J.; Macek, T. Phytoextraction of Heavy Metals: A Promising Tool for Clean-up of Polluted Environment? *Front. Plant Sci.* **2018,** *9,* 1476.

Tharmavaram, M.; Rawtani, D.; Pandey, G. Fabrication Routes for One-dimensional Nanostructures via Block Copolymers. *Nano Conv.* **2017,** *4* (1), 1–13.

The World Bank. *World Development Report 2018: Learning to Realize Education's Promise*, 2018.

Torimiro, N.; Daramola, O. B.; Oshibanjo, O. D.; Otuyelu, F. O.; Akinsanola, B. A.; Yusuf, O. O.; Ore, O. T.; Omole, R. K. Ecorestoration of Heavy Metals and Toxic Chemicals in Polluted Environment using Microbe-mediated Nanomaterials. *Int. J. Environ. Bioremed. Biodegrad.* **2021,** *9,* 8–21.

Ushani, U.; Lu, X.; Wang, J.; Zhang, Z.; Dai, J.; Tan, Y.; Wang, S.; Li, W.; Niu, C.; Cai, T.; Wang, N.; Zhen, G. Sulfate Radicals-based Advanced Oxidation Technology in Various Environmental Remediation: A State-of-the–art Review. *Chem. Eng. J.* **2020,** *402*. https://doi.org/10.1016/j.cej.2020.126232

Vijayan, S. R.; Santhiyagu, P.; Singamuthu, M.; Ahila, N. K.; Jayaraman, R.; Ethiraj, K. Synthesis and Characterization of Silver and Gold Nanoparticles using Aqueous Extract of Seaweed, *Turbinaria conoides*, and their Antimicrofouling Activity. *Sci. World J.* **2014,** *2014*.

Wang, L.; Zhang, C.; Gao, F.; Mailhot, G.; Pan, G. Algae Decorated TiO2/Ag Hybrid Nanofiber Membrane with Enhanced Photocatalytic Activity for Cr (VI) Removal under Visible Light. *Chem. Eng. J.* **2017,** *314,* 622–630.

Wang, P.-T.; Song, Y.-H.; Fan, H.-C.; Yu, L. Bioreduction of Azo Dyes was Enhanced by In-Situ Biogenic Palladium Nanoparticles. *Bioresour. Technol.* **2018,** *266,* 176–180.

Wang, R.; Wang, S.; Tai, Y.; Tao, R.; Dai, Y.; Guo, J.; Yang, Y.; Duan, S. Biogenic Manganese Oxides Generated by Green Algae Desmodesmus sp. WR1 to Improve Bisphenol A Removal. *J. Hazard. Mater.* **2017,** *339,* 310–319.

Wang, X.; Zhang, D.; Pan, X.; Lee, D.-J.; Al-Misned, F. A.; Mortuza, M. G.; Gadd, G. M. Aerobic and Anaerobic Biosynthesis of Nano-selenium for Remediation of Mercury Contaminated Soil. *Chemosphere* **2017,** *170,* 266–273.

Wang, X.; Li, Y.; Zhao, J.; Yao, H.; Chu, S.; Song, Z.; He, Z.; Zhang, W. Magnetotactic Bacteria: Characteristics and Environmental Applications. *Front. Environ. Sci. Eng.* **2020,** *14* (4). https://doi.org/10.1007/s11783-020-1235-z

Wu, R.; Wu, H.; Jiang, X.; Shen, J.; Faheem, M.; Sun, X.; Li, J.; Han, W.; Wang, L.; Liu, X. The Key Role of Biogenic Manganese Oxides in Enhanced Removal of Highly Recalcitrant 1, 2, 4-triazole from Bio-treated Chemical Industrial Wastewater. *Environ. Sci. Pollut. Res.* **2017,** *24* (11), 10570–10583.

Xie, Y.; He, Y.; Irwin, L. Jin. Tony and S. Xianming. *Appll. Environ. Microbiol.* **2011,** *77,* 2325.

Yan, F.-F, Wu, C.; Cheng, Y.-Y.; He, Y.-R.; Li, W.-W.; Yu, H.-Q. Carbon Nanotubes Promote Cr (VI) Reduction by Alginate-immobilized *Shewanella oneidensis* MR-1. *Biochem. Eng. J.* **2013,** *77,* 183–189.

Yousefi, N.; Emtyazjoo, M.; Sepehr, M. N.; Darzi, S. J.; Sepahy, A. A. Green Synthesis of *Pseudomonas aeruginosa* Immobilized Fe3O4-multiwalled Carbon Nanotubes Bio-Adsorbent for the Removal of 2,4,6-trinitrophenol from Aqueous Solution. *Environ. Technol. Innov.* **2020,** *20,* 101071.

Yilmaz, B.; Terekeci, H.; Sandal, S.; Kelestimur, F. Endocrine Disrupting Chemicals: Exposure, Effects on Human Health, Mechanism of Action, Models for Testing and Strategies for Prevention. *Rev. Endocr. Meta. Disord.* **2020,** *21* (1). https://doi.org/10.1007/s11154-019-09521-z

Zamora-Ledezma, C.; Negrete-Bolagay, D.; Figueroa, F.; Zamora-Ledezma, E.; Ni, M.; Alexis, F.; Guerrero, V. H. Heavy Metal Water Pollution: A Fresh Look about Hazards, Novel and Conventional Remediation Methods. *Environ. Technol. Innov.* **2021,** *22.* https://doi.org/10.1016/j.eti.2021.101504

Zeng, S.; Ma, J.; Yang, Y.; Zhang, S.; Liu, G. J.; Chen, F. Spatial Assessment of Farmland Soil Pollution and Its Potential Human Health Risks in China. *Sci. Total Environ.* **2019,** *687.* https://doi.org/10.1016/j.scitotenv.2019.05.291

Zhang, N.; Ishag, A.; Li, Y.; Wang, H.; Guo, H.; Mei, P.; Meng, Q.; Sun, Y. Recent Investigations and Progress in Environmental Remediation by Using Covalent Organic Framework-based Adsorption Method: A Review. *J. Clean. Prod.* **2020,** *277.* https://doi.org/10.1016/j.jclepro.2020.123360

Zhang, X.; Tong, S.; Huang, D.; Liu, Z.; Shao, B.; Liang, Q.; Wu, T.; Pan, Y.; Huang, J.; Liu, Y.; Cheng, M.; Chen, M. Recent Advances of Zr based Metal Organic Frameworks Photocatalysis: Energy Production and Environmental Remediation. *Coord. Chem. Re.* **2021,** 448. https://doi.org/10.1016/j.ccr.2021.214177

Zhou, H.; Fu, C. Manganese-oxidizing Microbes and Biogenic Manganese Oxides: Characterization, Mn(II) Oxidation Mechanism and Environmental Relevance. *Rev. Environ. Sci. Biotechnol.* **2020,** *19* (3). https://doi.org/10.1007/s11157-020-09541-1

Zhu, N.; Zhang, B.; Yu, Q. Genetic Engineering-facilitated Coassembly of Synthetic Bacterial Cells and Magnetic Nanoparticles for Efficient Heavy Metal Removal. *ACS Appl. Mater. Interfaces* **2020,** *12* (20), 22948–22957.

CHAPTER 5

Nanoparticles as Tools to Remediate Environmental Pollution

FERNANDA MARIA POLICARPO TONELLI[1],
FLÁVIA CRISTINA POLICARPO TONELLI[2], MOLINE SEVERINO LEMOS[3],
and VITÓRIA DE OLIVEIRA LOURENÇO[4]

[1]*Federal University of São João del Rei, Divinópolis, Minas Gerais, Brazil*

[2]*Estácio de Sá University, Divinópolis, Brazil*

[3]*Department of Cell Biology, ICB/UFMG, Belo Horizonte, Brazil*

[4]*Department of Biochemistry, UFSJ/CCO, Divinópolis, Brazil*

ABSTRACT

Environmental pollution concerns humans worldwide as it threatens the existence of life on Earth. Regarding water, for example, humans need to receive water of good quality from water-treatment plants; however, due to the increasing amount of pollution and also the increasing world's population, this task is not so easy to perform. The enormous array of organic and inorganic contaminants polluting surface water and groundwater makes even dictating quality standards regarding each pollutant a hard task to regulatory agencies. Innovative strategies to deal with pollution aiming to reduce the risk it represents to ecosystems are highly necessary and urgent. The nanotechnology field provides interesting solutions through structures carefully designed to present specific properties of interest. This chapter will be dedicated to exploring nanoparticles, generated by nongreen-based and through green-based synthesis protocols as tools to remediate polluted environments, addressing their potential, main characteristics, and future perspectives associated with them.

Sustainable Nanoremediation: Modern Technologies for a Clean Environment.
Fernanda Maria Policarpo Tonelli, Arpita Roy, and Khalid Rehman Hakeem (Eds.)

5.1 INTRODUCTION

Although the term nanotechnology was introduced in the 1970s by Norio Taniguchi (Taniguchi, 1974), materials in nanoscale for different applications are being developed since the 1950s and a large array of protocols regarding their synthesis and functionalization is available nowadays (Tonelli et al., 2022).

These structures possess, at least in one dimension, dimensions ranging from 1 to 100 nm. Quantum confinement, high surface-to-volume ratio, good chemical reactivity, high stability, and presence of catalytic activity are some important characteristics commonly found in nanomaterials (Xu et al., 2019; Bethi and Sonawane, 2018; Werkneh et al. 2018).

As environmental pollution is a worldwide problem that threatens living beings' survival, the properties of nanomaterials make them interesting options to remediate environmental contamination.

Environmental pollution is not a new issue; heavy metals, for example, since Bronze Age (Longman et al., 2018) present themselves as harmful to humans' health (beings poisons capable of impairing cognitive ability and causing diverse serious symptoms that can lead to death (Yamada et al., 2020)). However, what is especially concerning in a recent scenario is the number and high concentration of contaminants present in different environments. The process of industrialization and urbanization happened in a fast way increasing the threats to life on Earth and the negative impact over ecosystems (Chaoua et al., 2019; Zheng et al., 2020). Industrial wastewater, for example, can contain a large variety of contaminants of different chemical nature (from heavy metals to dyes, from pharmaceuticals to pesticides) and the majority of them can persistently pollute (Iloms et al., 2020).

Soil contamination should also receive attention once its degradation threatens agricultural production and food safety, as same as water pollution (Bakshi and Abhilash, 2020).

Low-cost and green-based nanomaterials receive increasing attention as tools to be applied in the environmental science field (Lofrano et al., 2017; Bolade et al., 2020; Nasrollahzadeh et al., 2021). Traditional methods of remediating environmental pollution present serious limitations that impair their success rate (Ahmed et al., 2021). Among these tools in nanoscale are materials that possess an inorganic chemical nature and others that are organic. Regarding morphology, they can be classified as nanoparticles (NPs), nanotubes (NTs), nanofibers (NFs), nanowires, and nanomembranes (Lu and Astruc, 2020).

This chapter aims to address NPs that are generated by different synthesis protocols as tools to remediate environmental pollution.

5.1.1 NANOPARTICLES FOR ENVIRONMENTAL POLLUTION REMEDIATION

There are diverse protocols available to synthesize nanomaterials improved in important characteristics to perform a specific task such as remediating environmental pollution (Khan and Malik, 2019).

When it comes to NPs, the main challenge is related to aggregation. They commonly aggregate as soon as they encounter water-indifferent extents that vary under the influence of the charge displayed on NP's surface due to synthesis or functionalization protocol or interactions established with other substances present on the medium such as organic molecules (Keller et al, 2010).

An important aspect that should also receive attention is safety of these materials for living beings. It is necessary to guarantee a security level that will not put in danger the health of people in contact with the nanomaterial (Pulizzi and Sun, 2018).

NPs can be organic or inorganic materials in nanoscale and are commonly studied due to the large array of applications they possess (Li et al., 2019). Organic NPs are commonly applied to deliver drugs as they degrade rapidly and avoid causing long-lasting toxicity in patients; chitosan NPs are an interesting example of this type of NPs (Mohammed et al., 2017). Inorganic NPs can also be designed to be highly biocompatible (Paul and Sharma, 2020) and are commonly more applied than organic NP when it comes to being used to remediate pollution. Metals and metalloids are popular materials to generate NPs aiming environmental remediation (Khin et al., 2012).

NPs can be synthesized through different strategies such as physical and chemical synthesis methods and recently green-based protocols are receiving increasing attention. The main reason for that is sustainability, trying to avoid the damages caused by toxicity reagents used in nongreen-based protocols. The biological synthesis can involve bacteria, fungi, algae, and plants in the process (Pal et al., 2019).

5.2 NONGREEN-BASED SYNTHESIZED NANOPARTICLES TO REMEDIATE ENVIRONMENTAL POLLUTION

A large array of examples of NPs generated through nongreen-based synthesis protocols can be found in literature. Some examples of protocol from the last 11 years are discussed here and summarized in Table 5.1.

TABLE 5.1 Examples of Nongreen-Based Nanoparticles to Remediate Environmental Pollution.

Nanoparticle	Functionalization/Chemical Modification	Pollutant(s) Remediated	References
Nanoscale zero-valent iron (nZVI)	–	U, Al, sulfates, As, Be, Cd, Cr, Cu, Ni, V, and Zn	Klimkova et al. (2011)
	–	Cr^{6+}	Gheju (2011), Guan et al. (2015)
	–	Cu^{2+}	Li et al. (2014)
	Coated with 4% bentonite	Cr^{6+}	Wang et al. (2018)
	–	Cu and Ni	Gil-Díaz et al. (2020)
	Sulfide-modified and associated with graphene aerogel composite	Trichloroethylene	Bin et al. (2020)
	Conjugated to biochar	Pb^{2+}	Li et al. (2020b)
		Trichloroethylene	Li et al. (2020c)
	Nonmodified, carboxymethyl cellulose (CMC)—Carbo-Iron® colloids (CICs), nano-Fe, and sodium alginate nZVI	Organic and inorganic pollutants	Zafar et al. (2021)
	Combined to potassium permanganate	Polycyclic aromatic hydrocarbons	Sun et al. (2022)
TiO2 nanoparticles		Methylene blue	Jain and Vaya (2017)
		As	Nazary et al. (2019)
	Hydrated and on a support of agricultural waste rice straw	Cu	Chen et al. (2020)
	Decorated with SnO2 quantum dots	Rhodamine B	Lee et al. (2014)
	Modified by fullerene	Methylene blue	Qi et al. (2016)
	Immobilized on polyethylene terephthalate nanofibers	Methylene blue	Yasin et al. (2020)
	Aeroxide® P25	Paraquat and acetaminophen	Nicosia et al. (2021)
	–	Cu^{2+}	Ezati et al. (2021)
	–	Ciprofloxacin	Usman et al. (2021)
Silver nanoparticles	–	Co^{2+} and Pb^{2+}	Attatsi and Nsiah (2020)
	Decorating, as same as titanium dioxide a nanosponge β-cyclodextrin polyurethane was modified by phosphorylated multiwalled carbon nanotubes	Trichloroethylene and Congo red	Taka et al. (2020)

TABLE 5.1 *(Continued)*

Nanoparticle	Functionalization/Chemical Modification	Pollutant(s) Remediated	References
Iron oxide nanoparticles	Associated with protocatechuate 3,4-dioxygenase isolated from *Rhizobium* sp. LMB-1	Organic pollutants	Zhang et al. (2017a)
	Associated with activated carbon	Methylene blue and brilliant green	Joshi and Garg (2019)
	–	Nitrate, nitrite, phosphate, ammonium	Hesni et al. (2020)
		As^{5+}	Zeng et al. (2020)
	With zeolite type 5A in nanocomposite	Pb^{2+}	Ramos-Guivar et al. (2020)
	Associated with biochar	Arsenite and arsenate	Priyadarshni et al. (2020)
	Coated with polyvinylpyrrolidone	Oil	Mirshahghassemi and Lead (2015)
		Cd, Cr, Ni, and Pb	Hong et al. (2020)
	Associated with grapheme	Methylene blue, Rhodamine B, Acid orange 7, and phenol	Hammad et al. (2021)
	–	Cu	Demangeat et al. (2021)
	–	Crystal violet dye	Samrot et al. (2021)
	–	Cr and PCB	Gil-Díaz et al. (2022)
MnO2 nanoparticles	Attached to reduced graphene oxide functionalized with functional groups of black cumin seeds	Methylene blue and arsenite	Tara et al. (2020)
	Deposited on magnetic halloysite nanotubes	Pb^{2+}	Afzali and Fayazi (2016)
	–	As^{3+}	Liu et al. (2020)
	Loaded on biochar	Cd^{2+}, Cu^{2+}, Zn^{2+}, and Pb^{2+}	Zhang et al. (2020)
	Associated with chitosan	Lead and lanthanum	Mahmoud et al. (2021b)
	–	Rhodamine B	Fang et al. (2021)

TABLE 5.1 *(Continued)*

Nanoparticle	Functionalization/Chemical Modification	Pollutant(s) Remediated	References
Palladium nanoparticles	Attached to chitosan–carbon nanotube	2-Nitroaniline, 4-nitrophenol, 4-nitro-*o*-phenylenediamine, 2,4-dinitrophenol, Methyl orange, Congo red, Methyl red, and methylene blue	Sargin et al. (2020)
	Decorating mesoporous polydopamine/bacterial nanocellulose membrane	Dyes	Derami et al. (2020)
	Attached to hierarchical flower-like TiO2 nanospheres	Cr^{6+}	Li et al. (2020a)
Zirconia nanoparticles	Modifying porous silicon	Phosphate	Anbia et al. (2018)
	Modifying Fe	Rhodamine B	Cheng et al. (2020)
	Associated with iron	Rhodamine B	Reddy et al. (2020)
	Loaded on pine biochar	As^{5+}	Liu et al. (2021)
Zeolite nanoparticles	In polysulfone membrane	Pharmaceuticals	Dong et al. (2016)
	Modified by (3-aminopropyl) triethoxy silane	Direct Red 23 (DR23) and Direct Red 80	Mahmoodi and Saffar-Dastgerdi (2019)
	Impregnated in polysulfone membrane	Heavy metals	Manap (2020)
Copper oxide nanoparticles		Methylene blue	Ghulam et al. (2013)
		Cd^{2+} and Ni^{2+}	Hassan et al. (2017)
	–	Ni^{2+}	Jain et al. (2021)
	In polyethersulfune membrane	Hydroquinone	Ghiasi et al. (2021)

TABLE 5.1 *(Continued)*

Nanoparticle	Functionalization/Chemical Modification	Pollutant(s) Remediated	References
Prussian blue-based nanoparticles	Associated with γ-Al2O3 spheres	Orange G dye	Doumic et al. (2015)
	Associated with polyacrylonitrile nanofibers	^{137}Cs	Kim et al. (2018)
	Associated with activated charcoal	Cs^+, Sr^{2+}, and Co^{2+}	Ali et al. (2020)
	Doped on mesoporous silica	Rubidium and cesium	Yuan et al. (2020)
		Cesium	Carniato et al. (2020)
WO3 nanoparticle	In an Au–WO3–C Janus micromotor	Sodium-2,6-dichloroindophenol and Rhodamine B	Zhang et al. (2017b)
		Methylene blue	Thilagavathi and Venugopal (2018)
	Polyaniline coated	Orange G dye	Hsini et al. (2021)

It is possible to start the discussion with nanoscale zero-valent iron (nZVI) particles. These nanoscale materials are versatile, being able to remediate a wide range of contaminants of different chemical nature (Jang et al., 2014). The high reactivity, low cost, low toxicity, and possibility to adapt protocols to generate bimetallic versions or chemically modified structures largely contribute to this feature (O'Carroll et al., 2013; Jiang et al., 2018). In polluted water, for example, they can reduce the availability of inorganic pollutants (Gil-Díaz et al., 2020). It can remediate U, Al, sulfates, As, Be, Cd, Cr, Cu, Ni, V, and Zn from acid-mine water (Klimkova et al., 2011), hexavalent chromium-contaminated wastewater (Gheju, 2011; Guan et al., 2015), Cu^{2+} from wastewater with high efficiency (Li et al., 2014). Modified versions of the nanomaterial can exhibit improved remediation capacity; nZVI could remove almost 80% of organic and inorganic pollutants from groundwater, but modified versions (carboxymethyl cellulose–Carbo-Iron® colloids, nano-Fe, and sodium alginate nZVI) offered a removal of almost 100% (Zafar et al., 2021). The presence of an iron oxide outer shell favors chemisorption and electrostatic interactions with heavy metals increasing the capacity to uptake of Cu^{2+}, Pb^{2+}, and Cd^{2+} (Tarekegn et al., 2021). nZVI coated with 4% bentonite could remediate hexavalent chromium offering better performance when compared to the uncoated version (Wang et al., 2018). Association to porous hydrophilic biochar increased the binding force of the complex offering the possibility of a remediation of Pb^{2+}'s pollution with higher efficiency and quality; the synergistic effect made it favorable to lead ions to perform diffusion and adsorption (Li et al., 2020b). Trichloroethylene from groundwater could also be removed by maize-derived biochars bound to nZVIs in a fast way by generating reactive oxygen species capable of promoting dichlorination of the pollutant (Li et al., 2020c). Sulfide-modified nZVI's associated with graphene aerogel composite could deal well with organic pollutants such as trichloroethylene in polluted water samples (Bin et al., 2020). In fact, modification involving S atom contribute to increasing selectivity for contaminants and elongating reactive lifetimes offering the possibility to performing defluorination reactions. Sulfidized nZVI allows in situ groundwater remediation, dealing well with chlorinated and fluorinated contaminants (Xu et al., 2021). nZVI can also be used to remediate polluted soil, but there are studies demonstrating that its capacity to promote water remediation is better; Cu and Ni ions, for example, are immobilized with higher efficiency by the nanomaterial in aqueous environment (Gil-Díaz et al., 2020). However, it is important to highlight that when it comes to remediating polluted soil using nZVI, a special attention must be directed to soil microbial communities. The nanomaterial can, as a second source of stress, negatively affect microbial quality of contaminated soil (Anza et al., 2019). Still, there are protocols successfully applying nZVI to remediate soil; the synergistic of the nanomaterial combined with potassium permanganate made it possible to deal with polycyclic aromatic hydrocarbons in contaminated soil. Sugar esters as dispersant offered the best result, generating products such as 9-methylene-9*H*-fluorene, fluoranthene, and 1,5-diphenyl-1,4-pentadiyn-3-one (Sun et al., 2022).

TiO_2 NPs are also efficient nanomaterials to be used in remediation protocols (Mustapha et al., 2020). These NPs, besides presenting high photocatalytic activity, are inexpensive and resistant to corrosion. They could, for example, successfully degrade the contaminant methylene blue, an organic dye (Jain and Vaya, 2017). Arsenic pollution (Zahra et al., 2020), organic co-contaminants (Hristovski et al., 2009), and cooper (Chen et al., 2020), for example, can also be remediated using these NPs. Some factors can influence on process's efficiency such as pollutant's and nanomaterial's concentration and pH. NPs could absorb 90% of arsenic and took 60 min to reach the level of drinking water standard. Keeping pollutant concentration and increasing TiO_2 NP concentration and also increasing pH offered an improved scenario (Nazary et al., 2019). Protocols applying hydrated version of the material are common, as same as protocols that immobilize this NP in different substrates. These NPs immobilized on the surface of polyethylene terephthalate NFs could, in aqueous samples, promote the photocatalytic degradation of methylene blue. The efficiency of the process reached 88% of the dye degraded at only 10 min (Yasin et al., 2020). TiO_2 is a very useful photocatalysts, specially to remediate organic environmental pollution (Ainali et al., 2021; Dlamini et al., 2021) and can also be chemically modified to enhance its potential. TiO_2 NPs decorated with SnO_2 quantum dots improved charge separation's potential and oxidation capacity which improved photocatalytic performance to degrade Rhodamine B (Lee et al., 2014). Nano-TiO_2's capacity to photocatalytic degrade methylene blue could be enhanced by fullerene by narrowing the band gap and also introducing an additional doping state between conduction band and valance. Charge separation was improved as same as light adsorption allowing optimized remediation performance (Qi et al., 2016). Aeroxide® P25 titanium dioxide NPs in thin films could in situ promote photodegradation of paraquat and acetaminophen efficiently (Nicosia et al., 2021). Nano-TiO_2 could be used as nanoadsorbent to remove at low-cost Cu^{2+} from polluted samples, and the efficiency of the process could be improved by an increase in pH and decrease in ionic strength (Ezati et al., 2021). Ciprofloxacin could be degraded by TiO_2 NPs and the process efficiency optimized at lighting for 5 h, pH 5.5 using 50 mg of the nanomaterial. The process generated degradation products such as *p*-fluoraniline, presenting low toxicity (Usman et al., 2021).

Silver nanoparticles (AgNPs) decorating the surface of phosphorylated multiwalled carbon NTs (modifying nanosponge β-cyclodextrin) together with titanium dioxide NPs offered a nanobiosorbent, capable of dealing with pollution caused by trichloroethylene and Congo red via an endothermic adsorption process (Taka et al., 2020). Cobalt and lead atoms could be removed by AgNPs from contaminated groundwater (Attatsi and Nsiah, 2020). As silver presents specific toxicity, the release of NPs involving this metal into the environment concerns; that is why there are studies also being developed to study in detail its accumulation in aquatic organisms and potential of transference along the food chain (Xiao et al., 2022) and also to develop protocols to promote AgNP's removal and recovery from wastewater-treatment plants that apply this technology to perform water remediation (Salih et al., 2019; Yin et al., 2020).

Iron oxide has drawn a lot of attention due to the fact that they are not only biocompatible and biodegradable but also magnetic. Their functionalization allows the design to optimize desirable features of interest (Lesiak et al., 2019). In soil contamination by Cu, magnetite (Fe_3O_4) NPs could influence the metal mobility especially at lower Cu concentrations not affecting growth of plants nearby (sunflower), enhancing their antioxidant enzymatic response (Demangeat et al., 2021). Pb^{2+} pollution could be remediated by a magnetic zeolite type 5A reusable nanocomposite containing iron NPs bringing contaminants concentration on aqueous samples to permissible levels of 10 μg/L (Ramos-Guivar et al., 2020). Levels of nitrate, nitrite, phosphate, and ammonium were diminished by iron oxide NPs applied on the remediation of the discharge burden of fish farms' effluents. The process was performed in a reactor and the efficiency was increased in the first 6 h (Hesni et al., 2020). Soils contaminated by mixed pollution involving metals/metalloids and organic compounds could be managed by iron NPs. Cr was adsorpted and reduced and polychlorinated biphenyls were also removed from polluted samples. However, the ability of these particles could not surpass the ability presented by nZVI and nZVI–Pd at pseudo-anaerobic condition to promote the recovery of soil co-contaminated with these pollutants (Gil-Díaz et al., 2022). Iron magnetic NPs, in a version coated with PVP (polyvinylpyrrolidone) could also remove oil-contaminating water (Mirshahghassemi and Lead, 2015) and metals (Cd, Cr, Ni, and Pb) from contaminated sea water and synthetic soft water (Hong et al., 2020). Crystal violet dye could be removed at rates superior than 90% by superparamagnetic iron oxide NPs at pH 9 and 2 h of contact (Samrot et al., 2021). Pollution caused by dyes such as methylene blue and brilliant green could successfully be remediated by Fe_3O_4/activated carbon NPs (Joshi and Garg, 2019). Sludge-containing iron can be used to generate these NPs and the nanomaterial proved to be capable of removing As^{5+} from polluted water (Zeng et al., 2020). In fact, arsenite and arsenate pollution could already be remediated by iron oxide NPs (Priyadarshni et al., 2020). A combination of nanostructures can also bring benefits to remediation promoted by iron NPs. Association to graphene could allow efficiencies of degradation superior to 99% when dealing with contaminants methylene blue, Rhodamine B, Acid orange 7, and phenol. Under UV-A irradiation, the nanocomposite took 60 min to achieve these results due to the large surface area and accelerated transfer of photo-generated electrons occurring between iron oxide and graphene (Hammad et al., 2021). Enzymes can also be associated with nanomaterials to improve remediation. From *Rhizobium* sp. LMB-1, the enzyme protocatechuate 3,4-dioxygenase could be isolated and immobilized on Fe_3O_4 NPs to deal with organic contaminants, not only enhancing the process's efficiency but also increasing proteins' stability and allowing reusability (Zhang et al., 2017a).

MnO2 NPs associated with reduced graphene offered an option to remediating methylene blue and arsenite from polluted water, efficiently (Tara et al., 2020). These particles when deposited on magnetic halloysite NTs through hydrothermal method generated a sorbent capable of removing Pb^{2+} by and spontaneous and endothermic process. The nanostructure proved to be stable and reusable to remediate aqueous solutions (Afzali and Fayazi, 2016). Manganese dioxide NPs can also be applied to deal with As^{3+} pollution; the efficiency of the process increases with a reduction in the value of zeta potential and is dependent also on pH and adsorbent dosage (Liu et al., 2020). MnO2 NPs loaded on biochar could promote the remediation of Cd^{2+}, Cu^{2+}, Zn^{2+}, and Pb^{2+} by promoting the adsorption of heavy metals to the structure. The process's efficiency suffered influence of pH from polluted water/wastewater (Zhang et al., 2020). Lead and lanthanum could be remediated from water/wastewater samples by manganese dioxide NPs associated with chitosan in a nanocomposite in a spontaneous and endothermic process at rates higher than 91% (Mahmoud et al., 2021b). MnO2 NP can also be used to remediate halogenated aromatic compounds such as the agrochemical Rhodamine B at pH $\leq$5.0 allowing recycling to new cycles of degradation (Fang et al., 2021).

Palladium NPs on chitosan–carbon NT could promote efficient degradation of different dyes such as Methyl orange, Congo red, Methyl red, and methylene blue. The nanocomposite could also perform the reduction of 2-nitroaniline, 4-nitrophenol, 4-nitro-*o*-phenylenediamine, and 2,4-dinitrophenol (Sargin et al., 2020). These NPs decorating a membrane composed of bacterial nanocellulose embedded with mesoporous polydopamine contributed to the high-efficiency removal of dyes from contaminated wastewater (Derami et al., 2020). A reusable catalyst to deal with Cr^{6+} pollution can also benefit from the presence of palladium NPs. These NPs attached to hierarchical flower-like TiO_2 nanospheres offered a reusable and efficient solution to remediate water contaminated by this heavy metal (Li et al., 2020a).

Zirconia NPs are also interesting option, for example, to remediate samples containing As^{5+}, contributing to the transformation of the contaminant into a stable form. Soil contaminated could be treated with these NPs loaded on pine biochar (Liu et al., 2021). Iron doped with NPs of Zr exhibited good photocatalytic and photoelectrocatalytic activities performing the photodegradation of the dye Rhodamine B (Reddy et al., 2020). Phosphate could be removed from wastewater, avoiding eutrophication, by an absorbent containing modified porous silicon associated with zirconium NPs. The composite offered a maximum adsorption of 47.7 mg P/g (at pH 4 and room temperature), being reusable (Anbia et al., 2018). Iron doped with zirconia NPs exhibited great photoelectrochemical properties and could reach a rate of 96% for degradation of Rhodamine B under visible light irradiation (Cheng et al., 2020).

Zeolite NPs proved to be useful to remediate pollution associated with pharmaceuticals. These NPs associated with a polysulfone support generated a membrane that presented high rejection to negatively charged pharmaceuticals possessing small molecular weight (Dong et al., 2016). Direct Red 23 (DR23) and Direct Red 80 from wastewater could be efficiently adsorbed by sodalite zeolite NPs modified by (3-aminopropyl) triethoxy silane; the best performance was observed at pH 2.1 through an exothermic and spontaneous reaction (Mahmoodi and Saffar-Dastgerdi, 2019). A polysulfone membrane impregnated with zeolite NPs could remediate pollution caused by lead and nickel cations in aqueous samples (Manap, 2020).

Cupric oxide NPs could remediate pollution caused by the dye methylene blue (Ghulam et al., 2013). Copper oxide NPs, via adsorption batch method, could remove Cd^{2+} and Ni^{2+} from its aqueous solutions (Hassan et al., 2017). Through the chemical precipitation method divalent nickel ions could also be removed from aqueous media by copper oxide NPs at an optimized way presenting as conditions pH 7.0, 90 min of contact time, approximately 26°C, 180 rpm, and NP dose of 0.2 g/L (Jain et al., 2021). Hydroquinone is another example of a pollutant that could be removed from wastewater by using copper oxide NPs; these nanomaterials in polyethersulfune membrane allowed the generation of an efficient membrane to nanofiltration aiming water pollution remediation (Ghiasi et al., 2021).

Prussian blue-based NPs also possess an important role when it comes to dealing with environmental pollution. The dye Orange G could be nanoadsorpted and suffered photocatalysis by a nanocomposite associating γ-Al_2O_3 spheres and Prussian blue NPs (Doumic et al., 2015). ^{137}Cs present in polluted water could be immobilized by polyacrylonitrile NFs associated with Prussian blue NPs (Kim et al., 2018). The high-efficiency adsorption of Cs^+, Sr^{2+}, and Co^{2+} could also be performed by these NPs associated with activated charcoal at pH 6.8 (Ali et al., 2020). Rubidium and cesium from complex matrixes, even at low concentrations, could be removed from polluted samples by these NPs doping mesoporous silica gel (Yuan et al., 2020). Radioactive Cesium-137's pollution could be remediated by Prussian blue NPs and the efficiency of the process could benefit from reduction of the particle size and increase in the specific surface area and pore volume. In 24 h, more than 98% of the contaminant could be removed from contaminated water (Carniato et al., 2020).

WO_3 NP associated with colloidal could be used to generate a light-driven Au–WO_3–C Janus micromotor capable of degrading, even in low concentration, sodium-2,6-dichloroindophenol and Rhodamine B (Zhang et al., 2017b). Methylene blue could be degraded through photocatalysis promoted by WO_3 NPs (Thilagavathi and Venugopal, 2018). Orange G dye could also be remediated by tungsten trioxide NPs in a version coated with polyaniline (Hsini et al., 2021).

5.3 NANOTECHNOLOGY FIELD AND GREEN SYNTHESIS

As attention directed to sustainability increases, in the field of nanotechnology new eco-friendly protocols of nanomaterials' synthesis are proposed. However, in the 1990s, green synthesized nanomaterials were already a reality, as a consequence of the development of Green chemistry (Anastas and Warner, 2000). As an alternative to conventional physicochemical methods, the green-based protocols reject the use of novice reagents, prevent waste production, apply renewable energy sources and/or materials, and consider in the first place offering the smallest negative impact possible to a living being and to the ecosystem (Nasrollahzadeh et al., 2021).

Green synthesis in the nanotechnology field involves the use of seaweeds, bacteria, cyanobacteria, yeasts, fungi, plants (living ones, biomass, extracts), and/or bio-derived products (Figure 5.1) to generate the nanomaterial (Bahrulolum et al., 2021).

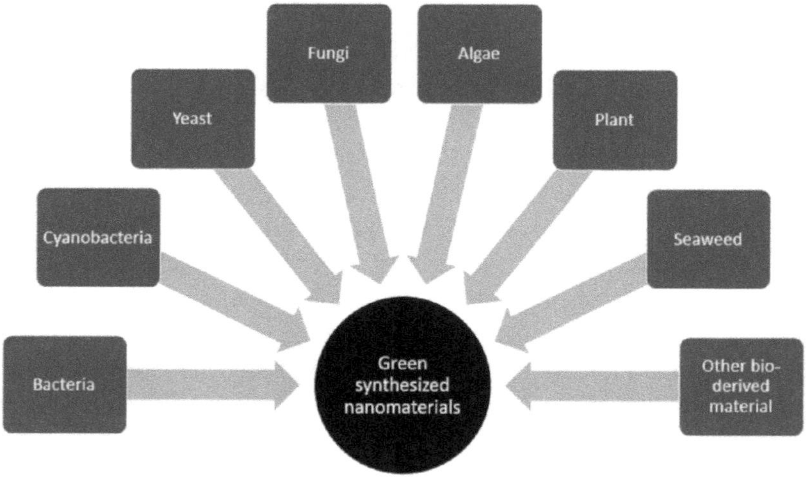

FIGURE 5.1 Green synthesis of nanomaterials involves the participation of bio-based materials.

5.3.1 GREEN-BASED SYNTHESIZED NANOPARTICLES TO REMEDIATE ENVIRONMENTAL POLLUTION

Green synthesized nanomaterials can be designed to remediate environmental pollution remediation. Table 5.2 presents some examples of this application of green-based NPs.

Platinum NPs could be generated to reduce the contaminants' aromatic nitrocompounds. To synthesize the NPs, the dried leaf powder of *Anacardium occidentale* was used (Sheny et al., 2013).

Gold NPs generated using the mushroom *Flammulina velutipes* could reduce through heterogeneous catalytic reactions of the contaminants methylene blue and 4-nitrophenol (Narayanan et al., 2015). When generated using *Sterculia acuminata* fruit extract, the NPs could degrade 4-nitrophenol, direct blue 24, methylene blue, and Methyl orange (Bogireddy et al., 2015). Fresh

TABLE 5.2 Examples of Green-Based Nanoparticles to Remediate Environmental Pollution.

Nanoparticle	Bio-based Material Applied on the Synthesis	Important Modification on Nanoparticle	Pollutant(s) Remediated	References
Platinum nanoparticles	Dried leaf powder of *Anacardium occidentale*		Aromatic nitrocompounds	Sheny et al. (2013)
Gold nanoparticles	Mushroom *Flammulina velutipes*		Methylene blue and 4-nitrophenol	Narayanan et al. (2015)
	Sterculia acuminata fruit extract		4-Nitrophenol, direct blue 24, methylene blue, and Methyl orange	Bogireddy et al. (2015)
	Fresh fruiting bodies of *Flammulina velutipes*		Methylene blue	Rabeea et al. (2020)
	Green tea	decorating TiO_2	Methylene blue	Perera et al. (2020)
Palladium nanoparticles	*Myrtus communis*'s leaf extract	Palladium/TiO_2 nanoparticle	Organic pollutants	Nasrollahzadeh and Sajadi (2016)
	Root extract from *Salvadora persica* L.		Aryl halides	Khan et al. (2016)
	Aqueous extract of agro-waste empty cotton boll peels		Congo red, Methyl orange, Sunset yellow, and Tartrazine	Narasaiah and Mandal (2020)
Silver nanoparticles	*Medicago polymorpha* extract		Methyl orange	Ismail et al. (2019)
	Zingiber officinale's rhizome extract		Methyl orange	Riaz et al. (2020)
	Leaf extract of *Pistacia atlantica*	Coated with nanomagnetic biochar dots	Dyes	Zahedifar et al. (2020)
	Peumo extract	Decorated on nanostructured porous silicon	Methylene blue	Naveas et al. (2022)

TABLE 5.2 *(Continued)*

Nanoparticle	Bio-based Material Applied on the Synthesis	Important Modification on Nanoparticle	Pollutant(s) Remediated	References
Iron oxide nanoparticles	Tea waste template		As^{3+} and As^{5+}	(Lunge et al., 2014)
	Extract of mortiño berry (*Vaccinium floribundum*)		Total petroleum hydrocarbons	Murgueitio et al. (2018)
	Crude latex of *Jatropha curcas* and leaf extract of *Cinnamomum tamala*	Coated	Organic/Inorganic pollutants	Das et al. (2020)
Copper nanoparticles	Pomegranate leaf extract	Deposition on vegetal activated carbon	Atrazine, caffeine, and diclofenac and nitrate	Peternela et al. (2018)
	Cynomorium coccineum		methylene blue	Sebeia et al. (2020)
	Citrofortunella macrocarpa		Rhodamine B	Rafique et al. (2020)
	Extract of mint leaves		Pb^{2+}, Ni^{2+}, and Cd^{2+}	Mahmoud et al. (2021a)
	Extract of orange peels		Pb^{2+}, Ni^{2+}, and Cd^{2+}	Mahmoud et al. (2021a)
Prussian blue nanoparticles	β-Cyclodextrin (surfactant) and ethyl alcohol (solvent)		Rhodamine B	Yang et al. (2021)
nZVI nanoparticles	Leaf extract of *Mentha piperita*		Organic and inorganic contaminants	Shad et al. (2020)
TiO2 nanoparticles	*Trianthema portulacastrum* extract		Cadmium	Irshad et al. (2022)
	Chenopodium quinoa extract		Cadmium	Irshad et al. (2022)
	Extract of *Mentha aquatica* leaves	Functionalized with carbon	Rhodamine B	Bathula et al. (2021)

fruiting bodies of *F. velutipes* could act as green reducers to the synthesis of gold NPs capable of degrading the organic pollutant methylene blue (Rabeea et al., 2020). Gold NPs decorating TiO_2 could be synthesized using green tea to remediate the pollution associated with the dye methylene blue through adsortion (Perera et al., 2020).

Palladium NPs to fast convert aryl halides into biphenyl could be generated using as bioreductant the root extract from *Salvadora persica* L. (Khan et al., 2016). Palladium/TiO_2 NPs synthesized using *Myrtus communis*'s leaf extract as a recyclable catalyst could degrade organic environmental pollutants (Nasrollahzadeh and Sajadi, 2016). Palladium NPs generated using the aqueous extract of agro-waste empty cotton boll peels (that provided reductants and stabilizers) could deal with pollution associated with Congo red, Methyl orange, Sunset yellow, and Tartrazine. Efficient degradation of toxic azo dyes was performed (Narasaiah and Mandal, 2020).

AgNPs synthesized using *Peumo* extract and deposited on nanostructured porous silicon thin films and could efficiently promote the reduction of organic dye methylene blue in aqueous samples (Naveas et al., 2022). Using *Medicago polymorpha* extract AgNPs could be generated and used to promote remediation of Methyl orange through reduction; these particles also presented antibacterial activity against *Escherichia coli* and *Staphylococcus aureus* (Ismail et al., 2019). The ones generated applying *Zingiber officinale*'s rhizome extract could reduce the dye Methyl orange and also presented antioxidant activity and capacity to act as antimicrobial against *Pseudomonas aeruginosa*, methicillin-resistant *S. aureus*, and *Candida albicans* (Riaz et al., 2020). AgNPs in nanomagnetic biochar dots coated version could be generated using leaf extract of *Pistacia atlantica*. The resulting nanocomposite could remediate the pollution caused by dyes in aqueous samples (Zahedifar et al., 2020).

Iron NPs generated using the extract of *Vaccinium floribundum* could remediate total petroleum hydrocarbons from water and soil, exhibiting a performance of more than 80% reduction of pollutants (Murgueitio et al., 2018). The magnetic ones generated using tea waste template proved to be useful to remediate As^{3+} and As^{5+} by promoting pollutants' adsorption (Lunge et al., 2014). Magnetite NPs in coated version generated using crude latex of *Jatropha curcas* and leaf extract of *Cinnamomum tamala* could remediate organic/inorganic pollution. The particles were generated at low costs and also presented the potential to serve as antioxidant and antibacterial agents (Das et al., 2020).

Copper NPs could be generated by using *Cynomorium coccineum*. The nanomaterial generated by this green-based protocol could deal well with pollution caused by the dye methylene blue (Sebeia et al., 2020). CuO NPs were generated using pomegranate leaf extract on vegetal activated carbon. The nanomaterials proved to be able to remediate pollution related to atrazine, caffeine, diclofenac, and nitrate (Peternela et al., 2018). *Citrofortunella macrocarpa* allowed the production of CuO NPs to remediate Rhodamine B by promoting dye's photocatalytic degradation under UV light (Rafique et al., 2020). Copper oxide NPs prepared using extracts of mint leaves and orange peels, separately, could act as nanosorbent to deal with pollution of Pb^{2+}, Ni^{2+}, and Cd^{2+} in this order of efficiency (Mahmoud et al., 2021a).

Prussian blue NPs synthesized applying β-cyclodextrin (surfactant) and ethyl alcohol (solvent) could remediate pollution associated with Rhodamine B; the dye could be degraded efficiently by the NP under visible light (Yang et al., 2021).

nZVI NPs synthesized using the leaf extract of *Mentha piperita* (reducing agent) could remediate pollution associated with organic and inorganic contaminants (Shad et al., 2020).

Titanium dioxide NPs could be synthesized using *Trianthema portulacastrum* and *Chenopodium quinoa* extracts separately. The nanomaterials proved to be able to remediate pollution associated with the heavy metal cadmium from aqueous samples with a contact time of 2 h. At a dosage 0.7 g/L the NPs (adsorbent) could remove 46 mg Cd/g of TiO_2 NPs and 44 mg Cd/g of the nanomaterial, respectively (Irshad et al., 2022). TiO_2 NPs generated using the extract of *Mentha Aquatica* leaves could, after being functionalized with carbon, remediate pollution associated with Rhodamine B. Photocatalytic oxidation took place in an efficient way and the low-cost eco-friendly solution proved also to be reusable (Bathula et al., 2021).

5.4 CONCERNING ASPECTS ON NANOPARTICLES' NANOREMEDIATION OF ENVIRONMENTAL POLLUTANTS

It is crucial to direct attention to NPs' toxicity and capacity to accumulate on living beings before suggesting to apply them on environmental pollution remediation. Sorbent NPs can be released in water that goes out of wastewater-treatment plants and soil, for example, possibly reaching the food chain. So, it is necessary to consider how harmful they can be specially to humans (that are at the end of the food chain). Nanomaterials for this application should be biocompatible avoiding toxicity or undesirable response from the organism in contact with them (Ferdous and Nemmar, 2020).

A large number of characteristics on NP can influence on how they will be noticed by organisms and how they will react to their presence, for example, purity, size, shape, and charge (Tonelli et al., 2020).

To select the methodology to access nanomaterial's cytotoxicity. Some conventional cytotoxicity assays are dependent on spectrophotometry and there are nanomaterials that present bands that may impair a fair result to be obtained, interfering on the results. Methodologies that access electrical impedance of attached cells are recommended to be performed in these situations (Lee et al., 2017).

5.5 FUTURE PROSPECTS

Researches on using NP to remediate environmental pollution are expected to continue to develop rapidly also filling important gaps that need to receive attention and are expected to be studied in details, for example, biochemical interactions established by these nanomaterials in living being organisms and their accumulation pattern. Protocols are expected to become even more secure, generating biodegradable materials, or materials that can be reused but easy to be removed from treatment areas avoiding contact to living beings. Biocompatibility, safety, and possibility of riskless use on a large scale are important aspects to be considered. There is also a large array of materials to be explored as reagents/stabilizers to be used in synthesis protocols. However, environmental awareness should be adopted worldwide as same as concern for sustainable development, remediating pollution without reducing or stopping its production, if possible, would result in useless effort.

5.6 CONCLUSIONS

Human actions ignoring sustainability principles have been causing serious damage to ecosystems worldwide and turned environmental pollution a serious threat to life on Earth.

The nanotechnology field is advancing in a fast pace and has been presenting various interesting solution to deal with this problem in an elegant and successful way. NPs deserve to be highlighted as a useful option and have yet an enormous potential to be explored by researchers through methods of synthesis green and nongreen. New protocols are continuously being developed to be more successful, inexpensive, eco-friendly, and secure. There are still some gaps to be filled but undoubtedly NPs are a very important, useful, and efficient solution to remediate environmental pollution.

KEYWORDS

- **nanoparticles**
- **green nanoparticles**
- **nanoremediation**
- **nanotechnology**
- **environmental pollution**

REFERENCES

Afzali, D.; Fayazi, M. Deposition of MnO2 Nanoparticles on the Magnetic Halloysite Nanotubes by Hydrothermal Method for Lead(II) Removal from Aqueous Solutions. *J. Taiwan Inst. Chem. Eng.* **2016,** *63*, 421–429. DOI: 10.1016/j.jtice.2016.02.025.

Ahmed, S. F.; Mofijur, M.; Nuzhat, S.; Chowdhury, A. T.; Rafa, N.; Uddin, M. A.; Inayat, A.; Mahlia, T. M. I.; Ong, H. C.; Chia, W. Y.; Show, P. L. Recent Developments in Physical, Biological, Chemical, and Hybrid Treatment Techniques for Removing Emerging Contaminants from Wastewater. *J. Hazard. Mater.* **2021,** *416*, 125912. DOI: 10.1016/j.jhazmat.2021.125912.

Ainali, N. M.; Kalaronis, D.; Evgenidou, E.; Bikiaris, D. N.; Lambropoulou, D. A. Insights into Biodegradable Polymer-Supported Titanium Dioxide Photocatalysts for Environmental Remediation. *Macromol* **2021,** *1* (3), 201–233. DOI: 10.3390/macromol1030015.

Ali, M. M. S.; Sami, N. M.; El-Sayed, A. A. Removal of Cs^+, Sr^{2+} and Co^{2+} by Activated Charcoal Modified with Prussian Blue Nanoparticle (PBNP) from Aqueous Media: Kinetics and Equilibrium Studies. *J. Radioanal. Nucl. Chem.* **2020,** *324* (1), 189–201. DOI: 10.1007/s10967-020-07067-y.

Anastas, P. T.; Warner, J. C. *Green Chemistry: Theory and Practice*. Oxford University Press: New York, NY, 2000.

Anbia, M.; Khodashenas, Z.; Kamel, L. Modified Porous Silicon with Zirconium Nanoparticles as a Selective Adsorbent for Phosphate Removal from Aqueous Systems. *Iran. J. Energy Environ.* **2018,** *9*, 182–190.

Anza, M.; Salazar, O.; Epelde, L.; Alkorta, I.; Garbisu, C. The Application of Nanoscale Zero-valent Iron Promotes Soil Remediation While Negatively Affecting Soil Microbial Biomass and Activity. *Front. Environ. Sci.* **2019,** *7*, 19–23. DOI: 10.3389/fenvs.2019.00019.

Attatsi, I. K.; Nsiah, F. Application of Silver Nanoparticles Toward Co(II) and Pb(II) Ions Contaminant Removal in Groundwater. *Appl. Water Sci.* **2020,** *10* (6), 152–162. DOI: 10.1007/s13201-020-01240-0.

Bahrulolum, H.; Nooraei, S.; Javanshir, N.; Tarrahimofrad, H.; Mirbagheri, V. S.; Easton, A. J.; Ahmadian, G. Green Synthesis of Metal Nanoparticles Using Microorganisms and Their Application in the Agrifood Sector. *J. Nanobiotechnol.* **2021,** *19* (1), 86. DOI: 10.1186/s12951-021-00834-3.

Bakshi, M.; Abhilash, P. C. Nanotechnology for Soil Remediation: Revitalizing the Tarnished Resource. In *Nano-materials as Photocatalysts for Degradation of Environmental Pollutants*; Elsevier: Amsterdam, 2020; pp 345–370.

Bathula, C.; Rabani, I.; Sekar, S.; Youi, H. K.; Choy, J. Y.; Kadam, A.; Shretha, N. K.; Seo, Y. S.; Kim, H. S. Enhanced Removal of Organic Dye by Activated Carbon Decorated TiO2 Nanoparticles from *Mentha aquatica* Leaves via Ultrasonic Approach. *Ceram. Int.* **2021,** *47* (6), 8732–8739. DOI: 10.1016/j.ceramint.2020.12.282.

Bethi, B.; Sonawane, S. H. Nanomaterials and Its Application for Clean Environment. In *Nanomaterials for Green Energy*; Elsevier: Amsterdam, 2018; pp. 385–409.

Bin, Q.; Lin, B.; Zhu, K.; Shen, Y.; Man, Y.; Wang, B.; Lai, C.; Chen, W. Superior Trichloroethylene Removal from Water by Sulfide-Modified Nanoscale Zero-Valent Iron/Graphene Aerogel Composite. *J. Environ. Sci. (China)* **2020,** *88*, 90–102. DOI: 10.1016/j.jes.2019.08.011.

Bogireddy, N. K. R.; Hoskote Anand, K. K. H.; Mandal, B. K. Gold Nanoparticles—Synthesis by *Sterculia acuminata* Extract and Its Catalytic Efficiency in Alleviating Different Organic Dyes. *J. Mol. Liq.* **2015,** *211*, 868–875. DOI: 10.1016/j.molliq.2015.07.027.

Bolade, O. P.; Williams, A. B.; Benson, N. U. Green Synthesis of Iron-Based Nanomaterials for Environmental Remediation: A Review. *Environ. Nanotechnol. Monit. Manage.* **2020,** *13*, 100279. DOI: 10.1016/j.enmm.2019.100279.

Carniato, F.; Gatti, G.; Vittoni, C.; Katsev, A. M.; Guidotti, M.; Evangelisti, C.; Bisio, C. More Efficient Prussian Blue Nanoparticles for an Improved Caesium Decontamination from Aqueous Solutions and Biological Fluids. *Molecules* **2020,** *25* (15), 3447. DOI: 10.3390/molecules25153447.

Chaoua, S.; Boussaa, S.; El Gharmali, A.; Boumezzough, A. Impact of Irrigation with Wastewater on Accumulation of Heavy Metals in Soil and Crops in the Region of Marrakech in Morocco. *J. Saudi Soc. Agric. Sci.* **2019,** *18* (4), 429–436. DOI: 10.1016/j.jssas.2018.02.003.

Chen, Y.; Shi, H.; Guo, H.; Ling, C.; Yuan, X.; Li, P. Hydrated Titanium Oxide Nanoparticles Supported on Natural Rice Straw for Cu(II) Removal from Water. *Environ. Technol. Innov.* **2020,** *20*, 101143. DOI: 10.1016/j.eti.2020.101143.

Cheng, Q.; Yang, W.; Chen, Q.; Zhu, J.; Li, D.; Fu, L.; Zhou, L. Fe-Doped Zirconia Nanoparticles with Highly Negative Conduction Band Potential for Enhancing Visible Light Photocatalytic Performance. *Appl. Surf. Sci.* **2020,** *530*, 147291. DOI: 10.1016/j.apsusc.2020.147291.

Das, C.; Sen, S.; Singh, T.; Ghosh, T.; Paul, S. S.; Kim, T. W.; Jeon, S.; Maiti, D. K.; Im, J.; Biswas, G. Green Synthesis, Characterization and Application of Natural Product Coated Magnetite Nanoparticles for Wastewater Treatment. *Nanomaterials (Basel)* **2020,** *10* (8), 1615. DOI: 10.3390/nano10081615.

Demangeat, E.; Pédrot, M.; Dia, A.; Bouhnik-Le-Coz, M.; Roperch, P.; Compaoré, G.; Cabello-Hurtado, F. Investigating the Remediation Potential of Iron Oxide Nanoparticles in Cu-Polluted Soil–Plant Systems: Coupled Geochemical, Geophysical and Biological Approaches. *Nanoscale Adv.* **2021,** *3* (7), 2017–2029. DOI: 10.1039/d0na00825g.

Derami, H. G.; Gupta, P.; Gupta, R.; Rathi, P.; Morrissey, J. J.; Singamaneni, S. Palladium Nanoparticle-Decorated Mesoporous Polydopamine/Bacterial Nanocellulose as a Catalytically Active Universal Dye Removal Ultrafiltration Membrane. *ACS Appl. Nano Mater.* **2020,** *3* (6), 5437–5448. DOI: 10.1021/acsanm.0c00787.

Dlamini, M. C.; Maubane-Nkadimeng, M. S.; Moma, J. A. The Use of TiO2/Clay Heterostructures in the Photocatalytic Remediation of Water Containing Organic Pollutants: A Review. *J. Environ. Chem. Eng.* **2021,** *9* (6), 106546. DOI: 10.1016/j.jece.2021.106546.

Dong, L. X.; Huang, X. C.; Wang, Z.; Yang, Z.; Wang, X. M.; Tang, C. Y. A Thin-Film Nanocomposite Nanofiltration Membrane Prepared on a Support with In Situ Embedded Zeolite Nanoparticles. *Sep. Purif. Technol.* **2016,** *166*, 230–239. DOI: 10.1016/j.seppur.2016.04.043.

Doumic, L.; Salierno, G.; Cassanello, M.; Haure, P.; Ayude, M. Efficient Removal of Orange G Using Prussian Blue Nanoparticles Supported over Alumina. *Cat. Today* **2015,** *240*, 67–72. DOI: 10.1016/j.cattod.2014.03.064.

Ezati, F.; Sepehr, E.; Ahmadi, F. The Efficiency of Nano-TiO2 and γ-Al2O3 in Copper Removal from Aqueous Solution by Characterization and Adsorption Study. *Sci. Rep.* **2021,** *11* (1), 18831. DOI: 10.1038/s41598-021-98051-3.

Fang, C.; Gujarati, H.; Osinaga, F.; Hsia, V.; Cheney, M. A.; Kharel, M. K. Optimization of the Catalytic Activity of Manganese Dioxide (MnO2) Nanoparticles for Degradation of Environmental Pollutants. *Res. Chem. Intermed.* **2021,** *47* (9), 3673–3690. DOI: 10.1007/s11164-021-04494-8.

Ferdous, Z.; Nemmar, A. Health Impact of Silver Nanoparticles: A Review of the Biodistribution and Toxicity Following Various Routes of Exposure. *Int. J. Mol. Sci.* **2020,** *21* (7), 2375. DOI: 10.3390/ijms21072375.

Gheju, M. Hexavalent Chromium Reduction with Zero-Valent Iron (ZVI) in Aquatic Systems. *Water Air Soil Pollut.* **2011,** *222* (1–4), 103–148. DOI: 10.1007/s11270-011-0812-y.

Ghiasi, S.; Roshanfekr, H.; Peyman, H. Preparation of Polyethersulfune Membrane with Copper Oxide Nanoparticle for Hydroquinone Removal from Waste Water. *Adv. Nanochem.* **2021,** *3*, 71–79.

Ghulam, M.; Hajira, T.; Muhammad, S.; Nasir, A. Synthesis and Characterization of Cupric Oxide (CuO) Nanoparticles and Their Application for the Removal of Dyes. *Afr. J. Biotechnol.* **2013,** *12* (47), 6650–6660. DOI: 10.5897/AJB2013.13058.

Gil-Díaz, M.; Álvarez, M. A.; Alonso, J.; Lobo, M. C. Effectiveness of Nanoscale Zero-Valent Iron for the Immobilization of Cu and/or Ni in Water and Soil Samples. *Sci. Rep.* **2020,** *10* (1), 15927. DOI: 10.1038/s41598-020-73144-7.

Gil-Díaz, M.; Pérez, R. A.; Alonso, J.; Miguel, E.; Diez-Pascual, S.; Lobo, M. C. Iron Nanoparticles to Recover a Co-contaminated Soil with Cr and PCBs. *Sci. Rep.* **2022,** *12* (1), 3541. DOI: 10.1038/s41598-022-07558-w.

Guan, X.; Sun, Y.; Qin, H.; Li, J.; Lo, I. M.; He, D.; Dong, H. The Limitations of Applying Zero-Valent Iron Technology in Contaminants Sequestration and the Corresponding Countermeasures: The Development in Zero-Valent Iron Technology in the Last Two Decades (1994–2014). *Water Res.* **2015,** *75*, 224–248. DOI: 10.1016/j.watres.2015.02.034.

Hammad, M.; Fortugno, P.; Hardt, S.; Kim, C.; Salamon, S.; Schmidt, T. C.; Wende, H.; Schulz, C.; Wiggers, H. Large-Scale Synthesis of Iron Oxide/Graphene Hybrid Materials as Highly Efficient Photo-Fenton Catalyst for Water Remediation. *Environ. Technol. Innov.* **2021,** *21*, 101239. DOI: 10.1016/j.eti.2020.101239.

Hassan, K. H.; Jarullah, A. A.; Saadi, S. K. Synthesis of Copper Oxide Nanoparticle as an Adsorbent for Removal of Cd(II) and Ni(II) Ions from Binary System. *Int. J. Appl. Environ. Sci.* **2017,** *12*, 1841–1861.

Hesni, M. A.; Hedayati, S. A.; Qadermarzi, A.; Pouladi, M.; ZangiAbadi, S.; Naghshbandi, N. Application of Iron Oxide Nanoparticles in the Reactor for Treatment of Effluent from Fish Farms. *Iran. J. Fish. Sci.* **2020,** *19*, 1319–1328.

Hong, J.; Xie, J.; Mirshahghassemi, S.; Lead, J. Metal (Cd, Cr, Ni, Pb) Removal from Environmentally Relevant Waters Using Polyvinylpyrrolidone-Coated Magnetite Nanoparticles. *RSC Adv.* **2020,** *10* (6), 3266–3276. DOI: 10.1039/c9ra10104g.

Hristovski, K. D.; Nguyen, H.; Westerhoff, P. K. Removal of Arsenate and 17α-Ethinyl Estradiol (EE2) by Iron (Hydro)Oxide Modified Activated Carbon Fibers. *J. Environ. Sci. Health A* **2009,** *44* (4), 354–361. DOI: 10.1080/10934520802659695.

Hsini, A.; Naciri, Y.; Bouziani, A.; Aarab, N.; Essekri, A.; Imgharn, A.; Laabd, M.; Navío, J. A.; Puga, F.; Lakhmiri, R.; Albourine, A. Polyaniline Coated Tungsten Trioxide as an Effective Adsorbent for the Removal of Orange G Dye from Aqueous Media. *RSC Adv.* **2021,** *11* (50), 31272–31283. DOI: 10.1039/d1ra04135e.

Iloms, E.; Ololade, O. O.; Ogola, H. J. O.; Selvarajan, R. Investigating Industrial Effluent Impact on Municipal Wastewater Treatment Plant in Vaal, South Africa. *Int. J. Environ. Res. Public Health* **2020,** *17* (3), 1096. DOI: 10.3390/ijerph17031096.

Irshad, M. A.; Shakoor, M. B.; Nawaz, R.; Yasmeen, T.; Arif, M. S.; Rizwan, M.; Rehman, M. Z.; Ahmad, S. R.; Latif, M.; Nasim, I.; Ali, S. Green and Eco-friendly Synthesis of TiO2 Nanoparticles and Their Application for Removal of Cadmium from Wastewater: Reaction Kinetics Study. *Z. Phys. Chem.* **2022,** 236 (5), 637–657. DOI: 10.1515/zpch-2021-3171.

Ismail, M.; Gul, S.; Khan, M. I.; Khan, M. A.; Asiri, A. M.; Khan, S. B. *Medicago polymorpha*-Mediated Antibacterial Silver Nanoparticles in the Reduction of Methyl Orange. *Green Process. Synth.* **2019,** *8* (1), 118–127. DOI: 10.1515/gps-2018-0030.

Jain, A.; Vaya, D. Photocatalytic Activity of TiO2 Nanomaterial. *J. Chil. Chem. Soc.* **2017,** *62* (4), 3683–3690. DOI: 10.4067/s0717-97072017000403683.

Jain, M.; Yadav, M.; Chaudhry, S. Copper Oxide Nanoparticles for the Removal of Divalent Nickel Ions from Aqueous Solution. *Toxicol. Rev.* **2021,** *40* (4), 872–885. DOI: 10.1080/15569543.2020.1799407.

Jang, M. H.; Lim, M.; Hwang, Y. S. Potential Environmental Implications of Nanoscale Zero-Valent Iron Particles for Environmental Remediation. *Environ. Health Toxicol.* **2014,** *29*, e2014022. DOI: 10.5620/eht.e2014022.

Jiang, D.; Zeng, G.; Huang, D.; Chen, M.; Zhang, C.; Huang, C.; Wan, J. Remediation of Contaminated Soils by Enhanced Nanoscale Zero Valent Iron. *Environ. Res.* **2018,** *163*, 217–227. DOI: 10.1016/j.envres.2018.01.030.

Joshi, S.; Garg, V. K.; Kataria, N.; Kadirvelu, K. Applications of Fe3O4@AC Nanoparticles for Dye Removal from Simulated Wastewater. *Chemosphere* **2019,** *236*. DOI: 10.1016/j.chemosphere.2019.07.011.

Keller, A. A.; Wang, H.; Zhou, D.; Lenihan, H. S.; Cherr, G.; Cardinale, B. J.; Miller, R.; Ji, Z. Stability and Aggregation of Metal Oxide Nanoparticles in Natural Aqueous Matrices. *Environ. Sci. Technol.* **2010,** *44* (6), 1962–1967. DOI: 10.1021/es902987d.

Khan, S. T.; Malik, A. Engineered Nanomaterials for Water Decontamination and Purification: From Lab to Products. *J. Hazard. Mater.* **2019,** *363*, 295–308. DOI: 10.1016/j.jhazmat.2018.09.091.

Khan, M.; Albalawi, G. H.; Shaik, M. R.; Khan, M.; Adil, S. F.; Kuniyil, M.; Alkhathlan, H. Z.; Al-Warthan, A.; Siddiqui, M. R. H. Miswak Mediated Green Synthesized Palladium Nanoparticles as Effective Catalysts for the Suzuki Coupling Reactions in Aqueous Media. *J. Saudi Chem. Soc.* **2017,** *21* (4), 450–457. DOI: 10.1016/j.jscs.2016.03.008.

Khin, M. M.; Nair, A. S.; Babu, V. J.; Murugan, R.; Ramakrishna, S. A Review on Nanomaterials for Environmental Remediation. *Energy Environ. Sci.* **2012,** *5* (8), 8075–8109. DOI: 10.1039/c2ee21818f.

Kim, H.; Kim, M.; Lee, W.; Kim, S. Rapid Removal of Radioactive Cesium by Polyacrylonitrile Nanofibers Containing Prussian Blue. *J. Hazard. Mater.* **2018,** *347*, 106–113. DOI: 10.1016/j.jhazmat.2017.12.050.

Klimkova, S.; Cernik, M.; Lacinova, L.; Filip, J.; Jancik, D.; Zboril, R. Zero-Valent Iron Nanoparticles in Treatment of Acid Mine Water from In Situ Uranium Leaching. *Chemosphere* **2011,** *82* (8), 1178–1184. DOI: 10.1016/j.chemosphere.2010.11.075.

Lee, K. T.; Lin, C. H.; Lu, S. Y. SnO2 Quantum Dots Synthesized with a Carrier Solvent Assisted Interfacial Reaction for Band-Structure Engineering of TiO2 Photocatalysts. *J. Phys. Chem. C* **2014,** *118* (26), 14457–14463. DOI: 10.1021/jp5045749.

Lee, Y. H.; Chuang, S. M.; Huang, S. C.; Tan, X.; Liang, R. Y.; Yang, G. C. C.; Chueh, P. J. Biocompatibility Assessment of Nanomaterials for Environmental Safety Screening. *Environ. Toxicol.* **2017,** *32* (4), 1170–1182. DOI: 10.1002/tox.22313.

Lesiak, B.; Rangam, N.; Jiricek, P.; Gordeev, I.; Tóth, J.; Kövér, L.; Mohai, M.; Borowicz, P. Surface Study of Fe3O4 Nanoparticles Functionalized with Biocompatible Adsorbed Molecules. *Front. Chem.* **2019,** *7*, 642. DOI: 10.3389/fchem.2019.00642.

Li, S.; Wang, W.; Yan, W.; Zhang, W. X. Nanoscale Zero-Valent Iron (nZVI) for the Treatment of Concentrated Cu(II) Wastewater: A Field Demonstration. *Environ. Sci. Process. Impacts* **2014,** *16* (3), 524–533. DOI: 10.1039/c3em00578j.

Li, W.; Cao, Z.; Liu, R.; Liu, L.; Li, H.; Li, X.; Chen, Y.; Lu, C.; Liu, Y. AuNPs as an Important Inorganic Nanoparticle Applied in Drug Carrier Systems. *Artif. Cells Nanomed. Biotechnol.* **2019,** *47* (1), 4222–4233. DOI: 10.1080/21691401.2019.1687501.

Li, H.; Gao, M.; Gao, Q.; Wang, H.; Han, B.; Xia, K.; Zhou, C. Palladium Nanoparticles Uniformly and Firmly Supported on Hierarchical Flower-Like TiO2 Nanospheres as a Highly Active and Reusable Catalyst for Detoxification of Cr(VI)-Contaminated Water. *Appl. Nanosci.* **2020a,** *10* (2), 359–369. DOI: 10.1007/s13204-019-01164-5.

Li, S.; Yang, F.; Li, J.; Cheng, K. Porous Biochar-Nanoscale Zero-Valent Iron Composites: Synthesis, Characterization and Application for Lead Ion Removal. *Sci. Total Environ.* **2020b,** *746*, 141037. DOI: 10.1016/j.scitotenv.2020.141037.

Li, Z.; Sun, Y.; Yang, Y.; Han, Y.; Wang, T.; Chen, J.; Tsang, D. C. W. Biochar-Supported Nanoscale Zero-Valent Iron as an Efficient Catalyst for Organic Degradation in Groundwater. *J. Hazard. Mater.* **2020c,** *383*, 121240. DOI: 10.1016/j.jhazmat.2019.121240.

Liu, C.; Ding, R.; Xie, F. Facile Synthesis of Manganese Dioxide Nanoparticles for Efficient Removal of Aqueous As(III). *J. Chem. Eng. Data* **2020,** *65* (8), 3988–3997. DOI: 10.1021/acs.jced.0c00311.

Liu, M.; Sun, F.; Lv, Y.; Xu, Y.; Li, M.; Wang, Y.; Yin, X.; Jiang, H. Remediation of Arsenic-Contaminated Soil by Nano-zirconia Modified Biochar. *Environ. Sci. Pollut. Res. Int.* **2021,** *28* (48), 68792–68803. DOI: 10.1007/s11356-021-15362-8.

Lofrano, G.; Libralato, G.; Brown, J. *Nanotechnologies for Environmental Remediation: Applications and Implications*; Springer: New York, NY, 2017; 325 p.

Longman, J.; Veres, D.; Finsinger, W.; Ersek, V. Exceptionally High Levels of Lead Pollution in the Balkans from the Early Bronze Age to the Industrial Revolution. *Proc. Natl. Acad. Sci. USA* **2018,** *115* (25), E5661–E5668. DOI: 10.1073/pnas.1721546115.

Lu, F.; Astruc, D. Nanocatalysts and Other Nanomaterials for Water Remediation from Organic Pollutants. *Coord. Chem. Rev.* **2020,** *408*. DOI: 10.1016/j.ccr.2020.213180.

Lunge, S.; Singh, S.; Sinha, A. Magnetic Iron Oxide (Fe3O4) Nanoparticles from Tea Waste for Arsenic Removal. *J. Magn. Magn. Mater.* **2014,** *356*, 21–31. DOI: 10.1016/j.jmmm.2013.12.008.

Mahmoodi, N. M.; Saffar-Dastgerdi, M. H. Zeolite Nanoparticle as a Superior Adsorbent with High Capacity: Synthesis, Surface Modification and Pollutant Adsorption Ability from Wastewater. *Microchem. J.* **2019,** *145*, 74–83. DOI: 10.1016/j.microc.2018.10.018.

Mahmoud, A. E. D.; Al-Qahtani, K. M.; Alflaij, S. O.; Al-Qahtani, S. F.; Alsamhan, F. A. Green Copper Oxide Nanoparticles for Lead, Nickel, and Cadmium Removal from Contaminated Water. *Sci. Rep.* **2021a,** *11* (1), 12547. DOI: 10.1038/s41598-021-91093-7.

Mahmoud, M. E.; Ibrahim, G. A. A.; Abdelwahab, M. S. Manganese Dioxide Nanoparticles Decorated with Chitosan for Effective Removal of Lead and Lanthanum Ions from Water by Microwave Sorption Technique. *Mater. Sci. Eng. B* **2021b,** *267*, 115091. DOI: 10.1016/j.mseb.2021.115091.

Manap, A. M. M. Application of Zeolite Nanofiltration for Removal of Heavy Metals from Urban Wastewater. *Civ. Eng. Beyond Limits* **2020,** *4*, 7–12.

Mirshahghassemi, S.; Lead, J. R. Oil Recovery from Water Under Environmentally Relevant Conditions Using Magnetic Nanoparticles. *Environ. Sci. Technol.* **2015,** *49* (19), 11729–11736. DOI: 10.1021/acs.est.5b02687.

Mohammed, M. A.; Syeda, J. T. M.; Wasan, K. M.; Wasan, E. K. An Overview of Chitosan Nanoparticles and Its Application in Non-parenteral Drug Delivery. *Pharmaceutics* **2017,** *9* (4), 53. DOI: 10.3390/pharmaceutics9040053.

Murgueitio, E.; Cumbal, L.; Abril, M.; Izquierdo, A.; Debut, A.; Tinoco, O. Green Synthesis of Iron Nanoparticles: Application on the Removal of Petroleum Oil from Contaminated Water and Soils. *J. Nanotechnol.* **2018,** *2018*, 1–8. DOI: 10.1155/2018/4184769.

Mustapha, S.; Ndamitso, M. M.; Abdulkareem, A. S.; Tijani, J. O.; Shuaib, D. T.; Ajala, A. O.; Mohammed, A. K. Application of TiO2 and ZnO Nanoparticles Immobilized on Clay in Wastewater Treatment: A Review. *Appl. Water Sci.* **2020,** *10* (1), 49. DOI: 10.1007/s13201-019-1138-y.

Nasrollahzadeh, M.; Sajadi, S. M. Green Synthesis, Characterization and Catalytic Activity of the Pd/TiO2 Nanoparticles for the Ligand-Free Suzuki–Miyaura Coupling Reaction. *J. Colloid Interface Sci.* **2016,** *465*, 121–127. DOI: 10.1016/j.jcis.2015.11.038.

Nasrollahzadeh, M.; Sajjadi, M.; Iravani, S.; Varma, R. S. Green-Synthesized Nanocatalysts and Nanomaterials for Water Treatment: Current Challenges and Future Perspectives. *J. Hazard. Mater.* **2021,** *401*, 123401. DOI: 10.1016/j.jhazmat.2020.123401.

Narasaiah, B. P.; Mandal, B. K. Remediation of Azo-Dyes Based Toxicity by Agro-Waste Cotton Boll Peels Mediated Palladium Nanoparticles. *J. Saudi Chem. Soc.* **2020,** *24* (2), 267–281. DOI: 10.1016/j.jscs.2019.11.003.

Narayanan, K. B.; Park, H. H.; Han, S. S. Synthesis and Characterization of Biomatrixed-Gold Nanoparticles by the Mushroom *Flammulina velutipes* and Its Heterogeneous Catalytic Potential. *Chemosphere* **2015,** *141*, 169–175. DOI: 10.1016/j.chemosphere.2015.06.101.

Naveas, N.; Manso-Silván, M.; Carmona, E.; Garrido, K.; Hernández-Montelongo, J.; Recio-Sánchez, G. Green Synthesized Silver Nanoparticles Decorated on Nanostructured Porous

Silicon as an Efficient Platform for the Removal of Organic Dye Methylene Blue. *Green Chem. Lett. Rev.* **2022,** *15* (1), 108–115. DOI: 10.1080/17518253.2021.2024609.

Nazary, A.; Nakhaei, M.; Yari, A. R. Arsenic Removal from Aqueous Solution Using Titanium Dioxide Nanoparticles (Anatase). *Qom Univ. Med. Sci. J.* **2019,** *13* (8), 62–72. DOI: 10.29252/qums.13.8.62.

Nicosia, A.; Vento, F.; Di Mari, G. M.; D'Urso, L.; Mineo, P. G. TiO2-Based Nanocomposites Thin Film Having Boosted Photocatalytic Activity for Xenobiotics Water Pollution Remediation. *Nanomaterials (Basel)* **2021,** *11* (2), 400. DOI: 10.3390/nano11020400.

O'Carroll, D.; Sleep, B.; Krol, M.; Boparai, H.; Kocur, C. Nanoscale Zero Valent Iron and Bimetallic Particles for Contaminated Site Remediation. *Adv. Water Resour.* **2013,** *51*, 104–122. DOI: 10.1016/j.advwatres.2012.02.005.

Pal, G.; Rai, P.; Pandey, A. Green Synthesis of Nanoparticles: A Greener Approach for a Cleaner Future. *Green Synthesis, Characterization and Applications of Nanoparticles* Elsevier: Amsterdam, 2019; pp 1–26.

Paul, W.; Sharma, C. P. Inorganic Nanoparticles for Targeted Drug Delivery. *Biointegration of Medical Implant Materials*; Woodhead Publishing: Sawston, UK, 2020; pp 333–373.

Perera, M.; Wijenayaka, L. A.; Siriwardana, K.; Dahanayake, D.; Nalin de Silva, K. M. N. Gold Nanoparticle Decorated titania for Sustainable Environmental Remediation: Green Synthesis, Enhanced Surface Adsorption and Synergistic Photocatalysis. *RSC Adv.* **2020,** *10* (49), 29594–29602. DOI: 10.1039/d0ra05607c.

Peternela, J.; Silva, M. F.; Vieira, M. F.; Bergamasco, R.; Vieira, A. M. S. Synthesis and Impregnation of Copper Oxide Nanoparticles on Activated Carbon through Green Synthesis for Water Pollutant Removal. *Mater. Res.* **2018,** *21* (1), 20160460. DOI: 10.1590/1980-5373-mr-2016-0460.

Priyadarshni, N.; Nath, P.; Nagahanumaiah; Chanda, N. Sustainable Removal of Arsenate, Arsenite and Bacterial Contamination from Water Using Biochar Stabilized Iron and Copper Oxide Nanoparticles and Associated Mechanism of the Remediation Process. *J. Water Process Eng.* **2020,** *37*, 101495. DOI: 10.1016/j.jwpe.2020.101495.

Pulizzi, F.; Sun, W. Treating Water with Nano. *Nat. Nanotechnol.* **2018,** *13* (8), 633. DOI: 10.1038/s41565-018-0238-4.

Qi, K.; Selvaraj, R.; Al Fahdi, T.; Al-Kindy, S.; Kim, Y.; Wang, G.; Tai, C. W.; Sillanpää, M. Enhanced Photocatalytic Activity of Anatase-TiO2 Nanoparticles by Fullerene Modification: A Theoretical and Experimental Study. *Appl. Surf. Sci.* **2016,** *387*, 750–758. DOI: 10.1016/j.apsusc.2016.06.134.

Rabeea, M. A.; Owaid, M. N.; Aziz, A. A.; Jameel, M. S.; Dheyab, M. A. Mycosynthesis of Gold Nanoparticles Using the Extract of *Flammulina velutipes*, Physalacriaceae, and Their Efficacy for Decolorization of Methylene Blue. *J. Environ. Chem. Eng.* **2020,** *8* (3), 103841. DOI: 10.1016/j.jece.2020.103841.

Rafique, M.; Shafiq, F.; Ali Gillani, S. S. A.; Shakil, M.; Tahir, M. B.; Sadaf, I. Eco-friendly Green and Biosynthesis of Copper Oxide Nanoparticles Using Citrofortunella microcarpa Leaves Extract for Efficient Photocatalytic Degradation of Rhodamine B Dye Form Textile Wastewater. *Optik* **2020,** *208*, 164053. DOI: 10.1016/j.ijleo.2019.164053.

Ramos-Guivar, J. A.; Taipe, K.; Schettino Jr., M. A.; Silva, E.; Morales Torres, M. A. M.; Passamani, E. C.; Litterst, F. J. Improved Removal Capacity and Equilibrium Time of Maghemite Nanoparticles Growth in Zeolite Type 5A for Pb(II) Adsorption. *Nanomaterials (Basel)* **2020,** *10* (9), 1668. DOI: 10.3390/nano10091668.

Reddy, C. V.; Reddy, I. N.; Harish, V. V. N.; Reddy, K. R.; Shetti, N. P.; Shim, J.; Aminabhavi, T. M. Efficient Removal of Toxic Organic Dyes and Photoelectrochemical Properties of Iron-Doped Zirconia Nanoparticles. *Chemosphere* **2020,** *239*, 124766. DOI: 10.1016/j.chemosphere.2019.124766.

Riaz, M.; Ismail, M.; Ahmad, B.; Zahid, N.; Jabbour, G.; Khan, M. S.; Mutreja, V.; Sareen, S.; Rafiq, A.; Faheem, M.; Shah, M. M.; Khan, M. I.; Bukhari, S. A. I.; Park, J. Characterizations and Analysis of the Antioxidant, Antimicrobial, and Dye Reduction Ability of Green Synthesized Silver Nanoparticles. *Green Process. Synth.* **2020,** *9* (1), 693–705. DOI: 10.1515/gps-2020-0064.

Salih, H. H. M.; El Badawy, A. M.; Tolaymat, T. M.; Patterson, C. L. Removal of Stabilized Silver Nanoparticles from Surface Water by Conventional Treatment Processes. *Adv. Nanopart.* **2019,** *8* (2), 21–35. DOI: 10.4236/anp.2019.82002.

Samrot, A. V.; Ali, H. H.; Selvarani A, J.; Faradjeva, E.; , Raji, P.; Prakash, P.; Kumar S, S. Adsorption Efficiency of Chemically Synthesized Superparamagnetic Iron Oxide Nanoparticles (SPIONs) on Crystal Violet Dye. *Curr. Res. Green Sustain. Chem.* **2021,** *4*, 100066. DOI: 10.1016/j.crgsc.2021.100066.

Sargin, I.; Baran, T.; Arslan, G. Environmental Remediation by Chitosan–Carbon Nanotube Supported Palladium Nanoparticles: Conversion of Toxic Nitroarenes into Aromatic Amines, Degradation of Dye Pollutants and Green Synthesis of Biaryls. *Sep. Purif. Technol.* **2020,** *247*, 116987. DOI: 10.1016/j.seppur.2020.116987.

Sebeia, N.; Jabli, M.; Ghith, A.; Saleh, T. A. Eco-friendly Synthesis of *Cynomorium coccineum* Extract for Controlled Production of Copper Nanoparticles for Sorption of Methylene Blue Dye. *Arab. J. Chem.* **2020,** *13* (2), 4263–4274. DOI: 10.1016/j.arabjc.2019.07.007.

Shad, S.; Belinga-Desaunay-Nault, M. A.; Sohail; Bashir, N.; Lynch, I. Removal of Contaminants from Canal Water Using Microwave Synthesized Zero Valent Iron Nanoparticles. *Environ. Sci.: Water Res. Technol.* **2020,** *6* (11), 3057–3065. DOI: 10.1039/D0EW00157K.

Sheny, D. S.; Philip, D.; Mathew, J. Synthesis of Platinum Nanoparticles Using Dried *Anacardium occidentale* Leaf and Its Catalytic and Thermal Applications. *Spectrochim. Acta A: Mol. Biomol. Spectrosc.* **2013,** *114*, 267–271. DOI: 10.1016/j.saa.2013.05.028.

Sun, X.; Ran, Z.; Wu, Y.; Zhong, C.; Zhu, W.; Hllah, H.; Yu, J. Optimization of PAHs Oxidation from Contaminated Soil Using Modified Nanoscale Zero-Valent Iron Combined with Potassium Permanganate. *RSC Adv.* **2022,** *12* (12), 7335–7346. DOI: 10.1039/d1ra09106a.

Taka, A. L.; Fosso-Kankeu, E.; Pillay, K.; Yangkou Mbianda, X. Metal Nanoparticles Decorated Phosphorylated Carbon Nanotube/Cyclodextrin Nanosponge for Trichloroethylene and Congo Red Dye Adsorption from Wastewater. *J. Environ. Chem. Eng.* **2020,** *8* (3), 103602. DOI: 10.1016/j.jece.2019.103602.

Taniguchi, N. On the Basic Concept of Nanotechnology. In *Proceedings of the International Conference on Production Engineering*, 1974; pp 18–23.

Tara, N.; Siddiqui, S. I.; Bach, Q. V.; Chaudhry, S. A. Reduce Graphene Oxide-Manganese Oxide-Black Cumin Based Hybrid Composite (rGO-MnO2/BC): A Novel Material for Water Remediation. *Mater. Today Commun.* **2020,** *25*, 101560. DOI: 10.1016/j.mtcomm.2020.101560.

Tarekegn, M. M.; Hiruy, A. M.; Dekebo, A. H. Nano Zero Valent Iron (nZVI) Particles for the Removal of Heavy Metals (CD^{2+}, Cu^{2+} and Pb^{2+}) from Aqueous Solutions. *RSC Adv.* **2021,** *11* (30), 18539–18551. DOI: 10.1039/d1ra01427g.

Thilagavathi, T.; Venugopal, D. Synthesis and Characterization of WO3 Nanoparticles for Photocatalytic Degradation of Methylene Blue Dye. *International Journal for Research in Engineering Application & Management* **2018,** *4,* 317–320.

Tonelli, F. M. P.; Tonelli, F. C. P.; Ferreira, D. R. C.; da Silva, K. E.; Cordeiro, H. G.; Ouchida, A. T.; Nunes, N. A. M. Biocompatibility and Functionalization of Nanomaterials. In *Intelligent Nanomaterials for Drug Delivery Applications*; Ahmad, N., Gopinath, P., Eds.; Elsevier: Amsterdam, 2020; 204 p.

Tonelli, F. M. P.; Tonelli, F. C. P.; Lemos, M. S.; Ferreira, D. R. C. Nanomaterials to Remediate Water Pollution. In *Nanotechnology for Environmental Pollution Decontamination. Tools, Methods, and Approaches for Detection and Remediation*; Tonelli, F. M. P., Bhat, R. A., Dar, G. H., Eds.; CRC Press: Boca Raton, FL, 2022.

Usman, M. R.; Prasasti, A.; Islamiah, S.; Firdaus, A. N.; Marita, A. W.; Fajriyah, S.; Noviyanti, A. R.; Eddy, D. R. Degradation of Ciprofloxacin by Titanium Dioxide (TiO2) Nanoparticles: Optimization of Conditions, Toxicity, and Degradation Pathway. *Bull. Chem. React. Eng. Catal.* **2021,** *16* (4), 752–762. DOI: 10.9767/bcrec.16.4.11355.752-762.

Wang, F.; Yang, W.; Zheng, F.; Sun, Y. Removal of Cr(VI) from Simulated and Leachate Wastewaters by Bentonite-Supported Zero-Valent Iron Nanoparticles. *Int. J. Environ. Res. Public Health* **2018,** *15* (10), 2162. DOI: 10.3390/ijerph15102162.

Werkneh, A. A.; Rene, R. R.; Lens, P. N. L. Application of Nanomaterials in Food, Cosmetics and Other Process Industries. In *Nanotoxicology: Toxicity Evaluation, Risk Assessment and Management*; Vol. 1. Kumar, V., Dasgupta, N., Ranjan, S., Eds.; CRC Press Taylor & Francis Group, 2018, pp 63–79.

Xiao, B.; Yang, R.; Chen, P.; Yang, J.; Sun, B.; Wang, K.; Zhang, T.; Zhu, L. Insights into the Lower Trophic Transfer of Silver Ions than Silver Containing Nanoparticles along an Aquatic Food Chain. *Sci. Total Environ.* **2022,** *804,* 150228. DOI: 10.1016/j.scitotenv.2021.150228.

Xu, C.; Nasrollahzadeh, M.; Sajjadi, M.; Maham, M.; Luque, R.; Puente-Santiago, A. R. Benign-by-Design Nature-Inspired Nanosystems in Biofuels Production and Catalytic Applications. *Renew. Sustain. Energy Rev.* **2019,** *112,* 195–252. DOI: 10.1016/j.rser.2019.03.062.

Xu, J.; Li, H.; Lowry, G. V. Sulfidized Nanoscale Zero-Valent Iron: Tuning the Properties of This Complex Material for Efficient Groundwater Remediation. *Acc. Mater. Res.* **2021,** *2* (6), 420–431. DOI: 10.1021/accountsmr.1c00037.

Yamada, D.; Hiwatari, M.; Hangoma, P.; Narita, D.; Mphuka, C.; Chitah, B.; Yabe, J.; Nakayama, S. M. M.; Nakata, H.; Choongo, K.; Ishizuka, M. Assessing the Population-Wide Exposure to Lead Pollution in Kabwe, Zambia: An Econometric Estimation Based on Survey Data. *Sci. Rep.* **2020,** *10* (1), 15092. DOI: 10.1038/s41598-020-71998-5.

Yang, J.; Huang, Z.; Yang, B.; Lin, H.; Qin, L.; Nie, M.; Li, Q. Green Route to Prussian Blue Nanoparticles with High Degradation Efficiency of RhB under Visible Light. *J. Mater. Sci.* **2021,** *56* (4), 3268–3279. DOI: 10.1007/s10853-020-05406-8.

Yasin, S. A.; Abbas, J. A.; Ali, M. M.; Saeed, I. A.; Ahmed, I. H. Methylene Blue Photocatalytic Degradation by TiO2 Nanoparticles Supported on PET Nanofibres. *Mater. Today Proc.* **2020,** *20,* 482–487. DOI: 10.1016/j.matpr.2019.09.174.

Yin, W.; Liu, M.; Zhao, T. L.; Qian, F. J.; Li, H.; Yao, Q. Z.; Fu, S. Q.; Zhou, G. T. Removal and Recovery of Silver Nanoparticles by Hierarchical Mesoporous Calcite: Performance, Mechanism, and Sustainable Application. *Environ. Res.* **2020,** *187,* 109699. DOI: 10.1016/j.envres.2020.109699.

Yuan, T.; Chen, Q.; Shen, X. Adsorption of Cesium Using Mesoporous Silica Gel Evenly Doped by Prussian Blue Nanoparticles. *Chin. Chem. Lett.* **2020,** *31* (10), 2835–2838. DOI: 10.1016/j.cclet.2020.03.035.

Zafar, A. M.; Javed, M. A.; Hassan, A. A.; Mohamed, M. M. Groundwater Remediation Using Zero-Valent Iron Nanoparticles (nZVI). *Groundwater Sustain. Dev.* **2021,** *15*, 100694. DOI: 10.1016/j.gsd.2021.100694.

Zahedifar, M.; Seyedi, N.; Salajeghe, M.; Shafiei, S. Nanomagnetic Biochar Dots Coated Silver NPs (BCDs-Ag/MNPs): A Highly Efficient Catalyst for Reduction of Organic Dyes. *Mater. Chem. Phys.* **2020,** *246*, 122789. DOI: 10.1016/j.matchemphys.2020.122789.

Zahra, Z.; Habib, Z.; Chung, S.; Badshah, M. A. Exposure Route of TiO2 NPs from Industrial Applications to Wastewater Treatment and Their Impacts on the Agro-environment. *Nanomaterials (Basel)* **2020,** *10* (8), 1469. DOI: 10.3390/nano10081469.

Zeng, H.; Zhai, L.; Qiao, T.; Yu, Y.; Zhang, J.; Li, D. Efficient Removal of As(V) from Aqueous Media by Magnetic Nanoparticles Prepared with Iron-Containing Water Treatment Residuals. *Sci. Rep.* **2020,** *10* (1), 9335. DOI: 10.1038/s41598-020-65840-1.

Zhang, L. S.; Fang, Y.; Zhou, Y.; Ye, B. C. Improvement of the Stabilization and Activity of Protocatechuate 3,4-Dioxygenase Isolated from *Rhizobium* sp. LMB-1 and Immobilized on Fe3O4 Nanoparticles. *Appl. Biochem. Biotechnol.* **2017a,** *183* (3), 1035–1048. DOI: 10.1007/s12010-017-2481-9.

Zhang, Q.; Dong, R.; Wu, Y.; Gao, W.; He, Z.; Ren, B. Light-Driven Au-WO3@C Janus Micromotors for Rapid Photodegradation of Dye Pollutants. *ACS Appl. Mater. Interfaces* **2017b,** *9* (5), 4674–4683. DOI: 10.1021/acsami.6b12081.

Zhang, H.; Xu, F.; Xue, J.; Chen, S.; Wang, J.; Yang, Y. Enhanced Removal of Heavy Metal Ions from Aqueous Solution Using Manganese Dioxide-Loaded Biochar: Behavior and Mechanism. *Sci. Rep.* **2020,** *10* (1), 6067. DOI: 10.1038/s41598-020-63000-z.

Zheng, H.; Zhang, C.; Liu, B.; Liu, G.; Zhao, M.; Xu, G.; Luo, X.; Li, F.; Xing, B. Biochar for Water and Soil Remediation: Production, Characterization, and Application. In *A New Paradigm for Environmental Chemistry and Toxicology*; Jiang, G., Li, X., Eds.; Springer: Singapore, 2020; pp 153–196.

CHAPTER 6

Nanoremediation of the Environment: Nanoparticles' Characteristics, Advantages, and Applications

YASHODHARA GOSWAMI and ISHANI CHAKRABARTTY

Department of Applied Biology, School of Biological Sciences, University of Science and Technology Meghalaya (USTM), Ri-Bhoi, Baridua, Meghalaya, India

ABSTRACT

Excellent features of nanoparticles have been brought about their fruitful consequences in upgrading human lifestyle by intriguing the concept of environmental remediation with the help of nanotechnology. High surface area, magnetic properties, optical activity, sensitivity, adsorption capacity, shape, catalytic behavior, low solubility, recyclability, and many more—these all characteristics of nanoparticles are interrelated and altogether contribute to their environmental and other scientific applications. In this book chapter, the influence of various advantageous properties in bioremediation and simple remediation of soil, water, and gas phases are briefly discussed, followed by the explanation of main categories of nanoparticles used in remediation processes and how the unique features imply their applications.

6.1 INTRODUCTION

Nanoparticles—the name itself depicts the significance of those materials which are simply dwarf in size, ranging from 1–100 nm in diameter. The term "nano" was coined by Norio Taniguchi in 1974 (Wang et al., 2020). They

Sustainable Nanoremediation: Modern Technologies for a Clean Environment.
Fernanda Maria Policarpo Tonelli, Arpita Roy, and Khalid Rehman Hakeem (Eds.)

show several unique characteristics in contrast to the large-sized macroparticles, which become advantageous to varieties of scientific research areas, including agricultural and environmental science, biomedical, chemical, and pharmaceutical industries, etc. For crystalline particles, periodic boundary conditions get damaged with the approach of the size of the particles toward the nanoscale. Again, when the size becomes closer to or smaller than the wavelength of light (or else, de Broglie wavelength), atomic density changes on the surface of amorphous particles (Guo et al., 2014). Nanoparticles can be categorized as zero-dimensional, which differs from the ones having one or two dimensions in macroscale. The zero-dimensional nanostructures contain higher quality properties such as better conductivity, optical stability, and special optical properties (Wang et al., 2020), cellular permeability, catalytic activity, chemical inertness, and wide range of adaptability and biocompatibility, etc. Therefore, they are valued as the harbinger of nanotechnology and its applications in sustainable as well as technical areas of science.

An environment is the surroundings of an individual including all the living and non-living components and the circumstances created by them. In the present moment, a number of contaminants have been entering and altering the conditions in an adverse manner, leading to a major problem of society called environmental pollution. Scientists and researchers have explored and implemented many novel technologies to remove those contaminants and purify the crucial components of the environment—air, water, and soil, in turn remediating the biosphere. When different technologies are used for the degradation of contaminants, it is important to note that they are able to overcome the challenges such as high volatility and complexity of the mixture of contaminant particles, as well as the employed materials are themselves non-pollutant and harmless for the environment too, so that they can offer eco-friendly, safer and more efficient solutions for the problem of environmental pollution. Adsorption, filtration, photocatalysis, and chemical reactions are some of the basic techniques utilized for the remediation of environmental pollution (Guerra et al., 2018).

Nanoparticles and nanoscale technologies have gained a lot of attention in the recent years in different areas of applied science which also include environmental science. The new remediation technologies focus on properties of target-specific capture, recyclability, cost-effectiveness, non-toxicity, etc. Since the nanoparticles possess a number of characteristics advantageous to environmental remediation, therefore the concept of nanoremediation has been developed intensively. Nanoremediation is the treatment process of contaminated water (basically groundwater), soil, or atmosphere using nanoparticles, to restore land and water resources. The basic reactions involved in nanoremediation are redox reactions, sorption processes, and some combinations of the mentioned chemical mechanisms (Koschitzky et al., 2017). Nanoremediation has become an emerging concern as the need for immediate treatment of hazardous contamination in landfills, industrial and manufacturing sites has arisen since it started affecting well-being of human health and civilization. Nanoremediation is an economically and ecologically sustainable method that allows in situ, less chemical-based environmental purification, comparatively faster in rate. Nanoremediation may further have a great impact on ecosafety, industrial competitiveness, and innovations of nanostructured materials (Corsi et al., 2018).

6.2 ADVANTAGEOUS CHARACTERISTICS TO NANOREMEDIATION

Nanoremediation is based on a number of advantageous characteristics of nanoparticles which can be briefly discussed in this content. Some general properties of nanoparticles can be commonly advantageous to various fields of applications (Sajid and Wasylka, 2020).

The size of nanoparticles has a great impact on their target applications. The optical properties of nanoparticles are directly affected by the size which is resulted by synthetic and reaction parameters. Gold nanoparticles of less than 10 nm size are evaluated for their optical properties. Again, silver nanoparticles and iron oxide nanoparticles have shown effects on bandgap and bacterial strains depending on their size (Sajid and Wasylka, 2020). The shape of nanomaterials is a major factor that affects their applications in different fields. It has a great impact on photocatalytic activity. The structure is particularly related to the fabrication method. Selenium nanoparticles are found in nanowires, nanotubes, nanospheres and many other shapes.

Nanoparticles possess high surface areas. This specific characteristic results in better catalytic activity, adsorption capacity, and chemical reactivity of nanoparticles. High surface area nanoparticles exhibit reasonably high adsorption capacities and excellence in photocatalytic within a waterbody. However, the ecotoxicity of several nanoparticles, such as of Silicon dioxide to the green alga *Pseudokirchneriella subcapitata* is dependent on the surface area of the nanomaterial. The surface area and shape of nanoparticles also determine the electrocatalytic activity of various electrochemical reactions (Sajid and Wasylka, 2020).

Magnetic properties should be distinctively specific for every type of application, as they are widely used in an extensive range of industries such as for therapeutic agents, materials for data storage, and removal of environmental pollutants. While the stability of the magnetic behavior of nanoparticles is crucial for biomedical applications, it is essential for a magnetic nanoparticle to be easily separable with the help of an external magnetic field so that it can be implemented as a great sorbent in wastewater treatment. Agglomeration may cause loss of magnetic activity and surface coating can be a worthy solution to this problem. Amalgamation of magnetic characteristics along with Surface Enhanced Raman Spectroscopy, provides high enrichment factor and enhancement of sensitivity toward analytes. Nanoparticles can be converted to superparamagnetic forms which may show zero remanence in external magnetic fields (Sajid and Wasylka, 2020; Guo et al., 2014).

Optical characteristics are in direct charge of aggregation and size of nanoparticles. Their optical properties are due to the surface plasmon resonance phenomenon. Gold nanoparticles are exploited to manufacture colorimetric sensors. Particle size and interparticle spacing are the two factors that are deployed in this context as the maximum absorption and emission wavelengths vary with them, ranging from infrared (IR) to ultraviolet (UV) region (Sajid and Wasylka, 2020).

Sensitivity and catalytic behavior are two other characteristics having significance in nanoremediation processes. The random distribution of active sites in high surface area in turn results in greater catalysis and adsorption. Nanofilters are the membranes that can selectively separate components of water effluents. Most importantly, Recyclability, a characteristic of paramount importance for the application of anything in environmental remediation and sustainable development, can be a pillar factor for the use of nanoparticles in nanoremediation processes. The hydrophobic nature of fullerene and the ferromagnetic behavior of iron nanoparticles lead to their potential in recyclability and adsorption properties. Some metal-based nanoparticles have Low solubility, leading to less or zero secondary pollution and adverse effects on the environment. High surface area to volume ratio of nanoparticles is the key factor of their Sensitive detectability and Specific affinity which empowers their application in heavy metal and water remediation prospects (Sajid and Wasylka, 2020). A detailed understanding of mechanical properties and applications of some classic theories are undoubtedly crucial for gaining knowledge about the importance of a handful of characteristics (elastic modulus, friction, movement law, and many more) in the field of environmental science (Guo et al., 2014).

Nanobioremediation is the technique of removal of pollutants from soil, water, and other incorporated natural toxins, by increasing microbial activities with the help of nanoparticles that are the results of plant, fungal, or microbial activities, along with the nanotechnological analysis. Enzyme-mediated nanoremediation is very much preferable since the nanoparticles make provision for an inert but biologically active and compatible micro-environment, with the least interference. Nonfertilizer, green nanooxidizers, and nanominerals are the newly explored and developed terms that will have greater significance along with the exploitation of other nanoparticles in the field of nanoremediation, in upcoming days (Benjamin et al., 2019).

6.3 NANOREMEDIATION OF ENVIRONMENT BY UTILISING ADVANTAGEOUS CHARACTERISTICS OF NANOPARTICLES

Different classes of nanoparticles are used for the remediation of environmental components such as air, soil, and water. Their ability to diffuse in small spaces, the interaction of nanofilter membranes with water effluents, and selective permeability influence the applications of nanotechnology for environmental remediation. In the nanoremediation process, nanoparticles can perform distinct types of functions—one group can act as an electron donor (for less toxic contaminants), while the other group can be used as a sorbent (for more toxic fixations). Again, a few nanoparticles are utilized as precipitant or co-precipitant. Chemical sequestration may also be done in two ways—either by the encapsulation of toxic materials in interfaces of clusters of nanoparticles, or by the fixation on the surface (Prado-Audelo et al., 2021) (Fig. 6.1)

6.4 NANOREMEDIATION OF AIR

The presence of several physical, chemical, and biological molecules or particulates in the atmosphere can alter natural characteristics and have a direct impact on human health and the climate of the surroundings. Nanotechnology can be used to prevent air pollution, to detect and sense the pollution factors and parameters as well as in remediation and control treatment. The major advantageous properties of various nanoparticles employed to overcome the problem of greenhouse gases with the help of nanoremediation technique are such as ordered and tunable porous structure, large pore

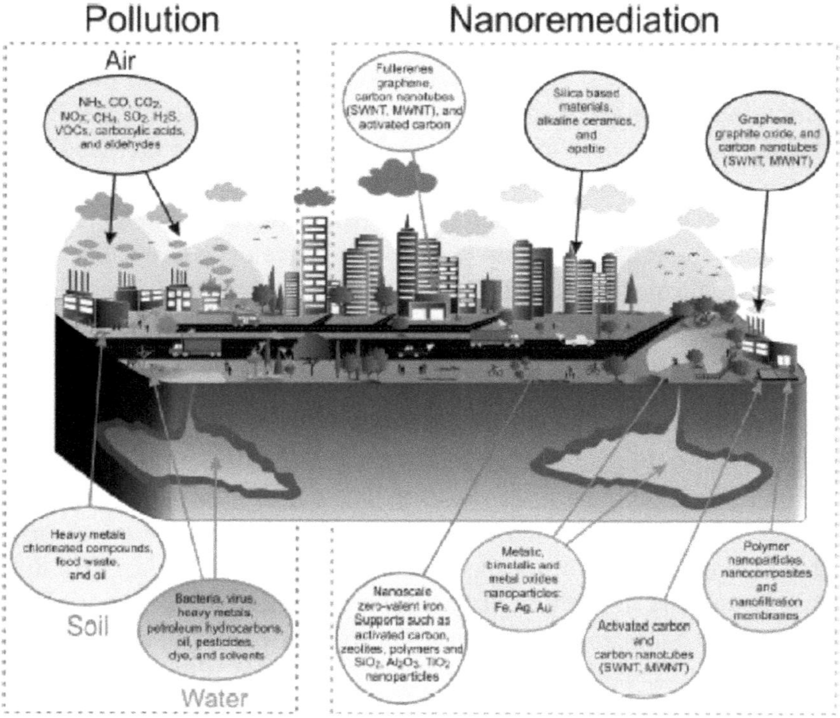

FIGURE 6.1 Application of different nanomaterials in remediation of the environment.

Source: Reprinted with permission from Del Prado-Audelo et al., 2021). https://creativecommons.org/licenses/by/4.0/

volume, high surface area, highly reactive sites on the surface, thermal stability, etc. (Prado-Audelo et al., 2021).

6.5 RECENT ADVANCEMENT IN SOIL AND GROUNDWATER NANOREMEDIATION

Based on the place where remediation takes place, the technique may be classified as either in situ or ex situ remediation. The main aim of soil and groundwater remediation is to reduce pollution up to a safe and recommendable extent by a chemical, physical, or biological manner. At the molecular level, the mechanisms involved in nanoremediation are chemical reduction, adsorption, and catalysis. Many engineered nanoparticles possess innovative surface coating, extensive site cleaning, diffusion, and penetration capability into tiny subsurface spaces, and larger spatial distribution, which make in situ nanoremediation preferable over the ex situ techniques. Due to their small size and quantified density, the gravity sedimentation factor will not affect their movement. Their physical movement in groundwater is mostly affected by Brownian movement. Therefore, direct injection treatment is functional for subsurface-level nanoremediation (Alazaiza et al., 2021).

6.6 NANOREMEDIATION OF SOIL

The basic causes of soil pollution include erroneous waste management, oil spills, industrial activities, etc. Expanding industrialization and application of pesticides, PAHs, and herbicides, along with mining, and road transport also exert influence on soil toxicity or sediment contamination. Engineered nanoscale materials are potentially effective in transforming complex contaminants into simpler, harmless compounds such as degradation of chlorinated organic pollutants, and the conversion of heavy metals to less toxic forms. They are applied as fruitful catalytic agents for in situ cleaning. Nanotechnology-based remediation works by reducing metalloid toxicity (of soil impurities) using the principle of immobilization. Nanoremediation techniques also focus on immediate practical solutions and long-term anticipation (Corsi et al., 2018). Characteristics such as high reactivity with great adsorption capacity, high surface area to volume ratio, surface functionalization, physical property modifications (morphology, porosity, and size), selectivity, efficiency, and economic feasibility are some of the explored qualities important for soil remediating agents (Prado-Audelo et al., 2021). With significantly fewer requirements and higher nutritional effectiveness, nanofertilizers play a major role in the soil remediation process in nanotechnological manner. They amplify soil aggregation, and moisture content retention. Encapsulation of nanofertilizers can be done in several ways, such as (a) polymer film coating; (b) encapsulation within a nanoporous capsule; (c) delivery to the field as nanoscale emulsions or particles, etc. The efficiency of nanoparticles can be improved by adding nanoclays and nanoporous zeolite to the list. This nano-based fertilizer is superior to conventional fertilizers since it prevents the leaching of excess nutrients.

6.7 NANOREMEDIATION OF WATER (WASTEWATER AND GROUND WATER)

Water sources get polluted by a number of emerging factors including pharmaceuticals, heavy metals and radioactive ions, toxic petroleum hydrocarbons, etc. Nanoremediation of water and wastewater is based on selectivity to chemical impurities and the adsorption capacity of several inorganic, polymeric, and carbon origin nanoparticles (Prado-Audelo et al., 2021). As a result of the eutrophication of seawater, the whole marine ecosystem faces dissolved oxygen depletion, resulting in water quality deterioration. Polypyrrole is an easily synthesizable, cheap, nontoxic, and highly stable, conducting polymer. When these beneficial properties are assisted with an antimicrobial capacity of zinc oxide and oxidant $AgNO_3$ nanoparticles, forming a highly efficient nanocomposite, it gives additive consequences in the field of wastewater bioremediation. The nanocomposite fabrication with a novel heating technique focuses on advantages such as homogenicity in thermal transmission, shorter reaction period, repeatability, purity, and environmental friendliness. This process is dependent on synergistic interaction for the bioremediation of wastewater in both single and double pollution systems and acts as an antimicrobial agent against several fungi and Gram-positive and negative bacteria (Mohamed et al., 2021). Nanoparticles of Fe_2O_3, Al_2O_3, and SiO_2 for a number of anions have strong affinity and therefore are used in water remediation in the global platform. For the water purification processes, mostly nanoparticles considered to be benign are utilized. Not only the high surface area is necessary for the adsorption of impurities but also the understanding of the transforming process of nanoparticles during and after use is important (Hamers, 2017) (Fig. 6.2).

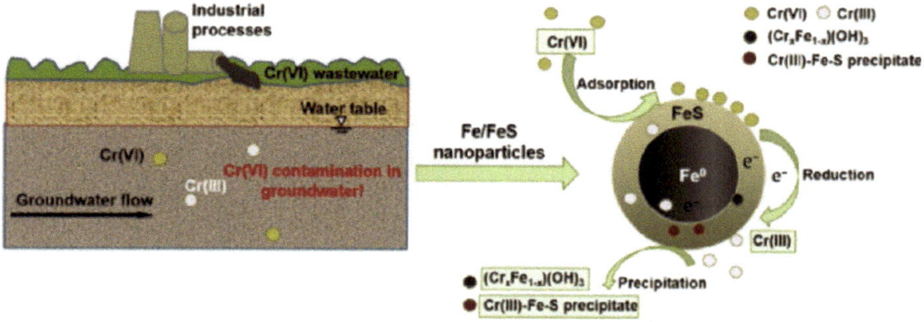

FIGURE 6.2 Groundwater nanoremediation using Fe/FeS nanomaterials.

Source: Reprinted with permission from Alazaiza et al., 2021. https://creativecommons.org/licenses/by/4.0/

Nanofiltered membrane is a technique widely used to purify huge amounts of water all over the world. To filter out pesticide transformation products

(PTP) from contaminated groundwater, this nanomembrane structure is utilized. Due to the smaller size of particles, low-pressure reverse osmosis can be applied for further effective removal of pesticide-related products (Benjamin et al., 2019).

6.8 NANOPARTICLE CHARACTERISTICS: APPLICATIONS IN THE FIELD OF IMMOBILIZATION AND BIOREMEDIATION

Various nanoparticles can facilitate the bioremediation of hydrocarbons with the help of oxidation. This is dependent on the reduction of toxicity and enhanced microbial growth as the result of their high adsorption property. Nanocalcium peroxide is an excellent oxygen-releasing material that intensifies the dissolution rate of hydrocarbons. Organophilic clay formation by the replacement of inorganic cations is possible with the use of surfactants such as quaternary ammonium cations that amplify the hydrophobic properties of the clay. Zinc oxide and Fe_2O_3 nanoparticles are beneficial for the adsorption of non-biodegradable, hydrophobic components of crude oil (Kumari and Singh, 2016). Titanium oxide nanoparticles in the presence of UV light, can easily degrade up to 78% organic pollutants from petroleum wastewater, even at a weak concentration of the photocatalyst. Cobalt nanoparticles and manganese nanoparticles encourage the reduction of PAHs. The presence of nanoiron particles at specific concentrations can influence their oxidation, energy which is essential for the growth of some microorganisms, applied for the decrease of duration in the lag phase and increase in the duration of stationary and log phase. Magnetic nanoparticles are the most preferable materials for the immobilization of enzymes due to their high surface area, biocompatibility, low mass transfer resistance, efficient reaction with substrates, higher binding capacity, easy separation ability, etc. Nanoparticles in combination with Chitosan will have high mechanical strength, lower vulnerability to chemical changes, film forming capacity, resulting it to be an effective immobilization matrix for enzymes. The enzyme immobilization helps in dealing with their kinetic limitations based on stability, activity, specificity, selectivity, purity, reuse competence, etc. Zhubanova and Mansurov (2008) designed biopreparation by immobilizing the microbial cells on nanoparticles. They utilized carbonized rice husk nanoparticles for crude oil degradation in soil. Hybrid nanoparticles lead to the removal of hydrophobic hydrocarbons, acting as the sequestering agents from crude oil.

With the help of processes such as adsorption, absorption, filtration, and chemical reactions, varieties of inorganic, carbon-based, and polymer-type nanoparticles follow the basic mechanism of nanoremediation. Based on these classes, their characteristics will be different which can be the basis of their applications in nanoremediation techniques. The use of biodegradable nanoparticles gives the outcome of zero waste production. Off-targeting allows low efficiency. Moreover, a decrease in monodispersity, stability, and enhancement in cluster formation are some of the challenges that can be cured by nanoremediation.

6.9 INORGANIC NANOPARTICLES IN ENVIRONMENTAL REMEDIATION

Metal and metal oxide nanoparticles exhibit high flexibility, fast kinetics, and high adsorption capacity. Therefore they are applied for in situ as well as ex situ wastewater treatments and heavy metal and chlorine-based chemical nanoremediation in aqueous systems (Guerra et al., 2018).

Titanium oxides show characteristics such as semiconducting nature, electronic and gas sensing, photocatalytic (activation by light), and energy conversion behavior, which are extensively useful for water remediation and removal of contaminants from various media. Titanium oxides produce highly reactive hydroxyl radicals, acting as adequate disinfectants. Moreover, titanate nanotubes are formulated with basic, acidic, and neutral pH media and loaded with manganese oxides to yield Mn-doped nanosheets, nanorods, nanotubes, etc. Neutrally fabricated manganese nanotubes are the best in catalytic reduction. Mixed oxide nanomaterials of titanium and silica origin have a greater ability for the mineralization of a wide variety of pollutants. Titanium oxides are doped with other transition metal ions, such as silver, to enhance the photocatalytic degradation performance. The use of more than one transition metal also was performed in the development of stable and monodispersed lecithin-nano-Ni/Fe that are highly efficient in the rapid removal of PCB77 as a target contaminant. In the case of bimetallic nanoparticles, the inclusion of a second metal can prevent agglomeration and enhance the solution stability, which in turn can speed up the contaminant degradation rate by expanding the capacity and efficiency (Guerra et al., 2018).

Nanozero-valent iron (nZVI) shows remarkable features such as high mobility, high reactivity, low reduction potential, and flexibility; therefore, reducing the pollutants by getting oxidized from Fe^0 to Fe^{2+} and then to Fe^{3+} (Ganie et al., 2021). Through redox transformation strategies sequestration, nZVI shows operative remediation of chlorine-based pesticides, polychlorinated biphenyls, DDT, nitrate, lead, cadmium, etc. Other noble metals are mixed with this particular class of nanomaterials to store their function. Providing ZVI nanoparticles along with acids influences the lessening of environmental risks by chromium and other transition metals. The efficiency is higher for nZVI in the case of water, compared to soil remediation. Zhu et al. used green tea to synthesize nZVI/Cu for groundwater nanoremediation. Numerous studies and experiments suggested that zeolite-supported nZVI is beneficial for the long-term remediation of metal-contaminated soil. Using nZVI combined with CNTs resulted in high removal efficiency for cobalt and selenium metal pollutants. Biochar-nZVI promises effective remediation of halogenated polluted groundwater (Prado-Audelo et al., 2021).

ZVI allows two methods for the remediation of chlorinated contaminants – (a) hydrogenolysis, replacing chlorine atoms by hydrogen atoms; (b) dehalogenation, forming new carbon–carbon bonds without using hydrogen atoms. In the soil remediation process by ZVI, the reaction of dehalogenation is preferable over hydrogenolysis. The chlorinated hydrocarbons may be qualitatively remediated using such nZVIs, in a low-cost and efficient manner (Ganie et al., 2021).

Nanocrystallites possessing super magnetic behavior can help in the decontamination related to organic toxic metals such as lead, copper, and arsenic. The nanoadsorbents follow second-order kinetics and showcase recyclability with 96.2% desorption capacity, depend on temperature, pH, and other physical and chemical factors.

Silica-based nanoparticles are composed of adjustable pore size, easily adjusting surfaces, and various other versatile features. Recent advancements have proved that silica nanoparticles can easily purify the environmental air from lead contaminants. Amino-modified silicates are able to diminish the climate change phenomenon, deleting organic ketone and aldehyde-related contaminants from the air (Prado-Audelo et al., 2021). Formation of an imine is required in the capture of these classes of organic materials. Carboxylic acid can be bound to dyes, pollutants, and metal ions by hydrogen bonds and this is the basis of the functionalization of mesoporous silica with the carboxyl group. The materials sometimes become good adsorbents under specific basic pH conditions (Guerra et al., 2018).

6.10 CARBON-BASED NANOPARTICLES IN NANOREMEDIATION

6.10.1 NANOREMEDIATION APPLICATIONS OF CARBON NANOTUBES

Carbon nanotubes (CNTs) are outstanding adsorbing nanomaterials for numerous heavy metals including chromium, cadmium, cobalt, lead, mercury, nickel, and many more, from aqueous systems. The regeneration capacity of the composite nanoparticle of multiwalled carbon nanotube (MWCNT), iron oxides, and β-cyclodextrin also supports cost-effective applications in wastewater remediation. Modifications in CNTs are done by coating with appropriately charged functional groups containing reactive materials to magnify their adsorption capacity. The exposure area and functional groups on the surface determine the adsorption capacity. In the atmosphere, they are implemented for the detection of gases such as H_2S, SO_2, VOCs, CO, CO_2, NO_x, etc. CNTs can be incorporated into the remediation of hexachlorocyclohexane and other hydrocarbons, intensification of microbial and plant growth in response to the aqueous dispersion, membrane filtration, etc. MWCNTs are deployed for the adsorption of hexavalent chromium, and gasoline removal in wastewater and groundwater treatments (Benjamin et al., 2019). Both single-walled carbon nanotubes (SWCNTs) and MWCNATs are of paramount importance for environmental nanoremediation. The feature of high microporosity of CNTs conveys additional advantages to nanoremediation processes (Guerra et al., 2018).

SWCNTs generally have hexagonal configuration, attaining a heterogeneous porous structure, that shows better adsorption capacity. The adsorption may take place either on the external surface or in between two neighboring tubes/inside a single tube, allowing higher adsorption energy usage. Oxygen content also regulates the characteristic adsorption of CNTs. The physical and chemical properties of adsorbate material such as the electric charge on it, dipole moment, and molecular weight also stimulate the activity of CNTs (Guerra et al., 2018). Other two factors influencing the activity of CNTs are sediment–sorbent contact period and dosage. MWCNTs are responsible for magnifying microbial growth in crude oil-contaminated soil, leading to faster and efficient degradation of hydrocarbons, and remediation of DDT and HCH from sediments (Alazaiza et al., 2021).

6.11 NANOREMEDIATION BY GRAPHENE MATERIALS

Varieties of applications of graphene nanomaterials in the aspect of environmental remediation are based on a few effective characteristics. Pristine graphene is used to eliminate fluoride from water sources by using efficient adsorption capacity. An important thing to note is that the surface modifications of graphene particles alter aggregation of the layers as well as potent surface area gets increased. Oxygen gets attached in the form of its functional groups to the carbon surface of graphene in the modified graphene oxides (GOs). GOs are functionally more attracted to the cationic adsorption. Metal oxides or organic groups may be the factors influencing the deletion of anionic metals (Guerra et al., 2018).

At an optimum graphene concentration in titanium oxide–graphene composites, photocatalytic activity is maximum for the nanoparticle. In the case of benzene remediation, it was found to be the composition of 0.5% graphene ratio by weight. Zinc oxide–graphene was synthesized to perform photocatalytic reduction under UV radiation. The primary photoactive component of cadmium sulfide nanoparticles-(CdS NPs)-graphene nanocomposite (CdS-graphene) is CdS and its lower ratio along with the low intensity of light can directly impact the photocatalysis of this nanocomposite. The addition of other metals and active compounds to the graphene widens up the range of impurities that can be removed by the nanoparticles in the environmental remediation procedure (Guerra et al., 2018).

6.12 ROLE OF ENGINEERED POLYMERIC NANOPARTICLES IN NANOREMEDIATION

To remediate silver ions from polluted water and improvisation of PHEN released from contaminated aqueous sources are the main objectives of using engineered nanopolyners in environmental remediation. They amplify the functional solubility of the organic pollutants (Benjamin et al., 2019).

These are the materials that use a host material as a matrix for the other nanoparticles to be attached to the entire polymer. These are more thermally stable than the pure or composite metal and carbon-based nanoparticles. Also, they have higher durability and mechanical strength. They can detect and eliminate a wide array of contaminants such as pharmaceuticals, organics, and chemicals too (Guerra et al., 2018).

Dendrimers remediate wastewater samples. They carry carboxylate, primary amine groups. They act as chelating agents. They are able to bind with and encapsulate cationic metal ions such as Cu^{2+}, Fe^{2+}, Zn^{2+}, U^{6+}, etc., and play a major role in wastewater treatment. The dendritic nanopolymers have a globular shape and lower polydispersity; therefore, they are bound to contaminants and make them pass through nanofilters. These polymers are recycled after the recovery of pure water and their low passing tendency through the pores helps in these applications up to a large extent (Guerra et al., 2018). The use of titanium oxide porous ceramic filters with immobilized pores gives rise to higher mechanical strength and high surface area containing hybrid filters (Benjamin et al., 2019).

Some specially designed nanopolymers exhibit rapid, target-specific capture of gas pollutants. Some robust capturing technology uses the basis of physical bond formation between the targeted group and the amine moiety (Guerra et al., 2018).

Following several mechanisms, polymeric nanocomposites remove impurities, depending on two key factors—biodegradability and biocompatibility. The presence of magnetic hydrogel in a hybrid green nanocomposite of wheat xylan can eliminate methylene blue dye up to 90%. Silver nanoparticles when embedded in cellulose acetate fibers, the hybrid possess antibacterial activity. SiO_2 in acrylamide hydrogel in combination with grafted copolymerized gum karaya helps in methylene blue dye removal from wastewater. Not only the size, but also the shape of silver nanoparticles, accompanied metal oxides or polymeric nanoparticles, affect the overall functionality of the nanocomposite in the field of nanoremediation (Guerra et al., 2018).

6.13 NANOREMEDIATION: DRAWBACKS AND SOLUTIONS

Ecological implications are produced by some nanoparticles, but exploitation of alternative green nanoparticles from plants may reduce the resultant toxicity. Nanofactories (fungi), and other bionanoparticles are safer alternatives (Sharma, 2019). Synthesis of well-defined greener, monodispersed bionanoparticles uses fungi including, *Aspergillus fumigates*, *fusarium semitectum*, *fusarium oxysporum*, etc. (Patil et al., 2016).

High concentrations of nZVI can form reactive oxygen species, in turn damaging the cell membrane, decreasing transpiration and translocation rates, genotoxicity, oxidative stress, pulmonary diseases, and ultimately leading to the death of organisms. The solution to all these drawbacks is achieved by encapsulating the iron nanoparticles in an oil membrane that is biodegradable in nature, known as emulsified zero-valent iron (E-ZVI) (Sharma, 2019). These membrane-bound nanoparticles can soak up insoluble impurities. The toxic contaminants get broken down into simpler, harmless forms. This sustainable approach also protects the nZVI from groundwater components, which may deteriorate the target-specific reduction potential of the iron particles. E-ZVIs form micelles with dense nonaqueous phase liquids (DNAPLs), which are generally heavier than water and slightly soluble in water, in situ, by the help of their unique miscible behavior. The basic mechanism involved here is abiotic reductive dechlorination. Within a short time period, the US have been successfully remediated halogenated volatile compounds from contaminated groundwater and soil (Patil et al., 2016).

6.14 CONCLUSION

To clean up the environmental components such as air, water, and soil in a time-effective, cost-effective manner, without the production of harmful reaction intermediates, the development and utilization of nanotechnology is the best option during the recent time period. The advantageous characteristics such as surface structures, chemical stability, adsorption capacity, robustness, etc. can be easily manipulated and improvised according to our own needs which become beneficial to the global sustainability (Ganie et al., 2021).

Nanoremediation has become an indispensable part of human welfare and commercial processes. Since they are very much susceptible to the manipulations; therefore, some rules should be designed for their molecular-level incorporations and proper understanding of how they interact with our surroundings (Hamers, 2017).

We have got to know about the prospective role of nanoparticles and their basic physicochemical features in the advancement of nanoremediation processes in combination with a few other technologies. A huge number of challenges have been overcome by the availability of nanotechnology, but some are left to be solved. Even the drawbacks of natural and engineered nanoparticles have been resolved by the invention and designing of novel compositions of nanoscale materials. It has been proved that nanoparticles have brought a new era to a number of sub-fields of science and technology. Not only the environmental issues have been solved, but also human health, medical, energy, agriculture-related questions are answered in satisfactory manner with the help of nanotechnological inventions. We can hope for greater evolution and maturation of the nanoremediation concept with time which will ensure further sustainable consequences for our environment and human lifestyle in the global platform.

KEYWORDS

- **nanotechnology**
- **nanoparticles**
- **nanoremediation**
- **nanoparticles' advantageous features**
- **nanoparticles' applications**

REFERENCES

Alazaiza, M. Y.; Albahnasawi, A.; Ali, G. A.; Bashir, M. J.; Copty, N. K.; Amr, S. S.; et al. Recent Advances of Nanoremediation Technologies for Soil and Groundwater Remediation: A Review. *Water* **2021,** *13,* 2186.

Benjamin, S. R.; Lima, F. d.; Tramontina Florean, E. O.; Guedes, M. I. Current Trends in Nanotechnology for Bioremediation. *Environ. Pollut.* **2019,** *22.*

Corsi, I.; Nielsen, M. W.; Sethi, R.; Punta, C.; Torre, C. D.; Libralato, G.; et al. Ecofriendly nanotechnologies and Nanomaterials for Environmental Applications: Key Issue and Consensus Recommendations for Sustainable and Ecosafe Nanoremediation. *Ecotoxicol. Environ. Safety* **2018,** 237–244.

Del Prado-Audelo, M. L.; Karden, L. G.; Escutia-Guadarrama, L.; Reyna-Gonzalez, J. M.; Magana, J. J.; Leyva-Gomez, G. Nanoremediation: Nanomaterials and Naotechnologies for Environmental Cleanup. *Front. Environ. Sci.* **2021,** *9.*

Ganie, A. S.; Bano, S.; Khan, N.; Sultana, S.; Rehman, Z.; Rahman, M. M.; et al. Nanoremediation Technologies for Sustainable Remediation of Contaminated Environments: Recent Advances and Challenges. *Chemosphere* **2021,** *20.*

Guerra, F. D.; Attia, M. F.; Whitehead, D. C.; Alexis, F. Nanotechnology for Environmental Remediation: Materials and Applications. *Molecules* **2018,** *23.*

Guo, D.; Xie, G.; Luo, J. Mechanical Properties of Nanoparticles: Basics and Applications. *J. Phys. D: Appl. Phys.* **2014,** *25.*

Hamers, R. J. Nanomaterials and Global Sustainability. *Acc. Chem. Res.* **2017,** *5.*

Koschitzky, H.-P.; Gens, A.; Roos, J. Nano-Rem: Nanoremediation for Soil and Groundwater Clean-up—Possibilities and Future Trends. *Adv. Fut. Perspect. Nanoremediat.* **2017,** *10.*

Kumari, B.; Singh, D. P. A Review on Multifaceted Application of Nanoparticles in the Field of Bioremediation of Petroleum Hydrocarbons. *Ecol. Eng.* **2016,** *8.*

Mohamed, F.; Allah, A. E.; Abu Al-Ola, K. A.; Shaban, M. Design and Characterization of a Novel ZnO-Ag/Polypyrrole Core-Shell Nanocomposite for Water Bioremediation. *Nanomaterials* **2021,** *21.*

Patil, S. S.; Shedbalkar, U. U.; Truskewycz, A.; Chopade, B. A.; Ball, A. S. Nanoparticles for Environmental Clean-up: A Review of Potential Risks and Emerging Solutions. *Environ. Technol. Innov.* **2016,** *12.*

Prado-Audelo, M. L.; Kerdan, I. G.; Escutia-Guadarrama, L.; Reyna-Gonzalez, J. M.; Magana, J. J.; Leyva-Gomez, G. Nanoremediation: Nanomaterials and Nanotechnologies for Environmental Cleanup. *Front. Environ. Sci.* **2021,** *7.*

Sajid, M.; Wasylka, J. P. Nanoparticles: Synthesis, Characteristics, and Applications in Analytical and Other Sciences. *Microchem. J.* **2020,** *53.*

Sharma, D. Nanoremediation. *Int. J. Life Sci. Technol.* **2019,** *12* (1), 1–6.

Wang, Z.; Hu, T.; Liang, R.; Wei, M. Applications of Zero-Dimensional Nanomaterials in Biosensing. *Front. Chem.* **2020,** *8*, 19.

CHAPTER 7

A Brief History of the Green Synthesis of Nanomaterials

ATIF HUSAINA and MALIK NASIBULLAHA

[c]*Department of Chemistry, Integral University, Dasauli, Lucknow, India*

ABSTRACT

Over the last few decades, green and sustainable processes have undergone a rapid metamorphosis into key strategic methods to address the majority of the drawbacks of conventional synthesis. Continual progression in the field of green synthesis has contributed vastly to the development of greener pathways leading to the establishment of newer, efficient, cleaner pathways at no cost of environmental sustainability. To overcome these, a search for a reliant and environmentally friendly pathway by minimizing the use of potentially hazardous substances and adherence to "green synthesis", processes is the need of the hour. In establishing these pathways in the exploitation of environmentally friendly synthesis approaches, there is a brighter scope in the nurturance of remedial measures for the environment paving the way for nanoremediation. This necessity has become a driving factor in the creation of new research achievements in the same sector in creating a new, eco-friendly, and regenerative platform for both nanoremediation and green chemistry. Despite the tremendous development of this sector over the last 10 years, it is continuously growing with more effect. Because nanotechnology is multidisciplinary and interdisciplinary, the relationship between green chemistry and sustainability has shown to be a good element. Thus, the confluence of three major domains of research, namely Green Chemistry, Nanotechnology, and Sustainability, produces enhanced, unique, and much better necessary areas of study to cope with a decrease in environmental repercussions.

Sustainable Nanoremediation: Modern Technologies for a Clean Environment.
Fernanda Maria Policarpo Tonelli, Arpita Roy, and Khalid Rehman Hakeem (Eds.)

7.1 INTRODUCTION

A greener and more sustainable route has been pursued over the last few decades through ongoing efforts within greener creation of nanomaterials for nanoremediation. This has resulted in significant developments in the creation of innovative nanomaterials and their environmentally benign synthetic processes, offering a fresh, renewable avenue both for nanoremediation and green nanomaterial synthesis. The choice of nanoparticles is based on their unique qualities, such as adaptability, minuscule size, high surface-to-volume ratio, good magnetic characteristics, and, most importantly, high biocompatibility. The manufacturing of nanoparticles should be commercially viable because of their selective properties., proceeding through green pathways, environmentally sustainable with fewer or no by-products (Saif et al., 2016). This chapter will present a brief overview of the green synthesis of nanomaterials, and the strategic principles applied in the "Greener" synthesis of nanomaterials toward remedial measures. The present and future challenges persist in carrying out green synthesis of nanomaterials in exploring remediation measures of the same pertinent to green synthesis. The leading examples of green synthesis of nanomaterials in various applications such as environmental pollution are herein discussed.

In the present scenario of environmental crisis where the necessity of sustainability is the need of the hour, the greener pathways of nanomaterials synthesis are simple, cost-effective, and ultimately environmentally friendly. By following the green pathways, the parameter of efforts and environmental damage is significantly reduced.

7.2 A GLIMPSE INTO NANOMATERIALS AND NANOTECHNOLOGY

Nanoparticles are defined by their diameters, which are smaller than 100 nm. A nanometer is one billion a meter (10?^{9} m), or ten times the size of a solitary atom at the nanoscale. This tiny size is ultimately accountable for the favored microparticles' unique physical, chemical, and biological capabilities in presenting numerous phenomena. Because of their small size, predicting their characteristics and applications is more difficult than predicting those found at larger sizes (Eckelman et al., 2008).

Nanotechnology, on the contrary, gives an in-depth look into the depths and information of materials and gadgets that will touch numerous sectors and have an impact on many preceding domains. By using nanoscale characteristics as an adjustable physical variable, it is now feasible to significantly improve the performance of current materials and chemicals (Wang, 2010).

Thus, the creation of select nanomaterials has emerged as one of the most important and active fields in nanotechnology. In the current context, there are several ways and tactics created on a "greener" scale to synthesize nanomaterials with specific traits and qualities for new prospective uses. Despite significant progress in the same sector, there are still several tough issues in the synthesis of nanomaterials, including the fabrication of nanoparticles. Despite significant progress in the same sector, there are still numerous tough issues in the synthesis of nanomaterials, including the production of selected structures with desired and regulated features in large quantities for diverse industrial applications (Das and Marsili, 2009).

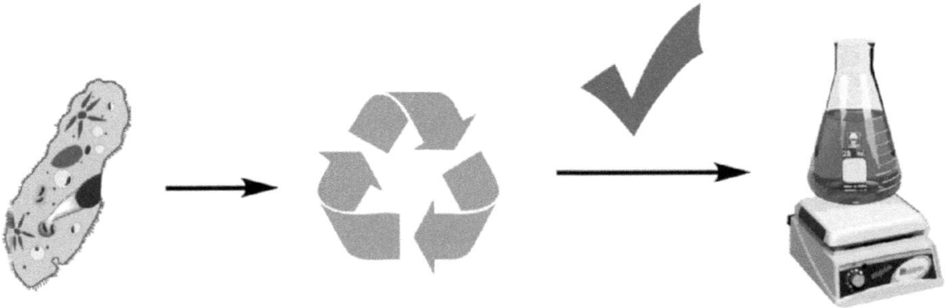

FIGURE 7.1 A typical process flow sheet for the "green synthesis" of nanoparticles from microbes.

As a result, throughout the last several decades, there has been a search for dependable, convenient, yet environmentally friendly synthesis procedures to produce newer and better nanomaterials while gradually reducing the usage of hazardous chemicals. The most effective technique to construct these "green" processes is to use nontoxic precursors and mild process conditions (Iravani, 2011).

7.3 A BRIEF ACCOUNT OF GREEN SYNTHESIS TOWARD SUSTAINABILITY

In the last two centuries, chemists and chemical engineers have aimed to continuously enhance chemical synthesis and transformation. Green chemistry offers a broad-spectrum framework for this goal. These green synthesis techniques are defined by 12 essential green chemistry concepts (Anstas and Warner, 1998). Only in the past 25 years has the term "green chemistry" been used, as well as its definition and guiding principles. The current green chemical movement was started in 1998 with address the intention of addressing the scientific difficulties associated with attempting to balance the effects on the environment and human health (Nasir Baig and Varma, 2013). The development of nanoparticles for environmental cleanup and other nano remedial techniques using green synthesis in nanotechnology is crucial.

This field of green nanotechnology encourages collaborations with other similar goal-oriented sciences creating a vision advancing sustainability fundamentally via scientific research. This can be done by making nano-manufacturing processes energy efficient. The fundamental ideas of green chemistry may be applied to nanotechnology through many means, such as (a) the use of affordable, nontoxic precursors; (b) the reduction in carcinogenic reagents and solvents; (c) the avoidance of potentially hazardous, acidic or basic analogs by using aqueous solvents; (d) the use of relatively few reagents to promote atom economy; and (e) the reduction of experiments involving pyrophoric chemicals or unstable precursors. (f) development of techniques to produce high-purity materials with few, if any, byproducts using high-yield methods; (g) atmospheric temperatures and pressure synthesis, if at all feasible, to avoid the requirement either for pressure or high-temperature processes (Binnig et al., 1983; Diallo and Brinker, 2011).

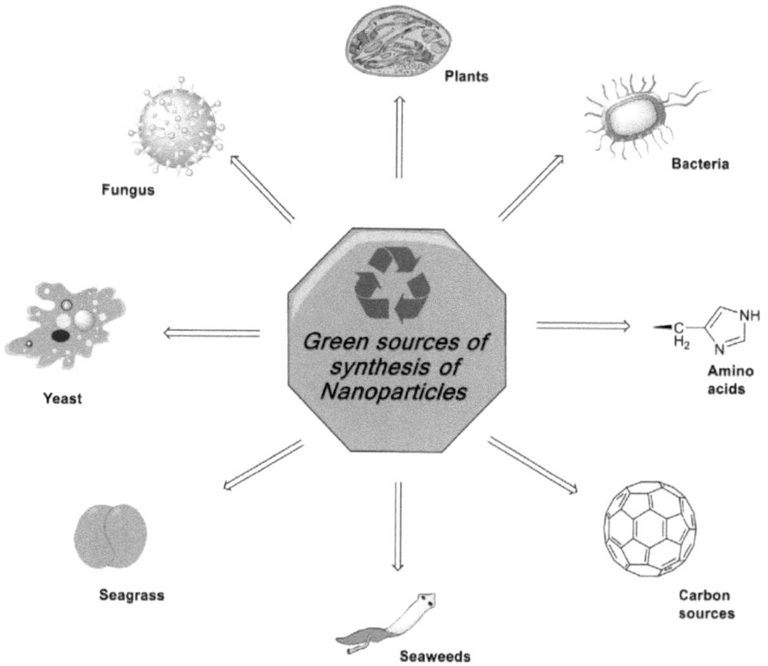

FIGURE 7.2 Some naturally occurring sources for the green synthesis of nanomaterials.

Recent research has revealed that throughout the last several decades, the discipline of green synthesis has placed a high priority on the synthesis, characterization, and use of nanomaterials in the pursuit of a sustainable future. Due to the widespread issues related to environmental deterioration, "greener" ecologically friendly procedures in chemical science and technology are becoming more and more popular and essential. In the modern world, eco-friendly solutions to these problems are becoming more and more common. By using a sustainable method, sustainability aims to provide an environmentally beneficial end result that is environmentally beneficial. Recent advances in green nanomaterial synthesis have been made to mitigate these limitations by supplying the environment with naturally biodegradable materials needed for their manufacture (Machado et al., 2013; Huang et al., 2014; Luo et al., 2016).

7.4 PARADIGM OF NANOREMEDIATION TOWARD GREEN CHEMISTRY

The successful employment of nanomaterials toward remedial measures or nanoremediation has seen a swift transition over the last few decades. The importance of green synthesis in the development of these selective structures cannot be overstated. The following are the main findings that may be condensed into quick recommendations for green nanomaterial synthesis: (1) Hazardous reagents and organic solvents must be avoided; hence, "greener" methods using safe lowering and capping agents are constantly being developed; (2) "green" synthesis generates few or no hazardous by-products; (3) "green" synthesis offers greater control over metal nanoparticle forms, sizes, and dispersivity; and (4) "green synthesis" does away with the requirement for purification of the nanoparticles as-prepared, which in turn removes the need for further processing (Jebali et al., 2011; Kirubaharan et al., 2012; Mandal et al., 2006).

Potential applications in Antimicrobial and environmental remediation have seen an abrupt increase over the past few years. Numerous studies have been conducted on the production of nanoparticles and their application in semiconductor technology due to their powerful photocatalytic activity when using semiconductor nanomaterials such as TiO_2 and ZnO (Armon et al., 2004; Keller et al., 2005). This environmental remediation is also extended for organic pollutant degradation in water and air, as well as antimicrobial agents (Pathakoti et al., 2014). These nanomaterials constructed with "green synthesis" by using eco-friendly and sustainable methods offer much more advantages than the traditional synthetic methods. Furthermore, the development of different green methods including bioinspired methods that mimic those of natural materials or living matter are taken into consideration for new remedial applications. Numerous plants and organisms have been utilized for antimicrobial applications, including bacteria because of their ease of handling and genetic modification (Jayaseelan et al., 2012), fungi because of their environmental adaptability with high metal resistance (Valix et al., 2001; Jain et al., 2013), seagrasses because of their abundance of biologically active substances (Mohamed et al., 2012), and phytochemical constituents with rich phytochemical content (Malarkodi et al., 2013; Thema et al., 2015; Sangeetha et al., 2011). Other natural reagents that are employed as "green agents" in addition to these bioinspired materials have also been used for the environmentally friendly synthesis of nanomaterials. In a recent work conducted within the last 10 years, ZnO nanoparticles with a size range of 50–110 nm were produced by decomposing a precursor called Zn(II) Schiff base complex at 400, 500, and 600°C. L-Alanine, zinc acetate, and sodium salicylaldehyde-5-sulfonate are all components of the complex that is created. The same technique was utilized to create nanomaterials that measured 38 3 nm (Gharagozlou et al., 2015; Vigneshwaran et al., 2006). However, the capacity to better regulate the size and form of the produced nanoparticles was not provided by these natural reagents.

7.5 PRINCIPLES OF GREEN SYNTHESIS APPLIED IN NANOTECHNOLOGY TOWARD REMEDIAL MEASURES

The "Green synthesis of nanomaterials for nanoremediation" can only be improved and stabilized by the development of new strategic ways to address the majority of the green chemistry principles. This can be achieved by the production of selective functional chemicals for the construction of desired nanomaterials with the utmost level of waste minimization and byproduct generation. The development of cleaner and sustainable processes such as step- and atom economy (Trost, 1991) and E-factor (Sheldon, 2016) provides a green alternative to traditional synthetic procedures and transformations. Additionally, Nanotechnology can also supply us with information about new materials and gadgets. We can achieve our aim of gradually improving the range of effectiveness of current chemicals and materials by using nanoscale characteristics as an adjustable physical variable. Furthermore, self-assembled monolayers (alignment with simple molecules in such an ordered array on a substrate surface) can operate as a modern trend for chemical and biological sensors (Wang, 2010). Green Nanotechnology is centered on reducing the risk of utilizing nanomaterials, limiting the risk of manufacturing nanomaterials, and using nanomaterials to reduce the danger of producing undesirable chemical intermediates and end products (Binnig et al., 1983). Green Nanotechnology is committed to rigor in applications connected to sustainability, as well as raising awareness of and preventing any detrimental implications of successful implementation. Indeed, the stated purpose of all of these evaluations is to guarantee that nanomaterials are designed in a smart way that eliminates dangers and provides a sustainability pathway throughout the materials' life cycle, possibly eliminating negative environmental consequences (National Nanotechnology Initiative Strategic Plan 2011).

The orientation of nanoparticle structures makes it possible for their surfaces to be functionalized with different polymers, metals, metal oxides, small molecules, surfactants, and other substances that are required to create the desired composition and are thus suitable for the intended application (Christian et al., 2008). In a nanoparticle, there are three layers: A surface that is simple to functionalize, a shell that can be added based on the required qualities of the finished goods, and a core that can be produced using a variety of techniques, reaction settings, and precursors are the first two requirements (Huczko, 2000). When the surface chemistry of these nanoparticles is examined, it is probable that the core and shell characteristics, rather than the core alone, will prevail in the case of nanoparticle destiny and transit in the environment. The dangers connected with the presence of nanoparticles in the environment, on the contrary, always are linked to the surface, core, and shell of the nanoparticle without stressing each of the components of the nanoparticle.

The creation of novel exopolysaccharides was suggested as a potential innovative, sustainable, and renewable approach toward bioremediation or green synthesis of nanomaterials (Pradip et al., 2009). These changes are just a result of heavy metal and dye-related environmental contamination, which over time grew more severe with brisk industrial and economic growth and created significant environmental issues.

7.6 PERSISTENT CHALLENGES TOWARD "GREENER" SYNTHESIS OF NANOMATERIALS

The turbulent pathway that leads to the green synthesis of nanomaterials is one of the major limiting factors in the construction of desired nanomaterials for new potential applications. To overcome these challenges a green approach

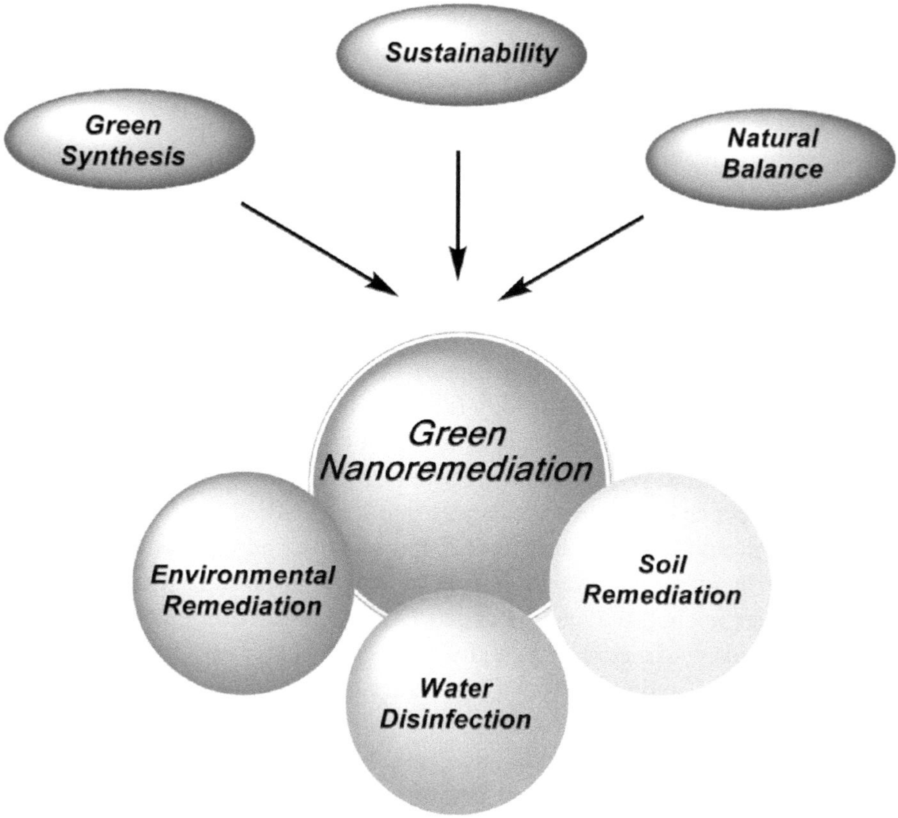

FIGURE 1.3 A graph visualizing the connection between sustainability and green synthesis for nanoremediation technologies.

to nanomaterial synthesis to environmentally sustain the ecosystem with available biodegradable matter needs to be developed (Elumalai et al., 2015). Some other challenges that need to be addressed to successfully manufacture nanomaterials and nanodevice are to solve the process of standardization and difficulties with other surface phenomena. For instance, the particle's outer layer atoms could differ from the remainder of the particle's atoms in terms of composition and chemistry. Additionally, that nanoparticle surface can be vulnerable to alterations in environmental factors such as redox state, pH, ionic strength, as well as the presence of microbes and organic debris from the environment. Additionally, the nanoparticles' tiny size and high surface-to-volume ratios may affect their physical and chemical characteristics, such as their melting point, thermal conductivity, and optical absorption, including mechanical, biological, and sterically catalytic activity (Huczko,

2000). Aside from this, one of the most challenging issues in nanomaterial synthesis is producing structures with desired and controllable qualities in large quantities for varied commercial purposes (Wang, 2010).

One of the most significant components of a synthesis of nanomaterials for a specific application is the functionalization of the nanoparticle surface. For example, strong chemical activity with a large surface is generally the main reason for aggregation, which is an irreversible and extremely undesired process since aggregation greatly reduces reactivity through the lowered specific surface area as well as the interfacial free energy (Pomogailo and Kestelman, 2005). As a result, nanoparticles should be functionalized with specific chemicals or minerals to boost their stability during storage, transit, and application to avoid aggregation. Addressing these issues clears the way for "green synthesis" of nanomaterials; however, it is still much more important for emerging researchers to resolve some of these issues, such as (1) identifying the risks associated with being exposed to these nanomaterials, (2) developing even more "green" synthesis approaches on the sustainable scale by eliminating power processes and increasing the use of various wastes, and (3) developing various analogs. Thus, despite current advancements and good outcomes, more research is required to enhance present methods and procedures, which will bring new possibilities and challenges for the scientific community and the general public in the years to come (Mandal et al., 2006; Narayanan and Sakthivel, 2010; Mukherjee et al., 2001).

Flow chemistry is yet another step introduced to better and meet the obstacles faced in the application of green chemistry, for the same diverse flow chemistry instruments are also introduced. This system drive option has proven its worth by providing improved mass/heat transfer possibilities, waste minimization, the use of little or less solvent, and the restriction of the use of hazardous and reactive chemicals with round-the-clock unmanned operations (Gholap et al., 2005)

7.7 LEADING EXAMPLES OF GREEN SYNTHESIS OF NANOMATERIALS

Over the past several years, a huge variety of nanomaterials have been created by adhering to the basic tenets of green chemistry. The usage of green synthesis techniques is one contemporary trend that is growing daily to lessen the effects of toxins and other negative characteristics. In terms of ease of use and safety, one green technique implemented by employing plant materials, plants, and microbes is applied for the synthesis of metal oxide nanoparticles (Makarov et al., 2014). The initial development of the NP core is aided by phytochemicals, which are another important class of precursors used in the production of nanomaterials. They and other phytochemicals serve as a coating material to stabilize the newly created nanoparticles. Humans have learned during the recent decades about phytochemical qualities and their accompanying uses, including their use in medicine. For instance, the Chinese have long employed traditional Chinese medicine and have been using herbs to treat numerous disorders. On the other hand, the understanding of inorganic nanoparticles and phytochemicals combined has undoubtedly created new avenues for obtaining them with specialized features using procedures that are safe for the environment (Qian et al., 2015). A further significant obstacle to sustainability is the choice of certain phytochemicals with predetermined features for the creation of nanomaterials.

Apart from plants, other major sources for the synthesis of inorganic materials are microorganisms such as magnetotactic bacteria which leads to the formation of magnetite nanoparticles (Philipse and Maas, 2002). This approach is not as cost-effective as the previous method of using plant extract. Another example of green synthesis of TiO_2 nanoparticles using microorganisms was synthesized by using bacteria and fungi. Microorganisms such as magnetotactic bacteria, which generate magnetite nanoparticles, are important sources for the production of inorganic minerals besides plants (Philipse and Maas, 2002). This strategy is less economical than the earlier way of employing plant extract. Another instance of environmentally friendly TiO_2 nanoparticle production was carried out employing bacteria and fungus. It was discovered that the size of the nanoparticles produced by the bacteria Lactobacillus varied depending on the source. Jha and Prasad (2010), for instance, analyzed various sources between yogurt and probiotic spore tablets and discovered that the particle size distribution from yogurt is broader (15–70 nm) than that of probiotic spore tablets (10–25 nm). Fungi are another important source, and their capacity to release enormous amounts of enzymes suggests that they are actively engaged in nanoparticle creation. Jha et al., for example, employed yeast to create spherically, 12.57 nm TiO_2 NPs (Jha et al., 2009). Rajakumar et al. (2012) used *Aspergillus flavus* to create TiO_2 nanoparticles with sizes ranging from 62 to 74 nm X-ray diffraction (XRD) analysis. Following their examination, Fourier transform infrared (FTIR) revealed peaks related to TiO_2. It also demonstrated the presence of amino acids on the NPs' surfaces. Bansal et al. produced NPs at 27°C using fungal mycelia and K2TiF6 at pH 3.5 in 2005. As indicated in the conventional transmission Electron Microscopy (TEM) (Bansal et al., 2005), the particles recovered are spherical with a size range of 6–13 nm.

7.8 CONCLUSIONS AND FUTURE PROSPECTS

In recent times, the green synthesis of nanomaterials pertaining with green synthesis and sustainability is emerging at a faster pace. This growth is witnessed due to new emerging potential applications in the same field striving the green measures and sustainable approaches with fewer or lesser harmful byproducts. In this chapter, we have mainly focused on the internal and external domains of "green synthesis" and the relevant fields of study with the aim of presenting a brief review of remedial measures toward nanoremediation. The principles of green synthesis for these remedial measures and the potential applications developing due to the constant effort of researchers working in multidisciplinary fields and the leading examples of green synthesis such as bioinspired materials, plant extracts, and microorganisms are also discussed. Because the process of nanomaterial production is not well understood at the moment, future research should focus just on the formation mechanism to regulate size and shape to make further gains in the area. Although green synthesis has some advantages over traditional approaches, it still has challenges with particle size, shape, and repeatability of a synthesis process. To further investigate novel nanomaterials, it is also critical to uncover new possible green sources functioning as antecedents for new nanomaterial production from secondary metabolites of plants or chemical ingredients.

KEYWORDS

- **green synthesis of nanomaterials**
- **green nanotechnology**
- **green chemistry**
- **nanotechnology**
- **sustainability**

REFERENCES

Anstas, P. T.; Warner, J. C. *Green Chemistry: Theory and Practice*; Oxford University Press, Inc; New York, 1998.

Armon, R.; Weltch-Cohen, G.; Bettane, P. G. Disinfection of Bacillus spp. Spores in drinking water by TiO2 photocatalysis as a model for Bacillus anthracis. *Water Sci. Technol,* **2004,** *4,* 7–14.

Bansal, V.; Rautaray, D.; Bharde, A.; Ahire, K.; Sanyal, A.; Ahmad, A.; Sastry M. Fungusmediated Biosynthesis of Silica and Titania Particles. *J. Mater. Chem.* **2005,** *15,* 2583–2589.

Binnig, G.; Rohrer, H.; Gerber, C.; Weibel, E. *Phys. Rev. Lett.* **1983,** *50,* 120–123.

Binnig, G.; Rohrer, H.; Gerber, C.; Weibel, E. *Phys. Rev. Lett.* **1983**, *50,* 120–123.

Christian, P.; Von der Kammer, F.; Baalousha, M.; Hofmann, T. Nanoparticles: Structure, Properties, Preparation and Behaviour in Environmental Media. *Ecotoxicology* **2008,** *17* (5), 326–343.

Das, S. K.; Marsili, E. A Green Chemical Approach for the Synthesis of Gold Nanoparticles: Characterization and Mechanistic Aspect. *Rev. Environ. Sci. Bio/Technol.* **2009**, *9,* 199–204.

Diallo, M.; Brinker, C. J. Nanotechnology for Sustainability: Environment, Water, Food, and Climate. In *Nanotechnology Research Directions for Societal Needs in 2020: Retrospective and Outlook*; Roco, M., Mirkin, C. A., Hersam, M. C., Eds.; Springer Publishing: New York, 2011; Chapter 5, pp 157−187.

Eckelman, M. J.; Zimmerman, J. B.; Anastas, P. T. Toward Green Nano. *J. Ind. Ecol.* **2008,** *12* (3), 316–328.

Elumalai, K.; Velmurugan, S. Green Synthesis, Characterization and Antimicrobial Activities of Zinc Oxide Nanoparticles from the Leaf Extract of Azadirachta Indica (L.). *Appl. Surf. Sci.* **2015,** *345,* 329–336.

Gharagozlou, M.; Baradaran, Z.; Bayati R. A Green Chemical Method for Synthesis of ZnO Nanoparticles from Solid-State Decomposition of Schiff-bases Derived from Amino Acid Alanine Complexes. *Ceram. Int.* **2015,** *41,* 8382–8387.

Gholap, A. R.; Venkatesan, K.; Pasricha, R.; Daniel, T.; Lahoti, R. J.; Srinivasan, K. V. Copperand ligand-free Sonogashira reaction Catalyzed by Pd(0) Nanoparticles at Ambient Conditions under Ultrasound Irradiation. *J. Org. Chem.* **2005**, *70,* 4869–4872.

Huang, L.; Weng, X.; Chen, Z.; Megharaj, M.; Naidu, R. Green Synthesis of Iron Nanoparticles by Various Tea Extracts: Comparative Study of the Reactivity. *Spectrochim. Acta A Mol. Biomol. Spectrosc.* **2014,** *130,* 295–301.

Huczko, A. Template-based Synthesis of Nanomaterials. *Appl. Phys. A: Mater. Sci. Process.* **2000,** *70* (4), 365–376.

Huczko, A. Template-based Synthesis of Nanomaterials. *Appl. Phys. A: Mater. Sci. Process.* **2000**, *70* (4), 365–376.

Iravani, S. Green Synthesis of Metal Nanoparticles Using Plants. *Green Chem.* **2011,** *13* (10).

Jain, N.; Bhargava, A.; Tarafdar, J.; Singh, S. K.; Panwar J. A Biomimetic Approach towards Synthesis of Zinc Oxide Nanoparticles. *Appl. Microbiol. Biotechnol.* **2013,** *97,* 859–869.

Jayaseelan, C.; Rahuman, A. A.; Kirthi, A. V.; Marimuthu, S. T.; Santhoshkumar, T.; Bagavan, A.; Gaurav, K.; Karthik, L.; Rao, K. V. B. Novel Microbial Route to Synthesize ZnO Nanoparticles Using Aeromonas Hydrophila and their Activity Against Pathogenic Bacteria and Fungi. *Spectrochim Acta A Mol. Biomol. Spectrosc.* **2012,** *90,* 78–84.

Jebali, A.; Ramezani, F.; Kazemi, B. Biosynthesis of Silver Nanoparticles by Geotricum. sp. *J. Cluster Sci.* **2011,** *22* (2), 225–232.

Jha, A. K.; Prasad K. Biosynthesis of Metal and Oxide Nanoparticles Using Lactobacilli from Yoghurt and Probiotic Spore Tablets. *Biotechnol. J.* **2010,** *5,* 285–291.

Jha, A. K.; Prasad, K.; Kulkarni, A. R. Synthesis of TiO2 Nanoparticles Using Microorganisms. *Colloids Surf. B: Biointerfaces* **2009,** *71,* 226–229.

Keller, V.; Keller, N.; Ledoux, M. J.; Lett MC. Biological Agent Inactivation in a Flowing Air Stream by Photocatalysis. *Chem. Commun.* **2005,** *23,* 2918–2920.

Kirubaharan, C. J.; Kalpana, D.; Lee, Y. S.; Kim, A. R.; Yoo, D. J.; Nahm, K. S.; Kumar, G. G. Biomediated Silver Nanoparticles for the Highly Selective Copper(II) Ion Sensor Applications. *Ind. Eng. Chem. Res.* **2012,** *51* (21), 7441–7446.

Luo, F.; Yang, D.; Chen, Z.; Megharaj, M.; Naidu, R. One-step Green Synthesis of Bimetallic Fe/ Pd Nanoparticles Used to Degrade Orange II. *J. Hazard. Mater.* **2016,** *303,* 145–153.

Machado, S.; Pinto, S. L.; Grosso, J. P.; Nouws, H. P. A.; Albergaria, J. T.; Delerue-Matos, C. Green Production of Zero-valent Iron Nanoparticles using Tree Leaf Extracts. *Sci. Total Environ.* **2013,** *445,* 1–8.

Makarov, V. V.; Love, A. J.; Sinitsyna, O. V.; Makarova, S. S.; Yaminsky, I. V.; Taliansky, M. E.; Kalinina, N. O. "Green" Nanotechnologies: Synthesis of Metal Nanoparticles Using Plants. *Acta Nat.* **2014**, *6,* 35–44.

Malarkodi, C.; Chitra, K.; Rajeshkumar, S.; Gnanajobitha, G.; Paulkumar, K.; Vanaja Mand Annadurai, G. Novel Eco-friendly Synthesis of Titanium Oxide Nanoparticles by Using Planomicrobium sp. and Its Antimicrobial Evaluation. *Der Pharmacia Sinica* **2013,** *4,* 59–66.

Mandal, D.; Bolander, M.; Mukhopadhyay, D.; Sarkar, G.; Mukherjee, P. The Use of Microorganisms for the Formation of Metal Nanoparticles and their Application. *Appl. Microbiol. Biotechnol.* **2006,** *69* (5), 485–492.

Mandal, D.; Bolander, M.; Mukhopadhyay, D.; Sarkar, G.; Mukherjee, P. The Use of Microorganisms for the Formation of Metal Nanoparticles and Their Application. *Appl. Microbiol. Biotechnol.* **2006,** *69* (5), 485–492.

Mohamed, S.; Hashim, S. N.; Rahman, H. A. Seaweeds: A Sustainable Functional Food Forcomplementary and Alternative Therapy. *Trends Food Sci. Technol.* **2012,** *23,* 83–96.

Mukherjee, P.; Ahmad, A.; Mandal, D.; Senapati, S.; Sainkar, S. R.; Khan, M. I.; Parishcha, R.; Ajaykumar, P. V.; Alam, M.; Kumar, R.; Sastry, M. Fungus-mediated Synthesis of Silver Nanoparticles and Their Immobilization in the Mycelial Matrix: A Novel Biological Approach to Nanoparticle Synthesis. *Nano Lett.* **2001,** *1* (10), 515–519.

Narayanan, K. B.; Sakthivel, N. Biological Synthesis of Metal Nanoparticles by Microbes. *Adv. Colloid Interface Sci.* **2010,** *156* (1−2), 1–13.

Nasir Baig, R. B.; Varma, R. S. Solvent-free synthesis. In *An Introduction to Green Chemistry Methods*; Luque, R., Colmenares, J. C., Eds.; Future Science: London, UK, 2013. http://www.futuremedicine.com/doi/book/10.4155/9781909453104

National Nanotechnology Initiative Strategic Plan, National Science and Technology Council Committee on Technology, Subcommittee on Nanoscale Science, Engineering, and Technology: Washington, DC, February **2011;** pp 29−32.

Pathakoti, K.; Huang, M.-J.; Watts, J. D.; He, X.; Hwang, H.-M. Using Experimental Data of Escherichia Coli to Develop a QSAR Model for Predicting the Photo-induced Cytotoxicity of Metal Oxide Nanoparticles. *J. Photochem. Photobiol. B Biol.* **2014,** *130* (Suppl. C), 234–240.

Philipse, A. P.; Maas D. Magnetic Colloids from Magnetotactic Bacteria: Chain Formation and Colloidal Stability. *Langmuir* **2002,** *18,* 9977–9984.

Pomogailo, A. D.; Kestelman, V. N. *Metallopolymer Nanocomposites*; Springer: Berlin, Heidelberg, New York, 2005; Vol 81, p 563.

Pradip, B.; Singh, R.; Kumar, M.; Prajapati, J.; Singh, R. Exopolysaccharides of Lactic Acid Bacteria: A Review. *J. Food Sci. Technol.* **2009**, *46,* 1–11.

Qian, Y.; Yao, J.; Russel, M.; Chen, K.; Wang X. Characterization of Green Synthesized Nanoformulation (ZnO–A. vera) and their Antibacterial Activity Against Pathogens. *Environ. Toxicol. Pharmacol.* **2015,** *39,* 736–46.

Rajakumar, G.; Rahuman, A. A.; Roopan, S. M.; Khanna, M. G.; Elango, G.; Kamaraj, C.; Zahir, A. A.; Velayutham K. Fungus-mediated Biosynthesis and Characterization of TiO2 Nanoparticles and their Activity Against Pathogenic Bacteria. *Spectrochim. Acta A Mol. Biomol. Spectrosc.* **2012,** *91,* 23–29.

Saif, S.; Tahir, A.; Chen, Y. Green Synthesis of Iron Nanoparticles and their Environmental Applications and Implications. *Nanomaterials* **2016,** *6* (11), 209.

Sangeetha, G.; Rajeshwari, S.; Venckatesh, R. Green Synthesis of Zinc Oxide Nanoparticles by Aloe Barbadensis Miller Leaf Extract: Structure and Optical Properties. *Mater. Res. Bull.* **2011,** *46,* 2560–2566.

Sheldon, R. A. Green Chemistry and Resource Efficiency: Towards a Green Economy. *Green Chem.* **2016**, *18,* 3180–3183 and references cited therein.

Thema, F. T.; Manikandan, E.; Dhlamini, M. S.; Maaza, M. Green Synthesis of ZnO Nanoparticles via Agathosma Betulina Natural Extract. *Mater. Lett.* **2015,** *161,* 124–127.

Trost, B. M. The atom Economy—A Search for Synthetic Efficiency. *Science* **1991**, *254,* 1471–1477.

Valix, M.; Tang, J. Y.; Malik, R. Heavy Metal Tolerance of Fungi. *Miner. Eng.* **2001,** *14,* 499–505.

Vigneshwaran, N.; Kumar, S.; Kathe, A. A.; Varadarajan, P. V.; Prasad V. Functional Finishing of Cotton Fabrics Using Zinc Oxide–Soluble Starch Nanocomposites. *Nanotechnology* **2006,** *17,* 5087–5095.

Wang, Z. L. Toward Self-powered Sensor Networks. *Nanotoday* **2010,** *5,* 512–514.

CHAPTER 8

Green Methods to Synthesize Nanomaterials to Remediate Polluted Environs

LINKON BHARALI and SIDDHARTHA S. DHAR

Department of Chemistry, National Institute of Technology, Silchar, Assam, India

ABSTRACT

Nanomaterials with particle size 1–100 nm have a broad range of applications in several fields. Various methods are employed for the synthesis of nanomaterials. Because of its sustainability, reliability, and environmental friendliness, green synthesis methods for synthesizing nanomaterials have gained a great deal of interest. The green synthesis is considered a significant technique for reducing the adverse effects of standard nanomaterial synthesis processes applied in laboratories and industries. The synthesis approach avoids employing highly toxic chemicals. Numerous organic and inorganic contaminants in wastewater have become a major source of concern in the aquatic environment. The green-synthesized nanomaretials can be used for the removal of such pollutants from water bodies. This chapter describes green methods of synthesizing several nanomaterials. This chapter also focuses on the applications of such nanomaterials in the field of environmental cleanliness.

8.1 INTRODUCTION

Evolution of nanotechnology and nanoscience has been proven to be a key factor in technological advancement from the last few decades (Kolahalam

Sustainable Nanoremediation: Modern Technologies for a Clean Environment.
Fernanda Maria Policarpo Tonelli, Arpita Roy, and Khalid Rehman Hakeem (Eds.)

et al., 2019). Nanotechnology is a broad and multidisciplinary area of research that grabbed the interest of chemists, biologists, physicists, and engineers (Aarthye and Sureshkumar, 2021). The researchers synthesize various nanomaterials with superior qualities to bulk materials. Nanomaterials have been developed as a fascinating class of materials that includes a wide range of samples with at least one dimension in the nanoscale from 1 to 100 nm (Manasreh, 2011). Some typical examples of naturally occurring nanoparticles in the environment include iron oxides, sulfides, gold, and silver (Sharma et al., 2015). In addition to these metal and metal oxides, naturally produced nanoparticles can also be found in ocean spray, fine sand, volcanic ash, small microorganisms like viruses etc. Again, certain nanoparticles are manufactured artificially. Prepared or engineered nanoparticles (EN) are gaining a lot of attention because they are frequently used in various commercial products and processes.

In general, there are two categories of nanomaterials: nanostructured materials and nanostructured elements. Its structural dimensions are on the nanoscale in nanostructured materials (Dingreville et al., 2005). One or more structural components in nanostructured elements have outside dimensions that are in the nanometer range (Koutsawa et al., 2017). Depending upon the dimension of nanoscale, i.e., less than 100 nm the nanomaterials are classified (Kolahalam et al., 2019) as follows:

a. Zero-dimensional nanomaterials (0D): All three dimensions of the nanomaterials in 0D are within the nanoscale range. The examples of such nanomaterials are quantum dots, smaller nanoparticles (GaP, ZnO etc.)
b. One-dimensional nanomaterials (1D): Here, one dimension will fall within the nanoscale range, while the other two dimensions are beyond this range. Nanowire, nanorods, and nanotubes are examples of 1D nanomaterials.
c. Two-dimensional nanomaterials (2D): In 2D nanomaterials, any two dimensions exist in nanoscale range and the others exist outside nanoscale. Nonocoating, nanolayers are two dimensional nanomaterials.
d. Three-dimensional nanomaterials (3D): These nanoparticles are not nanoscale in any dimension. This indicates that they are greater than 100 nm scale in three arbitrary dimensions. Such nanomaterials are also known as bulk nanomaterials. This class of nanomaterials include core shell nanomaterials, nanocomposites,

multi-nanolayers etc (Asghari et al., 2016; Aversa et al., 2018; Shiau et al., 2018).

Nanomaterials are further classified depending upon morphology, state, aspect ratios, and chemical compositions (Saleh, 2020). Metal nanomaterials (MNs) are one of these classifications of nanomaterials which includes metal nanoparticles such as iron, silver, gold, platinum, palladium, and other MNs. Such nanomaterials have specific electric, optical, chemical properties that are different from the bulk substances (Asghari et al., 2016). For example while considering the optical properties of these nanomaterials it can be observed that ruthenium, platinum, palladium, and nickel MNs absorb light radiation through intra-band transition. Again for copper, aluminum, silver, and gold intra-band transitions are seen (Kim and Lee, 2018). Another classification of nanomaterials are bimetallic nanomaterials (BNs) which contain two metals in their compositions. Au/Pd, Ag-Au, and Fe-Cu are some of the bimetallic nanomaterials (Chen et al., 2019; Lozhkomoev et al., 2019; Oh et al., 2018). From literatures it was found that several parameters, including the miscibility of the two metals and the conditions of fabrication, have a significant impact on the nanostructure of BNs (Saha et al., 2019). The nanomaterials such as SnO_2, Al_2O_3, Fe_2O_3, and ZnO are known as metal oxide nanomaterials (MONs). The common methods of synthesizing MONs include hydrothermal method and sol-gel method. A metal oxide offers a substantial advantage in various applications, including catalysis, semiconductors, and chemical sensors due to the modifications in surface characteristics, which influence the band gap energy of such materials (Saleh and Fadillah, 2019). A different class of nanomaterials known as nanocomposites exists in addition to these MONs. Solid materials known as nanocomposites are composed of many phases, at least one of which has dimensions smaller than 100 nm (Lozhkomoev et al., 2019). Composites may also include structures with nanoscale repetition gaps in between the phases (Oh et al., 2018). High surface-to-volume ratio of nanocomposites separates them from other composites. Various types of nanocomposites are there such as PMNC, polymer matrix nanocomposites, CMNC, ceramic matrix nanocomposites and MMNC, metal matrix nanocomposites. Nanocomposites have a wide range of applications in different fields. For example, K.C. Das, S.S. Dhar, D.G. Thakurata et al. reported the novel nanocomposite, $NiFe_2O_4$@HAp-Sn^{2+}, that performed excellent photocatalytic activity in the degradation of organic dye (Das et al., 2021). Also, from the literature it was found that the Box-Behnken method can be employed to produce chitosan–titanium oxide-based nanocomposites, which require minimum energy, chemicals that are applied in orange dye adsorption reaction (Saha et al., 2019). Semiconductor nanomaterials are another class of nanomaterials having both metallic and nonmetallic characteristics. Such materials are used in catalysis and electronic instruments. Group II–IV and Group III–V are some classes of semiconductor materials. The examples of the former class include CdSe (Muhammad et al., 2019), ZnO (Kolahalam et al., 2019), CdS (Ngamdee and Ngeontae, 2018), and ZnS (Rajabi et al., 2018), while InAs (Lee and McLaurin, 2018) and GaP (Yue et al., 2006) belong to the latter class. Carbon-based nanomaterials are the next class of nanomaterials whose primary constituent is carbon. There are different solid-state allotropes of carbon, including graphite, diamond, and amorphous carbon (Li et al., 2019). Materials can be categorized according to the form of carbon-based nanomaterials (CNs). Two such CNs are (a) Fullerene, the carbon allotrope, which has 60 or more number of carbon atoms packed together in buckyball form, also known as buckminsterfullerene. Fullerene derivatives have special characteristics that can neutralize reactive species like oxygen and nitrogen (Sumi and Chitra, 2019). (b) Graphene sheets, the CNs can form hexagonal lattice. They possess various exclusive properties such as excellent stability, higher electrical conductivity, and greater surface area. The exfoliation method is used in the isolation of graphene from graphite (Silva et al., 2018). In addition to the nanomaterials described above, there are other categories of nanomaterials. Some of them are zeolite-based nanomaterial (Liu et al., 2021), silica-based nanomaterials (Zeb et al., 2020), polymeric nanomaterials (Shi et al., 2017), ceramic nanomaterials (Zinatloo-Ajabshir et al., 2021), lipid-based nanomaterials, metal organic framework-based nanomaterials (Wen et al., 2019), and so on. These nanomaterials are used in many fields and beneficial to humanity.

Nanomaterials are fascinating because they exhibit novel optical, magnetic, electrical, and other properties. Technology, medicine, and other disciplines could be greatly impacted by these emerging features. The use of nanomaterials is influenced by a variety of aspects, such as their physical characteristics and higher surface area-to-volume ratios which provide the significant opportunities to enhance their usefulness (Gleiter, 2000). When considering the electronic properties it was found that various surface properties of nanomaterials can be enhanced by the addition of organic substances like ligands. Modification of nanomaterials such as ZnO, Al_2O_3 with organic can alter the electronic properties (Henkel et al., 2020). It was obtained from the literature that electric characteristics of polymer nanocomposites were enhanced by including inorganic components in the polymer structures (Hassan et al., 2019). Mechanical properties of materials may vary from materials to materials. Because of the surface area, volume, also quantum effects of nanoparticles, nanomaterials exhibit exceptional mechanical properties. When nanoparticles are added to an ordinary material, the nanoparticles will improve the grain to some point; as a result inter- or intra-granular structures are formed, which enhance the grain boundary and also increase the mechanical behavior of such nanomaterials (Wang and Meng, 2014; Wang et al., 2011; Zou et al., 2006). For instance, by introducing the metal oxide SnO_2, mechanical characteristics of acrylic polyurethane can be increased, specifically the adhesive nature, hardness, and also abrasion resistance (Bui et al., 2020). Applications in the medical and environmental domains use magnetic nanoparticles. Nanomaterials have significant magnetic characteristics. Several literature showed that the particle size smaller than 35 nm performed excellent properties (Lamouri et al., 2020; Shrimali et al., 2020). Various applications, including photocatalysts and optoelectronic devices, depend on the optical characteristics of the nanomaterials. The optical parameters of nanomaterials can be measured using UV-visible spectrophotometer by applying the optical formulae as well as the Beer–Lambert Law (Kalita et al., 2022). Thermal properties of nanomaterials are better than their fluid counterpart. It is due to the fact that they have huge surface area which allows direct heat flow on the surface. By increasing the amount of metal oxides in between the nanomaterials and polymer matrix, the thermal properties of the nanomaterials can be improved (Nomai and Schlarb, 2019).

The synthesis processes for nanomaterials have been covered in this chapter. The chapter primarily focuses on environment-friendly ways, i.e., green methods to produce nanomaterials and their use in environmental remediation.

8.2 SYNTHESIS METHODS OF NANOMATERIALS

There are various conventional methods of synthesis of nanomaterials. The synthesis techniques can be broadly classified into two categories: Top Down and Bottom Up Approach.

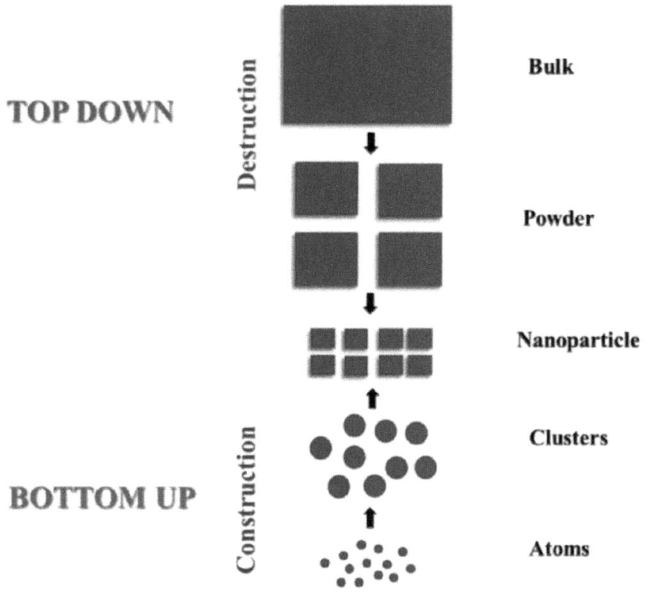

FIGURE 8.1 Schematic representation of top-down and bottom-up approaches of synthesizing nanoparticles.

The top-down approach of synthesizing nanomaterials involves breakdown of bulk materials to produce nanoparticles. Top-down techniques are electrospinning, ball milling, laser ablation, arc discharge method etc. (Jenkner et al., 1992). Electrospinning is one of the basic top-down methods for manufacturing nanomaterials. This technique is used for nanofibers from larger polymer materials (Ostermann et al., 2011). Coaxial electrospinning was a significant advancement in the field of electrospinning. It is a proficient and elementary top-down method for producing core-shell thin nanofibers on a wide scale. The technique involves the preparation of organic, inorganic, and hybrid materials (Kumar et al., 2014). The next top-down method of nanomaterial synthesis is ball milling. The method utilized mechanical abrasion in the production of nanomaterials. This process involves transfer of kinetic energy (KE) from the grinding form to the sample that undergoes reduction (Arole and Munde, 2014). Carbon nanomaterials synthesized by employing the ball milling technique are regarded as a new class of nanomaterials that hold potential to act as materials used for environmental remediation and also energy storage and conversion (Lyu et al., 2017). Again, by hitting the target substance or material with a strong laser beam, the laser ablation technique produces nanomaterials. Using this method a wide range of nanomaterials can be prepared easily which includes carbon-based nanomaterials (Chrzanowska et al., 2015; Ismail et al., 2019), metal nanomaterials (J. Zhang et al., 2017), ceramic nanomaterials (Brabazon et al., 2017), and oxide nanocomposites (Duque et al., 2019). Another high-energy top-down method for nanomaterials production is arc discharge method. From the beginning, this method was employed to prepare carbon nanotubes. It also involves production of fullerene, layered graphene, carbon nanohorns etc. (D. Zhang et al., 2019).

Besides the above-mentioned top-down methods, we can also utilize bottom-up approaches for the production of nanomaterials. The bottom-up approach begins with atoms or molecules for the preparation of nanomaterials. Chemical vapor deposition (CVD), solvothermal method, hydrothermal method, sol-gel method, and reverse micelle method are some of the bottom-up methods. CVD gains more importance for the fabrication of carbon-based nanostructured materials. In the CVD method, chemical reaction of vapor phase precursor takes place which results in the construction of thin film on the surface of the substrate. The precursors used in this process should have certain features such as greater stability, high purity, volatility, economic, nontoxic and also should have long shelf-life time (Jones and Hitchman, 2009). CVD is one of excellent methods of synthesizing nanomaterials with superior qualities (Machac et al., 2020). It is also the commonly used method for the production of two-dimensional (2D) nanomaterials (Q. Wu et al., 2016). The solvothermal and hydrothermal methods are some of the most commonly employed preparation methods for nanomaterial production (Cao et al., 2016; Wu et al., 2011). The solvothermal process involves utilization of solvent under required condition of pressure and temperature that eases the interaction of precursors during synthesis. Different solvents can be used in this method. When the solvent is water it is simply known as hydrothermal method. Various geometries such as powders, thin films, nanocrystals, and single crystals can be prepared using this method (Kitamura et al., 2007). Besides these, wet chemical methods like the sol-gel method are frequently utilized to develop nanomaterials. Sol-gel method is widely used to prepare glass materials, metal oxides, ceramics etc. (Cheng et al., 1995; Mani, 2014). The precursors used in this process are organometallic salts or metal alkoxides. Again, a useful way for producing nanomaterials with the required shapes and sizes is the reverse micelle method. Here, reverse micelles work as nanoreactors in the nanoparticle synthesis process (Malik et al., 2012). The nanomaterials prepared by this micelle method result in the development of fine particles as well as monodispersed in character (Nguyen, 2013).

The most widely used techniques for developing nanomaterials are those described in the paragraphs above. However, some environmentally benign approaches to nanomaterial preparation are becoming increasingly significant. Green methods of synthesizing nanomaterials are discussed in the proceeding section.

8.3 GREEN SYNTHESIS OF NANOMATERIALS

Conventional techniques have been used for a long time in the preparation of nanomaterials. But from various studies it was found that green approaches are more effective for synthesizing nanoparticles or nanomaterials because of easy characterization, more economic as well as facile preparation with less failure (Abdelghany et al., 2018).

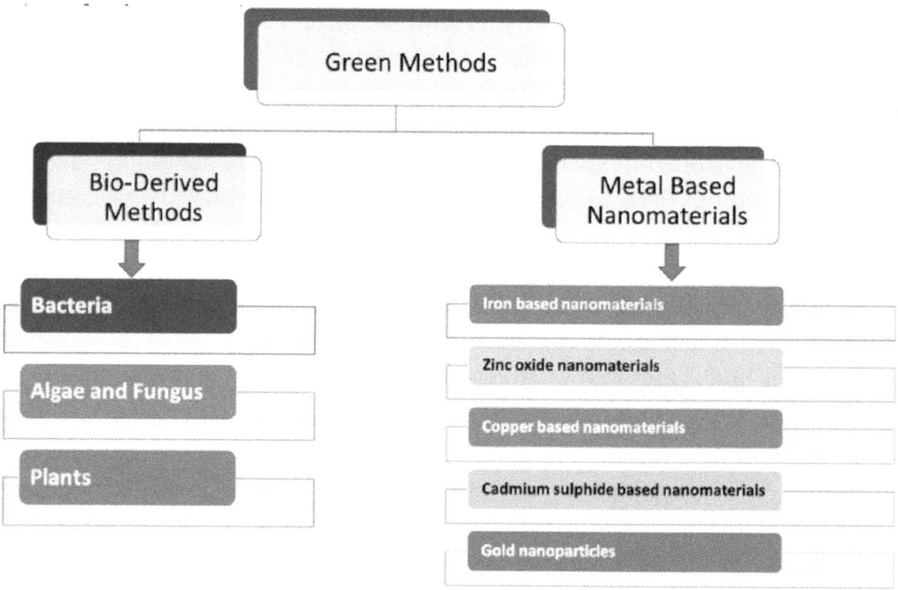

FIGURE 8.2 Green methods for synthesizing of nanomaterials.

Green methods gain more attraction due to their potential to decrease the toxic effects of the nanomaterials (NMs). There are several green methods for manufacturing NMs. They are discussed below:

8.3.1 BIO-DERIVED METHODS

Biological methods act as an alternative to the physical and chemical methods for preparing nanomaterials which results in the production of greatly stable, improved, and intact NMs. Some examples of bio-derived NMs are described as follows:

8.3.1.1 BACTERIA

Due to their defensive machinery, microbes like bacteria can be employed for the preparation of nanoparticles. Pseudomonas stutzeri AG259, a bacterial strain, was used in the synthesis of crystalline silver nanoparticles with changing compositions (Menon et al., 2017). At room temperature, TiO_2 nanoparticles can be prepared using isolated Bacillus mycoides from $TiO(OH)_2$ (Órdenes-Aenishanslins et al., 2014). TiO_2 nanoparticles can also be synthesized from *Aspergillus flavus* which acts as both reducing and capping agent. Such NMs show excellent antimicrobial properties against various other bacteria such as *E. coli*, *B. subtilis* etc. Furthermore, stability of Ag nanoparticles in dark was investigated by a research team (Shivaji et al., 2011). They found that Ag nanoparticles remain stable till 6–8 months in dark under a cell-free environment resuspended of different bacteria such as *Pseudomonas meridian*, *Bacillus indicus*, *Bacillus cecembensis*, *Arthrobacter kerguelensis*, and *Pseudomonas antarctica* etc.

8.3.1.2 ALGAE AND FUNGUS

The development of algae and fungus-based nanomaterials has led to their widespread use in different fields. Utilization of *Sargassum wightii* for the development of stable gold nanoparticles was investigated by a research group (Singaravelu et al., 2007). *Fusarium oxysporum* can be used for the transformation of $AgNO_3$ to well-dispersed silver oxide nanoparticles (Ahmed et al., 2018). After conjoining Se nanoparticles with chitosan and fungus as a reducing agent, a synergic interaction between the fungal *C. albicans* and Se nanoparticle was observed (Lara et al., 2018). Konishi et al. investigated that gold and platinum nanomaterials were found to be stored near the cytoplasmic region of Shewanella algae cells. The team also found that along with the deposition location, pH plays a key role in the fabrication of gold (Au) nanoparticles having desired morphologies and the sizes (Lara et al., 2018).

8.3.1.3 PLANTS

The preparation of various nanomaterials such as nano metal, nanometal oxides, polymeric nanoparticles etc. by green methods involves the utilization of renewable resources such as various antioxidants from coffee/tea, plant extracts, biodegradable polymers, and reducing sugars (Mallikarjuna and Varma, 2007; Nadagouda and Varma, 2007b, 2008). Kou and Varma (2012) reported a green and easy preparation of silver nanoparticles by microwave radiation employing beet juice that acts as a reducing agent. The synthesized nanomaterial exhibits excellent photocatalytic activity in dye degradation (Kou and Varma, 2012). Shankar et al. (2003) revealed possible reduction of gold ions into 20–40 nm nanoparticles by pelargonium graveolens extract (Shankar et al., 2003). Likewise, Trigonella foenum-graecum extract can be used in the reduction of Au ions into Au nanoparticles with size 15–25 nm (Aswathy Aromal and Philip, 2012). Also *Cymbopogan citratus*, *Ficus carica*, *Premna herbacea*, and *Ziziphora tenuior* leaf extracts are useful for the growth of silver nanoparticles with preferred sizes (Gupta et al., 2021; Jeeva et al., 2014; Sadeghi and Gholamhoseinpoor, 2015; Ulug et al., 2015).

8.3.2 GREEN SYNTHESIS OF METAL-BASED NANOMATERIALS

8.3.2.1 IRON, FE-BASED NANOMATERIALS

several literatures, it was found that iron-based nanoparticles can be easily prepared by employing green routes. Such nanomaterials are applied in a variety of fields. Prabhakar et al. reported the fabrication of Fe nanoparticles applying extract from terrestrial and aquatic weeds and used them in the removal of phosphates and nitrates from wastewater (Prabhakar and Samadder, 2017). In order to stabilize bimetallic Fe or Pd a research group employed water-soluble starch (He and Zhao, 2008). They also observed that starch has a prominent role in the stabilization of iron NMs. Again, Gao et al. investigated the preparation of iron oxide nanoparticles using sodium alginate, which is a biopolymer. Such nanomaterials were discovered to have regular, spherical shapes with an average diameter of 26–27 nm (Gao et al., 2008). Moreover, the synthesis of iron nanomaterials using vitamin C (i.e., ascorbic acid) was explored by Nadagouda et al. It showed the formation of core-shell iron using aqueous ascorbic acid and by it, the transition metal salts were subsequently reduced to the corresponding nanostructures (Nadagouda and Varma, 2007a). The production of amine functionalized iron NMs by employing the wet chemical route was reported by Krishna and coworkers (Krishna et al., 2012). l-lysine amino acid used for this method resulted in the production of highly crystallized iron nanoparticles. Green synthesis of Fe nanoparticles also involves use of bio-precursors in the synthesis process. Hemoglobin and myoglobin are iron-containing bio-precursors which were utilized for the one-pot preparative method of iron nanoparticles (Sayyad et al., 2012). Furthermore, utilizing *Terminalia chebula* fruit extract Mohan Kumar et al. developed iron and palladium nanoparticles (Mohan Kumar et al., 2013). The neem leaf (Azadirachta indica) extract was used by Pattanayak et al. (2013) as a cost-effective reductant in the manufacturing of nanoscale zero-valent iron under atmospheric environment (Pattanayak et al., 2013). The above-mentioned methods are some of the green methods for the production of iron nanomaterials with desired yields and applicability (Saif et al., 2016).

8.3.2.2 ZINC OXIDE, ZNO NANOMATERIALS

Zn oxide-based nanomaterials can be easily prepared by greener methods. In order to unite ZnO nanoparticles with typical size ranges of approximately 112–282 nm, fluid arrangement of Zn(NO3)2 was treated with the extract from *Cassia auriculata* blossoms (Naseer et al., 2020). *Zingiber officinale*, *Ficus benghalensis*, and *Parthenium hysterophorus* are recent examples of plants that have successfully produced ZnO-nanomaterials through the plant-mediated biosynthetic pathway (Shekhawat et al., 2015). Again, using Sedum alfredii Hance plant origin, zinc oxide nanoparticles can be easily synthesized that has applications in nanoelectronics (Qu et al., 2011).

8.3.2.3 COPPER, CU-BASED NANOMATERIALS

Green synthesis of copper (Cu)-based nanomaterials involves use of various plant extracts (Ghidan et al., 2016). Copper oxide nanoparticles have been manufactured using the *Punica granatum* peel extract. In this process, brown color solution indicated the formation of Cu oxide nanoparticles (Waris et al., 2021). Gunalan et al. introduced an important medicinal plant called Aloe barbadensis Miller which was utilized to synthesize copper oxide nanoparticles (Gunalan et al., 2012). Besides these, flower extracts are also utilized to prepare Cu nanoparticles. One of such flowers is *Cordia sebestena* from which copper oxide nanomaterials can be manufactured (Prakash et al., 2018). Moreover, copper oxide can be prepared from utilizing fungus strain Trichoderma (Saravanakumar et al., 2019). Using *Cissus vitiginea*, Cu nanoparticles can be synthesized and such copper-based nanomaterials have been used as an antioxidant and antibacterial agent (Wu et al., 2020). CuO nanoparticles are produced via a colloidal heat combination technique. The CuO that had been added underwent decontamination and drying in order to produce CuO nanoparticles of various sizes (Padi and Cernik, 2013). In a method described by Manoj et al. (2018), a biosensor for the detection of nitrite ions in the medium is produced by using carboxymethyl cellulase as a substrate to create a greatly stable and also sensitive copper nanocomposite (Manoj et al., 2018).

8.3.2.4 CADMIUM SULFIDE, CDS-BASED NANOMATERIALS

Cadmium sulfide, CdS nanoparticles can be synthesized using different greener approaches. Trichoderma harzianum, a typical biofungicide, was used to manufacture CdS nanoparticles using a microwave aided technique (Bhadwal et al., 2014). Wei et al. showed that using starch as a capping agent, CdS nanoparticles can easily be manufactured (Wei et al., 2004). Moreover, applying the green method it is possible to develop cadmium sulfide quantum dots. In order to manufacture cadmium sulfide quantum dots, *Fusarium oxysporum* biomass was used. It produced dots with diameters ranging from 2 to 6 nm (Sandoval-Cárdenas et al., 2017). *Klebsiella pneumonia* and *E. coli* were found to be capable of producing CdS nanoparticles. Such microorganisms can be utilized for the generation of nanomaterials which can be applied in environmental remediation. Some of them can easily degrade harmful pollutants from water surface and also be utilized for the adsorption of heavy metals (Abd Elsalam et al., 2018).

8.3.2.5 GOLD, AU-NANOPARTICLES

Gold is one of the most valuable metals and is utilized in various fields such as sensing, medicine, catalysis etc. Green method of synthesizing Au nanoparticles gains more importance than conventional synthesis methods as it has a positive impact on the environment. Francis et al. synthesized gold nanoparticle by *M. Glabrata* leaf extract from its metal salt precursor, employing the microwave assistance process. The group also investigated and found that such nanomaterials have water-purifying nature due to their excellent antimicrobial activity. Such materials can inhibit different pathogenic organisms *Staphylococcus aureus*, *E. coli*, *Penicillium chrysogenum* etc (Francis et al., 2017). Again, preparation of gold nanoparticles from concentrations of Cucurbitapepo L. was reported by Gonnelli et al (2015). From the literature survey it was found that the medicinally significant tree *Couroupita guianensis* can biosynthesize gold nanoparticles. Such gold nanoparticles have shown anticancer properties (Geetha et al., 2013).

Thus, the section discusses various green methods of synthesizing nanomaterials. The future of nanomaterials depends on the use of green chemistry. This field of nanoscience should lead to the formation of safe, environmentally acceptable nanoparticles and gain widespread recognition in the nanotechnology field (Gour and Jain, 2019).

8.4 CONCLUSION

In recent decades, green nanomaterial synthesis has become more significant. The advantages of the green methodologies for the production of nanoparticles over the conventionally used processes include the following points:

- Use of naturally occurring materials that are nontoxic.
- The green processes are easy and cost-effective.
- Nonhazardous by-products are formed.
- Green nanomaterials are highly stable, simply stored, and easily transportable.
- The materials may be manufactured more easily worldwide by utilizing regular and frequently biodegradable ingredients.

However, "green" nanomaterial fabrication based on bio constituent-derived materials is expected to be employed extensively in environmental cleanliness as well as in other important fields like the food, cosmetics, pharmaceutical etc. Organic, inorganic, pharmaceutical, and also heavy metal contaminants can be successfully and economically removed from aqueous streams using green-synthesized nanomaterials and biogenic nano catalysts. Therefore, due to their effective potential, the applications of nanomaterials are expected to increase significantly in the near future and it will execute an important role in a sustainable improvement.

KEYWORDS

- **nanomaterials**
- **green synthesis**
- **environmental friendly**
- **natural resources**
- **catalyst**

REFERENCES

Aarthye, P.; Sureshkumar, M. Green Synthesis of Nanomaterials: An Overview. *Mater. Today: Proc.* **2021,** *47,* 907–913.

Abd Elsalam, S. S.; Taha, R. H.; Tawfeik, A. M. Antimicrobial Activity of Bio and Chemical Synthesized Cadmium Sulfide Nanoparticles. *Egypt. J. Hosp. Med.* **2018,** *70* (9), 1494–1507.

Abdelghany, T. M.; Al-Rajhi, A. M. H.; Al Abboud, M. A.; Alawlaqi, M. M.; Ganash Magdah, A.; Helmy, E. A. M.; Mabrouk, A. S.. Recent Advances in Green Synthesis of Silver Nanoparticles and their Applications: About Future Directions. A Review. *Bionanoscience* **2018,** *8* (1), 5–16.

Ahmed, A. A.; Hamzah, H.; Maaroof, M. Analyzing Formation of Silver Nanoparticles from the Filamentous Fungus Fusarium Oxysporum and their Antimicrobial Activity. *Turk. J. Biol.* **2018,** *42* (1), 54–62.

Arole, V. M.; Munde, S. V. Fabrication of Nanomaterials by Top-Down and Bottom-up Approaches—An Overview. *JAAST: Mater. Sci.* **2014,** *1* (2), 2–89.

Asghari, F.; Jahanshiri, Z.; Imani, M.; Shams-Ghahfarokhi, M.; Razzaghi-Abyaneh, M. Antifungal Nanomaterials: Synthesis, Properties, and Applications. In *Nanobiomaterials in Antimicrobial Therapy: Applications of Nanobiomaterials*; Elsevier Inc., 2016.

Aswathy Aromal, S.; Philip, D. Green Synthesis of Gold Nanoparticles Using Trigonella Foenum-Graecum and its Size-Dependent Catalytic Activity. *Spectrochim. Acta A Mol. Biomol. Spectrosc.* **2012,** *97,* 1–5.

Aversa, R.; Modarres, M. H., Cozzini, S.; Ciancio, R.; Chiusole, A. Data Descriptor: The First Annotated Set of Scanning Electron Microscopy Images for Nanoscience. *Sci. Data* **2018,** *5*, 180172.

Bhadwal, A. S.; Tripathi, R. M.; Gupta, R. K.; Kumar, N.; Singh, R. P.; Shrivastav, A. Biogenic Synthesis and Photocatalytic Activity of CdS Nanoparticles. *RSC Adv.* **2014,** *4* (19), 9484–9490.

Brabazon, D.; Pellicer, E.; Zivic, F.; Sort, J.; Baró, M. D.; Grujovic, N.; Choy, K. L. Commercialization of Nanotechnologies—A Case Study Approach. In *Commercialization of Nanotechnologies—A Case Study Approach*; 2017, pp 1–315.

Bui, T. M. A.; Nguyen, T. V.; Nguyen, T. M.; Hoang, T. H.; Nguyen, T. T. H.; Lai, T. H.; Tran, T. N.; Nguyen, V. H.; Hoang, V. H.; Le, T. L.; Tran, D. L.; Dang, T. C.; Vu, Q. T.; Nguyen-Tri, P. Investigation of Crosslinking, Mechanical Properties and Weathering

Stability of Acrylic Polyurethane Coating Reinforced by SiO2 Nanoparticles Issued from Rice Husk Ash. *Mater. Chem. Phys.* **2020,** *241* (August 2019), 122445.

Cao, S.; Zhao, C.; Han, T.; Peng, L. Hydrothermal Synthesis, Characterization and Gas Sensing Properties of the WO3 Nanofibers. *Mater. Lett.* **2016,** *169,* 17–20.

Chen, X.; Shen, Y.; Zhou, P.; Zhong, X.; Li, G.; Han, C.; Wei, D.; Li, S. Bimetallic Au/Pd Nanoparticles Decorated ZnO Nanowires For NO2 Detection. *Sens. Actuators B Chem.* **2019,** *289* (2), 160–168.

Cheng, H.; Ma, J.; Zhao, Z.; Qi, L. Hydrothermal Preparation of Uniform Nanosize Rutile and Anatase Particles. *Chem. Mater.* **1995,** *7* (4), 663–671.

Chrzanowska, J.; Hoffman, J.; Małolepszy, A.; Mazurkiewicz, M.; Kowalewski, T. A.; Szymanski, Z.; Stobinski, L. Synthesis of Carbon Nanotubes by the Laser Ablation Method: Effect of Laser Wavelength. *Phys. Status Solidi B Basic Res.* **2015,** *252* (8), 1860–1867.

Das, K. C.; Dhar, S. S.; Thakurata, D. G.; Das, J. Sn(II) Inserted On Hydroxyapatite Encapsulated Nickel Ferrite (NiFe2O4@HAp-Sn2+): A Novel Nanocomposite for the Effective Photo-Degradation of Rhodamine B Dye. *J. Clean. Prod.* **2021,** *290,* 125172.

Dingreville, R.; Qu, J.; Mohammed Cherkaoui. Surface Free Energy and its Effect on the Elastic Behavior of Nano-Sized Particles, Wires and Films. *J. Mech. Phys. Solids* **2005,** *53*(8), 1827–1854.

Duque, J. S.; Madrigal, B. M.; Riascos, H.; Avila, Y. P. Colloidal Metal Oxide Nanoparticles Prepared by Laser Ablation Technique and their Antibacterial Test. *Colloids Interfaces* **2019,** *3* (1).

Francis, S.; Joseph, S.; Koshy, E. P.; Mathew, B. Green Synthesis and Characterization of Gold and Silver Nanoparticles Using Mussaenda Glabrata Leaf Extract and Their Environmental Applications to Dye Degradation. *Env. Sci. Pollut. Res.* **2017,** *24* (21), 17347–17357.

Gao, S.; Shi, Y.; Zhang, S.; Jiang, K.; Yang, S.; Li, Z.; Takayama-Muromachi, E. Biopolymer-Assisted Green Synthesis of Iron Oxide Nanoparticles and their Magnetic Properties. *J. Phys. Chem. C* **2008,** *112* (28), 10398–10401.

Geetha, R.; Ashokkumar, T.; Tamilselvan, S.; Govindaraju, K.; Sadiq, M.; Singaravelu, G. Green Synthesis of Gold Nanoparticles and their Anticancer Activity. *Cancer Nanotechnol.* **2013,** *4* (4–5), 91–98.

Ghidan, A. Y.; Al-Antary, T. M.; Awwad, A. M. Green Synthesis of Copper Oxide Nanoparticles Using Punica Granatum Peels Extract: Effect on Green Peach Aphid. *Environ. Nanotechnol. Monit. Manag.* **2016,** *6,* 95–98.

Gleiter, H. Nanostructured Materials: Basic Concepts and Microstructure. *Acta Mater.* **2000,** *48* (1), 1–29.

Gonnelli, C.; Cacioppo, F.; Giordano, C.; Capozzoli, L.; Salvatici, M. C.; Colzi, I.; Del Bubba, M.; Ancillotti, C.; Ristori, S. Cucurbita Pepo L. Extracts as a Versatile Hydrotropic Source for the Synthesis of Gold Nanoparticles with Different Shapes. *Green Chem. Lett. Rev.* **2015,** *8* (1), 39–47.

Gour, A.; Jain, N. K. Advances in Green Synthesis of Nanoparticles. *Artif. Cells Nanomed. Biotechnol.* **2019,** *47* (1), 844–851.

Gunalan, S.; Sivaraj, R.; Rajendran, V. Green Synthesized ZnO Nanoparticles Against Bacterial and Fungal Pathogens. *Prog. Nat. Sci.: Mater. Int.* **2012,** *22* (6), 693–700.

Gupta, P.; Kour, J.; Bakshi, M.; Kalsi, R. Flavonoids. In *Nutraceuticals and Health Care*; 2021, Vol. 2013, pp 105–113.

Hassan, M. L.; Ali, A. F.; Salama, A. H.; Abdel-Karim, A. M. Novel Cellulose Nanofibers/Barium Titanate Nanoparticles Nanocomposites and their Electrical Properties. *J. Phys. Org. Chem.* **2019,** *32* (2), 1–13.

He, F.; Zhao, D. Response to Comment on Manipulating the Size and Dispersibility of Zerovalent Iron Nanoparticles by Use of Carboxymethyl Cellulose Stabilizers. *Environ. Sci. Technol.* **2008,** *42* (9), 3480.

Henkel, C.; Wittmann, J. E.; Träg, J.; Will, J.; Stiegler, L. M. S.; Strohriegl, P.; Hirsch, A.; Unruh, T.; Zahn, D.; Halik, M.; Guldi, D. M. Mixed Organic Ligand Shells: Controlling the Nanoparticle Surface Morphology Toward Tuning the Optoelectronic Properties. *Small* **2020,** *16* (2).

Ismail, R. A.; Mohsin, M. H.; Ali, A. K.; Hassoon, K. I.; Erten-Ela, S. Preparation and Characterization of Carbon Nanotubes by Pulsed Laser Ablation in Water for Optoelectronic Application. *Phys. E: Low-Dimens. Syst. Nanostructures* **2020,** *119* (May 2019), 113997.

Jeeva, K.; Thiyagarajan, M.; Elangovan, V.; Geetha, N.; Venkatachalam, P. Caesalpinia Coriaria Leaf Extracts Mediated Biosynthesis of Metallic Silver Nanoparticles and their Antibacterial Activity Against Clinically Isolated Pathogens. *Ind. Crops Prod.* **2014,** *52,* 714–720.

Jenkner, P. K.; Spielberger, A.; Hengge, E. Synthesis and Characterization of Functional Cyclopentasilanes. *Heteroat. Chem.* **1992,** *3* (4), 443–447.

Jones, A. C.; ; Hitchman, M. L. Overview of Chemical Vapour Deposition. In *Royal Society of Chemistry*; 2009.

Kalita, J.; Das, B.; Dhar, S. S. Synergistic Effect of Iron and Copper in Hydroxyapatite Nanorods for Fenton-Like Oxidation of Organic Dye. *Colloids Surf. A: Physicochem. Eng. Asp.* **2022,** *643* (February), 128750.

Kim, C.; Lee, H. Light-Assisted Surface Reactions on Metal Nanoparticles. *Catal. Sci. Technol.* **2018,** *8* (15), 3718–3727.

Kitamura, Y.; Okinaka, N.; Shibayama, T.; Mahaney, O. O. P.; Kusano, D.; Ohtani, B.; Akiyama, T. Combustion Synthesis of TiO2 Nanoparticles as Photocatalyst. *Powder Technol.* **2007,** *176* (2–3), 93–98.

Kolahalam, L. A.; Kasi Viswanath, I. V.; Diwakar, B. S.; Govindh, B.; Reddy, V.; Murthy, Y. L. N. Review on Nanomaterials: Synthesis and Applications. *Mater. Today: Proc.* **2019,** *18,* 2182–2190.

Kou, J.; Varma, R. S. Beet Juice-Induced Green Fabrication of Plasmonic AgCl/Ag Nanoparticles. *Chemsuschem* **2012,** *5* (12), 2435–2441.

Koutsawa, Y.; Tiem, S.; Yu, W.; Addiego, F.; Giunta, G. A Micromechanics Approach for Effective Elastic Properties of Nano-Composites with Energetic Surfaces/Interfaces. *Compos. Struct.* **2017,** *159,* 278–287.

Krishna, R.; Titus, E.; Krishna, R.; Bardhan, N.; Bahadur, D.; Gracio, J. Wet-Chemical Green Synthesis of L-Lysine Amino Acid Stabilized Biocompatible Iron-Oxide Magnetic Nanoparticles. *J. Nanosci. Nanotechnol.* **2012,** *12* (8), 6645–6651.

Kumar, P. S.; Sundaramurthy, J.; Sundarrajan, S.; Babu, V. J.; Singh, G.; Allakhverdiev, S. I.; Ramakrishna, S. Hierarchical Electrospun Nanofibers for Energy Harvesting, Production and Environmental Remediation. *Energy Environ. Sci.* **2014,** *7* (10), 3192–3222.

Lamouri, R.; Mounkachi, O.; Salmani, E.; Hamedoun, M.; Benyoussef, A.; Ez-Zahraouy, H. Size Effect on the Magnetic Properties of CoFe2O4 Nanoparticles: A Monte Carlo Study. *Ceram. Int.* **2020,** *46* (6), 8092–8096.

Lara, H. H.; Guisbiers, G.; Mendoza, J.; Mimun, L. C.; Vincent, B. A.; Lopez-Ribot, J. L.; Nash, K. L. Synergistic Antifungal Effect of Chitosan-Stabilized Selenium Nanoparticles Synthesized by Pulsed Laser Ablation in Liquids Against Candida Albicans Biofilms. *Int. J. Nanomed.* **2018,** *13,* 2697–2708.

Lee, S. K.; Mclaurin, E. J. Recent Advances in Colloidal Indium Phosphide Quantum Dot Production. *Curr. Opin. Green Sustain. Chem.* **2018,** *12,* 76–82.

Li, Z.; Wang, L.; Li, Y.; Feng, Y.; Feng, W. Carbon-Based Functional Nanomaterials: Preparation, Properties and Applications. *Comp. Sci. Technol.* **2019,** *179* (April), 10–40.

Liu, Y.; Pang, H.; Wang, X.; Yu, S.; Chen, Z.; Zhang, P.; Chen, L.; Song, G.; Saleh Alharbi, N.; Omar Rabah, S.; Wang, X. Zeolitic Imidazolate Framework-Based Nanomaterials for the Capture of Heavy Metal Ions and Radionuclides: A Review. *Chem. Eng. J.* **2021,** *406,* 127139.

Lozhkomoev, A. S.; Pervikov, A. V.; Chumaevsky, A. V.; Dvilis, E. S.; Paygin, V. D.; Khasanov, O. L.; Lerner, M. I. Fabrication of Fe-Cu Composites From Electroexplosive Bimetallic Nanoparticles by Spark Plasma Sintering. *Vacuum* **2019,** *170* (September), 108980.

Lyu, H.; Gao, B.; He, F.; Ding, C.; Tang, J.; Crittenden, J. C. Ball-Milled Carbon Nanomaterials for Energy and Environmental Applications. *ACS Sustain. Chem. Eng.* **2017,** *5* (11), 9568–9585.

Machac, P.; Cichon, S.; Lapcak, L.; Fekete, L. Graphene Prepared by Chemical Vapour Deposition Process. *Graph. Technol.* **2020,** *5* (1–2), 9–17.

Malik, M. A.; Wani, M. Y.; Hashim, M. A. Microemulsion Method: A Novel Route to Synthesize Organic and Inorganic Nanomaterials. 1st Nano Update. *Arab. J. Chem.* **2012,** *5* (4), 397–417.

Mallikarjuna, N. N.; ; Varma, R. S. Microwave-Assisted Shape-Controlled Bulk Synthesis of Noble Nanocrystals and their Catalytic Properties. *Cryst. Growth Des.* **2007,** *7* (4), 686–690.

Manasreh, O. *Introduction to Nanomaterials and Devices*; 2011.

Mani, A. D. *Combustion Synthesis of Nanomaterials For Photocatalytic And Photovoltaic Applications*; 2014.

Manoj, D.; Saravanan, R.; Santhanalakshmi, J.; Agarwal, S.; Gupta, V. K.; Boukherroub, R. Towards Green Synthesis of Monodisperse Cu Nanoparticles: an Efficient and High Sensitive Electrochemical Nitrite Sensor. *Sens. Actuators B Chem.* **2018,** *266,* 873–882.

Menon, S. S.; R.; S.; V. K. (2017). A Review On Biogenic Synthesis of Gold Nanoparticles, Characterization, and its Applications. *Resource-Efficient Technol.* **2017,** *3* (4), 516–527.

Mohan Kumar, K.; Mandal, B. K.; Siva Kumar, K.; Sreedhara Reddy, P.; Sreedhar, B. Biobased Green Method to Synthesise Palladium and Iron Nanoparticles Using Terminalia Chebula Aqueous Extract. *Spectrochimica Acta—Part A: Molecular and Biomolecular Spectroscopy*, **2013,** *102,* 128–133.

Muhammad, F.; Tahir, M.; Zeb, M.; Wahab, F.; Kalasad, M. N.; Khan, D. N.; Karimov, K. S. Cadmium Selenide Quantum Dots: Synthesis, Characterization and their Humidity and Temperature Sensing Properties with Poly-(Dioctylfluorene). *Sens. Actuators B Chem.* **2019,** *285* (November 2018), 504–512.

Nadagouda, M. N.; Varma, R. S. A Greener Synthesis of Core (Fe, Cu)-Shell (Au, Pt, Pd, and Ag) Nanocrystals Using Aqueous Vitamin C. *Cryst. Growth Des.* **2007a,** *7* (12), 2582–2587.

Nadagouda, M. N.; Varma, R. S. Synthesis of Thermally Stable Carboxymethyl Cellulose/Metal Biodegradable Nanocomposites for Potential Biological Applications. *Biomacromolecules* **2007b,** *8* (9), 2762–2767.

Nadagouda, M. N.; Varma, R. S. Green Synthesis of Silver and Palladium Nanoparticles at Room Temperature Using Coffee and Tea Extract. *Green Chemistry* **2008,** *10* (8), 859–886.

Naseer, M.; Aslam, U.; Khalid, B.; Chen, B. Green Route To Synthesize Zinc Oxide Nanoparticles Using Leaf Extracts of Cassia Fistula and Melia Azadarach and their Antibacterial Potential. *Sci. Rep.* **2020,** *10* (1), 1–10.

Ngamdee, K.; Ngeontae, W. Circular Dichroism Glucose Biosensor Based on Chiral Cadmium Sulfide Quantum Dots. *Sens. Actuat. B: Chem.* **2018,** *274* (April), 402–411.

Nguyen, T. D. From Formation Mechanisms to Synthetic Methods Toward Shape-Controlled Oxide Nanoparticles. *Nanoscale* **2013,** *5* (20), 9455–9482.

Nomai, J.; Schlarb, A. K. Effects of Nanoparticle Size and Concentration on Optical, Toughness, and Thermal Properties of Polycarbonate. *J. Appl. Polym. Sci.* **2019,** *136* (23), 1–7.

Oh, Y.; Lee, J.; Lee, M. Fabrication of Ag-Au Bimetallic Nanoparticles by Laser-Induced Dewetting of Bilayer Films. *Appl. Surf. Sci.* **2018,** *434,* 1293–1299.

Órdenes-Aenishanslins, N. A.; Saona, L. A.; Durán-Toro, V. M.; Monrás, J. P.; Bravo, D. M.; Pérez-Donoso, J. M. Use of Titanium Dioxide Nanoparticles Biosynthesized by Bacillus Mycoides in Quantum Dot Sensitized Solar Cells. *Microb. Cell Factories* **2014,** *13* (1), 1–10.

Ostermann, R.; Cravillon, J.; Weidmann, C.; Wiebcke, M.; Smarsly, B. M. Metal-Organic Framework Nanofibers via Electrospinning. *Chem. Commun.* **2011,** *47* (1), 442–444.

Padi, V. V. T.; Cernik, M. Green Synthesis of Copper Oxide Nanoparticles Using Gum Karaya as a Biotemplate and their Antibacterial Application. *Int. J. Nanomed.* **2013,** *8,* 889–898.

Pattanayak, M.; Mohapatra, D.; Nayak, P. L. Green Synthesis and Characterization of Zero Valent Iron Nanoparticles from the Leaf Extract of Syzygium Aromaticum (Clove). *Middle East J. Sci. Res.* **2013,** *18* (5), 623–626.

Prabhakar, R.; Samadder, S. R.; Jyotsana. Aquatic and Terrestrial Weed Mediated Synthesis of Iron Nanoparticles for Possible Application in Wastewater Remediation. *J. Clean. Prod.* **2017,** *168,* 1201–1210.

Prakash, S.; Elavarasan, N.; Venkatesan, A.; Subashini, K.; Sowndharya, M.; Sujatha, V. Green Synthesis of Copper Oxide Nanoparticles and its Effective Applications in Biginelli Reaction, BTB Photodegradation and Antibacterial Activity. *Adv. Powder Technol.* **2018,** *29* (12), 3315–3326.

Qu, J.; Luo, C.; Hou, J. Synthesis of ZnO Nanoparticles from Zn-Hyperaccumulator (Sedum Alfredii Hance) Plants. *Micro Nano Lett.* **2011,** *6* (3), 174–176.

Rajabi, H. R.; Karimi, F.; Kazemdehdashti, H.; Kavoshi, L. Fast Sonochemically-Assisted Synthesis of Pure and Doped Zinc Sulfide Quantum Dots and Their Applicability in Organic Dye Removal from Aqueous Media. *J. Photochem. Photobiol. B: Biol.* **2018,** *181* (November 2017), 98–105.

Sadeghi, B.; Gholamhoseinpoor, F. A Study on the Stability and Green Synthesis of Silver Nanoparticles Using Ziziphora Tenuior (Zt) Extract at Room Temperature. *Spectrochim. Acta A Mol. Biomol. Spectrosc.* **2015,** *134,* 310–315.

Saha, R.; Arunprasath, D.; Sekar, G. Surface Enriched Palladium on Palladium-Copper Bimetallic Nanoparticles as Catalyst for Polycyclic Triazoles Synthesis. *J. Catal.* **2019,** *377,* 673–683.

Saif, S.; Tahir, A.; Chen, Y. Green Synthesis of Iron Nanoparticles and their Environmental Applications and Implications. *Nanomaterials* **2016,** *6* (11), 1–26.

Saleh, T. A. Nanomaterials: Classification, Properties, and Environmental Toxicities. *Environ. Technol. Innov.* **2020,** *20,* 101067.

Saleh, T. A.; Fadillah, G. Recent Trends in the Design of Chemical Sensors Based on Graphene–Metal Oxide Nanocomposites for the Analysis of Toxic Species and Biomolecules. *TrAC, Trends Anal. Chem.* **2019,** *120.*

Sandoval-Cárdenas, D. I.; Gómez-Ramírez, M.; Rojas-Avelizapa, N. G.; Vidales-Hurtado, M. A. Synthesis of Cadmium Sulfide Nanoparticles by Biomass of Fusarium Oxysporum f. sp. Lycopersici. *J. Nano Res.* **2017,** *46,* 179–191.

Saravanakumar, K.; Shanmugam, S.; Varukattu, N. B.; Mubarakali, D.; Kathiresan, K.; Wang, M. H. Biosynthesis and Characterization of Copper Oxide Nanoparticles from Indigenous Fungi and its Effect of Photothermolysis on Human Lung Carcinoma. *J. Photochem. Photobiol. B: Biol.* **2019,** *190* (October 2018), 103–109.

Sayyad, A. S.; Balakrishnan, K.; Ci, L.; Kabbani, A. T.; Vajtai, R.; Ajayan, P. M. Synthesis of Iron Nanoparticles From Hemoglobin and Myoglobin. *Nanotechnology* **2012,** *23* (5).

Shankar, S. S.; Ahmad, A.; Sastry, M. Geranium Leaf Assisted Biosynthesis of Silver Nanoparticles. *Biotechnol. Prog.* **2003,** *19* (6), 1627–1631.

Sharma, V. K.; Filip, J.; Zboril, R.; Varma, R. S. Natural Inorganic Nanoparticles-Formation, Fate, and Toxicity in the Environment. *Chem. Soc. Rev.* **2015,** *44* (23), 8410–8423.

Shekhawat, M. S.; Ravindran, C. P.; Manokari, M. An Ecofriendly Method for the Synthesis of Zinc Oxide Nanoparticles Using Lawsonia Inermis L. Aqueous Extracts. *Int. J. Innovat.* **2015,** *5* (1), 1.

Shi, Y.; Liu, M.; Deng, F.; Zeng, G.; Wan, Q.; Zhang, X.; Wei, Y. Recent Progress and Development on Polymeric Nanomaterials for Photothermal Therapy: A Brief Overview. *J. Mater. Chem. B,* **2017,** *5* (2), 194–206.

Shiau, B. W.; Lin, C. H.; Liao, Y. Y.; Lee, Y. R.; Liu, S. H.; Ding, W. C.; Lee, J. R. The Characteristics and Mechanisms of Au Nanoparticles Processed by Functional Centrifugal Procedures. *J. Phys. Chem. Solids* **2018,** *116* (62), 161–167.

Shivaji, S.; Madhu, S.; Singh, S. Extracellular Synthesis of Antibacterial Silver Nanoparticles Using Psychrophilic Bacteria. *Proc. Biochem.* **2011,** *46* (9), 1800–1807.

Shrimali, V. G.; Gadani, K.; Rajyaguru, B.; Gohil, H.; Chudasama, D. K.; Dhruv, D.; Joshi, A. D.; Pandya, D. D.; Solanki, P. S.; Shah, N. A. Size Dependent Dielectric, Magnetic, Transport and Magnetodielectric Properties of BiFe0.98Co0.02O3 Nanoparticles. *J. Alloys Compound.* **2020,** *817,* 152685.

Silva, A. A.; Pinheiro, R. A.; Rodrigues, A. C.; Baldan, M. R.; Trava-Airoldi, V. J.; Corat, E. J. Graphene Sheets Produced by Carbon Nanotubes Unzipping and their Performance as Supercapacitor. *Appl. Surf. Sci.* **2018,** *446,* 201–208.

Singaravelu, G.; Arockiamary, J. S.; Kumar, V. G.; Govindaraju, K. A Novel Extracellular Synthesis of Monodisperse Gold Nanoparticles Using Marine Alga, Sargassum Wightii Greville. *Colloids Surf. B: Biointerfaces* **2007,** *57* (1), 97–101.

Sumi, N.; Chitra, K. C. Fullerene C60 Nanomaterial Induced Oxidative Imbalance in Gonads of the Freshwater Fish, Anabas Testudineus (Bloch, 1792). *Aquat. Toxicol.* **2019,** *210* (January), 196–206.

Ulug, B.; Haluk Turkdemir, M.; Cicek, A.; Mete, A. Role of Irradiation in the Green Synthesis of Silver Nanoparticles Mediated by Fig (Ficus Carica) Leaf Extract. *Spectrochim. Acta A Mol. Biomol. Spectrosc.* **2015,** *135,* 153–161.

Wang, J. L.; Meng, L. J. Influence of Carbon Nano-Fiber on Mechanical Property of PALC. *Appl. Mech. Mater.* **2014,** *535,* 785–787.

Wang, X.; Xu, C.; Yi, M.; Zhang, H. Effects of Nano-ZrO2 on the Microstructure and Mechanical Properties of Ti(C, N)-Based Cermet Die Materials. *Adv. Mater. Res.* **2011,** *154–155,* 1319–1323.

Waris, A.; Din, M.; Ali, A.; Ali, M.; Afridi, S.; Baset, A.; Ullah Khan, A. A Comprehensive Review of Green Synthesis of Copper Oxide Nanoparticles and their Diverse Biomedical Applications. *Inorg. Chem. Commun.* **2021,** *123* (December 2020), 108369.

Wei, Q.; Kang, S. Z.; Mu, J. "Green" Synthesis of Starch Capped CdS Nanoparticles. *Colloids Surf. A: Physicochem. Eng. Asp.* **2004,** *247* (1–3), 125–127.

Wen, M.; Li, G.; Liu, H.; Chen, J.; An, T.; Yamashita, H. Metal-Organic Framework-Based Nanomaterials for Adsorption and Photocatalytic Degradation of Gaseous Pollutants: Recent Progress and Challenges. *Environ. Sci. Nano* **2019,** *6* (4), 1006–1025.

Wu, Q.; Wongwiriyapan, W.; Park, J. H.; Park, S.; Jung, S. J.; Jeong, T.; Lee, S.; Lee, Y. H.; Song, Y. J. In Situ Chemical Vapor Deposition of Graphene and Hexagonal Boron Nitride Heterostructures. *Curr. Appl. Phys.* **2016,** *16* (9), 1175–1191.

Wu, S.; Rajeshkumar, S.; Madasamy, M.; Mahendran, V. Green Synthesis of Copper Nanoparticles Using Cissus Vitiginea and its Antioxidant and Antibacterial Activity Against Urinary Tract Infection Pathogens. *Artif. Cells Nanomed. Biotechnol.* **2020,** *48* (1), 1153–1158.

Wu, X.; Lu, G. Q.; Wang, L. Shell-In-Shell TiO2 Hollow Spheres Synthesized by One-Pot Hydrothermal Method for Dye-Sensitized Solar Cell Application. *Energy Environ. Sci.* **2011,** *4* (9), 3565–3572.

Yue, L. Y.; Zhang, Z. C.; Chen, X. Absorption and Fluorescence Spectra of Gallium Phosphide (Gap) Nanoparticles. *Trans. Nonferrous Met. Soc. China (English Edition)*, **2006,** 16 (4), 863–867.

Zeb, S.; Ali, N.; Ali, Z.; Bilal, M.; Adalat, B.; Hussain, S.; Gul, S.; Ali, F.; Ahmad, R.; Khan, S.; Iqbal, H. M. N. Silica-Based Nanomaterials as Designer Adsorbents to Mitigate Emerging Organic Contaminants from Water Matrices. *J. Water Proc. Eng.* **2020,** *38* (September), 101675.

Zhang, D.; Ye, K.; Yao, Y.; Liang, F.; Qu, T.; Ma, W.; Yang, B.; Dai, Y.; Watanabe, T. Controllable Synthesis of Carbon Nanomaterials by Direct Current Arc Discharge from the Inner Wall of the Chamber. *Carbon* **2019,** *142,* 278–284.

Zhang, J.; Chaker, M.; Ma, D. Pulsed Laser Ablation Based Synthesis of Colloidal Metal Nanoparticles for Catalytic Applications. *J. Colloid Interface Sci.* **2017,** *489*, 138–149.

Zinatloo-Ajabshir, S.; Baladi, M.; Salavati-Niasari, M. Sono-Synthesis of MnWO4 Ceramic Nanomaterials as Highly Efficient Photocatalysts for the Decomposition of Toxic Pollutants. *Ceram. Int.* **2021,** *47* (21), 30178–30187.

Zou, B.; Huang, C. Z.; Wang, J.; Liu, B. Q. (2006). Effect of Nano-Scale Tin on the Mechanical Properties and Microstructure of Si3N4 Based Ceramic Tool Materials. *Key Eng. Mater.* **2006,** *315–316,* 154–158.

CHAPTER 9

Green-Synthesized Nanomaterials' Potential for Environmental Pollution Remediation

SUMEL ASHIQUE[1], AFZAL HUSSAIN[2], NEERAJ MISHRA[3], ARSHAD FARID[4], MOHD. RIHAN[5], and ASHISH GARG[6]

[1]*Department of Pharmaceutics, Bharat Institute of Technology (BIT), School of Pharmacy, Meerut, Uttar Pradesh, India*

[2]*Department of Pharmaceutics, College of Pharmacy, King Saud University, Riyadh, Saudi Arabia*

[3]*Department of Pharmaceutics, Amity Institute of Pharmacy, Amity University, Gwalior, Madhya Pradesh, India*

[4]*Gomal Center of Biochemistry and Biotechnology, Gomal University, Dera Ismail Khan, Pakistan*

[5]*Department of Pharmacology, National Institute of Pharmaceutical Education and Research (NIPER), S.A.S. Nagar, Punjab, India*

[6]*Department of P.G. Studies and Research in Chemistry and Pharmacy, Rani Durgavati University, Jabalpur, Madhya Pradesh, India*

ABSTRACT

Environmental remediation is the process of eliminating harmful pollutants from different environmental sources (water, soil, air), which includes all the remediation approaches for diminishing the risks of environmental pollution. Currently, nanotechnology has become a promising approach for the successful removal of environment-associated contamination. Although

Sustainable Nanoremediation: Modern Technologies for a Clean Environment.
Fernanda Maria Policarpo Tonelli, Arpita Roy, and Khalid Rehman Hakeem (Eds.)

these advances in nanotechnology have numerous benefits, various nanomaterial development methods have a negative impact on the environment as well as on human health. Therefore, researchers are now considering the eco-friendly plant-based synthesis of nanomaterials for various environmental remediation. Recently, the scientific community has been considering the utilization of various biomaterials for synthesizing nanoparticles (NPs) and affirmed that this method is safe, less toxic, and eco-friendly which is known as green nanotechnology. The use of nanotechnology-based green nanomaterials for environmental remediation plays a significant role due to some unique beneficial features of plant-based nanoparticles. This chapter describes several environmental contaminants, their resources, various green-based nanomaterials, and the application of those nanomaterials for environmental remediation.

9.1 INTRODUCTION

The world is on the verge of a serious environmental catastrophe that will bankrupt us. The existing state of the ecosystem is constantly declining. Global environmental problems are becoming more and more urgent, and we must act accordingly. We must adopt a fresh viewpoint and address disasters with fresh ideas and approaches, as well as with full awareness and seriousness. Nature takes millions of years to clean up pollution in the air, water, and soil. The primary causes of the majority of environmental pollution are industry and automotive exhaust emissions (Adeleye et al., 2016). Particles that pollute the environment are referred to as pollutants. Life can be harmed by exposure to these toxins, and the effects on people and other animals are well known. Pollutants can enter the environment in a variety of ways, both naturally and as a result of human activity (C.L et al., 2020). The soil, water, and air all contain mixtures of the majority of pollutants. Nanotechnology offers a range of capabilities and technologies to improve the current environment in this way. Innovative approaches to the synthesis, characterization, and application of nanomaterials in society have been made possible by nanotechnology and the science of miniaturization. Because NPs serve as a link between atomic or molecular systems and bulk material assemblies, there has been an increase in scientific interest in these objects. Many well-characterized bulk materials have intriguing nanoscale characteristics. Compared to most materials, NPs have higher reactivity and efficacy due to their high aspect ratio. Over time, researchers have proven their abilities and created unique nano-based materials and complementary composites (Liu et al., 2011; Sinha et al., 2018; Nassar, 2010). AC adsorption, condensation, flocculation, froth flotation, and filtering are some of the long-used techniques. These, however, have their own drawbacks, including ineffective metal ion scraping, excessive energy input, and the creation of nonrecyclable compounds. Nanotechnology looks to be an emerging solution to these issues (Roy et al., 2021). The use of green nanotechnology (Fig 9.1) could greatly contribute to the creation of cleaner, greener technologies with considerable benefits for the sake of the planet and human health (Ahmed et al, 2021). As well as being explored for their potential to improve the effectiveness of conventional environmental cleanup approaches, nanotechnology techniques may also offer options for pollution management and mitigation (Marlon et al., 2019). By using less energy throughout the production and manufacturing processes, enabling products to be recycled after use, and creating and utilizing environmentally benign materials, nanotechnology can benefit the environment (Kaur and Roy, 2021). Currently, nanotechnology shows great promise for tackling sustainability challenges, but we also need to consider the risks to the environment and public health (Khan, 2020). Therefore, the significance of numerous environmentally friendly nanomaterials that can remove contaminants has been covered in this review. This environmentally friendly synthesis not only creates extremely effective nanocarriers but also completes its intended task without endangering any living things or the environment. This chapter focuses on current research projects that deal with environmentally friendly NP synthesis, which has advantages over conventional methods that involve environmentally harmful chemical agents. The following chapter examines traditional synthetic techniques with a focus on recent innovations in more environmentally friendly ways to produce metal, metal oxide, and other significant NPs.

9.2 SEVERAL CATEGORIES OF ENVIRONMENTAL POLLUTANTS

In the current situation, the world is full of progressive industries, and our environment is becoming contaminated with several categories of harmful contaminants released from human functions as well as from industrial sources. Some examples of these contaminants include particles, hydrocarbons, nitrogen oxides, organic compounds (such as dioxins and volatile organic compounds), sulfur dioxide, and heavy metals (such as arsenic, chromium, lead, cadmium, mercury, and zinc). Natural gas, coal, and oil combustion by humans have the ability to significantly alter emissions from those sources (Yedjou et al., 2012). Additionally, water contamination is a problem, just like air pollution, and is brought on by a number of variables such as trash disposal, oil spills, fertilizer, herbicide, and pesticide leaks, byproducts of industrial operations, combustion, and the mining of fossil fuels (Krantzberg et al., 2010). These harmful contaminants are mixed with the environment by following various mechanisms (naturally and by humankind) (Seewagen, 2020). The several kinds of contaminants, such as insecticides, poly aromatic hydrocarbons, heavy metals, and dyes, are mentioned in Table 9.1 and Figure 9.1.

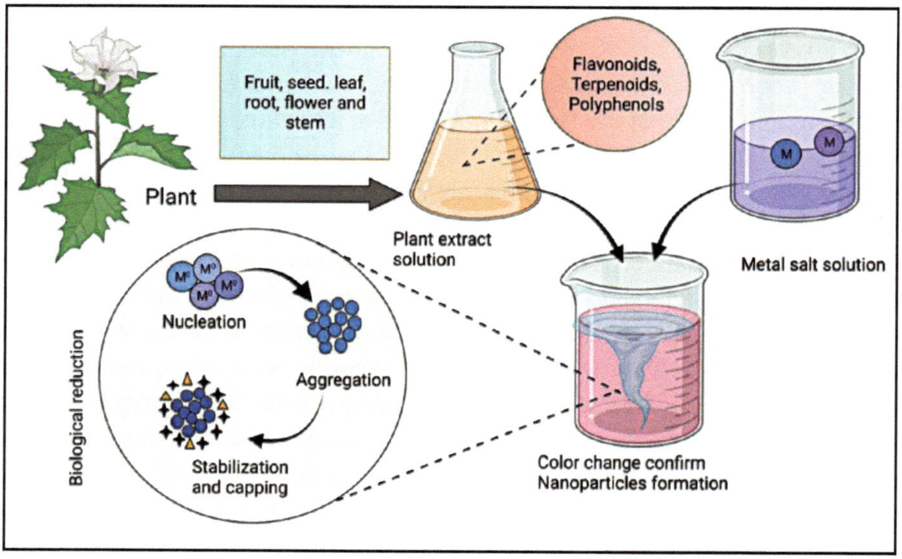

FIGURE 9.1 Schematic representation for plant-mediated biosynthesis of nanoparticles.
Source: Created with BioRender.com

9.3 ENVIRONMENTAL POLLUTION REMOVAL TECHNIQUES USED IN THE PAST AND THEIR NEGATIVE IMPACTS

A combination of classic techniques like electrochemical and biological processes, coagulation, adsorption, and sophisticated oxidation processes (Table 9.2) have been applied prior to the introduction of nanobiotechnology. Mechanisms like traditional methods of coagulation and flocculation approach to polluted water remediation. They are also applied for the pretreatment of biologicals, increasing the rate of biodegradability of waste and micropollutants (Amuda and Amoo, 2007). Some agents like ferric

TABLE 9.1 Various Categories of Harmful Environmental Pollutants.

Pollutants	Description	Ref
Dyes	It is a synthetic organic chemical utilized in various industries (such as textiles). At the time of manufacturing and while fabric dyeing, they cause environmental pollution. The synthetic dyes lead to a reduction of light penetration into rivers, changing the photosynthetic efficacy of aquatic vegetation. Thus, the food supply of aquatic creatures gets hampered.	Kant, 2012
	Biochemical oxygen demand (BOD) gets enhanced because of industrial dye effluents. These persistent organic pollutants (POPs) are xenobiotic in nature.	Pereira and Alves, 2012
	Dyes such as Azure-B can alter the helix DNA and duplex RNA structures seen in living things. The colors Orange 1 and Disperse Red 1 are mutagenic in nature.	Lellis et al., 2019
Heavy Metal	Heavy metal contaminants are having a negative impact on several fields all over the world. Commonly used heavy metals are cobalt (Co), arsenic (As), molybdenum (Mo), mercury (Hg), and lead (Pb).	Briffa et al., 2020
	Toxic heavy metals are mainly present on the Earth's surface. Due to the significant increase in their use, there will soon be an influx of these metallic agents affecting both terrestrial and aquatic resources. Heavy metal contamination is becoming an insignificant issue in anthropogenic function, commonly because of metal mining, steelmaking, functions like foundries and other sectors relying on metal.	Tchounwou et al., 2012
Pesticides	Applied to eliminate undesirable organisms from gardens, farms, and other public areas. They occasionally pose a risk to all living beings, the environment, and food safety.	Hassaan and El Nemr, 2020
	Most pesticides (like DDT) react with the receptor present in the adipose fat tissue and accumulate in lipids. DDT is lipophilic by nature, so when humans consume any aquatic animal that has consumed it along with food, it too accumulates in the body.	Z.-Q. Jia et al., 2020
	These pesticides lead to inhibition of reproduction, alteration of the hormonal system occurs, inhibition of the immune system, chances of malignancies, and impact on teratogens, DNA, and cell apoptosis. This happens due to these pollutants.	Hwang et al., 2018
Polyaromatic hydrocarbons	As they are highly toxic, mutagenic, and carcinogenic, they are classified as persistent environmental pollutants, which are dangerous to people. Low molecular weight polyaromatic hydrocarbons are freely soluble in water. The Environmental Protection Agency has identified such compounds as naphthalene, phenanthrene ($C4H10$), anthracene ($(C6H4CH)2$), fluoranthene ($C16H10$), pyrene ($C16H10$), and chrysene ($C18H12$), which must be kept in low concentration with high consideration.	Masih et al., 2010

chloride and polymers are mixed, which decompose the colloidal particles and make the small particles aggregate into large flocs which can easily precipitate. It is a physicochemical method by which metal contaminants can be removed from wastewater by removing the soluble components by generating flocs (Iwuozor, 2019). Additionally, it aids in the removal of organic materials and suspended solids (SS). For waste sanitary landfill leachates, coagulation and ozonation are used (Poznyak et al., 2008), whereas adsorption is a surface mechanism that can remove inorganic contaminants (Zambianchi et al., 2017). The adsorption mechanism is associated with activated carbon, which is used to remove a multitude of contaminants. In comparison to other compounds created for contamination removal, natural zeolites have a high abundance and offer superior selectivity for heavy metals including cadmium, copper, and zinc. Large pores, increased surface area, thermal stability, low cost, chemical inertness, and easily modifiable surface functional groups provide them with the ability to exhibit good selectivity toward pollutants (El-Geundi, 1997; Diagboya and Dikio, 2018). Complete decomposition of organic pollutants with mineralization is gained by progressive oxidation approaches. It includes mechanisms including ozonation, the Fenton reaction, UV light, and managing peroxide, and their mixtures. Fenton and solar photo-Fentons, which are able to treat recurring organic compounds, can cure polyphenols. By using a mixture of NaClO and hydrogen peroxide, singlet oxygen and hydroxide ions can be eliminated, creating singlet oxygen and hydroxide ions (Suty et al., 2004). Photolysis (decomposition by light) mixed with hydrogen peroxide is also applied for the decomposition of pesticides by using trichloromethane and carbon tetrachloride as solvents (James et al., 2014). Chedeville et al. (2009) found that ozonation is dependent on the oxidation and/or decomposition of pesticides as a means for greater inhibition in chemical oxygen demand and potent removal of aromatic pollutants (Chedeville, et al., 2009). The Electrochemical Advanced Oxidation mechanism has made it possible to completely mineralize pollutants while also removing them using an environmentally benign process. Even in ideal current and catalyst conditions, bulk hydroxyl ions form (Dirany et al., 2012). For the cleanup of low permeability matrices and the removal and remediation of polluted environments, both organic and inorganic, electrokinetic remediation is a key technique (Hassan et al., 2015). Electrochemical oxidation is a widely considered mechanism for avoiding organic contaminants (Ferro et al., 2006). Biological approaches are being applied as they are less toxic and cost-effective for removing pollutants (Zhang et al., 2021; Ganzenko et al., 2014). They are mixed with other methods such as ozonation and moist air oxidation for their improved effectiveness in removing contaminants that are hazardous and refractory in nature (Jeworski and Heinzle, 2000).

TABLE 9.2 Adverse Events of Applying Traditional Approaches for Harmful Contaminant Removal.

Process to remove contaminants	Adverse effects	Ref
Biological processes	It is unable to treat toxic or refractory organic contaminants and shows certain disadvantages.	Ganzenko et al., 2014
Zeolite adsorption	However, the surface may be very hydrophilic after template removal due to greater surface silanol density, resulting in reduced adsorption for hydrophobic contaminants.	Diagboya and Dikio, 2018
Photocatalysis	Has a greater price for higher mineralization levels. Photocatalysis's post-reaction chemicals are still dangerous.	Y. Pi et al., 2018
Electro kinetics	Inorganic pollutants cannot be removed because of ionic mobility, the crystallization of metal salts, and dehydration.	D. Wen et al., 2021
Electrochemical advanced oxidation methods	Till the last mineralization of pollutants, high cost, greatest energy usage	Ganzenko et al., 2014
Progressed oxidation approach	Mainly influenced by pollutant nature, category, and concentration of oxidants and catalyst toxicity.	Ribeiro et al., 2015
Electrocoagulation	A myriad of designs for reactor development	Mitchell et al., 1999
Ozonation	Challenging process of mineralization because of hydroxyl ion scavengers.	K. Jayaswal et al., 2018
Classical Fenton methods.	Use of hydrogen peroxide for bulk treatments and unsafe methods of transportation and storage	Fernandes et al., 2015

9.4 NANOPARTICLES AND ENVIRONMENTAL REMEDIATION

Environmental contamination has grown into a serious issue that needs to be resolved right away. Surprisingly, there are a remarkable number of polluted sites in the private sector, industry, military, oil fields, landfills, and industrial sectors all over the world. Water aquifers with high chloride concentrations and low dissolved oxygen levels (0.75 mg/L), along with a trace amount of pesticides, oils, greases, heavy metals, and other harmful components, are now cited by Jadhav et al. (2013). Previous research was done by Patil et al. (2014a) and Adetutu et al. (2015), who described the efficiency of various approaches for the green-based treatment strategies for groundwater plumes from chloroethene-polluted places in Victoria, Australia. By considering the trend of applying biological approaches such as bioremediation, phytoremediation, and rhizoremediation, nanotechnology is currently widely accepted for environmental remediation (also known as nanoremediation). Nanomaterial-based remediation currently provides a novel and safe approach for remediation environmental detection and monitoring (Rajan, 2011). Nanomaterials provide significant advantages over other methods for site remediation, and they also increase the efficiency of the overall remediation steps. The existing approaches (thermal treatment, chemical oxidation, and surfactant cosolvent flushing) for environmental remediation are very costly and challenging, so they have been avoided (Patil et al., 2016; Löffler and Edwards, 2006) with the introduction of the quick, low-cost, and effective nanoremediation techniques for the elimination and detoxification of impurities, including bimetallic nanoparticles, nanoscale zeolites, metal oxides, carbon nanotubes, fibers, etc. The most favored among these is nanoscale zero-valent iron (nZVI) (Garner and Keller, 2014). A substantial agent that helps with the removal or breakdown of contaminants from soils and water is produced as a result of the greater surface area involved with their greater function. Low standard inhibition of potential, advantageous quantum size properties, and a large improvement in transport efficiency via groundwater's subsurface matrix are additional factors that have been taken into account for nZVIs (Tosco et al., 2014). These NMs have the characteristics that make it possible to remove a number of dangerous contaminants, including heavy metal ions, polychlorinated biphenyls, organochlorine pesticides, chlorinated compounds, and inorganic anions, by chemical reduction as well as catalysis (Karn et al., 2009). E-nano is becoming a generally acknowledged strategy for environmental applications such as soil, water, and groundwater remediation, oil spill cleanup, wastewater remediation, and water purification by adhering to the benefits offered by this technique (Ghasemzadeh et al., 2014). Ag/Fe, Ni/Fe, and Cu/Fe bimetallic nanoparticles were used to clean water bodies of dangerous contaminants such as heavy metals, hexachlorobenzene, chlorinated chemicals, and explosive substances (Nie et al., 2013).

Carbon nanotubes (CNTs) act as a promising approach for the adsorption of contaminants from water bodies because of their unique adsorption features and are capable of fusing with functional groups, enhancing the affinity toward the target agent (Savage and Diallo, 2005). Activated sludge and wastewater rinsed with organic compounds from dyes, insecticides, medicines, and therapies can be transformed by single-walled or multi-walled carbon nanotubes (CNTs) (Yu et al., 2014). Additionally, the removal of metalloids like arsenic constituents, organic biological adulteration, volatile organic compounds, and dioxins as well as heavy metals like Cr^{3+}, Pb^{2+}, and Zn^{2+} has also been done using CNTs (Li et al., 2003).

9.5 MECHANISM OF NANOTECHNOLOGY ON ENVIRONMENTAL REMEDIATION

Three categories, including (i) environmentally friendly and/or sustainable agents (green chemistry or pollution reduction), (ii) remediation of materials contaminated with hazardous agents, and (iii) sensors for environmental agents, are the main foundations for nanotechnology-based applications in the environmental field (Karn et al., 2011). It must be remembered that these three categories also apply to microbial agents and biological materials even though they are mostly made up of chemical components or nonbiological substances. Recently, nanotechnology-based approaches are playing a promising role in advancing detection techniques and decontamination mechanisms of deleterious biological components (Koper et al., 2002), which are connected, either directly or indirectly, to environmental pollution. The effectiveness of nanotechnology-based use for the remediation of environmental pollutants demonstrates the range of "nonnano" methods for hazardous contaminant cleanup. Adsorptive versus reactive and in-situ versus ex-situ are two key contrasts that define the category of conventional remediation methods and also try nanotechnologies for cleanup. While reactive procedures effect the degradation of hazardous pollutants, sometimes even leading to decontaminated agents, absorptive remediation involves the removal of metal contaminants through sequestration (e.g., CO_2 and H_2O for organic pollutants purposes). Ex situ procedures prefer treatment after relocating the polluting material to a more suitable location, whereas in-situ approaches focus on treating pollutants where they are already located (e.g., pumping polluted groundwater to the surface and treatment in aboveground reactors).

9.6 THE ADVANCEMENT OF SYNTHETIC NANOMATERIALS

The advancement of science and technology has made significant advancements in the creation of nanomaterials, which have several distinctive properties and are advantageous over other bulk materials. The nanoparticles are synthesized by following various methods, including chemical, physical, and biological (green synthesis) (Fig 9.2), and out of these three approaches, the green synthesis process has become a very promising way to overcome the development of hazardous by-products generated by chemical and physical methods, and this method is also more eco-friendly than other methods. For the green synthesis of nanoparticles, the main factors involved are suitable solvent systems and natural sources. The world is progressing toward a quick headway via advanced modern technologies and innovations, and the utilization of green resources has increased rapidly, with valid concerns. Currently, researchers are focusing more on the application of nanotechnology for environmental remediation. The major factors behind choosing for better affinity of nanoparticles toward environmental contaminants, enhanced reactivity, and ease of disposal augment their positive outcomes in environmental remediation (Yunus et al., 2012).

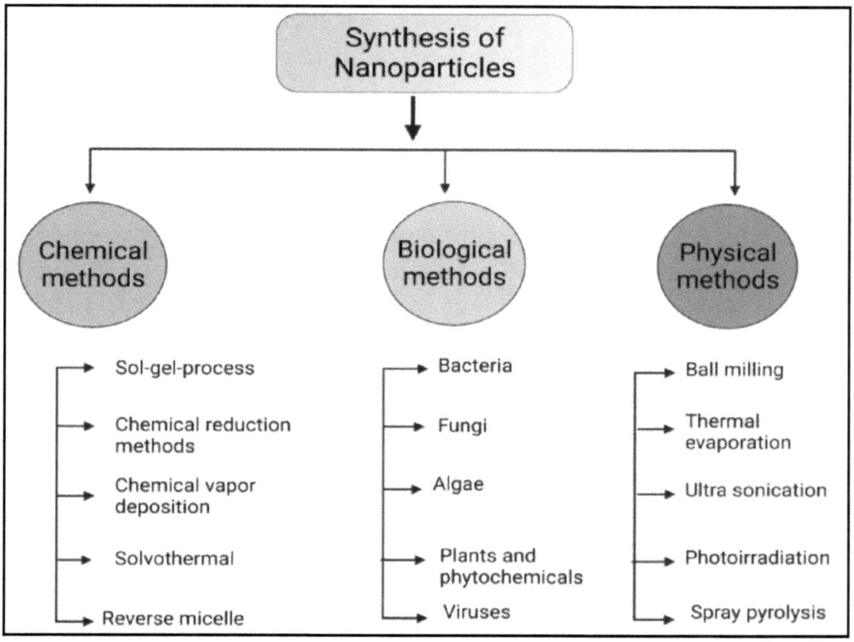

FIGURE 9.2 Schematic representation of various methods adopted for NP synthesis.

Source: Created with BioRender.com

TABLE 9.3 Green Nanoparticle Production Utilizing Various Plants and Their Application in Environmental Remediation.

Plants (species)	Portion of plant	Green (NPs)	Uses	Reference
Acalypha Indica	Leaf	Ag	Anti-bacterial	Krishnaraj et al., 2010
Apiin	Leaf	Ag	Reducing agent	Kasthur et al., 2009
Black Tea	Leaf	Ag and Au		Begum et al., 2009
Alfalfa (Medicago Sativa)	Leaf	Ti-Ni		Schabes-Retchkiman et al., 2006
Mentha Piperita (peppermint)	Leaf	Au	Antibacterial activity.	Mubarak Ali et al., 2011
Jatropha Curcas	Latex	Ag	Both a reducing and capping agent	Bar et al., 2009
Pear fruit	Peel	Au		Ghodake et al., 2010
Sedum Alfredii Hance	Shoots	ZnO		Qu et al., 2011
Medicago Sativa	Leaf	FeO		Herrera-Becerra et al., 2008
Green, oolong, and black tea	Leaf	nZVI	Removing monochlorobenzene (MCB) from wastewaters	Kuang et al., 2013
Tea	Leaf	nZVI	Hexavalent chromium reduction	Mystrioti et al., 2016
Eucalyptus	Leaf	nZVI	Treatment of eutrophic wastewater	Wang et al., 2014
Terminalia Chebula	Fruit	Ag	Reduction of anthropogenic pollutant	Edison and Sethuraman, 2012
Vitis rotundifolia	Fruit	CoO	Decay of the acid blue dye	Samue et al., 2020
Lactuca serriols	Seed	NiO	Degradation of dye	Ali et al., 2021
Punica granatum	Peel	ZnO	Antibacterial property	Abdelmigid et al., 2022
Ficus carica	Leaf	Fe3O4	Antioxidant activity	Üstün et al., 2022
Moringa oleifera	Seed	Fe	Nitrate is removed from water, and there is antibacterial activity	Katata-Seru et al., 2018
Eucalyptus sp	Leaf	Fe	Treatment of eutrophic water	Zhao et al., 2019
Citrus Sinensis	Peel	Ag	Cadmium and chromium absorption activities	Tavker et al., 2021

9.6.1 GREEN NANOPARTICLE PRODUCTION FROM PLANT SOURCES

It is quite simple to create nanoparticles from plant materials, and when making plant-based nanomaterials, a range of reducing and capping agents are used since they are economical, safe, biocompatible, and environmentally benign. The presence of biomolecules and functional groups in plant extracts aids in the development of nanoparticles since they serve as capping and reducing agents during the creation of the particles. The creation of nanoparticles is aided by the powerful components found in plant leaf extracts. The extract and metal precursor solutions are combined at a number of experimental features (Mittal et al., 2013) and the featured parameters are recognized to have an impact on the pace of nanoparticle development along with the stability of the particles (Dwivedi and Gopal, 2010). Green nanoparticles (NPs) bioreduction involves a number of active ingredients, including flavones, terpenoids, sugars, ketones, aldehydes, carboxylic acids, and amides (Mathew et al., 2010). According to Ahmad et al. (2010), the enol-to-keto-transformation tracks the emergence of green silver (Ag) nanoparticles in sweet basil extracts. Panigrahi et al. (2010) also noted that the presence of sugars (glucose and fructose) in plant extracts promotes the synthesis of metallic nanoparticles (Panigrahi et al., 2004). In order to create Ag-NPs, silver nitrate ($AgNO_3$) was combined with a variety of plant extracts, including Alternanthera dentate leaf extract, Acorus calamus rhizome, tea extract, Vitis vinifera fruit extract, Salvadora persica stem extract, Vasaka (Justicia adhatoda L.) leaf extract, Vasaka (Justicia adhatoda L.) extract, and beetroot extract (Bindhu and Umadevi, 2015). Coleus amboinicus, Dillenia indica fruit extract, Dioscorea bulbifera tuber extract, Euphorbia hirta leaf extract, Zingiber officinale, Mentha piperita, and similarly zinc oxide were used to create gold nanoparticles (Au-NPs). For the successful synthesis of NPs, additional plant resources (Calatropis gigantean, Plectranthus amboinicus, Agathosma betulina, Vitex negundo, Nephelium lappaceum, Azadirachta indica, Moringa oleifera, and Plectranthus amboinicus) are also utilized (Agarwal et al., 2017; Kumar et al., 2017; Nadeem et al., 2017).

9.6.2 BACTERIAL SOURCE

Several bacterial species are used in various commercial biotechnological processes like environmental remediation, genetic alteration, and biosorption (Gericke, 2006). Bacteria help in the reduction of metal ions and also aid the synthesis of NPs (Iravani, 2011). Because of its ability to provide both economic effectiveness and viability, the bacteria-based NP synthesis process has been a widely accepted approach for the development of metallic NPs (Zhang et al., 2020). Bacteria such as Delftia acidovorans are able to produce pure Au-NPs (Johnston et al., 2013). The nonribosomal peptide (NRP) delftibactin is commonly used in the synthesis of Au-NPs due to its capacity to affect ion resistance to poisonous Au. The transition metal gold demonstrated no toxicity toward bacteria due to the creation of inert Au-NPs linked to delftibactin (Pantidos and Horsfall, 2014). It was found that Bacillus licheniformis synthesized Ag-NPs within its cells (Kalimuthu et al., 2008).

9.6.3 FUNGI AS MEDIUM

Numerous species of fungi were successfully mentioned in previous studies, and findings showed that fungal cells can aid in fabricating the monodispersed NPs as fungal cells are a rich source of enzymes, which assist in the development of NPs. In between bacteria and fungi, fungal species are more capable of producing NPs because they release a greater amount of protein, which causes more synthesis of nanoparticles. Green nanoparticle synthesis by using fungi is also a very quick process as fungi are a rich source of enzymes that initiate the synthesis of metal nanoparticles (Chen et al., 2009). Rather than bacteria, some specific fungi are more capable of synthesizing a greater volume of nanomaterials from green sources (Mohanpuria et al., 2008). For example, *Trichoderma reesei* provides more advantages than any other fungi, produces a bulk quantity of enzymes, and leads to an enhancement of nanoparticle synthesis rate (Vahab et al., 2011). Neurospora crassa is utilized to synthesize platinum nanoparticles (Pt-NPs) (Castro et al., 2013).

9.6.4 ALGAE AS MEDIUM

Algae is a widely accepted medium for synthesizing nanomaterials as it can aggregate several heavy metals. It provides significant effects in various fields as a unique and Fucoidans, a polymer released by algae's cell walls, have many benefits because of some unique characteristics like being an anticancerous, antiviral, and slow aging agent, and anti-inflammatory. There are metabolites in the algae cell walls that act as stabilizing and reducing agents for maintaining the nanosize of NPs. Several reports have reported that many algal species are found to be potent in synthesizing various nanoparticles. Heavy metals get aggregated in algae, which can be utilized in the biogenic generation of metallic NPs (Luangpipat et al., 2011). The dried unicellular alga Chlorella Vulgaris would synthesize tetrahedral, decahedral, and icosahedral NPs, which accumulate near the surface. Some specific proteins are found in the algae extract, which act as stabilizers, reducing agents, and shape-control modifiers (Kahzad and Salehzadeh, 2020). Some algae named *Sargassum wightii*, marine algae, are capable of generating extracellular Ag and Au NPs (Govindaraju et al., 2009; Namvar et al., 2015). The Au-NPs can be synthesized from living Euglena gracilis microalga cells which are cultivated under a specific mixotrophic or autotrophic environment (Dahoumane et al., 2016).

9.7 GREEN SYNTHESIZED NANOPARTICLES FOR ENVIRONMENTAL REMEDIATION

Advanced nanotechnology-based approaches are used to inhibit the existing environment-associated issues. Novel nanomaterials sometimes result in an increase in the risk of the environment for repeated contamination and, therefore, to human health. Thus, the synthesis of nanomaterials by following green-based approaches has become significant and has thus increased its acceptance in the scientific field. The application of nanomaterials for environmental restoration has a promising role due to the beneficial features provided by the NPs (such as electronic, optical, magnetic, and catalytic properties). It was found that the role of nanomaterials for environmental remediation is greater than bulk due to the vast surface area and, as a result, their great multifunctional properties (Singh et al., 2010) (Fig. 9.3)

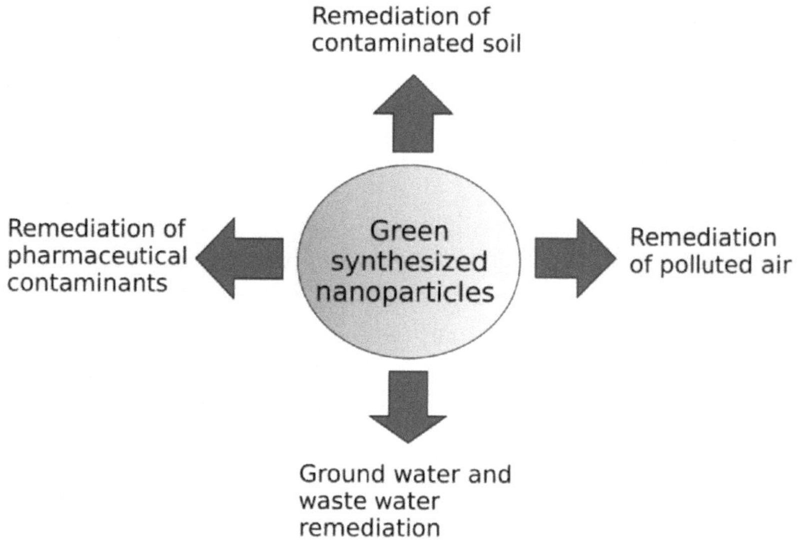

FIGURE 9.3 Application of green synthesized nanomaterials in environmental remediation.

Source: Created with BioRender.com

9.7.1 APPLICATIONS OF ENVIRONMENTALLY FRIENDLY NANOMATERIALS FOR WASTEWATER AND WATER FILTRATION

To reduce or get rid of the dangerous polluted materials in water resources, sewage systems, treatment facilities, and membrane bioreactors may be examined for remediation using green-synthesized and biogenic NPs, and other cutting-edge water purification devices. However, size control, stability, aggregation, and sedimentation problems continue to plague commercial applications of biogenic NPs in the treatment of effluents. Due to their high efficiency and large surface area, nano-adsorbents, nanocatalysts, and nano-films have been used to degrade hazardous dyes like methyl orange (MO), Congo red (CR), Eosin Y (EY), rhodamine B (RhB), methylene blue (MB), and others as well as to remove heavy metals from inorganic, organic, radioactive, and pharmaceutical pollutants (Lapworth et al., 2012; Yadav et al., 2015; Gautam et al., 2015).

9.7.1.1 REMOVAL OF ORGANIC AND INORGANIC CONTAMINANTS

The generation of Au NPs on Al2O3 substrates in a single step at ambient temperature is shown by polyphenols from the bayberry tannin (BT) plant (Huang et al., 2011b); as a reducing and stabilizing agent, BT initially reduced AuCl4 to create Al2O3/BT/Au NPs. This resulted in the formation of a bridged structure when glutaraldehyde-assisted self-crosslinking was performed on porous Al2O3. The activity of well-dispersed, spherically-shaped gold nanoparticles on Al2O3 with a size of 23 nm allowed for the reduction of 4-NP. This reduction was aided by NaBH4-mediated reduction. For environmental applications, particularly wastewater treatment, biogenic metal NPs from natural sources, including algae, bacteria, plants, and fungi, have demonstrated exceptional capabilities. In these remediation efforts, earth-abundant materials should be taken into first consideration in these remediation efforts. Green tea extract in aqueous form catalyzed the degradation of malachite green in iron nanoparticles (Weng et al., 2013; Plachtová et al, 2018) and their effects on ecotoxicology have been assessed. Reports on the creation, characterization, and biocompatibility of silver nanoparticles made from green tea are also available (Markova et al., 2014). In a different work, aqueous lemongrass (Cymbopogon citratus) extract was utilized to produce anatase TiO2 NPs doped with iron in an environmentally benign manner (Solano et al., 2019). These Fe-TiO2 NPs showed potential for organic pollutant degradation and wastewater photocatalytic treatment. Additionally, it has been suggested that the highly dangerous chemical trichloroethane may be removed from water through a reductive degradation mechanism using tea extract-facilitated bioproduction of Fe and Fe/Pd bimetallic NPs (20–30 nm) (Smuleac et al., 2011). Reports on the creation, characterization, and biocompatibility of silver nanoparticles made from green tea are also available (Markova et al., 2014). In a different work, aqueous lemongrass (Cymbopogon citratus) extract was utilized to produce anatase TiO2 NPs doped with iron in an environmentally benign manner (Solano et al., 2019). These Fe-TiO2 NPs showed potential for organic pollutant degradation and wastewater photocatalytic treatment. Additionally, it has been suggested that the highly dangerous chemical trichloroethane may be removed from water through a reductive degradation mechanism using tea extract-facilitated bioproduction of Fe and Fe/Pd bimetallic NPs (20–30 nm). In one work, eucalyptus leaf extracts were used to spread the greener iron oxide NPs attached to zeolite, where they were used to simultaneously remove phosphate and ammonia from aqueous solutions (Xu et al., 2020).

9.7.1.2 GETTING RID OF PHARMACEUTICAL TOXINS

Pharmaceutical substances, particularly antibiotics in natural water systems, present several difficulties and endanger human health. The use of biogenic nanoparticles for cleanup is possible. One of the most popular antibiotics for both human and veterinary use is tetracycline, which can be eliminated using nano zero-valent technology-based techniques. Tetracycline elimination was improved by using a bentonite-supported composite to address colloidal instability, and it did so with a noticeably lower NP loading (Yi et al., 2018; Gopal et al., 2020). In a different study, pomegranate (Punica granatum) peel extract, which is high in polyphenols, was used to create an environmentally friendly nickel–iron nanocomposite. To create nanocomposite beads (GS-NiFe beads), nickel–iron was subsequently adsorbed onto biocompatible and biodegradable alginate (Ravikumar et al., 2020). With the use of aqueous Tilia extract residues, green-synthesized Cu NPs were produced in a hurry (Husein et al., 2019), and three distinct pharmaceutical drugs were extracted from wastewater samples using the ensuing biogenic NPs. Another attempt was made to produce a unique $Fe3O_4$ nano sorbent utilizing a sustainable technique and plant extracts from black grapes, cucumbers, and lemons (Citrus limon) (Vitis vinifera) (Stan et al., 2017).

9.7.1.3 MEMBRANE-BASED WATER TREATMENT

Due to their numerous positive attributes, including as their readily available low-cost raw materials, comparatively nontoxic metal sources with appropriate biocompatibility, and desired physicochemical features, MOFs can be employed as potential options for environmental cleanup (such as semiconductor properties, high porosity, and framework flexibility) (Hou et al., 2018). Interesting to note is the use of MOF particles as a green template to produce a porous matrix membrane (PMM) utilizing an eco-friendly process. Despite being insoluble in polar organic solvents, these particles are quickly removed by water. These systems can be utilized for pressure-driven membrane processes, such as Lee et al. (2014) use carbonaceous nanofiber membranes to extract and separate NPs with astonishing selectivity, or osmotically-driven membrane processes.

9.8 REMEDIATION OF CONTAMINATED SOIL BY GREEN SYNTHESIZED NANOMATERIALS

Global soil contamination is serious, harming human health as well as the functions of natural ecosystems. Ex situ treatments, such as thermal desorption, dirt washing, or excavation followed by disposal, are common practices for treating polluted soils. However, these techniques can be time-consuming, and expensive to utilize, and they can generate a lot of secondary pollutants. As a result, low-impact in situ therapeutic methods like NP injection are becoming more and more popular. Many designed NPs are said to have strong reactivity toward a variety of contaminants, such as organic pollutants and heavy metals. Their rapid mobility, particular surface area, and reactivity are the causes of this. These advantageous characteristics of green NPs also make them appropriate for soil remediation. For instance, the production of nZVI has utilized green tea soil remediation as a reducing agent. Green tea nZVI was examined in silica sand and sand that had been treated with aluminum hydroxide, two granular media. They discovered that the amount of the heavy metal Cr(VI) might be decreased by 30–66%. According to reports, the soil's pH dropped, which caused metal leaching (Cr, Fe, and Pb). The pesticide fipronil, also known as phenyl pyrazole, is suspected to be neurotoxic, carcinogenic, and endocrine disruptive (Romeh, 2018). As a result, efforts have been made to use green Ag-NPs produced from Brassica alba and Plantago major plant extracts to repair soil contaminated with fipronil. The elimination efficacy of the Brassica alba NPs was 68.8% while that of the NPs generated using Plantago major leaf extract was only 10.1 percent. Research on the use of NPs for soil remediation has reported the outcomes of several green stabilizing agents. For instance, Hoch et al. (2008) described a carbon-supported nZVI product stabilized with CMC and polyacrylate (Table 9.3).

9.9 NANOMATERIALS IN THE REMEDIATION OF AIR POLLUTION

The huge specific surface areas and high reactivates of green nanomaterials make them excellent adsorbents, catalysts, and sensing materials. The nanomaterials have the capacity to absorb a lot of air pollutants or accelerate reactions significantly more quickly as a result of their enormous surface area and high surface energy. As a result, the energy required for degradation is decreased, and it may even help prevent the release of pollutants. In several ways, nanotechnology can aid in air purification. One approach is to utilize nanocatalysts with a bigger surface area for gaseous reactions. Catalysts quicken chemical reactions that change dangerous fumes from vehicles and factories into harmless gasses. The ground-based reduction of air pollution is ineffective, and consequently, disposal of the filtered substances is challenging (R.D et al., 2012). Filters are helpful in the fight against air pollution. A filter's constrained porosity structure keeps particles in their proper places while allowing gas to pass through. The membrane filter operates in three ways: surface contact with the filter structure, inertial force provided by moving gas, and particle and filter-generated electricity design (Shao et al., 2016). The main problem with filters is that they result in a pressure decrease, which demands a significant amount of energy. Nano-filters are more successful than traditional filters at eliminating a range of organic contaminants and micro-organisms because their pores have a size of between 1 and 10 nm. Carbon nanotube-based membranes work better in separating CO2 from other gasses (Skoulidas et al., 2006). In a variety of industries, nanomembrane technology can be used to divide and clean gasses and polluting vapors as well as to stop their release into the environment (Zhang et al., 2015).

9.10 BENEFITS AND LIMITATIONS OF GREEN NANOTECHNOLOGY

Although nanotechnology-based approaches are widely accepted for future purposes, they are providing greater benefits in different fields. Although nothing is always advantageous, at some point, there are pros and cons.

TABLE 9.4 Examples of Green Nanoparticles (NPs) Used for Contaminated Soil, Groundwater, and Wastewater Remediation.

Green NPs	NP size (nm)	Media treated	Removal mechanism	Contaminant type	References
Fe3O4	38~78	Soil	Surface adsorption	Heavy meta	Es'haghi et al., 2016
nZVI	15~45	Soil (sandy)	Chemical reduction	Pharmaceutical	Machado et al., 2013
Carbon-nZVI	20~150	Soil	Chemical reduction	Heavy metal	Hoch et al., 2008
Fe (clay supported)	40~80	Soil	Chemical reduction	Heavy metal	Soliemanzadeh and Fekri, 2017
Fe/Cu	60~120	Groundwater	Catalytic reduction	Heavy metal	Zhu et al., 2017
Fe/Pd	20~30	Groundwater	Chemical reduction	Chlorinated solvent	Smuleac et al., 2011
Fe	40~50	Wastewater	Adsorption (spontaneous and exothermic)	Pharmaceutical	Ali et al., 2016b
Fe3O4	38~42	Wastewater	Adsorption	Dye	Buazar et al., 2016
Fe	100~200	Wastewater	Chemical reduction and adsorption	Domestic waste water	Devatha et al., 2016
Ag	25	Wastewater	Electron transfer mediator (redox catalyst)	Dye	Edison and Sethuraman, 2012
Pb	47	Wastewater	UV-induced photocatalytic activity	Dye	Elango and Roopan, 2015
Ag	30	Wastewater	Chemical reduction	Organic compound	Edison and Sethuraman, 2013
ZnO	14~29	Wastewater	UV-induced photocatalytic activity	Petroleum refinery wastewater	El-Gendy et al., 2016
Fe	100	Wastewater	Adsorption	Heavy metal	Fazlzadeh et al., 2017
NiO	400~600	Wastewater	Adsorption	Dye	Hu et al., 2016
Fe/Pd	10~100	Wastewater	Adsorption and catalytic reduction	Dye	Luo et al., 2016
C nanodots	5~10	Wastewater	UV photocatalytic reactor	Dye	Mahajan et al., 2017
C-nZVI	27	Wastewater	Adsorption and chemical reduction	Chlorinated organic compounds	Liu and Zhang, 2010
TiO2	10~20	Surface water	Photocatalyst involving H2O2 generation	Bacterial	Dhandapani et al., 2012

The advantages and limitations of nanotechnology are easy to describe. The application fields are estimated by following their scale and scope by market forecasting generations and estimating potential toxicities and challenges (Manso and Castro-Gomes, 2015). The main advantages are that green-nanotechnology follows renewable resources, which means that future generations will surely get advantages from them also without causing any adverse effects on the planet. Management of waste generation brings techniques for waste removal and recycling and decreases the chances of global warming by reducing CO_2 emissions. Additionally, nanotechnology also provides economic conveniences to specific fields (farming) by enhancing the rate of productivity. Above all the beneficial effects of nanotechnology, it also has some limitations. Added initiatives are being conducted to find ways of measuring the effect of nanotechnology on particular purposes, like green growth. Nanotechnology is a very time-consuming, costly, and challenging task as it sometimes causes critical toxicities to the environment. The risks of utilizing new green nanotechnologies must be considered in relation to the dangers of applying current technologies and measured compared to the costs of ineffectiveness on people and the environment, identifying significant global limitations (de la Guardia, 2014).

9.11 FUTURE CROSSTALK

Nanotechnology has been widely accepted for future purposes as it reduces many undesirable adverse effects in many fields, like nanotechnology-based green nanomaterials, which provide several beneficial effects like enhanced catalytic efficacy and selectivity, and cost-effectiveness, and are linked to less toxic waste disposal. Green synthesized nanoparticles also have various advantages like antimicrobial nanomaterial combinations, which are used for the application of local plant extracts without using any toxic chemical reducing reagents, along with additional use in antibacterial bandages. Not only the beneficial window but also nanotechnology's future challenges and current perspectives associated with plant-based nanoparticle synthesis should be considered, converting lab-based conformance to a realizable industrial standard focusing on the current/past problems in terms of the effects on human health and the environment. Furthermore, plant-based techniques for nanomaterial synthesis are required for successful environmental remediation and, additionally, in various fields like the food, cosmetics, and pharmaceutical industries. Since various biomaterials synthesized from marine plants and algae resources remain undiscovered today, it is hoped that in the near future, numerous great establishments of plant-based synthesis approaches can be a promising tool. To enhance the bulk production of green nanomaterials, scientific research laboratories, and pharmaceutical industries must collaborate. Above all these advanced nanoparticles somehow generate components or by-products that can cause environmental pollution, so protective concerns should be investigated further (Samuel et al., 2022).

9.12 CONCLUSIONS

Concerns regarding the NPs' potential side effects and a growing need for more environmentally friendly solutions have emerged in the field of green and sustainable cleanup. Approving green synthesis has the prospect of lowering manufacturing costs in addition to preventing secondary conservation pollution. To help the profession advance, however, there are still research gaps that need to be filled. NPs with a variety of geometric shapes were created as part of the present green synthesis investigation, but further composite forms with more intricate surface areas still need to be created utilizing various methods. The development of green synthesized NPs has received wide acceptance as it associates no harmful agents in its synthesis process. In order to investigate the positive effects of plant-based NPs on plant health and the environment, plant-based NPs have thus emerged as a potential method. Green synthesis has thus far concentrated on minimizing secondary consequences during NP production rather than taking into account the impacts that arise during their use. It is advised that more investigation be done into the broader life-cycle effects of green NPs and how best to improve green synthesis methods to lessen these effects. New opportunities in health, energy, and environmental research materials are expected as the nanotechnology field presents an enormous potential yet to be explored.

ACKNOWLEDGMENT

We are very much thankful to our respective Institutions.

CONFLICT OF INTEREST

All the authors declare no conflict of interest

FUNDING

Not applicable

KEYWORDS

- **green nanoparticles**
- **synthesis**
- **environmental remediation**
- **future perspectives**

REFERENCES

Abdelmigid, H. M.; Hussien, N. A.; Alyamani, A. A.; Morsi, M. M.; Alsufyani, N. M.; Kadi, H. A. Green Synthesis of Zinc Oxide Nanoparticles Using Pomegranate Fruit Peel and Solid Coffee Grounds vs. Chemical Method of Synthesis, with Their Biocompatibility and Antibacterial Properties Investigation. *Molecules* **2022,** *27,* 1236.

Adeleye, A. S.; Conway, J. R.; Garner, K.; Huang, Y.; Su, Y.; Keller, A. A. Engineered Nanomaterials for Water Treatment and Remediation: Costs, Benefits, and Applicability. *Chem. Eng. J.* **2016,** *286,* 640–662.

Adetutu, E. M.; Gundry, T. D.; Patil, S. S.; Golneshin, A.; Adigun, J.; Bhaskarla, V.; Aleer, S.; Shahsavari, E.; Ross, E.; Ball, A. S. Exploiting the Intrinsic Microbial Degradative Potential for Field-Based In Situ Dechlorination of Trichloroethene Contaminated Groundwater*J. Hazard. Mater.* **2015**, *300,* 48–57.

Agarwal, H.; Kumar, S. V.; Rajeshkumar, S. A Review on Green Synthesis of Zinc Oxide Nanoparticles—An Eco-Friendly Approach. *Resour.-Efficient Technol.* **2017,** *3* (4), 406–413.

Ahmad, N.; Sharma, S.; Alam, M. K.; Singh, V. N.; Shamsi, S. F.; Mehta, B. R.; Fatma, A. Rapid Synthesis of Silver Nanoparticles Using Dried Medicinal Plant of Basil. *Colloids Surf. B: Biointerfaces* **2010,** *81* (1), 81–86.

Ahmed, H. M.; Roy, A.; Wahab, M.; Ahmed, M.; Othman-Qadir, G.; Elesawy, B. H.; Khandaker, M. U.; Islam, M. N.; Emran, T. B. Applications of Nanomaterials in Agrifood and Pharmaceutical Industry. *J. Nanomater.* **2021**.

Ali, I.; AL-Othman, Z. A.; Alwarthan, A. Synthesis of Composite Iron Nano Adsorbent and Removal of Ibuprofen Drug Residue from Water. *J. Mol. Liq.* **2016,** *219,* 858–864.

Ali, T.; Warsi, M. F.; Zulfiqar, S.; Sami, A.; Ullah, S.; Rasheed, A.; Alsafari, I. A.; Agboola, P. O.; Shakir, I.; Baig, M. M. Green Nickel/Nickel Oxide Nanoparticles for Prospective Antibacterial and Environmental Remediation Applications. *Ceram. Int.* **2021,** *48,* 8331–8340.

Amuda, O. S.; Amoo, I. A. Coagulation/Flocculation Process and Sludge Conditioning in Beverage Industrial Wastewater Treatment. *J. Hazard. Mater.* **2007,** *141* (3), 778–783.

Bar, H.; Bhui, D. K.; Sahoo, G. P.; Sarkar, P.; De, S. P.; Misra, A. Green Synthesis of Silver Nanoparticles Using Latex of Jatropha Curcas. *Colloids Surf. A: Physicochem. Eng. Asp.* **2009,** *339* (1–3), 134–139.
Begum, N. A.; Mondal, S.; Basu, S.; Laskar, R. A.; Mandal, D. Biogenic Synthesis of Au and Ag Nanoparticles Using Aqueous Solutions of Black Tea Leaf Extracts. *Colloids Surf. B: Biointerfaces* **2009,** *71* (1), 113–118.
Bindhu, M. R.; Umadevi, M. Antibacterial and Catalytic Activities of Green Synthesized Silver Nanoparticles. Spectrochim. Acta A Mol. Biomol. Spectrosc. **2015,** *135,* 373–378.
Bose, D.; Chatterjee, S. Antibacterial Activity of Green Synthesized Silver Nanoparticles Using Vasaka (Justicia Adhatoda L.) Leaf Extract. *Indian J. Microbiol.* **2015,** *55* (2), 163–167.
Briffa, J.; Sinagra, E.; Blundell, R. Heavy Metal Pollution in the Environment and their Toxicological Effects on Humans. *Heliyon* **2020,** *6* (9), e04691.
Buazar, F.; Baghlani-Nejazd, M. H.; Badri, M.; Kashisaz, M.; Khaledi-Nasab, A.; Kroushawi, F. Facile One-Pot Phytosynthesis of Magnetic Nanoparticles Using Potato Extract and Their Catalytic Activity. *Starch Stärke.* **2016,** *68* (7–8), 796–804.
Castro, L.; Blázquez, M. L.; Muñoz, J. A.; González, F.; Ballester, A. Biological Synthesis of Metallic Nanoparticles Using Algae. *IET Nanobiotechnol.* **2013,** *7* (3), 109–116.
Chedeville, O.; Debacq, M.; Porte, C. Removal of Phenolic Compounds Present in Olive Mill Wastewaters by Ozonation. *Desalination* **2009,** *249* (2), 865–869.
Chen, Y. L.; Tuan, H. Y.; Tien, C. W.; Lo, W. H.; Liang, H. C.; Hu, Y. C. Augmented Biosynthesis of Cadmium Sulfide Nanoparticles by Genetically Engineered Escherichia coli. *Biotechnol. Prog.* **2009,** *25* (5), 1260–1266.
Chrysochoou, M.; Johnston, C. P.; Dahal, G. A Comparative Evaluation of Hexavalent Chromium Treatment in Contaminated Soil by Calcium Polysulfide and Green-tea Nanoscale Zero-valent Iron*J. Hazard. Mater.* **2012,** *201,* 33–42.
Dahoumane, S. A.; Yéprémian, C.; Djédiat, C.; Couté A, Fiévet, F.; Coradin, T.; Brayner, R. Improvement of Kinetics, Yield, and Colloidal Stability of Biogenic Gold Nanoparticles Using Living Cells of Euglena Gracilis Microalga. *J. Nanopart. Res.* **2016,** *18* (3), 1–2.
de la Guardia, M. The Challenges of Green Nanotechnology. *BioImpacts* **2014,** *4* (1), 1.
Devatha, C. P.; Thalla, A. K.; Katte, S. Y. Green Synthesis of Iron Nanoparticles Using Different Leaf Extracts for Treatment of Domestic Waste Water. *J. Clean. Prod.* **2016,** *139,* 1425–1435.
Dhandapani, P.; Maruthamuthu, S.; Rajagopal, G. Bio-mediated Synthesis of TiO2 Nanoparticles and Its Photocatalytic Effect on Aquatic Biofilm. *J. Photochem. Photobiol. B: Biol.* **2012,** *11,* 43–49.
Diagboya, P. N.; Dikio, E. D. Silica-based Mesoporous Materials; Emerging Designer Adsorbents for Aqueous Pollutants Removal and Water Treatment. *Microporous Mesoporous Mater..* **2018,** *266,* 252–267.
Dirany, A.; Sires, I.; Oturan, N.; Özcan, A.; Oturan, M. A. Electrochemical Treatment of the Antibiotic Sulfachloropyridazine: Kinetics, Reaction Pathways, and Toxicity Evolution. *Environ. Sci. Technol.* **2012,** *46* (7), 4074–4082.
Dwivedi, A. D.; Gopal, K. Biosynthesis of Silver and Gold Nanoparticles Using Chenopodium Album Leaf Extract. *Colloids Surf. A: Physicochem. Eng. Asp.* **2010,** *369* (1–3), 27–33.
Edison, T. J.; Sethuraman, M. G. Biogenic Robust Synthesis of Silver Nanoparticles Using Punica Granatum Peel and Its Application as a Green Catalyst for the Reduction of an

Anthropogenic Pollutant 4-Nitrophenol. *Spectrochimica Acta A: Mol. Biomol. Spectrosc.* **2013,** *104,* 262–264.

Edison, T. J.; Sethuraman, M. G. Instant Green Synthesis of Silver Nanoparticles Using Terminalia Chebula Fruit Extract and Evaluation of their Catalytic Activity on Reduction of Methylene Blue. *Process Biochem.* **2012,** *47* (9), 1351–1357.

El-Gendy, N. S.; El-Salamony, R. A.; Younis, S. A. Green Synthesis of Fluorapatite from Waste Animal Bones and the Photo-Catalytic Degradation Activity of a New ZnO/Green Biocatalyst Nano-Composite for Removal of Chlorophenols. *J. Water Proc. Eng.* **2016,** *12,* 8–19.

El-Geundi, M. S. Adsorbents for Industrial Pollution Control. *Adsorp. Sci. Technol.* **1997,** *15* (10), 777–787.

Es'haghi, Z.; Vafaeinezhad, F.; Hooshmand, S. Green Synthesis of Magnetic Iron Nanoparticles Coated by Olive Oil and Verifying Its Efficiency in Extraction of Nickel from Environmental Samples via UV–vis Spectrophotometry. *Proc. Safety Environ. Protec.* **2016,** *102,* 403–409.

Fazlzadeh, M.; Rahmani, K.; Zarei, A.; Abdoallahzadeh, H.; Nasiri, F.; Khosravi, R. A Novel Green Synthesis of Zero Valent Iron Nanoparticles (NZVI) Using Three Plant Extracts and Their Efficient Application for Removal of Cr (VI) from Aqueous Solutions. *Adv. Powder Technol.* **2017,** *28* (1), 122–130.

Fernandes, A.; Pacheco, M. J.; Ciríaco, L.; Lopes, A. J. Review on the Electrochemical Processes for the Treatment of Sanitary Landfill Leachates: Present and Future. *Appl. Catal. B: Environ.* **2015**, *176,* 183–200.

Ganzenko, O.; Huguenot, D.; Van Hullebusch, E. D.; Esposito, G.; Oturan, M. A. Electrochemical Advanced Oxidation and Biological Processes for Wastewater Treatment: A Review of the Combined Approaches. *Environ. Sci. Pollut. Res.* **2014,** *21* (14), 8493–8524.

Garner, K. L.; Keller, A. A. Emerging Patterns for Engineered Nanomaterials in the Environment: A Review of Fate and Toxicity Studies. *J. Nanopart. Res.* **2014,** *16* (8), 1–28.

Gautam, P. K.; Gautam, R. K.; Banerjee, S.; Lofrano, G.; Sanroman, M. A.; Chattopadhyaya, M. C.; Pandey, J. D. Preparation of Activated Carbon from Alligator Weed (Alternenthera philoxeroids) and Its Application for Tartrazine Removal: Isotherm, Kinetics and Spectroscopic Analysis. *J. Environ. Chem. Eng.* **2015,** *3* (4), 2560–2568.

Gericke, M.; Pinches, A. Microbial Production of Gold Nanoparticles. *Gold Bull.* **2006,** *39* (1), 22–28.

Ghasemzadeh, G.; Momenpour, M.; Omidi, F.; Hosseini, M. R.; Ahani, M.; Barzegari, A. Applications of Nanomaterials in Water Treatment and Environmental Remediation. *Front. Environ. Sci. Eng.* **2014,** *8* (4), 471–482.

Ghodake, G. S.; Deshpande, N. G.; Lee, Y. P.; Jin, E. S. Pear Fruit Extract-Assisted Room-Temperature Biosynthesis of Gold Nanoplates. *Colloids Surf. B: Biointerfaces* **2010,** *75* (2), 584–589.

Gnanajobitha, G.; Paulkumar, K.; Vanaja, M.; Rajeshkumar, S.; Malarkodi, C.; Annadurai, G.; Kannan, C. Fruit-Mediated Synthesis of Silver Nanoparticles Using Vitis Vinifera and Evaluation of their Antimicrobial Efficacy. *J. Nanostructure Chem.* **2013,** *3* (1), 1–6.

Gopal, G.; Sankar, H.; Natarajan, C.; Mukherjee, A. Tetracycline Removal Using Green Synthesized Bimetallic nZVI-Cu and Bentonite Supported Green nZVI-Cu Nanocomposite: A Comparative Study. *J. Environ. Manag.* **2020,** *254,* 109812.

Govindaraju, K.; Kiruthiga, V.; Kumar, V. G.; Singaravelu, G. Extracellular Synthesis of Silver Nanoparticles by a Marine Alga, Sargassum Wightii Grevilli and Their Antibacterial Effects. *J. Nanosci. Nanotechnol.* **2009,** *9* (9), 5497–5501.

Hassaan, M. A.; El Nemr, A. Pesticides Pollution: Classifications, Human Health Impact, Extraction and Treatment Techniques. *Egypt. J. Aquat. Res.* **2020,** *46* (3), 207–220.

Hassan, I.; Mohamedelhassan, E.; Yanful, E. K. Solar Powered Electrokinetic Remediation of Cu Polluted Soil Using a Novel Anode Configuration. *Electrochim. Acta* **2015,** *181,* 58–67.

Herrera-Becerra, R.; Zorrilla, C.; Rius, J. L.; Ascencio, J. A. Electron Microscopy Characterization of Biosynthesized Iron Oxide Nanoparticles. *Appl. Phys. A* **2008,** *91* (2), 241–246.

Hoch, L. B.; Mack, E. J.; Hydutsky, B. W.; Hershman, J. M.; Skluzacek, J. M.; Mallouk, T. E. Carbothermal Synthesis of Carbon-Supported Nanoscale Zero-Valent Iron Particles for the Remediation of Hexavalent Chromium. *Environ. Sci. Technol.* **2008,** *42* (7), 2600–2605.

Holt, P.; Barton, G.; Mitchell, C. Electrocoagulation as a Wastewater Treatment. *Third Annu. Aust. Environ. Eng. Res. Event.* **1999,** *1000,* 41–46.

Hou, S.; Wu, Y. N.; Feng, L.; Chen, W.; Wang, Y.; Morlay, C.; Li, F. Green Synthesis and Evaluation of an Iron-based Metal–organic Framework MIL-88B for Efficient Decontamination of Arsenate from Water. *Dalton Trans.* **2018,** *47* (7), 2222–2231.

Hu, H.; Chen, G.; Deng, C.; Qian, Y.; Wang, M.; Zheng, Q. Green Microwave-Assisted Synthesis of Hierarchical NiO Architectures Displaying a Fast and High Adsorption Behavior for Congo Red. *Mater. Lett.* **2016,** *170,* 139–141.

Huang, X.; Liao, X.; Shi, B. Synthesis of Highly Active and Reusable Supported Gold Nanoparticles and Their Catalytic Applications to 4-Nitrophenol Reduction. *Green Chem.* **2011,** *13* (10), 2801–2805.

Husein, D. Z.; Hassanien, R.; Al-Hakkani, M. F. Green-synthesized Copper Nano-adsorbent for the Removal of Pharmaceutical Pollutants from Real Wastewater Samples. *Heliyon* **2019,** *5* (8), e02339.

Hwang, J. I.; Zimmerman, A. R.; Kim, J. E. Bioconcentration Factor-based Management of Soil Pesticide Residues: Endosulfan Uptake by Carrot and Potato Plants. *Sci. Total Environ.* **2018,** *627,* 514–522.

Iravani, S. Bacteria in Nanoparticle Synthesis: Current Status and Future Prospects. *Int. Schol. Res. Not.* **2014,** 2014.

Iravani, S. Green Synthesis of Metal Nanoparticles Using Plants. *Green Chem.* **2011,** *13* (10):2638–2650.

Iwuozor, K. O. Prospects and Challenges of Using Coagulation-flocculation Method in the Treatment of Effluents. *Adv. J. Chem. Sect. A* **2019,** *2* (2), 105–127.

Jadhav, S. D.; Jadhav, M. S.; Jawale, R. W. Study of Chloride and Nitrate Concentration of MulaMutha River in Pune City (Maharashtra). *Int. J. Chem. Life Sci.* **2013**, *2,* 1140–1142.

Jain, C. K.; Singh, R. D. Technological Options for the Removal of Arsenic with Special Reference to South East Asia. *J. Environ. Manage.* **2012,** *10,* 1–8.

James, C. P.; Germain, E.; Judd, S. Micropollutant Removal by Advanced Oxidation of Microfiltered Secondary Effluent for Water Reuse. *Sep. Purif. Technol.* **2014,** *127,* 77–83.

Jayaswal, K.; Sahu, V.; Gurjar, B. R. Water Pollution, Human Health and Remediation. In *Water Remediation*; Springer, 2018; pp 11–27.

Jeworski, M.; Heinzle. E. *Combined Chemical-Biological Treatment of Wastewater Containing Refractory Pollutants*. 2000; pp 163–196.

Jia, Z. Q.; Zhang, Y. C.; Huang, Q. T.; Jones, A. K.; Han, Z. J.; Zhao, C. Q. Acute Toxicity, Bioconcentration, Elimination, Action Mode and Detoxification Metabolism of Broflanilide in Zebrafish, Danio Rerio. *J. Hazard. Mater.* **2020,** *394,* 122521.

Johnston, C. W.; Wyatt, M. A.; Li, X.; Ibrahim, A.; Shuster, J.; Southam, G.; Magarvey, N. A. Gold Biomineralization by a Metallophore from a Gold-associated Microbe. *Nat. Chem. Biol.* **2013,** *9* (4), 241–243.

Kahzad, N.; Salehzadeh, A. Green Synthesis of CuFe2O4@ Ag Nanocomposite Using the Chlorella Vulgaris and Evaluation of Its Effect on the Expression of norA Efflux Pump Gene Among Staphylococcus Aureus Strains. *Biol. Trace Elem. Res.* **2020,** *198* (1), 359–370.

Kalimuthu, K.; Babu, R. S.; Venkataraman, D.; Bilal, M.; Gurunathan, S. Biosynthesis of Silver Nanocrystals by Bacillus Licheniformis. *Colloids Surf. B: Biointerfaces* **2008,** *65* (1), 150–153.

Kant, R. Textile Dyeing Industry an Environmental Hazard. *Nat. Sci.* **2012,** *4* (1), pp 22–26.

Karn, B.; Kuiken, T.; Otto, M. Nanotechnology and In Situ Remediation: A Review of the Benefits and Potential Risks. *Environ. Health Perspect.* **2009,** *117* (12), 1813–1831.

Karn, B.; Kuiken, T.; Otto, M. Nanotechnology and in Situ Remediation: A Review of the Benefits and Potential Risks. *Ciência Saúde Coletiva* **2011,** *16,* 165–178.

Karthik, C.; Swathi, N.; Caroline, D. G. Green Synthesized rGO-AgNP Hybrid Nanocomposite—An Effective Antibacterial Adsorbent for Photocatalytic Removal of DB-14 Dye from Aqueous Solution. *J. Environ. Chem. Eng.* **2020,** *8* (1), 103577.

Kasthuri, J.; Veerapandian, S.; Rajendiran, N. Biological Synthesis of Silver and Gold Nanoparticles Using Apiin as Reducing Agent. *Colloids Surf. B: Biointerfaces* **2009,** *68* (1), 55–60.

Katata-Seru, L.; Moremedi, T.; Aremu, O. S.; Bahadur, I. Green Synthesis of Iron Nanoparticles Using Moringa Oleifera Extracts and Their Applications: Removal of Nitrate from Water and Antibacterial Activity Against Escherichia coli. *J. Mol. Liq.* **2018,** *256,* 296–304.

Kaur, S.; Roy A. Bioremediation of Heavy Metals from Wastewater Using Nanomaterials. *Environ. Dev. Sustain.* **2021,** *23* (7), 9617–9640.

Khan, S. H. Green Nanotechnology for the Environment and Sustainable Development. In *Green Materials for Wastewater Treatment*; Springer: Cham, **2020,**; pp 13–46.

Koper, O. B.; Klabunde, J. S.; Marchin, G. L.; Klabunde, K. J.; Stoimenov, P.; Bohra, L. Nanoscale Powders and Formulations with Biocidal Activity Toward Spores and Vegetative Cells of Bacillus Species, Viruses, and Toxins. *Curr. Microbiol.* **2002,** *44* (1), 49–55.

Krantzberg, G.; Tanik, A.; do Carmo, J. S.; Indarto, A.; Ekdal, A.; Gurel, M.; Pehlivanoglu Mantas, E.; Wang, Z.; Wang, G.; Zhao, C.; Machiwal D. *Advances in Water Quality Control*; Scientific Research Publishing, Inc.: USA; 2010.

Krishnaraj, C.; Jagan, E. G.; Rajasekar, S.; Selvakumar, P.; Kalaichelvan, P. T.; Mohan, N. J. Synthesis of Silver Nanoparticles Using Acalypha Indica Leaf Extracts and its Antibacterial Activity Against Water Borne Pathogens. *Colloids Surf. B: Biointerfaces* **2010,** *76* (1), 50–56.

Kuang, Y.; Wang, Q.; Chen, Z.; Megharaj, M.; Naidu, R. Heterogeneous Fenton-Like Oxidation of Monochlorobenzene Using Green Synthesis of Iron Nanoparticles. *J. Colloid Interface Sci.* **2013,** *410,* 67–73.

Kumar, B.; Smita, K.; Cumbal, L.; Debut, A. Green Synthesis of Silver Nanoparticles Using Andean Blackberry Fruit Extract. *Saudi J. Biol. Sci.* **2017,** *24* (1), 45–50.

Kumar, D. A.; Palanichamy, V.; Roopan, S. M. Green Synthesis of Silver Nanoparticles Using Alternanthera Dentata Leaf Extract at Room Temperature and Their Antimicrobial Activity. *Spectrochim. Acta A Mol. Biomol. Spectrosc.* **2014,** *127,* 168–171.

Lapworth, D. J.; Baran, N.; Stuart, M. E.; Ward, R. S. Emerging Organic Contaminants in Groundwater: A Review of Sources, Fate and Occurrence. *Environ. Pollut.* **2012,** *163,* 287–303.

Lee, J. Y.; Tang, C. Y.; Huo, F. Fabrication of Porous Matrix Membrane (PMM) Using Metal-Organic Framework as Green Template for Water Treatment. *Sci. Rep.* **2014,** *4* (1), 1–5.

Lellis, B.; Fávaro-Polonio, C. Z.; Pamphile, J. A.; Polonio, J. C. Effects of Textile Dyes on Health and the Environment and Bioremediation Potential of Living Organisms. *Biotechnol. Res. Innov.* **2019,** *3* (2), 275–290.

Li, Y. H.; Ding, J.; Luan, Z.; Di, Z.; Zhu, Y.; Xu, C.; Wu, D.; Wei, B. Competitive Adsorption of Pb2+, Cu2+ and Cd2+ Ions from Aqueous Solutions by Multiwalled Carbon Nanotubes. *Carbon* **2003,** *41* (14), 2787–2792.

Liu, J.; Qiao, S. Z.; Hu, Q. H.; Lu, G. Q. Magnetic Nanocomposites with Mesoporous Structures: Synthesis and Applications. *Small* **2011,** *7,* 425–443.

Liu, R.; Zhao, D. Reducing Leachability and Bioaccessibility of Lead in Soils Using a New Class of Stabilized Iron Phosphate Nanoparticles. *Water Res.* **2007,** *41* (1), 2491–2502.

Liu, Z.; Zhang, F. S. Nano-zerovalent Iron Contained Porous Carbons Developed from Waste Biomass for the Adsorption and Dechlorination of PCBs. *Bioresour. Technol.* **2010,** *101* (7), 2562–2564.

Löffler, F. E.; Edwards, E. A. Harnessing Microbial Activities for Environmental Cleanup. *Curr. Opin. Biotechnol.* **2006,** 17 (3), 274–84.

Luangpipat, T.; Beattie, I. R.; Chisti, Y.; Haverkamp, R. G. Gold Nanoparticles Produced in a Microalga. *J. Nanopart. Res.* **2011,** *13* (12), 6439–6445.

Luo, F.; Yang, D.; Chen, Z.; Megharaj, M.; Naidu, R. One-step Green Synthesis of Bimetallic Fe/Pd Nanoparticles Used to Degrade Orange, II. *J. Hazard. Mater.* **2016,** *303,* 145–153.

Machado, S.; Stawiński, W.; Slonina, P.; Pinto, A. R.; Grosso, J. P.; Nouws, H. P.; Albergaria, J. T.; Delerue-Matos, C. Application of Green Zero-Valent Iron Nanoparticles to the Remediation of Soils Contaminated with Ibuprofen. *Sci. Total Environ.* **2013,** *461,* 323–329.

Mahajan, R.; Bhadwal, A. S.; Kumar, N.; Madhusudanan, M.; Pudake, R. N.; Tripathi, R. M. Green Synthesis of Highly Stable Carbon Nanodots and their Photocatalytic Performance. *IET Nanobiotechnol.* **2017,** *11* (4), 360–364.

Manso, M.; Castro-Gomes, J. Green Wall Systems: A Review of their Characteristics. *Renew. Sustain. Energy Rev.* **2015,** *41,* 863–871.

Markova, Z.; Novak, P.; Kaslik, J.; Plachtova, P.; Brazdova, M.; Jancula, D.; Siskova, K. M.; Machala, L.; Marsalek, B.; Zboril, R.; Varma, R. Iron (II, III)–Polyphenol Complex Nanoparticles Derived from Green Tea with Remarkable Ecotoxicological Impact. *ACS Sustain. Chem. Eng.* **2014,** *2* (7), 1674–1680.

Marlon, J. R.; Bloodhart, B.; Ballew, M. T.; Rolfe-Redding, J.; Roser-Renouf, C.; Leiserowitz, A.; Maibach E. How Hope and Doubt Affect Climate Change Mobilization. *Front. Commun.* **2019,** *20.*

Martínez-Cabanas, M.; López-García, M.; Barriada, J. L.; Herrero, R.; de Vicente, M. E. Green Synthesis of Iron Oxide Nanoparticles. Development of Magnetic Hybrid Materials for Efficient as (V) Removal. *Chem. Eng. J.* **2016,** *301,* 83–91.

Martinez-Huitle, C. A.; Ferro, S. Electrochemical Oxidation of Organic Pollutants for the Wastewater Treatment: Direct and Indirect Processes. *Chem. Soci. Rev.* **2006,** *35* (12), 1324–1340.

Masih, J.; Masih, A.; Kulshrestha, A.; Singhvi, R.; Taneja, A. Characteristics of Polycyclic Aromatic Hydrocarbons in Indoor and Outdoor Atmosphere in the North Central Part of India. *J. Hazard. Mater.* **2010,** *177* (1–3), 190–198.

Mathew, L.; Chandrasekaran, N.; Mukherjee, A. Biomimetic Synthesis of Nanoparticles: Science, Technology & Applicability. In *Biomimetics Learning From Nature*; 2010.

Mittal, A. K.; Chisti, Y.; Banerjee, U. C. Synthesis of Metallic Nanoparticles Using Plant Extracts. *Biotechnol. Adv.* **2013,** *31* (2), 346–356.

Mohanpuria, P.; Rana, N. K.; Yadav, S. K. Biosynthesis of Nanoparticles: Technological Concepts and Future Applications. *J. Nanopart. Res.* **2008,** *10* (3), 507–517.

Mubarakali, D.; Thajuddin, N.; Jeganathan, K.; Gunasekaran, M. Plant Extract Mediated Synthesis of Silver and Gold Nanoparticles and its Antibacterial Activity Against Clinically Isolated Pathogens. *Colloids Surf. B: Biointerfaces* **2011,** *85* (2), 360–365.

Mystrioti, C.; Xanthopoulou, T. D.; Tsakiridis, P.; Papassiopi, N.; Xenidis, A. Comparative Evaluation of Five Plant Extracts and Juices for Nanoiron Synthesis and Application For Hexavalent Chromium Reduction. *Sci. Total Environ.* **2016,** *539,* 105–113.

Nadeem, M.; Abbasi, B. H.; Younas, M.; Ahmad, W.; Khan, T. A Review of the Green Syntheses and Anti-microbial Applications of Gold Nanoparticles. *Green Chem. Lett. Rev.* **2017,** *10* (4), 216–227.

Nakkala, J. R.; Mata, R.; Gupta, A. K.; Sadras, S. R. Biological Activities of Green Silver Nanoparticles Synthesized with Acorous Calamus Rhizome Extract. *Eur. J. Med. Chem.* **2014,** *85,* 784–794.

Nassar, N. N. Rapid Removal and Recovery of Pb(II) from Wastewater by Magnetic Nanoadsorbents. *J. Hazard. Mater.* **2010,** *184,* 538–546.

Nie, X.; Liu, J.; Zeng, X.; Yue, D. Rapid Degradation of Hexachlorobenzene by Micron Ag/Fe Bimetal Particles. *J. Environ. Sci.* **2013,** *25* (3), 473–478.

Panigrahi, S.; Kundu, S.; Ghosh, S.; Nath, S.; Pal, T. General Method of Synthesis for Metal Nanoparticles. *J. Nanopart. Res.* **2004,** *6* (4), 411–414.

Pantidos, N.; Horsfall, L. E. Biological Synthesis of Metallic Nanoparticles by Bacteria, Fungi and Plants. *J. Nanomed. Nanotechnol.* **2014,** *5* (5), 1.

Patil, S. S.; Adetutu, E. M.; Sheppard, P. J.; Morrison, P.; Menz, I. R.; Ball, A. S. Site-specific Pre-evaluation of Bioremediation Technologies for Chloroethene Degradation. *Int. J. Environ. Sci. Technol.* **2014,** *11* (7), 1869–1880.

Patil, S. S.; Shedbalkar, U. U.; Truskewycz, A.; Chopade, B. A.; Ball, A. S. Nanoparticles for Environmental Clean-up: A Review of Potential Risks and Emerging Solutions. *Environ. Technol. Innov.* **2016,** *5,* 10–21.

Pereira, L.; Alves, M. Dyes-environmental Impact and Remediation. In *Environmental Protection Strategies for Sustainable Development*; Springer; Dordrecht, **2012,**; pp 111–162.

Pi, Y.; Li, X.; Xia, Q.; Wu, J.; Li, Y.; Xiao, J.; Li, Z. Adsorptive and Photocatalytic Removal of Persistent Organic Pollutants (POPs) in Water by Metal-organic Frameworks (MOFs). *Chem. Eng. J.* **2018,** *337,* 351–371.

Plachtová. P.; Medrikova, Z.; Zboril, R.; Tucek, J.; Varma, R. S.; Maršálek, B. Iron and Iron Oxide Nanoparticles Synthesized with Green Tea Extract: Differences in Ecotoxicological Profile and Ability to Degrade Malachite Green. *ACS Sustain. Chem. Eng.* **2018,** *6* (7), 8679–8687.

Poznyak, T.; Bautista, G. L.; Chaírez, I.; Córdova, R. I.; Ríos, L. E. Decomposition of Toxic Pollutants in Landfill Leachate by Ozone after Coagulation Treatment. *J. Hazard. Mater.* **2008,** *152* (3), 1108–1114.

Qu, J.; Luo, C.; Hou, J. Synthesis of ZnO Nanoparticles From Zn-Hyperaccumulator (Sedum Alfredii Hance) Plants. *Micro Nano Lett.* **2011,** *6* (3), 174–176.

Rajan, C. S. Nanotechnology in Groundwater Remediation. *Int. J. Environ. Sci. Dev.* **2011,** *2* (3), 182.

Ravikumar, K. V.; Kubendiran, H.; Ramesh, K.; Rani, S.; Mandal, T. K.; Pulimi, M.; Natarajan, C.; Mukherjee, A. Batch and Column Study on Tetracycline Removal Using Green Synthesized NiFe Nanoparticles Immobilized Alginate Beads. *Environ. Technol. Innov.* **2020,** *17,* 100520.

Ribeiro, A. R.; Nunes, O. C.; Pereira, M. F.; Silva, A. M. An Overview on the Advanced Oxidation Processes Applied for the Treatment of Water Pollutants Defined in the Recently Launched Directive 2013/39/EU. *Environ. Int.* **2015,** *75,* 33–51.

Romeh, A. A. Green Silver Nanoparticles for Enhancing the Phytoremediation of Soil and Water Contaminated by Fipronil and Degradation Products. *Water Air Soil Pollut.* **2018,** *229* (5), 1–3.

Roy, A.; Elzaki, A.; Tirth, V.; Kajoak, S.; Osman, H.; Algahtani, A.; Islam, S.; Faizo, N. L.; Khandaker, M. U.; Islam, M. N.; Emran, T. B. Biological Synthesis of Nanocatalysts and Their Applications. *Catalysts* **2021,** *11* (12), 1494.

Samuel, M. S.; Ravikumar, M.; John J, A.; Selvarajan, E.; Patel, H.; Chander, P. S.; Soundarya, J.; Vuppala, S.; Balaji, R.; Chandrasekar, N. A Review on Green Synthesis of Nanoparticles and Their Diverse Biomedical and Environmental Applications. *Catalysts.* **2022,** *12* (5), 459.

Samuel, M. S.; Selvarajan, E.; Mathimani, T.; Santhanam, N.; Phuong, T. N.; Brindhadevi, K.; Pugazhendhi, A. Green Synthesis of Cobalt-Oxide Nanoparticle Using Jumbo Muscadine (Vitis Rotundifolia): Characterization and Photo-Catalytic Activity of Acid Blue-74. *J. Photochem. Photobiol. B: Biol.* **2020,** *211,* 112011.

Savage, N.; Diallo, M. S. Nanomaterials and Water Purification: Opportunities and Challenges. *J. Nanopart. Res.* **2005,** *7* (4), 331–342.

Schabes-Retchkiman, P. S.; Canizal, G.; Herrera-Becerra, R.; Zorrilla, C.; Liu, H. B.; Ascencio, J. A. Biosynthesis and Characterization of Ti/Ni Bimetallic Nanoparticles. *Opt. Mater.* **2006,** *29* (1), 95–99.

Seewagen, C. L. The Threat of Global Mercury Pollution to Bird Migration: Potential Mechanisms and Current Evidence. *Ecotoxicology* **2020,** *29* (8), 1254–1267.

Shankar, S. S.; Ahmad, A.; Pasricha, R.; Sastry, M. Bioreduction of Chloroaurate Ions by Geranium Leaves and Its Endophytic Fungus Yields Gold Nanoparticles of Different Shapes. *J. Mater. Chem.* **2003,** *13* (7), 1822–1826.

Singh, A. K.; Talat, M.; Singh, D. P.; Srivastava, O. N. Biosynthesis of Gold and Silver Nanoparticles by Natural Precursor Clove and their Functionalization with Amine Group. *J. Nanopart. Res.* **2010,** *12* (5), 1667–1675.

Singh, V. V. Green Nanotechnology for Environmental Remediation. In *Sustainable Nanotechnology for Environmental Remediation*; Elsevier, 2022; pp 31–61.

Sinha, A.; Cha, B. G.; Kim, J. Three-Dimensional Macroporous Alginate Scaffolds Embedded with Akaganeite Nanorods for the Filter-Based High-Speed Preparation of Arsenic-Free Drinking Water. *ACS Appl. Nano Mater.* **2018,** *1,* 1940–1948.

Skoulidas, A. I.; Sholl, D. S.; Johnson, J. K. Adsorption and Diffusion of Carbon Dioxide and Nitrogen Through Single-Walled Carbon Nanotube Membranes. *J. Chem. Phys.* **2006,** *124* (5), 054708.

Smuleac, V.; Varma, R.; Sikdar, S.; Bhattacharyya, D. Green Synthesis of Fe and Fe/Pd Bimetallic Nanoparticles in Membranes for Reductive Degradation of Chlorinated Organics. *J. Membr. Sci.* **2011,** *379* (1–2), 131–137.

Solano, R. A.; Herrera, A. P.; Maestre, D.; Cremades, A. Fe-TiO2 Nanoparticles Synthesized by Green Chemistry for Potential Application in Waste Water Photocatalytic Treatment. *J. Nanotechnol.* **2019,** 2019.

Soliemanzadeh, A.; Fekri, M. The Application of Green Tea Extract to Prepare Bentonite-supported Nanoscale Zero-Valent Iron and Its Performance on Removal of Cr (VI): Effect of Relative Parameters and Soil Experiments. *Microporous Mesoporous Mater.* **2017,** *239,* 60–69.

Stan, M.; Lung, I.; Soran, M. L.; Leostean, C.; Popa, A.; Stefan, M.; Lazar, M. D.; Opris, O.; Silipas, T. D.; Porav, A. S. Removal of Antibiotics from Aqueous Solutions by Green Synthesized Magnetite Nanoparticles with Selected Agro-Waste Extracts. *Process Saf. Environ. Prot.* **2017,** *107,* 357–372.

Sun, Q.; Cai, X.; Li, J.; Zheng, M.; Chen, Z.; Yu, C. P. Green Synthesis of Silver Nanoparticles Using Tea Leaf Extract and Evaluation of Their Stability and Antibacterial Activity. *Colloids Surf. A: Physicochem. Eng. Asp.* **2014,** *444,* 226–231.

Suty, H.; De Traversay, C.; Cost, M. Applications of Advanced Oxidation Processes: Present and Future. *Water Sci. Technol.* **2004,** *49* (4), 227–233.

Tahir, K.; Nazir, S.; Li, B.; Khan, A. U.; Khan, Z. U.; Ahmad, A.; Khan, F. U. An Efficient Photo Catalytic Activity of Green Synthesized Silver Nanoparticles Using Salvadora Persica Stem Extract. *Sep. Purif. Technol.* **2015,** *150,* 316–324.

Tavker, N.; Yadav, V. K.; Yadav, K. K.; Cabral-Pinto, M.; Alam, J.; Shukla, A. K.; Ali, F. A.; Alhoshan, M. Removal of Cadmium and Chromium by Mixture of Silver Nanoparticles and Nano-Fibrillated Cellulose Isolated from Waste Peels of Citrus Sinensis. *Polymers* **2021,** *13* (2), 234.

Tchounwou, P. B.; Yedjou, C. G.; Patlolla, A. K.; Sutton, D. J. Heavy Metal Toxicity and the Environment. *Mol. Clin. Environ. Toxicol.* **2012,** 133–164.

Tosco, T.; Papini, M. P.; Viggi, C. C.; Sethi, R. Nanoscale Zerovalent Iron Particles for Groundwater Remediation: A Review. *J. Clean. Prod.* **2014,** 77, 10–21.

Üstün, E.; Önbaş, S. C.; Çelik, S. K.; Ayvaz, M. Ç.; Sahin, N. Green Synthesis of Iron Oxide Nanoparticles by Using Ficus Carica Leaf Extract and Its Antioxidant Activity. *Biointerface Res. Appl. Chem.* **2022,** *2021,* 2108–2116.

Vahabi, K.; Mansoori, G. A.; Karimi, S. Biosynthesis of Silver Nanoparticles by Fungus Trichoderma Reesei (A Route for Large-Scale Production of AgNPs). *Insciences, J.* **2011,** *1* (1), 65–79.

Wang, T.; Jin, X.; Chen, Z.; Megharaj, M.; Naidu, R. Green Synthesis of Fe Nanoparticles Using Eucalyptus Leaf Extracts for Treatment of Eutrophic Wastewater. *Sci. Total Environ.* **2014,** *466,* 210–213.

Wen, D.; Fu, R.; Li, Q. Removal of Inorganic Contaminants in Soil by Electrokinetic Remediation Technologies: A Review. *J. Hazard. Mater.* **2021,** *401,* 123345.

Weng, X.; Huang, L.; Chen, Z.; Megharaj, M.; Naidu, R. Synthesis of Iron-Based Nanoparticles by Green Tea Extract and Their Degradation of Malachite. *Ind. Crop. Prod.* **2013,** *51,* 342–347.

Xu, Q.; Li, W.; Ma, L.; Cao, D.; Owens, G.; Chen, Z. Simultaneous Removal of Ammonia and Phosphate Using Green Synthesized Iron Oxide Nanoparticles Dispersed onto Zeolite. *Sci. Total Environ.* **2020,** *703,* 135002.

Yadav, I. C.; Devi, N. L.; Syed, J. H.; Cheng, Z.; Li, J.; Zhang, G.; Jones, K. C. Current Status of Persistent Organic Pesticides Residues in Air, Water, and Soil, and Their Possible Effect on Neighboring Countries: A Comprehensive Review of India. *Sci. Total Environ.* **2015,** *511,* 123–137.

Yi, H.; Jiang, M.; Huang, D.; Zeng, G.; Lai, C.; Qin, L.; Zhou, C.; Li, B.; Liu, X.; Cheng, M.; Xue, W. Advanced Photocatalytic Fenton-like Process Over Biomimetic Hemin-Bi2WO6 with Enhanced pH. *J. Taiwan Inst. Chem. Eng.* **2018,** *93,* 184–192.

Yu, J. G.; Zhao, X. H.; Yang, H.; Chen, X. H.; Yang, Q.; Yu, L. Y.; Jiang, J. H.; Chen, X. Q. Aqueous Adsorption and Removal of Organic Contaminants by Carbon Nanotubes. *Sci. Total Environ.* **2014,** *482,* 241–251.

Yunus, I. S.; Harwin, Kurniawan, A.; Adityawarman, D.; Indarto, A. Nanotechnologies in Water and Air Pollution Treatment. *Environ. Technol. Rev.* **2012,** *1* (1), 136–148.

Zambianchi, M.; Durso, M.; Liscio, A.; Treossi, E.; Bettini, C.; Capobianco, M. L.; Aluigi, A.; Kovtun, A.; Ruani, G.; Corticelli, F.; Brucale, M. Graphene Oxide Doped Polysulfone Membrane Adsorbers for the Removal of Organic Contaminants from Water. *Chem. Eng. J.* **2017,** *326,* 130–140.

Zhang, C.; Li, Y.; Shen, H.; Shuai, D. Simultaneous Coupling of Photocatalytic and Biological Processes: A Promising Synergistic Alternative for Enhancing Decontamination of Recalcitrant Compounds in Water. *Chem. Eng. J.* **2021,** *403,* 126365.

Zhang, D.; Ma, X. L.; Gu, Y.; Huang, H.; Zhang, G. W. Green Synthesis of Metallic Nanoparticles and Their Potential Applications to Treat Cancer. *Front. Chem.* **2020,** *8,* 799.

Zhang, L.; Zhao, B.; Jiang, C.; Yang, J.; Zheng, G. Preparation and Transport Performances of High-density, Aligned Carbon Nanotube Membranes. *Nanosc. Res. Lett.* **2015,** *10* (1), 1–8.

Zhao, W.; Zheng, Z.; Zhang, J.; Roger, S. F.; Luo, X. Evaluation of the Use of Eucalyptus to Control Algae Bloom and Improve Water Quality. *Sci. Total Environ.* **2019,** *667,* 412–418.

Zhu, M.; Han, J.; Wang, F.; Shao, W.; Xiong, R.; Zhang, Q.; Pan, H.; Yang, Y.; Samal, S. K.; Zhang, F.; Huang, C. Electrospun Nanofibers Membranes for Effective Air Filtration. *Macromol. Mater. Eng.* **2017,** *302* (1), 1600353.

CHAPTER 10

Eco-Friendly Synthesis of Silver Nanoparticles: Principles and Their Antimicrobial Characteristics

SALIM ALBUKHATY[1,2], ALI ATHAFAH TOMAH[3,4]
YASMINA KHANE5,[6], HASSAN AL-KARAGOLY[7],
NISREEN JAWAD KADHIM[8],
FARES FENNICHE[9], and DJABER AOUF[9]

[1]*Department of Chemistry, College of Science, University of Misan, Maysan, Iraq*

[2]*College of Medicine, University of Warith Al-Anbiyaa, Karbala, Iraq*

[3]*Department of Plant Protection, College of Agriculture, University of Misan, Maysan, Iraq*

[4]*State Key Laboratory of Rice Biology, Institute of Biotechnology, Zhejiang University, Hangzhou, China*

[5]*Université de Ghardaia, Ghardaia, Algeria*

[6]*Laboratory of Applied Chemistry (LAC), University of Bouchaib Belhadj, Ain Temouchent, Algeria*

[7]*Department of Internal and Preventive Medicine, Veterinary Medicine College, University of Al-Qadisiyah, Al-Diwaniyah, Iraq*

[8]*Department of Biomedicine, Faculty of Medicine, Warith Al-Anbiyaa University, Iraq*

[9]*Laboratory of Dynamic Interactions and Reactivity of Systems, University of Kasdi Merbah, Ouargla, Algeria*

Sustainable Nanoremediation: Modern Technologies for a Clean Environment.
Fernanda Maria Policarpo Tonelli, Arpita Roy, and Khalid Rehman Hakeem (Eds.)

ABSTRACT

The advancement of eco-friendly techniques for creating nanoparticles is a crucial step in the progress of nanotechnology as a dependable, environmentally sustainable method for producing a large variety of biocompatible materials and nanomaterials, including metal/metal oxide nanomaterials and nanocomposites.Therefore, green synthesis is viewed as a crucial technique to minimize the adverse effects related to the conventional methods of synthesis for nanoparticles that are frequently used in laboratories and industry. We used silver nanoparticles (AgNPs) as an example when describing the fundamentals of "ecofriendly" synthesis approaches in this chapter due to their excellent antibacterial properties and capacity to combat pathogens that cause infections, including gram-positive and Gram-negative bacteria as well as multidrug-resistant pathogens. We focused on the functionality of bio- elements as reducing agents. Additionally, covered are nanoparticle stability, photodegradation, toxicity, and related surface engineering methods for improving biocompatibility. We finished by discussing antibacterial properties for use in biomedical applications.

10.1 INTRODUCTION

The phrase "nanotechnology" refers to the fabrication of materials at the nanoscale (between 1 and 100 nm), their formulation, modification, and use, where these newly generated nanomaterials display novel features in comparison to the same materials at their bulk size (Bayda et al., 2020; Whitesides, 2005). In general, nanoparticles (NPs) can be divided into inorganic and organic NPs. Organic NPs include carbon nanoparticles like carbon nanotubes, fullerenes, and quantum dots, while inorganic NPs include gold (Au), silver (Ag), copper (Cu), and other magnetic nanoparticles (MNPs). Au and Ag NPs, for example, are becoming more attractive due to their outstanding properties and practical adaptability (Khan et al., 2017; Sanzari et al., 2019; Jabir et al., 2020; Albukhaty et al., 2020; Yazdi et al., 2020). Nanobiotechnology provides essential tools and knowledge of nanotechnology to detect and treatment of different diseases. These include the application of nanoscale materials, such as biocompatible nanoparticles in life sciences include, including medicines, biotechnology, gene, drug delivery, and tissue engineering (Jihad et al., 2021; Albukhaty et al., 2018, 2021; Al-Kaabi et al., 2021).

Recent advances in nanobiotechnology have drawn a lot of attention because they address the sustainability concerns raised by traditional synthesis techniques (Castillo-Henríquez et al., 2020).

One of the most significant metallic nanoparticles is Ag NPs, which have elevated surface area-to-volume ratio as well as the ability to alter their biological, physical, and chemical characteristics (Pirtarighat et al., 2019; Akter et al., 2018; Marassi et al., 2018), so that these superior nanoparticles have been exploited for multiple scientific types of research and different applications such as medicine, food, industrial materials, and pharmacy (Akter et al., 2018, Marassi et al., 2018; Kumar et al., 2018; Fortunati et al., 2013; Tavaf et al., 2017).

There are different approaches for obtaining high quantity and good control of nanoparticles by different synthesis methods covering chemical, physical, and biological procedures that are relevant to their use (Wu et al., 2013; Khashan et al., 2021; Safat et al., 2021). Interestingly, biosynthesis of Ag NPs exhibits good results, solubility, and bio-stability (Prasher et al., 2020). Additionally, under the ideal circumstances for applied research, these methods ensure to be rapid, low-toxic, simple to prepare, reliable, and well-produced in terms of size and shape (Long et al., 2022; Tian et al., 2014). Therefore, it appears that biological methods for producing Ag NPs hold great promise. The purpose of this chapter is on recently developed approaches for the eco-friendly synthesis of Ag NPs using different natural elements such as herbal extracts, bacteria, and fungi and their antimicrobial and photocatalytic activities with a proposal of their mechanism.

10.2 SILVER NANOPARTICLES SYNTHESIS

Two general methods usually have been performed to fabricate nanomaterials and components on a nanoscale; these approaches include the bottom-up method, which is frequently a great option for creating uniformly sized nanoparticles covering a uniform framework in which catalysts (enzymes and reducing agent) synthesize and are controlled by the catalyst from molecular components of their own. In contrast, the top-down method uses bigger (macroscopic) starting structures that can have their geometry externally altered when developing nanostructures (Begum et al., 2022). (Fig. 10.1a).

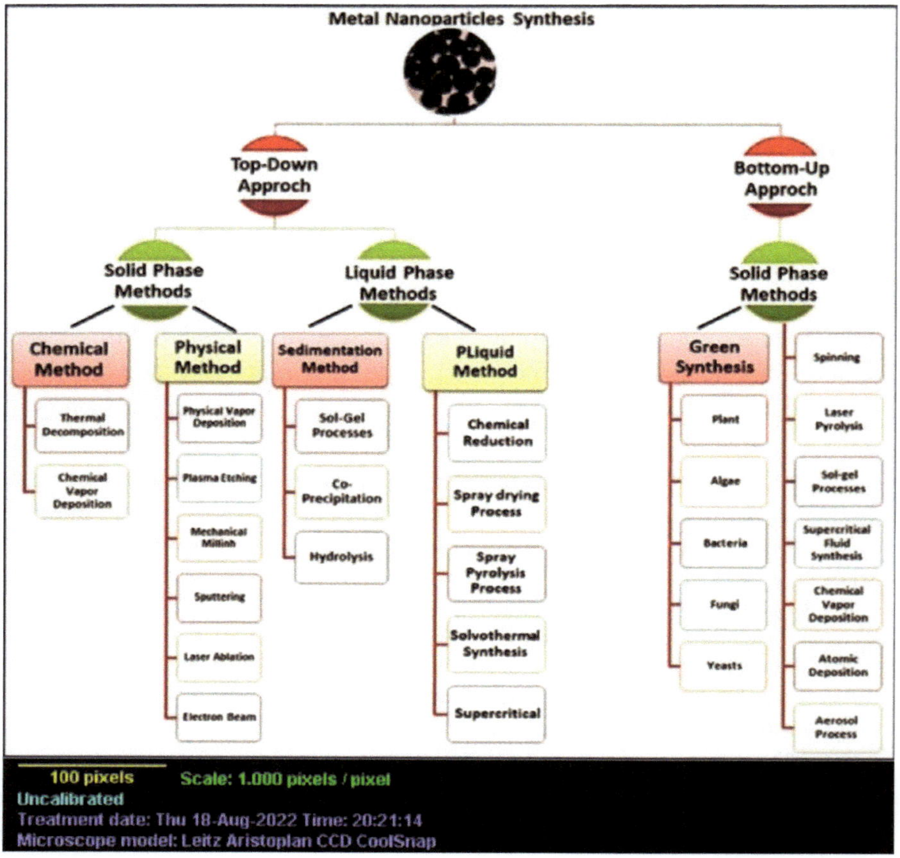

FIGURE 10.1a Schematic diagram of the various methods for creating nanoparticles.

Various approaches are obtainable for fabricating Ag NPs including chemical (Modan and Plăiaşu, 2020; Rashid et al., 2013; Slepička et al., 2020), physical (Sarkar et al., 2009; Mafune et al., 2000), and biosynthesis methodologies (Roy et al., 2019; Okafor et al., 2013). The chemical reduction, electrochemical, irradiation-assisted chemical, and pyrolysis processes are the main types of chemical processes for manufactured Ag NPs (Nakamura et al., 2019). These methods include arc-discharge (Tien et al., 2008), organic and inorganic reducing chemicals are used to reduce silver nanoparticles, which is the most used method of synthesis (Garcia-Barrasa et al., 2011). For the reduction of Ag^+ ions, a number of reducing agents could be utilized, including hydrogen (sodium ascorbate, borohydride, and citrate), as well as Tollens reagent. As a result of the reduction, metallic Ag ions are produced, which then aggregate to form oligomeric clusters. Eventually, these clusters cause the emergence of metallic colloidal silver particles (Garcia-Barrasa et al., 2011). To prevent AgNPs from agglomerating, stabilizing agents must be used during AgNP synthesis (Restrepo and Villa, 2021). It is important to remember that polymeric materials work well as protective agents to keep nanoparticles stable (Mahendia et al., 2013; Cunha et al., 2017). The physical methods commonly used for the production of Ag NPs are laser ablation, ultra-sonication, microwave (MW) irradiation, evaporation-condensation, gamma irradiation, and lithography. In these methods, NPS is integrated by evaporation condensation using a tube heater operating at barometric weight. The most important physical methods are laser removal and condensation evaporation. Without the use of chemical reagents, laser ablation is one of the most effective ways to create Ag NPs. By varying the number of laser pulses, it is possible to regulate the particle size of colloids (Pyatenko et al., 2004). However, this wide range of chemical, physical, or biological techniques used to synthesize silver nanoparticles can produce AgNPs of various sizes, homogeneity, and shape (Khodashenas and Ghorbani, 2015). Based on their morphology and shape, silver nanostructures can be divided into different groups (Loiseau et al., 2019). For example, cubes such as silver nanocubes with sharp (Zhou et al., 2016), spheres as chicory-silver nanospheres (Khatami et al., 2018), triangles as saponin-capped silver nanotriangles (Debnath and Das, 2019), prisms as citrate-capped silver nanoprisms (Haber and Sokolov, 2017), sheets as silver nanosheets using glucose and aluminum nitrate (Chen et al., 2006), disks as silver nanodisks (Kim and Lee, 2015), rods as tannic acid-silver nanorods (Patarroyo et al., 2016), bars as silver nano bars (Zhang et al., 2012), wires as silver nanowires from AgCl seeds (Puchana-Rosero et al., 2016), and flower-like (Lu et al., 2016), as shown in Figure 10.1b.

10.3 GREEN SYNTHESIS

There are numerous physical and chemical processes now being employed to produce nanoparticles. However, biogenic reduction of metal precursors to produce equivalent NPs is more cost-effective, chemical free, and environmentally benign for use in biological and medicinal applications. A scientific revolution has been initiated by the utilization of biological agents to produce silver nanoparticles because of their availability, usability, and environmentally friendly nature. Ag NPs can be biosynthesized using biological components including bacterial, fungal, yeast, algal, and plant

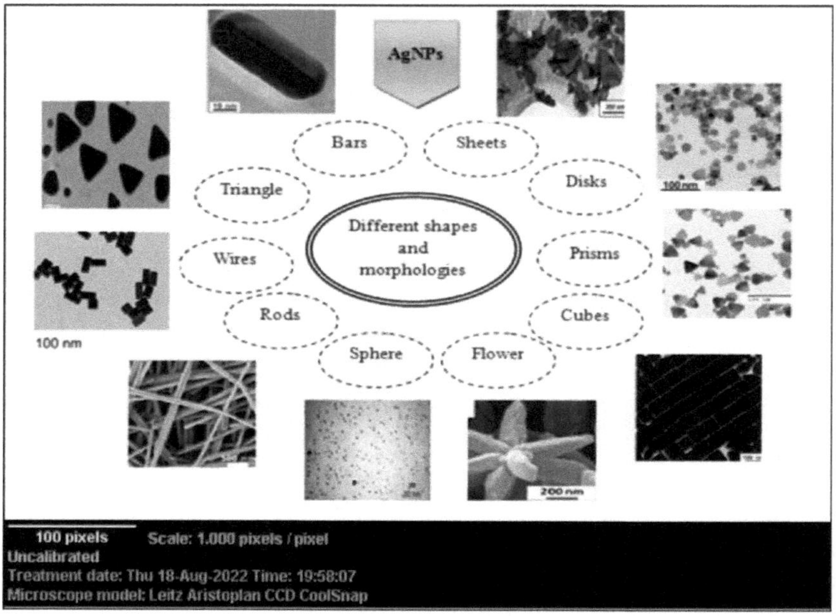

FIGURE 10.1b Images from transmission electron microscopy (TEM) of various silver nanoparticle morphologies and forms.

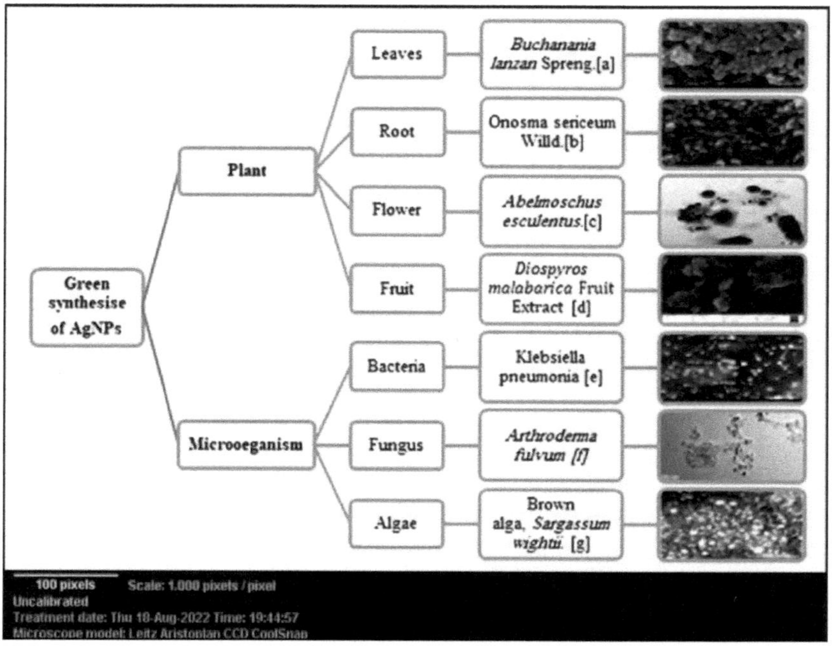

FIGURE 10.2 Detailed generally, biological components are utilized to biosynthesize Ag NPs.

extracts (Eid et al., 2020; Vahabi et al., 2011; Shu et al., 2020; Abdel-Raouf et al., 2019; Purohit et al., 2022; Çalhan and GÜndoĞan, 2020; Devanesan and AlSalhi, 2021; Xue et al., 2016), as shown in Figure 10.2.

Recent research has concentrated on the potential of biogenesis of Ag NPs using different strains of bacteria, including *Pseudomonas stutzeri* (Bachii and Abd-Al Sahib), Lactobacillus strains (Mohd Yusof et al., 2020), *Bacillus licheniformis* (Tan et al., 2021), *Escherichia coli* (Birla et al., 2009), *Brevibacterium casei* (Kalishwaralal et al., 2010), fungi including *Fusarium oxysporum* (Yin et al., 2016), Ganoderma neo-japonicum Imazeki (Gurunathan et al., 2013; Khane et al., 2022; Ali et al., 2016; Mirzaie et al., 2022), plant. Several biomolecules, including biopolymers (Leung et al., 2010), starch (Kumar et al., 2014), fibrinolytic enzyme (Deepak et al., 2011), and amino acids (Shankar and Rhim, 2015), were also used, as schematically illustrated in Figure 10.3, which depicts the methodology for producing silver nanoparticles using a variety of biological components. According to recent research, three general factors must be considered in green synthesis: the solvent medium, eco-friendly reducing reagents, and nanoparticle stabilizer (Peralta-Videa et al., 2016). Green synthesis may be superior to traditional synthesis because the ZnO NPs manufactured seem pure and devoid of nasty chemicals, be pure and devoid of nasty chemicals, thereby making them suitable for biological, therapeutic approaches, and pharmaceutical purposes (Gur et al., 2022; Abdelmigid et al., 2022). Several other researchers use plant extracts or bacterial proteins as reductants to manipulate nanoparticle size, shape, and monodispersity (Rajeshkumar and Bharath, 2017).

The generated nanoparticles are rapidly water-soluble, high density and stable, of which are extra benefits of green fabrication techniques (Karagoly et al., 2022).

AgNPs' biological activity is determined by their morphology and structure, which in turn is determined by the particles' sizes and shapes (Raza et al., 2016; Alshareef et al., 2017). AgNPs synthesized from algae, and bacterial sources have a size range of 100 nm or more, whereas AgNPs synthesized from algae, algal plants, pteridophytes, gymnosperms, and biopolymer sources have a size range of less than 50 nm. Instead of using plant extracts, the authors of this study used microorganisms and bio-polymers to synthesize AgNP, and they agitated the reaction mixture constantly to prevent agglomeration. In some cases, the production of nanoparticles can be sped up by mechanically agitating the reaction mixture. Aging, the synthesized AgNP solution morphed the nanospheres into flower-like frameworks (Table 10.1).

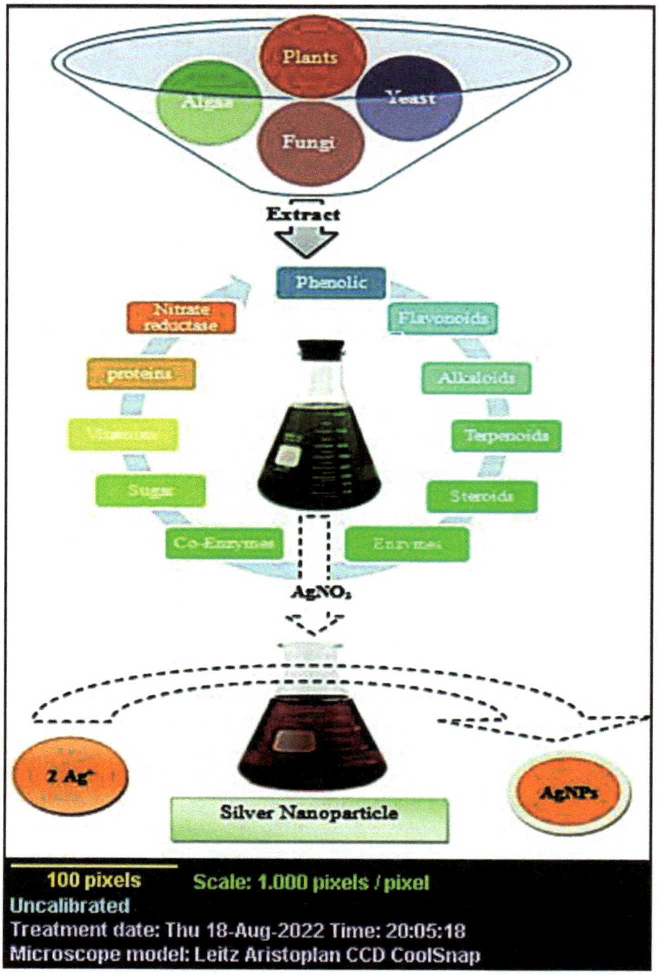

FIGURE 10.3 A schematic illustration of the method for producing silver nanoparticles in a sustainable manner using different biological entities.

When comparing different nanoparticles, those with smaller sizes and truncated-triangular shapes appear to be the most effective. Many studies have been able to synthesize AgNPs in a variety of shapes and sizes, but there are still some restrictions. For the synthesis of monodisperse and uniformly sized silver colloids, an abundance of a potent reducing agent like sodium borohydride ($NaBH_4$) was used to enhance the efficiency of production when it comes to morphology and structure (Hong and Lee, 2018). By optimizing the synthesis methods—including the number of byproducts, temp, pH, and

TABLE 10.1 Displayed Green Synthetic Ag NPs Produced Using Several Living Things, Including Plants, Bacteria, Fungi, Yeasts, and Algae.

Reducing factor	Size and shape	Characterization	Application	Ref no.
		Bacteria		
Culture Supernatant of *Shewanella sp.* ARY1	38 nm spherical	UV, FTIR, XRD, EDX, TEM, and SEM.	Antibacterial activity	(Mondal et al., 2020)
Thermophilic *Bacillus Sp.* AZ1	~7–31 nm spherical	UV, SEM, EDX and TEM	Antimicrobial activity	(Deljou and Goudarzi, 2016)
		Fungi and yeast		
Enteromorpha flexuosa	2–32 nm circular	UV, XRD, EDX, and TEM.	Antibacterial and antifungal activity against *E. coli, S. cerevisiae, C. Albicans, S. aureus, S. epidermidis, B. subtilis, B. pumulis, E. faecalis*	(Yousefzadi et al., 2014)
The culture supernatants of *Aspergillus sydowii*	1 and 24 nm spherical	UV, TEM and XRD.	Antifungal/antiproliferative activities	(Wang et al., 2021)
		Algal extracts		
Yeast strain MKY3	2–5 nm	UV, TEM, XRD, and XRP	Antifungal/antiproliferative activities	(Kowshik et al., 2003)
Aqueous filtrate of *Sargassum wightii*	18.45–41.59 nm spherical	UV, FTIR, XRD, FESEM, EDX, Z-P, and DLS.	Enzyme inhibitory, antibacterial, and antioxidant	(Deepak et al., 2018 Sep 18)
		Plant extracts		
leaf extract of *Prunus dulcis L.* (almond tree)	14.6 nm spherical	FTIR,XRD, UV, SEM, TEM, and Z-P.	Antioxidant and Antimicrobial Activity	(Aktepe and Baran, 2021)

TABLE 10.1 *(Continued)*

Reducing factor	Size and shape	Characterization	Application	Ref no.
Syzygium cumini fruit extract	~47 nm nearly spherical	UV, FTIR, XRD, and SEM.	Antibacterial, antioxidant, and anti-inflammatory	(Chakravarty et al., 2022)
Extract of *Jasminum officinal* L. Leaves	9.22 nm spherical	UV, FTIR, TEM, Z-P and XRD.	Cytotoxic Activity against Bladder and Breast Cancer (MCF-7) Cell Lines	(Elhawary et al., 2020)
Diospyros malabarica aqueous fruit extract	17.4 nm spherical	UV, XRD, FTIR, DLS, Z-P, FESEM, EDX, TEM and P-L.	Antibacterial, anticancer effect	(Bharadwaj et al., 2021)
Citrus limon Zest Extract	23 nm spherical	UV, Z-P, DLS, SEM, EDX, XRD, and TEM.	Antioxidant and antimicrobial properties	(Khane et al., 2022)
Root extract of *Duchesnea indica*	20.49 nm spherical	UV, TEM, SEM, XRD, EDX, FTIR, and Z-P.	Antimicrobial, anti-inflammatory, analgesic, and muscle relaxant	(HsanIlahi et al., 2021)
Root Extracts of *Rubus ellipticus* Sm.	13.85–34.30 nm spherical	UV, FTIR, XRD, FESEM, TEM and EDX.	Antioxidant and antibacterial activity	(Khanal et al., 2022)
Malva Sylvestris Flower Extract	20–30 nm spherical	UV, FTIR, SEM, TEM, and XRD	Antibacterial and catalytic activity	(Mehdizadeh et al., 2021)
Plumeria rubra Flower Extract	20–80 nm Spherical	UV, FTIR and TEM	Antimicrobial activities	(Mandal, 2018)

the quantity of limiting and stabilizing factors—biological processes make it easier to control the shape, size, and dispersion of the generated nanoparticles (Uzair et al., 2020; Tiburu et al., 2017).

10.4 CHARACTERIZATION

The behavior, bio-distribution, safety, and effectiveness of nanoparticles are largely dependent on their physicochemical qualities. In order to assess the functional characteristics of the produced nanoparticles, it is crucial to characterize AgNPs. According to Shin et al. (2015) size distribution, electrostatics, surface area, general form, and aggregation are a few physiochemical characteristics that may have a significant effect on the physiological interactions between nanomaterials and their intended biological targets.

The identification of AgNPs is thus considered a critical factor in assessing the performance properties of the related synthesized particles. The physiological interactions that take place between nanomaterials and their respected biological targets may be significantly impacted by specific physiochemical properties, such as particles size, electrostatics, surface area, particular shape, and aggregation, according to a study by Shin et al. (2015). Numerous analytical techniques, such as UV, FTIR, X-RD, XPS, DLS, SEM, TEM, and AFM analysis, have been used to characterize the materials (Gurunathan et al., 2015).

The ideas and use of several analytical techniques for the characterization of AgNPs have been discussed in a number of reputable books and reviews; however, the fundamentals of the key approaches are summarized below for clarity, optimization of the synthesis processes, taking into account the quantity of precursors, temperature, pH, and reducing and stabilizing agents used (Zhang et al., 2016).

10.4.1 UV-VISIBLE SPECTROSCOPY

An essential and effective technique for the comprehensive categorization of synthesized nanoparticles is UV-vis spectroscopy. It is also used to examine AgNPs' stability and manufacturing process (Ajitha et al., 2018; Fayaz et al., 2010). Due to the distinct optical characteristics of AgNPs, they strongly interact with specific light wavelengths (Ren et al., 2020). Also, UV-vis spectroscopy is said to be quick, easy, and simple to use. It is quick to measure different types of NPs and is both sensitive and selective for them.

Additionally, colloidal suspension particle characterization does not require calibration (Reddy et al., 2019).

According to the reports, since the valence band and conduction band are so close together in AgNPs, electrons can move freely between them. An SPR absorption band is formed when the electrons of silver nanoparticles collectively oscillate in resonance with the light wave, which is caused by the free electrons discussed here (Das et al., 2010).

The chemical environment, particle size, and dielectric medium all play a role in the absorption of AgNPs (Fayaz et al., 2010; Reddy et al., 2019). This peak, attributed to a surface plasmon, has been observed for a wide range of metal nanoparticles, from 2 nm to 100 nm (Salem et al., 2020). Furthermore, using UV-vis spectroscopy, an SPR peak at the same wavelength was identified, and the stability of AgNPs extracted via biological methods was established for more than a year.

10.4.2 X-RAY DIFFRACTION (XRD)

Depending on the constituent amount's intensity, there may be an increase or decrease. With information on unit cell size and shape derived from peak positions and information on electron density within the unit cell or atomic placement derived from peak intensities, this method has been used to represent the metallic nature of particles (Bunaciu et al., 2015; Thakar et al., 2022; Heera and Shanmugam, 2015).

10.4.3 DYNAMIC LIGHT SCATTERING (DLS)

The emotional light scattering technique (DLS) is used to measure the size of nanoparticles in suspensions and to determine the degree to which they have aggregated.

10.5 MECHANISM OF GREEN SYNTHESIS OF SILVER NANOPARTICLE AND SURFACE MODIFICATION

Lessening, stabilizing, and capping agents available through chemical methods are more limited in scope. Many plethoras of biological and eco-friendly AgNP synthesis techniques have been discovered through recent literature searches and experiments (Kaabipour and Hemmati, 2021). Researchers have used green, cost-effective, and environmentally friendly methods for the establishment of sustainable silver nanoparticles (Kumar et al., 2016; Bhardwaj et al., 2020), including microbial-taken compounds from (bacteria, fungi, yeasts, and actinomycetes) (Kiran et al., 2014; Daphne et al., 2018; Srivastava et al., 2019; Bayram et al., 2018; Abdel-Raouf et al., 2019) and also using microbial-taken combinations from (bacteria, fungi, yeasts, and actinomycetes) (Rajeshkumar and Bharath, 2017; Ranoszek-Soliwoda et al., 2019). In light of this, it is clear that biological agents in this approach represent a wide range of reagents that restore chemical reagents and incorporate the use of bioactive compounds like alkaloids, phenolic compounds, terpinoids, enzymes, proteins, sugars, saccharides, and vitamins (Mukherjee et al., 2018; May and Oluwafemi, 2016; Nadagouda et al., 2008) that are thought to show promise in the reduction and stabilization of metal salts into nanoparticles that in turn act as reducing agents.

Furthermore, this biogenic technique is classified under bottom-up synthesis methods (Rafique et al., 2017) one-step procedure, and this method appears to be more atomically efficient since it does not need the protection/deprotection steps often employed in organic synthesis in order to construct particles (Albrecht and Evans, 2006). Follow the principles of green chemistry by using less energy for synthesis by operating at near-ambient temperature, pressure, and pH (Parveen et al., 2016). This method can improve nanoparticles in a number of ways, including crystallinity, stability, size, and shape. Biomolecules (large macromolecules like proteins, polysaccharides, lipids, and nucleic acids, and small natural molecules like primary and secondary metabolites) produced by living organisms can convert the positive oxidation state of silver ion Ag^+ into the zero oxidation state of atoms Ag^0 and form the nanosilver particles in the biosynthesis of colloidal dispersions of AgNPs (Mittal et al., 2013).

In one of the studies reported by Liu and Hurt (2010), it was demonstrated that when AgNPs were exposed to the aquatic environment, silver ions would be released, decreasing the stability of the AgNPs. Additionally, Lee et al. (2012) hypothesized that first-order kinetics controls the release of Ag ions. When assessing these synthesized NPs, it is important to take into account a number of variables that impact the release rates of Ag ions. These variables primarily include particle size, and environmental variables including pH, temperature, time, radiation, and dissolved oxygen (Zhang et al., 2011; Liu et al., 2010) and capping agents (Ortega-Arroyo et al., 2013).

However, in most cases, the bioreduction of silver ions to create silver nanocrust by bio compounds derived from a biological source of plant or organism is rarely achieved.

Generally, Flavonoids, membrane proteins, NAD (P)$^+$ reductases, dehydrogenases, citric acid, polyphenols, and secondary metabolites are examples of reducing agents (Karatoprak et al., 2017), whereas extracellular proteins, enzymes, peptides, and tannic acids are examples of capping agents (Akhtar et al., 2013). However, there have been no published results on the identification of proteins involved in plant-mediated AgNPs production. Earlier research by de Barros et al. (2018) on protein role in the synthesis of AgNPs by fungus was elucidated; it involves nitrogen and sulfur containing groups.

10.6 MICROBIAL-MEDIATED SYNTHESIS OF AGNPS

Microorganisms' ability to synthesize metal nanoparticles is being considered a potential biological source. There are two distinct steps involved in creating AgNPs by microbes: intracellular formulation and extracellular fabrication (Roy et al., 2019). Methods that use enzymes and other biomolecules already present in microbial cells to convert Ag ions to NPs are called "intracellular" (Rana et al., 2020), and the concentration of Ag inside the cell leads to the nucleation of the formed AgNPs. This process is carried out continuously as the microorganisms proliferate. The living cells are extracted once the cells have grown to their maximum potential. Additionally, unique handling techniques are used to release the manufactured NPs from the retrieved cells (Roy et al., 2019). However, not all species have been shown to have the capability to transform metals into their nano forms.

Both an intracellular and an extracellular method exist for the microbial production of silver nanoparticles (Das et al., 2014). Intracellular silver buildup initiates silver nanoparticle production and furthers microbial development. Upon completion of the optimal period for bacterial development, nanoparticles containing viable cells are extracted.

Therefore, the generated nanoparticles need to be released from the collected cells in a certain way. Bacterial populations' extracellular secretions are extracted and employed as raw materials in the extracellular manufacturing process. In general, microorganisms are resistant to silver ions and efficient producers of AgNPs. It is hypothesized that the extracts' biomolecular mixtures (including enzymes/proteins, amino acids, polysaccharides, vitamins, etc.) are responsible for the reduction of Ag^+ ions. However, the precise process is not yet known. The bacterial strain *Streptomyces sp.* LK3 was used by Karthik et al. (2014) to show that AgNPs may be synthesized outside of cells. During the synthesis process the reduction of nitrate to nitrite can be performed by nitrate reductase. Generally, the presence of the enzyme nitrate reductase is the cornerstone of the widely accepted process to explain the mechanism of biogenic mediated silver nanoparticle (Ahmed and Ikram, 2016; Roy et al., 2013; Kalimuthu et al., 2008), as illustrated in Fig. 10.4.

Nitrate (NO_3?) is changed into nitrite (NO_2?) by the enzyme. After the nitrate ion (NO_3) is reduced to nitrate (NO_2), the silver ion (Ag^+) receives the electron and is then reduced to metallic silver (Ag^0). Producing silver nanoparticles using pure nitrate reductase from the fungus *F. oxysporum* (Kumar et al., 2007) provided a clear proof of this. The reduction of Ag^+ to Ag^0 can also occur in the absence of nitrate reductase due to functional groups on the bacterial cell wall (Kalimuthu et al., 2008).

When compared to bacteria, the high concentration of enzymes, proteins, and reducing components on the cell surface of fungus makes them a superior choice for the creation of nanoparticles. Fungal systems contain enzymes (such as naphthoquinones and anthraquinones) that can reduce Ag^+ that has been trapped on cell surfaces by the fungi (Kumar et al., 2007; Slawson et al., 1992; Mukherjee et al., 2001). The synthesis of AgNPs using yeast was unsuccessful. Nonetheless, there are very few reports to be found. It is believed that the combinational impact of bio-molecules in the extract is responsible for the formation of silver nanoparticles. Algae are no exception to this rule. In the presence of enzymes, actinomycetes are able to manufacture silver nanoparticles. It is believed that intracellular interactions lead to metal ion reduction (Mukherjee et al., 2001; Narayanan and Sakthivel, 2011).

10.7 PLANT-MEDIATED SYNTHESIS OF AGNPS

Plant extracts are favored over microbial creation of nanoparticles because of their low synthesis rate, rapidity, high efficiency, and viability. Synthesis of AgNPs from plant extract is likely to occur by an enzymatic process, the same to that used by microbes. Due to the presence of a complex of diverse antioxidant metabolites in plant cells that prevent oxidation and damage to biological components (Jain et al., 2021; Al-Musawi et al., 2022), the chemicals used for the stabilization and final capping of the nanoparticles must be distinct from those used for microbes. This means that biomolecules including enzymes, glycosides, and saponins can help stabilize the nanoparticle (Shang et al., 2019; Tripathi et al., 2013). Literature suggests that when metal salts are added to a herb extract, silver ions bind to water-soluble chemicals through --OH and --COOH groups.

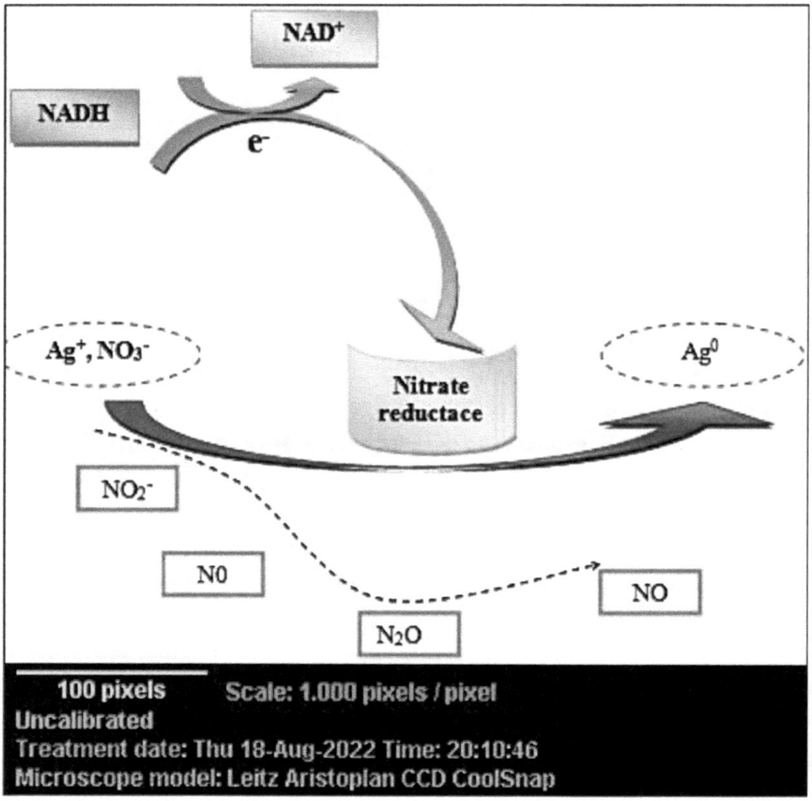

FIGURE 10.4 Biosynthesis of AgNPs using nitrate reductase.

This causes structural reforms in the protein complex that make it easier for the metal ion to change into a silver nanoparticle (Huang et al., 2015; Mikhailova, 2020). Proteins contribute to the reduction of silver and the production of AgNPs through their amino groups and cysteine residues (Kasithevar et al., 2017). Silver nanoparticles can be formed with the help of "capping" agents such as alkanes, amines, phenols, polyphenols, alkaloids, lignans, terpenoids, and flavonoids (Mandal et al., 2005; Elavazhagan and Arunachalam, 2011). Because of their potent antioxidant action, flavonoids are of special importance here. Nanoparticles are colloid-stable in water due to the hydrophilic functional groups of surrounding molecules (Ebrahiminezhad et al., 2017). Garlic extract's sucrose and fructose are very fascinating since they function as reducing and stabilizing chemicals (Von White et al., 2012). Additionally, polyols are in charge of the mechanism that converts Ag^+ into silver nanoparticles (Ghosh et al., 2012). Terpenoids, which are surface-active substances, are anticipated to adsorb on the surface of AgNPs to stabilize the nanoparticles and stop AgNPs from clumping together (Parlinska-Wojtan et al., 2016). One way that terpenoids might help turn Ag^{+} ions into Ag0 nanoparticles is by changing the C--O group of the terpenes into the --C--O group (Shankar et al., 2004).

It is thought that terpenoids help reduce metal ions by turning aldehydic groups in their molecules into carboxylic acids (Mahendia et al., 2013).

Li et al. (2007) have said that *Capsicum annuum* extract can be used to make AgNPs. Research has shown that to understand how AgNPs grow, you need to know how recognition-reduction-limited nucleation is related to growth. The first step in the recognition process was for proteins in the *C. annuum* extract and Ag ions to interact electrostatically. Extract proteins then reduce the number of Ag ions. This makes silver nuclei, which change the structure of proteins in a second step. Also, these nuclei grew because silver ions were taken away and then more atoms were added. By Ostwald ripening, the polycrystalline phase changed into the single crystalline phase as the AgNPs aged for longer (Jain et al., 2021). This made the AgNPs bigger. Bar et al. (2009) report that the Jatropha curcas seed extract can be used to make green silver NPs. As a result, it was found that curcain, which is an enzyme, curcacycline A, which is a cyclic octapeptide, and curcacycline B, which is a cyclic nonapeptide, could be used as reducing and capping agents. Characterized by XRD, UV-Vis, and TEM spectroscopy, AgNPs were found to have two different distributions: some spherical particles with a diameter of 20–40 nm and other irregular particles that were bigger. Further research showed that the enzyme curcain stabilized the larger NPs, while the cavities in the cyclic peptides stabilized the smaller ones (curcacycline A and curcacycline B). There is evidence that supports that carbonyl groups in amino acid residues or protein peptides are the ones responsible for presenting high affinity for metals (Lin et al., 2005).

Therefore, the protein can act as an encapsulating agent and preserve the NPs by inhibiting their aggregation. The cyclic protein curcacycline A or curcacycline B was thought to have first trapped the Ag ions in their core structure. Under the right processing circumstances, the amide groups of the host peptide reduced and stabilized AgNPs in situ. It was thought that the cyclic peptide cavity supported the smaller AgNPs since the radius of the majority of the generated AgNPs was equivalent to its cavity, but the enzyme curcain's massive, folded protein structure was thought to stabilize the irregularly sized AgNPs. The AgNPs produced by curcain latex was also shown to be stable even after one month, according to studies (Jain et al., 2021; Jain et al., 2021). According to the proposed mechanism of biosynthesized AgNPs using the *Phyllanthus amarus* leaf extract by Ajitha et al. (2018), the leaf extract itself served as a capping and reducing agent, significantly affecting the nanoparticles' size and form as illustrated in Fig. 10.5.

This mechanism suggests that the antioxidant phytochemicals of the leaf extract reduced the positively charged metal ions (Ag^+ ions) to a neutral state (Ag^0 atoms) and to small Ag^0 particles by electron donation to the ions with the help of the bioreducing agents in the first stage. The Ag0 atoms then expanded into petal shapes, which, with the help of bio-capping agents, coalesced into nanostructures resembling flowers. It is usual for reduced metals to undergo nucleation and cluster formation, with the resulting clusters assembling into particles that are stabilized by complexing with or making electrostatic interactions with oxygen-containing groups like --COO and --OH (Wu et al., 2014), The phytochemicals that caused the AgNPs to form also stopped them from further aggregating, in accordance with our suggested mechanism, by coating them and providing capping agents rich in electron-rich hydroxyl groups onto the particles.

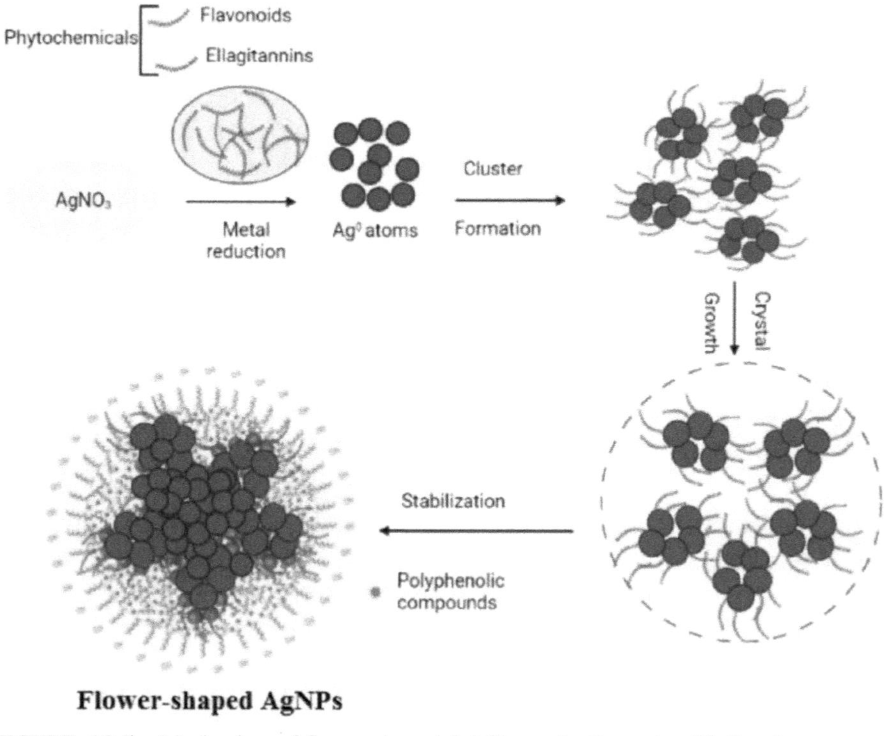

FIGURE. 10.5 Mechanism of flower-shaped AgNPs production using *Phyllanthus amarus* leaf extract.

To produce efficient engineered nanoparticles a deeper comprehension of nanoparticle structure is required. There are three main parts to a nanoparticle's structure: the core, the shell, and the surface (Christian et al., 2008). Different techniques, reaction conditions, and precursors can be used to synthesize the nanoparticle's core, while a shell can be added if required. However, when a single salt is used to create a nanoparticle, the shell and core often both consist of the same substance. The last and maybe most important consideration is the surface, which can be considered functional in several ways depending on the setting. The goal should be either to keep nanoparticles from clumping together or to endow them with a distinct role. The functionalization of the nanoparticle surface is the result of temporary van der Waals interactions between the chemicals and the nanoparticle surface (Roy et al., 2019).

Biological molecules operate as an in situ reducing and capping agent during green nanoparticle formation, reducing metal salts and covering the formed nanoparticles. This capping has advantages over others because (i) it stops the nanoparticles from aggregating together, (ii) it lessens the toxicity, and (iii) it boosts the antimicrobial effect (Roy et al., 2019).

The dose of biochemical reagents found in the extract has a considerable impact on the size and size distribution of the metallic nanoparticles as summarized in Table 10.2. Fast reaction rates and favorable conditions for the creation of smaller nanoparticles are made possible due to the presence of a powerful reductant in the concentrate. If the biomolecules quickly remove the salt, resulting in the constant generation of new nuclei or secondary nuclei, then the size distribution will be quite narrow. As a backup strategy, biomolecules can build a monolayer on the nanoparticle surface to inhibit aggregation (Restrepo and Villa, 2021). In contrast, recent studies about antibiotic-mediated production of nanoparticles highlight the importance of surface functionalization with antibiotics increases their biocidal action (Rai et al., 2010). Since extract-mediated nanoparticles have biomolecules attached to their surfaces, their antibacterial activity is greatly increased.

It has been suggested that "capping" agents can alter the relative surface area of a nanocrystal by attaching selectively to distinct types of facets with differing surface free energy (Xia et al., 2015). Thus, nanoparticle "capping" can serve several purposes, including preventing nanoparticle agglomeration, lessening toxicity, and boosting antibacterial characteristics. Furthermore, these compounds can increase the affiliation probability and activity of AgNPs on the bacterial cells (Roy et al., 2019; El-Rafie et al., 2011). Interestingly, the "capping" compounds found in plants often have their own antibacterial action, which can boost the efficacy of AgNPs (Mikhailova, 2020).

TABLE 10.2 Size, morphology, and antimicrobial activity of green synthesized AgNPs using various plant, microbial, and algal biomolecule

Biological materials	Size and morphology	Antibacterial activity against	Ref.
		Plants	
leaf extract of *elephantopus scaber*	37.86 nm spherical	*B. subtilis, L. lactis, P. fluorescens, P. aeruginosa, A. flavus*, and *A. penicillioides*	(Francis et al., 2017)
Acacia rigidula extract	8–66 nm spherical	*E. coli, P.s aeruginosa, B. subtilis*	(Escárcega-González et al., 2018)
Phyllanthus amarus Seeds	8–66 nm spherical	*E. coli, B. subtilis, Klebsiella pneumoniae,* and *Staphylococcus aureus*	(Joseph et al., 2021)
Phyllanthus niruri leaf extracts	8–66 nm spherical	*Streptococcus* spp., *E. coli, E. faecalis, E. clocae, Citrobacter freundii, Burkholderia cepacia* complex, *S. typhi*, carbapenem-resistant Enterobacteriaceae, MRSA, and VRE	(Kumar et al., 2020)
Phyllanthus amarus leaf extract	30–42 nm flower-shaped	*E. coli, Pseudomonas* spp., *Bacillus* spp., *Staphylococcus* spp., *A. niger, A. flavus*, and *Penicillium* spp.	(Singh et al., 2014)
Nelumbo nucifera leaf extract	30–40 nm spherical	*B. subtilis, E. coli, S. typhi*, and *V.cholerae*	(Premanand et al., 2016)
		Algae	
aqueous extract of seaweed *Enteromorpha compressa*	4 and 24 nm Spherical shaped	Bacterial : *E. coli, S. aureus, K. pneumoniae*, *Pseudomonas* sp., and *S. paratyphi* Fungal : *A. flavus, A. niger, A. ochraceus, A. terreus,* and *F. moniliforme*	(Ramkumar et al., 2017)
Ulva compressa Cladophora glomerata	66.3 nm /81.8 nm; cubic	*K. pneumoniae, P. aeruginosa, E. coli, E. faecium,* and *S. aureus*	(Minhas et al., 2018)
Spirogyra varians	~35 nm uniform and quasi-spherical	*S.aureus, B. cereus, S. typhimurium, E. coli*, and *Listeria monocytogenes*	(Salari et al., 2016)
Spatoglossum asperum	20–46 nm spherical	*K. pneumoniae*	(Ravichandran et al., 2018)
Spyridia fusiformis	5–50 nm spherical	*S. aureus, K. pneumoniae*	(Murugesan et al., 2017)
Nostoc sp.	51–100 nm; spherical	*R. solanacearum, X. campestris, A. niger*, and *T. harzianum*	(Sonker et al., 2017)

TABLE 10.2 *(Continued)*

Biological materials	Size and morphology	Antibacterial activity against	Ref.
		Fungi and yeasts	
Mushroom *Pleurotus ostreatus*	10–40 nm spherical	*B. subtilis*, *B. cereus*, *S. aureus*, *E. coli*, and *P. aeruginosa*	(Al-Bahrani et al., 2017)
Arthroderma fulvum	15.5 nm spherical	*Candida* spp., *Aspergillus* spp., and *Fusarium* spp.	(Xue et al., 2016)
Penicillium polonicum	10–15 nm spherical	*A. baumanii*	(Neethu et al., 2018)
Endophytic fungus *Curvularialunata*	10–50 nm spherical	*E. coli*, *P. aeruginosa*, *S.paratyphi*, *B. subtillis*, *S. aureus*, and *B.cereus*	(Ramalingmam et al., 2015)
		Bacteria	
Pseudomonas deceptionensis DC5	10–30 nm spherical	*V. parahaemolyticus*, *C. albicans*, *S. aureus*, *S. enterica*, and *B. anthracis*	(Jo et al., 2015)
Halotolerant *Bacillus endophyticus* SCU-	~5.1 nm spherical	*C. albicans*, *E. coli*, *S. typhi*, and *S. aureus*	(Gan et al., 2018)
Bacillus thuringiensis	10–30 nm spherical	*E. coli*	(Nayak et al., 2016)
Phenerochaete chrysosporium (MTCC-787)	34–90 nm spherical and oval	*P. aeruginosa*, *K. pneumoniae*, *S. aureus*, and *S. epidermidis*	(Suresh et al., 2014)
		Actinomycetes	
Streptacidiphilus durhamensis	~8–48 nm. spherical	*S. aureus*, *B. subtilis*, *E. coli*, *P. aeruginosa*, *K. pneumoniae*, and *P. mirabilis*	(Buszewski et al., 2018)
Nocardiopsis sp. MBRC-1	30–90 nm spherical	*B. subtilis*, *P. aeruginosa*, and *Candida albicans*	(Manivasagan et al., 2013)
Streptomyces sp.	20–45 nm; various	*C. albicans*, *C. Tropicalis*, and *C. krusei*	(Sanjenbam et al., 2014)
Streptomyces rochei	22–85 nm spherical	*B. subtilis*, *S. aureus*, *P. aeruginosa*, *B. cereus*, *S. typhimurium*, *E. coli*, *V. fluvialis*, and *V. damsela*	(Abdelnaby et al., 2016)

10.8 ANTIBACTERIAL ACTIVITY OF AGNPS

Silver nanostructures have proven remarkable potential in several industrial, environmental (Kaabipour and Hemmati, 2021), and medicinal applications (Wei et al., 2015). Silver's antibacterial properties, found via experimentation in the ancient world, perhaps contributed to its widespread usage in eating and drinking utensils made of silver (Srikar et al., 2016). It was generally accepted that the silver was employed in aqueous solutions, coins, and plates for aseptic and antibacterial purposes, and to preserve food and dairy articles until the 19th century because it kept them fresh longer than other containers (Kremer et al., 1996). The synthesis or modification of antimicrobial compounds, as well as alternative therapeutics, is required due to the dramatic growth in microbial infections and the quick development in drug resistance to existing antibiotics (Shankar and Rhim, 2017). In vitro studies indicate that silver nanoparticles (AgNPs) are effective antibacterial agents that can circumvent bacterial resistance to conventional antibiotics. As a result, the creation of AgNPs as antibacterial agents is considered necessary. The high surface-to-volume ratio and crystalline surface structure of AgNPs suggest they may be effective antibacterial agents, making them one of the many promising nanomaterials. The buildup of AgNPs in the cell wall and the creation of "pits" in the bacterial cell walls, ultimately leading to cell death, were observed in the seminal study reported by Sondi and Salopek-Sondi (2004), proving the antibacterial action of AgNPs against *E. coli*. Antibacterial activity against the same *E. coli* strainpathoginic bacteria was enhanced when the particles were smaller and had a higher surface-to-volume ratio (Baker et al., 2005).

Furthermore, antimicrobial activity of AgNPs is also influenced by the size and shape of the particle (Bistarelli et al., 2018). Bactericidal action against Gram-positive and Gram-negative bacteria, including extremely multi-resistant strains like methicillin-resistant *S. aureus*, was demonstrated by AgNPs manufactured by four distinct kinds of saccharides with an average size of 25 nm. As mentioned previously, the size is not the sole factor in determining efficacy but AgNPs also interact with the Gram-negative bacterium *E. coli* in a way that is shape-dependent (Raza et al., 2016). Additional research was conducted to examine the effectiveness of AgNPs as an antibacterial agent against yeast, *E. coli*, and *S. aureus*. Complete growth suppression was seen in yeast and *E. coli* at low doses of AgNPs, but *S. aureus* was very slightly affected (Chouhan and Guleria, 2020).

The effectiveness of different antibiotics, including penicillin G, amoxicillin, erythromycin, clindamycin, and vancomycin against *S. aureus* and

E. coli, was found to be enhanced in the presence of biologically produced AgNPs from the culture supernatants of *K. pneumoniae* (Jijie et al., 2017).

Silver nanoparticles (AgNPs) are effective against *E. coli*, but hydrogel-AgNP nanocomposites are more effective. Chitosan-Ag-nanoparticle composite was confirmed to have more potent antibacterial activity than its components at their respective concentrations (Sanpui et al., 2008).

This is likely due to the fact that one-pot synthesis encourages the creation of tiny AgNPs connected to the polymer, which can be dispersed in fluids of pH 6.3. When AgNPs were biologically produced using *S. aureus* culture supernatants, they showed significant antimicrobial activity against methicillin-resistant *S. aureus*, methicillin-resistant *S. epidermidis*, and *Streptococcus pyogenes* (Thomas et al., 2014), but only moderate antimicrobial activity against *S. typhi* and *K. pneumoniae*. In *E. coli*, the leakage of reducing sugars and proteins revealed the mechanisms of AgNP-induced cell death. Furthermore, AgNPs can damage bacterial cell membrane structure by destroying its permeability by forming numerous pits and holes in its surface (Vazquez-Muñoz et al., 2019). The silver nanocrystalline chlorhexidine (AgCHX) complex was highly effective against Gram-positive/negative bacteria and methicillin-resistant *S. aureus* (MRSA). It was found that the MICs of nanocrystalline Ag (III) CHX were significantly lower than those of the ligand (CHX), silver nitrate ($AgNO_3$), and gold (Au (III)). Several applications of silver nanoparticles (AgNPs) are depicted in Figure 10.6.

Nanoparticles immobilized on an amine-functionalized silica surface are more effective than colloidal AgNPs and have a higher concentration of silver ions in solution, as reported by Agnihotri et al. (2013). Silver/polyrhodanine-decorated silica nanoparticles form a nanocomposite that shows promising and improved antibacterial activity against *E. coli* and *S. aureus* (Song et al., 2013). In addition, the physical and surface properties of silver were tested against *S. aureus*, *B. megaterium*, *P. vulgaris*, and *S. sonnei* (Khurana et al., 2014). Particles with a hydrodynamic size of 59 nm were found to be more effective against bacteria than particles with a size of 83 nm. Gurunathan et al. (2013) also showed that antibiotics' antibacterial and anti-biofilm action was effective against significant pathogens like *P. aeruginosa*, *Shigella flexneri*, *S. aureus*, and *S. pneumoniae*.

These results suggest that at the lowest concentrations of antibiotics and AgNPs, the combination has significant antibacterial and anti-biofilm effects. Nanocomposite spheres composed of AgNPs coated on polymer colloids demonstrated impressive antibacterial activity (Chen et al., 2014; Jain et al., 2021).

Much attention has recently been paid to graphene and silver nanoparticle (AgNPs) nanocomposites killing bacteria. Compared to AgNPs, the graphene oxide (GO)-Ag nanocomposite was much more effective at killing *E. coli* bacteria than AgNPs. Based on the results of a study, scientists thought that silver nanoparticles might kill bacteria. AgNPs can release Ag^{+} ions that can stick to the walls and membranes of bacterial cells and even get into the cytoplasm. Antimicrobial activity is caused by the reactive oxygen species (ROS) made inside the cell by Ag+ ions.

This causes a chain of events: (1) DNA synthesis is stopped, (2) mRNA synthesis is stopped, (3) the cell is destroyed, and the cell contents leak out, (4) protein synthesis is stopped, (5) cell-wall synthesis is stopped, (6) mitochondria are damaged, and (7) the electron transport chain is stopped. When all of these things happen, cells die. It has been shown that AgNPs are antimicrobial on their own, in addition to releasing silver ions.

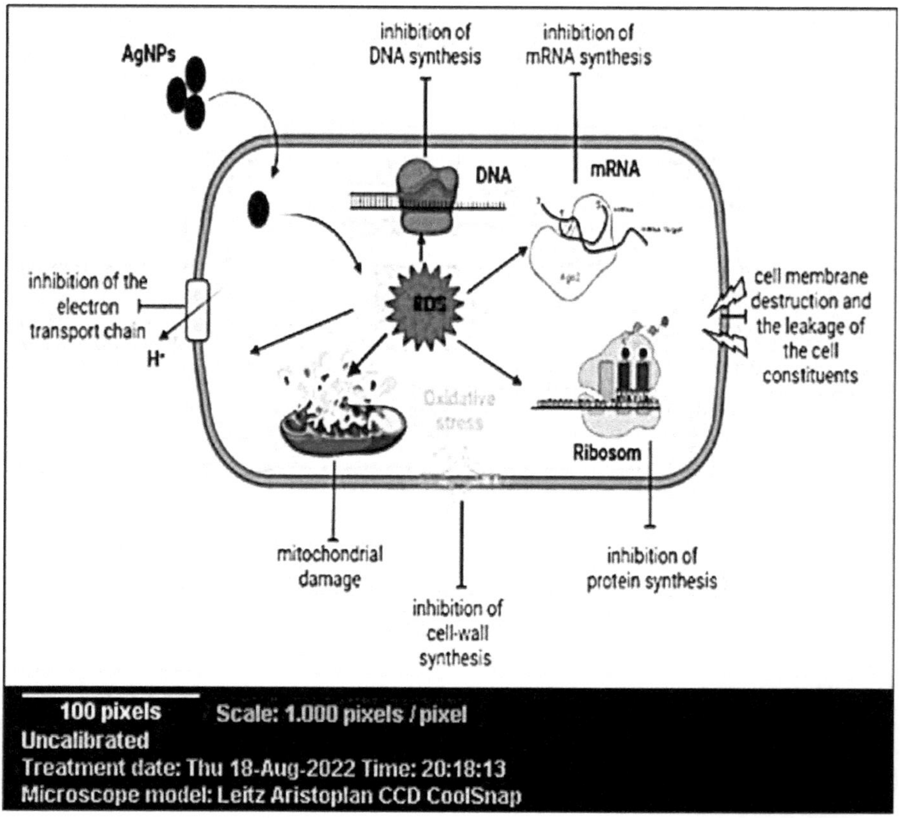

FIGURE 10.6 The suggested mechanism of antibacterial activity of silver nanoparticles' on bacterial cells.

Numerous studies, including those using a wide variety of plant extracts, suggest that AgNPs may physically interact with the cell surfaces of various bacteria. The biological effects of AgNP occur through several different pathways. Intracellular adhesion, penetration, and disruption of organelles and biomolecules (Dakal et al., 2016), as well as oxidative stress, are essential regulators of many bacterial functions. AgNPs were found attached to and deposited on the cell surface of bacteria, especially Gram-negative bacteria. AgNPs can enter Gram-negative bacteria cells through a water-filled channel called a porin in the outer membrane. The primary role of porins is the passive transport across the membrane of hydrophilic molecules of varying sizes and charges. Because silver ions are less likely to penetrate the thicker cell wall of Gram-positive bacteria, their action from AgNPs is more pronounced in Gram-negative bacteria (Chauhan et al., 2016). Since the negative charge of lipopolysaccharides encourages AgNP adherence, it is likely that they also strengthen the cell walls of Gram-negative bacteria, making them more vulnerable to silver nanoparticles (Pal et al., 2007). Some studies have suggested that the ability of silver nanoparticles to adhere to the bacterial cell wall is due to an electrostatic interaction between the positively charged silver ions and the negatively charged surface of the cell membrane caused by the carboxyl, phosphate, and amino groups in the cell wall. Because of their electrostatic attraction to one another, silver nanoparticles can cross the cell membrane and alter its molecular composition and permeability. This causes the proton motive force (PMF) to dissipate and the membrane to degrade (Netala et al., 2015; Rashid et al., 2017).

AgNPs could also be used as a way to get more Ag+ into bacterial cells. The proton motive force would lower the pH of the environment, which would cause more Ag^+ to be released (Ovais et al., 2016; McQuillan et al., 2012). Also, it is thought that when silver nanoparticles come in contact with bacteria, they make free radicals that damage the cell membrane and make it porous (Mikhailova, 2020). Some researchers say that AgNPs can damage DNA while inside a bacterial cell because they stick to the surface of bacteria and change the way their membranes work (Rajeshkumar and Bharath, 2017; Rajeshkumar and Bharath, 2017). For example, the review by MaQuillan et al. (2012) shows that the main way silver nanoparticles work is by breaking down cell membranes. Also, as the silver nanoparticles break down, silver ions that kill bacteria are released. These silver ions can interact with proteins in the cell wall that contain thiols and change how they work. When nanoparticles of silver attach to proteins and come in contact with the outer membrane, they can form complexes with electronic donors like oxygen, phosphorus, nitrogen, or sulfur. The most detailed explanation of how thiol groups work is found in the literature. Silver nanoparticles stop enzymes and proteins that are bound to a membrane from working because they break disulfide bonds and block active sites (Cakić et al., 2016).

Reports say that AgNPs may change the ratio of trans to cis unsaturated fatty acids in membranes. This could change how fluid the membrane is and how the lipid bilayer is made. It could cause changes to the structure of the membrane that stop it from working, making the membrane more permeable and weakening its integrity (Mikhailova, 2020).

AgNP accumulation on the cellular membranes results in problems with bilayer integrity and fissures, and AgNP penetration into the cell and interactions with vital proteins induce cell death. Ag ions have the potential to interact with thiol groups and the disulfide bonds of cellular metabolism enzymes, inactivating them and resulting in oxidative stress in microorganisms (Rashid et al., 2017; McQuillan et al., 2012). Thus, the antibacterial impact of nanoparticles interacting with *P. aeruginosa*'s cell membrane and generating ROS was evaluated (Yan et al., 2018). ROS oxidizes fatty acid double bonds in the membrane, generating free radicals that damage the cell membrane (Gamboa et al., 2019).

According to the discussion by Mikhailova (2020), the AgNPs were shown to have a catalytic activity to generate disulfide connections between oxygen molecules in cells and hydrogen atoms in thiol groups. Disulfide bonds, which silver catalyzes the creation of, alter the shape and structure of enzymes in cells and hence their function. Cells exposed to a 900 ppb Ag^+ solution showed changes in the expression of several proteins and enzymes, including 30S ribosomal subunit, succinyl coenzyme A synthetase, maltose transporter (MalK), and fructose bisphosphate aldolase. By forming a complex with the 30S ribosomal subunit, silver ions inactivate the ribosome complex and halt protein synthesis. The influence of Ag NPs on on ribosomal, transcriptional, and translation processes leads to the production of unfinished, membrane-forming precursor proteins, which ultimately leads to cell death. Disruption of cellular metabolism is caused by AgNPs because of their effect on succinyl coenzyme-a-synthetase, an enzyme critical to the tricarboxylic acid cycle (Rai et al., 2012).

Interruptions of bacterial RNA transcription as well as purines, pyrimidines, and fatty acids were related to the bactericidal effects of AgNPs. Silver nanoparticles induce oxidative stress by suppressing a wide range of cellular metabolic processes; this includes impeding nutrient uptake, altering gene expression, disrupting ATP production, and preventing cytochrome oxidase and NADH-succinate dehydrogenase from doing their jobs in the microbial respiratory chain. When AgNPs come into contact with DNA in a cell, it can cause replication errors (Bao et al., 2015). Ag^+ ion was discovered to form complexes with nucleic acids, which were then shown to disrupt the H-bonds between antiparallel base pairs.

AgNPs can also cause a change in the state of DNA molecules, from relaxed to compact, which reduces their capacity to replicate. Intercalation of AgNPs into the DNA helix has been shown to inhibit the transcription process in microorganisms (Morones et al., 2005). The phosphorylation cycle and the dephosphorylation cascade represent the relaying signal pathway needed for microbial activity and cellular function in microbial cells.

10.9 CONCLUSION

Over the past ten years, there has been a lot of interest in the biosynthesis of nanoparticles using reducing agents sourced from biological sources such plants, plant extracts, and microorganisms. The employment of no harmful or destructive components makes the green synthesis of NPs extremely cost-effective and a potential solution to the issue of antibiotic-resistant bacteria. When trying to create better treatments for bacterial infections, knowing how an antibacterial system works is a crucial first step. In general, nanoparticles have proven to be an effective replacement for antibiotic therapy and combinational therapy due to their astounding potential to address the issue of antibiotic-resistant microorganisms. In conclusion, AgNPs are a product with potential medical and hygiene applications, and their environmentally friendly synthesis can pave the way for these uses.

KEYWORDS

- **silver nanoparticles**
- **green-synthesis**
- **antibacterial activity**
- **photocatalytic properties**

REFERENCES

Abdelmigid, H. M.; Morsi, M. M.; Hussien, N. A.; Alyamani, A. A.; Alhuthal, N. A.; Albukhaty, S. Green Synthesis of Phosphorous-Containing Hydroxyapatite Nanoparticles (Nhap) as a Novel Nano-Fertilizer: Preliminary Assessment on Pomegranate (*Punica Granatum* L.). *Nanomaterials* **2022,** *12,* 1527. https://doi.org/10.3390/nano12091527

Abdelnaby, H.; Abo-Elala, G. M.; Abdel-Raouf, U. M.; Hamed, M. Antibacterial and Anticancer Activity of Extracellular Synthesized Silver Nanoparticles from Marine *Streptomyces Rochei* MHM13. *Egypt. J. Aquat. Res.* **2016,** *42,* 301–312.

Abdel-Raouf, N.; Al-Enazi, N. M.; Ibraheem, I. B.; Alharbi, R. M.; Alkhulaifi, M. M. Biosynthesis of Silver Nanoparticles by Using of the Marine Brown Alga Padina Pavonia and their Characterization. *Saudi J. Biol. Sci.* **2019,** *26* (6), 1207–1215.

Abdel-Raouf, N.; Al-Enazi, N. M.; Ibraheem, I. B.; Alharbi, R. M.; Alkhulaifi, M. M. Biosynthesis of Silver Nanoparticles by Using of the Marine Brown Alga Padina Pavonia and Their Characterization. *Saudi J. Biol. Sci.* **2019,** *26* (6), 1207–1215.

Agnihotri, S.; Mukherji, S.; Mukherji, S. Immobilized Silver Nanoparticles Enhance Contact Killing and Show Highest Efficacy: Elucidation of the Mechanism of Bactericidal Action of Silver. *Nanoscale.* **2013,** *5* (16), 7328–7340.

Ahmed, S.; Ikram, S. Biosynthesis of Gold Nanoparticles: A Green Approach. *J. Photochem. Photobiol. B: Biol.* **2016,** *161,* 141–153.

Ajitha, B.; Reddy, Y. A.; Jeon, H. J.; Ahn, C. W. Synthesis of Silver Nanoparticles in an Eco-Friendly Way Using Phyllanthus Amarus Leaf Extract: Antimicrobial and Catalytic Activity. *Adv. Powder Technol.* **2018,** *29* (1), 86–93.

Akhtar, M. S.; Panwar, J.; Yun, Y. S. Biogenic Synthesis of Metallic Nanoparticles by Plant Extracts. *ACS Sustai. Chem. Eng.* **2013,** *1* (6), 591–602.

Aktepe, N.; Baran, A. Fast and Low Cost Biosynthesis of Agnps with Almond Leaves: Medical Applications with Biocompatible Structures: Fast and Low Cost Biosynthesis of Agnps. *Progr. Nutr.* [Online] **2021,** *23* (3), E2021271. https://mattioli1885journals.com/index.php/progressinnutrition/article/view/11996 (accessed Jul 22, 2022).

Akter, M.; Sikder, M. T.; Rahman, M. M.; Ullah, A. K. M. A.; Hossain, K. F. B.; Banik, S.; Hosokawa, T.; Saito, T.; Kurasaki, M. A Systematic Review on Silver Nanoparticles-Induced Cytotoxicity: Physicochemical Properties and Perspectives. *J. Adv. Res.* **2018,** *9,* 1–16.

Al-Bahrani, R.; Raman, J.; Lakshmanan, H.; Hassan, A. A.; Sabaratnam, V. Green Synthesis of Silver Nanoparticles Using Tree Oyster Mushroom *Pleurotus Ostreatus* and Its Inhibitory Activity Against Pathogenic Bacteria. *Mater. Lett.* **2017,** *186,* 21–25.

Albrecht, M. A.; Evans, C. W.; Raston, C. L. Green Chemistry and the Health Implications of Nanoparticles. *Green Chem.* **2006,** *8* (5), 417–432.

Albukhaty, S.; Al-Bayati, L.; Al-Karagoly, H.; Al-Musawi, S. Preparation and Characterization of Titanium Dioxide Nanoparticles and In Vitro Investigation of their Cytotoxicity and Antibacterial Activity Against *Staphylococcus Aureus* and *Escherichia coli. Anim. Biotechnol.* **2020,** 1–7.

Albukhaty, S.; Al-Karagoly, H.; Allafchian, A. R.; Jalali, S. A. H.; Al-Kelabi, T.; Muhannad, M. Production and Characterization of Biocompatible Nanofibrous Scaffolds Made of B-Sitosterol Loaded Polyvinyl Alcohol/Tragacanth Gum Composites. *Nanotechnology* **2021,** *33,* 085102.

Albukhaty, S.; Naderi-Manesh, H.; Tiraihi, T.; Sakhi Jabir, M. Poly-L-Lysine-Coated Superparamagnetic Nanoparticles: A Novel Method for the Transfection of Pro-BDNF into Neural Stem Cells. *Artif. Cells Nanomed. Biotechnol.* **2018,** *46,* S125–S132.

Ali, M.; Kim, B.; Belfield, K. D.; Norman, D.; Brennan, M.; Ali, G. S. Green Synthesis and Characterization of Silver Nanoparticles Using Artemisia Absinthium Aqueous Extract—A Comprehensive Study. *Mater. Sci. Eng. C.* **2016** , *58,* 359–365.

Al-Kaabi, W. J.; Albukhaty, S.; Al-Fartosy, A. J. M.; Al-Karagoly, H. K.; Al-Musawi, S.; Sulaiman, G. M.; Dewir, Y. H.; Alwahibi, M. S.; Soliman, D. A. Development of *Inula Graveolens* (L.) Plant Extract Electrospun/Polycaprolactone Nanofibers: A Novel Material for Biomedical Application. *Appl. Sci.* **2021,** *11,* 828.

Al-Musawi, M. H.; Ibrahim, K. M.; Albukhaty S. *In Vitro* Study of Antioxidant, Antibacterial, and Cytotoxicity Properties of *Cordia Myxa* Fruit Extract. *Iran J. Microbiol.* **2022,** *14* (1), 97–103. doi: 10.18502/ijm.v14i1.8810.

Alshareef, A.; Laird, K.; Cross, R. B. Shape-Dependent Antibacterial Activity of Silver Nanoparticles on *Escherichia coli* and Enterococcus Faecium Bacterium. *Appl. Surf. Sci.* **2017,** *424,* 310–315.

Anil Kumar, S.; Abyaneh, M. K.; Gosavi, S. W.; Kulkarni, S. K.; Pasricha, R.; Ahmad, A.; Khan, M. I. Nitrate Reductase-Mediated Synthesis of Silver Nanoparticles From AgNo3. *Biotechnol. Lett.* **2007,** *29* (3), 439–445.

Chakravarty, A.; Ahmad, I.; Singh, P.; Sheikh, M. U. D.; Aalam, G.; Sagadevan, S.; Ikram, S. Green Synthesis of Silver Nanoparticles Using Fruits Extracts of *Syzygium Cumini* and their Bioactivity. *Chem. Phys. Lett.* **795,** *2022,* 139493. doi:10.1016/j.cplett.2022.139493.

Bachii, S. A.; Abd-Al Sahib, W. H. Preparation and Characterization of Silver Nanoparticles Biosynthesis by Pseudomonas Stutzeri Environmental Bacteria Isolated from Oil Fields and Their Antimicrobial Activity.

Baker, C.; Pradhan, A.; Pakstis, L.; Pochan, D. J.; Shah, S. I. Synthesis and Antibacterial Properties of Silver Nanoparticles. *J. Nanosci. Nanotechnol.* **2005,** 5 (2), 244–249.

Bao, H.; Yu, X.; Xu, C.; Li, X.; Li, Z.; Wei, D.; Liu Y. New Toxicity Mechanism of Silver Nanoparticles: Promoting Apoptosis and Inhibiting Proliferation. *PLoS ONE* **2015,** *10* (3), E0122535.

Bar, H.; Bhui, D. K.; Sahoo, G. P.; Sarkar, P.; De, S. P.; Misra A. Green Synthesis of Silver Nanoparticles Using Latex of Jatropha Curcas. *Colloids Surf. A: Physicochem. Eng. Asp.* **2009,** *339* (1–3), 134–139.

Bayda, S.; Adeel, M.; Tuccinardi, T.; Cordani, M.; Rizzolio, F. The History of Nanoscience and Nanotechnology: From Chemical–Physical Applications to Nanomedicine. *Molecules* **2020,** *25,* 112. https://doi.org/10.3390/molecules25010112

Bayram, S.; Zahr, O.; Re, J. D.; Blum, A. S. TMV Disk Scaffolds for Making Sub-30 Nm Silver Nanorings. In *Inprotein Scaffolds*; Humana Press: New York, NY, 2018; pp 109–118.

Begum, S. J. P.; Pratibha, S.; Rawat, J. M.; Venugopal, D.; Sahu, P.; Gowda, A.; Qureshi, K. A.; Jaremko, M. Recent Advances in Green Synthesis, Characterization, and Applications of Bioactive Metallic Nanoparticles. *Pharmaceuticals* **2022,** 1*5,* 455. https://doi.org/10.3390/ph15040455

Bharadwaj, K. K.; Rabha, B.; Pati, S.; Choudhury, B. K.; Sarkar, T.; Gogoi, S. K.; Kakati, N.; Baishya, D.; Kari, Z. A.; Edinur, H. A. Green Synthesis of Silver Nanoparticles Using *Diospyros Malabarica* Fruit Extract and Assessments of their Antimicrobial, Anticancer and Catalytic Reduction of 4-Nitrophenol (4-NP). *Nanomaterials (Basel).* **2021,** *11* (8), 1999. doi: 10.3390/nano11081999. PMID: 34443829; PMCID: PMC8401075.

Bhardwaj, B.; Singh, P.; Kumar, A.; Kumar, S.; Budhwar V. Eco-Friendly Greener Synthesis of Nanoparticles. *Adv. Pharmaceut. Bull.* **2020,** *10* (4), 566.

Birla, S. S.; Tiwari, V. V.; Gade, A. K.; Ingle, A. P.; Yadav, A. P.; Rai, M. K. Fabrication of Silver Nanoparticles by Phoma Glomerata and Its Combined Effect Against *Escherichia coli*, Pseudomonas Aeruginosa and Staphylococcus Aureus. *Lett. Appl. Microbiol.* **2009,** *48* (2), 173–179.

Bistarelli, S.; Rossi, F.; Santini, F. Not Only Size, but also Shape Counts: Abstract Argumentation Solvers are Benchmark-Sensitive. *J. Logic Comput.* **2018,** *28* (1), 85–117.

Bunaciu, A. A.; Udriştioiu, E. G.; Aboul-Enein, H. Y. X-Ray Diffraction: Instrumentation and Applications. *Crit. Rev. Anal. Chem.* **2015,** *45* (4), 289–299.

Buszewski, B.; Railean-Plugaru, V.; Pomastowski, P.; Rafińska, K.; Szultka-Młyńska, M.; Golinska, P.; Wypij, M.; Laskowski, D.; Dahm, H. Antimicrobial Activity of Biosilver Nanoparticles Produced by a Novel *Streptacidiphilus Durhamensis* Strain. *J. Microbiol. Immunol. Infect.* **2018,** *51,* 45–54.

Cakić, M.; Glišić, S.; Nikolić, G.; Nikolić, G. M.; Cakić, K.; Cvetinov, M. Synthesis, Characterization and Antimicrobial Activity of Dextran Sulphate Stabilized Silver Nanoparticles. *J. Mol. Struc.* **2016,** *1110,* 156–161.

Çalhan, S. D.; Gündoğan, M. Biosynthesis of Silver Nanoparticles Using Onosma Sericeum Willd. and Evaluation Oftheir Catalytic Properties and Antibacterial and Cytotoxic Activity. *Turk. J. Chem.* **2020,** *44* (6), 1587–600.

Castillo-Henríquez, L.; Alfaro-Aguilar, K.; Ugalde-Álvarez, J.; Vega-Fernández, L.; Montes De Oca-Vásquez, G.; Vega-Baudrit, J. R. Green Synthesis of Gold and Silver Nanoparticles from Plant Extracts and their Possible Applications as Antimicrobial Agents in the Agricultural Area. *Nanomaterials* **2020,** *10,* 1763. https://doi.org/10.3390/nano10091763

Chauhan, N.; Tyagi, A. K.; Kumar, P.; Malik, A. Antibacterial Potential of Jatropha Curcas Synthesized Silver Nanoparticles Against Food Borne Pathogens. *Front. Microbiol.* **2016,** *7,* 1748.

Chen, C. Y.; Hu, Y.; Lin, H. L. A Novel Low Temperature Sintering Process for Pmnn-PZT Ceramics. *Mater. Chem. Phys.* **2006,** *99* (1), 26–29.

Chen, G.; Lu, J.; Lam, C.; Yu, Y. A Novel Green Synthesis Approach for Polymer Nanocomposites Decorated with Silver Nanoparticles and their Antibacterial Activity. *Analyst* **2014,** *139* (22), 5793–5799.

Chouhan, S.; Guleria, S. Green Synthesis of Agnps Using Cannabis Sativa Leaf Extract: Characterization, Antibacterial, Anti-Yeast and A-Amylase Inhibitory Activity. *Mater. Sci. Energy Technol.* **2020** , *3,* 536–544.

Christian, P.; Von Der Kammer, F.; Baalousha, M.; Hofmann, T. Nanoparticles: Structure, Properties, Preparation and Behaviour in Environmental Media. *Ecotoxicology* **2008,** *17* (5), 326–343.

Cunha, M. N.; Felgueiras, H. P.; Gouveia, I.; Zille, A. Synergistically Enhanced Stability of Laccase Immobilized on Synthesized Silver Nanoparticles with Water-Soluble Polymers. *Colloids Surf. B: Biointerfaces.* **2017,** *154,* 210–220.

Dakal, T. C.; Kumar, A.; Majumdar, R. S.; Yadav, V. Mechanistic Basis of Antimicrobial Actions of Silver Nanoparticles. *Front. Microbiol.* **2016,** *7,* 1831.

Daphne, J.; Francis, A.; Mohanty, R.; Ojha, N.; Das, N. Green Synthesis of Antibacterial Silver Nanoparticles Using Yeast Isolates and Its Characterization. *Res. J. Pharm. Technol.* **2018,** *11* (1), 83–92.

Das, R.; Nath, S. S.; Chakdar, D.; Gope, G.; Bhattacharjee R. Synthesis of Silver Nanoparticles and their Optical Properties. *J. Exp. Nanosci.* **2010** , *5* (4), 357–362.

Das, V. L.; Thomas, R.; Varghese, R. T.; Soniya, E. V.; Mathew, J.; Radhakrishnan, E. K. Extracellular Synthesis of Silver Nanoparticles by the Bacillus Strain CS 11 Isolated from Industrialized Area. *3 Biotech* **2014,** *4* (2), 121–126.

De Barros, C. H.; Cruz, G. C.; Mayrink, W.; Tasic, L. Bio-Based Synthesis of Silver Nanoparticles from Orange Waste: Effects of Distinct Biomolecule Coatings on Size, Morphology, and Antimicrobial Activity. *Nanotechnol. Sci. Appl.* **2018,** *11,* 1.

Debnath, B.; Das, R. Controlled Synthesis of Saponin-Capped Silver Nanotriangles and their Optical Properties. *Plasmonics* **2019,** *14* (6), 1365–1375.

Deepak, P.; Amutha, V.; Birundha, R.; Sowmiya, R.; Kamaraj, C.; Balasubramanian, V.; Balasubramani, G.; Aiswarya, D.; Arul, D.; Perumal, P. Facile Green Synthesis of Nanoparticles from Brown Seaweed Sargassum Wightii and Its Biological Application Potential. *Adv. Nat. Sci.: Nanosci. Nanotechnol.* **2018,** *9* (3), 035019.

Deepak, V.; Umamaheshwaran, P. S.; Guhan, K.; Nanthini, R. A.; Krithiga, B.; Jaithoon, N. M.; Gurunathan, S. Synthesis of Gold and Silver Nanoparticles Using Purified URAK. *Colloid Surf. B* **2011,** *86,* 353–358.

Deljou, A.; Goudarzi S. Green Extracellular Synthesis of the Silver Nanoparticles Using Thermophilic *Bacillus Sp.* AZ1 and Its Antimicrobial Activity Against Several Human Pathogenetic Bacteria. *Iran. J. Biotechnol.* **2016,** *14* (2), 25.

Devanesan, S.; Alsalhi, M. S. Green Synthesis of Silver Nanoparticles Using the Flower Extract of Abelmoschus Esculentus for Cytotoxicity and Antimicrobial Studies. *Int. J. Nanomed.* **2021,** *16,* 3343.

Ebrahiminezhad, A.; Taghizadeh, S.; Ghasemi, Y. Green Synthesis of Silver Nanoparticles Using Mediterranean Cypress (Cupressus Sempervirens) Leaf Extract. *Am. J. Biochem. Biotechnol.* 2017.

Eid, A. M.; Fouda, A.; Niedbała, G.; Hassan, S. E.-D.; Salem, S. S.; Abdo, A. M.; Hetta, H. F.; Shaheen, T. I. Endophytic *Streptomyces Laurentii* Mediated Green Synthesis of Ag-NPs with Antibacterial and Anticancer Properties For Developing Functional Textile Fabric Properties. *Antibiotics* **2020,** *9,* 641. https://doi.org/10.3390/antibiotics9100641

Elavazhagan, T.; Arunachalam, K. D. Memecylon Edule Leaf Extract Mediated Green Synthesis of Silver and Gold Nanoparticles. *Int. J. Nanomed.* **2011,** *6,* 1265.

Elhawary, S.; El-Hefnawy, H.; Alzahraa, F. A.; Sobeh, M.; Mostafa, E.; Osman, S.; El-Raey, M. Green Synthesis of Silver Nanoparticles Using Extract of Jasminum Officinal L. Leaves and Evaluation of Cytotoxic Activity Towards Bladder (5637) and Breast Cancer (MCF-7) Cell Lines. *Int. J. Nanomed.* **2020,** *15,* 9771–9781 https://doi.org/10.2147/ijn.s269880

El-Rafie, M. H.; El-Naggar, M. E.; Ramadan, M. A.; Fouda, M. M.; Al-Deyab, S. S.; Hebeish, A. Environmental Synthesis of Silver Nanoparticles Using Hydroxypropyl Starch and their Characterization. *Carbohydr. Polym.* 2011, *86* (2), 630–635.

Escárcega-González, C. E.; Garza-Cervantes, J. A.; Vazquez-Rodríguez, A.; Montelongo-Peralta, L. Z.; Treviño-Gonzalez, M. T.; Díaz Barriga-Castro, E., Saucedo-Salazar, E. M.; Chávez Morales, R. M.; Regalado-Soto, D. I.; Treviño-González, F. M.; Carrazco Rosales, J. L.; Villalobos Cruz, R.; Morones-Ramirez, J. R. In Vivo Antimicrobial Activity of Silver Nanoparticles Produced Via a Green Chemistry Synthesis Using Acacia Rigidula as a Reducing and Capping Agent. *Int. J. Nanomed.* **2018,** *13,* 2349–2363, https://doi.org/10.2147/ijn.s160605

Fayaz, A. M.; Balaji, K.; Girilal, M.; Yadav, R.; Kalaichelvan, P. T.; Venketesan, R. Biogenic Synthesis of Silver Nanoparticles and their Synergistic Effect with Antibiotics: A Study Against Gram-Positive and Gram-Negative Bacteria. *Nanomedicine* **2010,** *6* (1), 103–109.

Fortunati, E.; Peltzer, M.; Armentano, I.; Jiménez, A.; Kenny, J. M. Combined Effects of Cellulose Nanocrystals and Silver Nanoparticles on the Barrier and Migration Properties of PLA Nano-Biocomposites. *J. Food Eng.* **2013,** *118,* 117–124.

Francis, S., Joseph, S., Koshy, E. P.; Mathew, B. Microwave Assisted Green Synthesis of Silver Nanoparticles Using Leaf Extract of Elephantopus Scaber and Its Environmental and Biological Applications. *Artif. Cells Nanomed. Biotechnol.* **2017,** *46* (4), 795–804. doi:10.1080/21691401.2017.1345921

Gamboa, S. M.; Rojas, E. R.; Martínez, V. V.; Vega-Baudrit, J. Synthesis and Characterization of Silver Nanoparticles and their Application as an Antibacterial Agent. *Int. J. Biosen. Bioelectron.* **2019,** *5,* 166–173.

Gan, L.; Zhang, S.; Zhang, Y.; He, S.; Tian, Y. Biosynthesis, Characterization and Antimicrobial Activity of Silver Nanoparticles by A Halotolerant *Bacillus Endophyticus* SCU-L. *Prep. Biochem. Biotechnol.* **2018,** *48,* 582–588.

Garcia-Barrasa, J.; Lopez-De-Luzuriaga, J. M.; Monge, M. Silver Nanoparticles: Synthesis Through Chemical Methods in Solution and Biomedical Applications. *Cent. Eur. J. Chem.* **2011,** *9,* 7–19.

Garcia-Barrasa, J.; Lopez-De-Luzuriaga, J. M.; Monge, M. Silver Nanoparticles: Synthesis Through Chemical Methods in Solution and Biomedical Applications. *Cent. Eur. J. Chem.* **2011,** *9,* 7–19.

Ghosh, S.; Patil, S.; Ahire, M.; Kitture, R.; Kale, S.; Pardesi, K.; Cameotra, S. S.; Bellare, J.; Dhavale, D. D.; Jabgunde, A.; Chopade, B. A. Synthesis of Silver Nanoparticles Using Dioscorea Bulbifera Tuber Extract and Evaluation of Its Synergistic Potential in Combination with Antimicrobial Agents. *Int. J. Nanomed.* **2012,** *7,* 483.

Gur, T.; Meydan, I.; Seckin, H.; Bekmezci, M.; Sen, F. Green Synthesis, Characterization and Bioactivity of Biogenic Zinc Oxide Nanoparticles. *Environ. Res.* **2022,** *204,* 111897.

Gurunathan, S.; Han, J. W.; Kim, E. S.; Park, J. H.; Kim, J. H. Reduction of Graphene Oxide by Resveratrol: A Novel and Simple Biological Method for the Synthesis of an Effective Anticancer Nanotherapeutic Molecule. *Int. J. Nanomed.* **2015,** *10,* 2951.

Gurunathan, S.; Raman, J.; Abd Malek, S. N.; John, P. A.; Vikineswary, S. Green Synthesis of Silver Nanoparticles Using Ganoderma Neo-Japonicum Imazeki: A Potential Cytotoxic Agent Against Breast Cancer Cells. *Int. J. Nanomed.* **2013,** *8,* 4399.

Gurunathan, S.; Han, J. W.; Eppakayala, V.; Jeyaraj, M.; Kim, J. H. Cytotoxicity of Biologically Synthesized Silver Nanoparticles in MDA-MB-231 Human Breast Cancer Cells. *Biomed. Res. Int.* **2013,** *2013,* 535796.

Haber, J.; Sokolov, K. Synthesis of Stable Citrate-Capped Silver Nanoprisms. *Langmuir* **2017,** *33,* 10525–10530.

Heera, P.; Shanmugam, S. Nanoparticle Characterization and Application: An Overview. *Appl. Sci.Int. J. Curr. Microbiol. Appl. Sci.* **2015,** *4* (8), 379–386.

Holt, K. B.; Bard, A. J. Interaction of Silver (I) Ions with the Respiratory Chain of Escherichia coli: An Electrochemical and Scanning Electrochemical Microscopy Study of the Antimicrobial Mechanism of Micromolar Ag+. *Biochemistry* **2005,** *44* (39), 13214–13223.

Hong, S. H.; Lee, W. G. Preparation and Characterization of Expanded Graphite/Ag Nanoparticle Composites for the Improvement of Thermal Diffusion. *Korean Chem. Eng. Res.* **2018,** *56* (3), 410–415. https://doi.org/10.9713/kcer.2018.56.3.410

Ilahi, I.; Khuda, F.; Sahibzada, M. U. K.; Alghamdi, S.; Ullah, R.; Zakiullah; Dablool, Anas S.; Alam, M.; Khan, A.; Khalil, A. A. K. Synthesis of Silver Nanoparticles Using Root Extract of Duchesnea Indica and Assessment of Its Biological Activities. *Arab. J. Chem.* **2021,** *14* (5), 103110, doi:10.1016/j.arabjc.2021.103110

Huang, J.; Lin, L.; Sun, D.; Chen, H.; Yang, D.; Li, Q. Bio-Inspired Synthesis of Metal Nanomaterials and Applications. *Chem. Soc. Rev.* **2015,** *44* (17), 6330–6374.

Jabir, M.; Sahib, U. I.; Taqi, Z.; Taha, A.; Sulaiman, G.; Albukhaty, S.; Al-Shammari, A.; Alwahibi, M.; Soliman, D.; Dewir, Y. H. Linalool-loaded Glutathione-Modified Gold Nanoparticles Conjugated with CALNN Peptide as Apoptosis Inducer and NF-κB Translocation Inhibitor in SKOV-3 Cell Line. *Int. J. Nanomed.* **2020,** *15,* 9025.

Jain, A. S.; Pawar, P. S.; Sarkar, A.; Junnuthula, V.; Dyawanapelly, S. Bionanofactories for Green Synthesis of Silver Nanoparticles: Toward Antimicrobial Applications. *Int. J. Mol. Sci.* **2021,** *22* (21), 11993.

Jain, A. S.; Pawar, P. S.; Sarkar, A.; Junnuthula, V.; Dyawanapelly, S. Bionanofactories for Green Synthesis of Silver Nanoparticles: Toward Antimicrobial Applications. *Int. J. Mol. Sci.* **2021,** *22,* 11993. https://doi.org/10.3390/ijms222111993

Jain, A. S.; Pawar, P. S.; Sarkar, A.; Junnuthula, V.; Dyawanapelly, S. Bionanofactories for Green Synthesis of Silver Nanoparticles: Toward Antimicrobial Applications. *Int. J. Mol. Sci.* **2021,** *22,* 11993. https://doi.org/10.3390/ijms222111993

Jain, A. S.; Pawar, P. S.; Sarkar, A.; Junnuthula, V.; Dyawanapelly, S. Bionanofactories for Green Synthesis of Silver Nanoparticles: Toward Antimicrobial Applications. *Int. J. Mol. Sci.* **2021,** *22,* 11993. https://doi.org/10.3390/ijms222111993

Jihad, M. A.; Noori, F. T. M.; Jabir, M. S.; Albukhaty, S.; Almalki, F. A.; Alyamani, A. A. Polyethylene Glycol Functionalized Graphene Oxide Nanoparticles Loaded with *Nigella Sativa* Extract: A Smart Antibacterial Therapeutic Drug Delivery System. *Molecules* **2021,** *26,* 3067.

Jijie, R.; Barras, A.; Teodorescu, F.; Boukherroub, R.; Szunerits, S. Advancements on the Molecular Design of Nanoantibiotics: Current Level of Development and Future Challenges. *Mol. Syst. Des. Eng.* **2017,** *2* (4), 349–369.

Jo, J. H.; Singh, P.; Kim, Y.J.; Wang, C.; Mathiyalagan, R.; Jin, C.-G.; Yang, D. C. *Pseudomonas Deceptionensis* DC5-Mediated Synthesis of Extracellular Silver Nanoparticles. *Artif. Cells Nanomed. Biotechnol.* **2015,** *44,* 1576–1581.

Joseph, J.; Deborah, K.; Raghavi, R.; Mary Shamya, A.; Aruni, W. Green Synthesis of Silver Nanoparticles Using *Phyllanthus Amarus* Seeds and their Antibacterial Activity Assessment. *Biomed. Biotechnol. Res. J.* **2021,** *5,* 35–38.

Kaabipour, S.; Hemmati, S. A Review on the Green and Sustainable Synthesis of Silver Nanoparticles and One-Dimensional Silver Nanostructures. *Beilstein J. Nanotechnol.* **2021,** *12,* 102–136. doi: 10.3762/bjnano.12.9. PMID: 33564607, PMCID: PMC7849236

Kaabipour, S.; Hemmati, S. A Review on the Green and Sustainable Synthesis of Silver Nanoparticles and One-Dimensional Silver Nanostructures. *Beilstein J. Nanotechnol.* **2021,** *12* (1), 102–136.

Kalimuthu, K.; Suresh Babu, R.; Venkataraman, D.; Bilal, M.; Gurunathan, S. Biosynthesis of Silver Nanocrystals by *Bacillus Licheniformis*. *Colloids Surf. B Biointerfaces* **2008,** *65* (1), 150–153. doi: 10.1016/j.colsurfb.2008.02.018

Kalishwaralal, K.; Deepak, V.; Pandian, S. R.; Kottaisamy, M.; Barathmanikanth, S.; Kartikeyan, B.; Gurunathan, S. Biosynthesis of Silver and Gold Nanoparticles Using Brevibacterium Casei. *Colloids Surf. B: Biointerfaces* **2010,** *77* (2), 257–262.

Karagoly, H.; Rhyaf, A.; Naji, H.; Albukhaty, S.; Almalki, F. A.; Alyamani, A. A.; Albaqami, J.; Aloufi, S. Green Synthesis, Characterization, Cytotoxicity, and Antimicrobial Activity of Iron Oxide Nanoparticles Using *Nigella Sativa* Seed Extract. *Green Process. Synth.* **2022,** *11,* 254–265.

Karatoprak, G. Ş.; Aydin, G.; Altinsoy, B.; Altinkaynak, C.; Koşar, M.; Ocsoy, I. The Effect of Pelargonium Endlicherianum Fenzl. Root Extracts on Formation of Nanoparticles and their Antimicrobial Activities. *Enzyme Microb. Technol.* **2017,** *97,* 21–26.

Karthik, L.; Kumar, G.; Kirthi, A. V.; Rahuman, A. A.; Bhaskara Rao, K. V. Streptomyces Sp. LK3 Mediated Synthesis of Silver Nanoparticles and Its Biomedical Application. *Bioprocess Biosyst. Eng.* **2014,** *37* (2), 261–267.

Kasithevar, M.; Saravanan, M.; Prakash, P.; Kumar, H.; Ovais, M.; Barabadi, H.; Shinwari, Z. K. Green Synthesis of Silver Nanoparticles Using Alysicarpus Monilifer Leaf Extract and Its Antibacterial Activity Against MRSA and Cons Isolates in HIV Patients. *J. Interdiscip. Nanomed.* **2017,***2* (2), 131–141.

Khan, I.; Saeed, K.; Khan, I. Nanoparticles: Properties, Applications and Toxicities. *Arab. J. Chem.* **2017,** *12,* 908–931.

Khane, Y.; Benouis, K.; Albukhaty, S.; Sulaiman, G. M.; Abomughaid, M. M.; Al Ali, A.; Aouf, D.; Fenniche, F.; Khane, S.; Chaibi, W.; Henni, A.; Bouras, H. D.; Dizge, N. Green Synthesis of Silver Nanoparticles Using Aqueous *Citrus Limon* Zest Extract: Characterization and Evaluation of their Antioxidant and Antimicrobial Properties. *Nanomaterials* **2022,** *12,* 2013. https://doi.org/10.3390/nano12122013

Khashan, K. S.; Sulaiman, G. M.; Abdulameer, F. A.; Albukhaty, S.; Ibrahem, M. A.; Al-Muhimeed, T.; Alobaid, A. A. Antibacterial Activity of TiO2 Nanoparticles Prepared by One-Step Laser Ablation in Liquid. *Appl. Sci.* **2021,** *11,* 4623.

Khatami, M.; Zafarnia, N.; Heydarpoor Bami, M.; Sharifi, I.; Singh, H. Antifungal and Antibacterial Activity of Densely Dispersed Silver Nanospheres with Homogeneity Size which Synthesized Using Chicory: An In Vitro Study. *J. Mycol. Med.* **2018,** *28,* 637–644.

Khodashenas, B.; Ghorbani, H. R. Synthesis of Silver Nanoparticles With Different Shapes. *Arab. J. Chem.* 2015, 1–16.

Khurana, C.; Vala, A. K.; Andhariya, N.; Pandey, O. P.; Chudasama, B. Antibacterial Activity of Silver: The Role of Hydrodynamic Particle Size at Nanoscale. *J. Biomed. Mater. Res. A* **2014,** *102,* 3361–3368.

Kim, B. H.; Lee, J. S. One-Pot Photochemical Synthesis of Silver Nanodisks Using A Conventional Metal-Halide Lamp. *Mater. Chem. Phys.* **2015,** *149,* 678–685.

Kiran, G. S.; Dhasayan, A.; Lipton, A. N.; Selvin, J.; Arasu, M. V.; Al-Dhabi, N. A. Melanin-Templated Rapid Synthesis of Silver Nanostructures. *J. Nanobiotechnol.* **2014,** *12* (1), 1–3.

Kowshik, M.; Ashtaputre, S.; Kulkani, S. K.; Parknikar, K. M. M. Extracellular Synthesis of Silver Nanoparticles by a Silver-Tolerant Yeast Strain MKY3. *Nanotechnology* **2003,** *14,* 95–100. doi: 10.1088/0957-4484/14/1/321

Kremer, J. R.; Mastronarde, D. N.; Mcintosh, J. R. Computer Visualization of Three-Dimensional Image Data Using IMOD. *J. Struc. Biol.* **1996,** *116* (1), 71–76.

Kumar, P.; Singh, P. K.; Hussain, M.; Kumar Das, A. Synthesis of Silver Metal Nanoparticles Through Electric Arc Discharge Method: A Review. *Adv. Sci. Lett.* **2016,** *22* (1), 3–7.

Kumar, B.; Smita, K.; Cumbal, L.; Debut, A.; Pathak, R. N. Sonochemical Synthesis of Silver Nanoparticles Using Starch: A Comparison. *Bioinorg. Chem. Appl.* **2014,** *2014,* 784268.

Kumar, S.; Shahid, M.; Khan, M. A.; Kumar, N.; Khan, H. M. Biosynthesis of Silver Nanoparticles From Phyllanthus Niruri Leaf Extracts and Its Antibacterial Activity

Against Antibiotics-Resistant Clinical Isolates. *Pathology* **2020,** *52,* S127. doi:10.1016/j.pathol.2020.01.432

Kumar, S. S. D.; Rajendran, N. K.; Houreld, N. N.; Abrahamse, H. Recent Advances on Silver Nanoparticle and Biopolymer-Based Biomaterials for Wound Healing Applications. *Int. J. Biol. Macromol.* **2018,** *115,* 165–175.

Lee, Y. J.; Kim, J.; Oh, J.; Bae, S.; Lee, S.; Hong, I. S.; Kim, S. H. Ion-Release Kinetics and Ecotoxicity Effects of Silver Nanoparticles. *Environ. Toxicol. Chem.* **2012,** *31* (1), 155–159.

Khanal, L. N.; Sharma, K. R.; Paudyal, H.; Parajuli, K.; Dahal, B.; Ganga, G. C.; Pokharel, Y. R.; Kalauni, S. K.; Green Synthesis of Silver Nanoparticles from Root Extracts of Rubus Ellipticus Sm. and Comparison of Antioxidant and Antibacterial Activity. *J. Nanomater.* **2022** 1832587. https://doi.org/10.1155/2022/183258

Leung, T. C.; Wong, C. K.; Xie, Y. Green Synthesis of Silver Nanoparticles Using Biopolymers, Carboxymethylated-Curdlan and Fucoidan. *Mater. Chem. Phys.* **2010,** *121,* 402–405.

Li, S.; Shen, Y.; Xie, A.; Yu, X.; Qiu, L.; Zhang, L.; Zhang, Q. Green Synthesis of Silver Nanoparticles Using Capsicum Annuum L. Extract. *Green Chem.* **2007,** *9* (8), 852–858.

Lin, Z.; Wu, J.; Xue, R.; Yang Y. Spectroscopic Characterization of Au3+ Biosorption by Waste Biomass of Saccharomyces Cerevisiae. *Spectrochim. Acta A Mol. Biomol. Spectrosc.* **2005,** *61* (4), 761–765.

Liu, J.; Hurt, R. H. Ion Release Kinetics and Particle Persistence in Aqueous Nano-Silver Colloids. *Environ. Sci. Technol.* **2010,** *44* (6), 2169–2175.

Liu, J.; Sonshine, D. A.; Shervani, S.; Hurt, R. H. Controlled Release of Biologically Active Silver from Nanosilver Surfaces. *ACS Nano* **2010,** *4* (11), 6903–6913.

Loiseau, A.; Asila, V.; Boitel-Aullen, G.; Lam, M.; Salmain, M.; Boujday, S. Silver-Based Plasmonic Nanoparticles for and their Use in Biosensing. *Biosensors* **2019,** *9,* 78. https://doi.org/10.3390/bios9020078

Long, Y.; Hu, S.; Lei, P.; Li, Y. Preparation of Green Silver Nanoparticles with High Antibacterial Ability Using N-Maleoyl Chitosan and Montmorillonite. *Mater. Lett.* **2022,** *316,* 132044.

Lu, Y.; Zhang, C. Y.; Zhang, D. J.; Hao, R.; Hao, Y. W.; Liu, Y. Q. Fabrication of Flower-Like Silver Nanoparticles for Surface-Enhanced Raman Scattering. *Chin. Chem. Lett.* **2016,** *27* (5), 689–692.

Mafune, F.; Kohno, J.-Y.; Takeda, Y.; Kondow, T.; Sawabe, H. Formation and Size Control of Silver Nanoparticles by Laser Ablation in Aqueous Solution. *J. Phys. Chem. B* **2000,** *104,* 9111–9117.

Mahendia, S.; Kumar Tomar, A.; Goyal, P. K.; Kumar, S. Tuning of Refractive Index of Poly (Vinyl Alcohol: Effect of Embedding Cu and Ag Nanoparticles. *J. Appl. Phys.* 2013, *113* (7), 073103.

Mahendia, S.; Kumar Tomar, A.; Goyal, P. K.; Kumar, S. Tuning of Refractive Index of Poly (Vinyl Alcohol: Effect of Embedding Cu and Ag Nanoparticles. *J. Appl. Phys.* **2013,** *113* (7), 073103.

Mandal, P.; Babu, S. S.; Mandal, N. C. Antimicrobial Activity of Saponins from Acacia Auriculiformis. *Fitoterapia* **2005,** *76* (5), 462–465.

Manivasagan, P.; Venkatesan, J.; Senthilkumar, K.; Sivakumar, K.; Kim, S. K. Biosynthesis, Antimicrobial and Cytotoxic Effect of Silver Nanoparticles Using A Novel *Nocardiopsis* sp. MBRC-1. *Biomed Res. Int.* **2013,** *2013,* 287638.

Marassi, V.; Di Cristo, L.; Smith, S. G. J.; Ortelli, S.; Blosi, M.; Costa, A. L.; Reschiglian, P.; Volkov, Y.; Prina-Mello, A. Silver Nanoparticles As a Medical Device In Healthcare

Settings: A Five-Step Approach for Candidate Screening of Coating Agents. *R. Soc. Open Sci.* **2018,** *5,* 171113.

May, B. M.; Oluwafemi, O. S. Sugar-Reduced Gelatin-Capped Silver Nanoparticles with High Selectivity for Colorimetric Sensing of Hg 2 and Fe 2 Ions in the Midst of Other Metal Ions in Aqueous Solutions. *Int. J. Electrochem. Sci.* **2016,** *11,* 8096–8108.

Mcquillan, J. S.; Groenaga Infante, H.; Stokes, E.; Shaw, A. M. Silver Nanoparticle Enhanced Silver Ion Stress Response in Escherichia coli K12. *Nanotoxicology* **2012,** *6* (8), 857–866.

Mikhailova, E. O. Silver Nanoparticles: Mechanism of Action and Probable Bio-Application. *J. Funct. Biomater.* **2020,** *11,* 84. https://doi.org/10.3390/jfb11040084

Mikhailova, E. O. Silver Nanoparticles: Mechanism of Action and Probable Bio-Application. *J. Funct. Biomater.* **2020,** *11,* 84. https://doi.org/10.3390/jfb11040084

Minhas, F. T.; Arslan, G.; Gubbuk, I. H.; Akkoz, C.; Ozturk, B. Y.; Asıkkutlu, B.; Arslan, U.; Ersoz, M. Evaluation of Antibacterial Properties on Polysulfone Composite Membranes Using Synthesized Biogenic Silver Nanoparticles with *Ulva Compressa* (L.) Kütz. and *Cladophora Glomerata* (L.) Kütz. Extracts. *Int. J. Biol. Macromol.* **2018,** *107,* 157–165.

Mirzaie, A.; Badmasti, F.; Dibah, H.; Hajrasouliha, S.; Yousefi, F.; Andalibi, R.; Kashtali, A. B.; Rezaei, A. H.; Bakhtiatri, R. Phyto-Fabrication of Silver Nanoparticles Using Typha Azerbaijanensis Aerial Part and Root Extracts. *Appl. Sci.Iran. J. Public Health.* **2022,** *51* (5), 1097.

Mittal, A. K.; Chisti, Y.; Banerjee, U. C. Synthesis of Metallic Nanoparticles Using Plant Extracts. *Biotechnol. Adv.* **2013,** *31* (2), 346–356. doi: 10.1016/j.biotechadv.2013.01.003

Modan, E. M.; Plăiașu, A. G. Advantages and Disadvantages of Chemical Methods in the Elaboration of Nanomaterials. *Ann. "Dunarea Jos" Univ. Galati. Fascicle IX Metall. Mater. Sci.* **2020,** *43,* 53–60.

Mohd Yusof, H.; Abdul Rahman, N. A.; Mohamad, R.; Zaidan, U. H. Microbial Mediated Synthesis of Silver Nanoparticles by Lactobacillus Plantarum TA4 and Its Antibacterial and Antioxidant Activity. *Appl. Sci.* **2020,** *10* (19), 6973.

Mondal, A. H.; Yadav, D.; Mitra, S.; Mukhopadhyay, K. Biosynthesis of Silver Nanoparticles Using Culture Supernatant of Shewanella sp. ARY1 and Their Antibacterial Activity. *Int. J. Nanomed.* **2020,** *15,* 8295.

Morones, J. R.; Elechiguerra, J. L.; Camacho, A.; Holt, K.; Kouri, J. B.; Ramírez, J. T.; Yacaman, M. J. The Bactericidal Effect of Silver Nanoparticles. *Nanotechnology* 2005, *16* (10), 2346.

Mukherjee, K.; Gupta, R.; Kumar, G.; Kumari, S.; Biswas, S.; Padmanabhan, P. Synthesis of Silver Nanoparticles by Bacillus Clausii and Computational Profiling of Nitrate Reductase Enzyme Involved in Production. *J. Genet. Eng. Biotechnol.* **2018,** *16* (2), 527–536.

Mukherjee, P.; Ahmad, A.; Mandal, D.; Senapati, S.; Sainkar, S. R.; Khan, M. I.; Parishcha, R.; Ajaykumar, P. V.; Alam, M.; Kumar, R.; Sastry, M. Fungus-Mediated Synthesis of Silver Nanoparticles and their Immobilization in the Mycelial Matrix: A Novel Biological Approach to Nanoparticle Synthesis. *Nano Lett.* 2001, *1* (10), 515–519.

Mukherjee, P.; Ahmad, A.; Mandal, D.; Senapati, S.; Sainkar, S. R.; Khan, M. I.; Parishcha, R.; Ajaykumar, P. V.; Alam, M.; Kumar, R.; Sastry, M. Fungus-Mediated Synthesis of Silver Nanoparticles and their Immobilization in the Mycelial Matrix: A Novel Biological Approach to Nanoparticle Synthesis. *Nano Lett.* 2001, *1* (10), 515–519.

Murugesan, S.; Bhuvaneswari, S.; Sivamurugan, V. Green Synthesis, Characterization of Silver Nanoparticles of A Marine Red Alga *Spyridia Fusiformis* and their Antibacterial Activity. *Int. J. Pharm. Pharm. Sci.* **2017,** *9,* 192–197.

Nadagouda, M. N.; Varma, R. S. Green Synthesis of Ag and Pd Nanospheres, Nanowires, and Nanorods Using Vitamin: Catalytic Polymerisation of Aniline and Pyrrole. *J. Nanomater.* **2008,** *2008.*

Nakamura, S.; Sato, M.; Sato, Y.; Ando, N.; Takayama, T.; Fujita, M.; Ishihara, M. Synthesis and Application of Silver Nanoparticles (Ag NPs) for the Prevention of Infection in Healthcare Workers. *Int. J. Mol. Sci.* **2019,** *20,* 3620. https://doi.org/10.3390/ijms20153620

Narayanan, K. B.; Sakthivel, N. Synthesis and Characterization of Nano-Gold Composite Using Cylindrocladium Floridanum and Its Heterogeneous Catalysis in the Degradation of 4-Nitrophenol. *J. Hazard. Mater.* **2011,** *189* (1–2), 519–525.

Nayak, P. S.; Arakha, M.; Kumar, A.; Asthana, S.; Mallick, B. C.; Jha, S. An Approach Towards Continuous Production of Silver Nanoparticles Using *Bacillus Thuringiensis*. *RSC Adv.* **2016,** *6,* 8232–8242.

Neethu, S.; Midhun, S. J.; Radhakrishnan, E.; Jyothis, M. Green Synthesized Silver Nanoparticles by Marine Endophytic Fungus *Penicillium Polonicum* and Its Antibacterial Efficacy Against Biofilm Forming, Multidrug-Resistant *Acinetobacter Baumanii*. *Microb. Pathog.* **2018,** *116,* 263–272.

Netala, V. R.; Kotakadi, V. S.; Nagam, V.; Bobbu, P.; Ghosh, S. B.; Tartte, V. First Report of Biomimetic Synthesis of Silver Nanoparticles Using Aqueous Callus Extract of Centella Asiatica and Their Antimicrobial Activity. *Appl. Nanosci.* **2015,** *5* (7), 801–807.

Okafor, F.; Janen, A.; Kukhtareva, T.; Edwards, V.; Curley, M. Green Synthesis of Silver Nanoparticles, Their Characterization, Application and Antibacterial Activity. *Int. J. Environ. Res. Public Health* **2013,** *10,* 5221–5238. https://doi.org/10.3390/ijerph10105221

Ortega-Arroyo, L.; Martin-Martinez, E. S.; Aguilar-Mendez, M. A.; Cruz-Orea, A.; Hernandez-Pérez, I.; Glorieux, C. Green Synthesis Method of Silver Nanoparticles Using Starch as Capping Agent Applied the Methodology of Surface Response. *Starch-Stärke.* **2013,** *65* (9–10), 814–821.

Ovais, M.; Khalil, A. T.; Raza, A.; Khan, M. A.; Ahmad, I.; Islam, N. U.; Saravanan, M.; Ubaid, M. F.; Ali, M.; Shinwari, Z. K. Green Synthesis of Silver Nanoparticles via Plant Extracts: Beginning A New Era in Cancer Theranostics. *Nanomedicine* **2016,** *12* (23), 3157–3177.

Pal, S.; Tak, Y. K. Y Song, J. M. Does the Antibacterial Activvity of Silver Nanoparticles Depend on the Shape of the Nanoparicle? A Study of the Gram-Negative Bacterium Escherichia coli. *Appl. Environ. Microbiol.* **2007,** *73* (6), 1712–1720.

Parlinska-Wojtan, M.; Kus-Liskiewicz, M.; Depciuch, J.; Sadik, O. Green Synthesis and Antibacterial Effects of Aqueous Colloidal Solutions of Silver Nanoparticles Using Camomile Terpenoids as a Combined Reducing and Capping Agent. *Bioprocess Biosyst. Eng.* **2016,** *39* (8), 1213–1223. doi: 10.1007/s00449-016-1599-4

Ramalingmam, P.; Muthukrishnan, S.; Thangaraj, P.; Biosynthesis of Silver Nanoparticles Using an Endophytic Fungus, Curvularialunata and its Antimicrobial Potential. *J. Nanosci. Nanoeng.* **2015,** *1* (4), 241–247.

Parveen, K.; Banse, V.; Ledwani, L. In *Green Synthesis of Nanoparticles: Their Advantages and Disadvantages*, AIP Conference Proceedings; 2016 Apr 13, AIP Publishing LLC, Vol. 1724 (1); p 020048.

Patarroyo, J.; Genç, A.; Arbiol, J.; Bastús, N. G.; Puntes, V. One-Pot Polyol Synthesis of Highly Monodisperse Short Green Silver Nanorods. *Chem. Commun.* **2016,** *52* (73), 10960–10963.

Peralta-Videa, J. R.; Huang, Y.; Parsons, J. G.; Zhao, L.; Lopez-Moreno, L.; Hernandez-Viezcas, J. A.; Gardea-Torresdey, J. L. Plant-Based Green Synthesis of Metallic Nanoparticles: Scientific Curiosity or a Realistic Alternative to Chemical Synthesis? *Nanotechnol. Environ. Eng.* **2016,** *1,* 4.

Mandal, P.; Biosynthesis of Silver Nanoparticles by Plumeria Rubra Flower Extract: Characterization and Their Antimicrobial Activities. *Int. J. Eng. Sci. Invent.* **2018,** *7* (1), 01–06.

Pirtarighat, S.; Ghannadnia, M.; Baghshahi, S. Green Synthesis of Silver Nanoparticles Using the Plant Extract of *Salvia Spinosa* Grown In Vitro and Their Antibacterial Activity Assessment. *J. Nanostructure Chem.* **2019,** *9,* 1–9.

Prasher, P.; Sharma, M.; Mudila, H.; Gupta, G.; Sharma, A. K.; Kumar, D.; Bakshi, H. A.; Negi, P.; Kapoor, D. N.; Chellappan, D. K.; et al. Emerging Trends in Clinical Implications of Bio-Conjugated Silver Nanoparticles in Drug Delivery. *Colloid Interface Sci. Commun.* **2020,** *35,* 100244.

Premanand, G., Shanmugam, N., Kannadasan, N. et al. Nelumbo Nucifera Leaf Extract Mediated Synthesis of Silver Nanoparticles and their Antimicrobial Properties Against Some Human Pathogens. *Appl. Nanosci.* **2016,** *6,* 409–415. https://doi.org/10.1007/s13204-015-0442-6

Puchana-Rosero, M. J.; Adebayo, M. A.; Lima, E. C.; Machado, F. M.; Thue, P. S.; Vaghetti, J. C.; Umpierres, C. S.; Gutterres, M. Microwave-Assisted Activated Carbon Obtained from the Sludge of Tannery-Treatment Effluent Plant for Removal of Leather Dyes. *Colloids Surf. A: Physicochem. Eng. Asp.* **2016,** *504,* 105–115.

Purohit, A.; Sharma, R.; Shiv Ramakrishnan, R.; Sharma, S.; Kumar, A.; Jain, D.; Kushwaha, H. S.; Maharjan, E. Biogenic Synthesis of Silver Nanoparticles (AgNPs) Using Aqueous Leaf Extract of Buchanania Lanzan Spreng and Evaluation of Their Antifungal Activity Against Phytopathogenic Fungi. *Bioinorg. Chem. Appl.* **2022,** 2022.

Pyatenko, A.; Shimokawa, K.; Yamaguchi, M.; Nishimura, O.; Suzuki, M. Synthesis of Silver Nanoparticles by Laser Ablation in Pure Water. *Appl. Phys. A* **2004,** *79* (4), 803–806.

Rafique, M.; Sadaf, I.; Rafique, M. S.; Tahir, M. B. A Review on Green Synthesis of Silver Nanoparticles and Their Applications. *Artif. Cells Nanomed. Biotechnol.* **2017,** *45* (7), 1272–1291.

Rai, A.; Prabhune, A.; Perry, C. C. Antibiotic Mediated Synthesis of Gold Nanoparticles with Potent Antimicrobial Activity and their Application in Antimicrobial Coatings. *J. Mater. Chem.* **2010,** *20* (32), 6789–6798.

Rai, M. K.; Deshmukh, S. D.; Ingle, A. P.; Gade, A. K. Silver Nanoparticles: The Powerful Nanoweapon Against Multidrug-Resistant Bacteria. *J. Appl. Microbiol.* **2012,** *112* (5), 841–852.

Rajeshkumar, S.; Bharath, L. V. Mechanism of Plant-Mediated Synthesis of Silver Nanoparticles—A Review on Biomolecules Involved, Characterisation and Antibacterial Activity. *Chem.-Biol. Interact.* **2017,** *273,* 219–227.

Rajeshkumar, S.; Bharath, L. V. Mechanism of Plant-Mediated Synthesis of Silver Nanoparticles—A Review on Biomolecules Involved, Characterisation and Antibacterial Activity. *Chem. Biol. Interact.* **2017,** *273,* 219–227. doi: 10.1016/j.cbi.2017.06.019

Ramkumar, V. S.; Pugazhendhi, A.; Gopalakrishnan, K.; Sivagurunathan, P.; Saratale, G. D.; Dung, T. N. B.; Kannapiran, E. Biofabrication and Characterization of Silver Nanoparticles Using Aqueous Extract of Seaweed Enteromorpha Compressa and Its Biomedical Properties. *Biotechnol. Rep.* **2017,** *14,* 1–7. doi:10.1016/j.btre.2017.02.001

Rana, A.; Yadav, K.; Jagadevan, S. A Comprehensive Review on Green Synthesis of Nature-Inspired Metal Nanoparticles: Mechanism, Application and Toxicity. *J. Clean. Prod.* **2020,** *272,* 122880.

Ranoszek-Soliwoda, K.; Tomaszewska, E.; Małek, K.; Celichowski, G.; Orlowski, P.; Krzyzowska, M.; Grobelny, J. The Synthesis of Monodisperse Silver Nanoparticles with Plant Extracts. *Colloids Surf. B: Biointerfaces* **2019,** *177,* 19–24.

Rashid, M.; Or, M.; Akhter, K. N.; Chowdhury, J. A.; Hossen, F.; Hussain, M.; Hossain, M. Characterization of Phytoconstituents and Evaluation of Antimicrobial Activity of Silver-Extract Nanoparticles Synthesized From Momordica Charantia Fruit Extract. *BMC Complement. Altern. Med.* **2017,** *17* (1), 1–7.

Rashid, M. U.; Bhuiyan, K. H.; Quayum, M. E. Synthesis of Silver Nano Particles(Ag-NPs) and Their Uses for Quantitative Analysis of Vitamin C Tablets. *Dhaka Univ. J. Pharm. Sci.* **2013,** *12,* 2933.

Ravichandran, A.; Subramanian, P.; Manoharan, V.; Muthu, T.; Periyannan, R.; Thangapandi, M.; Ponnuchamy, K.; Pandi, B.; Marimuthu, P. N. Phyto-Mediated Synthesis of Silver Nanoparticles Using Fucoidan Isolated From *Spatoglossum Asperum* and Assessment of Antibacterial Activities. *J. Photochem. Photobiol. B Biol.* **2018,** *185,* 117–125.

Raza, M. A.; Kanwal, Z.; Rauf, A.; Sabri, A. N.; Riaz, S.; Naseem, S. Size- and Shape-Dependent Antibacterial Studies of Silver Nanoparticles Synthesized by Wet Chemical Routes. *Nanomaterials* **2016,** *6,* 74. https://doi.org/10.3390/nano6040074

Raza, M. A.; Kanwal, Z.; Rauf, A.; Sabri, A. N.; Riaz, S.; Naseem, S. Size- and Shape-Dependent Antibacterial Studies of Silver Nanoparticles Synthesized by Wet Chemical Routes. *Nanomaterials* **2016,** *6,* 74. https://doi.org/10.3390/nano6040074

Reddy, J. M.; Anitha, R.; Rajeshkumar, S.; Lakshmi, T. Characterisation of Cumin Oil Mediated Silver Nanoparticles Using UV-Visible Spectrophotometer and TEM. *Res. J. Pharm. Technol.* **2019,** *12* (10), 4931–4933.

Ren, W.; Lin, G.; Clarke, C.; Zhou, J.; Jin, D. Optical Nanomaterials and Enabling Technologies for High-Security-Level Anticounterfeiting. *Adv. Mater.* **2020,** *32* (18), 1901430.

Restrepo, C. V.; Villa, C. C. Synthesis of Silver Nanoparticles, Influence of Capping Agents, and Dependence on Size and Shape: A Review. *Environ. Nanotechnol. Monit. Manag.* **2021,** *15,* 100428.

Restrepo, C. V.; Villa, C. C. Synthesis of Silver Nanoparticles, Influence of Capping Agents, and Dependence on Size and Shape: A Review. *Environ. Nanotechnol. Monit. Manag.* **2021,** *15,* 100428.

Roy, A.; Bulut, O.; Some, S.; Mandal, A. K.; Yilmaz, M. D. Green Synthesis of Silver Nanoparticles: Biomolecule-Nanoparticle Organizations Targeting Antimicrobial Activity. *RSC Adv.* **2019,** *9* (5), 2673–2702.

Roy, A.; Bulut, O.; Some, S.; Mandal, A. K.; Yilmaz, M. D. Green Synthesis of Silver Nanoparticles: Biomolecule-Nanoparticle Organizations Targeting Antimicrobial Activity. *RSC Adv.* **2019,** *9* (5), 2673–2702.

Roy, A.; Bulut, O.; Some, S.; Mandal, A. K.; Yilmaz, M. D. Green Synthesis of Silver Nanoparticles: Biomolecule-Nanoparticle Organizations Targeting Antimicrobial Activity. *RSC Adv.* **2019,** *9* (5), 2673–2702.

Roy, A.; Bulut, O.; Some, S.; Mandal, A. K.; Yilmaz, M. D. Green Synthesis of Silver Nanoparticles: Biomolecule-Nanoparticle Organizations Targeting Antimicrobial Activity. *RSC Adv.* **2019,** *9* (5), 2673–2702.

Roy, N.; Gaur, A.; Jain, A.; Bhattacharya, S.; Rani, V. Green Synthesis of Silver Nanoparticles: An Approach to Overcome Toxicity. *Environ. Toxicol. Pharmacol.* **2013,** *36* (3), 807–812

Safat, S.; Buazar, F.; Albukhaty, S.; Matroodi, S. Enhanced Sunlight Photocatalytic Activity and Biosafety of Marine-Driven Synthesized Cerium Oxide Nanoparticles. *Sci. Rep.* **2021,** *11,* 14734.

Salari, Z.; Danafar, F.; Dabaghi, S.; Ataei, S. A. Sustainable Synthesis of Silver Nanoparticles Using Macroalgae *Spirogyra Varians* and Analysis of Their Antibacterial Activity. *J. Saudi Chem. Soc.* **2016,** *20,* 459–464.

Salem, S. S.; EL-Belely, E. F.; Niedbała, G.; Alnoman, M. M.; Hassan, S. E.-D.; Eid, A. M.; Shaheen, T. I.; Elkelish, A.; Fouda, A. Bactericidal and In-Vitro Cytotoxic Efficacy of Silver Nanoparticles (Ag-NPs) Fabricated by Endophytic Actinomycetes and Their Use as Coating for the Textile Fabrics. *Nanomaterials* **2020,** *10,* 2082. https://doi.org/10.3390/nano10102082

Sanjenbam, P.; Gopal, J.; Kannabiran, K. Anticandidal Activity of Silver Nanoparticles Synthesized Using *Streptomyces* sp. VITPK1. *J. Mycol. Médicale* **2014,** *24,* 211–219.

Sanpui, P.; Murugadoss, A.; Prasad, P. D.; Ghosh, S. S.; Chattopadhyay, A. The Antibacterial Properties of A Novel Chitosan–Ag-Nanoparticle Composite. *Int. J. Food Microbiol.* **2008,** *124* (2), 142–146.

Sanzari, I.; Leone, A.; Ambrosone, A. Nanotechnology in Plant Science: To Make a Long Story Short. *Front. Bioeng. Biotechnol.* **2019,** *7,* 120.

Sarkar, P.; Bhui, D. K.; Bar, H.; Sahoo, G. P.; De, S. P.; Misra, A. Synthesis and Photophysical Study of Silver Nanoparticles Stabilized by Unsaturated Dicarboxylates. *J. Lumin.* **2009,** *129,* 704–709.

Shang, Y.; Hasan, M. K.; Ahammed, G. J.; Li, M.; Yin, H.; Zhou, J. Applications of Nanotechnology in Plant Growth and Crop Protection: A Review. *Molecules* **2019,** *24,* 2558. https://doi.org/10.3390/molecules24142558

Shankar, S.; Rhim, J. W. Preparation and Characterization of Agar/Lignin/Silver Nanoparticles Composite Films with Ultraviolet Light Barrier and Antibacterial Properties. *Food Hydrocoll.* **2017,** *71,* 76–84.

Shankar, S. S.; Rai, A.; Ahmad, A.; Sastry, M. Rapid Synthesis of Au, Ag, and Bimetallic Au Core–Ag Shell Nanoparticles Using Neem (Azadirachta Indica) Leaf Broth. *J. Colloid Interface Sci.* **2004,** *275* (2), 496–502.

Shankar, S.; Rhim, J. W. Amino Acid Mediated Synthesis of Silver Nanoparticles and Preparation of Antimicrobial Agar/Silver Nanoparticles Composite Films. *Carbohydr. Polym.* **2015,** *130,* 353–363.

Shin, S. W.; Song, I. H.; Um, S. H. Role of Physicochemical Properties in Nanoparticle Toxicity. *Nanomaterials* **2015,** *5,* 1351–1365. https://doi.org/10.3390/nano5031351

Shu, M.; He, F.; Li, Z.; Zhu, X.; Ma, Y.; Zhou, Z.; Yang, Z.; Gao, F.; Zeng, M. Biosynthesis and Antibacterial Activity of Silver Nanoparticles Using Yeast Extract as Reducing and Capping Agents. *Nanosc. Res. Lett.* **2020,** *15* (1), 1–9.

Mehdizadeh, S.; Ghasemi, N.; Ramezani, M.; Mahanpoor, K. Biosynthesis of Silver Nanoparticles Using Malva Sylvestris Flower Extract and Its Antibacterial and Catalytic Activity. *Chem. Methodol.* **2021,** *5* (4) 356–366. doi:10.22034/chemm.2021.132490

Singh, K.; Panghal, M.; Kadyan, S.; Chaudhary, U.; Yadav, J. P. Green Silver Nanoparticles of Phyllanthus Amarus: As an Antibacterial Agent Against Multi Drug Resistant Clinical Isolates of Pseudomonas Aeruginosa. *J. Nanobiotechnol.* **2014,** *12* (1). doi:10.1186/s12951-014-0040-x

Slawson, R. M.; Van Dyke, M. I.; Lee, H.; Trevors, J. T. Germanium and Silver Resistance, Accumulation, and Toxicity in Microorganisms. *Plasmid* **1992,** *27* (1), 72–79.

Slepička, P.; Slepičková Kasálková, N.; Siegel, J.; Kolská, Z.; Švorčík, V. Methods of Gold and Silver Nanoparticles Preparation. *Materials* **2020,** *13,* 1. https://doi.org/10.3390/ma13010001

Sondi; Salopek-Sondi, B. Silver Nanoparticles as Antimicrobial Agent: A Case Study On E. coli as a Model for Gram-Negative Bacteria. *J. Colloid Interf. Sci.* **2004,** *275* (1), 177–182.

Song, J.; Kim, H.; Jang, Y.; Jang, J. Enhanced Antibacterial Activity of Silver/Polyrhodanine-Composite-Decorated Silica Nanoparticles. *ACS Appl. Mater. Interf.* **2013,** *5,* 11563–11568.

Sonker, A. S.; Pathak, J.; Kannaujiya, V. K.; Sinha, R. P. Characterization and In Vitro Antitumor, Antibacterial and Antifungal Activities of Green Synthesized Silver Nanoparticles Using Cell Extract of *Nostoc* sp. Strain HKAR-2. *Can. J. Biotechnol.* **2017,** *1,* 26–37.

Srikar, S. K.; Giri, D. D.; Pal, D. B.; Mishra, P. K.; Upadhyay, S. N. Green Synthesis of Silver Nanoparticles: A Review. *Green Sustain. Chem.* **2016,** *6,* 34–56. doi: 10.4236/gsc.2016.61004.

Srivastava, S.; Bhargava, A.; Pathak, N.; Srivastava, P. Production, Characterization and Antibacterial Activity of Silver Nanoparticles Produced by Fusarium Oxysporum and Monitoring of Protein-Ligand Interaction Through In-Silico Approaches. *Microb. Pathogen.* **2019,** *129,* 136–145.

Suresh, G.; Gunasekar, P. H.; Kokila, D.; Prabhu, D.; Dinesh, D.; Ravichandran, N.; Ramesh, B.; Koodalingam, A.; Siva, G. V. Green Synthesis of Silver Nanoparticles Using *Delphinium Denudatum* Root Extract Exhibits Antibacterial and Mosquito Larvicidal Activities. *Spectrochim. Acta Part A Mol. Biomol. Spectrosc.* **2014,** *127,* 61–66.

Tan, L. V.; Tran, T.; Thi, V. D. Biosynthesis of Silver Nanoparticles From *Bacillus Licheniformis* TT01 Isolated from Quail Manure Collected In Vietnam. *Processes* **2021,** *9,* 584. https://doi.org/10.3390/pr9040584

Tavaf, Z.; Tabatabaei, M.; Khalafi-Nezhad, A.; Panahi, F. Evaluation of Antibacterial, Antibofilm and Antioxidant Activities of Synthesized Silver Nanoparticles (Agnps) and Casein Peptide Fragments Against Streptococcus Mutans. *Eur. J. Integr. Med.* **2017,** *12,* 163–171.

Thakar, M. A.; Jha, S. S.; Phasinam, K.; Manne, R.; Qureshi, Y.; Babu, V. H. X Ray Diffraction (XRD) Analysis and Evaluation of Antioxidant Activity of Copper Oxide Nanoparticles Synthesized From Leaf Extract of Cissus Vitiginea. *Mater. Today: Proc.* **2022,** *51,* 319–324.

Thomas, R.; Janardhanan, A.; Varghese, R. T.; Soniya, E. V.; Mathew, J.; Radhakrishnan, E. K. Antibacterial Properties of Silver Nanoparticles Synthesized by Marine Ochrobactrum sp. *Braz. J. Microbiol.* **2014,** *45,* 1221–1227.

Tian, Y.; Wang, F. L.; Liu, Y. X.; Pang, F.; Zhang, X. Green Synthesis of Silver Nanoparticles on Nitrogen-Doped Graphene for Hydrogen Peroxide Detection. *Electrochim. Acta* **2014,** *146,* 646–651.

Tiburu, E. K.; Salifu, A.; Aidoo, E. O.; Fleischer, H. N.; Manu, G.; Yaya, A.; Zhou, H.; Efavi, J. K. Formation of Chitosan Nanoparticles Using Deacetylated Chitin Isolated from Freshwater Algae and Locally Synthesized Zeolite A and Their Influence on Cancer Cell Growth. *Proc. J. Nano Res.* **2017,** *48,* 156–170.

Tien, D. C.; Tseng, K. H.; Liao, C. Y.; Huang, J. C.; Tsung, T. T. Discovery of Ionic Silver in Silver Nanoparticle Suspension Fabricated by Arc Discharge Method. *J. Alloys. Compd.* **2008,** *463,* 408–411.

Tripathi, R. M.; Ranac, D.; Shrivastav, A.; Singh, R. P.; Shrivastav, B. R. Biogenic Synthesis of Silver Nanoparticles Using Saraca Indica Leaf Extract and Evaluation of Their Antibacterial Activity. *Nano Biomed. Eng.* **2013,** *5* (1).

Uzair, B.; Liaqat, A.; Iqbal, H.; Menaa, B.; Razzaq, A.; Thiripuranathar, G.; Fatima Rana, N.; Menaa, F. Green and Cost-Effective Synthesis of Metallic Nanoparticles by Algae: Safe Methods For Translational Medicine. *Bioengineering* **2020,** *7,* 129. https://doi.org/10.3390/bioengineering7040129

Vahabi, K.; Mansoori, G. A.; Karimi, S. Biosynthesis of Silver Nanoparticles by Fungus Trichoderma Reesei (A Route for Large-Scale Production of Agnps). *Insciences J.* **2011,** *1* (1), 65–79.

Vazquez-Muñoz, R.; Meza-Villezcas, A.; Fournier, P. G.; Soria-Castro, E.; Juarez-Moreno, K.; Gallego-Hernández, A. L.; Bogdanchikova, N.; Vazquez-Duhalt, R.; Huerta-Saquero, A. Enhancement of Antibiotics Antimicrobial Activity due to the Silver Nanoparticles Impact on the Cell Membrane. *PLoS ONE.* **2019,** *14* (11), E0224904.

Von White, G.; Kerscher, P.; Brown, R. M.; Morella, J. D.; Mcallister, W.; Dean, D.; Kitchens, C. L. Green Synthesis of Robust, Biocompatible Silver Nanoparticles Using Garlic Extract. *J. Nanomater.* **2012,** *2012.*

Wang, D., Xue, B., Wang, L. et al. Fungus-Mediated Green Synthesis of Nano-Silver Using *Aspergillus Sydowii* and Its Antifungal/Antiproliferative Activities. *Sci. Rep.* **2021, 11,** *10356.* https://doi.org/10.1038/s41598-021-89854-5

Wei, L.; Lu, J.; Xu, H.; Patel, A.; Chen, Z. S.; Chen, G. Silver Nanoparticles: Synthesis, Properties, and Therapeutic Applications. *Drug Disc. Today* **2015,** *20* (5), 595–601.

Whitesides, G. M. Nanoscience, Nanotechnology, and Chemistry. *Small* **2005,** *1,* 172–179.

Wu, X.; Lu, C.; Zhou, Z.; Yuan, G.; Xiong, R.; Zhang, X. Green Synthesis and Formation Mechanism of Cellulose Nanocrystal-Supported Gold Nanoparticles with Enhanced Catalytic Performance. *Environ. Sci.: Nano* **2014,** *1* (1), 71–79.

Wu, S.-H.; Mou, C.-Y.; Lin, H.-P. Synthesis of Mesoporous Silica Nanoparticles. *Chem. Soc. Rev.* **2013,** *42,* 3862–3875.

Xia, Y.; Xia, X.; Peng, H. C. Shape-Controlled Synthesis of Colloidal Metal Nanocrystals: Thermodynamic Versus Kinetic Products. *J. Am. Chem. Soc.* **2015,** *137* (25), 7947–7966.

Xue, B.; He, D.; Gao, S.; Wang, D.; Yokoyama, K.; Wang, L. Biosynthesis of Silver Nanoparticles by the Fungus Arthroderma Fulvum and Its Antifungal Activity Against Genera of Candida, Aspergillus and Fusarium. *Int. J. Nanomed.* **2016,** *11,* 1899.

Xue, B.; He, D.; Gao, S.; Wang, D.; Yokoyama, K.; Wang, L. Biosynthesis of Silver Nanoparticles by the Fungus *Arthroderma Fulvum* and Its Antifungal Activity Against Genera of *Candida, Aspergillus* and *Fusarium. Int. J. Nanomed.* **2016,** *11,* 1899–1906.

Yan, X.; He, B.; Liu, L.; Qu, G.; Shi, J.; Hu, L.; Jiang, G. Antibacterial Mechanism of Silver Nanoparticles in Pseudomonas Aeruginosa: Proteomics Approach. *Metallomics* **2018,** *10* (4), 557–564.

Yazdi, M. E. T.; Darroudi, M.; Amiri, M. S.; Hosseini, H. A.; Nourbakhsh, F.; Mashreghi, M.; Farjadi, M.; Kouhi, S. M. M.; Mousavi, S. H. Anticancer, Antimicrobial, and Dye Degradation Activity of Biosynthesised Silver Nanoparticle Using Artemisia Kopetdaghensis. *Micro Nano Lett.* **2020,** *15,* 1046–1050.

Yin, Y.; Yang, X.; Hu, L.; Tan, Z.; Zhao, L.; Zhang, Z.; Liu, J.; Jiang, G. Superoxide-Mediated Extracellular Biosynthesis of Silver Nanoparticles by the Fungus Fusarium Oxysporum. *Environ. Sci. Technol. Lett.* **2016,** *3* (4), 160–165.

Yousefzadi, M.; Rahimi, Z.; Ghafori, V. The Green Synthesis, Characterization and Antimicrobial Activities of Silver Nanoparticles Synthesized from Green Alga *Enteromorpha Flexuosa* (Wulfen) *J. Agardh. Mater Let.* **2014,** *137,* 1–4. doi: 10.1016/j.matlet.2014.08.110

Zhang, Q.; Moran, C. H.; Xia, X.; Rycenga, M.; Li, N.; Xia, Y. Synthesis of Ag Nanobars in the Presence of Single-Crystal Seeds and a Bromide Compound, and their Surface-Enhanced Raman Scattering (SERS) Properties. *Langmuir* **2012,** *28* (24), 9047–9054.

Zhang, W.; Yao, Y.; Sullivan, N.; Chen, Y. Modeling the Primary Size Effects of Citrate-Coated Silver Nanoparticles on their Ion Release Kinetics. *Environ. Sci. Technol.* **2011,** *45* (10), 4422–4428.

Zhang, X.-F.; Liu, Z.-G.; Shen, W.; Gurunathan, S. Silver Nanoparticles: Synthesis, Characterization, Properties, Applications, and Therapeutic Approaches. *Int. J. Mol. Sci.* **2016,** *17,* 1534. https://doi.org/10.3390/ijms17091534

Zhou, S.; Li, J.; Gilroy, K. D.; Tao, J.; Zhu, C.; Yang, X.; Sun, X.; Xia, Y. Facile Synthesis of Silver Nanocubes with Sharp Corners and Edges in an Aqueous Solution. *Acs Nano.* **2016,** *10* (11), 9861–9870. doi:10.1021/acsnano.6b05776

CHAPTER 11

Green Synthesis of Nanomaterials from Algae Materials to Remediate Environmental Pollution

VANI TIWARI[1], VEDHA SAHITHYA[1], NILESH WAGH[1], and JAYA LAKKAKULA[1,2]

[1]*Amity Institute of Biotechnology, Amity University Maharashtra, Mumbai, Maharashtra, India*

[2]*Centre for Computational Biology and Translational Research, Amity Institute of Biotechnology, Amity University Maharashtra, Mumbai, Maharashtra, India*

ABSTRACT

In this study, a convenient, easy-to-perform, and green method has been employed for the synthesis of silver nanoparticles (AgNPs), ZnNPs, Cu-NPs, and FeNPs using different types of algae for environmental nanoremediation. It is thought that the moderate alga's phytochemicals, which are also non-toxic, act as both reducing and stabilizing agents. Using algae, contaminants can be removed from the environment or changed into harmless forms. Since the biomass produced by algae is used to produce biogas and biofuel, algal-based bioremediation is strongly favored. In order to completely remove and clean up polluted soil and water bodies and improve environmental health, nanobioremediation has emerged as a new discipline. In this study, by using spectroscopic and microscopic studies, the as-prepared nanoparticles were characterized. X-ray diffraction, UV–Vis spectroscopy, transmission electron microscopy (TEM), and Fourier-transform infrared spectroscopy (FTIR) were used to characterize the production of silver, zinc, iron, and

Sustainable Nanoremediation: Modern Technologies for a Clean Environment.
Fernanda Maria Policarpo Tonelli, Arpita Roy, and Khalid Rehman Hakeem (Eds.)

copper nanoparticles. Subsequently, cytotoxicity, photocatalytic, antifungal, and antioxidant properties of these nanoparticles were assessed. The results revealed that the four biosynthesized nanoparticles exhibited impressive potential for nanoremediation.

11.1 INTRODUCTION

Although natural occurrences like wildfires and erupting volcanoes can cause pollution, artificial activities have increased over the past few decades, leading to an increase in environmental contamination. Toxins that affect the environment are introduced, which is referred to as pollution. Pollution is the main environmental factor that contributes to sickness and early mortality (Bayda et al., 2020). Protecting our environment is one of our primary responsibilities and a natural way for us to take care of ourselves and our future generations. Following the creation of a pollution control program or strategy, an extensive operations plan should be established (Almansoori and Nazir, 2021). A variety of businesses are becoming more and more interested in the relatively new field of nanotechnology. This phrase describes the areas of science and engineering that leverage nanoscale phenomena in the design, evaluation, production, and application of novel compounds, structures, tools, and systems (Manisalidis et al., 2020). The typical customer already regularly utilizes products that contain nanotechnology. Current sunscreens, for instance, contain titanium or zinc nanoparticles that actively block UV radiation, while hand soaps, bandages, and socks contain silver nanoparticles (AgNPs) that have antibacterial properties (Smijs and Pavel, 2011).

An appealing and effective cleansing method to get rid of harmful waste from a damaged environment is bioremediation. Bioremediation strategies can offer eradication, degradation, detoxification, and immobilization of different physically dangerous elements and chemical wastes from contaminated areas—one of its strategies is the use of microorganisms. Degrading and transforming pollutants into less-hazardous forms is the fundamental idea (Brar et al., 2006). The primary manufacturing sector that discharges the most waste into water is the textile industry, which employs more than 10,000 dyes and pigments, the bulk of which are non-biodegradable, in its shading operations. In the textile industry, methylene blue, methyl orange, rhodamine B, and tri-phenyl methane dye, often known as crystal violet, are the dyes that are most frequently employed (Islam et al., 2019). Therefore, in order to safeguard the environment, it must be taken out of wastewater before being disposed of (Goswami et al., 2021).

Metal nanoparticles have recently drawn a lot of interest because of their prospective applications in a number of industries, including catalysis, nonlinear optical devices, bioimaging, water purification, and chemical sensing. The applicability of nanomaterials has increased interest in new approaches to obtain nanoparticles. AgNPs, which are employed in cryogenic superconductors, electrical component, catalysts, nonlinear optics, antimicrobials, and biosensors are among the inorganic nanomaterials with excellent material properties and a variety of useful applications (Korbekandi and Iravani, 2012). Metal nanoparticles are typically difficult to make and take a long time to make using conventional physical and chemical procedures, which also use a lot of energy with minimal conversion yield. Such procedures are expensive, harmful to the environment, and rely on hazardous reducing agents, organic solvents, and stabilizing compounds that are not biodegradable (de Oliveira et al., 2020). Biological creatures like plants, sea algae, bacteria, viruses, fungi, and lichens can be used in synthesis to create nanoparticles in more environmentally friendly ways. The green synthesis of nanomaterials has also used algae, such as cyanobacteria (Singh et al., 2018). AgNPs, zinc nanoparticles, and iron nanoparticles are inorganic nanomaterials, which have outstanding material qualities and functional diversity, making them valuable as cryogenic superconductors, catalysts, composite fibers, antimicrobials, and more (Yaqoob et al., 2020).

The study of semiconductor nanoparticles, in particular cadmium sulphide (CdS) nanoparticles, has grown dramatically during the last few decades. This is due to their low toxicity and favorable optical properties; CdS nanoparticles produced through biosynthetic processes are of utmost importance for upcoming bioremediation. The popular nutraceutical *Spirulina platensis* is used to make biologically produced CdS nanoparticles. CdS nanoparticles have a size range of 8–12 nm, as per transmission electron microscopy (TEM) studies. The generated photoluminescent CdS nanoparticles degrade the toxic pigment pollutant malachite green (Mandal et al., 2016).

Magnesium oxide nanoparticles (MgONPs), which are among the metal oxide nanoparticles, have been employed to treat a variety of illnesses because of their biocompatibility and outstanding stability under adverse conditions. The marine brown algae *Sargassum wighitii* is used to produce MgONP. It was discovered that MgONP possesses a crystalline shape and a face-centered cubic structure using X-ray diffraction investigation. They can degrade dyes and have antibacterial and anticancer properties. The outcomes demonstrated that MgONPs from brown algae significantly and dose-dependently killed both Gram-positive and Gram-negative. MgONPs exhibited strong catalytic activity under both visible and UV illumination when compared to the control. They spontaneously crystallized with a typical particle size of 68.02 nanometer (Pugazhendhi et al., 2019).

Differently from other microorganisms, a few algae have already been tested for their ability to allow the performance of green synthesis protocols to generate nanomaterials. Green synthesis, which uses organisms such as bacteria, fungi, seaweeds, lichens, plants, and viruses to produce nanoparticles, offers an environmentally friendly alternative. These biological systems are now found to be optimal for nanoparticle production under spontaneously adjusted pH at room temperature. In addition to cyanobacteria and diatoms, various types of plants and algae are used, including *Chlamydomonas reinhardtii, Chlorella vulgaris, Sargassum Wigiti*, and sea lettuce (Jeffryes et al., 2015).

In this chapter, we are going to discuss about the different ways used in the green synthesis of many metal and other nanoparticles such as silver, zinc, copper, and iron nanoparticles using different species of algae and their role in environmental bioremediation (Table 11.1).

11.2 ALGAE-BASED NANOMATERIALS TO CONTAMINATED ENVIRONS NANOREMEDIATION

11.2.1 ALGAE-BASED SYNTHESIS OF SILVER NANOMATERIALS

Phycosynthesis of AgNP using brown algae *Sargassum myriocystum* is an innovative and eco-friendly process to generate nanoparticle with high purity. AgNPs are synthesized using silver nitrate aqueous solution and 100 mL of seaweeds extract mixed with 900 mL of 1 Mm $AgNO_3$ after incubation for 24 h under dark conditions. Phycomolecules involved in capping and reducing process of AgNPs that were coated into Kbr pellet was identified by Fourier-transform infrared spectroscopy (FTIR) technique. Mid-region wavelength of 4000–400 cm^{-1} was used to record IR spectrum at a resolution of 4.0 cm. AgNPs were assayed against Gram-negative bacteria (*K. pneumoniae, P. pseudomonas aeruginosa, P. vulgaris, E coli)* and positive bacteria (*S. aureus, S. epidermidis*). Sterilization of NA was done by autoclaving it on 121 °C for 15 min and then it was spilled into sterile plates for solidifying aqueous seaweed extracts. AgNPs were aliquoted and added to sterile discs. Inhibition of bacterial growth was seen around the disc. Phycosynthesized AgNPs potential was additionally tried against biofilm creating

TABLE 11.1 Examples of Algae-Based Nanoparticles Capable of Remediating Environmental Contaminants.

S.No	Type of algae	Name of algae	Size of nano algae material	Characterization	Contaminants	References
1	Blue green algae	*Anabaena variabilis* and *Spirulina Platensis*	17.9–24.6 nm	Silver nanoparticles	Malachite green	Ismail et al. (2021)
2	Blue green algae	*Spirulina*	200–800 nm	Iron nanoparticles	Crystal violet dye	Bhukal et al. (2022)
3	Blue green algae	*Spirulina platensis*	5.54 nm	Iron nanoparticles	CV dye	Shalaby et al. (2021)
4	Green algae	*Spirulina*	40–65 nm	Silver oxide nanoparticles	Human pathogen	Rajamanickam et al. (2013)
5		*Ulva fasciata*	28–41 nm	Silver nanoparticles	Xanthomonas campestris pv. malvacearum	Sathiyamoorthy et al. (2012)
6		*Ulva flexuosa*	4.93–6.70 nm	Silver nanoparticles	*E. coli, S aureus*	Dixit et al. (2018)
7		*Enteromorpha flexuosa*	2–32 nm	Silver nanoparticles	Bacteria	Yousefzadi et al. (2014)
6		*Cholrella ellipsoidea*	220.8–31.3 nm	Silver nanoparticles	Methylene blue and methyl orange	Borah et al. (2020)
8		*Caulerpa serrulata*	2–10 nm	Ag nanoparticles	Congo red dye	Aboelfetoh et al. (2017)
9		*Ulva lactuca*	48.59 nm	Silver nanoparticles	Methyl orange dye	Kumar et al. (2013a)
10		*Chlorella vulgaris and Dunaliella tertiolecta*	80–200 nm	Zno nanoparticles	Paraffine hydrocarbons	Salehi et al. (2019)
11		*Ulva lactuca*	10–50 nm	Zinc nanoparticles	Organic dye	Kumar et al. (2013a)
12		*Chlorella vulgaris*	20 nm	Zinc nanoparticles	Dibenzothiophene (DBT)	Khalafi et al. (2019)
13		*Chlamydomonas reinhardtii*	10 nm	Copper nanoparticles	Cu+2 ions	Žvab et al. (2021)
14	Red algae	*Gelidium amansii*	27–54 nm	Silver nanoparticles	Bacteria	Pugazhendhi et al. (2018)
15		*Gelidiella acerosa*	22 nm	Silver (AgNPs)	Fungi	Vivek et al. (n.d.)
16		*Gracilaria corticata*	18–46 nm	Silver nanoparticles	Candida spp.	Kumar (2013b)
17		*Corallina officinalis and Corallina mediterranea*	12.7–13.6 nm	Copper nanoparticles	Lyngbya majuscula	El-Kassas and Okbah (2017)

TABLE 11.1 *(Continued)*

S.No	Type of algae	Name of algae	Size of nano algae material	Characterization	Contaminants	References
18	Brown algae	*Sargassum wightii*	8–27 nm	Silver nanoparticles	Bacteria	Govindaraju et al. (2009)
19		*Sargassum myriocystum*	20–2.2 nm	Silver (AgNPs) nanoparticles	Methylene blue	Balaraman et al. (2020)
20		*Sargassum serratifolium*	27.84 and 7.18 nm	Silver nanoparticles	Methylene blue, rhodamine B and methyl orange	Kim et al. (2021)
21		*Sargassum Vulgare*	100 nm	Zno nanoparticles	Methylene blue	Karkhane et al. (2020)
21		*Sargassum muticum*	20–25 nm	Zno and CuO nanoparticles	Biofilms and corrosive materials	Sadek et al. (2019)
22		*Sargassum polycystum*	3.6 nm	Copper oxide nanoparticles	*E. coli*	Bhattacharya et al. (2019)
23		*Padina pavonica thivy and Sargassum acinarium*	10–19.5 nm and 21.6–27.4 nm	Iron nanoparticles	Lead	Hala et al. (n.d.)
24		*Petalonia fascia, Colpomenia sinuosa, Padina pavonica*	Average 9.42 ± 2.84 nm	Ironoxide nano particles	Harmful algae	El-Sheekh et al. (2021)

bacterial strains *S. epidermidis* and *P. aeruginosa*. The maximum biofilm inhibition 67.75% was at 50 mg/L for AgNPs treated with these bacteria. For the antibiofilm test, bacteria culture (1 mL) and nutrient broth (4 mL) were added to individual sterile glass tubes. Phycosynthesized AgNPs and extracts of *S. myriocystum* potential for larvicidal activity were tested against larvae of *C. quinquefasciatus* and *Ae.aegypti* fourth instar). These mosquitoes are responsible for causing dengue and filariasis. Maximum mortality percentage was seen in these mosquitoes when they were treated at high concentration for 48 h incubation. Phycosynthesized AgNPs were also used for photocatalytic dye degradation of methylene blue, which can be easily confirmed by color change in dye solution. About 98.22% of degradation was found in 60 min, which was achieved by AgNPs mediated aqueous solution. This shows its potential in degradation of toxic dyes. The distinctive absorption peaks were observed at 660 nm and the supernatant was used to calculate the dye-degradation activity (Balaraman et al., 2020).

Anabaena variabilis and *Spirulina platensis*, two species of blue green algae, were studied and tested for their ability to remove color during the synthesis of AgNPs. A UV–Vis spectrophotometer operating at 400–450 nm was used to identify the color change from colorless to yellowish-brown, which indicated the synthesis of AgNPs. Batch mode adsorption experiment (in Erlenmeyer flask) was used to assess the potential of MG's removal by generated AgNPs from solutions. The volume of each flask was completed to 20 mL by deionized H_2O. As a dye-removing agent, the AgNPs produced by *S. platensis* and *A. variabilis* at various doses were used. During the course of 90 min, the absorbance (at 5 105 mol/L) decreased. At pH 6, 30 °C, and 100 rpm, the reaction was carried out. These biosynthesized AgNPs' ability to remove MG was studied using kinetic models including pseudo-first and pseudo-second orders. These kinetic models' resultant rate constants were identified. Additionally, it was looked into how the two kinetic models' rate constants ko1 and ko2 decreased as dye concentrations rose. Following the decontamination of MG, the dye-AgNPs-free water effluents generated were shown to be harmless to *T. aestivum* seedlings. Application of AgNPs produced by cyanobacteria in the removal of MG from solutions is a successful and environmentally safe biological method for treating water to remove polluting dyes (Ismail et al., 2021).

The dried biomass of the green alga *Chlorella ellipsoidea* has been used to create AgNPs in an easy, practical, and environmentally friendly way. Several spectroscopic and microscopic methods were used to characterize the AgNPs as they were being synthesized. UV–Vis was used to track the bioreduction of Ag+ ion to Ag_0 (AgNPs), and a KBr was used to record the FT-IR spectrum. The presence of elements in the produced nanoparticles was verified using energy-dispersive X-ray spectroscopy. Two potentially harmful dyes, methyl blue and methylene orange, were degraded by the photocatalytic activity of the as-synthesized AgNPs under UV light. In a typical procedure, 1 mL of freshly made $NaBH_4$ solution and 25 mL of a dye's 104 M stock solutions were combined. A UV–Vis spectrophotometer was used to measure the dye decrease at regular intervals. In aqueous solution, methylene blue, an organic dye, typically displays n-* and -* transition-related absorption bands at 662/nm and a shoulder at 612/nm, respectively. The strength of the absorption band with a maximum at 662 nm gradually decreases as the catalytic degradation progresses, while a second, relatively sharp band with a maximum at 280 nm simultaneously forms and intensifies over time. When exposed to light, the azo dye methyl orange has a maximum absorption at 460 nm in the visible spectrum. The intensity was reduced by adding the produced AgNPs as a catalyst to MO after it had been treated with $NaBH_4$. According to a kinetic analysis, the degradation rates are pseudo-first order, with methylene blue degrading at a rate of 4.72 102 min and methyl orange degrading at a rate of 3.24 102 min. The AgNPs showed notable antibacterial efficacy against four different types of harmful bacteria (Borah et al., 2020).

The aqueous extract of the green marine alga *Caulerpa serrulata* was used as a potent reducing and stabilizing agent to create stable colloidal AgNPs. By measuring the mixture through the reaction in the UV–Visible range, the synthesis of AgNPs was estimated. The solution's color changed from yellow to reddish brown throughout the reduction of Ag+ to Ag_0 nanoparticles, depending on various factors including the extract volume and stirring time temperature. UV–Vis, XRD, FT-IR, and HR-TEM were used to characterize the synthesized AgNPs. The produced AgNPs showed a prominent surface-plasmon resonance band at 412 nm when exposed to the ideal conditions, that is, 20% (v/v) extract and 95 °C. The catalytic characteristics of organic nitric oxide were investigated using the model reaction of $NaBH_4$ and azo dye (AgNPs). At regular intervals, the reaction was observed using a spectrophotometer with a scanning range of 350–650 nm. The greatest band of CR dye's absorption in aqueous media for AgNPs is at 498 nm. By using a Bruker spectrometer running at 400 MHz (in DMSO-d6) and reporting chemical shifts in ppm, the UV–Vis spectra of AgNPs were measured. The MXP-18 diffractometer used Cuk radiation to create the XRD pattern. Utilizing a KBr pellet system, FT-IR estimations were made using a PerkinElmer 1430. The degradation rate of CR with AgNPs ramped up with an increase in either the $NaBH_4$ concentration or the catalytic dosage. Additionally, the AgNPs showed antibacterial activity against *Salmonella typhi*, *Escherichia coli*, *Shigella sp.*, *Pseudomonas aeruginosa*, and *Staphylococcus aureus* at lower concentrations, indicating that they might be a suitable alternative therapeutic approach (Aboelfetoh et al., 2017).

Dyes are frequently employed in the textile industry and are a subset of synthetic organic chemicals. For the synthesis of AgNPs author employed materials such as seaweed samples which were taken from the Mandapam shore, Gulf of Mannar, Tamil Nadu. Seaweed extract which was boiled for 30 min was used to filter it through Whatmann No. 1 filter paper. A 250 mL Erlenmeyer flask was filled with 250 mg of powdered seaweed to create aqueous seaweed broth (ASB). XRD and spectrometry were two of the microscopy methods used to analyze Ag nanoparticles. UV–Vis spectrophotometer readings were recorded at 200–800 nm and FT-IR spectrophotometry at 4000–400 cm^{-1}. Zeta potential measurements were also performed using Zetasizer Nano ZS90 and Philps X'Pert Pro X-rays. Biosynthesized Ag nanoparticles were used to assess the photocatalytic destruction of methyl orange. UV–Vis spectroscopy, or SPR, is a helpful technique used to determine the reduction of metal ions based on optical properties. The discovery of a symmetrical absorption band suggests the presence of spherical, polydisperse nanoparticles. By decreasing the silver ions at room temperature, the incubation period is crucial in the production of nanoparticles. At 3429, 1637, and 685 cm, intense FT-IR bands were seen, indicating the presence of molecular functional groups in charge of the reduction of silver ions. Using solar irradiation technique at various time intervals using biometrically produced silver nanocatalysts, the photocatalytic degradation of the methyl orange dye was studied (Fig. 11.1). The methyl orange solution's characteristic absorption peak was determined to be 460 nm. According to Kansal et al., sunlight was found to discolor methyl orange in the presence of metal more quickly than that of other irradiation techniques. Excitation of SPR, which is nothing more than an oscillation in charge density that can propagate at the interface between metal and a dielectric media, may be the reason for the visible light photocatalytic activities of Ag nanoparticles. It has been shown that the biometric generation of AgNPs from U. lactuca at room temperature is effective in getting rid of methyl orange solution. These results imply that dyes can be degraded by AgNPs produced by a simple process, opening the door to environmental bioremediation (Kumar et al., 2013a).

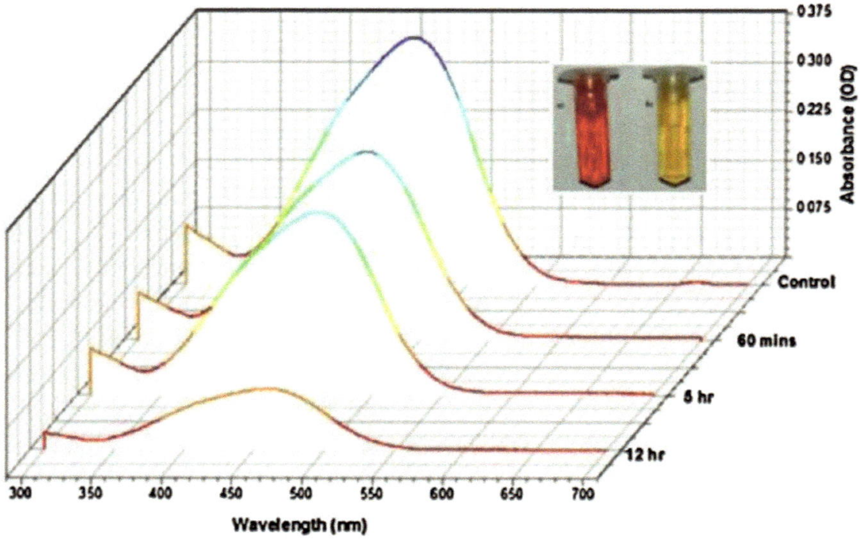

FIGURE 11.1 Photocatalytic degradation of methyl orange using the AgNPs synthesized from *Ulva lactuca*. Reprinted with permission from Kumar et al. (2013a).

Sargassum serratifolium, a marine algae extract, produces AgNPs. Utilizing techniques such as UV–Vis absorption spectroscopy, HR-TEM, selected area electron diffraction, dynamic light scattering, energy-dispersive X-ray spectroscopy, FTIS, and XRD, the characteristics of bio-synthesized SS-AgNPs were studied. Methylene blue (MB), methyl orange (MO), and rhodamine B (RB) were used to test the catalytic activity of biogenic AgNPs on the reduction of dyes at room temperature. The azo group (RN = NR0) in MO functions as a chromophore. MO has a vibrant orange color because of chromophore. Since MO has a maximal absorption band at 465 nm, it can be utilized as a pH indicator. MO was broken down into 4-aminobenzene-sulfonate and N, N-dimethyl-benzene-1, 4-diamine. A heterocyclic aromatic dye called methylene blue is used in the textile and healthcare industries. It works well as an antimalarial and treatment for methemoglobinemia. At 662 nm, reduced methylene blue, also known as leucomethylene blue, reaches its greatest UV–Vis spectrum peak. The textile and food sectors frequently use the water tracer fluorescent dye rhodamine B. At 553 nm, RB's highest absorption peak is located. The reduction products, amino compounds, are formed when RB is broken down. By using SS-AgNPs as the catalyst, all three dyes were reduced and changed into colorless solutions in a matter of only 18 min. The electrons of the green-synthesized nanoparticles were stimulated from the valence band to the conduction bands during photocatalytic reduction and interacted with oxygen and hydrogen from $NaBH_4$ to produce active radical species such the superoxide anion and OH radicals. The bonds of synthetic dyes were afterwards broken by free radicals (superoxide anion and OH radical) (Kim et al., 2021).

An easy and environmentally beneficial process for making hybrid bead AgNPs using *Ulva flexuosa* and water extracted from a natural, renewable source. According to this method, macroporosity methacrylic acid copolymer beads with Ag ions attached are reduced to AgNPs and then used as antibacterial agents for in vitro water disinfection. The nanoparticles were examined using EDXA, SEM, HR-TEM, UV–Vis, FTI spectroscopy, and inductively coupled plasma optical emission spectroscopy. Field-emission scanning electron microscopy (FESEM) was used to analyze the made AgNPs-bound copolymer beads. HR-TEM was used to analyse the microstructure of the nanoparticles. Atomic force microscopy (AFM) analysis and imaging were performed utilizing the NTMDT Ntegra Aura model (Moscow, Russia). The necessary bacterial stereomicroscopy-based AFM analysis and imaging preparation were completed. The immobilized AgNPs that exhibit high antibacterial properties and are stable at a concentration of 300 mg reduced germs by 100% contact time for all bacterial strains in hours within six days. As time goes on, there are more AgNPs-bound copolymer beads available (Dixit et al., 2018).

The UV–Vis spectroscopy, XRD, HR-TEM, and FT-IR techniques were used to characterize the AgNPs that were produced. The potential chemical traces left behind by nanoparticles may engage with biological systems. A growing demand to create environmentally friendly materials and synthesis methods has made biological synthesis of noble metal nanoparticles a hot topic. The bio reduction of Ag^+ ions in aqueous solution was monitored using regular (0.2 mL) sampling of aliquots of the suspension, dilution with 2 mL deionized water, and analysis of the UV–Vis spectra of the resultant diluents. Powders of the dry solutions containing biomass and Ag nanoparticles were centrifuged at 5000 rpm for 15 min, with suspensions as a result, was in sterile distilled water; redispersed, the clean pellets were examined by Thermo Nicolet Avatar 300 after being dried. The antibacterial efficiency of biologically produced AgNPs in water was evaluated using the agar well diffusion method. Wells were created using a cork borer and filled with different concentrations (20 L, 30 L, and 50 L) of chemically and biologically produced AgNPs. After 37 °C incubation, distilled water was used as the negative control and AgNPs as the positive control. For a full day, the clear zone was measured. It is almost certain that a new generation of bactericidal materials will be created by the preparation, surface modification, characterization, and functionalization of nanosized inorganic particles, as evidenced by the observed increased antibacterial activity of the seaweed (*Sargassum wightii*) mediated AgNPs (Govindaraju et al., 2009).

Green nanotechnology has recently sparked research interest in eco-friendly nanoparticle biosynthesis. This study combined the algal extract *Enteromorpha flexuosa* with silver nitrate. TEM, EDS, XRD, and UV–Vis were used to characterize the reduced AgNPs. In the in vitro studies, the produced *E. flexuosa* nanoparticles demonstrated potent antibacterial activity against Gram-positive bacteria and poor antibacterial activity against Gram-negative pathogens. Production of AgNPs using algae-based materials is relatively fast, inexpensive, and suitable for use in antimicrobial therapy. Ionizing radiation, photochemical processes, and biological strategies for chemical reduction of silver ions have been studied. In this experiment, seaweed extract was added to a solution of 1 mM $AgNO_3$ to reduce Ag to Ag_1 and form AgNPs. The colorless reaction mixture changed color. It shows the biotransformation of ions in a bright environment after 60 min of incubation in a dark brown solution. Silver becomes reduced silver because of the surface plasmon resonance phenomena. No color change was observed in the solution with controlled $AgNO_3$ without seaweed extract. The color change is due to the active ingredient in extract. Lastly, the silver metal ions were converted to AgNPs (Yousefzadi et al., 2014).

To date, many corrective strategies have been used to combat microfouling, but nanoparticles stand out among them. Green manufactured nanoparticles are currently in demand because of their eco-friendly, economical, and non-toxic properties, which have significantly enhanced their industrial applications (biomedicine, food, and textile). In this study, AgNPs were made using the marine red alga *Gelidium amansii*. On any manufactured or natural surface, microfouling is the buildup of microbial communities including bacteria, fungi, algae, etc. Following processes biofilm generation can induce surface's damage (due to the attachment of microbe and macro structures in additional layers). As they are antibacterial substances, AgNPs have received large attention among other types of nanoparticle. Red algae were applied to generate these Ag nanoparticles in the current investigation. *G. amansii* have the potential to be antibacterial agent, particularly against well-known microfouling bacteria. The detected absorption peak at 420 nm in the UV–Visible spectra further supported the conversion of silver nitrate to AgNPs (Fig. 11.2a). In order to pinpoint the putative biologically active molecules of the algal capping over the biosynthesized Ag NPs that were in charge of the reduction of the Ag+, the FTIR spectra was examined. Sharp absorption peaks at 2919, 2860, 1625, and 1515 cm^{-1} were visible in the FTIR analysis of GA-AgNPs (Fig. 11.2b). In conclusion, biosynthesis of nanoparticles opens up a straightforward and affordable synthesis approach using the machinery for algae reduction and also offers a practical way to lessen microfouling bacteria (Pugazhendhi et al., 2018).

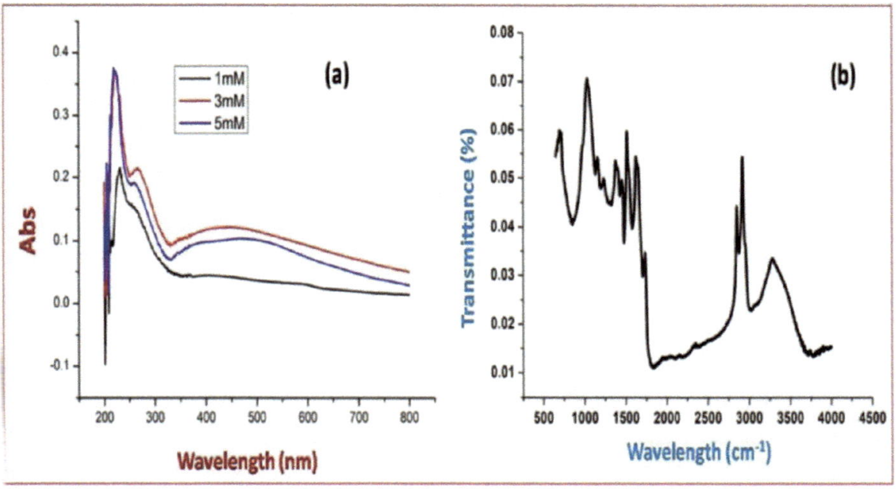

FIGURE 11.2 (a) UV–Vis spectroscopic analysis of GA-AgNPs; (b) FT-IR Spectrum of GA-AgNPs. Reprinted with permission from Pugazhendhi et al. (2018).

Utilizing bacteria isolated from Spirulina products, the production of AgNPs and its antibacterial property were examined. *MSK 3*, and *Brevundimonas sp*. Bacterial culture filtrate that contained $AgNO_3$ was used to create AgNPs. According to a FTIR study, and an energy-dispersive X-ray investigation identified silver as the main signal. X-ray diffraction and scanning electron microscopy were used to determine the structure of AgNPs. AgNPs with an average size of 40–65 nm exhibit antibacterial properties. Chemical techniques have been used to create nanoparticles without them aggregating, which is the simplest mode with the most benefit for industrial production. Most biological methods of producing nanoparticles used controlled size and shape production. The greater suppression of test pathogens by AgNPs may be caused by metabolites of the bacteria found in culture filtrate that are likely bound to the nanoparticles surface. Using a culture filtrate method, the ability of *MSK 1 Staphylococcus spp., MSK 2, Bacillus spp., MSK 3, and Brevundimonas spp. MSK 4* to synthesize the antibiotic AgNPs was examined. The isolates that were used to create the AgNPs were characterized using UV–Vis, EDAX, FTIR, XRD, and SEM, and their antibacterial activity was examined against a variety of test pathogens (Rajamanickam et al., 2013).

The *Gracilaria corticata* marine seaweed was used to quickly and easily produce AgNPs as a demonstration of the current investigation. By using the UV–Vis, FTIR, TEM, and dynamic light scattering studies, the produced Ag nanoparticles were thoroughly characterized. C. glabrata are effectively inhibited by biosynthesized Ag nanoparticles from G. corticate. Among many biological species, using plants to synthesize Ag nanoparticles is preferable to other biological techniques. When $AgNO_3$ was decreased in aqueous seaweed extract by heating at 60 °C, it resulted in an observable color change. As the solution was heated, the color gradually intensified, demonstrating the formation of Ag nanoparticles. The alteration in color is a result of Ag nanoparticle surface plasmon vibrations being excited. Notably, even 24 h later, there is still no discernible deepening of hue indicating reaction saturation which means there is good particle dispersion in the solution. Ag nanoparticles can be produced at a cost-effective rate from plants without the use of hazardous chemicals, high temperatures, or intense pressure. Before the experiment, SDA broth was prepared, inoculated, and incubated at 25 °C for three days with pure culture. The antifungal efficacy of Ag nanoparticles was assessed using the disc diffusion method. At a temperature of 60 °C, they were able to make biocompatible, stable Ag nanoparticles in 20 min. According to FTIR, the creation of Ag nanoparticles in seaweed extract is caused by the presence of functional bioactive chemicals. Ag nanoparticles were successfully used against *Candida spp.* as a potential antifungal agent (Kumar et al., 2013b).

Historically, hazardous and combustible substances have been utilized in wet chemical processes to create metallic nanoparticles. In this research, biosynthesis of AgNPs employing crude ethyl acetate extract from *Ulva fasciata* as a capping and reducing agent has been done. The bionanoparticles were characterized by EDX, FTIR, SEM, XRD, and UV–Vis. The nanoparticles are crystalline, spherical, polydispersed, and range in size from 28 to 41 nm, according to characterization. According to the study, crude ethyl acetate extract from *U. fasciata* could be utilized as a reducing agent for an easy, environmentally friendly way to make AgNPs. However, additional research is required before suggesting them for the treatment of disease. Due to its antibacterial qualities, silver has been utilized for a long time. Silver has undergone several shapes throughout history, including bulk silver, ionic silver, silver adsorbed on carrier materials, and the AgNPs we see today. We need to comprehend and control the behavior of nanoparticles in order to enhance nanotechnology for antimicrobial applications. Any material (insulator, conductor, or semiconductor) that has been precisely produced in the size range of 1–100 nm is referred to as a nanomaterial. The biosynthesized nanoparticles' size and form were evaluated using a SEM. Samples were created by drop-coating carbon copper grids with Ag nanoparticle solutions. Prior to measurement, the films on the grids were given time to dry. Energy-dispersive spectroscopy was used to confirm that there was elemental silver present. In this study, we reported the use of ethyl acetate extract from *U. fasciata* as a natural, affordable reducing and capping agent. Hazardous chemicals and solvents are not used in this technique. The biogenic nanosilver has shown encouraging antibacterial activity against the economically significant pathogen of the cotton plant, *X. campestris pv. malvacearum*, which significantly reduces the production across the globe's cotton-growing regions (Sathiyamoorthy et al., 2012).

An important area of nanotechnology is the synthesis, characterization, and application of biologically generated nanomaterials. The goal of the current study is to create AgNPs utilizing an aqueous extract of the red seaweed *Gelidiella acerosa* as a reducing agent. UV–Vis, XRD pattern, TEM, and SEM were used to confirm the creation of AgNPs. The produced AgNPs were polydispersed and mostly spherical in form. A study of the produced nano-Ag using FTI-R spectroscopy revealed that it was capped with bimolecular compounds that are in charge of reducing the silver ions. These nanoparticles were tested for their antifungal properties against the fungi *Humicola insolens, Mucor indicus, Fusarium dimerum*, and *Trichoderma reesei*. According to this study, AgNPs have significantly more antifungal activity than traditional antifungal medications. Because of their uses in the sciences of chemistry, physics, medicine, biology, green nanomaterial synthesis and characterization are a recent development in the field of nanotechnology. AgNPs produced through biological means have a diverse variety of uses due to their exceptional physical and chemical characteristics. In this paper, we outline a straightforward, one-step procedure for creating AgNPs by reducing aqueous Ag-ions with red seaweed extracts under direct sunlight. In conclusion, it has been shown that the aqueous extract of *Gelidiella acerosa* can bio-reduce aqueous $Ag+$ ions. This green chemical method for creating AgNPs has many benefits, including the simplicity of scaling up the procedure and its economic viability. Due to the applications for such nanoparticles in medicine and other fields, this technology may be employed for the mass manufacture of new inorganic nanomaterials. An entirely new class of antibacterial and antifungal medications has been made possible by AgNPs' toxicity study on human infections (Vivek et al., n.d.).

11.2.2 ALGAE-BASED SYNTHESIS OF ZINC, COPPER, AND IRON NANOMATERIALS

11.2.2.1 ZINC NANOMATERIALS

To produce more secure, stable, and pure ZnO NPs, a new environmentally friendly method was developed using microalgae *Chlorella* extract. The production of the ZnO NPs was carried out in ambient conditions using only zinc nitrate and *Chlorella* microalgae extract. According to XRD studies, the generated nanoparticles were revealed to have a high-crystalline hexagonal (Wurtzite) structure with an average size of about 19. 44 nm. And, according to TEM and SEM examinations, ZnO NPs are evenly distributed and have a hexagonal shape with an average size of 202.2 nm, respectively. This methodology starts green manufacturing processes using natural resources, allowing for the synthesis of nanoparticles for waste management, water treatment, renewable energy, and environmental rehabilitation. ZnO NPs are categorized as an essential semiconductor material due to their particular properties, which include their robust electron mobility, broad bandgap, and excellent transparency in the visual range. A home-made batch-type photo-reactor was used to investigate the photodegradation tests on DBT aqueous solution. Following UV light irradiation, the suspension is taken at regular intervals, and a gas chromatograph (GC) was used to analyze the breakdown of DBT. The reduction of DBT concentration was determined in GC by comparing the retention periods of the DBT sample generated with its final photocatalyzed solution by ZnO NPs. As-obtained ZnO NPs exhibit excellent photocatalytic activity toward DBT photodegradation as a hazardous pollutant in the aquatic environment, according to GC data obtained under optimum experimental conditions. The rate of DBT degradation remained almost same after five cycle cycles, indicating the increased durability and effectiveness of bio-assisted ZnO NPs. Due to the environmental benefits of biological biofabrication over conventional chemical and physical methods, additional investigations utilizing more potential natural sources are now being conducted (Khalafi et al., 2019).

Chlorella vulgaris and *Dunaliella tertiolecta*, two species of microalgae, were used as biological agents and ZnO nanoparticles as a photocatalyst to investigate the efficacy of eliminating hydrocarbons from oily water samples. Both of the investigated species were capable of removing hydrocarbons; however, *Chlorella vulgaris* was more effective at it. The obtained removal efficiencies for *D. tertiolecta*-ZnO and *C. vulgaris*-ZnO systems, respectively, were 100%, 78%, and 42% for initial hydrocarbon concentrations of 0. 05%, 0.1%, and 0. 5% (v/v). The utilization of secondary treatments, such as dissolved air floatation (DAF), advanced oxidation processes (AOPs), filtration, and biological treatments, is crucial as a result. It can be argued that because they consume less energy than the other methods mentioned, biological treatments and photocatalytic AOPs are more economically advantageous. Although TiO_2 is the most common photocatalyst used in photocatalytic reactions, ZnO can be used in its place because TiO_2 loses its effectiveness when exposed to visible light unaltered. The utilized nanoparticles in an AOP process should either be employed in an immobilized state or recovered from the discharged water because they can be harmful to living systems. FT-IR analysis was performed to look into how ZnO nanoparticles interact with algal cell surfaces. After being centrifuged with L-1 ZnO nanoparticles for seven days, the algal cells were extracted. In this study, a hybrid bio-photocatalyst was produced by combining biological processes using algae cells and AOPs using ZnO NPs. The results of experiments on the removal of hydrocarbons showed that the combination of algae and nanoparticles is capable of removing the hydrocarbon content of the medium, with the efficiency of removal depending on the type of algae, the quantity of hydrocarbons present, and the concentration of nanoparticles. Figures 11.3 and 11.4 illustrate how effectively *Dunaliella tertiolecta* and *Chlorella vulgaris* remove hydrocarbons. According to *D. tertiolecta* and *C. vulgaris*, the total removal efficiencies for samples with hydrocarbon contents of 0.1, 0.5, and 1% (v/v) were 92.9 0.1, 53.9 3.4, and 55.4 3.5, respectively. In most instances, increasing the hydrocarbon's molecular weight caused the clearance percentage to decline. This is brought on by the hydrocarbons' decreased biodegradability at greater molecular weights. Compared to Dunaliella tertiolecta, *Chlorella vulgaris* showed a stronger removal activity, according to the study's findings. (Salehi et al., 2019).

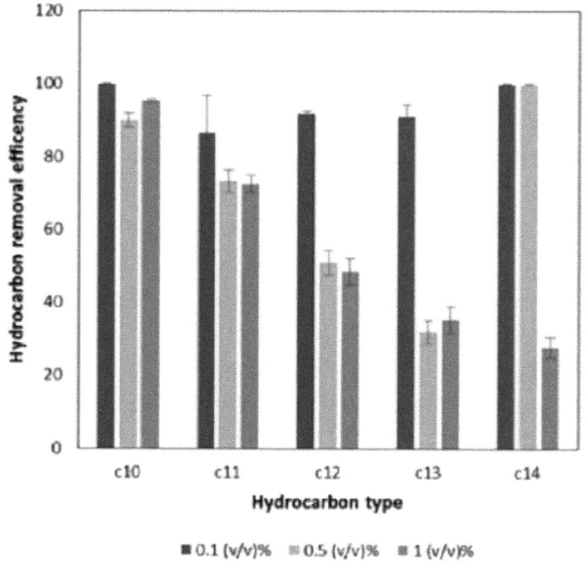

FIGURE 11.3 Removal percentage of different n-alkanes (containing C10–C14 alkanes) by Dunaliella tertiolecta for algal suspensions with 0.1, 0.5, and 1% (v/v) normal paraffines. The values are mean ± SD of three replicates. Reprinted with permission from Salehi et al. (2019).

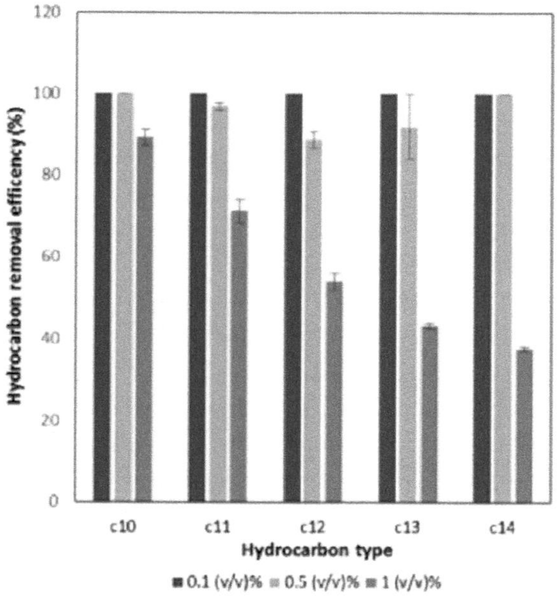

FIGURE 11.4 Removal percentage of n-alkanes (containing C10–C14 alkanes) by *Chlorella vulgaris* for algal suspensions with 0.1, 0.5, and 1% (v/v) normal paraffines. The values are mean ± SD of three replicates. Reprinted with permission from Salehi et al. (2019).

Chemical and biological corrosion in water pipes, which ruins expensive machinery and drives up maintenance costs, are the most prevalent problem in the industry. The brown alga *Sargassum muticum* extract was successfully used in the current work to form the nanocomposite of ZnO and CuO nanoparticles, which was then dispersed in polyethylene oxide polymer. Gamma radiation was used to shrink the produced nanoparticles` particle size. The effectiveness of the produced nanocomposites as anti-biofilm and anti-adherence agents was assessed after they had been characterized by UV–Vis, XRD, DLS, TEM, and FTIR analysis. The outcomes showed that ZnO and CuO nanoparticles' particle sizes were decreased by gamma radiation to 15 and 20 nm, respectively. Mirabilis induced the highest bio-corrosion rate of all the studied strains; 5 mpy on a coupon made of mild steel, and scanning electron microscope results showed that nanocomposite hindered its adherence (SEM). The research findings showed that the green nanocomposite was multifunctional and could be employed as a bio-corrosion inhibitor, an anti-biofilm, and anti-corrosion material. Sulfate-reducing bacteria (SRB) are more likely to proliferate in an open-air environment when a biofilm made by aerobic Pseudomonas aeruginosa is present. Due to the presence of numerous phenolic chemicals, steroids, fucoidans, terpenoids, and flavonoids that serve as reducing and stabilizing agents, brown algae called *Sargassum muticum* is employed for the green synthesis of nanoparticles. Natural reducing, capping, and stabilizing chemicals are used in this green synthesis process to cap produced nanoparticles. The nanocomposite was made utilizing a one-step green process, and DLS was used to gauge the particle size. The nanocomposite was transparent, included spherical nanoparticles, and was compatible with the DLS results, according to TEM images. This study successfully created a ZnO/CuO nanocomposite polymer that is both affordable and environmentally benign. By using EDX analysis, it was determined that the metal nanocomposite polymer had adhered to the metal surface and created a shield that prevented corrosion (Sadek et al., 2019).

The aqueous extract of *Sargassum vulgare* was served as a reducing and capping agent for the creation of ZnO-NPs. In order to characterize the biosynthesized ZnO-NPs, FTIR was recorded on an FTIR spectrometer, XRD was carried out on a Bruker high-resolution X-ray diffractometer system, and FESEM and EDS were acquired using model MIRA3 TESCAN-XMU. The functional groups of bioactive SVE compounds that are probably in charge of producing ZnO-NPs were looked into utilizing FT-IR analysis. The shape, size, and locality of the NPs were investigated using SEM and TEM. Utilizing X-ray diffraction, the crystalline phase of biosynthesized

ZnO-NPs was studied. ZnO-NPs were found to have a hydrodynamic diameter of about 200 nm and +12.7 mV according to the DLS and Zeta potential. The photocatalytic activity of the biosynthesized ZnO NPs was assessed using the degradation of methylene blue dye under UV light irradiation. The photocatalytic capacity of ZnO NPs was evaluated by comparing the areas under the curve of the spectra obtained from the un-reacted and reacted combinations. According to the research, the first 45 min of radiation caused around 50% of the MB dye to deteriorate. The reaction rate then decreased, and the leftover dye was broken down over the course of the next 75 min. Additionally, *Candida*, *Aspergillus* and *Saccharomyces cerevisiae* were all successfully eradicated by the ZnO-NPs' broad-spectrum antifungal effect. The antioxidant capacity of ZnO-NPs was found to be higher than that of SVE but around 65% lower than that of ascorbic acid and BHA. ZnO-NPs were discovered to be cytotoxic to the HepG2 cell line, with an estimated IC50 value of 128.0 g/mL. Given their strong antifungal, photocatalytic, and antioxidant properties, ZnO-NPs have a wide range of potential uses, including the biological control of pesticide residues and plant diseases (Karkhane et al., 2020).

A novel, straightforward green chemical method was employed to create ZnO NPs, using *Ulva lactuca* seaweed extract as a reducing and capping agent. The ZnO NPs produced by *U. lactuca* were characterized using a variety of techniques, including UV–Vis spectroscopic analysis, FTIR, XRD, SAED, EDAX, and TEM, which gives essential information for understanding the numerous physicochemical nano-features. The examination into the photocatalytic activity of these ZnO NPs used the deterioration of the MB's anionic and cationic character in the presence of strong sunlight. Additionally, the ability of the MB solution to decolorize the generated nanoparticles was roughly 90.4% after 120 min of exposure to direct sunlight. Visual examination was used to track the formation of ZnO NPs. Through the reaction, the solution's color changed from green to a light white hue, demonstrating the creation of ZnO NPs made by U. lactuca (Fig. 11.5a). After 4 h at 150 °C, the U. lactuca-fabricated ZnO NPs reached their UV–Visible absorption spectrum. ZnO nanoparticles made by U. lactuca displayed a distinctive UV absorption spectrum with an absorption wavelength of 325 nm, illuminating the fundamental band gap absorption of ZnO nanocrystals (Fig. 11.5b). Against both Gram-negative and positive bacteria, they demonstrated excellent bactericidal activity. High antibiofilm activity levels were investigated in both low- and high-light settings. Additionally, CLSM was used to assess the impact of a single ZnO NPs treatment on the composition of biofilms in both Gram-positive and Gram-negative bacteria. Additionally, at a dose of 50 g/mL, ZnO NPs completely eradicate all *A. aegypti* fourth instar larvae in under 24 h. Therefore, it is possible to advocate the multifunctional ZnO NPs made by *U. lactuca* as a new potential environmentally benign alternative to the chemical processes currently used for the nanosynthesis of antibiotics and mosquito larvicides (Kumar et al., 2013a).

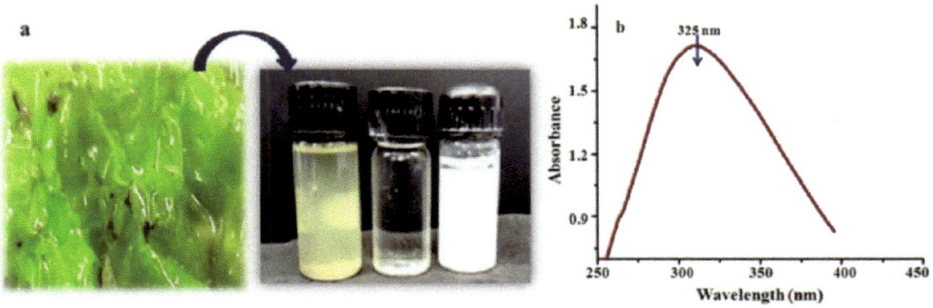

FIGURE 11.5 The green seaweed Ulva lactuca and the extract of seaweed before (left) and after (right) the synthesis of ZnO NPs, the central vial was zinc acetate: (a) UV–Visible absorption spectrum of Ulva lactuca-fabricated ZnO NPs showing a characteristic absorption peak at 325 nm and (b) XRD pattern of U. lactuca-fabricated ZnO NPs. Reprinted with permission from Kumar et al. (2013a).

11.2.2.2 COPPER NANOMATERIALS

Algal-mediated biosynthesis has the potential to reduce the cost of wastewater treatment and generate high-quality biosolids through the ecologically safe combination of cupric ion (Cu^{2+}) bioremediation and valuable copper-based chemicals in wastewater. The ability of a wild-type strain of *Chlamydomonas reinhardtii* to bioremediate free Cu^{2+} in nutrient media that are similar to sewage during copper biosynthesis of copper nanoparticles (Cu-NPs) was examined in this study. EDS and high-resolution TEM revealed spherical, polydisperse, and well-dispersed sub-10 nm Cu-NPs crystallites only in algae that had been heavily treated with copper. Cu-NPs' affinity for carbon-rich materials was also demonstrated by STEM and EDS using 10 mg L of diluted copper medium. The overall conclusion of this work is that Cu^{2+} bioremediation from the Cu-NPs is simultaneously produced by viable green microalga in a complicated nutritional medium resembling wastewater. After being received from CRC, algal cultures were kept for up to six months in a solid medium with a low photon flux of 5–10 mol^{-1} or in low light. A loopful of algae from a solid TAP medium was transferred into 50 mL of liquid

TAP media and cultivated to prepare 150 mL of culture, which was then grown to the late exponential phase in Erlenmeyer flasks at a photon flux density for fresh algal culture. With the addition of 10 mg/L Cu^{2+}, anaerobic TAP-S culture with high H_2 photoproduction underwent therapy and turned from blue to green. Samples including anaerobic TAP-S medium remained colorless and about 30% of Cu was taken away. The TAP-S no algae 10 Cu control sample showed a 5% reduction in CuT after being lit for seven days, demonstrating that Cu can sediment in irradiation media regardless of algae. Brown supernatants were produced by algae in aerobic Cu-polluted TAP-S medium (TAP-S a wt 10 Cu) and TAP medium (TAP a wt 10 Cu) (Fig. 11.6). Brown and green media supernatants were produced after *C. reinhardtii.* Algal elimination of hazardous free Cu^{2+} from contaminated media was demonstrated by cupric ISE treatment (Žvab et al., 2021).

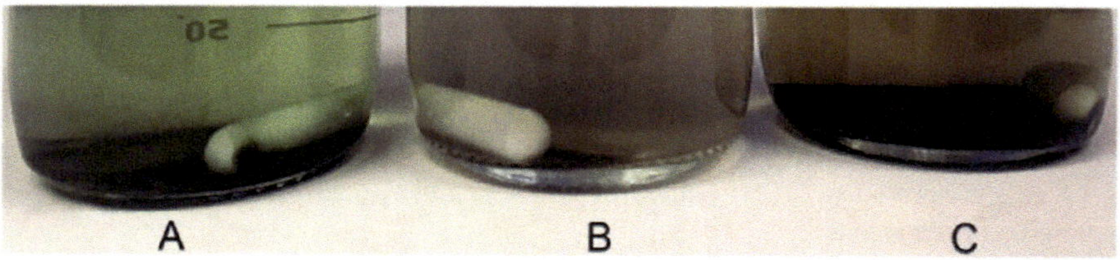

FIGURE 11.6 Colored supernatants of algae-treated Cu-polluted media. Green supernatant from TAP-S an H2 wt 10 Cu (A) and brown supernatants from TAP-S a wt 10 Cu (B) and TAP a wt 10 Cu (C). Reprinted with permission from Žvab et al. (2021).

In this study, Cu-NPs were produced using *Corallina officinalis.* According to the FTIR analyses, the phytochemicals present in the algal extracts, such as proteins, carboxylic acids, and complex polysaccharides, will act as encasing agents and change copper sulphate into Cu-NPs. EDX tests were used to confirm the synthesized Cu-NPs' copper content. The biosynthesized Cu-NPs, which likewise displayed time and concentration-dependent patterns, inhibited the development of *Lyngbya majuscula.* Chlorophyll *a* content of *L. majuscula* significantly decreased as Cu-NP concentrations increased. When Cu-NPs from C. officinalis were utilized, the reduction in chlorophyll *a* was 48% (4 lg/mL). The results of this investigation suggested that the inhibitory effects of Cu-NPs on *L. majuscula* are caused by the bioactive and allelopathic substances present in the two algal extracts spanning the (Cu^{2+} and (Cu^{2+}). In water treatment, Cu-NPs are used as a bactericidal substitute for nano-silver. The remarkable color changes that the extracts of the two *Corallina* species showed when exposed to copper sulphate suggested the creation of Cu-NPs which were characterized by TEM. According to TEM, the biosynthesized Cu-NPs using *C. officinalis* extract had a spherical morphology, were evenly scattered, and their size varied from 11.6 to 14.4 nm with an average of 12.7 nm. The purity and elemental composition of the Cu-NPs colloids had the most detrimental effects on L. majuscula when examined by EDX. At a concentration of 2 lg/mL, it reduced the dry weight of the algae by 85.4%. Cu-NPs have been produced by biosynthesis using the aqueous extracts of the red seaweed *C. officinalis Linnaeus* and *C. mediterranea*. For *C. officinalis* and *C. mediterranea*, the particles' typical diameters were 12.7 and 13.6 nm, respectively. Copper sulphate was changed into Cu-NPs by proteins, carboxylic acids, and complex polysaccharides acting as encapsulating agents. Copper-NPs made by *C. officinalis* have showed excellent harmful effects on *L. majuscula* in a dose-dependent manner. The findings also showed that as Cu-NP concentrations rose, *L. majuscula's* chlorophyll *a* content dropped significantly (El-Kassas and Okbah, 2017).

Surface water contamination is being caused by bacterial infestations from home and industrial wastes. Copper oxide nanoparticles (CuO NPs) with a particle size of 3.6 nm were created through green synthesis using algal extract. The produced particles demonstrated antibacterial efficacy against *Escherichia coli*, a gram-negative bacterium. In surface water samples, an optimized dose of 50 g/mL of NPs was found to be effective against the same bacteria. There are several different ways to treat wastewater produced by the paper and pulp industry, with the activated sludge process (ASP) being the most popular. Silver oxide nanoparticles were produced using a green method using *Sargassum longifolium*. The produced particles displayed good antifungal activity. CuO NPs were created using brown algae (*Sargassum polycystum*), and their antibacterial and anticancer properties were tested. *Pseudomonas aeruginosa* and *Aspergillus niger* both showed 96% antibacterial activity. The liquid was cooled, then passed through Whatman No. 1 filter paper and stored at 4 °C for use in the ensuing environmentally friendly nanoparticle synthesis. At pH 8.2, 1 mM Cu^{2+} concentration, 900 rpm stirring, and 5 g/100 mL algal extract were used to characterize the generated nanoparticles. The surface area of the nanoparticles was 23.4 m^2/g, and their average particle size was 3.6 nm. Customized particle sizes with efficient activity are provided by the environmentally friendly method of producing metal oxide nanoparticles. Green synthesis using algae offered a non-toxic and affordable way to create rod shaped nanoparticles engulfed in algal matrix. At a MIC value of 40 g/L, the nanoparticles showed antibacterial efficacy against *Escherichia coli*. Further toxicity investigations on human and animal cell lines have shown that CuO NPs are not hazardous to either type of cell up to a concentration of 2.5% (25,000 g/mL). The dose of CuO NPs suggested to have an antibacterial impact, that is, 40 g/mL, was therefore assumed to be non-toxic. As a result, the produced nanoparticles can be added to contaminated surface water at the necessary dosage to clean it up and make it safe for consumption (Bhattacharya et al., 2019).

11.2.2.3 IRON NANOMATERIALS

Due to their widespread industrial use, organic dyes make up a sizable portion of wastewater effluents around the world. In this study, cationic crystal violet (CV) and anionic methyl orange (MO) dyes were removed from their aqueous solutions using *Spirulina platensis* microalgae-based environmentally friendly iron oxide nanoparticle production (SP-IONPs). Overall, the findings demonstrated that the SP-IONPs sorbent was incredibly effective at removing significant dyes from aqueous medium. SP-IONPs have proven to be an excellent material for extracting color from actual dyeing effluents. The current study clarifies how the toxic dyes MO and CV can be controlled using iron oxide nanoparticles produced using green synthesis. The findings of the experiments showed that, when used under ideal circumstances, SP-IONPs had sorption efficiencies toward CV and MO dyes that were greater than 95%. The sorption data of CV and MO dyes onto SP-IONPs were fitted using a variety of kinetics and isotherm hypotheses. According to the modelling results of kinetics studies, it was revealed that the sorption of both CV and MO dyes was compatible with the pseudo-second-order model (Shalaby et al., 2021).

It was observed under TEM studies that the microalgae *Chlorella vulgaris* underwent morphological alterations, and the effects of generated nickel oxide nanoparticles on algal growth inhibition were evaluated. The valence state of nickel as well as the quantity of elements Ni and NiO inside algal cells were evaluated using samples of the treated algae and XRD and XPS analysis. The 32.28 mg NiO 72 h EC50 results proved that the NiO nanoparticles had a negative impact on the algae. NiO nanoparticle stress resulted in plasmolysis, cytomembrane rupture, and disruption of the thylakoid membrane in C. vulgaris cells. When NiO NPs were introduced to seawater, about 0.14% of ionic Ni was liberated. Aggregate adherence to algal cell surfaces and the presence of released ionic Ni most likely caused the negative effects. It's interesting to note that research using XRD and XPS showed that some NiO nanoparticles were changed into zero-valence nickel. The maximum ratios of nickel reduction were attained at 72 h after exposure, which suggests that some proteins from algae may be involved in the process, according to the time course of changes in the soluble protein content of treated *C. vulgaris*. The toxicity and bioavailability of NiO nanoparticles to marine algae are reduced by NiO aggregation and reduction. Marine algae therefore have the potential to be utilized in the bio-remediation of nano-pollution in aquatic systems (Gong et al., 2011).

Fe_3O_4-NPs were produced using a completely green biosynthetic method by reducing ferric chloride solution with brown seaweed water extracts. In this investigation, the two seaweeds *Sargassum acinarium Setchell* and *Padina pavonica Thivy* were employed. The reduction of $FeCl_3$ by the algal extract led to the phytosynthesis of Fe_3O_4-NPs. Iron nanoparticles are characterized by UV–Vis analysis to confirm the bio-synthesis of Fe_3O_4-NPs. Using TEM, the appearance and size of the phytogenic Fe_3O_4-NPs were examined. Using an X-ray micro-analyzer, the EDX spectrum was used to characterize the structure of Fe_3O_4-NPs. The biosynthesized Fe_3O_4-NPs were used in Pb adsorption tests after being trapped in calcium alginates beads. *S. acinarium* had a capacity of (78%) after 75 min., but *P. pavonica (Linnaeus) Thivy's* used for the biosynthesis of Fe_3O_4-NPs alginate beads shown a high capacity for bioremoval of Pb (91%). The values of the process variables for the maximum Pb removal effectiveness were computed (Hala et al., n.d.).

The removal of cationic crystal violet (CV) dye using ultrasonic waves is subsequently accomplished by employing green iron oxide nanoparticles (S-IONPs) produced from *spirulina*. Microalgae in powdered form, methanol, ethanol, and salts of CV dye were the chemicals employed for synthesis. Deionized water was used to create various solutions. Diluted HCl and NaOH were used to adjust the pH of all dye solutions and guarantee the even coloring. NPs in the algal extract solution underwent FTIS analysis to determine the effective groups of the active components. XRD was used to analyze the nano-crystalline adsorbent's makeup and phase. UV-Vis spectroscopy is used to analyze produced nano adsorbents in order to observe their optical properties. The iron oxide nano-adsorbent with Spirulina-mediated and CV dye-loaded FT-IR spectrum. By acting as nano-adsorbents, the NPs were able to remove toxic coloring from wastewater. Important details about the S-IONPs' absorption efficiency toward CV dye were also provided by the kinetics and adsorption isotherm. The ability of S-IONPs to adsorb the harmful dyes has been examined using various parameters. The dye removal efficacy of S-IONPs was demonstrated to be most effective at pH 12, 30 mg of nano-adsorbent, and 35 °C (Bhukal et al., 2022).

In this work, harmful algae and excess nutrients like nitrogen and phosphorus that impair the marine ecology are controlled using biosynthesized iron oxide nanoparticles which were produced using aqueous extracts of the brown seaweeds *Colpomenia sinuosa, Petalonia fascia,* and *Padina pavonica.* The extracts were utilized as ferric chloride reductants, and dark brown colors caused by the production of Fe_3O_4-NPs were seen. This suggests that the proteins in these extracts reduced $FeCl_3$, while the aromatic compounds stabilized the biogenic Fe_3O_4-NPs. An alternate, secure bioremediation method for wastewaters could involve applying nanoparticles using seaweed extracts. Fe_3O_4-NPs are now being used to reduce nitrogen and phosphorus and to prevent hazardous algae from blooming because there hasn't been much research on this subject. This study tracked optical density and Chl *a* concentration in the examined genuine seawater samples to report on the anti-algal effects of Fe_3O_4-NPs. In situ or ex situ nanoparticles hold a special potential for lengthy clean-up operations to restore such damaged environments with decreased cost and few unfavorable consequences. One of the main biological techniques for cleaning up polluted water is phytoremediation. In this study, Fe_3O_4-NPs were generated efficiently and sustainably using ferric chloride solution as an iron precursor and the aqueous extract of brown algae (*P. pavonica*) as reducing and stabilizing agents. The reduction of metal ions was done through the biosynthesis of Fe_3O_4-NPs utilizing the readily accessible aqueous extracts of brown algae. Algal-produced Fe_3O_4-NPs showed significant potential for dose-dependently lowering nitrogen and phosphorus levels simultaneously. The obtained results showed that there is a significant capacity for bioremoval of TN and TP using 4 mg/mL of Fe_3O_4-NPs created by *P. pavonica* extracts (El-Sheekh et al., 2021).

11.3 CONCLUSION

The developing discipline of nanoscience has been acknowledged as a gift for the improvement of study on a global scale. Despite the numerous studies conducted over the past ten years, it is still completely unknown what the biotechnological potential of green-synthesized nanoparticles is. Microalgae are the most typical organisms employed in environmental nanoremediation. Utilizing algae as a quick-growing, renewable bioresource, biocompatible silver, zinc, iron, and Cu-NPs were successfully produced using a straightforward, economical green method. The use of nanoparticles as adsorbents to remove heavy metals holds great promise. Due to their unique qualities, such as high selectivity and adsorption capacity, they are effective sorbents and frequently used to remove heavy metals from wastewater. CuO and ZnO nanoparticles have been found to be effective at removing metals like Cu^{2+} from wastewater. The generation of very pure and evenly dispersed Ag, CuO, ZnO, and FeO NPs is characterized and confirmed by visual color change, UV–Vis, TEM, SEM, XRD, and FTIR techniques. The clear zone of inhibition against both Gram-positive and Gram-negative bacteria and fungi revealed in this investigation served as evidence of AGNPs' antibacterial efficacy. AgNPs can also be used as a coating material over different surfaces to control micro-fouling. AgNPs are also demonstrated to be efficient for photocatalysis of methylene blue and methyl orange dye are used to degrade water-soluble contaminants. Thus, the current nanomaterial production methods are looking for safer, nontoxic, ecologically friendly options. The use of alga in the production of AgNPs could be extended to large-scale synthesis, as well as discover other metallic nanoparticles.

KEYWORDS

- **algae-based nanomaterials**
- **algae-based silver nanoparticles**
- **phycosynthesis of nanomaterials**
- **algae-based zinc nanoparticles**
- **algae-based cooper nanoparticles**
- **algae-based iron nanoparticles**

REFERENCES

Aboelfetoh, E. F.; El-Shenody, R. A.; Ghobara, M. M. Eco-Friendly Synthesis of Silver Nanoparticles Using Green Algae (*Caulerpa serrulata*): Reaction Optimization, Catalytic and Antibacterial Activities. *Environ. Monitor. Assess.* **2017,** *189* (7). https://doi.org/10.1007/s10661-017-6033-0

Almansoori, L.; Nazir, A. *Nanomaterials and Pollution Control*; 2021; pp 309–323. https://doi.org/10.4018/978-1-7998-8251-0.ch012

Balaraman, P.; Balasubramanian, B.; Kaliannan, D.; Durai, M.; Kamyab, H.; Park, S.; Chelliapan, S.; Lee, C. T.; Maluventhen, V.; Maruthupandian, A. Phyco-Synthesis of Silver Nanoparticles Mediated from Marine Algae *Sargassum myriocystum* and Its Potential Biological and Environmental Applications. *Waste Biomass Valoriz.* **2020,** *11* (10), 5255–5271. https://doi.org/10.1007/s12649-020-01083-5

Bayda, S.; Adeel, M.; Tuccinardi, T.; Cordani, M.; Rizzolio, F. The History of Nanoscience and Nanotechnology: From Chemical-Physical Applications to Nanomedicine. *Molecules* **2020,** *25* (1). https://doi.org/10.3390/molecules25010112

Bhattacharya, P.; Swarnakar, S.; Ghosh, S.; Majumdar, S.; Banerjee, S. Disinfection of Drinking Water via Algae Mediated Green Synthesized Copper Oxide Nanoparticles and Its Toxicity Evaluation. *J. Environ. Chem. Eng.* **2019,** *7* (1). https://doi.org/10.1016/j.jece.2018.102867

Bhukal, S.; Sharma, A.; Rishi, Divya, Kumar, S.; Deepak, B.; Pal, K.; Mona, S. Spirulina Based Iron Oxide Nanoparticles for Adsorptive Removal of Crystal Violet Dye. *Topics Catal.* **2020**. https://doi.org/10.1007/s11244-022-01640-3

Borah, D.; Das, N.; Das, N.; Bhattacharjee, A.; Sarmah, P.; Ghosh, K.; Chandel, M.; Rout, J.; Pandey, P.; Ghosh, N. N.; Bhattacharjee, C. R. Alga-Mediated Facile Green Synthesis of Silver Nanoparticles: Photophysical, Catalytic and Antibacterial Activity. *Appl. Organometal. Chem.* **2020,** *34* (5). https://doi.org/10.1002/aoc.5597

Brar, S. K.; Verma, M.; Surampalli, R. Y.; Misra, K.; Tyagi, R. D.; Meunier, N.; Blais, J. F. Bioremediation of Hazardous Wastes—A Review. *Pract. Period. Hazard. Toxic Radioactive Waste Manag.* **2006,** *10* (2), 59–72. https://doi.org/10.1061/ (ASCE)1090-025X (2006)10:2 (59)

de Oliveira, P. F. M.; Torresi, R. M.; Emmerling, F.; Camargo, P. H. C. Challenges and Opportunities in the Bottom-Up Mechanochemical Synthesis of Noble Metal Nanoparticles. *J. Mater. Chem. A* **2020,** *8* (32), 16114–16141. https://doi.org/10.1039/d0ta05183g

Dixit, D.; Gangadharan, D.; Popat, K. M.; Reddy, C. R. K.; Trivedi, M.; Gadhavi, D. K. Synthesis, Characterization and Application of Green Seaweed Mediated Silver Nanoparticles (AgNPs) as Antibacterial Agents for Water Disinfection. *Water Sci. Technol.* **2018,** *78* (1), 235–246. https://doi.org/10.2166/wst.2018.292

El-Kassas, H. Y.; Okbah, M. A. E. A. Phytotoxic Effects of Seaweed Mediated Copper Nanoparticles Against the Harmful Alga: Lyngbya Majuscula. *J. Genet. Eng. Biotechnol.* **2017,** *15* (1), 41–48. https://doi.org/10.1016/j.jgeb.2017.01.002

El-Sheekh, M. M.; El-Kassas, H. Y.; Shams El-Din, N. G.; Eissa, D. I.; El-Sherbiny, B. A. Green Synthesis, Characterization Applications of Iron Oxide Nanoparticles for Antialgal and Wastewater Bioremediation Using Three Brown Algae. *Int. J. Phytoremed.* **2021,** *23* (14), 1538–1552. https://doi.org/10.1080/15226514.2021.1915957

Gong, N.; Shao, K.; Feng, W.; Lin, Z.; Liang, C.; Sun, Y. Biotoxicity of Nickel Oxide Nanoparticles and Bio-Remediation by Microalgae *Chlorella vulgaris*. *Chemosphere* **2011,** *83* (4), 510–516. https://doi.org/10.1016/j.chemosphere.2010.12.059

Goswami, R. K.; Agrawal, K.; Shah, M. P.; Verma, P. Bioremediation of Heavy Metals from Wastewater: A Current Perspective on Microalgae-Based Future. *Lett. Appl. Microbiol.* **2021**. https://doi.org/10.1111/lam.13564

Govindaraju, K.; Kiruthiga, V.; Kumar, V. G.; Singaravelu, G. Extracellular Synthesis of Silver Nanoparticles by a Marine Alga, *Sargassum wightii grevilli* and Their Antibacterial Effects. *J. Nanosci. Nanotechnol.* **2009,** *9* (9), 5497–5501. https://doi.org/10.1166/jnn.2009.1199

Hala, E.-K.; Aly-Eldeen, M. A.; Gharib, S. M. Green Synthesis of Iron Oxide (Fe3O4) Nanoparticles Using Two Selected Brown Seaweeds: Characterization and Application for Lead Bioremediation (n.d.). https://doi.org/10.1007/s13131-016

Islam, M. A.; Ali, I.; Karim, S. M. A.; Hossain Firoz, M. S.; Chowdhury, A. N.; Morton, D. W.; Angove, M. J. Removal of Dye from Polluted Water Using Novel Nano Manganese Oxide-Based Materials. *J. Water Process Eng.* **2019,** *32*. https://doi.org/10.1016/j.jwpe.2019.100911

Ismail, G. A.; Allam, N. G.; El-Gemizy, W. M.; Salem, M. A. The Role of Silver Nanoparticles Biosynthesized by *Anabaena variabilis* and *Spirulina platensis* Cyanobacteria for Malachite Green Removal from Wastewater. *Environ. Technol.* **2021,** *42* (28), 4475–4489. https://doi.org/10.1080/09593330.2020.1766576

Jeffryes, C.; Agathos, S. N.; Rorrer, G. Biogenic Nanomaterials from Photosynthetic Microorganisms. *Curr. Opin. Biotechnol.* **2015,** *33*, 23–31. https://doi.org/10.1016/j.copbio.2014.10.005

Karkhane, M.; Lashgarian, H. E.; Mirzaei, S. Z.; Ghaffarizadeh, A.; cherghipour, K.; Sepahvand, A.; Marzban, A. Antifungal, Antioxidant and Photocatalytic Activities of Zinc Nanoparticles Synthesized by *Sargassum vulgare* Extract. *Biocatal. Agric. Biotechnol.* **2020,** *29*. https://doi.org/10.1016/j.bcab.2020.101791

Khalafi, T.; Buazar, F.; Ghanemi, K. Phycosynthesis and Enhanced Photocatalytic Activity of Zinc Oxide Nanoparticles Toward Organosulfur Pollutants. *Sci. Rep.* **2019,** *9* (1). https://doi.org/10.1038/s41598-019-43368-3

Kim, B.; Song, W. C.; Park, S. Y.; Park, G. Green Synthesis of Silver and Gold Nanoparticles via Sargassum Serratifolium Extract for Catalytic Reduction of Organic Dyes. *Catalysts* **2021,** *11* (3), 1–13. https://doi.org/10.3390/catal11030347

Korbekandi, H.; Iravani, S. Silver Nanoparticles. In: *The Delivery of Nanoparticles*; InTech, 2012. https://doi.org/10.5772/34157

Kumar, P.; Govindaraju, M.; Senthamilselvi, S.; Premkumar, K. Photocatalytic Degradation of Methyl Orange Dye Using Silver (Ag) Nanoparticles Synthesized from Ulva lactuca. *Colloids Surfaces B: Biointerf.* **2013a,** *103*, 658–661. https://doi.org/10.1016/j.colsurfb.2012.11.022

Kumar, P.; Senthamil Selvi, S.; Govindaraju, M. Seaweed-Mediated Biosynthesis of Silver Nanoparticles Using *Gracilaria corticata* for Its Antifungal Activity Against Candida spp. *Appl. Nanosci.* **2013b,** *3* (6), 495–500. https://doi.org/10.1007/s13204-012-0151-3

Mandal, R. P.; Sekh, S.; Sarkar, N. s.; Chattopadhyay, D.; De, S. Algae Mediated Synthesis of Cadmium Sulphide Nanoparticles and Their Application in Bioremediation. *Mater. Res. Express* **2016,** *3* (5). https://doi.org/10.1088/2053-1591/3/5/055007

Manisalidis, I.; Stavropoulou, E.; Stavropoulos, A.; Bezirtzoglou, E. Environmental and Health Impacts of Air Pollution: A Review. *Front. Public Health* **2020,** *8*. https://doi.org/10.3389/fpubh.2020.00014

Pugazhendhi, A.; Prabakar, D.; Jacob, J. M.; Karuppusamy, I.; Saratale, R. G. Synthesis and Characterization of Silver Nanoparticles Using *Gelidium amansii* and Its Antimicrobial Property Against Various Pathogenic Bacteria. *Microb. Pathogen.* **2018,** *114*, 41–45. https://doi.org/10.1016/j.micpath.2017.11.013

Pugazhendhi, A.; Prabhu, R.; Muruganantham, K.; Shanmuganathan, R.; Natarajan, S. Anticancer, Antimicrobial and Photocatalytic Activities of Green Synthesized Magnesium Oxide Nanoparticles (MgONPs) Using Aqueous Extract of Sargassum wightii. *J. Photochem. Photobiol. B: Biol.* **2019,** *190*, 86–97. https://doi.org/10.1016/j.jphotobiol.2018.11.014

Rajamanickam, K.; Sudha, S. S.; Francis, M.; Sowmya, T.; Rengaramanujam, J.; Sivalingam, P.; Prabakar, K. Microalgae associated Brevundimonas sp. MSK 4 as the Nano Particle Synthesizing Unit to Produce Antimicrobial Silver Nanoparticles. *Spectrochim. Acta—Part A: Mol. Biomol. Spectrosc.* **2013,** *113*, 10–14. https://doi.org/10.1016/j.saa.2013.04.083

Sadek, R. F.; Farrag, H. A.; Abdelsalam, S. M.; Keiralla, Z. M. H.; Raafat, A. I.; Araby, E. A Powerful Nanocomposite Polymer Prepared from Metal Oxide Nanoparticles Synthesized via Brown Algae as Anti-Corrosion and Anti-Biofilm. *Frontiers in Materials*, *6*. https://doi.org/10.3389/fmats.2019.00140

Salehi, M.; Biria, D.; Shariati, M.; Farhadian, M. Treatment of Normal Hydrocarbons Contaminated Water by Combined Microalgae—Photocatalytic Nanoparticles System. *J. Environ. Manag.* **2019,** *243*, 116–126. https://doi.org/10.1016/j.jenvman.2019.04.131

Sathiyamoorthy, R.; Rajesh, S.; Patric Raja, D.; Rathi, J. M.; Sahayaraj, K. Biosynthesis of Ag Nanoparticles Using *Ulva fasciata* (Delile) Ethyl Acetate Extract and Its Activity Against *Xanthomonas campestris* pv. Malvacearum, 2012. https://www.researchgate.net/publication/224858829

Shalaby, S. M.; Madkour, F. F.; El-Kassas, H. Y.; Mohamed, A. A.; Elgarahy, A. M. Green Synthesis of Recyclable Iron Oxide Nanoparticles Using *Spirulina platensis* Microalgae for Adsorptive Removal of Cationic and Anionic Dyes. *Environ. Sci. Pollut. Res.* **2021,** *28* (46), 65549–65572.

Singh, J.; Dutta, T.; Kim, K. H.; Rawat, M.; Samddar, P.; Kumar, P. "Green" Synthesis of Metals and Their Oxide Nanoparticles: Applications for Environmental Remediation. *J. Nanobiotechnol.* **2018,** *16* (1). https://doi.org/10.1186/s12951-018-0408-4

Smijs, T. G.; Pavel, S. Titanium Dioxide and Zinc Oxide Nanoparticles in Sunscreens: Focus on Their Safety and Effectiveness. *Nanotechnol. Sci. App.* **2011,** *4* (1), 95–112. https://doi.org/10.2147/nsa.s19419

Vivek, M.; Kumar, S.; Steffi, S.; Sudha, S. Biogenic Silver Nanoparticles by Gelidiella acerosa Extract and their Antifungal Effects, **n.d.,** *3* (3).

Yaqoob, A. A.; Ahmad, H.; Parveen, T.; Ahmad, A.; Oves, M.; Ismail, I. M. I.; Qari, H. A.; Umar, K.; Mohamad Ibrahim, M. N. Recent Advances in Metal Decorated Nanomaterials and Their Various Biological Applications: A Review. *Front. Chem.* **2020,** *8*. https://doi.org/10.3389/fchem.2020.00341

Yousefzadi, M.; Rahimi, Z.; Ghafori, V. The Green Synthesis, Characterization and Antimicrobial Activities of Silver Nanoparticles Synthesized from Green Alga *Enteromorpha flexuosa* (wulfen) J. Agardh. *Mater. Lett.* **2014,** *137*, 1–4. https://doi.org/10.1016/j.matlet.2014.08.110

Žvab, U.; Kukulin, D. S.; Fanetti, M.; Valant, M. Bioremediation of Copper Polluted Wastewater-Like Nutrient Media and Simultaneous Synthesis of Stable Copper Nanoparticles by a Viable Green Alga. *J. Water Process Eng.* **2021,** *42*. https://doi.org/10.1016/j.jwpe.2021.102123

CHAPTER 12

Algae-Based Nanoparticles for Contaminated Environs Nanoremediation

VENKATESH S. JOSHI[1], NITHIN GOWDA[1], SHILPA BOREHALLI MAYEGOWDA[2], and N. G. MANJULA[1]

[1]*School of Basic and Applied Sciences (SBAS), Dayananda Sagar University, Bengaluru, Karnataka, India*

[2]*Department of Psychology, Christ University, Kengeri Campus Mysore Road Kumbalgodu, Kanmanike, Bengaluru, Karnataka, India*

ABSTRACT

Currently, the rapidly growing human interference has increased the percentage of pollutants that include organic and inorganic and this has been threatening the ecosystems. Remediation by conventional physicochemical methods, bioremediation has gained immense acceptance due to their eco-friendly, economical, and sustainable approach. Microbial-based nanoparticles act as facilitators in remediating contaminants by microbial growth and immobilization of remediating agents, by inducing microbial remediating enzymes or enhanced biosurfactants that helps to improve solubility of hydrophobic hydrocarbons to create a conducive milieu for remediation. Algal-NPs can be produced easily using low-cost medium and simple scaling up process which is economically feasible. Silver nanoparticles (AgNPs) and gold nanoparticles (AuNPs) have been synthesized using *Nannochloropsis sps* (NN) and *Chlorella vulgaris* (CV), while, brown seaweeds *Petalonia fascia, Colpomenia sinuosa*, and *Padina pavonica* were used with iron oxide NPs along with their aqueous extracts. These applications have shown to be

Sustainable Nanoremediation: Modern Technologies for a Clean Environment.
Fernanda Maria Policarpo Tonelli, Arpita Roy, and Khalid Rehman Hakeem (Eds.)

promising alternative bioremediating methods that are safe. Algal-based NPs can act as a pollution abatement device that can help to effectively target the pollutants for efficient nanobioremediation and helps to promote environmental clean-up for eliminating heavy metals, dyes, and other organic and inorganic waste from the environment.

12.1 INTRODUCTION

Increased population has led to the modernization and industrialization consequently with degradation of agricultural lands causing damaged repercussions for feasible crop production and environment altogether. In order to achieve better crop protection and production, huge quantities of pesticides were used leading to resistance in pests. The overuse of chemical substances, fertilizers, and pesticides have caused substantial damage to the ecosystem leading to soil, air, and water pollution (Mukhopadhyay, 2014; Kar et al., 2016). Industrialization releases numerous pollutants that include nitrogen oxide, sulphur dioxide, organic compounds (dioxins), carbon monoxide (CO), chlorofluorocarbons (CFCs), various heavy metals, hydrocarbons, and particulate matter as effluents. The sources of air pollution include burning of coal, gas, and oil that has been responsible for the release of hazardous sulphur dioxide and nitrogen into the atmosphere. Water pollution include effluents from fertilizers leakage like pesticides, herbicides, oil spills, and sewage. Over the years, these pollutants have accumulated in large quantities and are one of the leading causes of soil infertility, contamination of water bodies, shrinkage of natural resources, and increased health risks in humans (Khalid et al., 2018) (Table 12.1).

The ill effects of these pollutants on humans and environment have led the researchers for pursuing sustainable solutions toward remediating these heavy metals pollutants from the milieu like air, soil, sediments, and water (ground or surface water) (Jaishankar et al., 2014). There have been different types of methods that has been used to achieve environmental remediation that includes physical remediation, chemical remediation, and biological remediation.

Physical remediation involves methods such as capping, soil washing, soil mixing, soil vapor extraction, land-farming solidification, and evacuation processes; however, increased cost and risk of contamination during shifting make these methods ineffective (Fei et al., 2022). Remediation involving chemical treatment transform the pollutants into less toxic forms

TABLE 12.1 Various Types of Pollutants, Toxicity with Permissible Limits (Fahimirad et al., 2017).

Heavy metals	Sources	Health hazards	Permitted levels in mg/L
As	Fungicides, pesticides	Toxicity, bronchitis, skin diseases	0.02
Cd	Fertilizers, pesticides, welding	Lung cancer, kidney damage, damage to bone marrow, renal disfunction	0.06
Pb	Paint, mining, pesticides, vehicle emission	Developmental defects, encephalopathy in infants, hazardous to kidneys and nervous system	0.1
Mn	Production of ferromanganese, welding	Damage to central nervous system	0.26
Hg	Paper and battery manufacturing units	Tremors, protoplasm toxicity, abortion	0.01
Zn	Metal plating, refineries	Dermatitis, injury to nervous system	15
Cr	Mineral sources	Irritation, injury to nervous system	0.05
Cu	Chemical industries, pesticides	Anaemia, liver and kidney misfunctioning	0.1

but are not absorbed by plants. Techniques used are chemical immobilization, actinide chelators, oxidation, critical fluid extraction, in situ catalyzed peroxide remediation, and photodegradation with certain metal retrieval that includes uranium. However, this method requires periodical treatment and operates on special instruments with skilled personnel. Furthermore, this method affects the physiological process and therefore are not applicable for large-scale remediation (Kaur et al., 2013; Gates and Siegrist, 1994; Ho et al., 1995). These conventional approaches are laborious, time consuming, and are highly expensive. Further, application of chemicals to degrade contaminants have proved to be harmful to the ecosystem. This becomes very challenging and is therefore considered not an effective method of remediation. Bioremediation is a natural or induced process that uses the metabolism of a microbe to remove the pollutants with the help of various techniques like bioventing, bioleaching, bioreactor, composting, rhizofiltration, biostimulation, bioaugmentation, and phytoremediation (Azubuike et al., 2016).

Phyco-remediation or algal-based remediation is an ecologically friendly and low-cost approach in removing contaminants from the environment. Algae belongs to a great group of photosynthetic microbes which comes in all sizes ranging from macroscopic macroalgae (seaweeds) or phytoplankton to microscopic microalgae. Being the primary producers of the aquatic

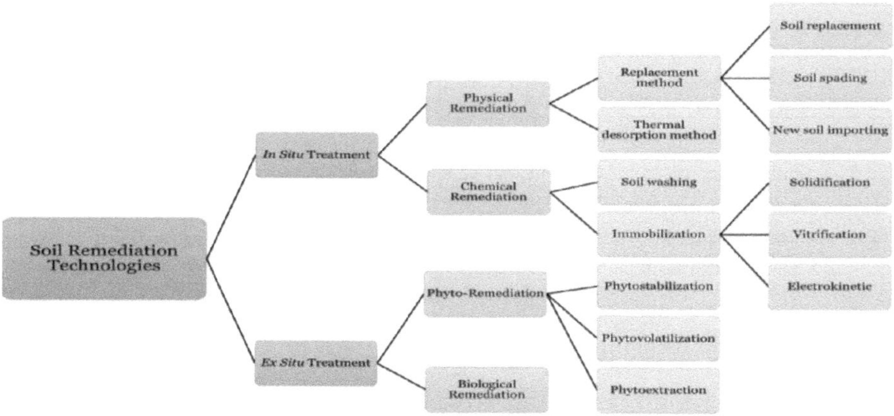

FIGURE 12.1 Different types of remediation involving physio-chemical and biological factors.

ecosystem, their ecological significance has been an utmost importance as they are variedly used as food, medicine, and agriculture. The application of algae for waste treatment dates back to the early civilization where water bodies were found to be useful disposal systems for human and animal wastes that provided nutrients for proliferation of microalgae which in turn supplied oxygen to the rest of the ecosystem (Olguín et al., 2003). However, with recent advancement in the field of science and technology, myco-remediation has been receiving attention for bioremedial applications.

Though bioremediation is an eco-friendly approach, the process is not spontaneous and takes time and often need to test for the appropriate environmental conditions for organism's growth (Pi et al., 2015). Moreover, due to small quantities of highly toxic contaminants, conventional techniques have been rendered useless and the essential methods that can monitor, recognize, and treat even small amounts of pollutants has become very much necessary. The application of nanotechnology in environment remediation have shown promising results to detect and reduce the effect of these atmosphere pollutants. Nanoparticles has been used in the process of redressing remediation known as nanoremediation (Guerra et al., 2018).

However, some nanoparticles which are chemically synthesized are harmful to certain physiological process of plants such as inhibitory effect on seed germination and root growth. In order to tackle this, scientists have turned toward green chemistry which involves biosynthesis of nanoparticles which have been proved to be eco-friendly in the remediation process and is called "nano-bioremediation," that involves the use of living organisms like algae, bacteria, and plants to synthesize nanoparticles and then apply to remove pollutants or contaminants from the ecosystem. This approach is found to reduce the threats posed by the contaminants to the ecosystem and also have been found beneficial to the environment (Pandey, 2018).

Lot of technologies are available for remediation like the biological systems of algal turfs or biomass and adsorption through nanoparticles. Numerous approaches have been used for the NPs synthesis, but algal-based approaches have been substantial and biodegradable. Bioremediating potential algal species has been efficient and successful to detoxify, transform, and volatize the substrates. NP synthesis using algal species has been established as a separate division called as phyco-nanotechnology. Algae has the ability to hyperaccumulate the heavy metals and convert them into pliable shapes. It has been observed that microalgae and macroalgae have tremendous applications in the biosynthesis of nanoparticles (Mukherjee et al., 2021). Nanoparticle-based phyco-remediation have shown promising results in eliminating the contaminants of the aquatic ecosystem that has been studied on microalgal species with silver (AgNPs), gold (AuNPs), iron (FeNPs), and palladium (PdNPs) nanoparticles (Aziz et al., 2015; Prasad et al., 2013; Subramaniyam et al., 2015; Momeni and Nabipour, 2015). For example, phyco-synthesis of Fe_30_4-NPs with green macroalga, *Ulva flexuosa*, has revealed high antibacterial activity but moderate antifungal activity (Mashjoor et al., 2018). In another study, AgNPs synthesised by *Chlorella vulgaris* has reported photocatalytic decolorization activity for degradation of methylene blue dye and also accumulated 1200–3300 μg/g of AgNPs when exposed to Ag salts (Rajkumar et al., 2021; Dhanker et al., 2021).

12.1.1 MICROALGAE IN PHYCO-REMEDIATION

Microalgae consist both prokaryotic cyanobacteria as well as eukaryotic algae. As the name suggests, they are microscopic compared to their macroalgae counterparts and are either unicellular or filamentous photosynthetic microbes (Masojidek and Prasil 2010). Microalgae that have been isolated from natural source, only a few of them are suitable for large-scale commercial use. Some of these microalgae are *Chlorella, Dunaliella, Haematococcus, Arthrospira, Nannochloropsis*, and *Tetraselmis* (Chisti, 2007).

Microalgae have been screened for their intrinsic characteristics before utilizing them for the phyco-remediation and their success rate or waste mitigation depends on factors like algal growth, nutrient efficiency, removal of pollutants, harvesting costs, extreme temperature tolerant, chemical composition, and sedimentation behavior, with a capacity to grow as mixotrophs (Olguin, 2003). Phyco-remediating microalgae can be used either in its free filamentous form or as immobilized cell on a varied array of resources like silica gel, glass beads, alginate, and carrageenan that has revealed a promising result for elimination of contaminants like phosphorous, nitrogen, and various heavy metals (Hameed and Ebrahim, 2007).

High-rate algae ponds (HRAP) is an open system, shallow ponds with paddlewheels and are used for growing microalgae for wastewater treatment. It is a suitable system for culturing microalgae in huge mass as in the tropical regions due to presence of abundant sunlight throughout the year. The action of *C. vulgaris* in HRAP is a boon that has been also used for end polishing the textile wastewater before being discharged in open systems (Lim et al., 2010). However, the capability depends upon the wastewater that has undergone with the varied stage of processing. Example, *Chlorella sps.* removed very low amount of phosphorous that was grown in activated sludge treated wastewater compared to their growth before and after primary settlement (Wang et al., 2010).

Phyco-remediation by microalgae is based on the mechanisms developed by them over the period of time in order to tolerate the concentrated pollutants by absorbing and degrading the pollutants or toxicants. Heavy metals such as cadmium and mercury are detoxified by transforming them into their respective sulphides having low solubility making them unavailable for biological activity. It is well noted that green algae *Chlamydomonas reinhaedtii,* red algae *Cyanidioschyzon merolae*, and cyanobacteria *Synechoccocus leopoliensis*, on exposure to cadmium, helps to detoxify into cadmium sulphide (Edwards et al., 2013). Acidic wastewater is remediated by the action of microalgae which induces alkalinity by nitrate assimilation resulting in the formation of metal hydroxides that has been studied on various fungal species like *Anabaena, Arthrospira, Oscillatoria, Cladophora, Scenedesmus, Chlorella, Spirulina* sp., and *Phaedoactylum tricornutum* that has been cultivated to induce alkalinity in water that was mixed in acid mine drainage (AMD), resulting in alkaline pH facilitating precipitation of metal hydroxide (Bwapwa, 2017). While, pesticides such as organophosphorus insecticides are degraded by microalgae and are utilized as phosphorous source. For instance, methyl parathion was degraded by *Anabaena sps.* and was transformed into O, O-dimethyl O-p nitroephenulthiophosphate further reducing to simpler substance (Megaraj et al., 1994; Barton et al., 2004). The effective bioremediation was found to be high with microalgae in combination with bacteria for pesticides, phenolics, tributyltin, naphthalene, oil, petroleum, and hydrocarbons making them potential bioremediating microbes (Olguín et al., 2003).

12.1.2 MACROALGAE IN PHYCO-REMEDIATION

Macroalgae are multicellular, macroscopic, eukaryotic, and autotrophic, and are commonly referred as seaweeds. They are broadly classified into three groups based on their color as Rhodophyta (red algae), Chlorophyta (green algae), and Ochrophyta–Phaeophyceae (brown algae). They have been used as bioindicator for water quality and in bioremediation by their mechanism of biosorption and bioaccumulation (Leandro et al., 2019). Phosphorus (P) and nitrogen (N) are important elements in agricultural fertilizers and wastewater but are of environmental concerns when washed out into the aquatic environment causing eutrophication. Different types of wastewaters are bioremediated to remove N and P by macroalgae. Most of the macroalgae grows submerged inside the water bodies and absorbs the dissolved nutrients through the entire surface area. *Ulva* (chlorophyta), an opportunistic green macroalgae having foliose morphology, efficiently uptake nutrients and have high growth rates making them to act fast upon favorable conditions (Sode et al., 2013).

12.1.3 NANO-BIOREMEDIATION

A process by which contaminants are converted into harmless products by the action of NPs in combination with biological source. Nanobioremediation involves the use of NPs synthesized using microbes such as bacteria, fungi, algae, etc., wherein several contaminants like heavy metals, organic, and inorganic pollutants are biodegraded (Enamala et al., 2019). These NPs are present widely in nature and some are synthesized artificially (also called as "anthropogenic NPs"). Natural NPs not only occur inside the living system but also in the biosphere and released because of the forest fire, volcanic eruption, weathering of rocks, etc. Nanoparticles are classified on the type of shape and dimension, phase composition, and nature of the materials (Jeevanandam et al., 2018).

Two types of process are used for the NPs synthesis:

TABLE 12.2 Various Microalgae and Macroalgae Used in Bioremediation Processes.

Organisms	Treatment of	Culture system	Pollutants removed	References
		MICROALGAE		
C. vulgaris (along with mixed bacteria)	Municipal wastewater	Photobioreactor	Removal of NH_4^+, phosphorus dissolved organic carbon (DOC)	He et al. (2013)
Chlorella zofingiensis	Pig farm wastewater	Tubular bubble column photobioreactor	Efficient in pollutant removal	Zhu et al. (2013)
C. vulgaris	Textile wastewater	Immobilized in alginate and carrageenan	Removal of dyes	Chu et al. (2009)
Anabaena sp.	Contaminated soil	–	Methyl parathion	Barton et al. (2004)
Arthrospira	Acidic wastewater	–	Heavy metals	Bwapwa, (2017)
Rhizocolonium sp.	Wastewater	–	Cadmium, arsenic	Saunders et al. (2012)
Isochrysis	Palm oil filled effluent (POME), inorganic fertilizer	Photobioreactor and outdoor cultures	NO_3, PO_4	Variappan and Ang, (2007)
		MACROALGAE		
Ulva lactuca	Metal contaminated water	Cd, Pb, Hg	–	Henriques et al. (2017)
Chlorogonium sp.	Spent water resource	NH3-N, NO3(-)-N, PO4(3-)-P	–	Lee et al. (2015)
Laminaria digitata	Aquatic ecosystem	Pesticides (diflubenzuron and lindane), Cd, Cu	–	Ancleto et al. (2017)
Enteromorpha prolifera (biochars)	Contaminated water	Polyclic aromatic hydrocarbons	–	Chen et al. (2020
Chara australis	Soil	Heavy metals and EDTA	–	Clabeaux et al. (2013)
Enteromorpha linza + *Bacillus subtilis*	Agricultural soil	Nitrates	–	Zhang et al. (2021)
Sargassum myriocystum	Wastewater	Dyes such as methylene blue		Balaraman et al. (2022)
Gracilaria lemaneiformis	Coastal water	Bisphenol A, nonylphenol	–	Zhang et al. (2021)

- **Top-down process:** It involves bulk materials that are broken down into nano scale objects. The techniques use microfabrication that uses external controlled tools. It is used to cut, mill, and shape materials into anticipated size and shape. In this method, metallic NPs are synthesized by mechanical milling, laser ablation, sputtering etching, and electro-explosion.
- **Bottom-up process:** It involves the joining and assembly of defined structures of atom-to-atom, molecule-by-molecule, and cluster or self-organizing. The properties of self-assembled single molecules are used to synthesize complex conformation at the nanoscales. Synthesis of NPs is by chemical reduction, laser pyrolysis, molecular condensation, and, utmost significant, by means of green synthesis (Khanna et al., 2019).

However, NPs produced by extreme physico-chemical conditions are potential threat to the environment. To prevent these risks, green synthesis of NPs is preferred, particularly by microbes as biosynthesis and bioprocessing are mostly sustainable, eco-friendly, and cost-effective (Dhanker et al., 2021). Currently, the use of algal species in the green synthesis of nanoparticles has been improved variedly due to their rich contents of proteins, phytochemicals, pigments, etc. They are righteous biofabricators for the synthesis of NPs with other advantages as they can grow fast, are cost-efficient, easy-to-handle, eco-friendly, and capable of hyper-accumulating heavy metals and metals changing them to NPs (AlNadhari et al., 2021).

12.2 ALGAL BIOSYNTHESIS OF NANOPARTICLES

Algae are considered as a perfect model microbe that fabricate bionanomaterials as they can hyperaccumulate heavy metals and modify them into further pliable forms. NPs biosynthesis is introduced by a precursor metal solution along with the extracts of algae-containing proteins, minerals, oil fats, carbohydrates, and PUFA along with the combination of bioactive components (antioxidants and pigments) like carotenoids, chlorophylls, and phycobilin. These compounds act as stabilizing and reducing agents in the process (Khanna et al., 2019).

These agents in the process help to reduce the charge of metal ion to valent zero state in the three-step way that includes: (a) activation phase—the metal ions are reduced and nucleation by algal enzymes takes place which is obvious from the change in color of the solution, (b) the growth phase—wherein the nucleated metals integrate with each other to form thermodynamically different size and shape of NPs that are stable, and (c) termination phase—the NPs are processed, acquiring final shape (Chaudhary et al., 2020).

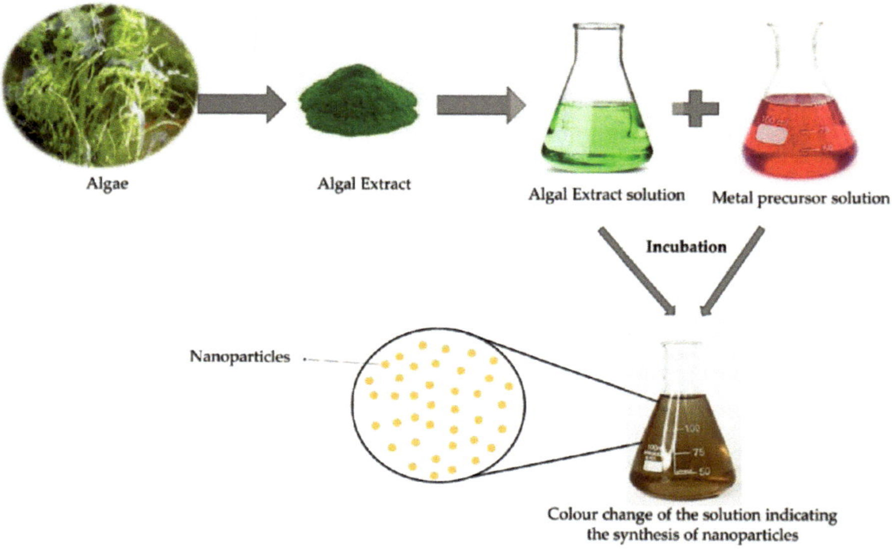

FIGURE 12.2 Biosynthesis of NPs using algal extracts.

12.3 ALGAL SYNTHESIS OF NPS OCCURS VIA TWO ROUTES

1. **Intracellular synthesis:** Biosynthesis of NPs occurs inside the cell of an organism. Algal pre-treatment is not required as the method is dependent involving the metabolic pathway like respiration, photosynthesis, and nitrogen fixation. NADPH- or NADPH-dependent reductase are the reducing agents that generate energy during photosynthesis or respiration by ETS at thylakoid membrane or cell wall respectively. *Rhizoclonium fontinales* and *Ulva intestinalis* change the thallus color from green to purple on treatment with chloroauric acid for 20 °C for 72 hours indicating the formation of AgNPs (Khanna et al., 2019). Algae are grown in a suitable liquid medium and incubated. After incubation, steady conditions are maintained to settle down the biomass. The supernatant is removed and cells are washed, centrifuged, and exposed to metal solutions of different dilutions and are incubated until visible color change is observed followed by purification and recovery process (Dahoumane et al., 2017).

2. **Extracellular synthesis:** Synthesis is outside the cell or at the cell surface, aided from the cell exudations ions, pigments, metabolites enzymes, proteins, etc. For example, the cell surface and cell exudates get attached to the various heavy metal ions that are reduced at the surface making the biosynthesis more convenient because the nanoparticles purification is easy. However, algal biomass being pre-treated is required like washing and blending. Algae is grown inside a suitable medium and after incubation the broth is subjected to centrifugation to obtain supernatant from which the NPs are isolated and transferred to a container having suitable metal ion concentration and incubated until visible color change is observed followed by appropriate purification and recovery process (Dahoumane et al., 2017).

FIGURE 12.3 Color change in algal biomass indicates the biosynthesis of Fe-NPs.

12.4 CHARACTERIZATION OF NANOPARTICLES

NPs are characterized to analyze the shape, structure, size, and various characteristics. Transmission electron microscopy (TEM) helps to denote orientation, exfoliation, and intercalation of NPs with high quality and powerful magnification. Scanning electron microscopy (SEM) is used for surface topography, composition, and 2-D structure (El-Sheekh et al., 2022). UV-Visible spectroscopy gives quick analysis of NPs size, shape, and material composition that are highly accurate and precise (Grand et al., 2019). X-ray diffraction (XRD) understandings on crystallinity, orientation of the crystallite, and phase composition (El-Sheekh et al., 2022). FTIR identifies functional groups which determine the reducing and capping agent responsible for synthesis and stability (Kalishwaralal et al., 2010).

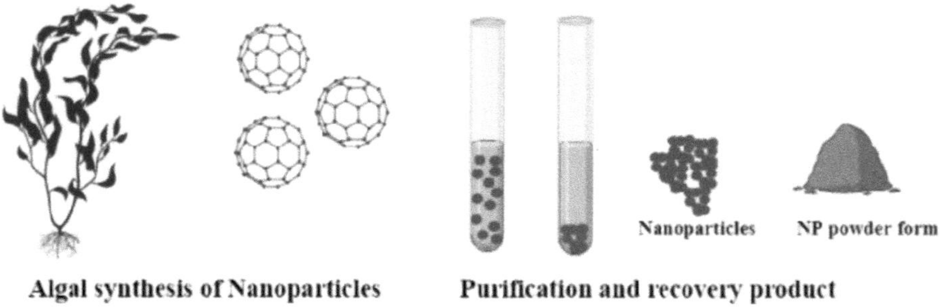

FIGURE 12.4 Flow chart outlining the overview of the mechanism of nano-bioremediation.

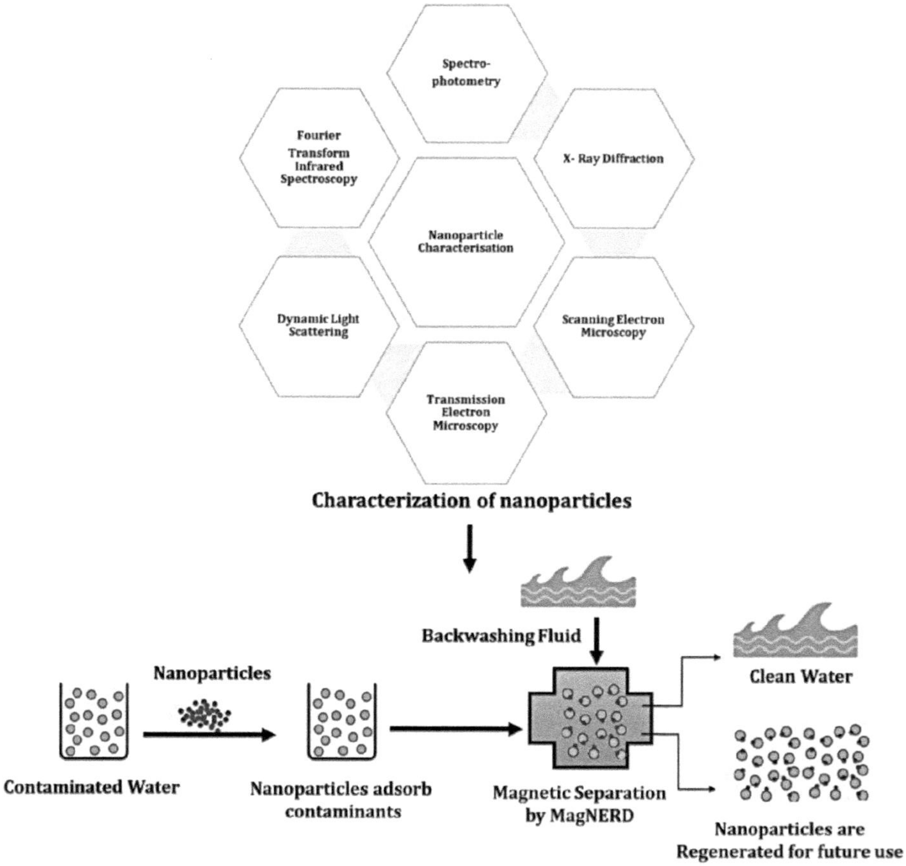

FIGURE 12.5 Mechanism of Algal Nano-Bioremediation.

To have a clear idea on how nano-bioremediation process is carried out, it is necessary to understand the mechanism behind it. There are two major mechanisms, biosorption and bioconcentration, by which microalgae degrade pollutants. Due to their large surface area, microalgae can absorb and metabolize heavy metals and hence play an important role in bioremediation. Bioconcentration accumulates the contaminants from the surrounding ecosystem into the microalgae, whereas, biosorption refers to absorption of toxicants on mainly dead biomass (Rajamani et al., 2007; Mehta and Gaur 2005). Several green algae, such as *Hydrodictyon sps, Oedogonium sps,* and *Rhizoclonium sps* have implemented the above-mentioned mechanisms to remove heavy metals like vanadium and arsenic from wastewater and have successfully treated it (Saunders et al., 2012). There has been effective role of mixotrophs for degradation of pollutants that include the combined action of *Chlorella pyrenoidosa, C. vulgaris*, and *Scenedesmus obliquus* that degraded inside the cells by means of adsorption and further hydrolyzing diclofopmethyl to diclofop (Cai et al., 2007).

The NPs synthesized by algal organisms follow three main routes for bioremediation of environmental contaminants which are adsorption, transformation, and catalysis (Bhatt et al., 2021). Adsorption involves the binding of contaminants on the surface of a nano-adsorbent and trapping the pollutants at its surface. The exothermic reaction involves ionic interaction with the contaminants and the surface of the algal-NPs. Enthalpy is negative due to the restricted movement of the nano-adsorbate during adsorption. At constant temperature and pressure, the adsorption increases spontaneously (Chen, 2012). Transformation involves oxidation and reduction of organic and inorganic pollutants. It reduces metal toxicity resulting in increased metal speciation. It was observed that zero valent iron-nanoparticles (ZVI-NPs) can be used as electron donor that reduce Cr present as contaminants (Singh et al., 2012). Catalysis involves the catalysts that are most commonly used in the production and emission control of CO, NO and hydrocarbons (Chaturvedi et al., 2012). The character of an active catalyst includes an enhancement in its number of sites so that it can expand its surface area along with the decrease of catalytic particle size. Biosynthesized NPs like iron oxide-magnetic NPs act as perfect catalysts in removal of magnetic impurities and heavy metals in the treatment of wastewater (Saif et al., 2016; Brar et al., 2022).

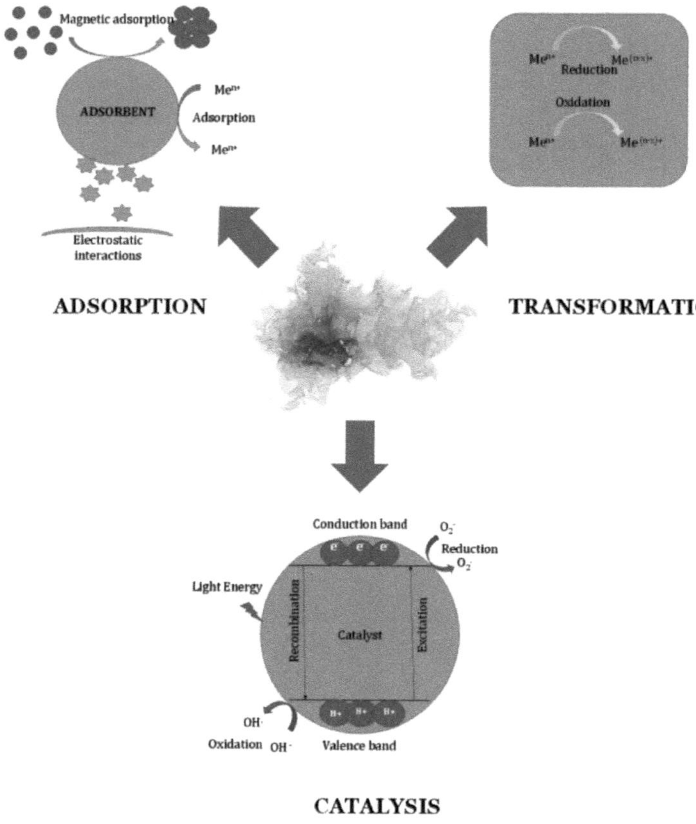

FIGURE 12.6 Phyco-remediation mechanism.

12.5 APPLICATIONS OF ALGAL-NPS FOR BIOREMEDIATION OF CHEMICAL AND BIOLOGICAL POLLUTANTS

12.5.1 HEAVY METAL PHYCO-REMEDIATION

Heavy metals like As, Cd, Cr, Pb, and Hg are carcinogenic that can be treated and used by microalgae species involving strategies against their toxicity by immobilization, gene regulation, antioxidant overexpression, and chelation. Further, Zn, Mg, Mo, Boron, cobalt, iron, and copper have been processed by the algae and used for their cellular components (Leong, et al., 2020). Algae can sequester dissolved metals through a two-step process that include—binding of metals passively to the cell surface and active transportation across the cell membrane and stored in the intracellular storage vacuoles. Once inside the algal cells, excess metal concentration is sequestered by metal-binding phyco-chelatins to form a metal protein complex that are stored in vacuoles to isolate metals from cellular process in an inert and detoxified form (Roberts et al., 2015). The free radicals released during the process are counteracted by the synthesis of catalase, peroxidase, superoxide dismutase (SOD), and glutathione reductase that are antioxidant enzymes and also involves the nonenzymatic antioxidants such as cysteine, proline, carotenoids, ascorbic acid (ASC), and glutathione (GSH). The first line of defence is SOD against the superoxide ions that are degraded to water and hydrogen peroxide, further acted upon by catalase to form oxygen and water molecules (Leong, et al., 2020). *Desmococcus oivaceus* has been used for the treatment of chromium rich sludge that has been waste from the electroplating industry (Sivasubramanian et al., 2010). Heavy metal chromium (IV) has been reduced from concentration by 42–100% using *Chlorella pyrenoidosa, C. vulgaris, Scenedesmus incrassatulus, Chlamydomonas reinhardtii,* and *Scenedesmus obliquus* through chromium reductase and a glutathione molecule (Yen et al., 2017).

Several leather tanning, textile, and metallurgy industries discharge heavy metals as pollutants such as chromium into the aquatic eco-system as a waste product. Chromium being highly toxic contaminants with two forms as Cr (III) and Cr (VI)—of these two, Cr (VI) being extremely toxic with carcinogenic properties (Xia et al., 2021). The ions of Cr (VI) infiltrate the cell membrane generating oxygen reactive radicals that oxidize the biomolecules (Nethaji et al., 2013; Bansal et al., 2009). It can be remediated the action of nanoscale zerovalent iron (nZVI) (Crane and Scott, 2012). These nanoparticles are reported to effectively remove organic, inorganic dyes, chlorinated compounds, and many heavy metals like Cu (II), As (V), Ni (II), Cr (VI) and Pb (II) (Barreto-Rodrigues et al., 2009; Shi et al., 2013; Xiao et al., 2011; Gheju, 2011). Leptolyngbya valderiana was used to biosynthesis nanoscale zerovalent iron utilizing both intracellular and extracellular mechanisms to produce nanosorbents (Banerjee et al., 2021). All three nanosorbent, intra and extracellular nano-iron (NI and NE), respectively, and biomass-loaded nano-iron (BN) were utilized in removal of Cr (VI) heavy metal ions. Among the three nanosorbents, BN showed maximum removal capacity and was the fastest among the three (Azeez et al., 2021).

12.5.1.1 PESTICIDE PHYCO-REMEDIATION

Increasing population has generated agriculture activities to help the need to cater living beings with an increased ground water contamination and surface water systems due to chemical pesticides (García-Galán et al., 2020). Microalgae has been useful in remediating these pollutants as they use them as carbon source (Priyadarshani et al., 2011). Degradation process is a multistep involving the transformation of hydrophilic compounds by reduction or oxidation of hydroxyl groups, hydrogenation and ring cleavage reactions, hydrolysis, and carboxylation with an initial process involving the cytochrome P450 enzymes (Ardal, 2014) followed by glutathione enzyme that protects cell against oxidation processes (Singhal et al., 2021, Rempel et al., 2021). Algae such as Chlorophyta (green algae), Rhodophyta (red algae), and Chromophyta reduce methyl parathion, malathion, etc., that are organophosphate-containing pesticides. However, microalgal-bacterial consortium is found to be more effective as seen in *C. vulgaris, Chlorella sorokiniana,* Coccomyxa subellipsoidea, and *Scenedesmus sps* (Nanda et al., 2021; Nicodemus et al., 2020; Avila et al., 2021).

12.5.1.2 ANTIBIOTIC PHYCO-REMEDIATION

Recent years have seen a drastic increase in the antibiotics with the potential harmful risk in the milieu. With excessive use, it has led to the antibiotic resistant genes in microbial systems (Xiong et al., 2016). Mechanism of algae degrading antibiotics is similar to the pesticide degradation. *Chlorella sps* are found to be highly effective in degrading cephalosporins, cephalexin, amoxicillin, ceftazindin, cephradine, azithromycin, levofloxacin, enrofloxacin, florfenicol, etc., by three-step process. The physiochemical interactions that take between the cell surface and antibiotics are rapid that help to adsorb passively and transfer antibiotic molecules across the cell membrane further followed by bioaccumulation and degradation (Yu et al., 2017).

12.5.1.3 POLYCYCLIC AROMATIC HYDROCARBON (PAH) PHYCO-REMEDIATION

These PAHs are xenobiotics and are carcinogenic in nature (Yu et al., 2013). Algae have been able to metabolize PAHs such as naphthalene, phenanthrene, anthracene, etc. (Ghosal et al., 2016). *Selenstrum capriconutum* has been a potential degradative for benzopyrene (EI-Sheekh et al., 2012). While, the phyco-remediation of fluoranthene (FLT) was confirmed by an enzymatic activity involving *C. vulgaris* and *Ochromonas danica* (Warshawsky et al., 1995).

12.5.1.4 BIOLOGICAL POLLUTANT PHYCO-REMEDIATION

The wastewater contains pathogenic bacteria that causes cholera, typhoid and tuberculosis, the virus including hepatitis and COVID-19, and parasites, too. The photosynthetic microalgae help in removal of these pathogens from the water bodies by increased DO that promotes the production of reactive oxygen species (ROS) that help to remove pathogens by damaging cellular components (Ugya et al., 2020). Algal proliferation decreases the nutrient availability promoting competition between the bacteria creating hypoosmotic shock causing bacterial cell lysis such as *E. coli, Enterococci, Clostridium perfringens*, and *Salmonella enterica* (Cerniglia et al., 1980). *C. vulgaris* and *Synechocystis sp* form toxins microsistin-LR, increasing pH that helps to lyse the pathogens (Chai et al., 2021). *Chlorella sorokiniana, Scenedesmus obliquus*, and *Galdieria sulphuraria* (extremophilic red microalgae) helps in wastewater treatment for eliminating microbial load (Gupta et al., 2016; Delanka-Pedige et al., 2019). Viral inactivation has been recently studied by the use of extracted polysaccharides from Porphyridium sps that has been able to inhibit high load of viruses including HIV 1 and 2 (Chai, et al., 2021; Talyshinsky et al., 2002). Microalgae has been used to eliminate Enterovirus and Norovirus in wastewater treatment (Delanka-Pedige et al., 2020a, 2020b). *Galdieria sulphuraria*, an acidophilic microalga, decreases pH of wastewater and coupled with high temperatures can be used to inactivate virus load (Templeton et al., 2008).

12.6 FUTURE PERSPECTIVES

Increasing pollutants with modernization and urbanization has led to the world's most serious environmental challenge for remediation. Microbes such as micro and macroalgae, and cyanobacteria are sustainable and eco-friendly accepted green technology for the synthesis of NPs for phyco-remediation process. Algae is an excellent source for the nanoparticle synthesis with ironic source containing secondary metabolites that are important source of reducing and capping agents for the process. Nevertheless, its application has been still in the initial stages and needs to be scaled up for the remediation purposes. There needs a knowledge regarding the synthesis mechanism involved with the use of algal-based NPs. Further research on the large-scale build-up of photobioreactors needs to be addressed with issues pertaining to kinetics, cell viability, yield, and physiochemical properties of algal-based

NPs. Future attempts need for more screening of algal strains for phyco-remediation and improvising by genetic engineering methods for algal tolerant against pollutants, accelerate the process, and identification of the various enzymes and proteins involved in the process of phyco-nanotechnology.

12.7 CONCLUSION

Algae-based NPs that are eco-friendly offers nontoxic method that are sustainable to the environment as a clean-up bioremediating process. NPs-phytoremediation has been a great potential for treatment of pollutants from the environment in wastewaters or soil with the traditional processes itself by applying them and being economically feasible. The algal cultivation and application for the reclamation of contaminated environment are a promising approach as they help to reduce eutrophication rates, increase dissolved oxygen concentrations, and generate pioneering products like pigments and enzymes. Algal-based NPs propagates an innovative, advanced technique to engineer and analyze various parameters that have tremendous vision in bioremediation. Further, these algal-NPs can be implemented and utilized on a large scale for commercial applications in remediation. In coming years, algal-based NPs can be useful in eradicating harmful contaminants from the environment to help maintain its sanctity along with cost-reduction, processing, and help to replenish the environment. Moreover, the advancement of algal-based NPs can be explored further for biosensors and nanocomposite applications, too. However, long-term effects of NPs' use need further proper assessment before counterfeiting ahead with commercial-scale implementation to curb future potential environmental effects.

KEYWORDS

- **environment**
- **nanoparticles**
- **algae**
- **heavy metals**
- **pollutantsbioremediation**

REFERENCES

AlNadhari, S.; Al-Enazi, N. M.; Alshehrei, F.; Ameen, F. A Review on Biogenic Synthesis of Metal Nanoparticles Using Marine Algae and Its Applications. *Environ. Res.* **2021,** *194*, 110672. DOI: 10.1016/j.envres.2020.110672.

Anacleto, P.; van den Heuvel, F. H. M.; Oliveira, C.; Rasmussen, R. R.; Fernandes, J. O.; Sloth, J. J.; Barbosa, V.; Alves, R. N.; Marques, A.; Cunha, S. C. Exploration of the Phycoremediation Potential of Laminaria Digitata towards Diflubenzuron, Lindane, Copper and Cadmium in a Multitrophic Pilot-Scale Experiment. *Food Chem. Toxicol.* **2017,** *104*, 95–108. DOI: 10.1016/j.fct.2017.01.030.

Avila, R.; Peris, A.; Eljarrat, E.; Vicent, T.; Blánquez, P. Biodegradation of Hydrophobic Pesticides by Microalgae: Transformation Products and Impact on Algae Biochemical Methane Potential. *Sci. Total Environ.* **2021,** *754*, 142114. DOI: 10.1016/j.scitotenv.2020.142114.

Azeez, N. A.; Dash, S. S.; Gummadi, S. N.; Deepa, V. S. Nano-Remediation of Toxic Heavy Metal Contamination: Hexavalent Chromium [Cr(VI)]. *Chemosphere* **2021,** *266*, 129204. DOI: 10.1016/j.chemosphere.2020.129204.

Aziz, N.; Faraz, M.; Pandey, R.; Shakir, M.; Fatma, T.; Varma, A.; Barman, I.; Prasad, R. Facile Algae-Derived Route to Biogenic Silver Nanoparticles: Synthesis, Antibacterial, and Photocatalytic Properties. *Langmuir* **2015,** *31* (42), 11605–11612. DOI: 10.1021/acs.langmuir.5b03081.

Azubuike, C. C.; Chikere, C. B.; Okpokwasili, G. C. Bioremediation Techniques–Classification Based on Site of Application: Principles, Advantages, Limitations and Prospects. *World J. Microbiol. Biotechnol.* **2016,** *32* (11), 180. DOI: 10.1007/s11274-016-2137-x.

Balaraman, P.; Balasubramanian, B.; Liu, W.-C.; Kaliannan, D.; Durai, M.; Kamyab, H.; Alwetaishi, M. et al. Sargassum Myriocystum-Mediated TiO2-Nanoparticles and Their Antimicrobial, Larvicidal Activities and Enhanced Photocatalytic Degradation of Various Dyes. *Environ. Res.* **2022,** *204*, 112278. DOI: 10.1016/j.envres.2021.112278.

Banerjee, S.; Banerjee, I.; Dutta, M.; Pal, R. Fabrication of Iron Nanoparticles Using Leptolyngbya Valderiana and Investigation of Its Cr (VI) Removal Potential in the Free and Biomass Associated Forms. *Algal Res.* **2021,** *58*, , 102373. DOI: 10.1016/j.algal.2021.102373.

Bansal, M.; Singh, S.; Garg, V. K. A Comparative Study for the Removal of Hexavalent Chromium from Aqueous Solution by Agriculture Wastes' Carbons. *J. Hazard. Mater.* **2009,** *171* (1–3), 83–92. DOI: 10.1016/j.jhazmat.2009.05.124.

Banwell, C. N. *Fundamentals of Molecular Spectroscopy*, 3rd ed.; McGraw-Hill: London ; New York, 1983.

Barreto-Rodrigues, M.; Silva, F. T.; Paiva, T. C. B. Optimization of Brazilian TNT Industry Wastewater Treatment Using Combined Zero-Valent Iron and Fenton Processes. *J. Hazard. Mater.* **2009,** *168* (2–3), 1065–1069. DOI: 10.1016/j.jhazmat.2009.02.172.

Barton, J. W., Kuritz, T.; O'Connor, L. E.; Ma, C. Y.; Maskarinec, M. P.; Davison, B. H. Reductive Transformation of Methyl Parathion by the Cyanobacterium Anabaena Sp. Strain PCC7120. *Appl. Microbiol. Biotechnol.* **2004,** *65* (3), 330–335. DOI: 10.1007/s00253-004-1557-y.

Bhatt, P.; Pandey, S. C.; Joshi, S.; Chaudhary, P.; Pathak, V. M.; Huang, Y.; Wu, X.; Zhou, Z.; Chen, S. Nanobioremediation: A Sustainable Approach for the Removal of Toxic

Pollutants from the Environment. *J. Hazard. Mater.* **2022,** *427*, 128033. DOI: 10.1016/j.jhazmat.2021.128033.

Brar, K. K.; Magdouli, S.; Othmani, A.; Ghanei, J.; Narisetty, V.; Sindhu, R.; Binod, P.; Pugazhendhi, A.; Awasthi, M. K.; Pandey, A. Green Route for Recycling of Low-Cost Waste Resources for the Biosynthesis of Nanoparticles (NPs) and Nanomaterials (NMs)-A Review. *Environ. Res.* **2022,** *207*, 112202. DOI: 10.1016/j.envres.2021.112202.

Bwapwa, J. K.; Jaiyeola, A. T.; Chetty, R. Bioremediation of Acid Mine Drainage Using Algae Strains: A Review. *South Afr. J. Chem. Eng.* **2017,** *24*, 62–70. DOI: 10.1016/j.sajce.2017.06.005.

Cai, X.; Liu, W.; Jin, M.; Lin, K. Relation of Diclofop-Methyl Toxicity and Degradation in Algae Cultures. *Environ. Toxicol. Chem.* **2007,** *26* (5), 970. DOI: 10.1897/06-440R.1.

Cerniglia, C. E., Van Baalen, C.; Gibson, D. T. Metabolism of Naphthalene by the Cyanobacterium Oscillatoria Sp., Strain JCM. *Microbiology* **1980,** *116* (2), 485–494. DOI: 10.1099/00221287-116-2-485.

Chai, W. S.; Tan, W. G.; Munawaroh, H. S. H.; Gupta, V. K.; Ho, S.-H.; Show, P. L. Multifaceted Roles of Microalgae in the Application of Wastewater Biotreatment: A Review. *Environ. Pollut.* **2021,** *269*, 116236. DOI: 10.1016/j.envpol.2020.116236.

Chaturvedi, S.; Dave, P. N.; Shah, N. K. Applications of Nano-Catalyst in New Era. *J. Saudi Chem. Soc.* **2012,** *16* (3), 307–325. DOI: 10.1016/j.jscs.2011.01.015.

Chaudhary, R.; Nawaz, K.; Khan, A. K.; Hano, C.; Abbasi, B. H.; Anjum, S. An Overview of the Algae-Mediated Biosynthesis of Nanoparticles and Their Biomedical Applications. *Biomolecules* **2020,** *10* (11), 1498. DOI: 10.3390/biom10111498.

Chen, L.; He, S.; He, B.-Y.; Wang, T.-J.; Su, C.-L.; Zhang, C.; Jin, Y. Synthesis of Iron-Doped Titanium Oxide Nanoadsorbent and Its Adsorption Characteristics for Fluoride in Drinking Water. *Ind. Eng. Chem. Res.* **2012,** *51* (40), 13150–13156. DOI: 10.1021/ie300102v.

Chen, M.; He, F.; Hu, D.; Bao, C.; Huang, Q. Broadened Operating PH Range for Adsorption/Reduction of Aqueous Cr(VI) Using Biochar from Directly Treated Jute (Corchorus Capsularis L.) Fibers by H3PO4. *Chem. Eng. J.* **2020,** *381*, 122739. DOI: 10.1016/j.cej.2019.122739.

Chisti, Y. Biodiesel from Microalgae. *Biotechnol. Adv.* **2007,** *25* (3), 294–306. DOI: 10.1016/j.biotechadv.2007.02.001.

Chu, W.-L.; See, Y.-S.; Phang, S.-M. Use of Immobilised *Chlorella vulgaris* for the Removal of Colour from Textile Dyes. *J. Appl. Phycol.* **2009,** *21* (6), 641–648. DOI: 10.1007/s10811-008-9396-3.

Clabeaux, B. L.; Navarro, D. A.; Aga, D. S.; Bisson, M. A. Combined Effects of Cadmium and Zinc on Growth, Tolerance, and Metal Accumulation in Chara Australis and Enhanced Phytoextraction Using EDTA. *Ecotoxicol. Environ. Saf.* **2013,** *98*, 236–243. DOI: 10.1016/j.ecoenv.2013.08.014.

Crane, R. A.; Scott, T. B. Nanoscale Zero-Valent Iron: Future Prospects for an Emerging Water Treatment Technology. *J. Hazard. Mater.* **2012,** *211–212*, 112–125. DOI: 10.1016/j.jhazmat.2011.11.073.

Dahoumane, S. A.; Mechouet, M.; Wijesekera, K.; Filipe, C. D. M.; Sicard, C.; Bazylinski, D. A.; Jeffryes, C. Algae-Mediated Biosynthesis of Inorganic Nanomaterials as a Promising Route in Nanobiotechnology—Ā Review. *Green Chem.* **2017,** *19* (3), 552–587. DOI: 10.1039/C6GC02346K.

Delanka-Pedige, H. M. K.; Cheng, X.; Munasinghe-Arachchige, S. P.; Abeysiriwardana-Arachchige, I. S. A.; Xu, J.; Nirmalakhandan, N.; Zhang, Y. Metagenomic Insights into

Virus Removal Performance of an Algal-Based Wastewater Treatment System Utilizing Galdieria Sulphuraria. *Algal Res.* **2020,** *47*, 101865. DOI: 10.1016/j.algal.2020.101865.
Delanka-Pedige, H. M. K.; Munasinghe-Arachchige, S. P.; Cornelius, J.; Henkanatte-Gedera, S. M.; Tchinda, D.; Zhang, Y.; Nirmalakhandan, N. Pathogen Reduction in an Algal-Based Wastewater Treatment System Employing Galdieria Sulphuraria. *Algal Res.* **2019,** *39*, 101423. DOI: 10.1016/j.algal.2019.101423.
Delanka-Pedige, H. M. K.; Munasinghe-Arachchige, S. P.; Zhang, Y.; Nirmalakhandan, N. Bacteria and Virus Reduction in Secondary Treatment: Potential for Minimizing Post Disinfectant Demand. *Water Res.* **2020,** *177*, 115802. DOI: 10.1016/j.watres.2020.115802.
Dhanker, R.; Hussain, T.; Tyagi, P.; Singh, K. J.; Kamble, S. S. The Emerging Trend of Bio-Engineering Approaches for Microbial Nanomaterial Synthesis and Its Applications. *Front. Microbiol.* **2021,** *12*, 638003. DOI: 10.3389/fmicb.2021.638003.
Edwards, C. D.; Beatty, J. C.; Loiselle, J. B. R.; Vlassov, K. V.; Lefebvre, D. D. Aerobic Transformation of Cadmium through Metal Sulfide Biosynthesis in Photosynthetic Microorganisms. *BMC Microbiol.* **2013,** *13* (1), 161. DOI: 10.1186/1471-2180-13-161.
El-Sheekh, M. M.; Ghareib, M. m.; Abou-EL-Souod, G. W. Biodegradation of Phenolic and Polycyclic Aromatic Compounds by Some Algae and Cyanobacteria. *J. Bioremed. Biodegrad.* **2012,** *03* (01). DOI: 10.4172/2155-6199.1000133.
El-Sheekh, M. M.; Morsi, H. H.; Hassan, L. H. S.; Ali, S. S. The Efficient Role of Algae as Green Factories for Nanotechnology and Their Vital Applications. *Microbiol. Res.* **2022,** *263*, 127111. DOI: 10.1016/j.micres.2022.127111.
Enamala, M. K.; Kolapalli, B.; Sruthi, P. D.; Sarkar, S.; Kuppam, C.; Chavali, M. Applications of Nanomaterials and Future Prospects for Nanobionics. In: *Plant Nanobionics*; Prasad, R., Ed.; Springer: Cham, 2019; pp 177–197. DOI: 10.1007/978-3-030-16379-2_6.
Fei, L.; Bilal, M.; Qamar, S. A.; Imran, H. M.; Riasat, A.; Jahangeer, M.; Ghafoor, M.; Ali, N.; Iqbal, H. M. N. Nano-Remediation Technologies for the Sustainable Mitigation of Persistent Organic Pollutants. *Environ. Res.* **2022,** *211*, 113060. DOI: 10.1016/j.envres.2022.113060.
García-Galán, M. J.; Monllor-Alcaraz, L. S.; Postigo, C.; Uggetti, E.; de Alda, M. L.; Díez-Montero, R.; García, J. Microalgae-Based Bioremediation of Water Contaminated by Pesticides in Peri-Urban Agricultural Areas. *Environ. Pollut.* **2020,** *265*, 114579. DOI: 10.1016/j.envpol.2020.114579.
Gates, D. D.; Siegrist, R. L. In Situ Chemical Oxidation of Trichloroethylene Using Hydrogen Peroxide. *J. Environ. Eng.* **1995,** *121* (9), 639–644. DOI: 10.1061/(ASCE)0733-9372(1995)121:9(639).
Gheju, M. Hexavalent Chromium Reduction with Zero-Valent Iron (ZVI) in Aquatic Systems. *Water Air Soil Pollut.* **2011,** *222* (1–4), 103–148. DOI: 10.1007/s11270-011-0812-y.
Ghosal, D.; Ghosh, S.; Dutta, T. K.; Ahn, Y. Current State of Knowledge in Microbial Degradation of Polycyclic Aromatic Hydrocarbons (PAHs): A Review. *Front. Microbiol.* **2016,** *7*. DOI: 10.3389/fmicb.2016.01369.
Grand, J.; Auguié, B.; Le Ru, E. C. Combined Extinction and Absorption UV–Visible Spectroscopy as a Method for Revealing Shape Imperfections of Metallic Nanoparticles. *Analy. Chem.* **2019,** *91* (22), 14639–14648. DOI: 10.1021/acs.analchem.9b03798.
Guerra, F.; Attia, M.; Whitehead, D.; Alexis, F. Nanotechnology for Environmental Remediation: Materials and Applications. *Molecules* **2018,** *23* (7), 1760. DOI: 10.3390/molecules23071760.
Gupta, S. K.; Ansari, F. A.; Shriwastav, A.; Sahoo, N. K.; Rawat, I.; Bux, F. Dual Role of Chlorella Sorokiniana and Scenedesmus Obliquus for Comprehensive Wastewater

Treatment and Biomass Production for Bio-Fuels. *J. Clean. Product.* **2016,** *115*, 255–264. DOI: 10.1016/j.jclepro.2015.12.040.

He, P.J., Mao, B.; Lü, F.; Shao, L. M.; Lee, D. J.; Chang, J. S. The Combined Effect of Bacteria and *Chlorella vulgaris* on the Treatment of Municipal Wastewaters. *Bioresour. Technol.* **2013,** *146*, 562–568. DOI: 10.1016/j.biortech.2013.07.111.

Henriques, B.; Rocha, L. S.; Lopes, C. B.; Figueira, P.; Duarte, A. C.; Vale, C.; Pardal, M. A.; Pereira, E. A Macroalgae-Based Biotechnology for Water Remediation: Simultaneous Removal of Cd, Pb and Hg by Living Ulva Lactuca. *J. Environ. Manag.* **2017,** *191*, 275–289. DOI: 10.1016/j.jenvman.2017.01.035.

Ho, C. L.; Shebl, M. A.-A.; Watts, R. J. Development of an Injection System for In Situ Catalyzed Peroxide Remediation of Contaminated Soil. *Hazard. Waste Hazard. Mater.* **1995,** *12* (1), 15–25. DOI: 10.1089/hwm.1995.12.15.

Jaishankar, M.; Tseten, T.; Anbalagan, N.; Mathew, B. B.; Beeregowda, K. N. Toxicity, Mechanism and Health Effects of Some Heavy Metals. *Interdiscipl. Toxicol.* **2014,** *7* (2), 60–72. DOI: 10.2478/intox-2014-0009.

Jeevanandam, J.; Barhoum, A.; Chan, Y. S.; Dufresne, A.; Danquah, M. K. Review on Nanoparticles and Nanostructured Materials: History, Sources, Toxicity and Regulations. *Beilstein J. Nanotechnol.* **2018,** *9*, 1050–1074. DOI: 10.3762/bjnano.9.98.

Kar, S.; Rathore, V. S.; Champati Ray, P. K.; Sharma, R.; Swain, S. K. Classification of River Water Pollution Using Hyperion Data. *J. Hydrol.* **2016,** *537*, 221–233. DOI: 10.1016/j.jhydrol.2016.03.047.

Kaur, M.; Zhang, H.; Martin, L.; Todd, T.; Qiang, Y. Conjugates of Magnetic Nanoparticle—Actinide Specific Chelator for Radioactive Waste Separation. *Environ. Sci. Technol.* **2013,** *47* (21), 11942–11959. DOI: 10.1021/es402205q.

Khalid, S.; Shahid, M.; Natasha, I. B.; Sarwar, T.; Shah, A.; Niazi, N. A Review of Environmental Contamination and Health Risk Assessment of Wastewater Use for Crop Irrigation with a Focus on Low and High-Income Countries. *Int. J. Environ. Res. Public Health* **2018,** *15* (5), 895. DOI: 10.3390/ijerph15050895.

Khanna, P.; Kaur, A.; Goyal, D. Algae-Based Metallic Nanoparticles: Synthesis, Characterization and Applications. *J. Microbiol. Methods* **2019,** 163 (August 2019): 105656. DOI: 10.1016/j.mimet.2019.105656.

Leandro, A.; Pereira, L.; Gonçalves, A. M. M. Diverse Applications of Marine Macroalgae. *Marine Drugs* **2019,** *18* (1), 17. DOI: 10.3390/md18010017.

Lee, K. Y.; Ng, T. W.; Li, G.; An, T.; Kwan, K. K.; Chan, K. M.; Huang, G.; Yip, H. Y.; Wong, P. K. Simultaneous Nutrient Removal, Optimised CO2 Mitigation and Biofuel Feedstock Production by Chlorogonium Sp. Grown in Secondary Treated Non-Sterile Saline Sewage Effluent. *J. Hazard. Mater.* **2015,** *297*, 241–250. DOI: 10.1016/j.jhazmat.2015.04.075.

Leong, Y. K.; Chang, J.-C. Bioremediation of Heavy Metals Using Microalgae: Recent Advances and Mechanisms. *Bioresour. Technol.* **2020,** *303*, 122886. DOI: 10.1016/j.biortech.2020.122886.

Lim, S.-L.; Chu, W.-L.; Phang, S.-M. Use of *Chlorella vulgaris* for Bioremediation of Textile Wastewater. *Bioresour. Technol.* **2020,** *101* (19), 7314–7322. DOI: 10.1016/j.biortech.2010.04.092.

Mallick,N.[NoTitleFound].*BioMetals***2002,***15*(4),377–390.DOI:10.1023/A:1020238520948.

Mashjoor, S.; Yousefzadi, M.; Zolgharnain, H.; Kamrani, E.; Alishahi, M. Organic and Inorganic Nano-Fe3O4: Alga Ulva Flexuosa-Based Synthesis, Antimicrobial Effects and

Acute Toxicity to Briny Water Rotifer Brachionus Rotundiformis. *Environ. Pollut.* **2018,** *237*, 50–64. DOI: 10.1016/j.envpol.2018.02.036.

Masojídek, J.; Prášil, O. The Development of Microalgal Biotechnology in the Czech Republic. *J. Ind. Microbiol. Biotechnol.* **2010,** *37* (12), 1307–1317. DOI: 10.1007/s10295-010-0802-x.

Megharaj, M.; Madhavi, D. R.; Sreenivasulu, C.; Umamaheswari, A.; Venkateswarlu, K. Biodegradation of Methyl Parathion by Soil Isolates of Microalgae and Cyanobacteria. *Bull. Environ. Contamin. Toxicol.* **1994,** *53* (2). DOI: 10.1007/BF00192047.

Mehta, S. K.; Gaur, J. P. Use of Algae for Removing Heavy Metal Ions from Wastewater: Progress and Prospects. *Crit. Rev. Biotechnol.* **2005,** *25* (3), 113–152. DOI: 10.1080/07388550500248571.

Microalgae Based Sustainable Bioremediation of Water Contaminated by Pesticides. *Biointerf. Res. Appl. Chem.* **2021,** *12* (1), 149–169. DOI: 10.33263/BRIAC121.149169.

Momeni, S.; Nabipour, I. A Simple Green Synthesis of Palladium Nanoparticles with Sargassum Alga and Their Electrocatalytic Activities Towards Hydrogen Peroxide. *Appl. Biochem. Biotechnol.* **2015,** *176* (7), 1937–1949. DOI: 10.1007/s12010-015-1690-3.

Mukherjee, A.; Sarkar, S.; Sasmal, S. A Review of Green Synthesis of Metal Nanoparticles Using Algae. *Front. Microbiol.* **2021,** *12*, 693899. DOI: 10.3389/fmicb.2021.693899.

Mukhopadhyay, S. S. Nanotechnology in Agriculture: Prospects and Constraints. *Nanotechnol. Sci. App.* **2014,** *63*. DOI: 10.2147/NSA.S39409.

Mystrioti, C.; Sparis, D.; Papasiopi, N.; Xenidis, A.; Dermatas, D.; Chrysochoou, M. Assessment of Polyphenol Coated Nano Zero Valent Iron for Hexavalent Chromium Removal from Contaminated Waters. *Bull. Environ. Contamin. Toxicol.* **2015,** *94* (3), 302–307. DOI: 10.1007/s00128-014-1442-z.

Nanda, M.; Kumar, V.; Fatima, N.; Pruthi, V.; Verma, M.; Chauhan, P. K.; Vlaskin, M. S.; Grigorenko, A. V. Detoxification Mechanism of Organophosphorus Pesticide via Carboxylestrase Pathway That Triggers de Novo TAG Biosynthesis in Oleaginous Microalgae. *Aqua. Toxicol.* **2019,** *209*, 49–55. DOI: 10.1016/j.aquatox.2019.01.019.

Nethaji, S.; Sivasamy, A.; Mandal, A. B. Preparation and Characterization of Corn Cob Activated Carbon Coated with Nano-Sized Magnetite Particles for the Removal of Cr(VI). *Bioresour. Technol.* **2013,** *134*, 94–100. DOI: 10.1016/j.biortech.2013.02.012.

Nicodemus, T. J.; DiRusso, C. C.; Wilson, M.; Black, P. N. Reactive Oxygen Species (ROS) Mediated Degradation of Organophosphate Pesticides by the Green Microalgae Coccomyxa Subellipsoidea. *Bioresour. Technol. Rep.* **2020,** *11*, 100461. DOI: 10.1016/j.biteb.2020.100461.

Olguín, E. Phycoremediation: Key Issues for Cost-Effective Nutrient Removal Processes. *Biotechnol. Adv.* **2003,** *22* (1–2), 81–91. DOI: 10.1016/S0734-9750(03)00130-7.

Pandey, G. Prospects of Nanobioremediation in Environmental Cleanup. *Orient. J. Chem.* **2018,** *34* (6), 2838–2850. DOI: 10.13005/ojc/340622.

Pi, Y.; Xu, N.; Bao, M.; Li, Y.; Lv, D.; Sun, P. Bioremediation of the Oil Spill Polluted Marine Intertidal Zone and Its Toxicity Effect on Microalgae. *Environ. Sci. Processes Impacts* **2015,** *17* (4), 877–885. DOI: 10.1039/C5EM00005J.

Prasad, T. N. V. K. V., Kambala, V. S. R.; Naidu, R. Phyconanotechnology: Synthesis of Silver Nanoparticles Using Brown Marine Algae Cystophora Moniliformis and Their Characterisation. *J. Appl. Phycol.* **2013,** *25* (1), 177–182. DOI: 10.1007/s10811-012-9851-z.

Rajamani, S.; Siripornadulsil, S.; Falcao, V.; Torres, M.; Colepicolo, P.; Sayre, R. Phycoremediation of Heavy Metals Using Transgenic Microalgae. In: *Transgenic*

Microalgae as Green Cell Factories; León, R., Galván, A., Fernández, E., Eds., Vol. 616; Springer: New York, 2007; pp 99–109. DOI: 10.1007/978-0-387-75532-8_9.

Rajkumar, R.; Ezhumalai, G.; Gnanadesigan, M. A Green Approach for the Synthesis of Silver Nanoparticles by *Chlorella vulgaris* and Its Application in Photocatalytic Dye Degradation Activity. *Environ. Technol. Innov.* **2021,** *21*, 101282. DOI: 10.1016/j.eti.2020.101282.

Rempel, A.; Gutkoski, J. P.; Nazari, M. T.; Biolchi, G. N.; Cavanhi, V. A. F.; Treichel, H.; Colla, L. M. Current Advances in Microalgae-Based Bioremediation and Other Technologies for Emerging Contaminants Treatment. *Sci. Total Environ.* **2021,** *772*, 144918. DOI: 10.1016/j.scitotenv.2020.144918.

Roberts, D. A.; Paul, N. A.; Bird, M. I.; de Nys, R. Bioremediation for Coal-Fired Power Stations Using Macroalgae. *J. Environ. Manag.* **2015,** *153*, 25–32. DOI: 10.1016/j.jenvman.2015.01.036.

Saif, S.; Tahir, A.; Chen, Y. Green Synthesis of Iron Nanoparticles and Their Environmental Applications and Implications. *Nanomaterials* **2016,** *6* (11), 209. DOI: 10.3390/nano6110209.

Saunders, R. J.; Paul, N. A.; Hu, Y.; de Nys, R. Sustainable Sources of Biomass for Bioremediation of Heavy Metals in Waste Water Derived from Coal-Fired Power Generation. Edited by Joshua L. Heazlewood. *PLoS ONE* **2012,** *7* (5), e36470. DOI: 10.1371/journal.pone.0036470.

Shams ul Hassan, S.; Jin, H.-z.; Abu-Izneid, T.; Rauf, A.; Ishaq, M.; Suleria, H. A. R. Stress-Driven Discovery in the Natural Products: A Gateway towards New Drugs. *Biomed. Pharmacother.* **2019,** *109*, 459–467. DOI: 10.1016/j.biopha.2018.10.173.

Shi, J.; Yi, S.; He, H.; Long, C.; Li, A. Preparation of Nanoscale Zero-Valent Iron Supported on Chelating Resin with Nitrogen Donor Atoms for Simultaneous Reduction of Pb2+ And. *Chem. Eng. J.* **2013,** *230*, 166–171. DOI: 10.1016/j.cej.2013.06.088.

Singh, R.; Misra, V.; Singh, R. P. Removal of Hexavalent Chromium from Contaminated Ground Water Using Zero-Valent Iron Nanoparticles. *Environ. Monitor. Assess.* **2012,** *184* (6), 3643–3651. DOI: 10.1007/s10661-011-2213-5.

Sode, S.; Bruhn, A.; Balsby, T. J. S.; Larsen, M. M.; Gotfredsen, A.; Rasmussen, M. B. Bioremediation of Reject Water from Anaerobically Digested Waste Water Sludge with Macroalgae (Ulva Lactuca, Chlorophyta). *Bioresour. Technol.* **2013,** *146*, 426–435. DOI: 10.1016/j.biortech.2013.06.062.

Subramaniyam, V.; Subashchandrabose, S. R.; Thavamani, P.; Megharaj, M.; Chen, Z.; Naidu, R. Chlorococcum Sp. MM11—a Novel Phyco-Nanofactory for the Synthesis of Iron Nanoparticles. *J. Appl. Phycol.* **2015,** *27* (5), 1861–1869. DOI: 10.1007/s10811-014-0492-2.

Talyshinsky, M. M.; Souprun, Y. Y.; Huleihel, M. M. [No Title Found]. *Cancer Cell Int.* **2002,** *2* (1), 8. DOI: 10.1186/1475-2867-2-8.

Ugya, A. Y.; Imam, T. S.; Li, A.; Ma, J.; Hua, X. Antioxidant Response Mechanism of Freshwater Microalgae Species to Reactive Oxygen Species Production: A Mini Review. *Chem. Ecol.* **2020,** *36* (2), 174–193. DOI: 10.1080/02757540.2019.1688308.

Vairappan, C. S.; Yen, A. M. Palm Oil Mill Effluent (POME) Cultured Marine Microalgae as Supplementary Diet for Rotifer Culture. *J. Appl. Phycol.* **2008,** *20* (5), 603–608. DOI: 10.1007/s10811-007-9305-1.

Wang, L.; Min, M.; Li, Y.; Chen, P.; Chen, Y.; Liu, Y.; Wang, Y.; Ruan, R. Cultivation of Green Algae Chlorella Sp. in Different Wastewaters from Municipal Wastewater Treatment Plant. *Appl. Biochem. Biotechnol.* **2010,** *162* (4), 1174–1186. DOI: 10.1007/s12010-009-8866-7.

Warshawsky, D.; Cody, T.; Radike, M.; Reilman, R.; Schumann, B.; LaDow, K.; Schneider, J. Biotransformation of Benzo[a]Pyrene and Other Polycyclic Aromatic Hydrocarbons and Heterocyclic Analogs by Several Green Algae and Other Algal Species under Gold and White Light. *Chem.-Biol. Interact.* **1995,** *97* (2), 131–148. DOI: 10.1016/0009-2797(95)03610-X.

Xia, X.; Wu, S.; Zhou, Z.; Wang, G. Microbial Cd(II) and Cr(VI) Resistance Mechanisms and Application in Bioremediation. *J. Hazard. Mater.* **2021,** *401*, 123685. DOI: 10.1016/j.jhazmat.2020.123685.

Xiao, S.; Ma, H.; Shen, M.; Wang, S.; Huang, Q.; Shi, X. Excellent Copper(II) Removal Using Zero-Valent Iron Nanoparticle-Immobilized Hybrid Electrospun Polymer Nanofibrous Mats. *Colloids Surf. A: Physicochem. Eng. Aspects* **2011,** *381* (1–3), 48–54. DOI: 10.1016/j.colsurfa.2011.03.005.

Xiong, Q.; Hu, L.-X.; Liu, Y.-S.; Zhao, J.-L.; He, L.-Y.; Ying, G.-G. Microalgae-Based Technology for Antibiotics Removal: From Mechanisms to Application of Innovational Hybrid Systems. *Environ. Int.* **2021,** *155*, 106594. DOI: 10.1016/j.envint.2021.106594.

Yen, H.-W.; Chen, P.-W.; Hsu, C.-Y.; Lee, L. The Use of Autotrophic *Chlorella vulgaris* in Chromium (VI) Reduction under Different Reduction Conditions. *J. Taiwan Inst. Chem. Eng.* **2017,** *74*, 1–6. DOI: 10.1016/j.jtice.2016.08.017.

Yu, F.; Ge, F.; Zhou, W.; Tao, N.; Liang, Z.; Zhu, L. Subcellular Distribution of Fluoranthene in *Chlorella vulgaris* with the Presence of Cetyltrimethylammonium Chloride. *Chemosphere* **2013,** *90* (3), 929–935. DOI: 10.1016/j.chemosphere.2012.06.032.

Yu, Y.; Zhou, Y.; Wang, Z.; Torres, O. L.; Guo, R.; Chen, J. Investigation of the Removal Mechanism of Antibiotic Ceftazidime by Green Algae and Subsequent Microbic Impact Assessment. *Sci. Rep.* **2017,** *7* (1), 4168. DOI: 10.1038/s41598-017-04128-3.

Zhang, C.; Lu, J.; Wu, J. Enhanced Removal of Phenolic Endocrine Disrupting Chemicals from Coastal Waters by Intertidal Macroalgae. *J. Hazard. Mater.* **2021,** *411*, 125105. DOI: 10.1016/j.jhazmat.2021.125105.

Zhang, H.; Chen, X.; Song, L.; Liu, S.; Li, P. The Mechanism by Which Enteromorpha Linza Polysaccharide Promotes Bacillus Subtilis Growth and Nitrate Removal. *Int. J. Biol. Macromol.* **2022,** *209*, 840–849. DOI: 10.1016/j.ijbiomac.2022.04.082.

Zhu, L.; Wang, Z.; Shu, Q.; Takala, J.; Hiltunen, E.; Feng, P.; Yuan, Z. Nutrient Removal and Biodiesel Production by Integration of Freshwater Algae Cultivation with Piggery Wastewater Treatment. *Water Res.* **2013,** *47* (13), 4294–4302. DOI: 10.1016/j.watres.2013.05.004.

CHAPTER 13

Plant-Based Nanomaterials for Contaminated Environs Nanoremediation

AHMED ALBAHNASAWI[1], MOTASEM Y.D. ALAZAIZA[2], ERCAN GÜRBULAK[1], and MURAT EYVAZ[1]

[1]*Department of Environmental Engineering, Gebze Technical University, Gebze-Kocaeli, Turkey*

[2]*Department of Civil and Environmental Engineering, A'Sharqiyah University, Ibra, Oman*

ABSTRACT

In the most recent technological-scientific revolution, nanotechnology has rapidly increased in appeal across various industries and academic fields, particularly environmental remediation. For the advancement of innovative nanotechnology, controlled, simple, affordable, and safe nanomaterial synthesis is looked for. Due to the toxic chemicals used or the byproducts created nanomaterial production methods like chemical synthesis are not safe or ecologically benign. A few nanoparticles are also produced as a byproduct of the washing process; these compounds could build up in the soil, water, and air systems and constitute a significant threat to plants, animals, and bacteria. In contrast, it has been shown that employing plants to synthesize nanomaterials is both cost-effective and environmentally friendly. Compared to many other types, plant-based nanoparticles are less toxic, making them intriguing candidates to prevent widespread harm to the microbiome caused by present methods. In the realm of environmental remediation, the usage of plant-based NPs has shown a variety of benefits and uses. Therefore, this

Sustainable Nanoremediation: Modern Technologies for a Clean Environment.
Fernanda Maria Policarpo Tonelli, Arpita Roy, and Khalid Rehman Hakeem (Eds.)

chapter aims to examine the current state of knowledge on products made from plant-based nanoparticles for environmental cleanup. The broad use of plant-NP interactions in environmental remediation will also be covered.

13.1 INTRODUCTION

Water makes up two-thirds of the earth's surface and is an element that all life forms require. Discharge of various inorganic or organic elements, viruses, heavy metals has varying physical, chemical, or biological qualities that make it detrimental to environment (Bora and Dutta, 2014). Additionally, the water on Earth is a precious and limited resource that is endlessly renewed through the water-cycle. As a result, the planning and management of water resources sustainably is urgently needed given that they are becoming more contaminated by various organic and inorganic contaminants, as well as urban and industrial wastewater discharges.

To deal with the absence of clean water, safe water for all habitats is the global challenge of this era. The scarcity of clean, treated water impacted our society's standard of living and reduced possibilities for everyone to succeed (Qadir et al, 2007). In 2012, the World Health Organization (WHO) reported that 780 million people worldwide still lacked access to a reliable source of safe drinking water (Qadir et al., 2007). Ninety percent of infections in the majority of developing nations are brought on by using contaminated water for drinking (WHO, 2012). The United Nations World Water Development Report (2017) states that during the last five years, the lack of clean water has significantly increased the risk to the human population.

About five hundred million people live in places where water demand is twice as high as treated water supplies in nations like India, Western South America, Australia, China, Western North America, and Central America (Mekonnen and Hoekstra, 2016). Around four billion persons around the world now have access to modern sanitation, while developing nations continue struggle with a shortage of sanitation infrastructure, according to reports from UNICEF and WHO in 2015 (Kjellén et al, 2011). Without a centralized facility to give connections to a water supply, some developing nations rely on their citizens or nongovernmental organization support.

According to Bora and Dutta (2014), wastewater is primarily divided into two categories: domestic wastewater (municipal wastewater) and nondomestic wastewater. Domestic wastewater is a major source of ecosystem contamination in developing countries (urban or industrial wastewater).

Wastewater is consisting of 1% impurities and 99% water, according to a 2015 UN-Water Report (Bora and Dutta, 2014).

Wastewater is thought to include highly harmful organic and inorganic (metals) components for both humans and the environment (Goutam et al., 2018; Goutam et al, 2020). However, the chemical oxygen demand (COD) and biological oxygen demand (BOD), the two primary water metrics, can be used to quantify the organic content of wastewater (Henze and Comeau, 2008). The standard of living and way of life of individuals have an impact on the volume of household wastewater (Chandrappa and Das, 2012). Additionally, industrial, agricultural, and commercial activities are the primary producers of nondomestic or urban wastewater, and their origins find the makeup of the wastewater. The integrated management planning of water sector is currently needed urgently if everyone can have access to a safe and clean water supply. There are currently a wide variety of physical, chemical, and biological treatment methods for water and wastewater remediation; however, these technologies are not always successful and necessitate a large financial investment, specific operational conditions, high maintenance costs, and energy requirements (Baruah and Dutta, 2013; Bora and Dutta, 2014).

The evolution of technologies for next-generation (waste) water treatment, on the other hand, is greatly facilitated by the use of nanotechnology, which has the potential to displace current wastewater treatment methods (Goutam et al., 2020). Due to nanoparticles' (NPs) high reactivity and functionalization, size-dependent characteristics, vast surface area, etc., NPs are more and more used for the purification and remediation of (waste) water (Gawande et al., 2011; Weng, 2020)

Due to their distinctive qualities, such as enhanced adsorption catalysis and catalysis properties and strong reactivity, nanomaterials have recently offered significant benefits in wastewater treatment (Gawande et al., 2011; Guo et al, 2013). Numerous nanoparticles, including metal oxide nanoparticles, zerovalent metal nanoparticles, nanocomposites, and carbon nanotubes, have been successfully used to treat water and wastewater (Hano and Abbasi, 2022).

The photocatalytic treatment of industrial wastewaters has made substantial use of NPs, particularly the chemically produced titanium dioxide (TiO_2). A highly efficient approach for the destruction and detoxification of resistant contaminants from industrial wastewaters is nano-sized TiO_2-based photocatalytic treatment (Goutam et al., 2020; Lu et al., 2016). Under specified conditions, the chemical preparation of NPs involves extremely expensive, poisonous, and dangerous substances, and the discharge of these compounds into the environment raises major ecotoxicological concerns (Kumari et al, 2021). However, the green synthesis, which uses biodegradable components like plant extracts, microorganisms, and enzymes to create NPs, is a straightforward, environmentally friendly, and less harmful method of doing so (Tiwari et al, 2008). However, the most practical method for creating NPs is to use plant extracts because it reduces the likelihood of related contamination, speeds up reaction times, and preserves cell structure (Siddiqi and Husen, 2020; Tiwari et al., 2008). Additionally, using green NPs to clean industrial wastewater could be a terrific way to combat environmental contamination.

Only a small number of research have, to date, documented the use of green synthetic NPs for wastewater treatment (Devi et al, 2021; Elsakhawy et al., 2022). As a result, the focus of this chapter is on the environmentally friendly production of nanomaterials using green, especially from plant extract. In addition, the applications of these nanomaterials in the remediation of (waste) water are discussed. In this chapter, the green preparation method for NPs, as well as their uses in wastewater dye degradation and water and wastewater treatment, are covered. Additionally, difficulties and problems with using green synthesized NPs for the treatment of (waste) water are also covered.

13.2 PLANT-BASED NPS SYNTHESIS AND CHARACTERIZATION

The idea of "Green Chemistry" for "Sustainable Development" has been extensively researched during the past ten years (Hano and Abbasi, 2022). Sustainable development aimed to afford present requirements of resources, and at the same time, to safe the right of the next generation from these resources that satisfy present requirements (Kates et al, 2012). Sustainable growth is crucial for many chemistry-based industries because it raises issues like pollution evidence and indiscriminate resource consumption (Omer, 2008).

The three most crucial issues for the green synthesis of NPs are the choice of a green or ecologically friendly solvent, a safe chemical for stabilization, and a suitable nontoxic reducing agent. In fact, a wide variety of synthetic processes have been employed to create nanoparticles, with the most common ones being physical, chemical, and biosynthetic. Chemical treatments typically cost too much money and require the use of dangerous and poisonous chemicals, which pose a number of ecosystem problems (Nath and Banerjee, 2013).

On the other hand, green synthesis provides a secure, environmentally responsible, and biocompatible green technique of creating NPs for a range of applications, containing biological ones (Razavi et al., 2015). This green synthesis has been performed by fungi, bacteria, algae, and plants. However, it has been frequently used to synthesize various NPs from parts of plants, such as fruits, roots, stems, leaves, and seeds (Narayanan and Sakthivel, 2011).

Indeed, NPs with specific shape, size, and content can be made using plant extracts. Additionally, a large variety of phytochemicals included in their extract may serve as organic stabilizing and/or reducing agents for the formation of NPs. It is widely acknowledged that plant-derived NPs have a lower risk of having negative side effects on people than chemically produced NPs do. They also have a high biological potential and can be used in a variety of fields including agriculture, food science and technology, cosmetics, bioengineering, human health protection, and nanomedicine.

To secure reproducibility in their biological activity, manufacture, and safety, these NPs must be accurately and fully described. To do this, the synthesized NPs are extremely carefully characterized using physicochemical techniques, such as Fourier transform infrared spectroscopy (FTIR), transmission electron microscopy (TEM), scanning electron microscopy (SEM), atomic force microscopy (AFM), field emission scanning electron microscopy (FE-SEM), X-ray diffractometer (XRD), X-ray photoelectron microscopy (XPS), energy dispersion analysis of X-ray (EDAX), thermal gravimetric differentiation, photoluminescence analysis (PL), Raman spectroscopy, transmission electron microscopy (TEM), dynamic light scattering (DLS), UV-visible diffuse reflectance spectroscopy (UV-DRS), emission scanning electron microscopy (FE-SEM), atomic force microscopy (AFM), scanning electron microscopy (SEM field), X-ray diffractometer (XRD), energy dispersion analysis of X-ray (EDAX), thermal gravimetric differential thermal analysis (TG-DTA), and X-ray photoelectron microscopy (XPS) (Khan et al., 2020; Mallikarjuna et al., 2021; Zaeem et al., 2020).

13.2.1 GREEN SYNTHESIS OF TIO2 NP

The secondary metabolites found in plants, such as saponins, alkaloids, flavonoids, terpenoids, steroids, tannins, and polyphenols, are typically found in high concentrations. Many of these metabolites are said to act as stabilizing and reducing agents to stop the agglomeration and aggregation of new metal NPs, according to a number of studies (Aslam et al, 2021; Goutam et al., 2018; Iravani, 2011).

There are typically three processes involved in the production of NPs with the help of metabolites. The activation procedure, in which metal ions are nucleated, is the first stage. Small surrounding NPs will mix to produce larger particles and increase in thermodynamic stability in the following step, which referred to as the growth phase. Through the termination reaction, the final step establishes the forms of the synthesized NPs (Nasrollahzadeh and Sajadi, 2015). Due to its attachment to the highly electronegative oxygen atom, every hydrogen atom in the hydroxyl group first obtains a partial charge. In addition, the titanium atom's hydroxide group functions similarly to how oxygen and hydrogen atoms do when they pick up positive charge and a partial negative, respectively (Aslam et al., 2021).

The titanium complex's oxygen atom, which has lone pair electrons, brings the positively charged hydrogen from the quercetin molecule, resulting in the formation of an intermediary structure. To make a positively charged intermediate called $TiO(OH)_2$, a water molecule is detached by the intermediate compound, leaving a positive charge on the titanium atom. The necessary TiO_2 NPs are created in the following stage by a similar electron transfer method. The generated NPs are then centrifuged, precipitated, and cleaned using the proper solvent to remove contaminants (Lee et al., 2011).

The active metabolites then decrease the metal precipitate to NPs, which are isolated using column chromatography before usage (Nasrollahzadeh and Sajadi, 2015). Polysaccharides, enzymes, organic acids, vitamins, amino acids, and proteins all of which are known to be often present in plant extracts—are bioactive molecules that can be used in the synthesis of NPs (Raut et al, 2014). The main functional groups that can reduce metal ions include those that can electrostatically bind to metal ions, such as hydroxyl, methoxide, carbonyl, and amino groups.

The type and concentration of plant extract and determine the form of the resulting NPs, while pH and temperature are primarily in charge of finding their size and growth. It has been proposed that the shape of the generated NPs can be significantly influenced by the regulation of metal salt content (El-Seedi et al., 2019). As a result, these factors can be changed to improve the synthesis procedure. Further research should be done on a variety of plant extracts that are suitable for the manufacture of TiO_2 NPs.

13.2.1.1 TiO_2 NPS SYNTHESIS FROM PLANT

Plant extracts are often made from the original plants' leaves, roots, blossoms, and stems. Herbs are mostly used to apply them in the medicinal industry. Because the active components in the extract are heat unstable, the extraction procedure typically takes place at a temperature of not more than 100°C (Jeetah et al, 2014).

The type of titanium precursor, reaction time, pH, and extract and precursor concentrations are just a few of the crucial synthesis parameters. Once the desired NPs have been produced, they are characterized using many instrumentation techniques, including XRD, FESEM, TEM, FTIR, and light scattering (LS), among others, to find key physical and chemical characteristics that can be linked to catalytic activity. In various studies, plant blossoms have been employed to make TiO_2 NPs, and the findings are promising (Ahmed et al, 2017).

Freshly obtained leaves (20 g) were used to make aqueous *Psidium guajava* leaf extract. The leaves were first washed with water, then they were submerged in 250 mL of double-distilled water at 60°C for 15 min. The filtered extract was then combined for 24 h at 25°C with 80 mL of $TiO(OH)_2$ to produce green TiO_2 NPs. The synthesis of TiO_2 NPs was aided by the presence of alcohol and primary and aromatic amines in the leaf extract. The XRD results of the produced TiO_2 NPs indicated the existence of both rutile and anatase forms. The synthetic TiO_2 NPs were also suitably crystalline in nature, according to the EDX investigations (Santhoshkumar et al., 2014). Other researchers, in the meantime, created irregularly shaped TiO_2 NPs with a size range of 25–110 nm using *Catharanthus roseus* leaf extract (Velayutham et al., 2012).

It was found that the NPs had size of 65 nm. Following the extraction of the compounds produced by the plant from the leaf, a broth solution was made using the *Catharanthus roseus* leaf extract. After filtering, the extract was mixed with the precursor TiO_2 while being stirred to create TiO_2 NPs. Following that, investigations using FTIR, SEM, XRD, and AFM were conducted to find the precise properties and shape of the produced TiO_2 NPs. Many researchers agree that plant extracts make excellent sources of capping and reducing agents for the biological production of NPs. More examples of leaf extract are *Eclipta prostrata L.*, also referred to as false daisy, which was used in the aqueous reduction process to create TiO_2 NPs. The successful production of TiO_2 NPs was confirmed by a color shift at room temperature in the prepared solution after it had been boiled, filtered, and combined with $TiO(OH)_2$. The created TiO_2 NPs were then put through a variety of tests, including FTIR, XRD, AFM, and FESEM tests. With an average NPs size of 49.5 nm and a size range of 36–68 nm, spherically shaped NPs were produced (Rajakumar et al., 2012). Like this, TiO_2 NPs were made with success using an extract of *Azadirachta indica* leaves. The prepared solution was made as fellows: The plant leaves were gathered, dried, and then extracted under experimental circumstances. Then, 5 mM of titanium isopropoxide solution was added to the aqueous leaf extract to begin a reaction at a pH of 1.5. At 50°C, the solution was agitated until a deep, intense brown color appeared, showing that TiO_2 NPs had successfully formed. The produced TiO_2 NPs were tightly bonded and had spherical shape, with particle size of 124 nm, according to the SEM and DLS studies (Sankar et al, 2015). The literature has not yet developed a full knowledge of the relationship between synthesis conditions and particle size. The production of metal oxide NPs does not just need plant materials like leaves and flowers. Additionally, they can be created using plant root extract. A reducing and stabilizing agent, *Euphorbia heteradena Jaub* root extract, has been seen to contribute to the production of TiO_2 NPs (Jalill et al., 2016).

Under precise experimental conditions, the $TiO(OH)_2$ then combined with aqueous root extract was mixed to produce the TiO_2 NPs. Confirmation of the creation of NPs came when the mixture's hue changed to a light gray. It was in fact an easy, affordable, nontoxic, and environmentally friendly approach to make TiO_2 rutile NPs (Jalill et al., 2016).

13.2.2 GREEN SYNTHESIS OF NANO ZERO VALENT IRON (NZVI) NANOPARTICLES USING PLANTS

In recent years, plant-mediated synthesis of nZVI has become increasingly popular as a viable alternative to traditional methods for assembling metal nanoparticles (Kozma et al, 2016). Biochemical molecules that are prevalent in biomass extracts, such as proteins, enzymes, amino acids, organic acids (citrates), vitamins, and polysaccharides, may operate as reducing agents as well as effective capping for the reduction of metal ions or can even be created on their own (Machado et al, 2015; Monga et al., 2020).

13.2.2.1 CURRENT SYNTHESIS OF NZVI USING PLANT EXTRACTS/ DERIVATIVES

The synthesis and stabilization of nZVI was conducted by using many kinds of plant extracts. The produced nZVI was assessed in terms of application for environmental remediation. In addition, the mechanism was determined (Huang et al, 2014a; Njagi et al., 2011; Plachtová et al., 2018).

Chrysochoou et al. (2012) investigated the performance of pure epigallocatechin gallate (EGCG) solution and green tea synthesized nZVI for the reduction of Cr(IV). The results showed that, compared to EGCG-nZVI, the green tea nZVI could reduce Cr(IV) two times more. Machado et al. (2014) produced nZVI by reducing iron (III) in aqueous solution using natural extracts from a variety of tree leaves (Machado et al, 2014). The extracts' ability to act as antioxidants was used to assess their quality (Machado et al., 2013). Using wastes from citrus juice (lemon, orange, mandarin, and lime), the same group created green nZVI, which might be used for a variety of purposes, including environmental restoration. The generation of nZVI from neem extract under air circumstances was established by Pattanayak and Nayak (2013). Although numerous studies on the environmentally friendly synthesis nZVI from plant extracts have been conducted, the highly active components in these procedures are mostly unknown (Zhang et al., 2020).

Madhavi et al. (2013) examined the production of Fe/Pd bimetallic NPs using grape leaf extract. The synthesis created particles with quasi-spherical shape and particle size of 2–20 nm in the Fe^0 state, according to SEM, TEM, XPS, and EDS examinations. Luo et al. (2016) used GC-MS to examine grape leaf extract in the methanolic solution in order to clarify the active biomolecules involved in the reduction and came to the conclusion that phytols, terpenoids, and antioxidants performed the dual functions of capping and reducing agents (Luo et al., 2016).

Xiao et al. (2016) investigated the production of nZVI using 15 different plant extracts and $FeCl_3$. In addition, the performance of produced nZVI in terms of Cr(VI) removal was examined. The results showed that three plant extracts, namely, *Alston (Syzygium jambos L.), Oolong tea,* and *Willd (Aleurites moluccana L.)*, could produce nZVI with a high ability to remove Cr(VI). In another study, Tandon et al. (2013) created clay-supported nZVI particles by combining tea liquor with ferric nitrate-impregnated montmorillonite K10. Similarly, Soliemanzadeh and Fekri (2017) created nZVI particles from green tea extract using a similar technique.

13.2.3 PLANT-MEDIATED SYNTHESIS

The NPs produced by microbes have slow production rate and low dispersion (Dhillon et al, 2012). According to a study by Kalaiarasi and Jayallakshmi (2010), the production of metallic nanoparticles using plants is simple, cost-effective, and highly repeatable and reproducible. Plant extracts have a proven track record of creating very stable metal nanoparticles at high rates and in significant amounts (Iravani, 2011). Plant-based nanomaterial production is recommended as naturally powerful biomolecular reducing agents can be extracted from many plants (Mondal et al, 2020; Mukunthan and Balaji, 2012).

Antioxidants are active phytochemical components found in high concentrations in the fruits, leaves, seeds, and stems of many herbs and plants. Due to the fact that biogenic synthesis through methods exploiting plant phytochemicals reduces environmental contamination during the manufacturing of nanomaterials, these green synthesis protocols were responsible for milestones associated to the obtainment of highly practical green nanotools that are sustainable and ecofriendly.

13.2.3.1 SYNTHESIS BY LEAF EXTRACT

Numerous plant leaf extracts have undergone significant research to find their polyphenolic and other medicinally valuable phytochemical components. During the synthesis of NPS, it is necessary to perfect a number of variables, including the precursor moiety, solution temperature, pH, solvent, biomass-to-solvent ratio, and extract amount. Huang et al. (2015) investigated the variables affecting how well Fe NPs made from green tea leaf extracts removed Malachite Green dye.

To figure out the ideal circumstances, the study is experimented with different amounts of tea extract, temperatures, and solution pH. It was found that increasing extract volume and solution pH had a detrimental effect on color removal. However, a rise in temperature enhanced the reactivity of nanoparticles. The produced Fe NPs achieved the highest dye removal efficiency of 90.56%. For the purpose of treating eutrophic wastewater, Wang et al. (2014) synthesized spheroidal Fe NPs with size of 20–80 nm, using an extract from dried eucalyptus leaf. Through investigation, it was reported that the effluent had around 84.5% of COD and 71.7% of total N eliminated.

Using leaf extracts from 26 different tree species, including *Apple, Avocado, Apricot, Cherry, Kiwi, Eucalyptus, Lemon, Medlar, Mandarin, Mulberry, Olive, Oak, Orange, Peach, Pear, Pine, Pomegranate, Passion fruit, Plum, Raspberry, Quince, Strawberry, Tea-Green, Vine, Tea-Black,* and *Walnut*. Machado et al. (2013) evaluated the viability of each extract's ability. The antioxidant content of dried leaf extracts was higher than of fresh leaf extracts. Moreover, each leaf had a different set of best extraction conditions. The best results were obtained at 80 °C using leaf extract from oak, green tea, and pomegranate. In a different investigation, Machado et al. (2013) investigated the application of green-synthesized ZVI NPs to eliminate ibuprofen, a common anti-inflammatory medicine, from sandy soils. NPs were used in aqueous solutions to remove ibuprofen for about 54–66% of that amount, while NPs were used as a catalyst in the Fenton process to degrade the drug for around 95% of that amount. The plant extract had an impact on the NPs' particle size. Prasad et al. (2017) reported that Murrayakoenigii leaf extract was used to create NPs with size of 12 nm, according to the authors. The leaves of *Sageretia thea* were used by Khalil et al. (2017) to create tetragonal crystalline-shaped iron oxide nanoparticles. It was discovered that the typical nanoparticle size was 29 nm. Deionized water was used to extract the bioactive components from the plant leaves after they had been ground up, and it also served as a reducing agent.

Pseudomonas aeruginosa was discovered to be the most vulnerable of the five human pathogenic bacterial strains when the antibacterial impact of the produced nanoparticles was studied. According to Borja et al. (2015), dried green tea can be used to create nano zerovalent iron. Green tea leaves were used to extract bioactive chemicals and abundant polyphenols using a microwave. The resulting nanoparticles, which had a size range of 8–23 nm, were spherical in shape. The bioactive components were extracted using ethanol as the extractant solvent.

The created nanoparticles were used to evaluate the removal of nitrate and were contrasted with iron NPs created chemically: *Lantana camara, Eichhornia crassipes*, and *Mimosa pudica.* Synthesized iron NPd demonstrated removal efficiencies of 71.12, 74.52, and 65.23% nitrate elimination, respectively, while chemically generated iron NPs show 85.27% removal. Sathya et al. (2017) produced iron oxide nanoparticles using *Coriandrum sativum* leaves. The artificial nanoparticles had a typical size between 20 and 90 nm and were spherical in shape.

Techniques for synthesis included ultrasonication and continuous stirring. The ultrasonication approach, which was also used for an antibacterial activity test, produced shorter synthesis times. With average diameters of 22 mm for *Micrococcus luteus* and 18 mm for *Staphylococcus aureus*, respectively, these two bacteria show greater antibacterial activity. From the extract of green tea leaves, Singh et al. (2017) created uniformly formed 103 nm iron oxide nanoparticles. The created iron oxide nanoparticles were magnetically saturated and superparamagnetic in nature. The created nanoparticle was also used to remove the methylene blue dye. Madhubala and Kalaivani (2018) developed iron NPs fixed on zinc oxide using neem leaf extract. It was discovered that the artificially created iron oxide nanoparticles had brick-like shapes and had an average particle size of 38 nm. Then, zinc oxide was added to the produced iron oxide nanoparticles using ultrasonication, and the mixture was then autoclaved for 18 h to reach temperature of 200°C. The prepared nanomaterial size and shape were determined (size (45–61) nm, and ellipsoid-shaped).

Both *Amaranthus dubius* leaf extract-mediated iron oxide and chemically generated nanoparticles were produced by Harshiny et al. (2015). Different phytochemicals properties were found in the leaf extract. For example, the produced zeta potentials were between 44 and 66 mV and had better stability than iron nanoparticles made through chemical synthesis. When it comes to particle size, green synthesized produced NPs presented a size of 43 nm, while chemically manufactured ones ranged in size from 70 to 700 nm. Additionally, both created nanoparticles were used to accelerate the breakdown of methyl orange (MO). Under the testing conditions of 20 mg of NPs and 20 ppm MO, it was discovered that the degradation efficiency was higher for chemically manufactured nanoparticles (90%), while greenly synthesized NPs had an efficiency of 81%.

According to Senthil and Ramesh (2012), the leaf's greater carbohydrate content made it an effective reducing agent. Iron oxide NPs with an uneven surface and an irregular spherical form were revealed by morphological investigation. XRD examination, which verified the crystalline size ranging from 80 to 100 nm, proved the crystalline structure of Fe_3O_4. The agar well diffusion method was used to evaluate the antibacterial activity of green produced iron oxide NPs against *Pseudomonas aeruginosa*, and the findings showed their potent antibacterial activity. Kiruba et al. (2013) investigated the production of Fe NPs using the *Dodonaea viscosa* leaf extract. The results showed that the *Dodonaea viscosa* leaf extract's high concentration of polyphenols, including tannins, santin, saponins, and Pendleton, was found to be helpful in the creation of nanoparticles with stable capping.

Lawsonia inermis (henna) and *Gardenia jasminoides* leaf extracts were used to create Fe NPs by Naseem and Farrukh (2015) utilizing a straightforward conventional heating process. *Lawsonia inermis* leaf extract produced Fe NPs with an average diameter of 21 nm, while Gardenia leaf extract produced Fe NPs with an average diameter of 32 nm. The experiment revealed that lawsone (2-hydroxy-1,4-naphthoquinone), which is made up of molecules including p-benzoquinone, benzene, and phenols, is the primary constituent of henna. As a result, coated Fe NPs with such bioactive components were more effectively stabilized. Fe NPs that had been synthesized by henna were discovered to be aggregated and to have a deformed hexagonal shape. In contrast, agglomerated rod-like Fe NPs were seen when Gardenia leaf extract was used.

Due to improved bioactive coatings, the produced Fe NPs were discovered to show superior antibacterial activity against human pathogens like *Escherichia coli* and *Staphylococcus aureus*. Additionally, Kumar et al. (2015) created nZVI NPs using *Emblica officinalis* leaf extract. Ascorbic acid and polyphenol in high concentrations both helped to minimize metal ions while also forming stabilizing agents on the top of nZVI NPs. The nZVI NPs had a spherical morphology with an average diameter of 22.6 nm, according to morphological examination. The elimination of Pb (II) was studied using nanoparticles with high zeta potential values. With an additional adsorbent dosage of 2 mL/L, a further maximum removal of 99% was reached for a 10 ppm Pb (II) solution after 24 h. Wang et al. (2014) made use of eucalyptus leaves to create iron oxide nanoparticles using a quick and easy method. In experimental studies, a 2:1 volume ratio of the leaf extract and 0.10 M $FeSO_4$ was mixed. The development of iron oxide NPs was shown by a yellow to black color change in the solution. The NPs were discovered to be spherical in shape and to have an average diameter between 20 and 80 nm. The COD and TN removal efficiencies using the produced NPs were 84.5 and 71.7%, respectively.

13.2.3.2 SYNTHESIS BY SEED EXTRACT

Researchers discovered that using the fruit seeds to produce green reducing agent for creating nanoparticles was appropriate. In order to create magnetic Fe_3O_4-NPs, Venkateswarlu et al. (2014) used the *Syzygium cumini* seed extract. The abundance of polyphenols and carbohydrates made it easier to create Fe_3O_4 nanoparticles by a straightforward reduction reaction, acting as both a capping and a reducing agent. With a saturation magnetism value of 13.6 emu/g, the produced nanomaterials displayed excellent ferromagnetic properties.

The creation of pure Fe_3O_4 devoid of any impurities was, however, confirmed by Raman spectroscopic analysis. The average particle diameter was from 9 to 20 nm, morphological features revealed a crystalline inverse spinel cubical structure. A surface area of 3.517 m^2/g was found to be the average. It was thought that such manufactured magnetic nanoparticles will find extensive use in the fields of biomedicine and separation aspects.

Additionally, the Silybum marianum L. seed extract was used by Sajadi et al. (2016) to create copper-supported iron oxide nanoparticles. Plant extract was combined with a combination that included copper (II) chloride and iron (III) chloride solutions as precursors to create Cu/Fe_3O^4 nanoparticles. The creation of crystalline spherical nanoparticles with an average particle diameter ranging from 8.5 to 60 nm was confirmed by morphological investigation. Additionally, the produced nanomaterial was used as a catalyst in the reduction of nitroarenes in an ethyl alcohol solution. According to an experimental investigation, the output peaked at 95% in the first cycle and dropped to 92% by the conclusion of the fifth recycle. This shows that manufactured nanomaterial can be reused for catalytic application reasons.

13.2.3.3 SYNTHESIS BY ROOT AND STOLON

A few studies in the literature discuss the use of plants stolen and roots to produce Fe_3O_4 NPs. Sajadi et al. (2016) used potato extract that was high in starch. The Fe_3O_4-NPs are reduced and stabilized by the starch content extract when it is combined with capping agents that primarily include carbohydrate groups. NaOH was added to the starch extract to begin the oxidation process, which further generated electrons to create FeO nanoparticles. With an average particle diameter of 40 2.2 nm and a face-centered cubic shape, the crystalline structure of magnetite NPs was created. The nanoparticle's superparamagnetic properties led to a saturation magnetization of 28.78 emu/g. Additionally, organic methylene blue pollutants were catalytically degraded by the produced magnetite NPs at ambient temperature.

Mimosa pudica root extract was used by Niraimathee et al. (2016) to produce Fe_3O_4-NPs. Through UV measurement, the produced iron oxide NPs revealed a distinct absorbance peak at 294 nm. The saturation magnetization value of the Fe_3O_4-NPs, which was discovered to be 55.40 emu/g, significantly above stated values, showed the superparamagnetic behavior of the material. The produced NPs had an average particle diameter of 67 nm and were spherical in shape, according to the morphological examination. The magnetic NPs' single-crystal domain, which denotes a single direction of magnetic moment, was observed. After the magnetic field was removed, it was discovered that the magnetization value dropped to zero from its plateau value.

13.2.3.4 SYNTHESIS BY MARINE PLANT

It was also thought that seaweed or algae, a type of marine plant, may be used as a source of Fe_3O_4-NPs. South-East Asia is home to an abundance of seaweed, which is important to marine life since it offers food and a habitat. It is regarded as a source of food due to the abundance of fats, minerals, and some vitamins. Furthermore, because it has a variety of phytochemicals, including proteins, polyphenols, and polysaccharides, it can also be used medicinally. Macroalgae and Microalgae are the main categories of marine algae. Most of the time, macroalgae are employed to create nanoparticles that are like plants. As metal-reducing and producers of capping agents, phytochemicals play a significant role in the production of better-stabilized nanoparticles. Brown seaweed was used to create Fe_3O_4-NPs by Madhavi et al. (2013). (*Sargassum muticum*). Fe_3O_4-NPs were created by simply combining a solution of FeCl3 with a brown seaweed extract, according to one description. A few minutes of mixing were needed to start the reaction, and then Fe_3O_4-NPs started to form right away.

13.2.3.5 SYNTHESIS BY FRUIT PEEL

Fruit flesh is shielded from the elements and insects by the peel or skin. Peels from fruits and vegetables, which are typically seen as waste in society, have been used as a natural fertilizer. Nevertheless, agricultural waste was used for the production of NPs. Fruit peel extract has been used extensively in research to synthesize NPs as a source. Fe_3O_4-NPs was produced by Venkateswarlu and Yoon (2015), who used fruit peel extract from *Citrullus lanatus* (watermelon), and Ananas comosus (pineapple). Both fruit peel extracts produced spherical Fe_3O_4-NPs with an average size of about 17 nm. The magnetic Fe_3O_4-NPs produced by pineapple peel extract had a saturation magnetization of 21.7 emu/g. For a 60 mg/L Cd (II) solution, batch adsorption tests showed that a dosage of around 0.1 mg/L adsorbed resulted in 96.1% elimination at pH 6. A 49.1 mg/g maximum adsorption capacity was reached. Similarly, it was discovered that watermelon peel extract-mediated Fe_3O_4-NPs had magnetic properties and had a saturation magnetization of 28.4 emu/g. The magnetic iron oxide nanoparticle that was created was used to remove Mercury (II).

Fruit peels that would have normally been abandoned as rubbish for landfills were instead used to reduce waste and conserve resources. Peels were combined with ultrapure water at a high temperature for 80 min to create the extract. The resulting extract was then membrane-filtered and centrifuged. The iron oxide NPs that were produced had spherical shapes with an average diameter of 93.8 nm. Batch tests were performed using 2 mL of the nanoparticle and 100 mg/L of Cr (VI) solution. Results showed that 99.29% of the Cr (VI) was removed after 90 min. Ehrampoush et al. (2015) used the tangerine peel extract for magnetic iron oxide NPs production. The created NPs were used to remove Cd (II) from an aqueous solution. A morphological investigation found 50-nm diameter, spherical-shaped nanoparticles. The maximum Cd(II) removal of 89.6% was achieved after 90 min, at pH of 4.

13.2.3.6 SYNTHESIS BY OTHER PLANT MATERIALS AND WASTES

Alfalfa biomass was used by Becerra et al. (2007) as the starting material for the biosynthesis of iron oxide NPs. Such recycling of waste items was made possible using waste biomass. The average diameter was 5 nm. Alfalfa biomass was first ground into a powder and added to a salt solution of ferrous ammonium sulfate. According to experimental findings, the average nanoparticle diameter at best pH 10–14 was 3.6 nm. Additionally, the produced nanoparticles underwent a thorough characterization. The characterization method EELS (electron energy loss spectroscopy) was used to show the existence of Fe and O; however, HRTEM pictures were carefully examined for atomistic distribution coupled with FFT spectra to study the structure and the associated iron oxide species. The magnetite and wustite kinds made up most of the clusters that were created (fcc crystalline). Such structures were thought to be the most stable for most materials at the quantum dot size. Under ideal circumstances, both magnetites coexisted with clusters resembling wustites but with smaller particle sizes.

Applications for Fe_3O_4-NPs may differ depending on the synthesis substrate. Due to the significant amounts of polyphenolic content, they have, natural wastes from daily life such as fruit, rice straw, vegetable peels, coffee grounds, and waste residues can be used as significant components in the green synthesis of nanoparticles. Tea residues were used by Lunge et al. (2014) to create Fe_3O_4-NPs and then for research on the removal of arsenic. Arsenic (V) and arsenic (III) elimination achieved maximum adsorption capacities of 153.8 and 188.69 mg/g, respectively, according to experimental tests.

In a different work, Khataee et al. (2017) used coffee waste to produce hydrochar coated with Fe_3O_4 nanoparticles and used for a study on the removal of Acid Red 17 (azo dye). It was discovered that removal efficiency decreased as initial dye concentration increased. The efficiency was reduced by coexisting Na_2SO_4 and NaCl salts from 100 to 85%, and 100 to 91%, respectively. These studies prove the excellent surface stabilizing qualities of natural residuals and show how they can be applied to a variety of environmental applications.

13.3 APPLICATIONS OF PLANT-DERIVED NPS

Due to their numerous uses in electronics, industries, energy, environment, and more specifically in the biomedical disciplines, NPs are currently in high demand commercially. NPs, like the most well-known Ag and Au NPs, have been extensively researched in this field and hold great promise for biological applications. In general, plant-derived green NPs have greater application potential than chemically produced ones and are less likely to have serious adverse effects on people. These include, but are not limited to:

- Bioengineering applications of food science and technology (storing, processes for processing, and packaging food) (photocatalysts, biocatalysts, biosensors, etc.) (Shafiq et al., 2020).
- Nanomedicine and the preservation of human health (antiparasitic, antimicrobial, antiproliferative, pro- or anti-oxidative, pro-apoptotic, depending on the context, anti-inflammatory actions, etc.) (Andleeb et al., 2021; Anjum et al., 2021).
- Cosmetics, such as anti-aging products, sunscreen, hair growth aids, nano-emulsions, and bioactive ingredient delivery systems (Abbasi et al., 2020).

In the current Special Issue, two review papers discussing CuNPs and algae-based NP synthesis provide light on two green plant-based nanotechnology tools and approaches that have not received much attention (Chaudhary et al., 2020). Because they are abundant in secondary metabolites that function as reducing and capping agents, algae are unquestionably the best possible candidates for the green synthesis of NPs.

13.3.1 DEGRADATION OF DYES

Hoag et al. (2009) used green tea as a catalyst to help hydrogen peroxide break down bromomethyl blue in an aqueous solution to synthesize iron nanoparticles. According to the study's findings, green synthesized nanoscale zero-valent iron produced increased catalytic activity for Fe-EDTA and Fe-EDDS (nZVI). The removal of bromomethyl blue was shown to increase with increasing doses of green produced nZVI due to improved hydrogen peroxide catalyzation.

One hundred percent removal of methyl orange (MO) and methylene blue (MB) was conducted with an adsorbent dose of 10–200 mg/L. According to an experimental investigation, MB was removed more quickly than MO dye. About 80% of the MB was eliminated in the first 5 min, while after 1 h, MO dye obtained 80% clearance. After 200 min and 350 min, the full removal of the MB and MO dyes, respectively, was conducted. According to studies, iron nanoparticles produced via a green method are more efficient than those created through chemical reduction acting as Fenton-like catalysts. Huang and colleagues synthesized iron nanoparticles using oolong tea extract (Huang et al. 2014b). The degradation of malachite green (MG) was done using the nanoparticles. According to the experimental findings, the polyphenols in oolong tea extract were used as capping and reducing agents during the synthesis. The high polyphenolic content allowed for increased reactivity and reduced agglomeration of the produced nanoparticles. When applied to a 50 mg/L MG solution, the nanoparticles removed dye from the solution by 75.5%. In a related study, Kuang et al. (2013) employed three distinct tea extracts, green tea (GT), oolong tea (OT), and black tea (BT)—individually to synthesize iron nanoparticles. The created nanoparticles were also used as a catalyst to break down monochlorobenzene (MCB) to undergo Fenton-like oxidation. The chosen extracts' greatest polyphenol content led to better-synthesized Fe NPs for expedited breakdown. In 180 min, MCB was degraded by 69% using iron nanoparticles with a GT basis, compared to 53 and 39% using the OT and BT bases. Wang et al. (2013) created iron–polyphenol complex-based nanoparticles to prevent color degradation. The produced nanoparticles were evaluated for adsorption–flocculation using acid black 194 dye and eucalyptus leaves as reducing and capping agents. At 25 °C, it was discovered that the produced nanoparticles had an adsorption capacity of 1.6 g of acid black 194 per gram. In addition, Wang et al. (2014) synthesized Fe NPs from three different plant sources, namely *R. officinalis, M. nesophila,* and E. *tereticornis* which were then used to decolorize acid black dye. The acid black dye and 87% of the total organic carbon were completely removed by the extract of E. tereticornis that it produced in the form of Fe nanoparticles (TOC). These Fe NPs outperformed the others in terms of their catalytic abilities. In an aqueous solution, the as-synthesized nanoparticles had a reduced average diameter and superior dispersion properties.

In a different study, Luo et al. (2014) investigated how iron nanoparticles produced by grape leaf extract could be used to remove the acid orange II color. According to the study, plant-mediated Fe-NPs had superior reactivity than Fe-NPs made from aqueous extract and grape leaf methanolic. Therefore, it was found that plant-mediated Fe nanoparticles were more effective at catalyzing the breakdown of dyes under diverse experimental conditions. Fe-NPs from plants worked as a Fenton catalyst in the traditional Fenton reaction, but they were found to be better suited for Fenton-like reactions.

Singh et al. (2017) investigated how iron oxide NPs that were synthesized using green tea might remove the pigment methylene blue. According to the findings, 95% of the degrading efficiency was reached in 16 min. Additionally, 99% of the Blue 235 dye was removed by iron oxide nanoparticles that were successfully generated using *Syzygium cumini* leaf extract after 4 h of reaction. Hematite quantum dots and nanorods were created by Ahmed et al. (2014) using microwave and microwave-hydrothermal methods, respectively. The generated nanoparticles were used to degrade an organic dye, and the outcomes were contrasted with hematite that is readily available on the market.

In order to degrade malachite green, Huang et al. (2015, 2014a, 2014b) obtained Fe nanoparticles using oolong tea extracts. The outcome showed a 75.5% clearance efficiency at a concentration of 50 mg/L. Additionally, to create Fe NPs, which were used to catalyze the oxidation of Mono chlorobenzene, Kuang et al. (2013) used the extracts of green, oolong, and black tea. The elimination efficiency of Fe NPs was 69% for green tea, 53% for oolong tea, and 39% for black tea after 180 min (Kuang et al., 2013). The utilization of grape leaf extracts in the synthesis of Fe NPs and the elimination of acid orange II was reported in Luo et al. (2016).

Both methanolic and water extracts of grape leaf were employed by the authors. Fe NPs from methanolic extracts were shown to be more reactive than those from water. When methanolic Fe NPs were employed in comparison to 4% from water Fe NPs, nearly 80% of the acid orange II was eliminated using an initial concentration of 10 mg/L (Liu et al, 2018; Luo et al., 2016).

13.3.2 REMOVAL OF HEAVY METALS

In Rao et al. (2013), chromium (VI) was removed using yeast cells and a bio Fe^0/Fe_3O_4 nanocomposite. These nanocomposite materials worked well as bio-sorbents to remove Cr (VI). Compared to yeast cells, the sorption capacity of the transformed yeast cells increased three times. Under ideal circumstances, the modified adsorbent had a higher adsorption capacity than the unmodified adsorbent, which had a capacity of 137.31 mg/g, for a 1000 mg/L solution of Cr (VI). Savasari et al. (2015) reported on stable and effective green nZVI produced using ascorbic acid. The produced nZVI was used in adsorption experiments to reduce Cd (II) in two distinct investigations.

According to Ponder et al. (2000), using environmentally friendly synthetic iron nanoparticles, heavy metals are eliminated through surface adsorption and redox reactions. The various oxidation states of iron are the primary determinant of iron nanoparticle reactivity. According to Tang and Lo (2013), the chemical properties and the mechanism of the reaction between metal and pollutants will differ depending on the oxidation state of any given metal. In a recent study, Xiao et al. (2016) created iron nanoparticles from diverse plant leaf extracts and used them to remove chromium. The reduced potential activity allowed the chosen plants to show good antioxidant qualities. The *S. jambos (L.) Alston (SJA), Oolong tea (OT),* and *A. moluccana (L.) Wild (AMW)* extracts are used in the investigation. The antioxidant properties of the leaf extracts are listed in a decreasing order. The results showed that 1 mL of SJA-Fe NPs solution eliminated Cr (VI) to 91.9% in the first 5 min due to its maximum antioxidant property; however, total removal efficiency was achieved at a reaction time of 60 min.

In a related investigation, *Syzygium jambos* were used. Xiao et al. (2017) synthesized nZVI using Alston leaf extract to remove hexavalent chromium ions from aqueous solution. The effectiveness of chromium removal is influenced by variables including solution pH, nanoparticle dosage, initial Cr (VI) content, and temperature. For a given chromium content, the availability of active sites rises with increasing nanoparticle dosage, which improves the removal rate. The results show that when temperature increased from 20°C to 60°C, the removal efficiency increased by 15%. The elimination of chromium (VI) was shown to be more effective at low pH because of the electrostatic interaction between the surface of protonated nanoparticles and hexavalent chromium at acidic pH.

The experimental findings showed that 100% chromium removal was achieved at an ideal pH of 5.5, using 0.5 mL of produced nanoparticles, and an initial chromium concentration of 50 mg/L. Similar to this, Sebastian et al. (2018) used the coconut husk extract to create magnetite nanoparticles, which they then used to adsorb modest quantities of calcium and cadmium. The results showed that at the conclusion of 120 min, cadmium and calcium elimination had increased by more than 40 and 50%, respectively. According to experiments, equilibrium was obtained once the maximum elimination was done within 30 min.

Ehrampoush et al. (2015) used the tangerine peel extract to produce iron oxide NPs and used them to eliminate trace amounts of cadmium from water bodies. According to experimental investigations, 90% of the maximum clearance could be removed after 90 min at pH 4. Like this, Prasad et al. created a graphene oxide-based iron oxide nanocomposite to remove Pb(II) from water. According to the experimental data, Pb (II) was removed approximately 96% at pH 5, where fast adsorption happened after 40 min, and equilibrium was set up after 80 min. It was discovered that the adsorption study was pH-dependent, and that the adsorbents could be reused by treating them with an acid of a specific concentration to desorb Pb (II) ions (Prasad et al., 2017).

It is important to highlight that desorption alters the electrostatic equilibrium between adsorbent and adsorbate at lower pH levels. As the concentration of nickel ions in the solution rises, the capacity for sorption increases, but the efficiency of adsorption falls because of a scarcity of active sites. The efficacy of adsorption rises as the size of nanoparticles decreases because smaller adsorbents have more surface area. In a report, Mahmoud et al. (2016) used a modified co-precipitation approach to synthesize magnetic iron oxide coated with amino-silica support. Using the synthesized nanocomposite heated by microwaves, the adsorption of heavy metal ions such as Cu^2, Pb^2, Hg^2, and Cd^2 was conducted. The dissolved metal ion and nanocomposite was heated using microwaves for a brief period. The metal ion absorbs onto the nanocomposite during the sorption process. Among the other ions, Pb^2 showed a greater capacity for sorption after 15 s. According to the experimental findings, there was no metal desorption as the temperature rose.

13.3.3 WASTEWATER TREATMENT

Chrysochoou et al. (2012) investigated the transport of iron NPs produced from green tea extract aluminum hydroxide coated sand and refined silica sand as media. On introducing green manufactured nZVI, a sharp fall in effluent pH was observed, going from 8.5 to 2. Due to hydrolysis processes and the existence of Fe^{3+} as residuary discharge in the solution, this phenomenon was found. Instead of Fe^0 being present in the produced nanoparticles, a redox potential rise of 150–550 mV was found. Such a phenomenon was noted because of green tea's polyphenol oxidation, which is a sign that green produced nZVI is being transported across the medium. Similar to this, He and Zhao (2005) used bimetallic Fe/Pd nanoparticles produced by starch to study the breakdown of trichloroethylene (TCE). When compared to nanoparticles made without a stabilizer, those coated with starch were less agglomerated and showed to be more effective at degrading TCE. When the process was run with 0.1 g/L of nanoparticles, a maximum degradation of 98% was recorded in 1 h.

In order to treat eutrophic wastewater, Wang et al. (2014) used bio-synthesized iron nanoparticles produced from eucalyptus leaf extract. This experiment was the first to describe using environmentally friendly nanoparticles to clean such wastewater. It was discovered that Fe NPs made from eucalyptus leaf extract had smaller average particle sizes than those made through chemical synthesis. After 21 days, researchers looked at the removal percentages of total nitrogen, total phosphorus, and COD and discovered that they were, respectively, 71.7, 30.4, and 84.5%. Little phosphorus is removed when precipitating substances like calcium or aluminum are absent.

According to Wang et al. (2014), iron oxide NPs were made from eucalyptus leaves independently and green tea was effective at removing nitrate from wastewater. In both cases, spheroidal particles were produced. In the study, nitrate removal from iron nanoparticles produced chemically and using plant materials is compared. The results showed that green tea and eucalyptus-mediated NPs produced nitrate removal rates of about 59.7 and 41.4%, respectively, while nZVI and Fe3O4 nanoparticles produced removal rates of 87.6 and 11.7%. Despite having a higher removal effectiveness, green produced nanoparticles were more stable and were chosen over nZVI.

A comparison of the green produced Fe NPs and nZVI's reactivity efficiency after two months of exposure was also done through experiments. Green synthesized NPs' reactivity efficiency was found to be kept at 40.7% (eucalyptus) and 51.7% (green tea), but nZVI generated by chemical synthesis increased its efficacy by 2.1-fold to 45.4%. According to Groiss et al. (2017), *Cynometra rami flora* leaf extract was used to create iron oxide nanoparticles, which were then used to catalyze the breakdown of the carcinogen Rhodamine B in the presence of hydrogen peroxide. When 1.11 mM nanoparticles were used with 2% H_2O_2, the degradation was discovered to be at its best within 15 min. A procedure like this was discovered to create the least amount of sludge, making it superior to other Fenton-catalyzed processes. Yellow color fading occurred with the formation of 4-aminophenol during the reaction in the presence of the nano-catalyst. However, when the nano catalyst was absent, no discernible decrease was found. The study also shows that Pd/RGO/Fe_3O_4 nanocomposite outperformed Pd/RGO/Fe_3O_4 nanocomposite in terms of catalytic characteristics, showing the palladium nanoparticles' ability to enhance reactions.

Iron nanoparticles made from three distinct plant extracts—Rosa damascene (RD), Thymus vulgaris (TV), and Urtica dioica—were also used by Fazlzadeh et al. (2017) to remove Cr(VI) (UD). Greater than 90% effectiveness was reached within 10 min, according to experimental research. In soil and water, eight prioritized polycyclic aromatic hydrocarbons (PAHs) were reported to have been degraded by Shanker et al. (2017). Initial concentrations of such high molecular weight PAHs range from 50 to 250 mg/L. Within 48 h, anthracene and phenanthrene degraded 80–90% under the ideal testing conditions at a pH of 7.0. For complete breakdown of an organo-phosphorus pesticide, Mehrotra et al. (2017) used protein-capped zero-valent iron NPs made from yeast (Dichlorvos). With 2000 mg/L of NPs and 1000 mL H_2O_2 in the ideal experimental setup, after 60-min reaction time, a maximum removal of 99.9% was reached. Additionally, Sajadi et al. (2016) created a copper-supported iron oxide nano-catalyst to investigate the reduction of nitroarenes in wastewater. Sodium borohydride was used as a reducing agent during the degradation process. According to experimental research, there was no degradation seen when $NaBH_4$ was present alone, but the produced nano-catalyst and $NaBH_4$ together achieved a maximum degradation of 90% in under 90 min.

13.3.4 REMOVAL OF OTHER CONTAMINANTS

Wang et al. (2014) produced Fe NPs using eucalyptus leaf extracts. The resulting Fe NP seemed spheroidal was attributed to the potential impact of various natural chemicals that can be obtained from leaf extracts. The produced Fe NPs were examined for its performance for on swine wastewater treatment, where reducing and capping agents were extracted for leaf. The test's findings showed that after 21 days, COD, phosphorus, and nitrogen were all eliminated with efficiencies of 71.7, 30.4, and 84.5%, respectively. The lack of calcium, magnesium, or aluminum in the aqueous solution, which could have bonded to the P and aided in its precipitation, was blamed for the low efficiency in the removal of P. Total petroleum hydrocarbons (TPHs) were removed from water and soil samples by Murgueitio et al. (2018) using Fe NP made from *Mortino berry* extracts. Fe NPs with a diameter of 5–25 nm were produced. After 720 s of TPH removal for two different concentration levels, an efficiency of at least 88.24% was recorded. Compared to the current bioremediation approach, this takes less time to remove TPHs from water.

13.4 ENVIRONMENTAL AND TOXICITY STUDY OF IRON NPS

Despite having several environmental applications, iron-based nanoparticles pose a significant environmental risk. Groundwater and soil environmental problems could result from improper handling of wastewater and other industrial wastes. While iron nanomaterial that has been created for specific remedial needs can be moved from one medium to another. Among other iron NPs, nZVI are thought to be the most reactive. It has been noted that nZVIs are transported through porous soil during the treatment of groundwater pollutions and go through transformation upon interaction with contaminants as well as the exposed environment.

Due to iron NPs' harmful effects on soil fauna and microorganisms, the ecosystem is either directly or indirectly affected. According to a study, soil microorganisms experience toxic effects and changes in microbial biomass as a result of the stress that nano-sized iron generates (Saccà et al. 2014). Due to their distinct morphological characteristics, surface energy, particle size, self-assembly, and production method, iron nanoparticles interact aggressively with metals and enzymes in the environment. However, it becomes a mildly toxic agent for the environment when dangerous compounds like hydrazine are used as capping agents in biosynthesized nanoparticles.

According to Vittori Antisari et al. (2013), the harmful effects of designed iron nanoparticles caused a shift in the soil's microbial mass. Additionally, the soil bacteria *Bacillus cereus* was examined by Fajardo et al. (2013) for its transcriptional and proteomic stress activities brought on by nano-zero-valent iron. In a recent study, Das et al. (2018) used the Desulfovibrio bacteria strain LS4 to in situ synthesize maghemite NPs (Fe_2O_3) from saltpan sediments. An average particle diameter of 18 nm was produced. Zebrafish embryos were also used to investigate the toxicity of produced nanoparticles at various dosages. The results verified the DNA damage and developmental impairment of the zebrafish embryo.

Growing nanoparticle concentration in Zebrafish eggs was associated with an increased death rate, slower beat, and delayed hatching rate. High concentrations of maghemite nanoparticles had a negative impact on larval development and were thought to be harmful to aquaculture. Through experimental experiments, Auffan et al. (2008) have shown a link between the cytotoxicity of iron nanoparticles and their state of oxidation. Different iron oxides, including nZVI, maghemite and magnetite, were evaluated on gram-negative bacteria to see how cytotoxic they were. According to the results of the studies, nZVI was more hazardous than other iron oxide NPs. The research concluded that the oxidation state of iron nanoparticles was crucial for toxicity.

Many investigations have revealed that stabilizing or capping compounds do not reduce the toxicity of iron oxide nanoparticles. Four distinct iron NPs were coated with ascorbate, citrate, poly-vinylpyrrolidone, and dextran; the Daphnia magna neo-nates were used for the toxicity experiments (water flea). Iron nanoparticles with ascorbate and dextran coatings had the highest levels of immobilization, but those with citrate coatings had lower levels. Citrate, dextran, and citrate coated iron NPs showed incomplete ecdysis when present in used dosages; however, polyvinylpyrrolidone coated iron NPs had a little detrimental impact. The maximum colloidal stability was shown by PVP-coated NPs for various functionalized iron nanoparticles (Soliemanzadeh and Fekri, 2017).

Further research revealed that while colloidal stability and ion release generate ROS in daphnids, stabilizing forces controlling hydrodynamic diameter had insignificant impact on toxicity. Since chlorella vulgaris serves as an ecological indicator and contributes to the wellbeing of the aquatic ecosystem, Barhoumi and Dewez (2013) studded the toxicity effect of iron oxide NPs on green alga. The study exposes Chlorella vulgaris cells to a suspension of iron oxide NPs with three different chemical concentrations. Investigation revealed that iron nanoparticles caused oxidative stress, inhibited cell division, and furthered the disruption of algal cells' photochemical activity. Similar to this, Markova et al. (2014) used sensitive organisms to investigate the impact of green produced iron NPs on cyanobacteria (*Synechococcus nidulans*), green algae (*Pseudokirchneriella subcapitata*), and invertebrates (*Daphnia magna*). According to experimental findings, the green tea extract was used to produce iron nanoparticles that were green in color. These nanoparticles had an adverse impact on selected toxicological assays. The literature on the toxicological impacts of green manufactured NPs on the environment is still lacking, despite research showing that they are biocompatible with living things.

The scientific community is becoming increasingly concerned about the toxicity of nanoparticles on individuals and the environment because of the field of nanotechnology's rapid advancements. Nanoparticles interact intensively with enzymes in living beings and metals in environmental media because of their distinctive morphological characteristics, surface energy, particle size, synthesis processes, and agglomeration/self-assembly. In contrast to physical and chemical procedures, the potential for toxicity is lower with biosynthesized nanoparticles. This is due to the seldom use of hazardous capping and stabilizing chemicals like hydrazine (Bolade et al, 2020).

13.5 FUTURE PROSPECTIVE OF PLANT-BASED NANOMETALS FOR ENVIRONMENTAL REMEDIATION

The peculiarity of green nanomaterial synthesis needs to be explored more, and there are many different resources that can be used to conduct this synthesis. Understanding the fundamental mechanism of such synthesis is crucial to finding new improvements and alterations. In addition, advances should move toward a method that is more environmentally friendly than traditional techniques, can be scaled up, and is economically competitive.

The focus of research should be on creating these eco-friendly nanomaterials to improve efficiency in removing toxins with the least amount of environmental harm. Again, more attention should be paid to discussing and managing the destiny, toxicological consequences, and risk management of chemically generated iron nanoparticles.

a. Distinct microorganisms, such as bacteria, fungus, or yeast, have different mechanisms of response for producing nanoparticles. The numerous interactions between metal ion precursors and other biological agents result in nanoparticles with distinct morphologies and properties. Extracellular synthesis happens when metal ions and enzymes transfer electrons, as opposed to intracellular synthesis, which includes bio-reduction of metal ions on the cell wall. Due to the complexity of these reactions, more research is needed to better understand the process mechanisms that can be used to control the shape and other features of produced nanoparticles.
b. In iron oxide nanoparticles, zero valent irons that are present, due to their higher oxidation state, are reactive in nature. Consequently, such nanoparticles can have substantial harmful effects on soil and the aquatic system during transfer from one medium to another, which can harm the ecosystem. Therefore, research should concentrate more on improving the green way for stabilizing manufactured nanomaterials and improving the management of nanomaterial waste.

c. Research should focus on exploiting agricultural waste for the environmentally friendly production of nanomaterials. Additionally, more attention should be paid to the green method of creating metal–metal nanocomposite materials, which can improve their efficacy for heavy metal removal and wastewater treatment applications.
d. This study has shown that different plant species create nanoparticles with various morphologies and physical–chemical characteristics. The primary active biomolecules in charge of capping and stabilizing the produced nanomaterials should be thoroughly studied by researchers.
e. Research into novel microbes is necessary for the manufacture of iron nanoparticles. According to earlier research, microorganism-related synthesis produces fewer stable nanomaterials with narrower morphological orientations, which have a lower yield value. The goal of new microbe research should be to explore various species to produce nanoparticles with uniform, crystalline, and monodisperse features.
f. There should be a standardized procedure for the synthesis to manage risk because of the hazardous nature of the created iron NPs. Since varied synthesis techniques used by researchers result in nanomaterials with different properties and characteristics, thorough screening of the nano-waste before discarding is disregarded. This leads to inconsistent research results and makes it challenging to forecast human health and estimate environmental danger.
g. Because the produced iron nanoparticles are dangerous, there should be a defined approach for the synthesis to limit risk. Nanomaterials with varying qualities and characteristics are produced by researchers using a variety of synthesis procedures; therefore, careful examination of the nano-waste before disposal is neglected. As a result, research findings are inconsistent, making it difficult to predict human health and gauge environmental threat.
h. Due to the extremely small bed void volume, using such NPs as an adsorbent in columns presents a significant pressure drop problem. Incorporating NPs over supports such carbon composites, adsorbent beads, and activated carbon could solve these problems. The best interaction between the adsorbent and adsorbate for such an operation should be chemisorption. Thus, the top goal would be to prepare and choose certain NPs with higher surface functional groups.

13.6 CONCLUSIONS

This chapter focuses on the different environmentally friendly processes for making plant-based nanomaterials as well as their potential use in cleaning up the environment. An effort is made to classify different raw materials under "green synthesis" and to examine how they are used in various applications that are relevant to the environment. The analysis presented here highlights intricate characteristics of the various NPs' size, shape, properties, and use as a nanomaterial. For creating plant-based nanomaterials and nanocomposites, a thorough summary of various plant sources, biocompatible chemicals, and energy-saving techniques has been provided. In this chapter, several uses of nanoparticles as catalysts, adsorbents, and their toxicological consequences were covered. Furthermore, it has been noted that bimetallic nanocomposite synthesis using such a green technique is currently developing. The use of several plant metabolites as effective adsorbents on their surfaces has led to comparisons between the plant-based synthesis and composites based on graphene. Additionally, a critical evaluation of the stability, size distribution, and other aspects of the various green synthesis pathways has been conducted. Finally, the difficulties and potential of green nanoparticle creation are highlighted. In conclusion, different green nanotechnology processes have been described in this chapter. The readers of this chapter would find it helpful to gain in-depth knowledge on the environmentally friendly synthesis of nanoparticles and their remarkable performance in a variety of environmental applications.

KEYWORDS

- **plant-derived nanoparticles**
- **environmental cleanup**
- **green technology**
- **toxicity**
- **public health.**

REFERENCES

Abbasi, B. H.; Fazal, H.; Ahmad, N.; Ali, M.; Giglioli-Guivarch, N.; Hano, C. Nanomaterials for Cosmeceuticals: Nanomaterials-Induced Advancement in Cosmetics, Challenges, and Opportunities. *Nanocosmetics* **2020,** 79–108. https://doi.org/10.1016/B978-0-12-822286-7.00005-X

Ahmed, F.; Arshi, N.; Anwar, M. S.; Danish, R.; Koo, B. H. Quantum-Confinement Induced Enhancement in Photocatalytic Properties of Iron Oxide Nanoparticles Prepared by Ionic Liquid. *Ceram. Int.* **2014,** *40* (10), 15743–15751. https://doi.org/10.1016/J.CERAMINT.2014.07.098

Ahmed, S.; Annu, Chaudhry, S. A.; Ikram, S. A Review on Biogenic Synthesis of ZnO Nanoparticles Using Plant Extracts and Microbes: A Prospect Towards Green Chemistry. *J. Photochem. Photobiol. B: Biol.* **2017,** *166*, 272–284. https://doi.org/10.1016/J.JPHOTOBIOL.2016.12.011

Andleeb, A.; Andleeb, A.; Asghar, S.; Zaman, G.; Tariq, M.; Mehmood, A. et al. A Systematic Review of Biosynthesized Metallic Nanoparticles as a Promising Anti-Cancer-Strategy. *Cancers* **2021,** *13* (11), 2818. https://doi.org/10.3390/CANCERS13112818

Anjum, S.; Ishaque, S.; Fatima, H.; Farooq, W.; Hano, C.; Abbasi, B. H.; Anjum, I. Emerging Applications of Nanotechnology in Healthcare Systems: Grand Challenges and Perspectives. *Pharmaceuticals* **2021,** *14* (8), 707. https://doi.org/10.3390/PH14080707

Aslam, M.; Abdullah, A. Z.; Rafatullah, M. Recent Development in the Green Synthesis of Titanium Dioxide Nanoparticles Using Plant-Based Biomolecules for Environmental and Antimicrobial Applications. *J. Ind. Eng. Chem.* **2021,** *98*, 1–16. https://doi.org/10.1016/j.jiec.2021.04.010

Auffan, M.; Achouak, W.; Rose, J.; Roncato, M. A.; Chanéac, C.; Waite, D. T. et al. Relation Between the Redox State of Iron-Based Nanoparticles and Their Cytotoxicity Toward *Escherichia coli*. *Environ. Sci. Technol.* **2008,** *42* (17), 6730–6735. https://doi.org/10.1021/ES800086F/SUPPL_FILE/ES800086F-FILE003.PDF

Barhoumi, L.; Dewez, D. Toxicity of Superparamagnetic Iron Oxide Nanoparticles on Green Alga *Chlorella vulgaris*. *BioMed Res. Int.* **2013,** *2013*. https://doi.org/10.1155/2013/647974

Baruah, S.; K. Pal, S.; Dutta, J. Nanostructured Zinc Oxide for Water Treatment. *Nanosci. Nanotechnol.-Asia* **2013,** *2* (2), 90–102. https://doi.org/10.2174/2210681211202020090

Bolade, O. P.; Williams, A. B.; Benson, N. U. Green Synthesis of Iron-Based Nanomaterials for Environmental Remediation: A Review. *Environ. Nanotechnol. Monitor. Manag.* **2020,** *13*, 100279. https://doi.org/10.1016/j.enmm.2019.100279

Bora, T.; Dutta, J. Applications of Nanotechnology in Wastewater Treatment: A Review. *J. Nanosci. Nanotechnol.* **2014,** *14* (1), 613–626. https://doi.org/10.1166/JNN.2014.8898

Borja, J. Q.; Ngo, M. A. S.; Saranglao, C. C.; Tiongco, R. P. M.; Roque, E. C.; Dugos, N. P. Synthesis of Green Zero-Valent Iron Using Polyphenols from Dried Green Tea Extract. *J. Eng. Sci. Technol.* **2015,** *10* (Spec.issue7), 22–31.

Chandrappa, R.; Das, D. B. Waste Quantities and Characteristics. *Environ. Sci. Eng.* **2012,** 47–63. https://doi.org/10.1007/978-3-642-28681-0_2

Chaudhary, R.; Nawaz, K.; Khan, A. K.; Hano, C.; Abbasi, B. H.; Anjum, S. An Overview of the Algae-Mediated Biosynthesis of Nanoparticles and Their Biomedical Applications. *Biomolecules* **2020,** *10* (11), 1498. https://doi.org/10.3390/BIOM10111498

Chrysochoou, M.; Johnston, C. P.; Dahal, G. A Comparative Evaluation of Hexavalent Chromium Treatment in Contaminated Soil by Calcium Polysulfide and Green-Tea

Nanoscale Zero-Valent Iron. *J. Hazard. Mater.* **2012,** *201–202*, 33–42. https://doi.org/10.1016/J.JHAZMAT.2011.11.003

Das, K. R.; Kowshik, M.; Praveen Kumar, M. K.; Kerkar, S.; Shyama, S. K.; Mishra, S. Native Hypersaline Sulphate Reducing Bacteria Contributes to Iron Nanoparticle Formation in Saltpan Sediment: A Concern for Aquaculture. *J. Environ. Manag.* **2018,** *206*, 556–564. https://doi.org/10.1016/J.JENVMAN.2017.10.078

Devi, W. K.; Meitei, M. D.; Prasad, M. N. V. Remediation of Wastewater Using Plant Based Nano Materials. In: *Handbook of Assisted and Amendment: Enhanced Sustainable Remediation Technology*, 2021; pp 583–599. https://doi.org/10.1002/9781119670391.ch28

Dhillon, G. S.; Brar, S. K.; Kaur, S.; Verma, M. Green approach for nanoparticle biosynthesis by Fungi: Current Trends and Applications. **2012,** *32* (1), 49–73. https://doi.org/10.3109/07388551.2010.550568

Ehrampoush, M. H.; Miria, M.; Salmani, M. H.; Mahvi, A. H. Cadmium Removal from Aqueous Solution by Green Synthesis Iron Oxide Nanoparticles with Tangerine Peel Extract. *J. Environ. Health Sci. Eng.* **2015,** *13* (1). https://doi.org/10.1186/S40201-015-0237-4

El-Seedi, H. R.; El-Shabasy, R. M.; Khalifa, S. A. M.; Saeed, A.; Shah, A.; Shah, R. et al. Metal nanoparticles Fabricated by Green Chemistry Using Natural Extracts: Biosynthesis, Mechanisms, and Applications. *RSC Adv.* **2019,** *9* (42), 24539–24559. https://doi.org/10.1039/C9RA02225B

Elsakhawy, T.; Omara, A. E. D.; Abowaly, M.; El-Ramady, H.; Badgar, K.; Llanaj, X. et al. Green Synthesis of Nanoparticles by Mushrooms: A Crucial Dimension for Sustainable Soil Management. *Sustainability*, *14* (7), 1–27. https://doi.org/10.3390/su14074328

Fajardo, C.; Saccà, M. L.; Martinez-Gomariz, M.; Costa, G.; Nande, M.; Martin, M. Transcriptional and Proteomic Stress Responses of a Soil Bacterium *Bacillus cereus* to Nanosized Zero-Valent Iron (nZVI) Particles. *Chemosphere* **2013,** *93* (6), 1077–1083. https://doi.org/10.1016/J.CHEMOSPHERE.2013.05.082

Fazlzadeh, M.; Rahmani, K.; Zarei, A.; Abdoallahzadeh, H.; Nasiri, F.; Khosravi, R. A novel Green Synthesis of Zero Valent Iron Nanoparticles (NZVI) Using Three Plant Extracts and Their Efficient Application for Removal of Cr (VI) from Aqueous Solutions. *Adv. Powder Technol.* **2017,** *28* (1), 122–130. https://doi.org/10.1016/J.APT.2016.09.003

Gawande, M. B.; Branco, P. S.; Parghi, K.; Shrikhande, J. J.; Pandey, R. K.; Ghumman, C. A. A. et al. Synthesis and Characterization of Versatile MgO–ZrO 2 Mixed Metal Oxide Nanoparticles and Their Applications. *Catal. Sci. Technol.* **2011,** *1* (9), 1653–1664. https://doi.org/10.1039/C1CY00259G

Goutam, S. P.; Saxena, G.; Roy, D.; Yadav, A. K.; Bharagava, R. N. Green Synthesis of Nanoparticles and Their Applications in Water and Wastewater Treatment. In: *Bioremediation of Industrial Waste for Environmental Safety*, 2020. https://doi.org/10.1007/978-981-13-1891-7_16

Goutam, S. P.; Saxena, G.; Singh, V.; Yadav, A. K.; Bharagava, R. N.; Thapa, K. B. Green Synthesis of TiO2 Nanoparticles Using Leaf Extract of *Jatropha curcas* L. for Photocatalytic Degradation of Tannery Wastewater. *Chem. Eng. J.* **2018,** *336*, 386–396. https://doi.org/10.1016/J.CEJ.2017.12.029

Groiss, S.; Selvaraj, R.; Varadavenkatesan, T.; Vinayagam, R. Structural Characterization, Antibacterial and Catalytic Effect of Iron Oxide Nanoparticles Synthesised Using the Leaf Extract of *Cynometra ramiflora*. *J. Mol. Struct.* **2017, 1128,** 572–578. https://doi.org/10.1016/J.MOLSTRUC.2016.09.031

Guo, D.; Xie, G.; Luo, J. Mechanical Properties of Nanoparticles: Basics and Applications. *J. Phys. D: Appl. Phys.* **2013,** *47* (1), 013001. https://doi.org/10.1088/0022-3727/47/1/013001

Hano, C.; Abbasi, B. H. Plant-Based Green Synthesis of Nanoparticles: Production, Characterization and Applications. *Biomolecules* **2022,** *12* (1), 1–9. https://doi.org/10.3390/biom12010031

Harshiny, M.; Iswarya, C. N.; Matheswaran, M. Biogenic Synthesis of Iron Nanoparticles Using *Amaranthus dubius* Leaf Extract as a Reducing Agent. *Powder Technol.* **2015,** *286,* 744–749. https://doi.org/10.1016/J.POWTEC.2015.09.021

He, F.; Zhao, D. Preparation and Characterization of a New Class of Starch-Stabilized Bimetallic Nanoparticles for Degradation of Chlorinated Hydrocarbons in Water. *Environ. Sci. Technol.* **2005,** *39* (9), 3314–3320. https://doi.org/10.1021/ES048743Y/ASSET/IMAGES/LARGE/ES048743YF00007.JPEG

Henze, M.; Comeau, Y. Wastewater Characterization. *Books.Google.Com*, 2008. https://www.google.com/books?hl=en&lr=&id=41JButufnm8C&oi=fnd&pg=PA33&dq=Henze+M,+Comeau+Y+ (2008)+Wastewater+characterization.+In:+Henze+M,+van+Loosdrecht+MCM,+Ekama+GA,+Brdjanovic+D+ (eds).+ISBN:+9781843391883Biological+wastewater+treatment:+prin-+ciples+

Herrera-Becerra, R.; Zorrilla, C.; Ascencio, J. A. Production of Iron Oxide Nanoparticles by a Biosynthesis Method: An Environmentally Friendly Route. *J. Phys. Chem. C* **2007,** *111* (44), 16147–16153. https://doi.org/10.1021/JP072259A/ASSET/IMAGES/MEDIUM/JP072259AE00003.GIF

Hoag, G. E.; Collins, J. B.; Holcomb, J. L.; Hoag, J. R.; Nadagouda, M. N.; Varma, R. S. Degradation of Bromothymol Blue by 'Greener' Nano-Scale Zero-Valent Iron Synthesized Using Tea Polyphenols. *J. Mater. Chem.* **2009,** *19* (45), 8671–8677. https://doi.org/10.1039/B909148C

Huang, L.; Luo, F.; Chen, Z.; Megharaj, M.; Naidu, R. Green Synthesized Conditions Impacting on the Reactivity of Fe NPs for the Degradation of Malachite Green. *Spectrochim. Acta Part A: Mol. Biomol. Spectrosc.* **2015,** *137*, 154–159. https://doi.org/10.1016/J.SAA.2014.08.116

Huang, L.; Weng, X.; Chen, Z.; Megharaj, M.; Naidu, R. Green Synthesis of Iron Nanoparticles by Various Tea Extracts: Comparative Study of the Reactivity. *Spectrochim. Acta Part A: Mol. Biomol. Spectrosc.* **2014a,** *130*, 295–301. https://doi.org/10.1016/J.SAA.2014.04.037

Huang, L.; Weng, X.; Chen, Z.; Megharaj, M.; Naidu, R. Synthesis of Iron-Based Nanoparticles Using Oolong Tea Extract for the Degradation of Malachite Green. *Spectrochim. Acta Part A: Mol. Biomol. Spectrosc.* **2014b,** *117*, 801–804. https://doi.org/10.1016/J.SAA.2013.09.054

Iravani, S. Green Synthesis of Metal Nanoparticles Using Plants. *Green Chem.* **2011,** *13* (10), 2638–2650. https://doi.org/10.1039/C1GC15386B

Jalill, R. D. A.; Nuaman, R. S.; Abd, A. N. Silver Nitrate and Zirconium Oxide Nanoparticles as Management of Wheat Damping-Off Caused by Fusarium Graminearum. *J. Genet. Environ. Resour. Conserv.* **2016.** https://www.researchgate.net/profile/Rasha-Saad-3/publication/332551232_Silver_nitrate_and_zirconium_oxide_nanoparticles_as_management_of_wheat_damping-off_caused_by_Fusarium_graminearum/links/5cbccd9d299bf12097765e3a/Silver-nitrate-and-zirconium-oxide-na

Jeetah, R.; Bhaw-Luximon, A.; Jhurry, D. Nanopharmaceutics: Phytochemical-Based Controlled or Sustained Drug-Delivery Systems for Cancer Treatment. *J. Biomed. Nanotechnol.* **2014,** *10* (9), 1810–1840. https://doi.org/10.1166/JBN.2014.1884

Kalaiarasi, R.; Jayallakshmi, N. Phytosynthesis of Nanoparticles and Its Applications. *Plant Cell Biotechnol.*. https://www.cabdirect.org/cabdirect/abstract/20133284851

Kates, R. W.; Parris, T. M.; Leiserowitz, A. A. What Is Sustainable Development? Goals, Indicators, Values, and Practice. **2012,** *47* (3), 8–21. https://doi.org/10.1080/00139157.2005.10524444

Khalil, A. T.; Ovais, M.; Ullah, I.; Ali, M.; Khan Shinwari, Z.; Maaza, M. Biosynthesis of Iron Oxide (Fe2O3) Nanoparticles via Aqueous Extracts of Sageretia Thea (Osbeck.) and Their Pharmacognostic Properties. **2017,** *10* (4), 186–201. https://doi.org/10.1080/17518253.2017.1339831

Khan, S. A.; Shahid, S.; Lee, C. S. Green Synthesis of Gold and Silver Nanoparticles Using Leaf Extract of Clerodendrum inerme; Characterization, Antimicrobial, and Antioxidant Activities. *Biomolecules* **2020,** *10* (6), 835. https://doi.org/10.3390/BIOM10060835

Khataee, A.; Kayan, B.; Kalderis, D.; Karimi, A.; Akay, S.; Konsolakis, M. Ultrasound-Assisted Removal of Acid Red 17 Using Nanosized Fe3O4-Loaded Coffee Waste Hydrochar. *Ultrasonics Sonochem.* **2017,** *35*, 72–80. https://doi.org/10.1016/J.ULTSONCH.2016.09.004

Kiruba Daniel, S. C. G.; Vinothini, G.; Subramanian, N.; Nehru, K.; Sivakumar, M. Biosynthesis of Cu, ZVI, and Ag Nanoparticles Using Dodonaea Viscosa Extract for Antibacterial Activity Against Human Pathogens. *J. Nanopart. Res.* **2013,** *15* (1), 1–10. https://doi.org/10.1007/S11051-012-1319-1/FIGURES/8

Kjellén, M.; Pensulo, C.; Nordqvist, P.; Madeleine, F. Global Review of Sanitation System Trends and Interactions with Menstrual Management Practices. *Stockholm Environment Institute, Project Report*, 2011.

Kozma, G.; Rónavári, A.; Kónya, Z.; Kukovecz, Á. Environmentally Benign Synthesis Methods of Zero-Valent Iron Nanoparticles. *ACS Sustain. Chem. Eng.* **2016,** *4* (1), 291–297. https://doi.org/10.1021/ACSSUSCHEMENG.5B01185/ASSET/IMAGES/LARGE/SC-2015-01185G_0007.JPEG

Kuang, Y.; Wang, Q.; Chen, Z.; Megharaj, M.; Naidu, R. Heterogeneous Fenton-Like Oxidation of Monochlorobenzene Using Green Synthesis of Iron Nanoparticles. *J. Colloid Interface Sci.* **2013,** *410*, 67–73. https://doi.org/10.1016/J.JCIS.2013.08.020

Kumar, R.; Singh, N.; Pandey, S. N. Potential of Green Synthesized Zero-Valent Iron Nanoparticles for Remediation of Lead-Contaminated Water. *Int. J. Environ. Sci. Technol.* **2015,** *12* (12), 3943–3950. https://doi.org/10.1007/S13762-015-0751-Z/FIGURES/6

Kumari, M.; Punia, S.; Lohchab, R. K.; Haritash, A. K. Green Synthesis of Nanoparticles and its Application in Sewage Treatment. **2021,** *4* (5), 861–866. https://doi.org/10.15680/IJMRSET.2021.0405034

Lee, J.; Kim, H. Y.; Zhou, H.; Hwang, S.; Koh, K.; Han, D. W.; Lee, J. Green Synthesis of Phytochemical-Stabilized Au Nanoparticles Under Ambient Conditions and Their Biocompatibility and Antioxidative Activity. *J. Mater. Chem.* **2011,** *21* (35), 13316–13326. https://doi.org/10.1039/C1JM11592H

Liu, Y.; Jin, X.; Chen, Z. The Formation of Iron Nanoparticles by Eucalyptus Leaf Extract and Used to Remove Cr (VI). *Sci. Total Environ.* **2018,** *627*, 470–479. https://doi.org/10.1016/J.SCITOTENV.2018.01.241

Lu, H.; Xue, Z.; Saikaly, P.; Nunes, S. P.; Bluver, T. R.; Liu, W. T. Membrane Biofouling in a Wastewater Nitrification Reactor: Microbial Succession from Autotrophic Colonization to Heterotrophic Domination. *Water Res.* **2016,** *88*, 337–345. https://doi.org/10.1016/J.WATRES.2015.10.013

Lunge, S.; Singh, S.; Sinha, A. Magnetic Iron Oxide (Fe3O4) Nanoparticles from Tea Waste for Arsenic Removal. *J. Magn. Magn. Mater.* **2014,** *356*, 21–31. https://doi.org/10.1016/J.JMMM.2013.12.008

Luo, F.; Chen, Z.; Megharaj, M.; Naidu, R. Biomolecules in Grape Leaf Extract Involved in One-Step Synthesis of Iron-Based Nanoparticles. *RSC Adv.* **2014,** *4* (96), 53467–53474. https://doi.org/10.1039/C4RA08808E

Luo, F.; Yang, D.; Chen, Z.; Megharaj, M.; Naidu, R. Characterization of Bimetallic Fe/Pd Nanoparticles by Grape Leaf Aqueous Extract and Identification of Active Biomolecules Involved in the Synthesis. *Sci. Total Environ.* **2016,** *562*, 526–532. https://doi.org/10.1016/J.SCITOTENV.2016.04.060

Machado, S.; Grosso, J. P.; Nouws, H. P. A.; Albergaria, J. T.; Delerue-Matos, C. Utilization of Food Industry Wastes for the Production of Zero-Valent Iron Nanoparticles. *Sci. Total Environ.* **2014,** *496*, 233–240. https://doi.org/10.1016/J.SCITOTENV.2014.07.058

Machado, S.; Pacheco, J. G.; Nouws, H. P. A.; Albergaria, J. T.; Delerue-Matos, C. Characterization of Green Zero-Valent Iron Nanoparticles Produced with Tree Leaf Extracts. *Sci. Total Environ.* **2015,** *533*, 76–81. https://doi.org/10.1016/J.SCITOTENV.2015.06.091

Machado, S.; Pinto, S. L.; Grosso, J. P.; Nouws, H. P. A.; Albergaria, J. T.; Delerue-Matos, C. Green Production of Zero-Valent Iron Nanoparticles Using Tree Leaf Extracts. *Sci. Total Environ.* **2013,** *445–446*, 1–8. https://doi.org/10.1016/J.SCITOTENV.2012.12.033

Machado, S.; Stawiński, W.; Slonina, P.; Pinto, A. R.; Grosso, J. P.; Nouws, H. P. A. et al. Application of Green Zero-Valent Iron Nanoparticles to the Remediation of Soils Contaminated with Ibuprofen. *Sci. Total Environ.* **2013,** *461–462*, 323–329. https://doi.org/10.1016/J.SCITOTENV.2013.05.016

Madhavi, V.; Prasad, T. N. V. K. V.; Reddy, A. V. B.; Ravindra Reddy, B.; Madhavi, G. Application of Phytogenic Zerovalent Iron Nanoparticles in the Adsorption of Hexavalent Chromium. *Spectrochim. Acta Part A: Mol. Biomol. Spectrosc.* **2013,** *116*, 17–25. https://doi.org/10.1016/J.SAA.2013.06.045

Madhubala, V.; Kalaivani, T. Phyto and Hydrothermal Synthesis of Fe3O4@ZnO Core-Shell Nanoparticles Using Azadirachta Indica and Its Cytotoxicity Studies. *Appl. Surf. Sci.* **2018,** *449*, 584–590. https://doi.org/10.1016/J.APSUSC.2017.12.105

Mahmoud, M. E.; Amira, M. F.; Zaghloul, A. A.; Ibrahim, G. A. A. Microwave-Enforced Sorption of Heavy Metals from Aqueous Solutions on the Surface of Magnetic Iron Oxide-Functionalized-3-Aminopropyltriethoxysilane. *Chem. Eng. J.* **2016,** *293*, 200–206. https://doi.org/10.1016/J.CEJ.2016.02.056

Mallikarjuna, K.; Nasif, O.; Alharbi, S. A.; Chinni, S. V.; Reddy, L. V.; Reddy, M. R. V.; Sreeramanan, S. Phytogenic Synthesis of Pd-Ag/rGO Nanostructures Using Stevia Leaf Extract for Photocatalytic H2 Production and Antibacterial Studies. *Biomolecules* **2021,** *11* (2), 190. https://doi.org/10.3390/BIOM11020190

Markova, Z.; Novak, P.; Kaslik, J.; Plachtova, P.; Brazdova, M.; Jancula, D. et al. Iron (II,III)-Polyphenol Complex Nanoparticles Derived from Green Tea with Remarkable Ecotoxicological Impact. *ACS Sustain. Chem. Eng.* **2014,** *2* (7), 1674–1680. https://doi.org/10.1021/SC5001435/ASSET/IMAGES/LARGE/SC-2014-001435_0006.JPEG

Mehrotra, N.; Tripathi, R. M.; Zafar, F.; Singh, M. P. Catalytic Degradation of Dichlorvos Using Biosynthesized Zero Valent Iron Nanoparticles. *IEEE Trans. Nanobiosci.* **2017,** *16* (4), 280–286. https://doi.org/10.1109/TNB.2017.2700232

Mekonnen, M. M.; Hoekstra, A. Y. Four Billion People Facing Severe Water Scarcity. *Sci. Adv.* **2016,** *2* (2). https://doi.org/10.1126/SCIADV.1500323

Mondal, P.; Anweshan, A.; Purkait, M. K. Green Synthesis and Environmental Application of Iron-Based Nanomaterials and Nanocomposite: A Review. *Chemosphere* **2020,** *259*, 127509. https://doi.org/10.1016/j.chemosphere.2020.127509

Monga, Y.; Kumar, P.; Sharma, R. K.; Filip, J.; Varma, R. S.; Zbořil, R.; Gawande, M. B. Sustainable Synthesis of Nanoscale Zerovalent Iron Particles for Environmental Remediation. *ChemSusChem* **2020,** *13* (13), 3288–3305. https://doi.org/10.1002/cssc.202000290

Mukunthan, K. S.; Balaji, S. Silver Nanoparticles Shoot Up from the Root of Daucus carrota (L.). *Int. J. Green Nanotechnol.* **2012,** *4* (1), 54–61. https://doi.org/10.1080/19430892.2012.654745

Murgueitio, E.; Cumbal, L.; Abril, M.; Izquierdo, A.; Debut, A.; Tinoco, O. Green Synthesis of Iron Nanoparticles: Application on the Removal of Petroleum Oil from Contaminated Water and Soils. *J. Nanotechnol.* **2018,** *2018*. https://doi.org/10.1155/2018/4184769

Narayanan, K. B.; Sakthivel, N. Green Synthesis of Biogenic Metal Nanoparticles by Terrestrial and Aquatic Phototrophic and Heterotrophic Eukaryotes and Biocompatible Agents. *Adv. Colloid Interf. Sci.* **2011,** *169* (2), 59–79. https://doi.org/10.1016/J.CIS.2011.08.004

Naseem, T.; Farrukh, M. A. Antibacterial Activity of Green Synthesis of Iron Nanoparticles Using Lawsonia Inermis and Gardenia Jasminoides Leaves Extract. *J. Chem.* **2015,** *2015*. https://doi.org/10.1155/2015/912342

Nasrollahzadeh, M.; Sajadi, S. M. Synthesis and Characterization of Titanium Dioxide Nanoparticles Using Euphorbia Heteradena Jaub Root Extract and Evaluation of Their Stability. *Ceram. Int.* **2015,** *41* (10), 14435–14439. https://doi.org/10.1016/J.CERAMINT.2015.07.079

Nath, D.; Banerjee, P. Green Nanotechnology—A New Hope for Medical Biology. *Environ. Toxicol. Pharmacol.* **2013,** *36* (3), 997–1014. https://doi.org/10.1016/J.ETAP.2013.09.002

Niraimathee, V. A.; Subha, V.; Ernest Ravindran, R. S.; Renganathan, S. Green Synthesis of Iron Oxide Nanoparticles from *Mimosa pudica* Root Extract. *Int. J. Environ. Sustain. Dev.* **2016,** *15* (3), 227–240. https://doi.org/10.1504/IJESD.2016.077370

Njagi, E. C.; Huang, H.; Stafford, L.; Genuino, H.; Galindo, H. M.; Collins, J. B. et al. Biosynthesis of Iron and Silver Nanoparticles at Room Temperature Using Aqueous Sorghum Bran Extracts. *Langmuir* **2011,** *27* (1), 264–271. https://doi.org/10.1021/LA103190N/ASSET/IMAGES/LARGE/LA-2010-03190N_0011.JPEG

Omer, A. M. Energy, Environment and Sustainable Development. *Renew. Sustain. Energy Rev.* **2008,** *12* (9), 2265–2300. https://doi.org/10.1016/J.RSER.2007.05.001

Pattanayak, M.; Nayak, P. Green Synthesis and Characterization of Zero Valent Iron Nanoparticles from the Leaf Extract of *Azadirachta indica* (Neem). *World J. Nano Sci.* **2013**. Retrieved from https://www.academia.edu/download/75838853/2.pdf

Plachtová, P.; Medříková, Z.; Zbořil, R.; Tuček, J.; Varma, R. S.; Maršálek, B. Iron and Iron Oxide Nanoparticles Synthesized with Green Tea Extract: Differences in Ecotoxicological Profile and Ability to Degrade Malachite Green. *ACS Sustain. Chem. Eng.* **2018,** *6* (7), 8679–8687. https://doi.org/10.1021/ACSSUSCHEMENG.8B00986/ASSET/IMAGES/LARGE/SC-2018-009863_0006.JPEG

Ponder, S. M.; Darab, J. G.; Mallouk, T. E. Remediation of Cr (VI) and Pb (II) Aqueous Solutions Using Supported, Nanoscale Zero-Valent Iron. *Environ. Sci. Technol.* **2000,** *34* (12), 2564–2569. https://doi.org/10.1021/ES9911420/ASSET/IMAGES/LARGE/ES9911420F00006.JPEG

Prasad, C.; Krishna Murthy, P.; Hari Krishna, R. H.; Sreenivasa Rao, R.; Suneetha, V.; Venkateswarlu, P. Bio-Inspired Green Synthesis of RGO/Fe3O4 Magnetic Nanoparticles Using Murrayakoenigii Leaves Extract and Its Application for Removal of Pb (II) from Aqueous Solution. *J. Environ. Chem. Eng.* **2017,** *5* (5), 4374–4380. https://doi.org/10.1016/J.JECE.2017.07.026

Qadir, M.; Sharma, B. R.; Bruggeman, A.; Choukr-Allah, R.; Karajeh, F. Non-Conventional Water Resources and Opportunities for Water Augmentation to Achieve Food Security in Water Scarce Countries. *Agric. Water Manag.* **2007,** *87* (1), 2–22. https://doi.org/10.1016/J.AGWAT.2006.03.018

Rajakumar, G.; Rahuman, A. A.; Priyamvada, B.; Khanna, V. G.; Kumar, D. K.; Sujin, P. J. Eclipta Prostrata Leaf Aqueous Extract Mediated Synthesis of Titanium Dioxide Nanoparticles. *Mater. Lett.* **2012,** *68*, 115–117. https://doi.org/10.1016/J.MATLET.2011.10.038

Rao, A.; Bankar, A.; Kumar, A. R.; Gosavi, S.; Zinjarde, S. Removal of Hexavalent Chromium Ions by *Yarrowia lipolytica* Cells Modified with Phyto-Inspired Fe0/Fe3O4 Nanoparticles. *J. Contamin. Hydrol.* **2013,** *146*, 63–73. https://doi.org/10.1016/J.JCONHYD.2012.12.008

Raut, R. W.; Mendhulkar, V. D.; Kashid, S. B. Photosensitized Synthesis of Silver Nanoparticles Using *Withania somnifera* Leaf Powder and Silver Nitrate. *J. Photochem. Photobiol. B: Biol.* **2014,** *132*, 45–55. https://doi.org/10.1016/J.JPHOTOBIOL.2014.02.001

Razavi, M.; Salahinejad, E.; Fahmy, M.; Yazdimamaghani, M.; Vashaee, D.; Tayebi, L. Green Chemical and Biological Synthesis of Nanoparticles and Their Biomedical Applications. In: *Green Processes for Nanotechnology: From Inorganic to Bioinspired Nanomaterials*; 2015; pp 207–235. https://doi.org/10.1007/978-3-319-15461-9_7/COVER

Saccà, M. L.; Fajardo, C.; Costa, G.; Lobo, C.; Nande, M.; Martin, M. Integrating Classical and Molecular Approaches to Evaluate the Impact of Nanosized Zero-Valent Iron (nZVI) on Soil Organisms. *Chemosphere* **2014,** *104*, 184–189. https://doi.org/10.1016/J.CHEMOSPHERE.2013.11.013

Sajadi, S. M.; Nasrollahzadeh, M.; Maham, M. Aqueous Extract from Seeds of *Silybum marianum* L. as a Green Material for Preparation of the Cu/Fe3O4 Nanoparticles: A Magnetically Recoverable and Reusable Catalyst for the Reduction of Nitroarenes. *J. Colloid Interf. Sci.* **2016,** *469*, 93–98. https://doi.org/10.1016/J.JCIS.2016.02.009

Sankar, R.; Rizwana, K.; Shivashangari, K. S.; Ravikumar, V. Ultra-Rapid Photocatalytic Activity of *Azadirachta indica* Engineered Colloidal Titanium Dioxide Nanoparticles. *Appl. Nanosci.* **2015,** *5* (6), 731–736. https://doi.org/10.1007/S13204-014-0369-3/FIGURES/5

Santhoshkumar, T.; Rahuman, A. A.; Jayaseelan, C.; Rajakumar, G.; Marimuthu, S.; Kirthi, A. V. et al. Green Synthesis of Titanium Dioxide Nanoparticles Using *Psidium guajava* Extract and Its Antibacterial and Antioxidant Properties. *Asian Pac. J. Trop. Med.* **2014,** *7* (12), 968–976. https://doi.org/10.1016/S1995-7645 (14)60171-1

Sathya, K.; Saravanathamizhan, R.; Baskar, G. Ultrasound Assisted Phytosynthesis of Iron Oxide Nanoparticle. *Ultrasonics Sonochem.* **2017,** *39*, 446–451. https://doi.org/10.1016/J.ULTSONCH.2017.05.017

Savasari, M.; Emadi, M.; Bahmanyar, M. A.; Biparva, P. Optimization of Cd (II) Removal from Aqueous Solution by Ascorbic Acid-Stabilized Zero Valent Iron Nanoparticles Using Response Surface Methodology. *J. Ind. Eng. Chem.* **2015,** *21*, 1403–1409. https://doi.org/10.1016/J.JIEC.2014.06.014

Sebastian, A.; Nangia, A.; Prasad, M. N. V. A Green Synthetic Route to Phenolics Fabricated Magnetite Nanoparticles from Coconut Husk Extract: Implications to Treat Metal

Contaminated Water and Heavy Metal Stress in *Oryza sativa* L. *J. Clean. Product.* **2018,** *174*, 355–366. https://doi.org/10.1016/J.JCLEPRO.2017.10.343

Senthil, M.; Ramesh, C. Biogenic Synthesis of Fe3O4 Nanoparticles Using Tridax Procumbens Leaf Extract and Its Antibacterial Activity on *Pseudomonas aeruginosa. Digest J. Nanomater. Biostruct.* **2012,** *7* (4), 1655–1661.

Shafiq, M.; Anjum, S.; Hano, C.; Anjum, I.; Abbasi, B. H. An Overview of the Applications of Nanomaterials and Nanodevices in the Food Industry. *Foods* **2020,** , *9* (2), 148. https://doi.org/10.3390/FOODS9020148

Shanker, U.; Jassal, V.; Rani, M. Green Synthesis of Iron Hexacyanoferrate Nanoparticles: Potential Candidate for the Degradation of Toxic PAHs. *J. Environ. Chem. Eng.* **2017,** *5* (4), 4108–4120. https://doi.org/10.1016/J.JECE.2017.07.042

Siddiqi, K. S.; Husen, A. Current Status of Plant Metabolite-Based Fabrication of Copper/Copper Oxide Nanoparticles and Their Applications: A Review. *Biomater. Res.* **2020,** *24* (1), 1–15. https://doi.org/10.1186/s40824-020-00188-1

Singh, K. K.; Senapati, K. K.; Sarma, K. C. Synthesis of Superparamagnetic Fe3O4 Nanoparticles Coated with Green Tea Polyphenols and Their Use for Removal of Dye Pollutant from Aqueous Solution. *J. Environ. Chem. Eng.* **2017,** *5* (3), 2214–2221. https://doi.org/10.1016/J.JECE.2017.04.022

Soliemanzadeh, A.; Fekri, M. The Application of Green Tea Extract to Prepare Bentonite-Supported Nanoscale Zero-Valent Iron and Its Performance on Removal of Cr (VI): Effect of Relative Parameters and Soil Experiments. *Microporous Mesoporous Mater.* **2017,** *239*, 60–69. https://doi.org/10.1016/j.micromeso.2016.09.050

Tandon, P. K.; Shukla, R. C.; Singh, S. B. Removal of Arsenic (III) from Water with Clay-Supported Zerovalent Iron Nanoparticles Synthesized with the Help of Tea Liquor. *Ind. Eng. Chem. Res.* **2013,** *52* (30), 10052–10058. https://doi.org/10.1021/IE400702K/SUPPL_FILE/IE400702K_SI_005.PDF

Tang, S. C. N.; Lo, I. M. C. Magnetic Nanoparticles: Essential Factors for Sustainable Environmental Applications. *Water Res.* **2013,** *47* (8), 2613–2632. https://doi.org/10.1016/J.WATRES.2013.02.039

Tiwari, D. K.; Tiwari, D. K.; Behari, J.; Sen, P. Application of Nanoparticles in Waste Water Treatment 1. *CiteSeer*. http://130.203.136.95/viewdoc/summary?doi=10.1.1.388.2845

Velayutham, K.; Rahuman, A. A.; Rajakumar, G.; Santhoshkumar, T.; Marimuthu, S.; Jayaseelan, C. et al. Evaluation of Catharanthus Roseus Leaf Extract-Mediated Biosynthesis of Titanium Dioxide Nanoparticles Against *Hippobosca maculata* and *Bovicola ovis. Parasitol. Res.* **2012,** *111* (6), 2329–2337. https://doi.org/10.1007/S00436-011-2676-X/FIGURES/4

Venkateswarlu, S.; Natesh Kumar, B.; Prasad, C. H.; Venkateswarlu, P.; Jyothi, N. V. V. Bio-Inspired Green Synthesis of Fe3O4 Spherical Magnetic Nanoparticles Using *Syzygium cumini* Seed Extract. *Physica B: Condensed Matter* **2014,** *449*, 67–71. https://doi.org/10.1016/J.PHYSB.2014.04.031

Venkateswarlu, S.; Yoon, M. Surfactant-Free Green Synthesis of Fe3O4 Nanoparticles Capped with 3,4-dihydroxyphenethylcarbamodithioate: Stable Recyclable Magnetic Nanoparticles for the Rapid and Efficient Removal of Hg (II) Ions from Water. *Dalton Trans.* **2015,** *44* (42), 18427–18437. https://doi.org/10.1039/C5DT03155A

Vittori Antisari, L.; Carbone, S.; Gatti, A.; Vianello, G.; Nannipieri, P. Toxicity of Metal Oxide (CeO2, Fe3O4, SnO2) Engineered Nanoparticles on Soil Microbial Biomass and

Their Distribution in Soil. *Soil Biol. Biochem.* **2013,** *60*, 87–94. https://doi.org/10.1016/J.SOILBIO.2013.01.016

Wang, T.; Jin, X.; Chen, Z.; Megharaj, M.; Naidu, R. Green Synthesis of Fe Nanoparticles Using Eucalyptus Leaf Extracts for Treatment of Eutrophic Wastewater. *Sci. Total Environ.* **2014,** *466–467*, 210–213. https://doi.org/10.1016/J.SCITOTENV.2013.07.022

Wang, T.; Lin, J.; Chen, Z.; Megharaj, M.; Naidu, R. Green Synthesized Iron Nanoparticles by Green Tea and Eucalyptus Leaves Extracts Used for Removal of Nitrate in Aqueous Solution. *J. Clean. Product.* **2014,** *83*, 413–419. https://doi.org/10.1016/J.JCLEPRO.2014.07.006

Wang, Z. Iron Complex Nanoparticles Synthesized by Eucalyptus Leaves. *ACS Sustain. Chem. Eng.* **2013,** *1* (12), 1551–1554. https://doi.org/10.1021/SC400174A/ASSET/IMAGES/LARGE/SC-2013-00174A_0008.JPEG

Weng, C. H. Water Pollution Prevention and State of the Art Treatment Technologies. *Environ. Sci. Pollut. Res.* **2020,** *27* (28), 34583–34585. https://doi.org/10.1007/s11356-020-09994-5

WHO. *Progress on Drinking Water and Sanitation: 2012 Update*, 2012.

Xiao, Z.; Yuan, M.; Yang, B.; Liu, Z.; Huang, J.; Sun, D. Plant-Mediated Synthesis of Highly Active Iron Nanoparticles for Cr (VI) Removal: Investigation of the Leading Biomolecules. *Chemosphere* **2016,** *150*, 357–364. https://doi.org/10.1016/J.CHEMOSPHERE.2016.02.056

Xiao, Z.; Zhang, H.; Xu, Y.; Yuan, M.; Jing, X.; Huang, J. et al. Ultra-Efficient Removal of Chromium from Aqueous Medium by Biogenic Iron Based Nanoparticles. *Sep. Purif. Technol.* **2017,** *174*, 466–473. https://doi.org/10.1016/J.SEPPUR.2016.10.047

Zaeem, A.; Drouet, S.; Anjum, S.; Khurshid, R.; Younas, M.; Blondeau, J. P. et al. Effects of Biogenic Zinc Oxide Nanoparticles on Growth and Oxidative Stress Response in Flax Seedlings vs. In Vitro Cultures: A Comparative Analysis. *Biomolecules* **2020,** *10* (6), 918. https://doi.org/10.3390/BIOM10060918

Zambre, A.; Upendran, A.; Shukla, R.; Chanda, N.; Katti, K. K.; Cutler, C. et al. Green Nanotechnology—A Sustainable Approach in the Nanorevolution. *R. Soc. Chem.* **2012,** 144–156. https://doi.org/10.1039/9781849735469-00144

Zhang, Y.; Jiao, X.; Liu, N.; Lv, J.; Chemosphere, Y. Y. *Enhanced Removal of Aqueous Cr (VI) by a Green Synthesized Nanoscale Zero-Valent Iron Supported on Oak Wood Biochar.* Elsevier, 2020. https://www.sciencedirect.com/science/article/pii/S0045653519327821

CHAPTER 14

Plant-Derived Nanocellulosic Material: A Promising Technology Application in Environmental Bioremediation

HEMAVATHI BRIJESH[1], BRIJESH AJJAPPALA[2], ANKITA CHATTERJEE[1], RAMYA MANJUNATH[1], HARISH BYRAPURA SIDDAPPA[3], and SHILPA BOREHALLI MAYEGOWDA[4]

[1]*Department of Biotechnology, School of Applied Sciences, REVA University, Bangalore, Karnataka, India*

[2]*Department of Plant Biotechnology, University of Agricultural Sciences, GKVK, Bangalore, Karnataka, India*

[3]*Department of Biotechnology, University of Horticultural Sciences, Bangalore, Karnataka, India*

[4]*Department of Psychology, Christ University, Bengaluru, Karnataka, India*

ABSTRACT

Nanocellulose (NC) polymers derived from plant sources are gaining enormous interest in environmental remediation owing to their low cost and potential for renewable adsorption. Plant-derived nanocellulose is applied in waste water treatment because of its unique features and functionality. The word "nanocellulose" refers to cellulosic materials having a dimension of nanoscopic scale/or nanoscale. One such nanomaterial is a cellulose-based material with a well-aligned nanocellulose composition indicating its structural hierarchy. Nanocellulose has been recognized as a remarkable natural biomaterial adsorbent which is obtained from renewable sources such as

Sustainable Nanoremediation: Modern Technologies for a Clean Environment.
Fernanda Maria Policarpo Tonelli, Arpita Roy, and Khalid Rehman Hakeem (Eds.)

wood, plants, fruit peel, can be found abundantly on earth, and biodegradable and can be easily used in the surface fabrication. Due to its increased surface area, nanocellulose has gained considerable advantage over conventional cellulose fibers. Application of nanocellulosic material in environmental remediation and wastewater treatment has recently emerged as a potential adsorbent generating, and aroused much attention in addressing the environmental issue. Nanocellulose may adsorb a wide range of contaminants, such as heavy metals, dissolved pollutants (organic), dyes, petroleum oil, and unwanted effluents. This review provides focus on the structure, properties, isolation, and adsorbent classes of nanocellulosic materials, as well as their applications in environmental remediation.

14.1 INTRODUCTION

In an effort for sustainable development, there has been a lot of attention directed to the growing usage of green, biodegradable, and renewable materials in place of nonrenewable resources. The nanomaterial technology has recently made it possible to detect and eliminate the toxic gases, organic/inorganic pollutants, and bio-based materials. They can perform as adsorbents, catalysts, or membranes, and can be available with different ranges of morphologies. The accessibility of the abundant plant resources has made it possible to produce green nanomaterials from a wide variety of plant extracts (Lu and Ozcan, 2015). Due to its extensive availability in the natural form and a renewable resource, cellulose is considered to be one of the best polymers present mainly in the form of lignocellulosic biomass which is generated from agricultural waste, oil crops, and forestry residues. The terminology "nanocellulose" refers to cellulose particles with one dimension at the nanoscale (1–100 nm). The effectiveness and broad applicability of cellulose offer a number of benefits, which include cost effectiveness, surface modification, sustainability, high aspect ratio, elevated adsorbent capacity, good mechanical strength and rigidity, strong chemical resistance, and the ability to selectively adsorb via appropriate surface functionalization. The nontoxic properties of nanocellulose materials, lack of environmental side effects, and low safety concerns are the most crucial requirements for using them (Bhatnagar et al., 2015). Cellulose nanocrystals (CNCs), cellulose nanofibers (CNFs), and bacterial nanocellulose (BNC) are the types of nanocellulose categories classified based on the source and extraction technique employed and hold tremendous potential in environmental cleanup due to its low cost and renewable adsorption (Tshikovhi et al., 2020). CNCs and CNFs have been used as construction materials in membranes applied for wastewater treatment, gas separation, removal of dyes, removal of heavy metals (Cruz-Tato et al., 2017; Ansaloni et al., 2017; Karim et al., 2014; Chitpong and Husson, 2017). This review emphasizes about the structure, properties, isolation of nanocellulose materials derived from plant-based lignocellulose, and recent progress in their application in environmental remediation (Fig. 14.1).

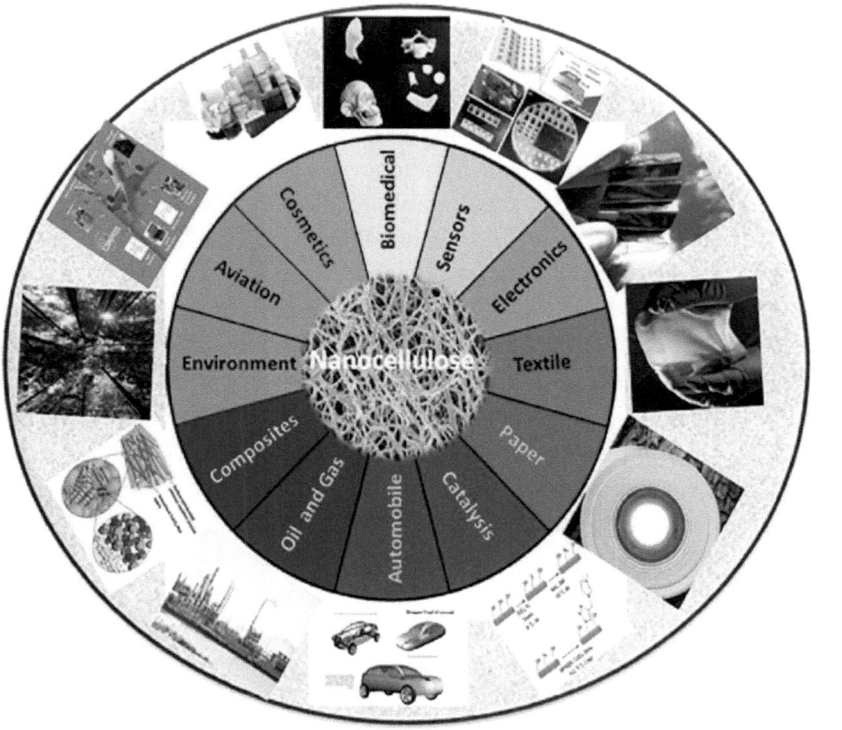

FIGURE 14.1 Smart materials based on nanocellulose.

Source: Reprinted with permission from Mishra et al., 2020. Copyright © 2020 Elsevier Inc.

14.2 STRUCTURE OF NANOCELLULOSIC MATERIALS

Nanocellulose $(C_6H_{10}O_5)n$ is the material obtained from cellulose and is constituted of a lengthy chain of homoglycans or homo-polysaccharides that comprise of dextrose (D-glucose) and repeated units of cellobiose $(C_{12}H_{22}O_{11})$ and the two glucose $(C_6H_{12}O_6)$ molecules that are connected by a β-1,4 glycosidic linkage (Klemm et al., 2005). Nearly 10,000 glucose/glucopyranose units make up the cellulose derived from wood products, whereas about

15,000 glucose units make up the cellulose derived from cotton. Several plant sources (corn, cotton, rice, husk, bamboo wood pulp) have been used to produce nanocellulosic polymers (CNFs and CNCs) (Singla et al., 2016). CNCs are rod-like crystalline structures which are made up of elongated, cylindrical, less flexible nanoparticles with a width of 4–70 nm, a length of 100–6000 nm. CNFs are elongated fibers with relatively tiny diameter in the nanoscale range that are produced by crushing cellulose under high pressure, which creates highly twisted fibril networks.

14.3 PROPERTIES OF NANOCELLULOSIC MATERIAL

The presence of nanosize filler in the nanocellulose reinforced composites will replace the conventional synthetic polymer composites (Gan et al., 2020). This reinforcing is accountable to its superior mechanical strength, high stiffness, and light weight. Nanocellulose also offers a great utility by surface modification which can be achieved either during the preparation step or post-production to improve the efficiency for the given application. Desired chemical can be incorporated to the surface to modify the reaction of nanocelluloses with foreign substances (Wei et al., 2017). According to the literature, CNCs are easy to be functionalized and reactive because of its increased surface area-to-volume ratio. The chemical modification of CNCs enhances their compatibility by stabilized negative and positive charges on the surface which on the other hand helps in better distribution of particles (Kaboorani and Riedl, 2015). Few authors have shown light on surface modification methods namely, nucleophilic substitution, silylation, polymer grafting, sulfonation, carbamation, esterification, etherification, and oxidation. Use of polymer coatings and adsorbing surfactants is also reported (Daud and Lee, 2017; Huang et al., 2020). Nanocellulose composites having tremendous features like high mechanical strength, nonporous character, tunable electrical and thermal conductivity, stretchability, photothermal and photodynamic activity, high adsorption capacity, and flexibility promise itself in a varied array of applications in water purification, storage and conversion, food packaging, construction of fire retardants, shape memory devices, and biomedical uses (Bacakova et al., 2020).

14.4 NANOCELLULOSIC MATERIALS DERIVED FROM PLANTS

Cellulosic fibers extracted from plants are one of the most thoroughly investigated and primarily utilized materials used for the fabrication of nanocellulose. Six different types of plant-derived fibers are categorized as the source of cellulose and they are: bast, core, grass and reed, leaf, and seed fibers (García et al., 2006). Wood pulps are the most extensively employed plant fiber for producing cellulose with high purity, robustness, and flexible networks, with excellent physical attributes when compared to other plant derivatives (Menon et al., 2017). Nanocellulose obtained from wood pulp has been used since past several years to modify the catalysts by fabrication for reducing 4-nitrophenol and membranes employed for removal of microorganisms like bacteria and virus including and heavy metals using microfiltration (Lin et al., 2011; Wang et al., 2013). Advances in the plant-based nanocellulosic technology have a lot to offer in the field of biomaterial sciences for environmental remediation.

14.4.1 ISOLATION OF CELLULOSE NANOFIBERS (CNFs) AND CELLULOSE NANOCRYSTALS (CNCs)

The length and diameter of CNCs have nano-dimensions, whereas CNFs are structurally in micro-dimensions length, and diameter in the nano-dimension. Strong acid treatment has the potential to hydrolyze the amorphous component of the cellulose microfibers, creating rod-like crystalline structures of CNCs (Fig. 14.2). Acid hydrolysis of cellulose microfibrils has been done using a variety of acids, including sulfuric acid, hydrochloric acid, oxalic acid, hydrobromic acid, and nitric acid (Maiti et al., 2013; Sonia et al., 2013; Lee et al., 2009). Cellulosic material subjected for acid hydrolysis and treated with sulfuric acid has been very effective and widely employed to extract CNCs. CNCs were resulted in the range of 100–300 nm length when extracted using acid hydrolysis a 60% (V/V) sulfuric acid solution from waste paper with constant stirring at 45°C (Danial et al., 2015). Another technique for extracting cellulose nanofibers is TEMPO (2,2,6,6-tetra methyl piperidine-1-oxyl), which entails altering the native cellulosic fibers from hydroxyl group to a carboxylate group while preserving their fibrous form and crystallinity (Saito et al., 2007). Raw dried cotton fibers were successfully fibrillated into single-cellulose nanofibers using chemical purification and combining high-speed blending with high-pressure homogenization (Chen et al., 2014). Similarly, CNFs extracted from poplar wood combined with explosive chemical pretreatment and high-intensity ultrasonic treatment (more than 1000 W) resulted in 5–20 nm width, and several microns lengthy CNFs (Chen et al., 2014). CNFs were successfully extracted from the banana plant by the autoclave-based steam explosion process and by chemical (alkaline treatment) and enzymatic treatment (xylanase) which resulted in diameter of 7.6 nm and a length of 2889.7 nm when the nanofibers were extracted from the enzymatic treatment (Tibolla et al., 2014) (Fig. 14.2).

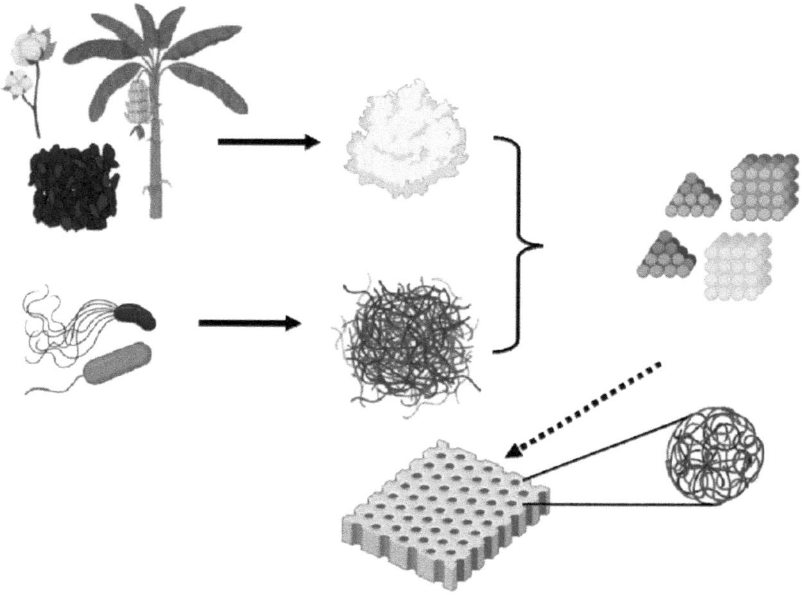

FIGURE 14.2 Cellulose nanomaterials are obtained through mechanical and chemical processing of cellulose feed stocks as well as direct production from living organisms. Acid hydrolysis of cellulose nanofibers (CNFs) to remove amorphous regions yields cellulose nanocrystals (CNCs).

14.5 ADSORBENT CLASSES OF NANOCELLULOSIC MATERIAL

Nanocellulosic material is considered to be an effective adsorbent due to their biodegradability, nontoxicity, and surface reactivity characteristics. Along with the adsorption properties, stability of these materials in water plays an important role in the reduction of biofouling in the environment (Habibi, 2014). Based on the dimension, shape, properties, function, and utilization, nanocellulose is categorized into many classes. Inter- and intramolecular interaction of hydrogen bonds are responsible for the formation of CNCs. The morphology of CNCs depends greatly on the cellulose source from which it is produced (Nechyporchuk et al., 2016). Cellulose nanofibers include the group of cellulosic fibrillar fines, nanofibrillar cellulose, nano-scale fibrillated cellulose, nanofibrillated cellulose, nanofibers, and nanofibril. Often, microfibrils are also included in the category of cellulose nanofibrils (Trache et al., 2017). Cellulose nanofibril, specifically, microfibrils was first reported with the isolation of cellulose microfibrils around 1983. Slurry of cellulose fibers of softwood pulp were first exposed to high-pressure homogenizer under extreme shear forces for the synthesis of the nanofibrils. This process is important to strengthen the entangled arrangements of the nanofibrils. Nanofibrils possess both crystalline and amorphous properties due to the extreme forces. On the completion of the process, viscous gel is formed which is required to be homogenized followed by drying to finally retrieve the network of synthesized nanofibrils (Table 14.1). Though at certain times, cellulose nanofibrils are also designated as microfibrils, however, it is considered to be misleading since it does not represent the accurate dimensions of the fibrils. The diameter of cellulose nanofibrils ranges between 5 and 50 nm (Khalil et al., 2012). A detailed description of cellulosic nanofibers synthesis is tabulated in Table 14.1. Aerogels are artificial porous material which has extremely low density. Although carbon and silica are often used for aerogels synthesis, at certain times, even nanocelluloses are used for producing the network structure. Nanocellulose aerogels are one of the unique mesoporous structures which are effectively advantageous when compared to inorganic aerogels. Nanocellulose-based aerogels are best produced using freeze–drying. In this case, the solvent is usually frozen followed by sublimation. The nanocellulose aerogels are environmental friendly due to their renewable sources of production. Flexibility, immense mechanical stability, ductility, and increased compression strength are few properties which result in the uniqueness of nanocellulose aerogels (Yuan et al., 2020) (Fig. 14.2).

14.6 APPLICATION OF NANOCELLULOSIC MATERIALS IN ENVIRONMENTAL REMEDIATION

14.6.1 NANOCELLULOSE-DERIVED MATERIALS AS ADSORBENTS

Adsorption is a process where a liquid solute or adsorbate accumulates on the adsorbent surface (Singh et al., 2018). Extensive industrialization has resulted in the accumulation of organic (organic solvents, dyes etc.) and inorganic water pollutants (heavy metals, salts etc.). Because of its high affinity

TABLE 14.1 Synthesis of Cellulose Nanofibers.

Name of the nanocellulose	Steps	Description	Methods
Cellulose nanofibers	Pretreatment		Enzyme hydrolysis: with the help of cellobiohydrolases and endoglucanase
			Application of alkali and acid: Sodium hydroxides and hydrochloric acid are preferred
			Utilization of organic salts
	Mechanical treatment	Mechanical treatment is carried out for defibrillation of cellulosic materials.	Homogenization: The cellulosic pulp is subjected to high pressure, velocity, and shear forces.
			Microfluidization: In this case, intensifier pump is applied to increase the pressure for better defibrillation.
			Grinding: Grinding involves passing of the pulp through stones where one stone rotates around one fixed stone causing a force for breakdown of the fibers.
			Cryocrushing: In this process, the cellulose is soaked in water and allowed to absorb water followed by immersing them in liquid nitrogen. This results in the solidification of the water crystals which can be crushed using mortar and pestle for the conversion of cellulose intro nanocellulose.
			High-intensity ultrasonication: Application of hydrodynamic ultrasound forces along with oscillation power is implemented to obtain the cellulose fibrils.
			Ball milling process: In this process, the suspension of cellulose is placed within a hollow cylinder which contains ceramic, zirconia, and metal balls. As the container rotates, due to the collision, cellulose fibers are obtained.
	Chemical hydrolysis	Chemical processes aid in broadening the spaces between hydroxyl groups, or breaking the hydrogen bonds and thereby alteration of crystalline characters.	Acid hydrolysis, preferably using hydrochloric and sulfuric acids.

adsorption for organic contaminants, nanocellulose has been considered a potent adsorbent derived from natural biomaterial for wastewater treatment (Fig. 14.3). One of the lignocellulose biomass resulted in the form of empty fruit bunch (EFB)-based nanocellulose was made functional with activated carbon which could be utilized for the effective water remediation process. Septevani et al. (2020) demonstrated the efficiency of selectivity with an 86% (24.94 mg/g) metal adsorption capacity within 3 min and was almost greater than that of the rice-straw nanocellulose by approximately two-fold. CNC-derived nanocellulose was successfully applied for the removal of dye (disperse yellow) from its aqueous solution (Mutar and Jasim, 2021). Another application examined the potential of nanocellulose isolated from cinnamon developed by hydrolysis with acid catalyst for the decolorization of azo dye methyl orange and its dye removal was found to be 40.64%, which was twice as high as that of an uncatalytic process at room temperature (Nematollahi et al., 2022). Recently, Wu et al. (2022) reported an interesting revelation pertaining to the remarkable adsorption ability (14.71 mg/g) of a nanocellulosic membrane under extreme acidic pH (pH 1.22) with high stability, when membrane was formulated using CNF along with chitosan using gel casting method (Fig. 14.4). The membrane stability against a highly acidic environment was responsible due to the bonding interactions between the hydrogen and ion groups of chitosan and CNF. Another study developed nanocellulose adsorbents for removing cadmium and nickel ions from polluted water by treating cotton microfibers with acid, followed by filtering (Taleb, 2016) (Fig. 14.4).

14.6.2 NANOCELLULOSIC MATERIALS AS FLOCCULANT

Enormous interest has been generated in developing flocculants derived from biopolymers from plants viz. chitosan, tannins, cellulose in bio-based waste water treatment considered to be an efficient alternative by many researchers. Natural bio-based flocculants are biodegradable in nature and belong to two different types (Zhu et al., 2015). One of the categories includes implanting natural polymers with polyacrylamide to develop semi-natural flocculants, such as hydroxypropyl methyl cellulose. The second kind is produced by altering natural polymers directly to produce improvised native version of nanocellulosic flocculants (dicarboxylic acid) (Das et al., 2013). The hydroxyl (-OH) groups on the cellulose allow for excellent functionalization of nanocellulose, providing it the desired functionality as an effective

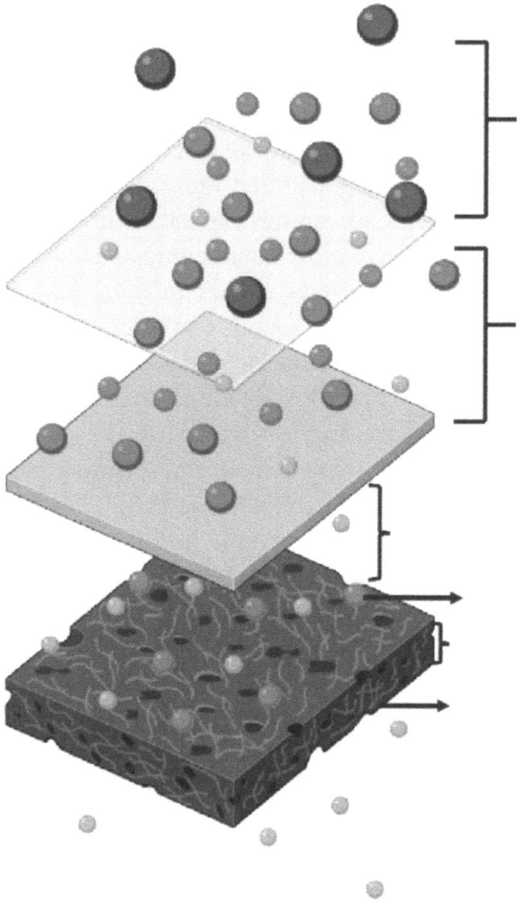

FIGURE 14.3 Application of nanocellulose as a potent adsorbent which is also obtained from various natural biomaterials, implicated in the treatment of wastewater and sewage.

flocculant (Eyley, 2015). Shak et al. (2018) have reviewed in detail about the possible applications of cellulosic derivative flocculants in wastewater treatment (Fig. 14.5). Previously, functionality of nanocellulose materials obtained from birch fibers of *Betula verrucosa* and *B. pendula* species that were bleached before the fabrication of dicarboxylic acid nanocellulose flocculant examined its performance in flocculation process when treating municipal wastewater along with anionic and cationic dialdehyde as water chemicals were tested (Suopajärvi et al., 2017). The observations demonstrated the potential utility of functionalized nanocellulose in wastewater treatment (Fig. 14.5).

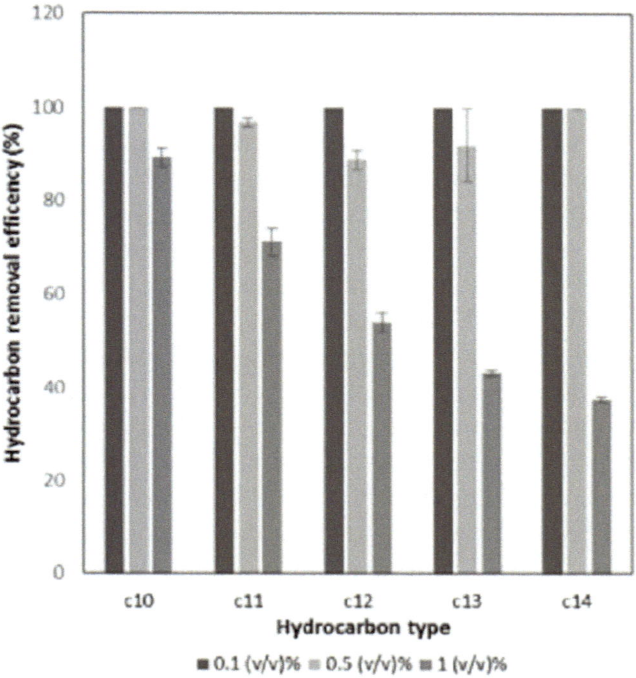

FIGURE 14.4 Advantages of nanocellulose over synthetic fibers.

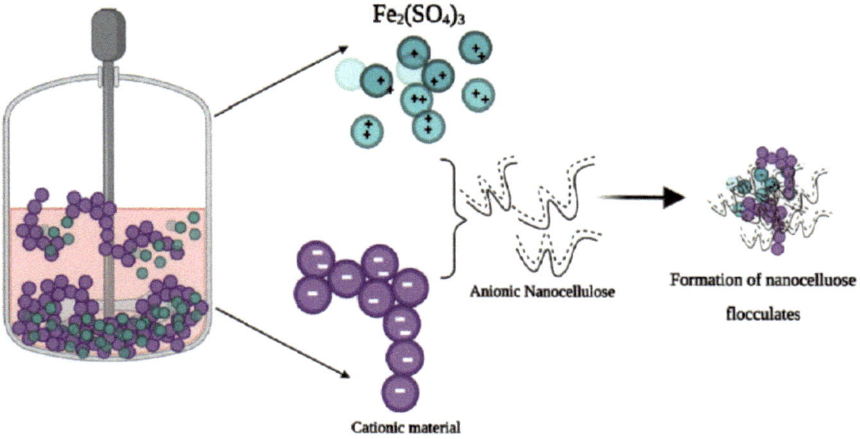

FIGURE 14.5 Natural bio-based flocculation using nanocellulose which poses hydroxypropyl methyl cellulose to fasten the process.

14.6.3 NANOCELLULOSIC MATERIALS AS PHOTOCATALYST

Photocatalytic destruction of different organic contaminants could be achieved using photocatalyst nanostructured with cellulose-based materials. Recently, a lot of interest has been generated for the possible utilization of fibrillated cellulose as a substrate or carrier for catalytic nanoparticles. Applying irradiation of UV and visible light, various research teams have exhibited photocatalytic wastewater treatment employing metal oxide nanostructures derived from cellulose which are made into thin films, membranes, fibers, and other hybrid materials. For example, degradation rate of organic pollutants was better when metal oxide composites such as TiO_2, ZnO, graphene oxide, and Fe_2O_3 coupled with nanocellulose when compared to nanocellulosic materials alone. The production of hydrogen generation by photocatalytic process was higher when using the nano-fibrillated cellulose/magnetite/titanium dioxide nanocomposites compared to the nanofibrillated cellulose/titanium dioxide sample (An et al., 2017). Due to its simplicity of collection and recycling, the use of nanocellulosic films and membranes in wastewater treatment demonstrated tremendous potential and also reaction system displayed absence of photocatalyst remnant. Additionally, it is possible to rapidly encapsulate small particle sized oxide (metal) nanoparticles and more surface area in the cellulosic thin film to produce a hybrid material (metal oxide nanoparticle/cellulose nanocomposite). Three photocatalysts, including commercial (TiO_2 P_{25}), laboratory constructed (TiO_2), and graphene oxide coated TiO_2 (GO-TiO_2), were fabricated on cellulose membranes (flat sheet) that demonstrated multi-functionality such as water filtration coupled with photodegradation of inorganic pollutant and dye (diphenhydramine and methyl orange). These studies revealed the efficacy of employing cellulosic film to provide support to the TiO_2 nanoparticle together with limiting the loss of TiO_2 during photocatalytic treatment of wastewater (Pastrana-Martínez et al., 2015). Another noteworthy application of nanocellulosic technology is gas separation, where CO2 gas produced as a result of various anthropogenic activities triggering the greenhouse effect can be effectively removed by nanocellulose-based membranes (Jahan et al., 2018). Nanocellulosic membranes are selectively allowed to pass CO2 gas from the feed side to the permeate side. Nanocellulose was used as an adjuvant component to the PVA membrane for bifurcation of mixture of gases (CO_2/N_2) in the study investigated by Torstensen et al. (2019) (Fig. 14.6).

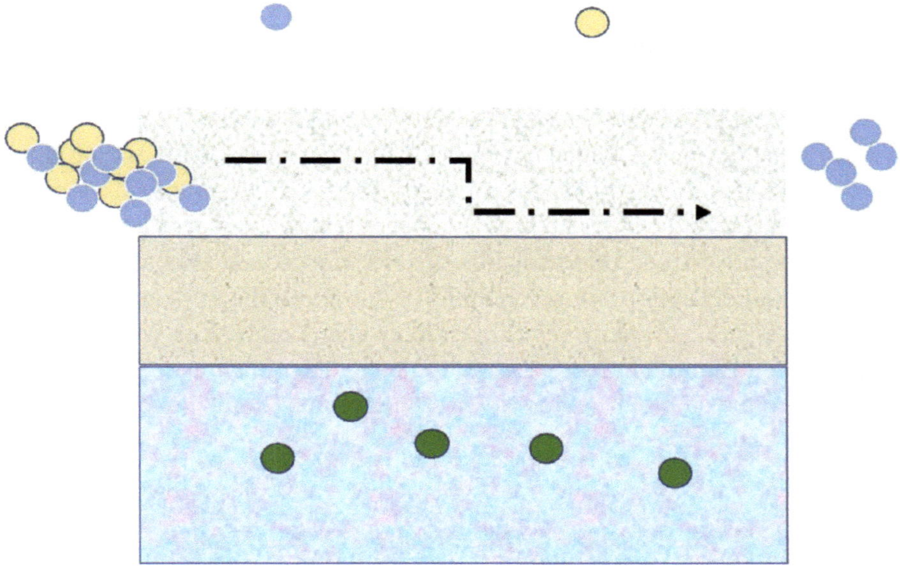

FIGURE 14.6 Mechanism of CO2/N2 separation using PVA/CNC as active layer.

14.6.4 NANOCELLULOSIC MATERIALS AS MEMBRANES

Membrane waste water treatment processes are by far the most prominent and dominant industrial methods due to their ease of scalability, low operational costs, versatility, ease of installation into existing technologies, low chemical footprint, efficiency, and excellent properties with enhanced adsorption capacity (Ray et al., 2020). Diverse nanocellulosic-derived composites have been studied for the removal of various pollutants in water, including cellulose/carbon nanotube (CNT) composites, cellulose/graphene nanocomposites, cellulose/metal nanocomposites, cellulose/ceramic nanocomposites, etc. By employing cellulose nanocrystals and chitosan, positively charged dyes have been eliminated using cross-linked and non-cross-linked processes that boosted the membrane's mechanical and dimensional stability, including efficiency of pollutant removal (Karim et al., 2014). In a different work, tissue paper (TP) was effectively transformed into an effective separation membrane using a unique superhydrophobic covering made from nanocellulose (Roy et al., 2018). As mentioned, cellulose nanofibers (CNFs) and octadecylamine were used in a straightforward one-step procedure coupled with glutaraldehyde (GA) in the presence of alcohol and deionized water to generate a unique, environmentally friendly, and benign superhydrophobic layer. It was observed that the modified TP was converted into a highly effective membrane for the partition of oil and water combinations when the coating was applied to the TP (Roy et al., 2018). An effective and cost-effective hydrophobic "nanowood" membrane was produced by Princeton University researchers with high porosity (89. 3%) (Hou et al., 2019). Water vapor transfer was facilitated based on the allocation of pore size of the native crystalline nanofibrils made from woody materials in conjunction with the xylem vessels and channels. The developed material is a possible substitute for membranes made of petroleum since it has an exceptional water flow and thermal efficacy of more than 60%. Contrary to the intricate fabrication procedures used to construct conventional membranes like polypropylene and polytetrafluoroethylene, it was revealed that this "nanowood" membrane was generated employing a top-down strategy which could be scalable easily for industrial purposes.

14.6.5 NANOCELLULOSIC MATERIALS AS POLLUTANT SENSORS

Technology advancements have produced sensors, a blooming and versatile functional material that responds to stimuli and is ideally used in applications ranging from the environment to health care. Sensors are signaling agents which detect and sense changes in their environment (physical, chemical, and biological) and transform them into analytical signals. Nanocellulosic material in native state is essential for developing new sensors, especially for creating a multidimensional architecture (Shahi et al., 2021). Cholesterol oxidase (ChOx) was immobilized on crystalline nanocellulose/ionic liquid/polyaniline to develop the electrochemical cholesterol biosensor. The sensors developed were extremely sensitive and could detect cholesterol as low as to 0.48 μM concentration (Abdi et al., 2019). Previous report suggests that CNF and (TEMPO)-oxidized cellulose nanofibers/glycerol sensors displayed effective gas-sensing ability against various inorganic chemicals such as acetone, ammonia, methane, and hydrogen sulfide (Shahi et al., 2021).The sensor was devised to detect cyanide in different water samples by inserting stable copper nanoparticles in the nanocellulosic structure (ECNPs-NC) by directly overlaying copper ions onto the nanocellulose by chemical reduction process with minimal quantity of sodium borohydride (NaBH4) (Pouzesh et al., 2019). This approach has a number of key advantages, including simplicity, the ability to sense in pure aqueous media, and the absence of any organic synthesis or solvent requirement. Recently, Jaffar et al. (2022) have mentioned various types of sensors employing nanocellulose and their application in the review. They have discussed various types of sensors used, and recent progress in the development and application of nanocellulose-based sensors viz, electrochemical, optical, colorimetric, fluorescent types, and also biosensors for the detection of pollutants including heavy metals, gases, minerals, and salts.

14.7 CONCLUSION

Nanocellulose derived from cellulosic waste products is considered to be efficient, promising, eco-friendly, and useful for minimizing carbon footprints. Research on their functional biomaterial property has expanded as a result of recent significant focus on the isolation and characterization of nanocellulose from various lignocellulosic feedstocks. The characteristics of nanocellulosic materials can be improved further by functionalization than by retaining them unaltered. These improvements include the maximal active binding sites, ion exchange properties, and the availability of active functional groups for large-scale industrial applications. Future applications for nanocellulose include the replacement of synthetic resin and regular petroleum fibers as well as the development of sophisticated materials for employment in building, aircraft, and fire retardants. Future developments will therefore facilitate the efficient and beneficial advancement of this "wonder" biomaterial for industrial applications by focusing on the economical and environmentally friendly processes of nanocellulose extraction and modification.

KEYWORDS

- **nanocellulose**
- **environmental remediation**
- **nanomaterials**
- **heavy metal**
- **adsorbent**

REFERENCES

Abdi, J.; Yahyanezhad, M.; Sakhaie, S.; Vossoughi, M.; Alemzadeh, I. Synthesis of Porous TiO2/ZrO2 Photocatalyst Derived from Zirconium Metal Organic Framework for Degradation of Organic Pollutants Under Visible Light Irradiation. *J. Environ. Chem. Eng.* **2019,** *7* (3), 103096.

An, X.; Cheng, D.; Dai, L.; Wang, B.; Ocampo, H.J.; Nasrallah, J.; Jia, X.; Zou, J.; Long, Y.; Ni, Y. Synthesis of Nano-Fibrillated Cellulose/Magnetite/Titanium Dioxide (NFC@ Fe3O4@ TNP) Nanocomposites and Their Application in the Photocatalytic Hydrogen Generation. *Appl. Catal. B: Environ.* **2017,** *206*, 53–64.

Ansaloni, L.; Salas-Gay, J.; Ligi, S.; Baschetti, M. G. Nanocellulose-Based Membranes for CO2 Capture. *J. Membr. Sci.* **2017,** *522*, 216–225.

Bacakova, L.; Pajorova, J.; Tomkova, M.; Matejka, R.; Broz, A.; Stepanovska, J.; Prazak, S.; Skogberg, A.; Siljander, S.; Kallio, P. Applications of Nanocellulose/Nanocarbon Composites: Focus on Biotechnology and Medicine. *Nanomater.* **2020,** *10* (2), 196.

Bhatnagar, A.; Sillanpää, M.; Witek-Krowiak, A. Agricultural Waste Peels as Versatile Biomass for Water Purification—A Review. *Chem. Eng. J.* **2015,** *270*, 244–271.

Chen, W.; Abe, K.; Uetani, K.; Yu, H.; Liu, Y.; Yano, H. Individual Cotton Cellulose Nanofibers: Pretreatment and Fibrillation Technique. *Cellulose* **2014,** *21* (3), 1517–1528.

Chitpong, N.; Husson, S. M. Polyacid Functionalized Cellulose Nanofiber Membranes for Removal of Heavy Metals from Impaired Waters. *J. Membr. Sci.* **2017,** *523*, 418–429.

Cruz-Tato, P.; Ortiz-Quiles, E. O.; Vega-Figueroa, K.; Santiago-Martoral, L.; Flynn, M.; Diaz-Vazquez, L. M.; Nicolau, E. Metalized Nanocellulose Composites as a Feasible Material for Membrane Supports: Design and Applications for Water Treatment. *Environ. Sci. Technol.* **2017,** *51* (8), 4585–4595.

Danial, W. H.; Majid, Z. A.; Muhid, M. N. M.; Triwahyono, S.; Bakar, M. B.; Ramli, Z. The Reuse of Wastepaper for the Extraction of Cellulose Nanocrystals. *Carbohydr. Polym.* **2015,** *118*, 165–169.

Das, R.; Ghorai, S.; Pal, S. Flocculation Characteristics of Polyacrylamide Grafted Hydroxypropyl Methyl Cellulose: An Efficient Biodegradable Flocculant. *Chem. Eng. J.* **2013,** *229*, 144–152.

Daud, J. B.; Lee, K.Y. Surface Modification of Nanocellulose. In *Handbook of Nanocellulose and Cellulose Nanocomposites*; Kargarzadeh, H., Ahmad, S. T., Dufresne, A., Eds.; Weinheim: Wiley, 2017; pp 101–122.

Eyley, S.; Vandamme, D.; Lama, S.; Van den Mooter, G.; Muylaert, K.; Thielemans, W. CO2 Controlled Flocculation of Microalgae Using pH Responsive Cellulose Nanocrystals. *Nanoscale* **2015,** *7* (34), 14413–14421.

Gan, P. G.; Sam, S. T.; Abdullah, M. F. B.; Omar, M. F. Thermal Properties of Nanocellulose-Reinforced Composites: A Review. *J. Appl. Polym. Sci.* **2020,** *137* (11), 48544.

García, A.; Gandini, A.; Labidi, J.; Belgacem, N.; Bras, J. Industrial and Crop Wastes: A New Source for Nanocellulose Biorefinery. *Ind. Crops Products* **2016,** 93, 26–38.

Habibi, Y. Key Advances in the Chemical Modification of Nanocelluloses. *Chem. Soc. Rev.* **2014,** *43* (5), 1519–1542.

Hou, D.; Li, T.; Chen, X.; He, S.; Dai, J.; Mofid, S.A.; Hou, D.; Iddya, A.; Jassby, D.; Yang, R.; Hu, L. Hydrophobic Nanostructured Wood Membrane for Thermally Efficient Distillation. *Sci. Adv.* **2019,** *5* (8), eaaw3203.

Huang, Z. X.; Cao, S. L.; Xu, P.; Wu, H.; Zong, M. H.; Lou, W. Y. Preparation of a Novel Nanobiocatalyst by Immobilizing Penicillin Acylase onto Magnetic Nanocrystalline Cellulose and Its Use for Efficient Synthesis of Cefaclor. *Chem. Eng. J.* **2018,** *346*, 361–368.

Jaffar, S. S.; Saallah, S.; Misson, M.; Siddiquee, S.; Roslan, J.; Saalah, S.; Lenggoro, W. Recent Development and Environmental Applications of Nanocellulose-Based Membranes. *Membranes* **2022,** *12* (3), 287.

Jahan, Z.; Niazi, M. B. K.; Hägg, M. B.; Gregersen, Ø. W. Decoupling the Effect of Membrane Thickness and CNC Concentration in PVA Based Nanocomposite Membranes for CO2/CH4 Separation. *Sep. Purif. Technol.* **2018,** *204*, 220–225.

Kaboorani, A.; Riedl, B. Surface Modification of Cellulose Nanocrystals (CNC) by a Cationic Surfactant. *Ind. Crops Products* **2015,** *65*, 45–55.

Karim, Z.; Mathew, A. P.; Grahn, M.; Mouzon, J.; Oksman, K. Nanoporous Membranes with Cellulose Nanocrystals as Functional Entity in Chitosan: Removal of Dyes from Water. *Carbohydr. Polym.* **2014,** *112*, 668–676.

Khalil, H. A.; Bhat, A. H.; Yusra, A. I. Green Composites from Sustainable Cellulose Nanofibrils: A Review. *Carbohydr. Polym.* **2012,** *87* (2), 963–979.

Klemm, D.; Heublein, B.; Fink, H. P.; Bohn, A. Cellulose: Fascinating Biopolymer and Sustainable Raw Material. *Angew. Chem. Int. Ed.* **2005,** *44* (22), 3358–3393.

Lee, S. Y.; Mohan, D. J.; Kang, I. A.; Doh, G. H.; Lee, S.; Han, S. O. Nanocellulose Reinforced PVA Composite Films: Effects of Acid Treatment and Filler Loading. *Fibers Polym.* **2009,** *10* (1), 77–82.

Lin, X.; Wu, M.; Wu, D.; Kuga, S.; Endo, T.; Huang, Y. Platinum nanoparticles using Wood Nanomaterials: Eco-Friendly Synthesis, Shape Control and Catalytic Activity for p-Nitrophenol Reduction. *Green Chem.* **2011,** *13* (2), 283–287.

Lu, Y.; Ozcan, S. Green Nanomaterials: On Track for a Sustainable Future. *Nano Today* **2015,** *10* (4), 417–420.

Maiti, S.; Jayaramudu, J.; Das, K.; Reddy, S. M.; Sadiku, R.; Ray, S. S.; Liu, D. Preparation and Characterization of Nano-Cellulose with New Shape from Different Precursor. *Carbohydr. Polym.* **2013,** *98* (1), 562–567.

Menon, M. P.; Selvakumar, R.; Kumar, P. S.; Ramakrishna, S. Extraction and Modification of Cellulose Nanofibers Derived from Biomass for Environmental Applications. *RSC Adv* **2017,** *7* (68), 42750–42773.

Mishra, D.; Shanker, K.; Khare, P. Nanocellulose-Mediated Fabrication of Sustainable Future Materials. In: *Sustainable Nanocellulose and Nanohydrogels from Natural Sources*; Elsevier, 2020; pp 217–236. DOI:10.1016/B978-0-12-816789-2.00010-9.

Mutar, H. R.; Jasim, K. K. Adsorption Study of Disperse Yellow Dye on Nanocellulose Surface. *Mater. Today: Proc.* **2021**.

Nechyporchuk, O.; Belgacem, M. N.; Bras, J. Production of Cellulose Nanofibrils: A Review of Recent Advances. *Ind. Crops Products* **2016,** *93*, 2–25.

Nematollahi, R.; Ziyadi, H.; Ghasemi, E.; Taheri, H. Cinnamon Nanocellulose as a Novel Catalyst to Remove Methyl Orange from Aqueous Solution. *Inorg. Chem. Commun.* **2022,** *137*, 109222.

Pastrana-Martinez, L. M.; Morales-Torres, S.; Figueiredo, J. L.; Faria, J. L.; Silva, A. M. Graphene Oxide Based Ultrafiltration Membranes for Photocatalytic Degradation of Organic Pollutants in Salty Water. *Water Res.* **2015,** *77*, 179–190.

Pouzesh, M.; Nekouei, S.; Ferdosi Zadeh, M. A.; Keshtpour, F.; Wang, S.; Nekouei, F. Fabrication of Stable Copper Nanoparticles Embedded in Nanocellulose Film as a

Bionanocomposite Plasmonic Sensor and Thereof for Optical Sensing of Cyanide Ion in Water Samples. *Cellulose* **2019,** *26* (8), 4945–4956.

Ray, S. S.; Iroegbu, A. O. C.; Bordado, J. C. Polymer-Based Membranes and Composites for Safe, Potable, and Usable Water: A Survey of Recent Advances. *Chem. Afr.* **2020,** *3* (3), 593–608.

Roy, S.; Zhai, L.; Van Hai, L.; Kim, J. W.; Park, J. H.; Kim, H. C.; Kim, J. One-Step Nanocellulose Coating Converts Tissue Paper into an Efficient Separation Membrane. *Cellulose* **2018,** *25* (9), 4871–4886.

Saito, T.; Kimura, S.; Nishiyama, Y.; Isogai, A. Cellulose Nanofibers Prepared by TEMPO-Mediated Oxidation of Native Cellulose. *Biomacromolecules* **2007,** *8* (8), 2485–2491.

Septevani, A. A.; Rifathin, A.; Sari, A. A.; Sampora, Y.; Ariani, G. N.; Sondari, D. Oil Palm Empty Fruit Bunch-Based Nanocellulose as a Super-Adsorbent for Water Remediation. *Carbohydr. Polym.* **2020,** *229*, 115433.

Shahi, N.; Lee, E.; Min, B.; Kim, D. J. Rice Husk-Derived Cellulose Nanofibers: A Potential Sensor for Water-Soluble Gases. *Sensors* **2021,** *21* (13), 4415.

Shak, K. P. Y.; Pang, Y. L.; Mah, S. K. Nanocellulose: Recent Advances and Its Prospects in Environmental Remediation. *Beilstein J. Nanotechnol.* **2018,** *9* (1), 2479–2498.

Singh, N. B.; Nagpal, G.; Agrawal, S. Water Purification by Using Adsorbents: A Review. *Environ. Technol. Innov.* **2018,** *11*, 187–240.

Singla, R.; Guliani, A.; Kumari, A.; Yadav, S. K. Metallic Nanoparticles, Toxicity Issues and Applications in Medicine. In: *Nanoscale Materials in Targeted Drug Delivery, Theragnosis and Tissue Regeneration*; Springer: Singapore, 2016; pp 41–80.

Sonia, A.; Dasan, K. P.; Alex, R. Celluloses Microfibres (CMF) Reinforced Poly (Ethylene-co-vinyl Acetate) (EVA) Composites: Dynamic Mechanical, Gamma and Thermal Ageing Studies. *Chem. Eng. J.* **2013,** *228*, 1214–1222.

Suopajärvi, T.; Sirviö, J. A.; Liimatainen, H. Nanofibrillation of Deep Eutectic Solvent-Treated Paper and Board Cellulose Pulps. *Carbohydr. Polym.* **2017,** *169*, 167–175.

Taleb, K.; Rusmirović, J.; Rančić, M.; Nikolić, J.; Drmanić, S.; Veličković, Z.; Marinković, A. Efficient Pollutants Removal by Amino-Modified Nanocellulose Impregnated with Iron Oxide. *J. Serbian Chem. Soc.* **2016,** *81* (10), 1199–1213.

Tibolla, H.; Pelissari, F. M.; Menegalli, F. C. Cellulose Nanofibers Produced from Banana Peel by Chemical and Enzymatic Treatment. *LWT* **2014,** *59* (2), 1311–1318.

Torstensen, J. Ø.; Helberg, R. M.; Deng, L.; Gregersen, Ø. W.; Syverud, K. PVA/Nanocellulose Nanocomposite Membranes for CO2 Separation from Flue Gas. *Int. J. Greenhouse Gas Control* **2019,** *81*, 93–102.

Trache, D.; Hussin, M. H.; Haafiz, M. M.; Thakur, V. K. Recent Progress in Cellulose Nanocrystals: Sources and Production. *Nanoscale* **2017,** *9* (5), 1763–1786.

Tshikovhi, A.; Mishra, S. B.; Mishra, A. K. Nanocellulose-Based Composites for the Removal of Contaminants from Wastewater. *Int. J. Biol. Macromol.* **2020,** *152*, 616–632.

Wang, R.; Guan, S.; Sato, A.; Wang, X.; Wang, Z.; Yang, R.; Chu, B. Nanofibrous Microfiltration Membranes Capable of Removing Bacteria, Viruses and Heavy Metal Ions. *J. Membr. Sci.* **2013,** *446*, 376–382.

Wei, L.; Agarwal, U. P.; Hirth, K. C.; Matuana, L. M.; Sabo, R. C.; Stark, N. M. Chemical Modification of Nanocellulose with Canola Oil Fatty Acid Methyl Ester. *Carbohydr. Polym.* **2017,** *169*, 108–116.

Wu, J.; Dong, Z.; Li, X.; Li, P.; Wei, J.; Hu, M.; Geng, L.; Peng, X. Constructing Acid-Resistant Chitosan/Cellulose Nanofibrils Composite Membrane for the Adsorption of Methylene Blue. *J. Environ. Chem. Eng.* **2022,** *10* (3), 107754.

Yuan, B.; Li, L.; Murugadoss, V.; Vupputuri, S.; Wang, J.; Alikhani, N.; Guo, Z. Nanocellulose-Based Composite Materials for Wastewater Treatment and Waste-Oil Remediation. *ES Food Agroforestry* **2020,** *1* (7), 41–52.

Zhu, H.; Zhang, Y.; Yang, X.; Liu, H.; Shao, L.; Zhang, X.; Yao, J. One-Step Green Synthesis of Non-Hazardous Dicarboxyl Cellulose Flocculant and Its Flocculation Activity Evaluation. *J. Hazard. Mater.* **2015,** *296*, 1–8.

CHAPTER 15

Green Iron Nanoparticles for Nanoremediation

SURESHKUMAR KEMPAHANUMAKKAGARI and T. RAMAKRISHNAPPA

Department of Chemistry, B.M.S. Institute of Technology and Management, Bengaluru, Karnataka, India

ABSTRACT

The iron nanoparticles (FeNPs) have been extensively used as nanoremediants for environmental protection from last two decades, because of its structural and chemical features. The FeNPs has been used for detoxifying organic (pesticides, insecticides, antibiotics, dyes, drugs, and phenolic compounds) and inorganic pollutants (toxic heavy metals like Cr^{6+},Pb^{2+},Hg^{2+}, Cd^{2+} As^{3+}, and As^{5+}) in water, soil, and sediments successfully. However, synthesis of FeNPs with good structural and functional features (stable homogeneous-sized particles with less aggregation, large surface area, and more reactivity) is still challenging. Numerous chemical and advanced technologies achieved the synthesis of FeNPs with the above-said properties. However, these technologies are not cost-effective, needs corrosive and environmentally toxic chemicals (as reducing and oxidising materials during synthesis), and also demand the sophisticated instrumentation. As an alternative, green approaches have emerged in the course of time for producing FeNPs. The approaches employed were non-corrosive, nontoxic, environmentally benign, as well as sustainable and chemicals or technologies were used. These approaches utilized various plant parts, microorganisms, and physical approaches (solvent less ball milling). However, producing FeNPs with necessary characteristics for nanoremidiation purpose is still not fully achieved using these greener technologies and hard efforts are in progress by research groups. This chapter comprehensively discusses the greener approaches used for FeNPs generation

Sustainable Nanoremediation: Modern Technologies for a Clean Environment.
Fernanda Maria Policarpo Tonelli, Arpita Roy, and Khalid Rehman Hakeem (Eds.)

for nanoremidiation applications. The challenges and future prospective regarding the same will be discussed in the final sections of this report.

15.1 INTRODUCTION

In recent years, concern regarding environmental pollution is tremendous and worldwide. This is to decrease deterioration of the air, water, and soil quality for future generations. The continuous circulation of the pollutants among air, water, and soil creates the source as well as sink. Initially, even pollutant concentration is less; however, continues exposure leads to accumulation and results in the toxicity. Various environmental protecting agencies have clearly set the threshold limit values (TLVs) of various organic and inorganic pollutants in order to decrease the toxicity to human beings. The organic pollutants obtained from petroleum, dye, drugs, and pesticides/insecticides include phenolic compounds, chlorides, and antibiotics. The inorganic pollutants especially heavy metal ions pose serious health effects to humans when exposed to environmental matric above TLVs. The health risks associated with the heavy metal ions include cancer, neurogenerative disorders, arthritis, and respiratory problems. The toxic heavy metals include Hg^{2+}, As^{3+}, As^{5+}, Pb^{2+}, Cr^{6+}, and Cd^{2+}, which will be generated by mining/industries and added to environmental flux. In order to clean up the above-mentioned pollutants in the environment, nanotechnology is most suitable through nanoremidiation. Nanotechnology was a celebrated field in the last decade in terms of financial support and research opportunities. It witnessed exponential increase in R&D activities in terms of scientific papers, patents, and projects in the last decade which provided sufficient knowledge regarding the development of new materials and the new possible applications of nanomaterials. Among various nanomaterials, iron nanoparticles (FeNPs) find application in numerous fields like catalysis, biotechnology, cosmetics, medical (drug delivery, cancer treatment, and anticancer drug treatment), space field, electronics, and environmental remediation (Wang et al., 2014; Xu et al., 2022). The environmental remediation involves the neutralization, removal by adsorption, and degradation of the toxic/hazardous components present in the environmental matrix (water and soil) using the FeNPs. The FeNPs percolate into the subsurface or small pores of the soil, or suspend in the water and react better for longer times, with high rate of dispersibility, and reach far into bulk materials (Machado et al., 2015). The FeNPs exists as metallic, bimetallic, trimetallic, or oxides. It is surprising that 16 different oxides of iron are possible. The FeNPs are used for nanoremediation, especially for environmental cleaning process. The cleaning process involves degradation of dyes, antibiotics, insecticides, pesticides from wastewater, removal of hazardous metals (by adsorption process), and reduction of toxic nitro-aromatics to non-toxic aromatic amines (catalytic reduction of nitro benzene to amino benzene). The special physical and chemical features of FeNPs, which includes large surface area, high reactivity, magnetic susceptibility, non-toxicity, and dual redox activity upon treating with water, make them useful in nanoremediation . The adsorption and catalytic activities of FeNPs are found to be 20 times higher than that of the bulk Fe-based materials. The techniques used for the synthesis of FeNPs can be broadly classified into two approaches, namely, top-down and bottom-up approach. The former technique involves grinding, mechanical milling, and machining, which breaks down larger particles to smaller one. Whereas, the latter approach involves process like sol-gel/chemical reduction strategies, which involves assembling of smaller particles to build larger structure. The top-down approaches demand specific instruments which are not cost-effective, whereas the bottom-up approaches are quite simple, in which the synthesis involves simple reactions, for example, Fe-based salt precursors will be chemically reduced using the reducing agents like $NaBH_4$ etc. However, these chemical methods are proved to be hazardous because of using reducing agents like $NaBH_4$ that has serious environmental concerns along with production of H_2 gas during reactions, which has to be carefully eliminated otherwise it would lead to explosion and therefore demands extra-sophisticated apparatus. This limited the chemical synthesis and search for alternative environmentally sustainable green protocols for the production of FeNPs. The green and sustainable technologies reported for synthesis of FeNPs includes use of biomaterials as reducing and capping agents. These methods utilize plant or microorganism extracts for production of FeNPs. The phytochemicals or enzymes present in these biomaterials (plant/animal extracts) acts as reducing and capping agents (Wang et al., 2014). These synthetic methods don't involve any toxic chemicals and, recently, some of the methods reported the synthesis at ambient conditions. These are perfect alternative to chemical methods as green and sustainable technologies for production of FeNPs without causing much damage to environment. The FeNPs obtained by greener approaches are mono-dispersed, non-toxic, less probability for agglomeration, and also the size and shape of the materials can be controlled by varying the concentration of plant extract/growth conditions of the microorganism. The added advantage of green synthesis of FeNPs will be the freely and abundantly available raw materials (plant and microorganisms). However, there are several concerns regarding the green protocols developed for nanomaterials, like they can't be scaled up to industrial level, and seldom green technologies has been commercialized. This report will summarize some of the recent green and sustainable protocols for FeNPs used for nanoremediation applications.

15.2 GREEN TECHNOLOGIES FOR FENPS SYNTHESIS

As mentioned in the introduction section, the chemical methods possess serious environmental pollution issues due to the use of toxic chemicals and toxic and flammable gases evolved during the synthesis process. Thus, physical methods like ultraviolet, aerosol, sonication, and thermal decomposition methods demand sophisticated instruments in order to achieve high temperatures (temperature of 2400 K necessary to form aerosol atomic droplets during aerosol nanoparticle synthesis) and pressures. So, alternative and environmentally sustainable methods are using biomaterials as reducing/capping agents for obtaining FeNPs. The advantages and environmental sustainability of these methods have been discussed in the introduction section. Generally, the green protocols utilize the various plant extracts (extracts of leaves, pericarp, roots, bark, flowers, and seeds) and microorganisms. In the case of plant extracts, phytochemicals, proteins, phenols, alkanes, aldehydes, vitamins, caffeine, amines, and enzymes will assist in FeNPs formation by acting as reducing or capping agents (Wang et al., 2014). In the case of microorganisms, the enzymes secreted will assist in nanoparticle formation in extracellular conditions. Some of the metal ions can penetrate into the cell membrane and nanoparticle formation occurs in intracellular conditions also. This section summaries some of the recent green protocols reported for synthesis of FeNPs for nanoremediation. The various bioresources used for production of Fe-based NPs and their nanoremediation activities have been illustrated in the Figure 15.1.

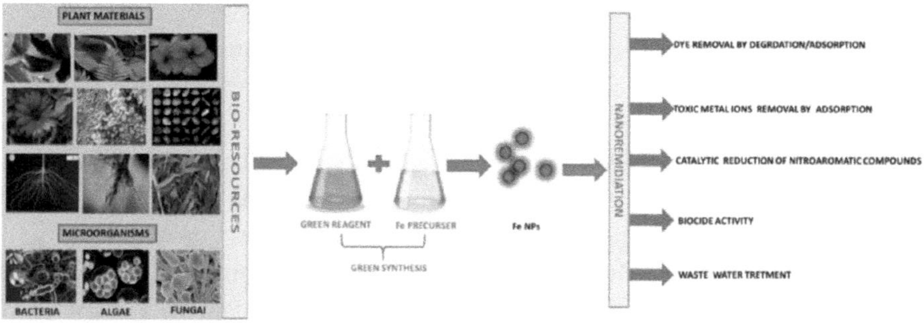

FIGURE 15.1 The bioresources used for FeNPs green synthesis and nanoremediation activities of green synthesized FeNPs.

15.2.1 FENPS SYNTHESIS USING PLANT MATERIALS

The green synthesis of zerovalent FeNPs was reported using leaf extracts of 26 different species of plants (Machado et al., 2015). The plant species used for this purpose includes apple, apricot, avocado, cherry, eucalyptus, kiwi, lemon, mandarin, medlar, mulberry, oak, olive, orange, passion fruit, peach, pear, pine, pomegranate, plum, quince, raspberry, strawberry, tea-black, tea-green, vine, and walnut (Machado et al., 2015). The leaves were collected and dried and powder was obtained by the physical milling process. The obtained powder was sieved to get fine particles. Green synthesis was performed by mixing 1 mL of aqueous extracts of leaf powder obtained in the above-discussed section with 250 μL of 0.1 M Fe (III) precursor solution followed by shaking. The nanoparticle formation was instantaneous, which was indicated by the darkening the color of the solution (Machado et al., 2015). Among all the leaf extracts, mulberry and pomegranate leaf extracts produced small zerovalent FeNPs with sizes ranging from 5 to 10 nm. Whereas the leaf extracts of pine, apple, and plum produced the larger size zerovalent FeNPs (nearly 100 nm) among all other leaf extracts (Machado et al., 2015). The tendency of agglomeration of green synthesized zerovalent FeNPs were studied. The zerovalent FeNPs synthesized using oak and eucalyptus started settling at concentrations of 4–9 ppm, whereas the zerovalent FeNPs synthesized from other leaf extracts didn't settled down even at concentrations greater than that of the 900 ppm (Machado et al., 2015). Another green synthetic approach reported for green synthesis of Fe_2O_3 NPs was using rose flower extract (Rosa Indica) (Prema et al., 2022). In this approach, the extract from rose flowers was prepared by boiling the petals of rose flower in hot water followed by filtration. The so-obtained rose flower petal extracts was mixed with 100 mM of $FeCl_3$ solution in 1:1 ratio. To this mixture, 1 M NaOH was added and sonicated for 1 h. The reaction yielded the pellet, which was washed with deionized water for several times in order to remove the contaminants (Prema et al., 2022). Finally, the pellet was desiccated at 80 °C for 3 h to yield the Fe_2O_3 NPs powder. The so-obtained Fe_2O_3 NPs were spherical and polydisperse with diameter of 70–120 nm. The Fe_2O_3 NPs exhibited mean average size of around 115.5 nm with polydispersity index (PDI) of 0.420 (Prema et al., 2022). The mixture of zerovalent FeNPs, FeO NPs, and Fe oxyhydroxide NPs were obtained using the jack fruit peel extract (Jain et al., 2021). First, jack fruit (*Artocarpus heterophyllus*) peel was collected and sundried, followed by being heated in hot air oven at 60 °C for 3 h. Then, the dried peel was grinded using pestle and mortar. The extract was prepared by heating the mixture of powder and water at 60 °C for 30 min. The obtained yellow color solution was cooled to room temperature, and filtered under vacuum. Then the extract and Fe precursor (0.1 M $FeCl_2$ solution) was mixed in 2:1 ratio and the solution pH was adjusted to 6.0 by adding the NaOH solution. The obtained black color solution was heated to evaporate the solvent and black powder was collected. The average size of the obtained FeNPs were found to be 33 nm (Jain et al., 2021). The eucalyptus leaf extracts were used to synthesize FeNPs, as hydrolysis of leaf extracts produces the organic acids like gallic acid, epicatechin, and quercetin-glucuronide. During hydrolysis, electrons generated will reduce Fe salt precursors to FeNPs (Wang et al., 2014). The FeNPs synthesis was also reported using watermelon rinds (Prasad et al., 2016). The Fourier-transform spectroscopic studies of obtained FeNPs revealed the surface functional groups responsible for Fe salt reduction to FeNPs. These include polyphenols, acid derivatives, and proteins (Prasad et al., 2016).The zerovalent FeNPs were also synthesized using the lemon and orange pulp waste extracts (Al-Juhaimi and Ghafoor 2013; Machado et al., 2014). The bimetallic nanoparticles also can be synthesized/stabilized using the green reagents. The bimetallic FeNPs like Fe-Ni, Ag-Fe_3O_4, and Fe-Cu were reported. The Fe-Ni bimetallic NPs had been synthesized using the eucalyptus leaves extract, which acts as reducing and the stabilizing agent. Similarly, Ag-Fe_3O_4 NPs were stabilized by carboxymethyl cellulose (CMC). Then Fe-Cu bimetallic nanoparticles were synthesized by using the chitosan as the stabilizing agent. The production of above three bimetallic FeNPs has been illustrated in Figure 15.2.

15.2.2 FENPS SYNTHESIS USING MICROORGANISMS

Alternative to plant materials for green nanoparticle synthesis, nature has provided us diverse microorganisms like bacteria, fungus, algae, yeast, and viruses. By using microorganisms, nanoparticles can be obtained by two approaches, that is, by extracellular or intracellular approach (Pereira et al., 2015). The nanoparticles can be synthesized by directly using the microorganisms itself or water-soluble molecules (DNA, enzymes), extracted or secreted by the microorganisms. Various biomolecules (reductase enzymes and other biomolecules) of afore-mentioned microorganism will assist in FeNPs formation by reduction of Fe salt precursors. The size as well as the shape of FeNPs obtained by microorganism-assisted synthesis can be controlled by controlling the factors affecting the microbial growth like concentration of the medium, temperature, and pH conditions. The intracellular as well as

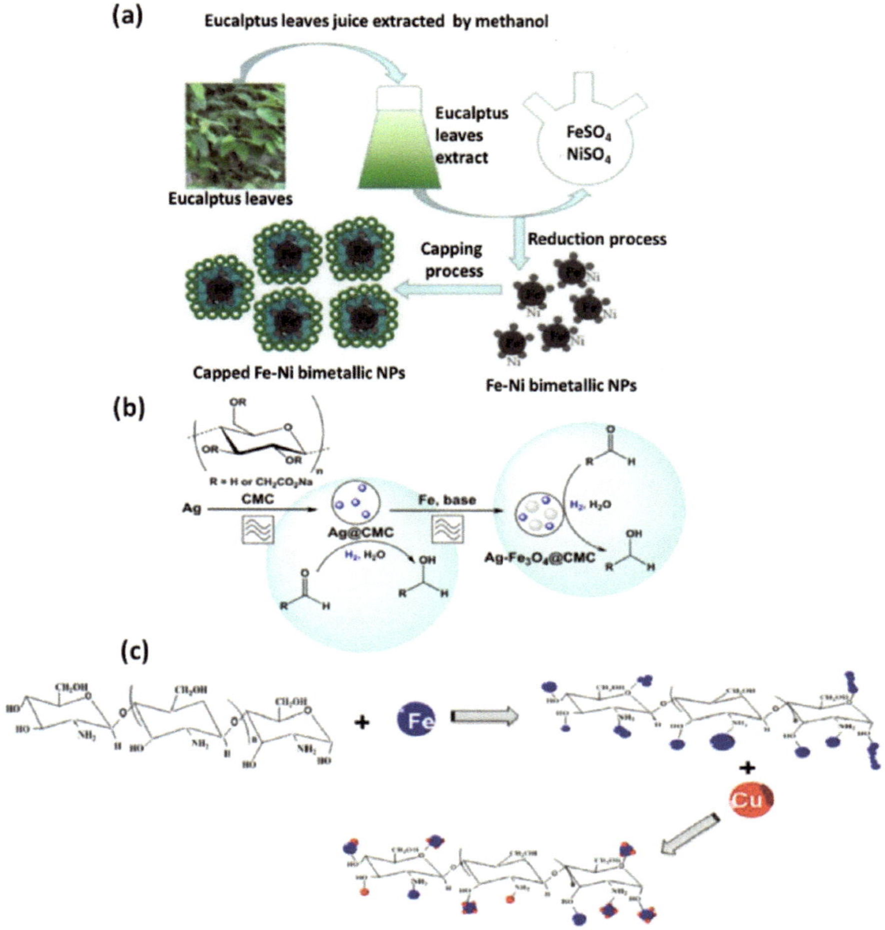

FIGURE 15.2 Illustrating the synthesis and stabilization of bimetallic nanoparticles: (a) Eucalyptus leaf extract assisted synthesis of Fe-Ni NPs, (b) CMC stabilized Ag-Fe NPs, and (c) Chitosan stabilized Fe-Cu NPs.

Source: a) Reprinted with permission from Weng et al., 2017. © 2016 Elsevier B.V. b) Reprinted with permission from Li et al., 2016. Copyright © 2016, American Chemical Society c) Reprinted with permission from Jiang et al., 2018. Copyright © 2018 Elsevier B.V.

extracellular synthesis of FeNPs using microorganisms has been illustrated in the Figure 15.3. The spherical Fe3O4 nanoparticles had been obtained by using *Aspergillus niger*. The *Aspergillus niger* decomposes FeCl2 and FeSO4 to Fe2O3 and FeS, respectively, through enzymatic action. The decomposed compounds react under supercritical ethanol environment to FeNPs (Abdeen et al., 2016; Dhillon et al., 2012).

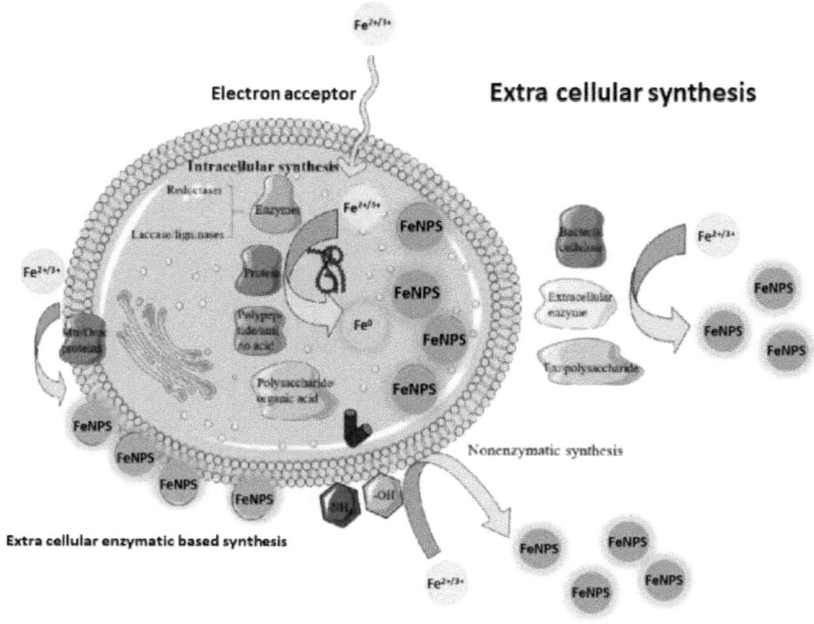

FIGURE 15.3 Illustrating possible mechanism involved in the FeNPs synthesis using microorganisms.

Source: Reprinted with permission from Yang et al., 2022. Copyright © 2022 Published by Elsevier Inc.

15.3 NANOREMEDIATION USING GREEN SYNTHESIZED FENPS

The Fe has been used for environmental remediation purposes since long. The bulk Fe has been used in reactive subsurface permeable barriers layers to remove chlorinated organic compounds in ground water. However, the advent of nanotechnology stopped using bulk Fe materials for environmental remediation and other purposes. This is because of more reactivity (FeNPs are proved to be 20 times more reactive than bulk Fe) of FeNPs than that of bulk Fe particles. Therefore, FeNPs have been extensively used in nanoremediation in order to clean up the polluted environment. The FeNPs can be directly poured into the contaminant sites in order to achieve effective in situ remediation with respect to removal/degradation of pollutants. Some studies reported that reducing agents can be directly injected into contaminated zones enriched with iron compounds in order to convert to zerovalent FE NPs. However, zerovalent Fe undergoes oxidation by exposure to air and moisture because of its electro donor nature (Guan et al., 2015). This affects the reactivity during nanoremediation and the yield of zerovalent Fe particles as well. Among various available Fe-based NPs, the zerovalent FeNPs, FeS NPs, and Fe_3O_4 NPs were the best for nanoremediation technologies. Bare FeNPs with zero oxidation state is ideally good for nanoremediation because of their larger surface areas and high reactivity. The zerovalent FeNPs under moisture are aerobic conditions, and undergo oxidation result in core–shell structures with core containing zerovalent FeNPs. The possible oxidation route of zerovalent FeNPs is given in eq 15.1 (Guan et al., 2015).

$$Fe^0 \rightarrow Fe(OH)_2 \rightarrow Fe_3O_4 \rightarrow \gamma\text{–}Fe_2O_3 \tag{15.1}$$

The formation of core–shell structure of FeNPs is given in eqs (15.2–15.11).

$$Fe^0 + 2H_2O \rightarrow Fe(OH)_2 + H_2\uparrow \tag{15.2}$$

$$2Fe^0 + 2H_2O + O_2 \rightarrow 2Fe^2 + 4OH^- \tag{15.3}$$

$$6Fe^2 + 6H_2O + O_2 \rightarrow 2Fe_3O_4\downarrow + 12H^+ \tag{15.4}$$

$$4Fe^{2+} + 6H_2O + O_2 \rightarrow 4Fe(OH)_3\downarrow + 8H^+ \tag{15.5}$$

$$Fe^2 + 2Fe(OH)_3 \rightarrow Fe_3O_4\downarrow + 2H_2O + 2H^+ \tag{15.6}$$

$$4Fe^2 + 6H_2O + O_2 \rightarrow 4Fe_3OOH\downarrow + 8H^+ \tag{15.7}$$

$$Fe^2 + 2FeOOH \rightarrow Fe_3O_4\downarrow + 2H^+ \tag{15.8}$$

$$Fe^0 + O_2 + 2H^+ \rightarrow Fe^{2+} + H_2O_2 \tag{15.9}$$

$$Fe^0 + H_2O_2 + 2H^+ \rightarrow Fe^{2+} + 2H_2O_2 \tag{15.10}$$

$$Fe^{2+} + H_2O_2 \rightarrow Fe^{3+} + \cdot OH + OH^- \tag{15.11}$$

15.3.1 NANOREMEDIATION BY ADSORPTION PATHWAY

Adsorption is the best, clean, and green approach for nanoremediation. The FeNPs are very good adsorbing agents for nanoremediation for environmental cleaning. The FeNPs with smaller size and larger surface areas are found to be effective materials for adsorption of pollutants. The FeNPs with core Fe^0

and shell ($Fe2O3$, $Fe2O3$, and $Fe(OH)3$) adsorbs pollutants (organic and inorganic) from the environmental matrix through shell $Fe2O3$ species by electrostatic and surface complexation. The inorganic pollutants (Hg^{2+}, As^{3+}, As^{5+}, Pb^{2+}, Cr^{6+}, and Cd^{2+}) with lower or slightly higher electrode potentials than Fe will form surface complexation with outer shell components of core–shell structures. The nanoremediation action of zerovalent FeNPs toward organic and inorganic pollutants was illustrated in Figure 15.4. Another added advantage of the green synthesized FeNPs for nanoremediation is their higher ζ- potentials. The higher ζ- potentials of FeNPs decreases their agglomeration and their reactivity toward nanoremediation will be preserved for longer times compared to that of chemically synthesized FeNPs. The pollutants form complexation with functional groups (amines, phytochemical, aldehydes, vitamins, enzymes, reductases, etc., of bioresource material used during the green synthesis of FeNPs) present on the surface of green synthesized FeNPs and reduce by oxidation of core Fe^0 present in core–shell structure.

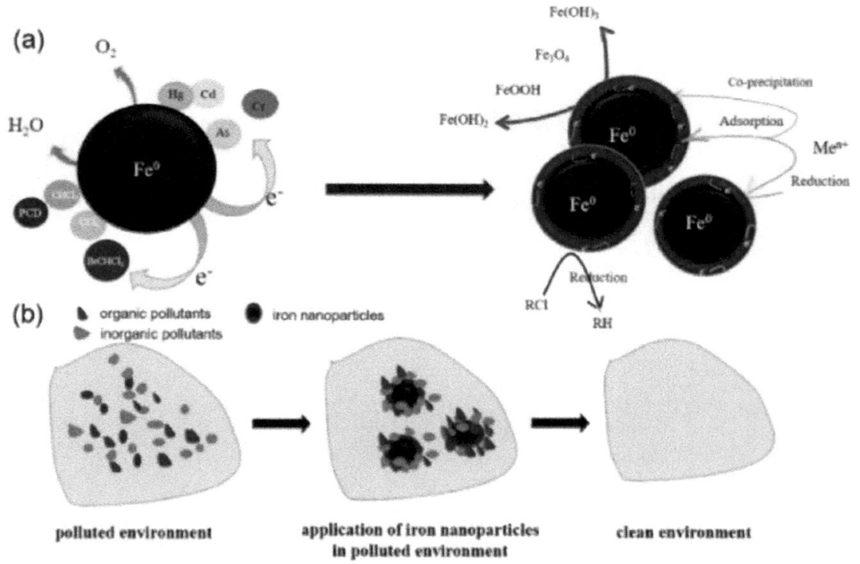

FIGURE 15.4 Illustrating the nanoremediation action of zerovalent FeNPs toward organic as well as inorganic pollutants.

Source: Reprinted with permission from Guan et al., 2015. Copyright © 2015 Elsevier Ltd.

15.3.2 NANOREMEDIATION BY REDUCTION PATHWAY

The nanoremediation using green synthesized FeNPs is found to be beneficial compared to the chemically synthesized one through reduction pathways.

The FeNPs synthesized using green tea extracts exhibited accelerated dye removal (bromophenol blue). This is because polyphenols and caffeine of green tea extract act as reducing and capping agents, which reduces the flocculation of small molecules and increases the rate of dye removal. In addition, the FeNPs synthesized using green tea extracts also catalyzed the H_2O_2, which assisted in degradation of organic dye (bromothymol blue) effectively. The natural polymers like humic acid and fluvic acid will assist in reduction of pollutants for FeNPs by acting as long-time electron donors for the anaerobic degradation of environmental pollutants. The large functional groups of humic acid has strong affinity toward the Fe(II) and Fe(III) species, and these groups facilitate the electron transfer between the Fe(0) and Fe(III) species resulting in the acceleration of active Fe(II) species concentration during pollutants remediation. The increase in Fe(II) species concentration during removal of heavy metals increases the removal efficiency to a large extent. Along with the functional organic molecules present in bioresources as mentioned in previous sections of discussion, the special functional groups that can have capability of electron shuttle and acts as co–resultants were present, which are nothing but the quinone moieties.

15.4 FUTURE PROSPECTIVE

The green synthesized FeNPs are proved to be best for nanoremediation compared to that of the chemically synthesized FeNPs because of their green and sustainable properties which help in preserving the environment from pollution. As an advantage, there is plenty of endless bioresources available in our mother Earth which can be explored for production of FeNPs with better stability, reactivity, and dispersibility toward environmental nanoremediation purpose. Many of the reported green approaches for FeNPs synthesis demonstrated their ability to neutralize/degrade the pollutants from environmental matrix. However, there is still some bolted necks that need to be overcome regarding green synthesis of nanomaterials. Studies regarding the active components of biomaterial responsible for nanoparticle formation and stabilization is seldom. The perfect docking and through spectroscopic analysis of biomaterial (plant or microorganism based) used for green synthesis is very much necessary. The fate of used FeNPs for nanoremediation is again a big question. An accurate and large quantum of research is left out with respect to green synthesis of nanomaterials especially used for environmental nanoremediation applications.

KEYWORDS

- **green iron nanoparticles**
- **nanoremediation**
- **iron-based nanoparticles**
- **eco-friendly remediation**
- **environmental pollution**

REFERENCES

Abdeen, M. I.; Sabry, S. A.; Ghozlan, H. A.; El-Gendy, A. A.; Carpenter, E. E. Microbial-Physical Synthesis of Fe and Fe3O4 Magnetic Nanoparticles Using *Aspergillus niger* YESM1 and Supercritical Condition of Ethanol. *J. Nanomater.* **2016,** *2016*, 9174891.

Al-Juhaimi, F. Y.; Ghafoor, K. Bioactive Compounds, Antioxidant and Physico-Chemical Properties of Juice From Lemon, Mandarin and Orange Fruits Cultivated in Saudi Arabia, 2013.

Dhillon, G. S.; Brar, S. K.; Kaur, S.; Verma, M. Green approach for Nanoparticle Biosynthesis by Fungi: Current Trends and Applications. *Crit. Rev. Biotechnol.* **2012,** *32* (1), 49–73.

Guan, X.; Sun, Y.; Qin, H.; Li, J.; Lo, I. M. C.; He, D.; Dong, H. The Limitations of Applying Zero-Valent Iron Technology in Contaminants Sequestration and the Corresponding Countermeasures: The Development in Zero-Valent Iron Technology in the Last Two Decades (1994–2014). *Water Res.* **2015,** *75*, 224–248.

Jain, R.; Mendiratta, S.; Kumar, L.; Srivastava, A. Green Synthesis of Iron Nanoparticles Using Artocarpus Heterophyllus Peel Extract and Their Application as a Heterogeneous Fenton-Like Catalyst for the Degradation of Fuchsin Basic Dye. *Curr. Res. Green Sustain. Chem.* **2021,** *4*, 100086.

Jiang, D.; Huang, D.; Lai, C.; Xu, P.; Zeng, G.; Wan, J.; Tang, L.; Dong, H.; Huang, B.; Hu, T. Difunctional Chitosan-Stabilized Fe/Cu Bimetallic Nanoparticles for Removal of Hexavalent Chromium Wastewater. *Sci. Total Environ.* **2018,** *644*, 1181–1189.

Machado, S.; Grosso, J. P.; Nouws, H. P. A.; Albergaria, J. T.; Delerue-Matos, C. Utilization of Food Industry Wastes for the Production of Zero-Valent Iron Nanoparticles. *Sci. Total Environ.* **2014,** *496*, 233–240.

Machado, S.; Pacheco, J. G.; Nouws, H. P. A.; Albergaria, J. T.; Delerue-Matos, C. Characterization of Green Zero-Valent Iron Nanoparticles Produced with Tree Leaf Extracts. *Sci. Total Environ.* **2015,** *533*, 76–81.

Pereira, L.; Mehboob, F.; Stams, A. J. M.; Mota, M. M.; Rijnaarts, H. H. M.; Alves, M. M. Metallic Nanoparticles: Microbial Synthesis and Unique Properties for Biotechnological Applications, Bioavailability and Biotransformation. *Crit. Rev. Biotechnol.* **2015,** *35* (1), 114–128.

Prasad, C.; Gangadhara, S.; Venkateswarlu, P. Bio-Inspired Green Synthesis of Fe3O4 Magnetic Nanoparticles Using Watermelon Rinds and Their Catalytic Activity. *Appl. Nanosci.* **2016,** *6* (6), 797–802.

Prema, P.; Nguyen, V.-H.; Venkatachalam, K.; Murugan, J. M.; Ali, H. M.; Salem, M. Z. M.; Ravindran, B.; Balaji, P. Hexavalent Chromium Removal from Aqueous Solutions Using Biogenic Iron Nanoparticles: Kinetics and Equilibrium Study. *Environ. Res.* **2022,** *205*, 112477.

Wang, Y.; Fang, Z.; Kang, Y.; Tsang, E. P. Immobilization and Phytotoxicity of Chromium in Contaminated Soil Remediated by CMC-Stabilized nZVI. *J. Hazard. Mater.* **2014,** *275*, 230–237.

Weng, X.; Guo, M.; Luo, F.; Chen, Z. One-Step Green Synthesis of Bimetallic Fe/Ni Nanoparticles by Eucalyptus Leaf Extract: Biomolecules Identification, Characterization and Catalytic Activity. *Chem. Eng. J.* **2017,** *308*, 904–911.

Xu, W.; Yang, T.; Liu, S.; Du, L.; Chen, Q.; Li, X.; Dong, J.; Zhang, Z.; Lu, S.; Gong, Y.; Zhou, L.; Liu, Y.; Tan, X. Insights into the Synthesis, Types and Application of Iron Nanoparticles: The Overlooked Significance of Environmental Effects. *Environ. Int.* **2022,** *158*, 106980.

CHAPTER 16

Green Silver Nanoparticles for Nanoremediation

VINICIUS MARX SILVA DELGADO[1], CHRISTOPHER SANTOS SILVA[1], VITÓRIA DE OLIVEIRA LOURENÇO[1], FLÁVIA CRISTINA POLICARPO TONELLI[2], and FERNANDA MARIA POLICARPO TONELLI[3]

[1]*Department of Biochemistry, UFSJ/CCO, Divinópolis, Brazil*

[2]*Estácio de Sá University, Divinópolis, Brazil*

[3]*Federal University of São João del Rei, Divinópolis, Minas Gerais, Brazil*

ABSTRACT

Environmental pollution has been increasing in level and diversity of pollutants as industrialization continues to occur at a fast pace. As a consequence, serious damage has been done to the health of living beings and to the ecosystem's balance. It is urgent to bring in changes in order to offer sustainability in each aspect of life possible, to avoid generating even more pollution. However, it is also necessary to remove the contaminants already present in air, water, and soil worldwide. Green silver nanoparticles (AgNPs) are interesting tools in this sense because they are eco-friendly nanomaterials capable of promoting remediation of polluted areas. This chapter will be dedicated to explore this important role of green AgNPs.

16.1 INTRODUCTION

Environmental pollution concerns the world's population once it affects ecosystems and threatens the survival of life on Earth.

Sustainable Nanoremediation: Modern Technologies for a Clean Environment.
Fernanda Maria Policarpo Tonelli, Arpita Roy, and Khalid Rehman Hakeem (Eds.)

16.1.1 ENVIRONMENTAL POLLUTION THREATENING HUMAN LIFE

Since the Bronze Age, heavy metals and relevant inorganic environmental pollutants have been doing damage to mankind (Longman et al., 2018). They can cause acute poisoning (e.g., impairing cognitive ability), and depending on the dose and exposure time they can even provoke death (Yamada et al., 2020).

However, recently the discussions on environmental pollution threats to human life have been receiving increasing attention. It is mainly due to the fact that the problem affects the whole world and the variety and level of environmental contaminants are enhancing in a fast way. Urbanization and industrialization largely contributed to this scenario, a consequence of promoting development ignoring sustainability principles (Zheng et al., 2020). And it is important to highlight that environmental pollution can cause ecological disruption and directly threatens not only living beings; alimentary safety is also at risk (Guowei et al., 2021).

Global premature deaths due to air pollution have reached 6.7 million, the ones caused by water pollution—1.4 million, and the ones related to toxic occupational hazards—870 000 in 2019. The number indicates an increase of 7% since 2015 and of more than 66% since 2000 (Beng et al., 2022).

Pollutants can be from inorganic chemical nature (such as heavy metals—presenting density greater than 5 g/cm^3 like titanium, vanadium, chromium, manganese, iron, cobalt, nickel, copper, zinc, arsenic, molybdenum, silver, cadmium, tin, platinum, gold, mercury, and lead) (Briffa et al., 2020), from biological nature (such as pathogenic microorganism) (Kumar et al., 2021), and from organic nature (such as pesticides and dyes) (Kumar et al., 2022) (Fig. 16.1).

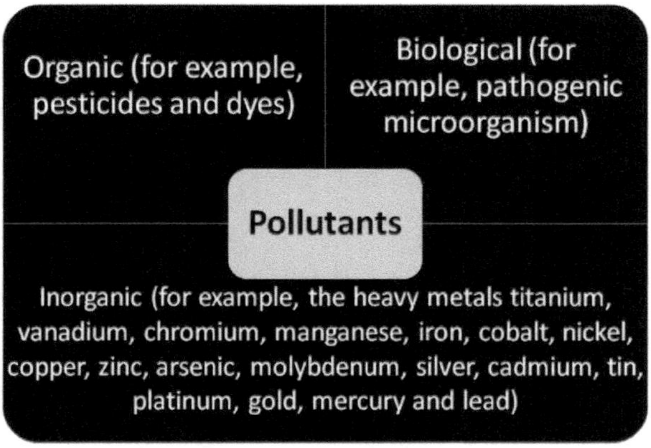

FIGURE 16.1 Different chemical nature from pollutants.

Living beings' exposure to heavy metals has increased due to their increased use in agriculture, domestic environment, and in industries and also actions related to inadequate disposal of waste, illegal refining, and artisanal mining (Anyanwu et al., 2018). Although some heavy metals are present in humans and other living forms as they are essential for important biochemical processes, exposure to values above the secure ones allows the toxicity mechanism induced by these substances to be triggered leading to serious damages that can end up causing death (Balali-Mood et al., 2021).

Among the main organic pollutants are dyes applied in different industrial environments such as textile industries that employ approximately 35 million workers (Desore and Narula, 2018). Basic, acid, direct, and reactive dyes are soluble organic compounds containing chromophoric groups and capable of coloring substrates. When polluting water bodies, they can cause deaths due to increase in biochemical and chemical oxygen demands and/or photosynthesis impairment, bioaccumulation, mutagenicity, and carcinogenicity (Lellis et al, 2019). Some dyes are phenolic chemicals, but not all phenolic substances are dyes. Phenol and its derivatives also deserve to be highlighted as relevant highly toxic environmental pollutants. They can be a by-product of industrial production (e.g., from petrochemical industries) and can also be present in pesticides, wood preservatives, and some polymers. These substances can cause severe hazards to living beings; chlorophenols, for example, are carcinogenic (El-Naggar et al., 2022).

So, it is urgent that efficient strategies to remediate polluted environs containing contaminants from one or various chemical natures start getting applied to protect the environment and the health of living beings. Bioremediation strategies and protocols that apply green nanomaterials are receiving increasing attention. Sustainability is very important to be promoted together with industrialization and urbanization to achieve balance on different ecosystems.

16.1.2 NANOMATERIALS AND NANOREMEDIATION

Norio Taniguchi, in the 1970s, used the term "nanotechnology" to refer to technologies applying material in nanoscale—presenting at least one dimension ranging from 1 to 100 nanometers (Taniguchi, 1974). Nanotechnologies can be generated by applying wet (using living beings or substances obtained from them such as enzymes), computational (applying simulations using computers), and/or dry (inorganic items are used) approaches (Chung et al., 2016).

The advantageous properties of nanomaterials that allow them to be efficiently used with different goals are being explored since the 1950s and among them the capacity to suffer specific surface chemical modification through various protocols, good chemical reactivity, high surface-to-volume ratio, and high stability have been identified (Bethi and Sonawane, 2018; Tonelli et al., 2020; Tonelli et al., 2022; Werkneh et al. 2018; Xu et al., 2019).

One of the important applications of nanomaterials is to promote nanoremediation, offering the possibility to restore polluted air, water, and soil. They allow dealing with organic and inorganic pollutants in an efficient way through different morphologies such as nanotubes, nanomembranes, and nanoparticles (Lu and Astruc, 2020).

However, it is still being debated about ecotoxicity and cytotoxicity associated with nanotechnology. Silver nanoparticles in some in vitro studies, for example, are related to toxicity associated with oxidative stress induction, DNA damage, and cytokine production regulation (Stensberg et al., 2011). Changes in water and soil microbial community composition were also observed (Tortella et al., 2020). However, negative effects are commonly observed after high concentration exposure and/or long-term exposure (Kakakhel et al., 2021).

In order to reduce negative impacts upon the environment and living beings that could be caused by harmful substances applied on nanoparticle's synthesis, green-based nanomaterials are interesting options. These materials can also offer advantages, besides being eco-friendly, for example, they can be generated at low costs (Fei et al., 2022; Nasrollahzadeh et al., 2021).

16.2 SILVER NANOPARTICLES ADVANTAGEOUS CHARACTERISTICS TO DIFFERENT APPLICATIONS

Silver nanoparticles (AgNPs) present interesting characteristics that attract interest regarding their use to serve specific functions with efficiency. Silver presents great antimicrobial activity and is also capable of offering desirable catalytic properties (Monti et al., 2021), high thermal and electrical conductivity (Coetzee et al., 2020), good chemical stability (Calderón-Jiménez et al., 2017), among other important features that allow AgNPs to perform efficiently diverse roles (Fig. 16.2).

FIGURE 16.2 Important silver features that make it possible a large array of applications for this metal.

AgNPs can, for example, serve as antimicrobial to treat diseases, be impregnated in textiles, or be applied in smart packages to inhibit adhesion and growth of fungi species (such as *Candida albicans*) and different bacteria (such as *Enterococcus faecalis*, *Escherichia coli*, *Pseudomonas aeruginosa*, *Streptococcus mutans*, and *Staphylococcus aureus*) (Rodrigues et al., 2019; Yin et al., 2020).

Organic reactions can also benefit from AgNPs as nanocatalysts of low cost and simple preparation. The material can be reused and offer the opportunity, for example, to increase velocity of synthesizing amines, nitriles, and heterocycles, performing coupling reactions, reducing nitro compounds, and performing hydration of nitriles to amides (Ardakani et al., 2021).

Silver nanoparticles are also useful to perform pollutants' removal (Rania et al., 2020). This application will be explored in the following section of this chapter, focusing on AgNPs generated through green protocols.

Formulations of sunscreen can apply AgNPs aiming protection against UVB-induced skin damage as the nanomaterial presents absorbance in the UV range (Ho et al., 2020).

When it comes to cancer therapy, AgNPs offer the possibility of personalized and patient-specific design optimizing targeting and treatment success and reducing side effects. These nanomaterials can also synergistically enhance other therapeutic approaches, such as chemotherapy (Kovács et al., 2022). Other treatments can also benefit from AgNPs, such as wound healing, favoring tissue regeneration, etc. (Mathur et al., 2018).

Early diagnosis has a significant influence on treatment success in a large array of diseases. AgNPs can also be applied in diagnosis as this nanomaterial presents plasmonic properties, high conductivity, and catalytic activity—necessary to the function of efficient biosensors. Silver nanoparticles can enhance biosensor's sensitivity being successfully used in electrochemical and optical biosensors to detect, for example, cancer cells (Miranda et al., 2022).

AgNPs can also perform delivery, for example, of drugs. Ciprofloxacin could be delivered by AgNPs releasing 98% of the substance within 750 min. The nanocomposite presented no toxicity when assayed in normal 3T3 cell line (Hussein-Al-Ali et al., 2022).

However, large-scale application of AgNPs faces challenges such as high-energy consumption required for the preparation of silver nanoparticles and the necessity of using harmful substances in conventional synthesis protocols.

Green synthesis presents itself as an alternative to biosynthesize the nanomaterial surpassing the barriers previously mentioned. Easier and eco-friendly protocols are capable of generating nanoparticles presenting optimized properties such as antimicrobial one (Zhao et al., 2022).

When it comes to environmental pollution, AgNPs can favor the detection of pollutants and/or their removal and/or degradation from polluted areas. In fact, prior to performing the remediation of any type of pollutant in a sample or in contaminated environments, it is necessary to detect the pollutant present to, for example, choose an adequate remediation methodology.

AgNPs can successfully and efficiently contribute in this aspect. Different studies have proved that silver nanoparticles can be effective in nanosensors to detect environmental contaminants, especially heavy metals. Nanoparticles make possible the synthesis of innovative, simple, low-cost, and easy-to-use devices to efficiently detect pollutants. Quantitative analysis of inorganic pollutants is crucial once legislations generally establish limits of concentration that must be observed specially in potable water. Expensive techniques are commonly applied with this purpose, demanding also qualified users to operate pieces of equipment difficult to remove for analysis in the field, for example (Annadhasan et al., 2014).

An eco-friendly approach to detect Hg^{2+} and Sn^{2+} in water was developed using garlic extract. The substance alliin allowed the generation of a colorimetric detection with high sensitivity to perform the dual detection of mentioned ions. The brown color of the solution becomes less intense in the presence of analytes (Paw et al., 2021).

The AgNPs generated from *Ganoderma lucidum* and mentioned previously in this chapter, could, beside exhibiting catalytic and antimicrobial activity, promote the colorimetric detection of Fe^{3+} in aqueous samples being suitable to be used not only in environmental remediation, but also in environmental monitoring (Nguyen et al., 2022).

When it comes to remediating environmental pollution using green nanoparticles, most of the attention is directed to remediating water contamination. Due to worldwide population explosion experienced in the last decades, it has been facing an enhanced necessity of providing an exponentially larger amount of potable water. Simultaneously with the population's increase, there has also been an increasing industrialization resulting in higher levels of water bodies' pollution (caused by organic, inorganic, and biological contaminants) and reduction of potable water availability (Santhosh et al., 2016). A large array of studies however has presented green AgNPs as efficient tools to promote the remediation of different environmental pollutants previously mentioned in this chapter. The next section will be dedicated to explore this important application of this nanomaterial.

16.3 GREEN SILVER NANOPARTICLES FOR NANOREMEDIATION

AgNPs can be synthesized to perform remediation of polluted samples by applying green protocols.

16.3.1 SYNTHESIS OF AGNPS

AgNPs can be synthesized by applying two types of methods: top-down or bottom-up. Top-down methods use as raw material structures possessing wider dimensions and technologies are applied to them aiming for size reduction until the final desired dimension is achieved. Ball-milling, laser ablation, sonication, electro explosion, and sputtering can be performed in this kind of approach (Mohammadlou et al., 2016). For example, Chen and

Yeh could generate AgNPs through a silver rod that underwent laser ablation in the presence of surfactants, SDS, and CTAB (Chen and Yeh, 2002). More recently, Hwang and coworkers proposed a methodology to obtain AgNPs from debris obtained as a result of laser ablation of silver nanowires in liquid media (Hwang et al., 2021). Another example of top-down technique to generate the silver nanoparticles is exploding wire (Mohammed and Naje, 2021)—used to synthesize not only AgNPs but also other metal nanoparticles (Sen et al., 2003).

When it comes to bottom-up synthesis, the idea is to construct a larger structure from smaller ones. Supercritical fluid synthesis, condensation, chemical vapor deposition, sol-gel, spinning, laser pyrolysis, flame synthesis, ultrasonic method, and green synthesis can be performed in this kind of approach (Mohammadlou et al., 2016). Silver salts can be used in these methodologies to obtain AgNPs with different functions, such as antiproliferative and apoptotic against human breast cell lines MCF-7, MCF-7/TAMR-1, and MCF-10A (Zulkifli et al., 2020).

When it comes to the nature of the synthesis method to generate these nanoparticles, it can be classified as physical (mechanical or vapor-based in approaches such as laser ablation, radiation, sinocation, and electro-deposition), chemical (reducing silver ions to atoms through a process involving steps of nucleation and growth—using sol-gel technologies, redox processes, or hydrothermal methods) or biological (green synthesis using, e.g., plants, bacteria, and fungi) (Shanmuganathan et al., 2019; Xu et al., 2020).

Among the above-mentioned chemical natures of the synthesis, chemical is the chosen one in the majority of standard protocols to obtain AgNPs. These protocols require the use of chemical reagents in order to generate nanoparticles of uniform size and presenting good stability and dispersion; these chemicals can act as stabilizers and/or agents to promote reduction and/or capping (Díaz-Núñez et al., 2017; Xu et al., 2020). However, these reagents are commonly cytotoxic, becoming, at the end of the synthesis, harmful pollutants capable of endangering ecosystems.

In order to avoid this negative impact of chemical protocols, green chemistry received attention as an interesting alternative, offering biological-based green synthesis protocols. These protocols also proved to be economical besides being eco-friendly and the raw green material can be provided, for example, by plants (Giri et al., 2022), bacteria (Naseer et al., 2022), fungi (Murillo-Rábago et al., 2022), and algae (Chugh et al., 2021). These raw materials provide substances capable of substituting the chemical reactants as stabilizers and capping agents, and also reducing silver salts (Xu et al., 2020).

In AgNPs standard synthesis, there are three steps—reduction, growth, and stabilization. In green synthesis protocols, commonly water is the solvent chosen to be applied and non-toxic agents to promote reduction in step one and stabilization (to prevent undesirable aggregation), step three comes from bio-based material, for example plant extract. In all steps, green synthesis protocols apply natural substances—eco-friendly aiming to reduce negative impacts to the environment (Mohammadlou et al., 2016).

This section will present brief subsections dedicated to explore plants, bacteria, fungi, and algae as raw material to be applied in green protocols to generate AgNPs. Green AgNPs can be produced to perform different tasks; however, in Table 16.1. examples of green protocols to generate AgNPs to perform nanoremediation of different environmental pollutants are presented.

16.3.1.1 PLANT-BASED SYNTHESIS OF AGNPS TO NANOREMEDIATION

The use of plant-based material has become a sustainable alternative to the obtainment of AgNPs. This alternative route is due to the presence of plant metabolites presenting different chemical properties, allowing the avoidance of harmful exogenous chemicals during synthesis, reducing damages to the environment and also offering a material with optimized stability, through a process with lower cost and higher yield. Plants also offer specific proteins capable of reducing Ag^{+} ions (e.g., NADH-dependent nitrate reductase) and protocols more simple to scale up when compared to protocols applying microorganisms. This type of synthesis does not require cell cultures to be maintained and allows large-scale synthesis of nanoparticles (Chung et al., 2016; Widatalla et al., 2022).

Plants can provide not only nutrients but also active compounds capable of benefiting living beings in various ways. Lipids, pharmaceutics, and phytochemicals, for example, can be extracted from vegetal species. When it comes to phytochemicals, organic acids, flavonoids, and alkaloids, differentially distributed in plant parts such as root, leaf, bark, and seeds can perform important functions to maintain the plant well established in a certain environment (Ralte et al., 2020). These substances can be obtained and applied in green synthesis protocols among other substances present in plant extracts.

Plant metabolites capable of participating efficiently in AgNPs synthesis must present antioxidant/reducing power when it comes to Ag ions, be

TABLE 16.1 Recent Examples of Green AgNP Performing Nanoremediation.

Biomaterial used in green synthesis		**Pollutant involved in experiments**	**References**
Plants	Punica granatum	4-Nitrophenol	Edison et al. (2013)
	Allivum sativum	Polycyclic aromatic hydrocarbons (PAHs)	Abbasi et al. (2013)
	Anacardium occidentale	Congo red and methyl orange	Edison et al. (2016)
	Mussaenda erythrophylla	Methyl orange	Varadavenkatesan et al. (2016)
	Lippia citriodora	*Escherichia coli, Salmonella typhi, Bacillus subtilis, Staphylococcus aureus* and *Candida albicans*	Elemike et al. (2017)
		Methylene blue	Elemike et al. (2017)
	Psidium guajava	*Alcaligenes faecalis, Escherichia coli* and *Aspergillus niger*	Wang et al. (2018)
		Methyl orange and Coomassie brilliant blue G-250	Wang et al. (2018)
	Ocimum sanctum	4-Nitrophenol	Khatoon et al. (2018)
	Ocimum sanctum and *Artemisia annua*	Reactive Blue 4, Reactive Orange 4 and Reactive Red 120	Khatoon et al. (2018)
	Trigonella foenum-graecum (TF)	RB19 and RY186 dyes	Singh et al. (2019)
	Syzygium cumini	*Pseudomonas aeruginosa* and *Escherichia coli*	Parmar et al. (2019)
		Methylene blue, crystal violet, amaranth, indigo carmine, acid red, vat red, malachite green, congo red, rhodamine B	Parmar et al. (2019)
	Zingiber officinale	4-Nitrophenol, 2-nitrophenol	Ismail et al. (2019)
		Rhodamine B, methyl red and congo red	Ismail et al. (2019)
		Methyl Orange	Riaz et al. (2020)
		Pseudomonas aeruginosa, Methicillin-resistant Staphylococcus aureus and *Candida albicans*	Riaz et al. (2020)
	Aloe barbendensis	Naphthalene	Abbas et al. (2020)

TABLE 16.1 *(Continued)*

Biomaterial used in green synthesis		Pollutant involved in experiments	References
	Chloroxylon swietenia	Congo red, coomassie blue and crystal violet	Nga et al. (2022)
		Staphylococcus nepalensis, Staphylococcus gallinarum, Bacillus subtilis, Enterococcous faecalis and Pseudomonas stuteria	Nga et al. (2022)
	Origanum majorana	Multidrug-resistant *Klebsiella pneumoniae*	Yassin et al. (2022)
	Smilax aspera	*Klebsiella pneumoniae, Streptococcus pneumoniae, Escherichia coli Pseudomonas aeruginosa* and *Staphylococcus aureus*	Negi et al. (2022)
		Methylene blue, rose bengal and acridine Orange	Negi et al. (2022)
	Catharanthus roseus	Chromium and cadmium	Verma and Bharadvaja (2022)
Bacteria	*Anabaena variabilis and Spirulina platensis*	malachite green	Ismail et al. (2021)
	Bacillus licheniformis M09	Methylene blue	Momin et al. (2019)
	Bacillus amyloliquefaciens MSR5	4-nitrophenol	Samuel et al. (2020)
Algae	*Marine red algae Gelidium amansii*	*Staphylococcus aureus, Bacillus pumilus, Escherichia coli, Pseudomonas aeruginosa, Vibrio parahaemolyticus, Aeromonas hydrophila*	Pugazhendhi et al. (2018)
Fungi	Cryptococcus laurentii and Rhodotorula glutinis	Fungi	Fernández et al. (2016)
	Rhizopus arrhizus, Trichoderma gamsii and Aspergillus niger	*Escherichia coli, Staphylococcus aureus* and *Pseudomonas aeruginosa*	Ottoni et al. (2017)
	Poria cocos	4-Nitrophenol	Doan et al. (2022)
	Penicillium oxalicum GRS-1	*Staphylococcus aureus, Escherichia coli* and *Salmonella typhimurium*	Rose et al. (2018)
	Penicillium polonicum	*Acinetobacter baumannii*	Neethu et al. (2018)
	Trichoderma spp.	*Pseudomonas aeruginosa* and *Escherichia coli*	Ramos et al. (2020)
	Species from Metschnikowia genus	*Escherichia coli, Pseudomonas aeruginosa* and *Staphylococcus aureus*	Liu et al. (2021)

present in sufficient molar concentration, and present the capacity to stabilize the particle generated. Polyphenol family, such as flavonoids and tannins, offers important substances to perform these roles (Filippi et al., 2017).

The phytochemical's profile from a vegetal can vary according to different circumstances to which the vegetal is exposed. Biotic and abiotic stress, for example, can impact this profile, as same as plant stage of development and age Phytochemicals of wheatgrass presenting antioxidant activity are significantly affected by seedling age and stress treatments (Jiang et al., 2021).

The combination of secondary metabolites present in plants will define how efficient the extract will be to generate the nanoparticles. Species such as *Coleus aromaticus, Syzygium cumini, Azadirachta indica, Citrus sinensis, Zingiber officinale, Ocimum sanctum, Glycyrrhiza glabra, Nigella arvensis, Dioscorea bulbifera, Lantana camara, Helianthus annuus, Rosmarinus officinalis, Mimusops elengi, Syzygium cumini, Ocimum sanctum, Acalypha indica, Gardenia jasminoides, Withania somnifera, Lippia citriodora, Decalepis hamiltonii, Achyranthes aspera, Hybanthus enneaspermus, Desmodium triflorum, Sesuvium portulacastrum, Solanum xanthocarpum, Mentha piperita, Anacardium occidentale, Dioscorea bulbifera, Iresine herbstii, Trianthema decandra, Morinda pubescens, Carica papaya, Annona squamosa, Trianthema decandra, Aegle marmelos, Rosa rugosa, Hibiscus rosa, Leonuri herba, Lonicera japonica, Mangifera indica, Eucalyptus sp., Alternanthera sessilis, Boswellia serrata, Plumeria rubra,* and *Piper betle* have already been applied to generate AgNPs and different secondary metabolites present in these species (mainly flavonoids but also proteins, ascorbic acid, alkaloids, glycosids, terpenoids, catechins, tannins, ketone, aldehydes...) generated different nanoparticles regarding size and other characteristics (Marslin et al., 2018).

Although AgNPs are among the easiest metal nanoparticles to prepare and almost all plants have the potential to be used to generate this nanomaterial, the selection of the plant to be used as green material in AgNPs' synthesis as redutor agent/stabilizer, for example, or the vegetal part to be used will influence on the results obtained from synthesis (Hano and Abbasi, 2022).

Pterodon emarginatus possesses on its phytochemical profile substances that have the potential to contribute to AgNPs' synthesis. Are examples: di-C-glycosylflavones, sesquiterpenes hydrocarbons, C,O-glycosylflavones, saponins, luteolin-7-O-rutinoside, pheophytin, steroids, and flavonoids. However, the presence of these substances can vary in plants' leaves according to conditions like the season. AgNPs generated from plant material collected during summer are different from the ones generated from material collected during winter. The yield of AgNPs generated, hydrodynamic diameters, and zeta potentials were different (Oliveira et al., 2019).

When it comes to using plant-based material to generate AgNPs to perform nanoremediation, multidrug-resistant *Klebsiella pneumoniae* strain, that can be emitted from human sources and is considered an environmental contaminant (Rocha et al., 2022), could be eliminated by green AgNPs. Biogenic AgNPs presenting antimicrobial effectiveness against this strain could be generated using the aqueous leaf extract of *Origanum majorana* (Yassin et al., 2022).

Ginger's (*Zingiber officinale*) rhizome extract was successfully used to promote the synthesis of AgNPs capable of exhibiting antimicrobial activity against *Pseudomonas aeruginosa*, *Methicillin-resistant Staphylococcus aureus*, and *Candida albicans* (Riaz et al., 2020).

Compounds such as 4-nitrophenol could be degraded by AgNPs synthesized from aqueous extracts of pomegranate (*Punica granatum*) peels. The nanoparticles exhibited high catalytic activity and also good stability during the chemical reduction (Edison et al., 2013).

Organic pollutants congo red and methyl orange, azo dyes, could be degraded by silver nanoparticles generated using *Anacardium occidentale*. Nanoparticles exhibited good stability as phytoconstituents of the plant acted as capping agents preventing aggregation (Edison et al., 2016).

Hydroalcoholic extract from leaves of *Cannabis sativa, Solanum nigrum, Achillea millefolium, Artemisia annua, Azadirachta indica, Saraca asoca, Aloe vera, Psidium guajava, Ocimum sanctum, Brassica juncea, Mentha, Murraya koenigii, Acacia arabica, Zizyphus jujuba, Lawsennia inermis, Cassia fistula, Eucalyptus, Syzygium cumini, Tamarindus indica, Tagetes*, and *Salvia officinalis* could be used to generate AgNPs. The ones produced from *Ocimum sanctum* and *Artemisia annua* exhibited capacity to adsorb textile dyes (Reactive Blue 4, Reactive Orange 4 and Reactive Red 120) from aqueous solutions, being also able to remove microbial and other contaminants from Yamuna River's water. *Ocimum sanctum*'s AgNPs could also reduce 63% of 4-nitrophenol's initial concentration in 20 min (Khatoon et al., 2018).

Aqueous leaf extract of *Lippia citriodora* could be successfully used in two different temperatures (50 and 90 °C) to generate AgNPs. The nanomaterial produced exhibited not only larvicidal effect (against *Culex quinquefasciatus*) but also the capacity to eliminate Gram-negative (*Escherichia coli*, and *Salmonella typhi*) and Grampositive (*Bacillus subtilis* and *Staphylococcus aureus*) bacteria and *Candida albicans*. The dye methylene blue, an organic pollutant present in industrial waste, could also be degraded due to photocatalytic properties of the green AgNPs (Elemike et al., 2017).

AgNPs could also be generated using ginger rhizome powder. The nanoparticle proved to be able to reduce the organic dye methyl orange, and also other pollutants such as 4-nitrophenol, 2-nitrophenol (2-NP), rhodamine B, methyl red, and congo red from water samples. Not only separately the carcinogenic dyes and the toxic nitrophenols could be degraded, but also collectively. Besides the catalytic properties, AgNPs also exhibited good stability and reusability (Ismail et al., 2019).

Fresh leaf extract from *Diospyros lotus* was used at an optimum pH of 8.6 to perform the synthesis of AgNPs capable of promoting decolorization (72.91%) of dyes present in industrial wastewater (Yasmin et al., 2020). Remediation of the pollution caused by the dye methyl orange was promoted by AgNPs generated using rhizome extract from Zingiber officinale; the AgNPs performed dye reduction (Riaz et al., 2020).

Aqueous extract obtained from the leaves of *Smilax aspera* allowed the generation of Ag-NPs capable of presenting different properties such as antibacterial activity against pathogens (*Klebsiella pneumoniae, Streptococcus pneumoniae, Escherichia coli Pseudomonas aeruginosa*, and *Staphylococcus aureus*), antioxidant potential, and antidiabetic activity. Besides that, AgNPs were also capable of remediating pollution caused by the dyes methylene blue, rose bengal, and acridine orange (Negi et al., 2022).

AgNPs produced using *Chloroxylon swietenia*'s leaves could degrade more than 90% of the dyes congo red, coomassie blue, and crystal violet. Besides being able to perform the degradation of these pollutants, the nanoparticles could be also used against food pathogens. Bactericidal activity was exhibited against *Staphylococcus nepalensis, Staphylococcus gallinarum, Bacillus subtilis, Enterococcous faecalis*, and *Pseudomonas stuteria* (Nga et al., 2022).

AgNPs generated using *Syzygium cumini* fruit extract exhibited antibacterial activity against *Pseudomonas aeruginosa* and *Escherichia coli.* They also proved to be capable of degrading the pollutant dyes methylene blue, crystal violet, amaranth, indigo carmine, acid red, vat red, malachite green, congo red, and rhodamine B (Parmar et al., 2019).

Green AgNPs synthetized from *Ganoderma lucidum*'s extract could convert more than 98% of the 4-nitrophenol into 4-aminophenol in 9 min. They also exhibited antimicrobial activity against *Bacillus subtilis, Bacillus cereus, escherichia coli, Staphylococcus aureus*, and *Pseudomonas aeruginosa* (Nguyen et al., 2022).

Phytochemicals present in *Catharanthus roseus* leaf extract promoted the synthesis of AgNPs by bioreducing silver salts. The nanomaterial offered chromium and cadmium removal of 47.84% and 5.68%, respectively (Verma and Bharadvaja, 2022).

AgNPs generated from *Psidium guajava* leaves' extract exhibited antimicrobial activity against *Alcaligenes faecalis*, *Escherichia coli*, and *Aspergillus niger* and were also capable of degrading the organic dyes methyl orange and Coomassie brilliant blue G-250 using UV or solar light (Wang et al., 2018).

16.3.1.2 FUNGI-BASED SYNTHESIS OF AGNPS TO NANOREMEDIATION

Due to the fact that they are able to secrete large amounts of proteins and possess a large array of enzymes and present a high growth rate, fungi stand out as green raw material to the biosynthesis of metallic nanoparticles. Due to these characteristics, species in the fungi kingdom have the ability to synthesize intracellularly and extracellularly large amounts of nanoparticles, with well-defined dimensions, high monodispersity, and good stability. They are an option of bio-based raw material that favors cost-effectiveness compared to other synthesis methods and protocols applying different types of microorganisms (Singh et al., 2018; Hamedi et al., 2013; Mohanpuria et al., 2007).

The intracellular synthesis of nanoparticles occurs through the association of the metal, due to electrostatic interactions, with the catalyst enzymes present in the fungal cell wall, causing the reduction of the metallic ions. On the other hand, extracellular synthesis occurs due to the interaction between metal ions and the reductase enzyme secreted by the fungus, causing the formation of nanoparticles (Garg et al., 2020; Zhao et al., 2017; Khan et al., 2017).

Furthermore, different species of fungi have been shown to be able to produce eco-friendly AgNPs through different protocols to perform the remediation of different environmental pollutants.

16.3.1.2.1 Filamentous Fungi and the Green Synthesis of AgNPs to Remediation

Ottoni et al. (2017) synthesized AgNPs by using the fungi's strains *Rhizopus arrhizus*, *Trichoderma gamsii*, and *Aspergillus niger*. Due to nitrate reductase activity exhibited the biosynthesis could occur successfully. The generated nanoparticles offered antimicrobial activities against *Escherichia coli*, *Staphylococcus aureus*, and *Pseudomonas aeruginosa*. Mushroom extracts (*Poria cocos*) were also used for the synthesis of AgNPs in order to degrade 4-Nitrophenol (4-NP), exhibiting high stability and catalytic potential during the conversion of 4-NP into 4-Aminophenol (4-AP). The nanomaterial could also be reused approximately five times without loss of catalytic capacity (Doan et al., 2022).

Neethu et al. (2018) synthesized AgNPs through aqueous filtrates the fungus *Penicillium polonicum*'s cells, demonstrating effective antibacterial property against the formation of biofilms of the gram-negative and highly resistant bacteria *Acinetobacter baumannii*. A sample of fungi of the genus *Trichoderma* spp., taken from trees of the Amazonia flora, was also able to biosynthesize AgNPs. The sample, treated at pH 5 for nine days, synthesized small nanoparticles with great polydispersity, presenting good antibacterial activity against Gram-negative species, such as *Pseudomonas aeruginosa* and *Escherichia coli* (Ramos et al., 2020).

Bacteria present in contaminated environs can cause severe health problems to humans. When it comes to this type of pollution, green AgNP are highly useful due to the previously mentioned property of silver to induce the death of these pathogens. Food-born pathogens *Staphylococcus aureus*, *Escherichia coli*, and *Salmonella typhimurium* were eliminated by green synthesized AgNPs at a minimum concentration of 32, 16 and 32 μg/mL, respectively. These nanoparticles were generated using the extract of *Penicillium oxalicum* GRS-1, which presents nitrate reductase activity (Rose et al., 2018).

16.3.1.2.2 Yeasts-Based Green AgNPs to Remediation

Yeasts are classified in the kingdom Fungi and are a relevant source of different enzyme types. They are easy to submit to mass production and present rapid growth, features that make them relevant microorganisms to produce nanoparticles. Yeast generally provide high-quality stabilization of nanoparticles and species such as Saccharomyces cerevisiae, Rhodotorula glutinis, and Geotrichum candidum, for example, were already used as green material to generate AgNPs (Roychoudhury et al., 2020).

AgNPs generated from the yeasts *Cryptococcus laurentii* and *Rhodotorula glutinis* (presenting nitrate reductase activity) were dispersed and stable and presented antifungal activity (Fernández et al., 2016).

Antibacterial effect (against *Escherichia coli, Pseudomonas aeruginosa,* and *Staphylococcus aureus*), as same as anticancer activity, was exhibited by AgNPs generated through two new species of yeasts from *Metschnikowia* genus (Liu et al., 2021).

16.3.1.3 GREEN AgNPs SYNTHESIZED USING BACTERIA TO PROMOTE NANOREMEDIATION

Bacteria can also be used to generate green silver nanoparticles capable of performing different functions such as remediating environmental pollution. Although using plants as green raw material can offer more rapid and simple synthesis protocol, as previously discussed the variation in phytochemicals due to plant parts, season, or other conditions may offer difficulties in reproducibility. Plant-based synthesis can also end up producing polydispersed nanoparticles due to involvement of different phytochemicals in the synthesis process (Ghosh et al., 2012; Singh et al., 2015).

When it comes to bacteria-based synthesis of green AgNPs, these variations are not the problem. The only additional care necessary is regarding sterile conditions to perform the microorganism culture and sterile conditions for nanoparticle synthesis is required (Salunkhe et al. 2014; Singh et al., 2015). Bacteria-based synthesis presents important advantages such as generating a product with increase stability (due to capping of the nanoparticle's surface by biomolecules from the microorganism) and offering a good yield due to the remarkable ability that various bacterium species possess to reduce metal ions (Ovais et al., 2018). Among the main reducing agents is the enzyme NADH-dependent nitrate reductase that can efficiently promote silver ions reduction (Tsekhmistrenko et al., 2020).

The cianobacteria *Anabaena variabilis* and *Spirulina platensis* were successfully applied in the synthesis of AgNPs capable of removing the dye malachite green from wastewater. Maximum removal capacity was observed at the rates of 93% for *S. platensis* and 82% for *A. variabilis* AgNPs. The treatment converted a toxic sample into a non-toxic one when assayed in *Triticum aestivum* L (Giza 171) seedlings (Ismail et al., 2021).

The cell-free supernatant generated by the mutant *Bacillus licheniformis* M09 allowed the production of green AgNPs capable of performing the photocatalytic degradation of methylene blue dye. Antimicrobial activity against foodborne pathogens was also demonstrated by the nanomaterial (Momin et al., 2019).

Cellfree supernatant of *Bacillus amyloliquefaciens* MSR5 was successfully used to generate AgNPs capable of degrading 98% of the 4-nitrophenol in the presence of $NaBH_4$ within 15 min (Samuel et al., 2020).

16.3.1.4 ALGAE-BASED SYNTHESIS OF AgNPs TO REMEDIATE

Material obtained from algae can also be applied in green synthesis protocols to generate silver nanoparticles. Algaes can be macroscopic (macroalgae) or macroscopic (microalgae) and they play important roles especially in medicine, pharmaceutical industry, aquaculture and forestry (Chugh et al., 2021).

Microalgae deserve to be highlighted as an interesting source of raw material once they can survive in extreme conditions being easy to grow in different environments (Jacob et al., 2021).

AgNPs can be generated by applying algae cell free extract, dry extract, living culture, and/or whole cells (Chugh et al., 2021).

AgNPs generated applying marine red algae *Gelidium amansii* could exhibit antibacterial activity against pathogenic Gram-positive (*Staphylococcus aureus, Bacillus pumilus*) and Gram-negative (*Escherichia coli, Pseudomonas aeruginosa, Vibrio parahaemolyticus*, and *Aeromonas hydrophila*) species. The nanomaterials could avoid biofilm formation by reducing bacterial growth (Pugazhendhi et al., 2018).

16.4 FUTURE PROSPECTS

It is expected that new eco-friendly protocols to generate AgNPs from different types of green raw material continue to be proposed. In the near future it is also expected to increase the number of studies evaluating green AgNPs through field experiments. These studies are relevant to understand interactions between that nanomaterial and microbiote of soil and/or water, for example. It is also necessary to study interaction of nanoparticles with tissues and organs of different living beings to understand not only accumulation pattern and biodistribution of these nanomaterials if they manage to enter their bodies but also toxicity profile in different cell types from these various species of living beings. These results will contribute to making the use of green AgNPs more safe. It is also desirable that attention is dedicated to developing efficient and cost-effective methods to remove the nanoremediators back from the environment and treat them to reuse the AgNPs in other remediation cycles.

16.5 CONCLUSIONS

Silver presents important features that allow the use of this metal to perform different functions and among them generation of nanoparticles capable of remediating environmental pollution. This application of AgNPs is interesting once environmental pollution provoked by development without sustainability is a worldwide problem, threatening the survival of life on Earth and diminishing life quality of living beings. However, the ideal scenario is producing this nanomaterial through an eco-friendly protocol, avoiding the use of harmful reagents. So, green synthesis of AgNPs deserve to be highlighted, offering methodologies to obtain the nanoparticles using bio-based raw material from different sources. Although it is still not a reality the large-scale use of green AgNPs produced by a large array of sustainable and low-cost protocols to remediate environmental pollution, they have exhibited the capacity to deal in a successful way with contaminants from various chemical nature (organic, inorganic, and biological).

KEYWORDS

- **green silver nanoparticles**
- **nanoremediation**
- **environmental pollution remediation**
- **green synthesis of AgNPs**
- **eco-friendly silver nanoparticles**

REFERENCES

Abbas, S.; Nasreen, S.; Haroon, A.; Asharf, M. A. Synhesis of Silver and Copper Nanoparticles from Plants and Application as Adsorbents for Naphthalene Decontamination. *Saudi J. Biol. Sci.* **2020,** 1016–1023.

Abbasi, M.; Saeed, F.; Rafique, U. Preparation of Silver Nanoparticles from Synthetic and Natural Sources: Remediation Model for PAHs. *IOP Conf. Ser.: Mater. Sci. Eng.* **2013,** 1–9.

Annadhasan, M.; Muthukumarasamyvel, T.; Babu, V. R. S.; Rajendiran, N. Green Synthesized Silver and Gold Nanoparticles for Colorimetric Detection of Hg2+, Pb2+, and Mn2+ in Aqueous Medium. *ACS Sustain. Chem. Eng.* **2014,** *2* (4), 887–896.

Anyanwu, B. O.; Ezejiofor, A. N.; Igweze, Z. N.; Orisakwe, O. E. Heavy Metal Mixture Exposure and Effects in Developing Nations: An Update. *Toxics* **2018,** *6* (4), 65.

Ardakani, L. S.; Surendar, A.; Thangavelu, L.; Mandal, T. Silver Nanoparticles (AgNPs) as Catalyst in Chemical Reactions. *Synth. Commun.* **2021,** *51* (10), 1516–1536.

Balali-Mood, M.; Naseri, K.; Tahergorabi, Z.; Khazdair, M. R.; Sadeghi, M. Toxic Mechanisms of Five Heavy Metals: Mercury, Lead, Chromium, Cadmium, and Arsenic. *Front. Pharmacol.* **2021,** *12*, 643972.

Beng, R. F.; Landrigan, P. J.; Balakrishnan, K.; Bathan, G.; Bose-O'Reilly, S.; Brauer, M.; Caravanos, J.; Chiles, T.; Cohen, A.; Corra, L. et al. Pollution and Health: A Progress Update. *Lancet: Planetary Health* **2022,** *6* (6), E535–E547.

Bethi, B.; Sonawane, S. H. Nanomaterials and Its Application for Clean Environment. In: *Nanomaterials for Green Energy*; Bhanvase, B. A.; Pawade, V. B.; Dhoble, S. J.; Sonawane, S. H.; Ashokkumar, M., Eds.; Elsevier: Amsterdam, 2018; pp 385–409.

Briffa, J.; Sinagra, E.; Blundell, R. Heavy Metal Pollution in the Environment and Their Toxicological Effects on Humans. *Heliyon* **2020,** *6* (9), e04691.

Chen, Y. H.; Yeh, C. S. Laser Ablation Method: Use of Surfactants to Form the Dispersed Ag Nanoparticles. *Colloids Surf. A: Physicochem. Eng. Aspects* **2002,** *197*, 133–139.

Chugh, D.; Viswamalya, V. S.; Das, B. Green Synthesis of Silver Nanoparticles with Algae and the Importance of Capping Agents in the Process. *J. Genet. Eng. Biotechnol.* **2021,** *19*, 126.

Chung, M.; Park, I.; Seung-Hyun, K.; Thiruvengadam, M.; Rajakumar, G. Plant-Mediated Synthesis of Silver Nanoparticles: Their Characteristic Properties and Therapeutic Applications. *Nanosc. Res. Lett.* **2016,** *11*, 40.

Coetzee, D.; Venkataraman, M.; Militky, J.; Petru, M. Influence of Nanoparticles on Thermal and Electrical Conductivity of Composites. *Polymers* **2020,** *12* (4), 742–750.

Doan, VD.; Phan, T. L.; Le, V. T.; Vasseghian, Y.; Evgenievna, L. O.; Tran, D. L.; Le, V. T. Efficient and Fast Degradation of 4-Nitrophenol and Detection of Fe (III) Ions by *Poria cocos* Extract Stabilized Silver Nanoparticles. *Chemosphere* **2022,** *286*, 131894.

Desore, A.; Narula, S. A. An Overview on Corporate Response Towards Sustainability Issues in Textile Industry. *Environ. Dev. Sustain.* **2018,** *20* (4), 1439–1459.

Díaz-Núñez, P.; González-Izquierdo, J.; González-Rubio, G.; Guerrero-Martínez, A.; Rivera, A.; Perlado, J.; Bañares, L.; Peña-Rodríguez, O. Effect of Organic Stabilizers on Silver Nanoparticles Fabricated by Femtosecond Pulsed Laser Ablation. *Appl. Sci.* **2017,** *7*, 793.

Edison, T. N. J. I.; Atchudan, R.; Sethuraman, M. G.; Lee, Y. R. Reductive-Degradation of Carcinogenic Azo Dyes Using *Anacardium occidentale* Testa Derived Silver Nanoparticles. *J. Photochem. Photobiol. B: Biol.* **2016,** *162*, 604–610.

Edison, T. J. I.; Sethuraman, M. G. Biogenic Robust Synthesis of Silver Nanoparticles Using *Punica granatum* Peel and Its Application as a Green Catalyst for the Reduction of an Anthropogenic Pollutant 4-Nitrophenol. *Spectrochim. Acta Part A: Mol. Biomol. Spectrosc.* **2013,** *104*, 262–264.

Elemike, E. E.; Onwudiwe, D. C.; Ekennia, A. C.; Ehiri, R. C.; Nnaji, N. J. Phytosynthesis of Silver Nanoparticles Using Aqueous Leaf Extracts of *Lippia citriodora*: Antimicrobial, Larvicidal and Photocatalytic Evaluations. *Mater. Sci. Eng.: C* **2017,** *75*, 980–989.

El-Naggar, N. A.; Moawad, M. N.; Ahmed, E. F. Toxic Phenolic Compounds in the Egyptian Coastal Waters of Alexandria: Spatial Distribution, Source Identification, and Ecological Risk Assessment. *Water Sci.* **2022,** *36* (1), 32–40.

Fei, L.; Bilal, M.; Qamar, S. A.; Imran, H. M.; Riasat, A.; Jahangeer, M.; Ghafoor, M.; Ali, N.; Iqbal, H. M. N. Nano-Remediation Technologies for the Sustainable Mitigation of Persistent Organic Pollutants. *Environ. Res.* **2022,** *211*, 113060.

Fernández, J. G.; Fernández-Baldo, M. A.; Berni, E.; Camí, G.; Durán, N.; Raba, J.; Sanz, M. I. Process Biochemistry Production of Silver Nanoparticles Using Yeasts and Evaluation of Their Antifungal Activity Against Phytopathogenic Fungi. *Process Biochem.* **2016,** *51* (9), 1306–1313.

Filippi, A.; Mattiello, A.; Musetti, R.; Petrussa, E.; Braidot, E.; Marchiol, L. Green Synthesis of Ag Nanoparticles Using Plant Metabolites. *AIP Conf. Proc.* **2017,** 1873, 020004.

Garg, D.; Sarkar, A.; Chand, P.; Bansal, P.; Gola, D.; Sharma, S.; Khantwal, S.; Surabhi; Mehrotra, R.; Chauhan, N. Synthesis of Silver Nanoparticles Utilizing Various Biological Systems: Mechanisms and Applications—A Review. *Progress Biomater.* **2020,** *9* (3), 81–95.

Ghosh, S.; Patil, S.; Ahire, M.; Kitture, R.; Kale, S.; Pardesi, K.; Cameotra, S. S.; Bellare, J.; Dhavale, D. D.; Jabgunde, A.; Chopade, B. A. Synthesis of Silver Nanoparticles Using *Dioscorea bulbifera* Tuber Extract and Evaluation of Its Synergistic Potential in Combination with Antimicrobial Agents. *Int. J. Nanomed.* **2012,** *7*, 483–496.

Giri, A. K.; Jena, B.; Biswal, B.; Pradhan, A. K.; Arakha, M.; Acharya, S.; Acharya, L. Green Synthesis and Characterization of Silver Nanoparticles Using *Eugenia roxburghii* DC. Extract and Activity Against Biofilm-Producing Bacteria. *Sci. Rep.* **2022,** *12*, 8383.

Guowei, Q.; Niu, Z.; Yu, J.; Li, Z.; Ma, J.; Xiang, P. Soil Heavy Metal Pollution and Food Safety in China: Effects, Sources and Removing Technology. *Chemosphere* **2021,** *267*, 129205.

Hamedi, S.; Shojaosadati, S. A.; Shokrollahzadeh, S.; Hashemi-Najafabadi, S. Extracellular Biosynthesis of Silver Nanoparticles Using a Novel and Non-Pathogenic Fungus, Neurospora Intermedia: Controlled Synthesis and Antibacterial Activity. *World J. Microbiol. Biotechnol.* **2013,** *30* (2), 693–704.

Hano, C.; Abbasi, B. H. Plant-Based Green Synthesis of Nanoparticles: Production, Characterization and Applications. *Biomolecules* **2022,** *12* (1), 31.

Ho, Y. Y.; Sun, D. S.; Chang, H. H. Silver Nanoparticles Protect Skin from Ultraviolet B-Induced Damage in Mice. *Int. J. Mol. Sci.* **2020,** *21* (19), 7082.

Hussein-Al-Ali, S. H.; Abudoleh, S. M.; Abualassal, Q. I. A.; Abudayeh, Z.; Aldalahmah, Y.; Hussein, M. Z. Preparation and Characterisation of Ciprofloxacin-Loaded Silver Nanoparticles for Drug Delivery. *IET Nanobiotechnol.* **2022,** *16* (3), 92–101.

Hwang, J. S.; Park, J. E.; Kim, G. W.; Nam, H.; Yu, S.; Jeon, J. S.; Kim, S.; Lee, H.; Yang, M. Recycling Silver Nanoparticle Debris from Laser Ablation of Silver Nanowire in Liquid Media Toward Minimum Material Waste. *Sci. Rep.* **2021,** *11* (2021), 2262–2271.

Ismail, G. A.; Allam, N. G.; El-Gemizy, W. M.; Salem, M. A. The Role of Silver Nanoparticles Biosynthesized by *Anabaena variabilis* and *Spirulina platensis* Cyanobacteria for Malachite Green Removal from Wastewater. *Environ. Technol.* **2021,** *42* (2), 4475–4489.

Ismail, M.; Khan, M. I.; Khan, M. A.; Akhtar, K.; Asiri, A. M.; Khan, S. B. Plant-Supported Silver Nanoparticles: Efficient, Economically Viable and Easily Recoverable Catalyst for the Reduction of Organic Pollutants. *Appl. Organ. Chem.* **2019,** *33* (8), e4971.

Jacob, J. M.; Ravindran, R.; Narayanan, M.; Samuel, S. M.; Pugazhendhi, A.; Kumar, G. Microalgae: A Prospective Low Cost Green Alternative for Nanoparticle Synthesis. *Curr. Opin. Environ. Sci.* **2021,** *20*, 100163.

Jiang, B.; Gao, G.; Ruan, M.; Bian, Y.; Geng, F.; Yan, W.; Xu, X.; Shen, M.; Wang, J.; Chang, R.; Xu, L.; Zhang, X.; Feng, F.; Chen, Q. Quantitative Assessment of Abiotic Stress on the Main Functional Phytochemicals and Antioxidant Capacity of Wheatgrass at Different Seedling Age. *Front. Nutr.* **2021,** *8*, 731555.

Kakakhel, M. A.; Wu, F.; Sajjad, W.; Qi Zhang, Q.; Khan, I.; Ullah, K.; Wang, W. Long-Term Exposure to High-Concentration Silver Nanoparticles Induced Toxicity, Fatality, Bioaccumulation, and Histological Alteration in Fish (*Cyprinus carpio*). *Environ. Sci. Europe* **2021,** *33*, 14.

Khan, A. U.; Malik, N.; Khan, M.; Cho, M. H.; Khan, M. M. Fungi-Assisted Silver Nanoparticle Synthesis and Their Applications. *Bioprocess Biosyst. Eng.* **2017,** *41* (1), 1–20.

Khatoon, N.; Alam, H.; Manzoor, N.; Sardar, M. Removal of Toxic Contaminants from Water by Sustainable Green Synthesised Non-Toxic Silver Nanoparticles. *IET Nanobiotechnol.* **2018,** *12* (8), 1090–1096.

Kovács, D.; Igaz, N.; Gopisetty, M. K.; Kiricsi, M. Cancer Therapy by Silver Nanoparticles: Fiction or Reality? *Int. J. Mol. Sci.* **2022,** *23* (2), 839.

Kumar, J. A.; Krithiga, T.; Sathish, S.; Renita, A. A.; Prabu, D.; Lokesh, S.; Geetha, R.; Namasivayam, S. K. R.; Sillanpaa, M. Persistent Organic Pollutants in Water Resources: Fate, Occurrence, Characterization and Risk Analysis. *Sci. Total Environ.* **2022,** *831*, 154808.

Kumar, P.; Kausar, M. A.; Singh, A. B.; Singh, R. Biological Contaminants in the Indoor Air Environment and Their Impacts on Human Health. *Air Qual. Atmos. Health* **2021,** *14*, 1723–1736.

Lellis, B.; Fávaro-Polonio, C. Z.; Pamphile, J. A., Polonio, J. C. Effects of Textile Dyes on Health and the Environment and Bioremediation Potential of Living Organisms. *Biotechnol. Res. Innov.* **2019,** *3* (2), 275–290.

Liu, X.; Chen, J. L.; Yang, W. Y.; Qian, Y. C.; Pan, J. Y.; Zhu, C. N.; Liu, L.; Ou, W. B.; Zhao, H. X.; Zhang, D. P. Biosynthesis of Silver Nanoparticles with Antimicrobial and Anticancer Properties Using Two Novel Yeasts. *Sci. Rep.* **2021,** *11*, 15795.

Longman, J.; Veres, D.; Finsinger, W.; Ersek, V. Exceptionally High Levels of Lead Pollution in the Balkans from the Early Bronze Age to the Industrial Revolution. *Proc. Natl. Acad. Sci.* **2018,** *115*, E5661–E5668.

Lu, F.; Astruc, D. Nanocatalysts and Other Nanomaterials for Water Remediation from Organic Pollutants. *Coordination Chem. Rev.* **2020,** 2020, 408.

Marslin, G.; Siram, K.; Maqbool, Q.; Selvakesavan, R. K.; Kruszka, D.; Kachlicki, P.; Franklin, G. Secondary Metabolites in the Green Synthesis of Metallic Nanoparticles. *Materials* **2018,** *11* (6), 940.

Mathur, P.; Jha, S.; Ramteke, S.; Jain, N. Pharmaceutical Aspects of Silver Nanoparticles. *Artif. Cells Nanomed. Biotechnol.* **2018,** *46*, 115–126.

Miranda, R. R.; Sampaio, I.; Zucolotto, V. Exploring Silver Nanoparticles for Cancer Therapy and Diagnosis. *Colloids Surfaces B: Biointerfaces* **2022,** *210*, 112254.

Mohammadlou, M.; Maghsoudi, H.; Jafarizadeh-Malmiri, H. A Review on Green Silver Nanoparticles Based on Plants: Synthesis, Potential Applications and Eco-Friendly Approach. *Int. Food Res. J.* **2016,** *23* (2), 446–463.

Mohammed, A. H.; Naje, A. N. Preparation and Characterization of Graphene Nanosheet Doped with Silver Nanoparticles. *J. Phys.: Conf. Ser.* **2021,** *2063*, 1–8.

Mohanpuria, P.; Rana, N. K.; Yadav, S. K. Biosynthesis of Nanoparticles: Technological Concepts and Future Applications. *J. Nanopart. Res.* **2007,** *10* (3), 507–517.

Momin, B.; Rahman, S.; Jha, N.; Annapure, U. S. Valorization of Mutant *Bacillus licheniformis* M09 Supernatant for Green Synthesis of Silver Nanoparticles: Photocatalytic

Dye Degradation, Antibacterial Activity, and Cytotoxicity. *Bioprocess Biosyst. Eng.* **2019,** *42*, 541–553.

Monti, G. A.; Correa, N. M.; Falcone, R. D.; Silbestri, G. F.; Moyano, F. New Insights into the Catalytic Activity and Reusability of Water-Soluble Silver Nanoparticles. *Chem. Select* **2021,** *6* (29), 7436–7442.

Murillo-Rábago, E. I.; Vilchis-Nestor, A. R.; Juarez-Moreno, K.; Garcia-Marin, L. E.; Quester, K.; Castro-Longoria, E. Optimized Synthesis of Small and Stable Silver Nanoparticles Using Intracellular and Extracellular Components of Fungi: An Alternative for Bacterial Inhibition. *Antibiotics* **2022,** *11*, 800.

Nasrollahzadeh, M.; Sajjadi, M.; Iravani, S.; Varma, R. S. Green-Synthesized Nanocatalysts and Nanomaterials for Water Treatment: Current Challenges and Future Perspectives. *J. Hazard. Mater.* **2021,** *401*, 123401.

Naseer, Q. A.; Xue, X.; Dang, W. S.; Din, S. U.; Kalsoom; Jamile, J. Synthesis of Silver Nanoparticles Using *Lactobacillus bulgaricus* and Assessment of Their Antibacterial Potential Síntese de nanopartículas de prata usando Lactobacillus bulgaricus e avaliação de seu potencial antibacteriano. *Braz. J. Biol.* **2022,** *82*, e232434.

Neethu, S.; Midhun, S. J.; Radhakrishnan, E. K.; Jyothis, M. Green Synthesized Silver Nanoparticles by Marine Endophytic Fungus *Penicillium polonicum* and Its Antibacterial Efficacy Against Biofilm Forming, Multidrug-Resistant *Acinetobacter baumanii. Microb. Pathog.* **2018,** *116*, 263–272.

Negi, A.; Rana, P.; Vishwakarma, R. K.; Negi, D. S. Multi Dye Degradation, Antibacterial, Antidiabetic and Antioxidant Assessment of Silver Nanoparticles (Ag-NPs) Derived via Leaves of *Smilax aspera. Inorg. Chem. Commun.* **2022,** *143*, 109703.

Nga, N. T. A.; Raghavendra, V. B.; Sindhu, R.; Alshiekheid,, M.; Sabour, A.; Krishnan, R.; Chi, N. T. L.; Pugazhendhi, A. Green Fabrication of Silver Nanoparticles Using *Chloroxylon swietenia* Leaves and Their Application Towards Dye Degradation and Food Borne Pathogens. *Food Chem. Toxicol.* **2022,** *165*, 113192.

Nguyen, T. H. A.; Nguyen, V. C.; Phan, T. N. H.; Le, V. T.; Vasseghian, Y.; Trubitsyn, M. A.; Nguyen, A. T.; Chau, T. P.; Doan, V. D. Novel Biogenic Silver and Gold Nanoparticles for Multifunctional Applications: Green Synthesis, Catalytic and Antibacterial Activity, and Colorimetric Detection of Fe (III) ions. *Chemosphere* **2022,** *287*, 132271–132271.

Oliveira, G. Z. S.; Lopes, C. A. P.; Sousa, M. H.; Silva, L. P. Synthesis of Silver Nanoparticles Using Aqueous Extracts of *Pterodon emarginatus* Leaves Collected in the Summer and Winter Seasons. *Int. Nano Lett.* **2019,** *9*, 109–117.

Ottoni, C. A.; Simões, M. F.; Fernandes, S.; Santos, J. G.; Silva, E. S.; Souza, R. F. B.; Maiorano, A. E. Screening of filamentous fungi for antimicrobial silver nanoparticles synthesis. *Amb Express,* **2017,** 7, 1.

Ovais, M.; Khalil, A. T.; Ayaz, M.; Ahmad, I.; Nethi, S. K.; Mukherjee, S. Biosynthesis of metal nanoparticles via microbial enzymes: a mechanistic approach. *Int. J. Mol. Sci.* **2018,** 19, 4100.

Parmar, A.; Kaur, G.; Kapil, S.; Sharma, V.; Choudhary, M. K.; Sharma, S. Novel Biogenic Silver Nanoparticles as Invigorated Catalytic and Antibacterial Tool: A Cleaner Approach Towards Environmental Remediation and Combating Bacterial Invasion. *Mater. Chem. Phys.* **2019,** *238*, 121861.

Paw, R.; Hazarika, M.; Boruah, P. K.; Kalita, A. J.; Guha, A. K.; Das, M. R.; Tamuly, C. Highly Sensitive and Selective Colorimetric Detection of Dual Metal Ions (Hg2+ and Sn2+) in Water: An Eco-Friendly Approach. *RSC Adv.* **2021,** *11* (24), 14700–14709.

Pugazhendhi, A.; Prabakar, D.; Jacob, J. M.; Karuppusamy, I.; Saratale, R. G. Synthesis and Characterization of Silver Nanoparticles Using *Gelidium amansii* and Its Antimicrobial Property Against Various Pathogenic Bacteria. *Microb. Pathog.* **2018,** *114*, 41–45.

Ralte, L.; Khiangte, L.; Thangjam, N. M.; Kumar, A.; Singh, Y. T. GC–MS and Molecular Docking Analyses of Phytochemicals from the Underutilized Plant, *Parkia timoriana* Revealed Candidate Anti-Cancerous and Anti-Inflammatory Agents. *Sci. Rep.* **2022,** *12*, 3395.

Ramos, M. M.; Morais, E. S.; Sena, I. S.; Lima, A. L.; Oliveira, F. R.; Freitas, C. M.; Fernandes, C. P.; Carvalho, J. C. T.; Ferreira, I. M. Silver Nanoparticle from Whole Cells of the Fungi Trichoderma spp. Isolated from Brazilian Amazon. *Biotechnol. Lett.* **2020,** *42* (5), 833–843.

Rania, P.; Kumar, V.; Singh, P. P.; Matharu, A. S.; Zhang, W.; Kim, K. H.; Singh, J.; Rawat, M. Highly Stable AgNPs Prepared via a Novel Green Approach for Catalytic and Photocatalytic Removal of Biological and Non-Biological Pollutants. *Environ. Int.* **2020,** *143*, 105924.

Riaz, M.; Ismail, M.; Ahmad, B.; Zahid, N.; Jabbour, G.; Khan, M. S.; Mutreja, V.; Sareen, S.; Rafiq, A.; Faheem, M.; Shah, M. M.; Khan, M. I.; Bukhari, S. A. I.; Park, J. Characterizations and Analysis of the Antioxidant, Antimicrobial, and Dye Reduction Ability of Green Synthesized Silver Nanoparticles. *Green Process. Synth.* **2020,** *9*, 64.

Rocha, J.; Henriques, I., Gomila, M.; Manaia, C. M. Common and Distinctive Genomic Features of *Klebsiella pneumoniae* Thriving in the Natural Environment or in Clinical Settings. *Sci. Rep.* **2022,** *12*, 10441.

Rodrigues, A. G.; Gonçalves, P. J. R. O.; Ottoni, C. A.; Morgano, M. A.; Araújo, W. L.; Melo, I. S.; Souza, A. O.; Ruiz, R. C. Functional Textiles Impregnated with Biogenic Silver Nanoparticles from *Bionectria ochroleuca* and Its Antimicrobial Activity. *Biomed. Microdevices* **2019,** *21* (3), 56.

Rose, G. K.; Soni, R.; Rishi, P.; Soni, S. K. Optimization of the Biological Synthesis of Silver Nanoparticles Using *Penicillium oxalicum* GRS-1 and Their Antimicrobial Effects Against Common Food-Borne Pathogens. *Green Process. Synth.* **2018,** *8*, 42.

Roychoudhury, A. Yeast-mediated Green Synthesis of Nanoparticles for Biological Applications. *Indian J. Pharm. Biol. Res.* **2020,** *8* (3), 26–31.

Samuel, M. S.; Jose, S.; Selvarajan, E.; Mathimani, T.; Pugazhendhi, A. Biosynthesized Silver Nanoparticles Using *Bacillus amyloliquefaciens*; Application for Cytotoxicity Effect on A549 Cell Line and Photocatalytic Degradation of p-Nitrophenol. *J. Photochem. Photobiol. B: Biology* **2020,** *202*, 111642.

Santhosh, C.; Velmurugan, V.; Jacob, G.; Jeong, S. K.; Grace, A. N.; Bhatnagar, A. Role of Nanomaterials in Water Treatment Applications: A Review. *Chem. Eng. J.* **2016,** *306*, 1116–1137.

Sen, P.; Ghosh, J.; Abdullah, A.; Kumar, P.; Vandana. Preparation of Cu, Ag, Fe and Al Nanoparticles by the Exploding Wire Technique. *J. Chem. Sci.* **2003,** *115* (5–6), 499–508.

Shanmuganathan, R.; Karuppusamy, I.; Saravanan, M.; Muthukumar, H.; Ponnuchamy, K.; Ramkumar, V. S.; Pugazhendhi, A. Synthesis of Silver Nanoparticles and Their Biomedical Applications—A Comprehensive Review. *Curr. Pharm. Des.* **2019,** *25*, 2650–2660.

Singh, J.; Dutta, T.; Kim, KH.; Rawat, M.; Samddar, P.; Kumar, P. 'Green' Synthesis of Metals and Their Oxide Nanoparticles: Applications for Environmental Remediation. *J. Nanobiotechnol.* **2018,** *16*, 1.

Singh, J.; Kumar, V.; Jolly, Sukhwinder Singh; Kim, Ki-Hyun; Rawat, M.; Kukkar, D.; Tsang, Yiu Fai. Biogenic Synthesis of Silver Nanoparticles and Its Photocatalytic Applications for Removal of Organic Pollutants in Water. *J. Ind. Eng. Chem.* **2019,** 247–257.

Singh, R.; Shedbalkar, U. U.; Wadhwani, S. A.; Chopade, B. A. Bacteriagenic Silver Nanoparticles: Synthesis, Mechanism, and Applications. *Appl. Microbiol. Biotechnol.* **2015,** *99* (11), 4579–4593.

Stensberg, M. C.; Wei, Q.; McLamore, E. S.; Porterfield, D. M.; Wei, A.; Sepúlveda, M. S. Toxicological Studies on Silver Nanoparticles: Challenges and Opportunities in Assessment, Monitoring and Imaging. *Nanomedicine* **2011,** *6* (5), 879–898.

Taniguchi, N. On the Basic Concept of Nanotechnology. *Proc. Int. Conf. Prod. Eng.* **1974,** *1974*, 18–23.

Tonelli, F. M. P.; Tonelli, F. C. P.; Ferreira, D. R. C.; da Silva, K. E.; Cordeiro, H. G.; Ouchida, A. T.; Nunes, N. A. M. Biocompatibility and Functionalization of Nanomaterials. In: *Intelligent Nanomaterials for Drug Delivery Applications*; Ahmad, N.; Gopinath, P., Eds.; Elsevier: New York, 2020; pp 85–103.

Tonelli, F. M. P.; Tonelli, F. C. P.; Lemos, M. S.; Ferreira, D. R. C. Nanomaterials to Remediate Water Pollution. In: *Nanotechnology for Environmental Pollution Decontamination: Tools, Methods, and Approaches for Detection and Remediation*; Tonelli, F. M. P.; Bhat, R. A.; Dar, G. H., Eds.; CRC Press: Palm Bay (in press).

Tortella, G. R.; Rubilar, O.; Durán, N.; Diez, M. C.; Martínez, M.; Parada, J.; Seabra, A. B. Silver Nanoparticles: Toxicity in Model Organisms as an Overview of Its Hazard for Human Health and the Environment. *J. Hazard. Mater.* **2020,** *390*, 121974.

Tsekhmistrenko, S. I.; Bityutskyy, V. S.; Tsekhmistrenko, O. S.; Horalskyi, L. P.; Tymoshok, N. O.; Spivak, M. Y. Bacterial Synthesis of Nanoparticles: A Green Approach. *Biosyst. Divers.* **2020,** *28*, 9–17.

Varadavenkatesan, T.; Selvaraj, R.; Vinayagam, R. Phyto-Synthesis of Silver Nanoparticles from Mussaenda Erythrophylla Leaf Extract and Their Application in Catalytic Degradation of Methyl Orange Dye. *J. Mol. Liquids* **2016,** 1063–1070.

Verma, A.; Bharadvaja, N. Plant-Mediated Synthesis and Characterization of Silver and Copper Oxide Nanoparticles: Antibacterial and Heavy Metal Removal Activity. *J. Clust. Sci.* **2022,** *33*, 1697–1712.

Wang, L.; Lu, F.; Liu, Y.; Wu, Y.; Wu, Z. Photocatalytic Degradation of Organic Dyes and Antimicrobial Activity of Silver Nanoparticles Fast Synthesized by Flavonoids Fraction of *Psidium guajava* L. Leaves. *J. Mol. Liquids* **2018,** *263*, 187–192.

Werkneh, A. A.; Rene, R. R.; Lens, P. N. L. Application of Nanomaterials in Food, Cosmetics and Other Process Industries. In: *Nanotoxicology: Toxicity Evaluation, Risk Assessment and Management*; Kumar, V.; Dasgupta, N.; Ranjan, S., Eds.; CRC Press: Palm Bay, 2018; pp 63–79.

Widatalla, H. A.; Yassin, L. F.; Alrasheid, A. A.; Ahmed, S. A. R.; Widdatallah, M. O.; Eltilib, S. H.; Mohamed, A. A. Green Synthesis of Silver Nanoparticles Using Green Tea Leaf Extract, Characterization and Evaluation of Antimicrobial Activity. *Nanoscale Adv.* **2022,** *4*, 911–915.

Xu, C.; Nasrollahzadeh, M.; Sajjadi, M.; Maham, M.; Luque, R.; Puente-Santiago, A. R. Benign-by-Design Nature-Inspired Nanosystems in Biofuels Production and Catalytic Applications. *Renew. Sustain. Energy Rev.* **2019,** *112*, 195–252.

Xu, L.; Wang, Y. Y.; Huang, J.; Chen, C. Y.; Wang, Z. X.; Xie, H. Silver Nanoparticles: Synthesis, Medical Applications and Biosafety. *Theranostics* **2020,** *10* (20), 8996–9031.

Yamada, D.; Hiwatari, M.; Hangoma, P.; Narita, D.; Mphuka, C.; Chitah, B.; Yabe, J.; Nakayama, S. M. M.; Nakata, H.; Choongo, K.; Ishizuka, M. Assessing the Population-Wide

Exposure to Lead Pollution in Kabwe, Zambia: An Econometric Estimation Based on Survey Data. *Sci. Rep.* **2020,** *10*, 15092.

Yasmin, S.; Nouren, S.; Bhatti, H. N.; Iqbal, D. N.; Iftikhar, S.; Majeed, J.; Mustafa, R.; Nisar, N.; Nisar, J.; Nazir, A.; Iqbal, M.; Rizvi, H. Green Synthesis, Characterization and Photocatalytic Applications of Silver Nanoparticles Using *Diospyros* Lotus. *Green Process. Synth.* **2020,** *9*, 87–96.

Yassin, M. T.; Mostafa, A. A.-F.; Al-Askar, A. A.; Al-Otibi, F. O. Facile Green Synthesis of Silver Nanoparticles Using Aqueous Leaf Extract of *Origanum majorana* with Potential Bioactivity Against Multidrug Resistant Bacterial Strains. *Crystals* **2022,** *12*, 603.

Yin, I. X.; Zhang, J.; Zhao, I. S.; Mei, M. L.; Li, Q. L.; Chu, C. H. The Antibacterial Mechanism of Silver Nanoparticles and Its Application in Dentistry. *Int. J. Nanomed.* **2020,** *15*, 2555–2562.

Zhao, X.; Xu, X.; Ai, C.; Yan, L.; Jiang, C.; Shi, J. Advantages of Silver Nanoparticles Synthesized by Microorganisms in Antibacterial Activity. In: *Green Synthesis of Silver Nanomaterials: Nanobiotechnology for Plant Protection*; Kamel A. Abd-Elsalam, Ed.; Elsevier: New York, 2022; pp 571–586.

Zhao, X.; Zhou, L.; Rajoka, M. S. R.; Yan, L.; Jiang, C.; Shao, D.; Zhu, J.; Shi, J.; Huang, Q.; Yang, H. Fungal Silver Nanoparticles: Synthesis, Application and Challenges. *Crit. Rev. Biotechnol.* **2017,** *38* (6), 817–835.

Zheng, H.; Zhang, C.; Liu, B.; Liu, G.; Zhao, M.; Xu, G.; Luo, X.; Li, F.; Xing, B. Biochar for Water and Soil Remediation: Production, Characterization, and Application. In: *A New Paradigm for Environmental Chemistry and Toxicology*; Jiang, G.; Li, X., Eds.; Springer: Singapore, 2020; pp 153–196.

Zulkifli, N. I.; Muhamad, M.; Zain, N. N. M.; Tan, W. N.; Yahaya, N.; Bustami, Y.; Aziz, A. A.; Kamal, N. N. S. M. N. A Bottom-Up Synthesis Approach to Silver Nanoparticles Induces Anti-Proliferative and Apoptotic Activities Against MCF-7, MCF-7/TAMR-1 and MCF-10A Human Breast Cell Lines. *Molecules* **2020,** *25* (18), 4332.

CHAPTER 17

Green Palladium Nanoparticles for Nanoremediation

JURI KALITA and SIDDHARTHA S. DHAR

Department of Chemistry, National Institute of Technology, Silchar, Cachar, Assam, India

ABSTRACT

Nanotechnology is concerned with matter having dimensions of 1–100 nm, which has a wide range of applications in medicines, industries, and environmental remediation. Developing nanoparticles (NPs) using green synthetic method is a growing research area in green nanotechnology in comparison to other conventional chemical and physical approaches since this methodology is non-toxic or less hazardous, efficient, environmentally friendly, and less expensive. One of the valuable metallic nanoparticles is palladium. Many researchers are interested in biogenic synthesis of palladium (Pd) nanoparticles from plants and bacteria because it is cost-effective, environmentally benign, and sustainable. Metal salts are reduced to metal nanoparticles by a variety of primary and secondary metabolites found in plants and their components. Temperature, pH, processing time, plant extract, and metal salt concentrations, all influence the morphology, size, and stability of Pd nanoparticles. The Pd nanomaterials have particular and selective therapeutic effects that are size and form dependant. This chapter examines the biogenic production of Pd nanoparticles and their prospective applications including catalysts, biosensors, therapeutics, and diagnostic tools. Furthermore, the chapter also covers the significance of the green synthesis approach, as well as the benefits, problems, and future prospects.

Sustainable Nanoremediation: Modern Technologies for a Clean Environment.
Fernanda Maria Policarpo Tonelli, Arpita Roy, and Khalid Rehman Hakeem (Eds.)

17.1 INTRODUCTION

Nanotechnology is the study and engineering of materials which includes modeling, imaging, measuring, and controlling of matter at dimensions between 1 and 100 nm. The concept of nanotechnology is the development, synthesis, characterization, and use of materials and devices by altering their size and shape at the nanoscale range, that is, in the range of 1–100 nm. This branch of study simply combines all the natural sciences with biology, physics, chemistry, materials science, engineering, and computer sciences to construct nanostructures (Shenton et al., 1999). Based on their morphology, size, and distribution, the nanostructures have various uses in every field of research and technology including chemical industries, drug delivery, energy science, environment, catalysis, cosmetics and health care, biomedical, electronics, space industries, mechanics and optics, single-electron transistors, non-linear optical devices, and photo-electrochemical applications. There are growing attempts to utilize nanotechnology as a green technology to conserve the environment through pollution control, treatment, and remediation of long-term concerns including contaminated sites (Innovations, n.d.). Nanotechnology has shown to be beneficial to technology and science comprising the domains of organic chemistry as catalyst (Shiju et al., 2009), electrochemical sensors (Kouvaris et al., 2012), electrical conductor (Rezaei et al., 2016), environmental remediation (Kalita et al., 2022), surface chemistry, molecular electronics, molecular engineering, medical diagnosis, etc. Due to their distinct physicochemical and thermodynamic characteristics, metal nanoparticles have received a lot of attention and are now a viable option for usage in a variety of industries, including electronics (Cui et al., 2012), catalysis (Ha et al., 2016; Bortolotto et al., 2015), biological applications (Anand et al., 2016b), and optics (Eryürek et al., 2015). One of the most remarkable aspects of modern nanoscience and nanotechnology is the fabrication of nanomaterials that meet the necessary quality requirements.

Nanoparticle synthesis can be carried out using chemical, physical, and biological techniques. In general, there are two approaches for the nanoparticle synthesis: the top-down strategy, which uses energy from chemical, physical, and biological sources to break down larger structures into smaller pieces; and the bottom-up strategy, which develops a large nanostructure from the atomic level through diverse physical, chemical, or biological processes (Das et al., 2017). However, the physical and chemical processes are more expensive and often results in harmful compounds. The use of toxic substances may result in risks such as carcinogenic effects, toxicity, and ecotoxicity (Gupta et al., 2018). Especially the chemical process, which is a widely used method to produce nanoparticles could be hazardous to the environment. Due to the use of potentially harmful compounds including organic solvents, reducing agents, and stabilisers, toxicity issues are extremely noticeable. The aggregation of colloids is prevented by these substances. Nanoparticles can be used in a variety of therapeutic and biomedical applications; however, their utilization is restricted by the use of hazardous solvents and chemical contamination (Hua et al., 2018). Consequently, it is necessary to use a dependable, clean, physiologically suitable, and environmentally benign method to synthesize nanoparticles. The fabrication of metallic nanoparticles using biological resources in an environmentally sustainable manner has received considerable attention. Some advantages of biological synthesis are the use of low pressure, low temperatures, and less hazardous substances. Also, the production of substantial amounts of nanoparticles can be done through biological synthesis (by bio extract), which is non-toxic, affordable, and environmental friendly. Extracts from living organisms can also be used in the synthesis of nanoparticles as both reducing and stabilizing agents (Momeni et al., 2015). According to researchers, nanoparticles made biologically have higher pharmacological efficacy than those made physicochemically. By adjusting various parameters including incubation time, pH, mixing ratio, and temperature, metallic nanoparticles (MNPs) made from plant extracts are stable and simple to monodisperse. Various plants including mango leaves (Mangifera indica), tulsi leaves (Ocimum tenuiflorum), curry leaves (Murraya koenigii), guava leaves (Psidium guajava), neem leaves (Azadirachta indica), aloe vera leaves, etc., have been used to synthesized green nanoparticles. Biosynthesis of some transition and noble metallic nanoparticles such as Au, Ag, Pt, Pd, Ru, Fe, Ni, and Co are commonly used in biocatalysis, anti-microbial, anti-cancer, and cosmetic products (Narayanan et al., 2005).

Palladium nanoparticle (Pd NP) catalysts have attracted a lot of attention because of its prospective applications in the domains of biotechnology, biomedicine, and medicine. Due to their high specific surface area and surface activity, Pd NP can be employed in both homogeneous and heterogeneous catalysis. Pd NPs have remarkable thermal and chemical stability, optical properties, and electrical characteristics. The development of Pd NPs has therefore assumed extraordinary relevance (Wang et al., 2014). The research of materials based on palladium is extremely significant and useful since palladium is among the most effective metals for catalysis. In earlier methods, the fabrication of Pd NPs involving reduction of Pd(II) ions in the addition of stabilizers using chemical-reducing agents including ascorbic acid, sodium citrate, hydrazine, and sodium borohydride to produce homogeneous suspensions. According to reports Pd NPs can be synthesized using biological sources such as Diopyros kaki. leaf (Yong et al., 2010), Cinnamom zeylanicum bark (Sathishkumar et al., 2009), C. Camphora leaf (Yang et al., 2010), Curcuma longa tuber (Sathishkumar et al., 2009), banana peel (Bankar et al., 2010), Hippophaerhamnoides Linn (Nasrollahzadeh et al., 2018), Pistacia atlantica kurdica gum (Veisi et al., 2015), Rosa canina fruit (Veisi et al., 2016), pectin (Khazaei et al., 2013), Stachys lavandulifolia (Veisi et al., 2014), and Oak gum (Veisi et al., 2016). These natural compounds are considered to have an effective function in the metal ion reduction as well as the precipitation of corresponding nanoparticles. Table 17.1 provides an overview of the characteristics of Pd NPs that were synthesized using various plant parts. Because to their high energy of surface and large ratio of surface-to-volume, they are employed as both homogeneous as well as heterogeneous catalysts (Narayanan et al., 2005).The Pd NPs have applications in heterogeneous catalysis for the pollutant dye degradation, wastewater treatment, antibacterial and antifungal properties, and anticancer properties. In fact, other metal oxide and metal nanoparticles are used in a variety of medicinal applications, such as antibacterial and anticancer. Ag, Au, Fe, CuO, and ZnO are some of these NPs. The heterogeneous catalytic efficiency of Pd NPs synthesized from herbal extracts in the Suzuki-Miyaura coupling process has also been investigated (Nasrollahzadeh et al., 2015). The noble Pd NPs, on the other hand, have demonstrated exceptional physicochemical characteristics, including outstanding photocatalytic activity, excellent chemical and thermal stability, and good electrical and optical properties (Saldan et al., 2015). Pd NPs catalyze numerous diverse reactions, such as oxidation processes and catalysis that result in the formation of C–C bonds. In comparison to other nanostructured materials like Ag and Au nanospheres, plasmonic Palladium nanospheres demonstrated a stronger reactivity in refractive index variations (Sugawa et al., 2015). Along with having a wide range of uses, Pd NPs can improve therapeutic treatment and biological diagnosis with the fewest possible side effects. Pd NPs, on the other hand, typically agglomerate due to thermodynamic instability, which restricts their potential biological applications. To do this, fabrication using an algal extract may stabilize the Pd NPs by effective algal biomolecule surface adsorption. Polysaccharides, proteins, fatty acids, phenolics, flavonoids, and other bioactive compounds found in algal extracts can effectively reduce Pd^{2+} ions (Compounds, 2018). After successful synthesis of Pd NP, they can be characterized using transmission electron microscopy (TEM), scanning electron microscopy (SEM), X-ray diffraction (XRD), Fourier-transform infrared (FTIR) spectroscopy, and UV-Vis technique.

This chapter explored the biogenic synthesis of Pd NPs using different biological sources as well as an assessment of their potential applications in the treatment of cancer, catalysis, and other biological sciences. Also, the possible mechanism of biosynthesis of Pd NP investigated.

TABLE 17.1 Synthesis of Pd Nanoparticles Using Various Plant Extract.

Plant	Shape	Part used	Size (nm)	References
Anogeissus latifolia	Spherical	Gum	4.8	Kora et al. (2018)
Cinnamomum camphora	–	Leaves	3.2–6.0	Yang et al. (2010)
Cinnamom zeylanicum	Crystalline	Bark	15–20	Sathishkumar et al. (2009)
Euphorbia granulate	–	Leaves	25–35	Nasrollahzadeh and Sajadi (2016)
Curcuma longa	Spherical	Tuber	10–15	Sathishkumar et al. (2009)
Gardenia jasminoides	–	Leaves	3–5	Jia et al. (2009)
Moringa oleifera	Spherical	Waste petal	10–50	Anand et al. (2016b)
Glycine max	Spherical	Leaves	15	Petla et al. (2012)
Moringa oleifera	Spherical	Peel extract	27 ± 2	Surendra et al. (2016)
Pulicaria glutinosa	Crystalline and spherical	Whole plant	20–25	Tahir (2014)
Musa paradisica	Crystalline and irregular	Peeled banana	50	Bankar et al. (2010)
Pinus resinosa	Crystalline	Bark	16–20	Links et al. (2012)
Solanum nigrum	Polydispersed and Spherical	Leaf extract	21.55	Devanesan et al. (2020)
Prunus x yedoensis	Spherical	Leaves	50–150	Manikandan et al. (2016)
Cissus quadrangularis	Spherical	Stem extract	12–26	Anjana et al. (2019)
Bauhinia variegata	Cylindrical Irregular shape	Bark extract	2–9	Vaghela et al. (2018)
Aspalathus linearis	Triangular, spherical, cubic, rectangular, and decahedron	Leaf extract	12.7	Ismail et al. (2017)
Camellia sinensis (Lahijan Black tea)	Spherical	Leaf extract	7	Lebaschi et al. (2017)
Cinnamomum zeylanicum	Crystalline	Bark extract	15–20	Sathishkumar et al. (2009)

TABLE 17.1 *(Continued)*

Plant	Shape	Part used	Size (nm)	References
Camellia sinensis (White tea)	Spherical	Powder	6–18	Azizi et al. (2017)
Curcuma longa	Spherical	Tuber extract	10–15	Sathishkumar et al. (2009)
Coffee and tea extract	Spherical	Powder	5–100	Varma (2008)
Cinnamomum camphora	Quasi-spherical and irregular	Leaf broth	3.2–6.0	Yang et al. (2010)
Dioscorea bulbifera	Spherical and blunt-ended cubes	Tuber extract	10–25	Press (2015)
Evolvulus alsinoides	Spherical	Leaf extract	5	Gurunathan et al. (2015)
Euphorbia granulate	–	Leaf extract	25–30	Nasrollahzadeh and Sajadi (2016)
Filicium decipiens	Spherical	Leaf extract	6.36	Sharmila et al. (2017)
Moringa oleifera	Spherical	Flower extract	10–50	Anand et al. (2016b)
Melia azedarach	Spherical	Leaf extract	10–20	Gopinath (2017)
Prunus × yedoensis	Spherical	Leaves extract	50–150	Manikandan et al. (2016)
Santalum album	Spherical	Leaf extract	10–40	Sharmila et al. (2017)
Rosa canina	Spherical	Fruit extract	10 ± 3	Veisi et al. (2016)
Sapium sebiferum	Spherical	Leaf extract	2–5	Tahir et al. (2016)
Rosmarinus officinalis	Semi-spherical and polyhedr	Leaf extract	15–90	Rabiee et al. (2020)
Glycine max (Soya bean)	Spherical	Leaf extract	~15	Petla et al. (2012)

17.2 BIOSYNTHESIS OF PD NPs

For the bioreduction of metals, plant extracts, algae, and microorganisms were employed. Biosynthesis of Pd NPs with a variety of sizes, morphologies, chemical, physical, and biological characteristics can be synthesized through a clean, non-hazardous, and environmentally responsive manner. Due to the abundance of microorganisms (particularly fungus) and their ability to bio-reduce Pd^{2+} ions through enzymatic activity and convert Pd^{2+} ions into Palladium atoms keeping sufficient regulations over the morphologies of the Pd NPs via bio-patterning, microorganisms have the potential to be used in the green synthesis. However, green synthetic method using microorganisms requires time-taking multiple series complex method. The microorganisms enzymes essential for the biological reduction to occur could be deformed by the thermal effects (e.g., heating with microwave) required for the Pd NP to be generated. Additionally, there are other factors that should be taken into account when developing Pd NPs utilizing microorganisms, such as carefully selecting the best microorganism since each interacts with metal ions in a different way depending on the enzymatic activity and biochemical processes that are inherent in it.

17.2.1 BIOSYNTHESIS USING PLANT EXTRACT

When Catharanthus roseus methanolic extract was mixed with an aqueous $[Pd(OAc)_2]$ solution for 1 h at 60 °C, a color change was observed. In the UV-Visible spectrum, it displayed an absorption peak in the 360–400 nm range, which is consistent with spherical Pd NPs that are 40 nm in size. The extract of *Catharanthus roseus* is a combination of eight chemicals with hydroxyl groups that convert metal ions into metal nanoparticles.

Pd (CH3COO)2 + Reducing extract → Pd + 2CH3COOH

Pd NPs have been reported to be synthesized, characterized, and used as photocatalytic agents (Kumar et al., 2015). Pd NPs has been investigated for their ability to degrade phenol red (Velmurugan et al., 2015). Sargassum bovinum alga was employed by Momeni and Nabipour (Momeni et al., 2015) to synthesize Pd NPs. With the help of UV-vis spectroscopy, the authors monitored the transformation of Pd^{2+} ions into metallic Pd between 300 and 800 nm (Fig. 17.2). As demonstrated by the color changes from yellow to dark brown Pd NPs had formed. By electrochemical reduction of H_2O_2, the 5–10 nm Pd NPs were examined for catalytic activity. They were thought to be stabilized by polysaccharides found in the algal extract since they remained stable for up to five months. Cyclic voltametry also provided further evidence of the H_2O_2 decrease. The bimetallic Au@Pd NPs have the appearance of a light-colored shell surrounding a dark inner core. Their face-centered cubic structure size ranged from 47.8 to 2.3 nm on average (Sun et al., 2014).

Recent research has shown that the green production of Pd nanoparticles onto Fe_3O_4 nanospheres from root extract of Euphorbia condylocarpa M. bieb has catalytic activity (Nasrollahzadeh and Sajadi, 2015). Flavonoids, which are present in the extract, provide the electrons required to reduce the metal ions. Fe_3O_4 is extremely sensitive to air, even though the Pd/Fe_3O_4 catalyst is effective and can be used continuously for Suzuki and Sonogashira coupling reactions without losing any activity. Due to their magnetic properties, Pd and Fe were retrieved by a magnet from the reaction mixture and reused numerous times for Sonogashira coupling reactions with hardly any activity loss.

Pd NPs were biosynthesized using the fruit extract of barberry on RGO (Reduced Graphene Oxide), and they were then employed as a heterogeneous catalyst to convert nitroarenes to amines at 50°C in a 1:2 alcohol-water solution (Nasrollahzadeh et al., 2016).The presence of considerable amounts of vitamin C in the extract suggests that the metal ions were likely reduced to nanoparticles. It was found that the average dimension of Pd nanoparticles is close to 18 nm. Using $NaBH_4$, the catalytic activity of nitrobenzene was evaluated by converting it to aniline.

$PdCl_2$ and C. zeylanicum bark extract were combined at 30 °C to generate Pd NPs (Sathishkumar et al., 2009). Even though the reaction occurred after 24 h, it was finished after 72 h. The polydispersed, spherical nanoparticles have a size between 15 and 20 nm. The increase in leaf extract concentration was necessary for their formation. The existence of crystalline palladium was confirmed by the XRD pattern. Although precipitation occurs above pH 5, the impact of pH on the production of NPs is negligible. However, it only minimally affects size of nanoparticle rather than their form.

It was demonstrated that when only 5 mL of an extract was being mixed with 50 mL of $PdCl_2$ (1 mM) at 30 °C, almost 60% $PdCl_2$ were transformed into Pd nanoparticles. Since the solution would normally contain both the palladium NPs and Pd^{2+} ions, a higher concentration biomaterial may reduce the remaining 40% $PdCl_2$. The Pd^{2+} are converted to Pd nanoparticles by the presence of several compounds in the C. zeylanicum bark extract, including eugenol, linalool, cinnamaldehyde, methyl chavicol, caryophyllene, and ethyl cinnamate.

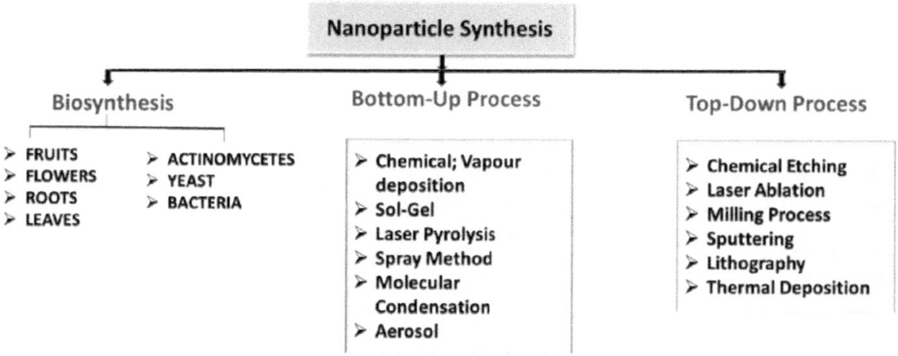

FIGURE 17.1 Different synthetic methods of nanoparticles.

17.2.2 *BIOSYNTHESIS USING OTHER BIOLOGICAL SYSTEMS*

The Pd^{2+} ions, after being treated ascorbic acid acting as a reducing agent and with sodium alginate, which acts as a brown alga stabilising agent, the Pd^{2+} ions were heated for 3 min at 850 W in a microwave oven. The change in color of the solution from light yellow to dark brown provided as a qualitative indicator of the formation of Pd NPs, and UV/Vis spectrophotometry used as a quantitative measurement (Ameri et al., 2020). The original absorption band at 441 nm for the Pd^{2+} ions was replaced with a wide absorption maximum at wavelength of 345 nm which demonstrates the biological reduction of Pd^{2+} ions to Pd NPs. The synthesized spherical Pd NPs ranged in size from 13 to 33 nm. In comparison to cisplatin and Pd (II) acetate, biosynthesized Pd NPs significantly inhibited the growth of A4549 lung cancer cells after 48 h of treatment. Additionally, the DPPH scavenging activity assay demonstrated that the generated Pd NPs had significantly stronger antioxidant property than Pd (II) acetate.

Another recent study described how the native Indian Khilar cow's urine was used to efficiently biosynthesize Pd NPs having adjustable size and morphology and lowered reaction time (Padvi et al., 2020). The NPs were generated by constantly stirring a solution of Pd (II) ions and cow urine at physiological pH and at 80 °C. The presence of a 410 nm UV/Vis peak indicates the production of Pd NPs. The formation of polydispersed, cylinder-shaped Pd NPs was observed using field emission scanning electron microscopy. The biosynthesized NPs shown effective antibacterial properties against *P. aeruginosa* (16 mm), S. typhi (16 mm), *E. coli* (15 mm), *Bacillus cereus* (15 mm), and *Bacillus thuringiensis* (15 mm) at a concentration of 200 g/mL. Pd NPs also shown excellent antioxidant properties, in compared to conventional ascorbic acid, which had a 28% radical scavenging ability as evaluated by the DPPH experiment.

17.3 MECHANISM OF BIOGENIC PD NP SYNTHESIS

The plant extract and metal salt solutions are typically mixed at room temperature to produce the Pd NPs. The nanoparticles are synthesized quickly and are visible as a color change in the reaction medium (Li et al., 2011). In order to control the morphology, crystallinity, and size of Pd NPs, it is necessary to optimize a number of variables, including the source and concentration of the biological extract as well as the concentration of solution of metal salts, the temperature and pH of the reaction mixture, the rate of addition of the extract, and the contact time.

Despite the fact that the precise mechanistic pathways for the biosynthesis of Pd NPs is uncertain and needs further study, a primary bottom-up mechanism with four crucial phases is proposed (Fig. 17.1). The first step in the biological synthesis of Pd NP involved the activation step which is the bioreduction of the metal ion and nucleation, and the reduction of metal ions into zero-oxidation states. In the second step, the small Pd NPs develop and aggregate into bulkier and thermodynamically more stable Pd NPs. Afterward, a terminating stage takes place when different Pd NPs morphologies are produced depending on the optimal conditions, including wires, rods, spheres, hexagons, pentagons, or triangles.

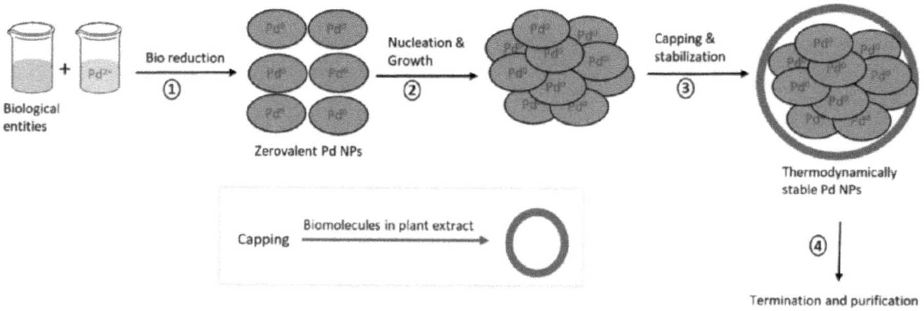

FIGURE 17.2 Mechanism of biosynthesis of Pd nanoparticles.

17.4 BIOMEDICAL APPLICATIONS OF PD NPs

Nanotechnology is growing very fast in a number of life science domains and nanomedicine is particularly interesting to the biogenic NPs because of its diversity and improved physico-chemical properties, which may help with the diagnosis and treatment of many diseases.

17.4.1 ANTIBACTERIAL APPLICATIONS

Recent research on the antibacterial properties of biogenic Pd NPs revealed that these nanoparticles had a lot of potential for use in antimicrobial applications. A Pd nanocrystal facet has the ability to damage microbial membranes. Reactive oxygen species (ROS) generated by Pd NP can damage bacterial cell membranes, DNA, induce protein denaturation, and disrupt electron transport which lead to bacterial mortality (Ullah et al., 2018) .The results of the experiments demonstrated that Pd NPs have superior anti-pathogenic activity to Pd^{2+} ions.

The antibacterial activity of Pd NPs is significantly influenced by their size and structure. Pd NPs with diameters ranging from 1 nm showed that small Palladium NPs were found to be more harmful to bacteria such as *E. coli* than larger palladium NPs while tiny palladium NPs exhibited an excellent antimicrobial property at relatively less concentrations (Karthik et al., 2014). Gram-negative and Gram-positive bacteria were examined for their susceptibility to Palladium nanocrystals with two different shapes: octahedrons and cubes. Pd nanocrystals have exceptional antibacterial properties, and their oxidase and peroxidase activities produce ROS.

The Moringa oleifera biomass waste used as a natural capping and reducing agent during the production of Pd nanoparticles showed strong antimicrobial property toward *E. faecalis* (Anand et al., 2016a). Pd NPs, which were generated in a green way using white tea extract (Pd@White tea NPs) also displayed antibacterial activity (Azizi et al., 2017). According to the experimental results, Pd@White tea NPs show superior antibacterial activities to white tea extract (on *E. coli* and Staphylococcus epidermidis). This is presumably because of the strong interaction between white tea extract and Pd@White tea NP. The antibacterial efficiency of Pd NPs made from Prunus yedoensis leaf extract and penicillin (used as a control) was evaluated against Gram-negative and Gram-positive bacteria. Palladium NPs demonstrated strong antibacterial efficacy against and Pseudomonas aeruginosa (5 mm) and Bacillus subtilis (6 mm) at a dosage of 250 g mL^{-1} (J. Ali et al., 2019). A novel composite with strong antibacterial properties could be produced by mixing PdNPs with other metals. For instance, Ag@ Pd /chitosan/HAp(hydroxyapatite) composites, which may be used as a treatment of dental scaffolds, have a significantly higher acidity after the addition of silver nanoparticles (Ag NPs) and Pd NPs (Ali et al., 2018).

17.4.2 BIOSENSING

Biosensors are instruments that identify target analytes with great specificity using biological procedures. A biosensor has two components—a bioreceptor that allows a specific analyte to attach to the sensor and a physical transducer that transforms the bio response into a quantifiable effect. DA (Dopamine) is one of the catecholamine neurotransmitters, was discovered in order to diagnose and treat several disorders, including Parkinson's, schizophrenia, and Alzheimer's.

Dopamine reacted in the Dopamine biosensor on adding Palladium NPs to NPG (nanoporous gold), an efficient electrochemically catalytic substance (Yi et al., 2017). With time, the synergistic effects of palladium electrochemical catalysis for Dopamine, 3D self-aiding, and nanoporous NPG wire significantly increased the current and the electrochemical signal. The biosensor is very efficient at detecting tracked Dopamine because to its high sensitivity, selectivity, and wide detection range (Soleymani et al., 2009). The generation of effective clinical diagnosis devices in biology and medicine depends greatly on nanoscale biosensors with remarkable performance.

17.4.3 DRUG AND GENE DELIVERY

The controlled drug delivery system is becoming more and more common in an effort to get beyond the inherent drawbacks of traditional pharmacological therapy, such as poor selectivity, quick excretion, and high toxicity (Yun et al., 2015). The nanoparticles may be engineered to target particular molecules and release substances at a controlled rate in addition to carrying drugs and Genetic material. Drug molecules may combine with NPs directly or indirectly through an associated molecule after being loaded into their pores. The ability of Pd NPs to penetrate makes them perfect for loading genes and medicines. Porous Pd NPs performed better in loading and releasing the gene than porous Au NPs when they were compared to porous Pd NPs (Kang et al., 2018). For instance, porous Pd NPs made from the chaga mushroom extract carried the cancer drug Dox via electrostatic association between the Dox and absorbed molecules on the surface of the NPs (Zeng et al., 2018). Dox was completely loaded after 6 h at room temperature. It is also possible to bind the drug to the Pd NPs indirectly using a connecting molecule.

In HeLa cells from patients with cervical cancer, Dox showed a pH-responsive release profile, which also had strong anticancer action in vivo against a HeLa tumor xenograft that was connected to PEGylated Pd NPs by a hydrazine bond (Shanthi et al., 2015). For specific cells/tumors, a detailed analysis of the cytotoxicity of both mono and dual mode in the cancer drug-loaded Pd NPs is required. The cytotoxicity of both mono and dual mode in the cancer medication loaded Pd NPs should be thoroughly investigated for certain cells/tumors. Pd NPs loaded with Dox killed 30% of the cells in the in vitro assay, while Pd NPs loaded with the extract of chaga mushroom destroyed 20% of the HeLa cells. To assure the efficiency of the drug/gene delivery system to the targeted tissues, it is necessary to investigate the concentration of Pd NPs that do not affect untargeted tissues.

17.4.4 ANTITUMOR THERAPY

Pd NPs have antibacterial properties in addition to being toxic to cancer cell lines. According to literature, the physicochemical interactions of Pd NPs with phosphate groups of DNA, nitrogen bases, functional groups of proteins, lactate dehydrogenase (LDH) leakage, free radical production, and cell cycle disruption are all part of the anticancer mechanism of Pd NPs. Interaction with Pd NPs may cause proteins and DNA to transition to inactive states. LDH is essential for the production of cellular energy. Several cancer cell types have higher LDH levels than healthy cells (Miao et al., 2013), and LDH leaking seriously affects the functions of cancer cell, free radical generation, which comprises both reactive nitrogen species (RNS) and ROS, causes damage to lipids, proteins, and DNA (Phaniendra et al., 2015).

The antitumor impact of Pd NPs has only been identified in a relatively few number of investigations until now. In 2014, Balbin et al. published the first assessment of antitumor efficacy of Pd NPs in human tumor cells (Balbín et al., 2014). The cytotoxicity of the PdNPs was examined using five distinct human cancer cell lines and PdNPs coated on mesoporous silica-based substrates. Pd NPs coated on mesoporous silica-based substrates, and five distinct human cancer cell lines were used to investigate the cytotoxicity of the PdNPs. Due to the lack of cytotoxicity of mesoporous silica, Pd NPs are crucial to the anticancer activities of hybrid composites. According to the in vitro research which revealed that the properties of hybrid materials that inhibit the growth of cancer cells are cell line-dependent, dose-dependent, and quality-dependent. The MTT test on peripheral lymphocyte normal cells and human lung cancer carcinoma cells (A549) revealed the potential for more anticancer research using tumor cell lines. The PdNPs with an IC50 value of only 41.37 g/mL had a powerful anti-proliferative effectiveness against MCF-7 cells. They were made utilizing a plant extract called Bauhinia variegate (Vaghela et al., 2018).

17.4.5 PD NPs FOR PHOTOTHERMAL THERAPY OF CANCER AND INFECTION

Photothermal treatment (PTT) is a technique that has the potential to treat both cancer and infections. The heat therapy used for bacteria and cancer cells in this procedure includes photo-absorbers, which convert photon energy into heat (Zou et al., 2016). PTT is a highly targeted and safe treatment since it uses external laser irradiation with customizable dosage and laser beam diameter. PTT can destroy original tumor or metastatic tumor of several different cancer cell types (Xiaohua Huang et al., 2011). The perfect photothermal agent need to possess the following crucial characteristics: water dispersion, small size, homogenous morphology, large NIR (near-infrared) absorption, excellent biocompatibility and photostability, targeting cancer cell and bacteria, and ease of renal cleaning. In order for the absorbed light to penetrate the tissue deeply, the photon absorbers must show significant absorption in the NIR (near IR) region, that is, in the range of 700–2500 nm, a "optical window" having the minimum amount of photon scattering and tissue absorption (Chen et al., 2019).

Because of their excellent photoconversion effectiveness, high stability, variety in form and size, and potent NIR absorptions, Pd NPs are now being explored as a potent photothermal activator for PTT. Huang et al. observed that the hexagonal Pd nanosheets that are very thin with controlled NIR absorption peaks exhibit effective photocatalytic conversion, excellent biocompatibility, and improved photostability, when compared to Ag or Au nanostructures (Huang et al., 2011). After being exposed to radiation for 10 min, the 30 ppm SPNS-containing solution increased its temperature from 17.0 to 52.3 °C. (808 nm, 1W; Tang et al., 2014).

Modern medicine has a serious challenge with multi-antibiotic resistance. Therefore, it is essential to develop novel therapies for the treatment of infections. Due to its excellent photothermal activity, Pd NPs were chosen as the photothermal chemical to encapsulate in the photothermally sensitive chitosan/PVA(polyvinyl alcohol) membrane. The fabricated hybrid membrane shows high moisture retention, high porosity, high swelling, good biocompatibility, and strong photothermal performance. Because of their significant absorption in the NIR region, photostability, biocompatibility, high photothermal conversion efficiency, variation in shape and size, and cost-effectiveness, Pd NPs can be regarded as potential candidates for photothermal therapy for cancer and infection.

17.5 CONCLUSION AND FUTURE PROSPECTIVE

Recent research has shown that Pd NPs offer promising therapeutic properties and biomedical applications such as antibacterial, anticancer, drug/gene delivery, PTT, and biosensor functions. Pd NPs have a good prospect in the development of highly advanced imaging-guided therapy, medication delivery, and surgery. Several investigations have shown that Pd NPs can occasionally outperform other metal nanoparticles for the same purposes. Pd NPs have unique optical characteristics that make them potential probes or diagnostic tools. Future medical trends might involve integrating many treatment modalities into a single system to maximize treatment effectiveness while minimizing unavoidable adverse effects. The diverse array of Pd NPs and their high porosity may promote the importance of multimodal therapy. By adjusting the ratio of particular metals and palladium NPs, different nanocomposites can be developed for particular biological purposes.

Future research is necessary to completely understand the metabolic pathways used in in vivo and in vitro evaluations, as well as the antibacterial or anticancer activities of Pd NPs at the molecular level. Before Pd NPs are developed into a theranostic drug for imaging-guided cancer treatment and diagnosis, they must undergo extensive research. In order to select the best PdNPs for the intended applications, it is essential to conduct comparative studies between other metal nanoparticles and Palladium NPs, as well as between Pd NPs made using different synthetic techniques and Pd NPs of various sizes and shapes. Future research is advised to include extensive animal investigations and clinical trials.

Recent reports have suggested that Pd NPs may have some hazardous effects on living things and the environment around them (Leso et al., 2018). However, more thorough research is required to fully understand the toxicity of Pd NPs. It is still necessary to do systematic and in-depth research on the pharmacokinetics of Pd NPs in order to gain a complete comprehend how the body responds to Pd NPs.

In order to promote the use of Pd NPs in pharmacological and clinical care and to make sure the successful transition of such ideas from study to reality, there seems to be a long path to go in dealing with many challenges. Pd NPs, however, have unique characteristics that could make them crucial nanoparticles in the field of nanomedicine in the future.

KEYWORDS

- **green silver nanoparticles**
- **nanoremediation**
- **environmental pollution remediation**
- **green synthesis of AgNPs**
- **eco-friendly silver nanoparticles**

REFERENCES

Ali, G. W.; El-Hotaby, W.; Hemdan, B.; Abdel-Fattah, W. I. Thermosensitive Chitosan/Phosphate Hydrogel-Composites Fortified with Ag versus Ag@Pd for Biomedical Applications. *Life Sci.* **2018,** *194*, 185–195.

Ali, J.; Ali, N.; Wang, L.; Waseem, H.; Pan, G. Revisiting the Mechanistic Pathways for Bacterial Mediated Synthesis of Noble Metal Nanoparticles. *J. Microbiol. Methods* **2019,** *159* (February), 18–25.

Ameri, A.; Shakibaie, M. Rapid and Facile Microwave-Assisted Synthesis of Palladium Nanoparticles and Evaluation of Their Antioxidant Properties and Cytotoxic Effects Against Fibroblast-Like (HSkMC) and Human Lung Carcinoma (A549) Cell Lines, 2020; pp 132–140.

Anand, K.; Tiloke, C.; Phulukdaree, A.; Ranjan, B.; Chuturgoon, A.; Singh, S.; Gengan, R. M. Biosynthesis of Palladium Nanoparticles by using *Moringa oleifera* Flower Extract and Their Catalytic and Biological Properties. *J. Photochem. Photobiol. B: Biol.* **2016**, *165*, 87–95.

Anjana, P. M.; Bindhu, M. R.; Umadevi, M.; Rakhi, R. B. Antibacterial and Electrochemical Activities of Silver, Gold, and Palladium Nanoparticles Dispersed Amorphous Carbon Composites. *Appl. Surf. Sci.* **2019,** *479* (Feb), 96–104.

Azizi, S.; Mahdavi, M.; Sulaiman, H.; Rahim, R. A.; Rasedee, A. Green Synthesis Palladium Nanoparticles Mediated by White Tea (Camellia sinensis) Extract with Antioxidant, Antibacterial, and Antiproliferative Activities Toward the Human Leukemia (MOLT-4) Cell Line, 2017; pp 8841–8853.

Balbín, A.; Gaballo, F.; Ceballos-Torres, J.; Prashar, S.; Fajardo, M.; Kaluderović, G. N.; Gómez-Ruiz, S. Dual Application of Pd Nanoparticles Supported on Mesoporous Silica SBA-15 and MSU-2: Supported Catalysts for C–C Coupling Reactions and Cytotoxic Agents Against Human Cancer Cell Lines. *RSC Adv.* **2014,** *4* (97), 54775–54787.

Bankar, A.; Joshi, B.; Ravi, A.; Zinjarde, S. Banana Peel Extract Mediated Novel Route for the Synthesis of Palladium Nanoparticles. *Mater. Lett.* **2010,** *64* (18), 1951–1953.

Bortolotto, T.; Elisa, S.; Gauna, S.; Ossig, A.; Liberato, C.; Vargas, J.; Endrigo, O.; Rodrigues, D.; Giacomelli, C.; Schmidt, V. Polymer-Coated Palladium Nanoparticle Catalysts for Suzuki Coupling Reactions. *J. Colloid Interface Sci.* **2015,** *439*, 154–161.

Chen, J.; Ning, C.; Zhou, Z.; Yu, P.; Zhu, Y.; Tan, G.; Mao, C. Nanomaterials as Photothermal Therapeutic Agents. *Progress Mater. Sci.* **2019,** *99* (July 2018), 1–26.

Compounds, S. B. Therapeutic Potential of Seaweed Bioactive Compounds Compounds, 2018.

Cui, Y.; Lieber, C. M. Functional Nanoscale Electronic Devices Assembled Using Silicon Nanowire Building Blocks. *Science* **2012,** *851* (2001).

Das, R. K.; Laxman, V.; Linson, P. Biological Synthesis of Metallic Nanoparticles: Plants, Animals and Microbial Aspects. *Nanotechnol. Environ. Eng.* **2017,** *2* (1), 1–21.

Devanesan, S.; Umadevi, M.; Aljaafreh, M. J.; Atif, M. Antimicrobial and Catalytic Activities of Biosynthesized Gold, Silver and Palladium Nanoparticles from *Solanum nigurum* Leaves. *J. Photochem. Photobiol. B: Biol.* **2020,** *202* (Oct 2019), 111713.

Eryürek, M.; Karadag, Y.; Tas, N.; Kılınc, N. Optical Sensor for Hydrogen Gas Based on a Palladium-Coated Polymer Microresonator. *Sens. Actuat. B: Chem.* **2015,** 212, 78–83.

Gopinath, K et al. Eco-Friendly Synthesis of Palladium Nanoparticles Using *Melia azedarach* Leaf Extract and Their Evaluation for Antimicrobial and Larvicidal Activities. *J. Cluster Sci.* **2017,** *28* (1), 463–476.

Gupta, R.; Xie, H.; Sciences, *H. HHS Public Access*. **2018,** *37* (3), 209–230.

Gurunathan, S.; Kim, E.; Han, J. W.; Park, J. H.; Kim, J. Green Chemistry Approach for Synthesis of Effective Anticancer Palladium Nanoparticles, 2015; pp 22476–22498.

Ha, C.; Zhu, N.; Shang, R.; Shi, C.; Cui, J.; Sohoo, I. Biorecovery of Palladium as Nanoparticles by *Enterococcus faecalis* and Its Catalysis for Chromate Reduction. *Chem. Eng. J.* **2016,** *288*, 246–254.

Hua, S.; Matos, M. B. C. De, Metselaar, J. M.; Storm, G.; Hua, S. Current Trends and Challenges in the Clinical Translation of Nanoparticulate Nanomedicines: Pathways for Translational Development and Commercialization. *Front. Pharmacol.* **2018,** *9* (July), 1–14.

Huang, X.; El-Sayed, M. A. Plasmonic Photo-Thermal Therapy (PPTT). *Alexandria J. Med.* **2011,** *47* (1), 1–9.

Huang, Xiaoqing, Tang, S.; Mu, X.; Dai, Y.; Chen, G.; Zhou, Z.; Ruan, F.; Yang, Z.; Zheng, N. Freestanding Palladium Nanosheets with Plasmonic and Catalytic Properties. *Nat. Nanotechnol.* **2011,** *6* (1), 28–32.

Innovations, S. (n.d.). Accepted Manuscript, pp 1–38.

Ismail, E.; Khenfouch, M.; Dhlamini, M.; Dube, S.; Maaza, M. Green Palladium and Palladium Oxide Nanoparticles Synthesized via *Aspalathus linearis* Natural Extract. *J. Alloys Compd.* **2017,** *695*, 3632–3638.

Jia, L.; Zhang, Q.; Li, Q.; Song, H. The Biosynthesis of Palladium Nanoparticles by Antioxidants in Gardenia jasminoides Ellis: Long Lifetime Nanocatalysts for p-Nitrotoluene Hydrogenation. *Nanotechnology* **2009,** 385601.

Kalita, J.; Das, B.; Dhar, S. S. Synergistic Effect of Iron and Copper in Hydroxyapatite Nanorods for Fenton-Like Oxidation of Organic Dye. *Coll. Surfaces A: Physicochem. Eng. Aspects* **2022,** *643* (Dec 2021), 128750.

Kang, S.; Shin, W.; Kang, K.; Choi, M. H.; Kim, Y. J.; Kim, Y. K.; Min, D. H.; Jang, H. Revisiting of Pd Nanoparticles in Cancer Treatment: All-Round Excellence of Porous Pd Nanoplates in Gene-Thermo Combinational Therapy. *ACS Appl. Mater. Interfaces* **2018,** *10* (16), 13819–13828.

Karthik, L.; Kumar, G.; Keswani, T.; Bhattacharyya, A.; Sarath Chandar, S.; Bhaskara Rao, K. V. Protease Inhibitors from Marine Actinobacteria as a Potential Source for Antimalarial Compound. *PLoS ONE*, **2014,** *9* (3).

Khazaei, A.; Rahmati, S.; Hekmatian, Z.; Saeednia, S. Journal of Molecular Catalysis A : Chemical A Green Approach for the Synthesis of Palladium Nanoparticles Supported on Pectin: Application as a Catalyst for Solvent-Free Mizoroki–Heck Reaction. *J. Mol. Catal. A Chem.* **2013,** *372*, 160–166.

Kora, A. J.; Rastogi, L. Green Synthesis of Palladium Nanoparticles Using Gum Ghatti (*Anogeissus latifolia*) and Its Application as an Antioxidant and Catalyst. *Arab. J. Chem.* **2018,** *11* (7), 1097–1106.

Kouvaris, P.; Delimitis, A.; Zaspalis, V.; Papadopoulos, D.; Tsipas, A.; Michailidis, N. Green Synthesis and Characterization of Silver Nanoparticles Produced Using Arbutus Unedo Leaf Extract. *Mater. Lett.* **2012,** *76*, 18–20.

Kumar, B.; Smita, K.; Cumbal, L.; Debut, A. Ultrasound Agitated Phytofabrication of Palladium Nanoparticles Using Andean Blackberry Leaf and Its Photocatalytic Activity. *J. Saudi Chem. Soc.* **2015,** *19* (5), 574–580.

Lebaschi, S.; Hekmati, M.; Veisi, H. Green Synthesis of Palladium Nanoparticles Mediated by Black Tea Leaves (*Camellia sinensis*) Extract: Catalytic Activity in the Reduction of

4-Nitrophenol and Suzuki-Miyaura Coupling Reaction Under Ligand-Free Co. *J. Colloid Interface Sci.* **2017,** *485*, 223–231.

Leso, V.; Iavicoli, I. Palladium Nanoparticles: Toxicological Effects and Potential Implications for Occupational Risk Assessment. *Int. J. Mol. Sci.* **2018,** *19* (2).

Li, X.; Xu, H.; Chen, Z.; Chen, G. Biosynthesis of Nanoparticles by Microorganisms and Their Applications **2011**. DOI: 10.1155/2011/270974

Links, D. A.; Coccia, F.; Tonucci, L.; Bosco, D.; Bressan, M. Green Chemistry and Their Catalytic Behaviour in Oxidation and Reduction Reactions. **2012**; pp 1073–1078.

Manikandan, V.; Velmurugan, P.; Park, J.; Lovanh, N.; Seo, S.; Jayanthi, P.; Park, Y.; Cho, M.; Oh, B. Synthesis and Antimicrobial Activity of Palladium Nanoparticles from Prunus Âyedoensis Leaf Extract. *Mater. Lett.* **2016,** *185*, 335–338.

Miao, P.; Sheng, S.; Sun, X.; Liu, J.; Huang, G. Lactate Dehydrogenase A in Cancer: A Promising Target for Diagnosis and Therapy. *IUBMB Life* **2013,** *65* (11), 904–910.

Momeni, S.; Nabipour, I. A Simple Green Synthesis of Palladium Nanoparticles with Sargassum Alga and Their Electrocatalytic Activities Towards Hydrogen Peroxide. 2015; pp 1937–1949.

Narayanan, R.; El-sayed, M. A. Catalysis with Transition Metal Nanoparticles in Colloidal Solution : Nanoparticle Shape. 2005; 12663–12676.

Nasrollahzadeh, M.; Sajadi, S. M. Journal of Molecular Catalysis A : Chemical Green synthesis of Pd/Fe3O4 Nanoparticles Using *Euphorbia condylocarpa* M. Bieb Root Extract and Their Catalytic Applications as Magnetically Recoverable and Stable Recyclable Catalysts for the Phosphine-fr. *J. Mol. Catal. A Chem.* **2015,** *396*, 31–39.

Nasrollahzadeh, M.; Sajadi, S. M. Pd Nanoparticles Synthesized In Situ with the Use of Euphorbia Granulate Leaf Extract: Catalytic Properties of the Resulting Particles. *J. Colloid Interface Sci.* **2016,** *462*, 243–251. DOI: 10.1016/j.jcis.2015.09.065

Nasrollahzadeh, M.; Sajadi, S. M.; Maham, M. Green Synthesis of Palladium Nanoparticles Using *Hippophae rhamnoides* Linn Leaf Extract and Their Catalytic Activity for the Suzuki–Miyaura Coupling in Water. *J. Mol. Catal. A Chem.* **2015,** *396*, 297–303.

Nasrollahzadeh, M.; Sajadi, S. M.; Maham, M.; Kohsari, I. Biosynthesis, Characterization and Catalytic Activity of the Pd/Bentonite Nanocomposite for Base- and Ligand-Free Oxidative Hydroxylation of Phenylboronic Acid and Reduction of Chromium (VI) and Nitro Compounds. *Micropor. Mesopor. Mater.* **2018,** *271* (April), 128–137.

Nasrollahzadeh, M.; Sajadi, S. M.; Rostami-vartooni, A.; Alizadeh, M. Green Synthesis of the Pd Nanoparticles Supported on Reduced Graphene Oxide Using Barberry Fruit Extract and Its Application as a Recyclable and Heterogeneous Catalyst for the Reduction of Nitroarenes. *J. Colloid Interface Sci.* **2016,** *466*, 360–368.

Padvi, S. R. P. M. N.; Shaikh, S. S. S. Y. I.; Samant, L. S. C. A. P.; Prasad, N. R. Bio-Inspired Synthesis of Catalytically and Biologically Active Palladium Nanoparticles Using Bos Taurus Urine. *SN Appl. Sci.* **2020,** *2* (4), 1–12.

Petla, R. K.; Vivekanandhan, S.; Misra, M. Soybean (Glycine max) Leaf Extract Based Green Synthesis of Palladium Nanoparticles. **2012,** *2012* (January), 14–19.

Phaniendra, A.; Jestadi, D. B.; Periyasamy, L. Free Radicals: Properties, Sources, Targets, and Their Implication in Various Diseases. *Indian J. Clin. Biochem.* **2015,** *30* (1), 11–26.

Press, D. Novel Platinum—Palladium Bimetallic Nanoparticles Synthesized by *Dioscorea bulbifera*: Anticancer and Antioxidant Activities. 2015; pp 7477–7490.

Rabiee, N.; Bagherzadeh, M.; Kiani, M.; Ghadiri, A. M. *Rosmarinus officinalis* Directed Palladium Nanoparticle Synthesis: Investigation of Potential Anti-Bacterial, Anti-Fungal and Mizoroki–Heck Catalytic Activities. *Adv. Powder Technol.* **2020,** *31* (4), 1402–1411.

Rezaei, B.; Shams-ghahfarokhi, L.; Havakeshian, E.; Ensa, A. A. An Electrochemical Biosensor Based on Nanoporous Stainless Steel Modified by Gold and Palladium Nanoparticles for Simultaneous Determination of Levodopa and Uric Acid. Talanta **2016,** *158*, 42–50.

Saldan, I.; Semenyuk, Y.; Marchuk, I. Chemical Synthesis and Application of Palladium Nanoparticles. 2015; pp 2337–2354.

Sathishkumar, M.; Forbes, E.; Sneha, K.; Yun, Y. Palladium Nanocrystal Synthesis Using *Curcuma longa* Tuber Extract. Jan 2009.

Sathishkumar, M.; Sneha, K.; Seob, I.; Mao, J.; Tripathy, S. J.; Yun, Y. Phyto-Crystallization of Palladium Through Reduction Process Using *Cinnamom zeylanicum* Bark Extract. **2009,** *171*, 400–404.

Shanthi, K.; Vimala, K.; Gopi, D.; Kannan, S. Fabrication of a pH Responsive DOX Conjugated PEGylated Palladium Nanoparticle Mediated Drug Delivery System: An In Vitro and In Vivo Evaluation. *RSC Adv.* **2015,** *5* (56), 44998–45014.

Sharmila, G.; Fathima, M. F.; Haries, S.; Geetha, S.; Kumar, N. M.; Muthukumaran, C. Green Synthesis, Characterization and Antibacterial Efficacy of Palladium Nanoparticles Synthesized Using *Filicium decipiens* Leaf Extract. *J. Mol. Struct.* **2017,** *1138*, 35–40.

Sharmila, G.; Haries, S.; Fathima, M. F.; Geetha, S.; Kumar, N. M.; Muthukumaran, C. Enhanced Catalytic and Antibacterial Activities of Phytosynthesized Palladium Nanoparticles Using Santalum Album Leaf Extract. *Powder Technol.* **2017,** *320*, 22–26. DOI: 10.1016/j.powtec.2017.07.026

Shenton, B. W.; Douglas, T.; Young, M.; Stubbs, G.; Mann, S. Inorganic ± Organic Nanotube Composites from Template Mineralization of Tobacco Mosaic Virus. **1999,** *3*, 253–256.

Shiju, N. R.; Guliants, V. V. Recent Developments in Catalysis Using Nanostructured Materials. *Appl. Catal. A: Gen.* **2009,** *356*, 1–17.

Soleymani, L.; Fang, Z.; Sargent, E. H.; Kelley, S. O. Programming the Detection Limits of Biosensors Through Controlled Nanostructuring. *Nat. Nanotechnol.* **2009,** *4* (12), 844–848.

Sugawa, K.; Tahara, H.; Yamashita, A.; Otsuki, J.; Sagara, T.; Harumoto, T.; Yanagida, S.. Refractive Index Susceptibility of the Plasmonic Palladium Nanoparticle: Potential as the Third Plasmonic. *Science* **2015,** *2*, 1895–1904.

Sun, D.; Zhang, G.; Huang, J.; Wang, H.; Li, Q. Plant-Mediated Fabrication and Surface Enhanced Raman Property of Flower-Like Au@Pd Nanoparticles. 2014; pp 1360–1369.

Surendra, T. V, Mohana, S.; Valan, M.; Al-dhabi, N. A.; Rayalu, G. M. RSM Optimized *Moringa oleifera* Peel Extract for Green Synthesis of *M. oleifera* Capped Palladium Nanoparticles with Antibacterial and Hemolytic Property. *J. Photochem. Photobiol. B: Biol.* **2016,** *162*, 550–557.

Tahir, K.; Nazir, S.; Li, B.; Ahmad, A.; Nasir, T.; Ullah, A.; Asim, S.; Shah, A.; Ul, Z.; Khan, H.; Yasin, G.; Usman, M. *Sapium sebiferum* Leaf Extract Mediated Synthesis of Palladium Nanoparticles and In Vitro Investigation of Their Bacterial and Photocatalytic Activities. *J. Photochem. Photobiol. B: Biol.* **2016,** *164*, 164–173.

Tahir, M. N. *Dalton Trans.* **2014,** *43* (24).

Tang, S.; Chen, M.; Zheng, N. Sub-10-nm Pd Nanosheets with Renal Clearance for Efficient Near-Infrared Photothermal Cancer Therapy. *Small* **2014,** *10* (15), 3139–3144.

Ullah, S.; Ahmad, A.; Khan, A.; Zhang, J.; Raza, M.; Rahman, A. ur, Tariq, M.; Ali khan, U.; Zada, S.; Yuan, Q. Palladium Nanoparticles Synthesis, Characterization Using Glucosamine as the Reductant and Stabilizing Agent to Explore Their Antibacterial & Catalytic Applications. *Microb. Pathog.* **2018,** *125*, 150–157.

Vaghela, H.; Shah, R.; Pathan, A. Palladium Nanoparticles Mediated Through, 2018; pp 168–177.

Varma, R. S. Green Synthesis of Silver and Palladium Nanoparticles at Room Temperature Using Coffee and Tea Extract, 2008; pp 859–862.

Veisi, H.; Ghorbani-vaghei, R.; Hemmati, S.; Haji, M. Green and Effective Route for the Synthesis of Monodispersed Palladium Nanoparticles Using Herbal Tea Extract (*Stachys lavandulifolia*) as Reductant, Stabilizer and Capping Agent, and Their Application as Homogeneous and Reusable Catalyst in Suzuki cou. Sept 2014.

Veisi, H.; Nasrabadi, N. H.; Mohammadi, P. Biosynthesis of Palladium Nanoparticles as a Heterogeneous and Reusable Nanocatalyst for Reduction of Nitroarenes and Suzuki Coupling Reactions, April 2016.

Veisi, H.; Rashtiani, A.; Barjasteh, V. Biosynthesis of Palladium Nanoparticles Using *Rosa canina* Fruit Extract and Their Use as a Heterogeneous and Recyclable Catalyst for Suzuki–Miyaura Coupling Reactions in Water, 2016 (Jan).

Veisi, H.; Reza, A.; Hemmati, S.; Gil, A. Green Synthesis of Palladium Nanoparticles Using Pistacia Atlantica Kurdica Gum and Their Catalytic Performance in Mizoroki–Heck and Suzuki–Miyaura Coupling Reactions in Aqueous Solutions, March, 2015.

Velmurugan, P.; Cho, M.; Lim, S.; Seo, S.; Myung, H.; Bang, K.; Sivakumar, S.; Cho, K.; Oh, B. Phytosynthesis of Silver Nanoparticles by *Prunus yedoensis* Leaf Extract and Their Antimicrobial Activity. *Mater. Lett.* **2015,** *138*, 272–275.

Wang, P.; Zhang, G.; Liu, L.; Jiao, H.; Deng, X. Palladium Supported on Mesoporous Silica via an In Situ Method as an Efficient Catalyst for Suzuki–Miyaura Coupling Reactions. *Mater. Res. Bull.* **2014,** *59*, 365–369.

Yang, X.; Li, Æ. Q.; Wang, Æ. H.; Huang, Æ. J.; Lin, Æ. L.; Hong, L.; Wang, Æ. Y.; He, Æ. N.; Jia, Æ. L. Green Synthesis of Palladium Nanoparticles Using Broth of *Cinnamomum camphora* Leaf. 2010; pp 1589–1598.

Yi, X.; Wu, Y.; Tan, G.; Yu, P.; Zhou, L.; Zhou, Z.; Chen, J.; Wang, Z.; Pang, J.; Ning, C. Palladium Nanoparticles Entrapped in a Self-Supporting Nanoporous Gold Wire as Sensitive Dopamine Biosensor. *Sci. Rep.* **2017,** *7* (1), 1–9.

Yong, J.; Kwon, S. Æ. E. Biological Synthesis of Platinum Nanoparticles Using *Diopyros kaki* Leaf Extract, 2010; pp 159–164.

Yun, Y. H.; Lee, B. K.; Park, K. Controlled Drug Delivery: Historical Perspective for the Next Generation. *J. Controll. Release* **2015,** *219*, 2–7.

Zeng, F. W.; Zhang, D.; Spicer, J. B. Palladium Nanoparticle Formation Processes in Fluoropolymers by Thermal Decomposition of Organometallic Precursors. *Phys. Chem. Chem. Phys.* **2018,** *20* (37), 24389–24398.

Zou, L.; Wang, H.; He, B.; Zeng, L.; Tan, T.; Cao, H.; He, X.; Zhang, Z.; Guo, S.; Li, Y. Current Approaches of Photothermal Therapy in Treating Cancer Metastasis with Nanotherapeutics. *Theranostics* **2016,** *6* (6), 762–772.

CHAPTER 18

Green Copper Nanoparticles for Nanoremediation

MADHURA ROY[1], ARPITA ROY[2], and SURESH GHOTEKAR[3]

[1]*Centre for Translational and Clinical Research, School of Chemical and Life Sciences, Jamia Hamdard University, New Delhi, India*

[2]*Department of Biotechnology, School of Engineering & Technology, Sharda University, Greater Noida, India*

[3]*Department of Chemistry, Smt. Devkiba Mohansinhji Chauhan College of Commerce and Science, University of Mumbai, Dadra and Nagar Haveli (UT), India*

ABSTRACT

Copper nanoparticles (Cu-NPs) are commonly employed because of their unique characteristics like distinguished strength, firmness, ductility, stiffness, large ratio of surface-to-volume, and flexibility. Cu-NPs demonstrate numerous uses as an anti-microbial, antioxidant, and anti-inflammatory agent. Nowadays, Cu-NPs have been more extensively used as compared to other metal NPs like silver or gold, because of its cost-effectiveness. The green synthesis of Cu-NPs has been suggested as it is relatively an easy process, economical, and environmental-friendly as compared to the chemical and physical synthesis process. The Cu-NPs obtained from plant extracts has been enormously used in multiple sectors and industries. Due to the increasing degradation of the environmental quality, various techniques are used for its recovery and restoration. Green-synthesized Cu-NPs are being studied due to their versatile properties and are considered for various applications of nanoremediation.

Sustainable Nanoremediation: Modern Technologies for a Clean Environment.
Fernanda Maria Policarpo Tonelli, Arpita Roy, and Khalid Rehman Hakeem (Eds.)

18.1 INTRODUCTION

The deterioration of the environment because of degradation of land, contamination of groundwater, and generation of wastewater has caused significant damage to the environmental quality in most parts of the globe. As a consequence, the global community faces the grave challenge of witnessing a clean environment. These threats are being addressed in the United Nations Sustainable Development Goals (UN SDGs)—"Life Below Water," "Life on Land," and "Clean Water and Sanitation" (UN, 2016). The bioaccumulation of harmful substances is accelerated by ongoing deposition in soil and aquatic habitats. Substances like fluoride, lead, arsenic, and cadmium, when exposed to the human body through the consumption of contaminated water and food, can cause serious problems related to health and can affect the lungs, brain, skin, kidneys, and other body parts (Schaefer et al., 2020; Wang et al., 2019). Due to their resistance and susceptibility to the physicochemical parameters of the natural atmosphere for disintegration, organic pollutants like synthetic dyes are challenging to manage (Lellis et al., 2019). The overuse of insecticides or pesticides and fertilizers has also polluted land and groundwater supplies globally (de Souza et al., 2020). The incomplete combustion of hydrocarbons during natural or anthropogenic processes like coal gasification, open burning, and coal-tar pitching can result in significant amounts of polycyclic aromatic hydrocarbons (PAHs) (Li et al., 2020). The contaminants generated from the personal care products, pharmaceutical industry, and healthcare sector are discharged in the wastewater and pose a great threat (Meyer et al., 2019). For decontaminating polluted land and groundwater, it is necessary to design products that are both affordable and environmentally safe. Nanotechnology has the remarkable opportunity to deliver pollutant levels to "almost zero," and it provides quick, affordable, and ecologically friendly alternatives (Bardos et al., 2018).

The method of using appropriate nanomaterials (NMs) to remove pollutants from the land, groundwater, and atmosphere is known as the nanoremediation of the ecosystem. Since the recovery process frequently occurs in situ, nanoremediation methods can reduce the need for extracting and transferring polluted soils (Cai et al., 2019; Fajardo et al., 2020). Additionally, a variety of techniques can be used to produce and repurpose NMs for purposes of pollutant remediation like extraction of metals from nanosorbents, and segregation of iron nanoparticles (Fe-NPs) using the magnetic technique (Mehta et al., 2015).

The size of NPs varies between 1 and 100 nanometers (nm). Their diminutive size and large surface area make them extremely dynamic and chemically active. It is also well recognized that the elements in metals and non-metals exhibit significant variations from the source bulk material at the nanoscale. These qualities allow them to be used in a plethora of different ways in a variety of industries and disciplines, including the healthcare sector, environmental remediation, etc. They have the capacity to control the levels of significant hazardous metals. The way that NPs react with numerous substances differs greatly from how the comparable metal interacts with substances of larger sizes. Due to its expanding uses, nanoparticle manufacturing is rising rapidly. The manufacturing of nanoparticles increased significantly during the course of the last few decades (PEN, 2013).

Due to their particular characteristics, such as strong electric conduction, high melting point, inexpensive cost, and limited electrochemical migration properties, copper nanoparticles (Cu-NPs) find use in industries, medicine, electronics, etc. Tanilvanan et al. (2014), reported that instead of employing noble metals, Cu-NPs can be utilized for biotransformation, inkjet printing, gas phase catalysis, electrocatalysis, and photocatalysis (Gawande et al., 2016).

Due to the biological uses of Cu-NPs, the reducing agent, solvent, and stabilizing agent utilized in their manufacture must be environment friendly. This indicates the bottom-up technique for the biosynthesis of Cu-NPs. Considering the various uses of Cu-NPs in the aforementioned domains as well as the difficulties in chemically synthesizing the aforementioned nanoparticles, this review highlights the importance of the synthesis process of Cu-NPs upon the major environmental pollutants, and applications of green Cu-NPs in nanoremediation.

18.2 NEED FOR GREEN COPPER SYNTHESIS—LIMITATIONS OF CHEMICALLY SYNTHESIZED NPS

The chemically synthesized NPs have been demonstrated to exhibit multiple issues. It has been reported that thermal processing has been employed for the development of certain NPs and this procedure can cause changes in the physicochemical characteristics of the NPs, like the occurrence of contaminants on the surface, density, and structure (Karakoti et al., 2013). The NPs that undergo chemical synthesis are also known as engineered nanomaterials (ENMs) and they may cause toxicity when utilized for biological activities. The NPs should be produced into the environment after appropriately treating them to be environmentally friendly. Karakoti et al. (2013) have reported that minute alterations in the temperature during preparation and storage can cause unforeseeable changes in the characteristics of the NPs that may not be recovered. Environment-friendly methods used for the generation of NPs are given importance because of the above-mentioned factors. The main purpose is to generate non-toxic and eco-friendly green synthesized NPs (Joerger et al., 2000; Chauhan et al., 2012).

The NPs are preferred to be prepared by the green synthesis procedure rather than the chemical synthesis method since NPs exhibit versatile characteristics, suitable dimensions, and enhanced stability and they can be prepared by a single-step process (Fig. 18.1). The unwanted factors that are produced during the chemical synthesis can be eradicated and during the green synthesis process utilizing biological substances like capping and reducing agents can be carried out under particular conditions like pH, temperature, and pressure in a cost-effective way (Ingale and Chaudhari, 2013; Noorjahan et al., 2015).

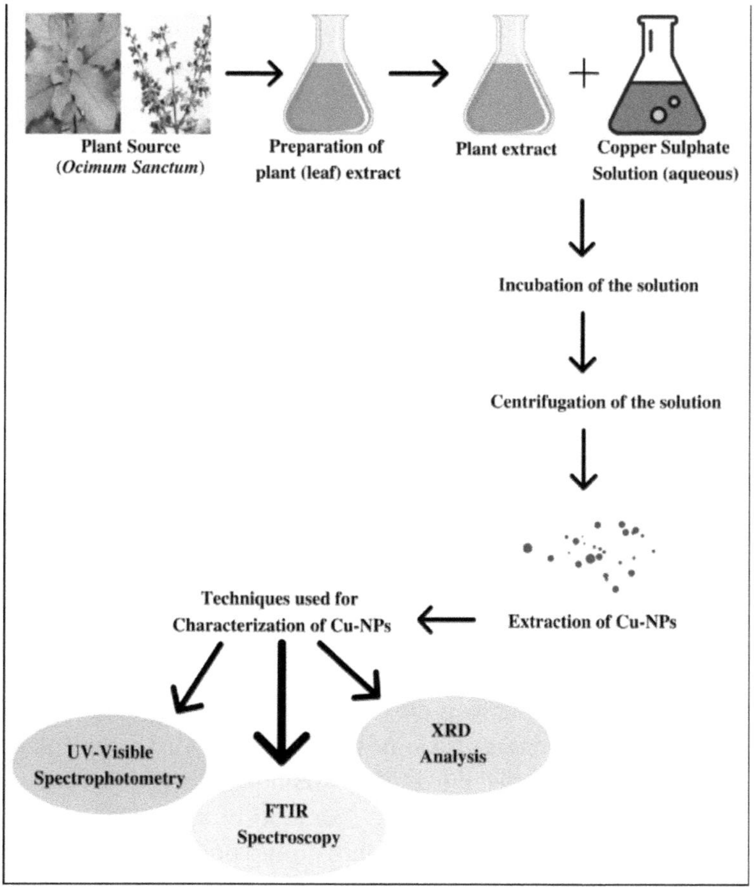

FIGURE 18.1 Schematic representation of synthesis and characterization of Cu-NPs.

As already discussed, the chemical, thermal, and physical procedures employed for the preparation of NPs have been demonstrated to be toxic, complex, and costly (Abd-Elkareem et al., 2016). The utilization of extract of plants for the green synthesized NPs provides a different opportunity to overcome the aforementioned problems. It is a simple, economical, and most importantly environment-friendly method. The plant extract is extremely affordable and resistant to various environmental stresses (Ocsoy et al., 2018).

Plant extracts demonstrate oxidation-reduction reactions, in which the ions of metals are reduced to synthesize NPs. Numerous compounds such as proteins, alkaloids, sugar, polyphenols, phenolic acids, and terpenoids reduce metal ions to synthesize NPs and maintain their stability (Shah et al., 2015; Kuppusamy et al., 2016). There have been evidences reported in which various extracts of plants like *Punica granatum* (Padma et al., 2018), *Moringa oleifera* (Das et al., 2020), *Solanum Lycopersicum* (Batoool et al., 2017), *Gnidia glauca* and *Plumbago zeylanica* (Jamdade et al., 2019), *Crataegus pontica L* (Długosz et al., 2020), *Eclipta prostrate* (Chung et al., 2017), *Camellia sinensis* (Ahmed et al., 2019), *Plantago asiatica* (Nasrollahzadeh et al., 2014), dates (Mohamed, 2020), and *Uncaria gambir Roxb* (Elisma et al., 2019) have been used for the preparation of Cu-NPs. Various types of plant extracts produce different properties of NPs.

The procedure followed for the preparation and characterization of Cu-NPs was as follows:

A. Preparation of the Plant (Leaf) Extract

The leaves of the *Ocimum sanctum* were utilized for the preparation of Cu-NPs. Around 50 mL of double distilled water was added to about 50 g of the leaf after it had been properly cleaned in distilled water and chopped into little fragments. For a period of 10–15 min, it was heated to 80 °C in the water bath. Whatman filter paper no. 1 was used to filter the end product, which was then placed in a refrigerator for storage.

B. Biosynthesis of Cu-NPs

A conical flask of 250 mL was filled with 100 mL of a 1 mM copper sulphate solution, and 25 mL of *Ocimum sanctum* leaf extract was added for the NP production process. Ten hours of incubation were performed on the solution. Three duplicates of the solution were positioned at three various temperatures—27, 40, and 80 °C. For 15 min, the solution had undergone centrifugation at 12,000 RPM.

The end product was cleaned using distilled water and the Cu-NPs were dried at 80 °C in a hot air oven. Additionally, the amount of Cu nanoparticles that were generated was determined at each of three distinct temperatures and periods. It was discovered that synthesis is temperature-dependent, peaking at around 80 °C.

C. Characterization of Cu-NPs

Through the use of characterization methods, we may reliably comprehend the measured values by accurately understanding the unique characteristics of the investigated nanocrystals. Various characterization techniques were applied to the synthesized NPs to know about their particular characteristics like size, morphology, composition, morphological, optical, and functional groups present. Three characterization techniques were employed for the evaluation of the green synthesized Cu-NPs namely X-ray diffraction (XRD), Fourier-Transform Infrared Spectroscopy (FTIR), and UV-Visible spectrophotometer.

1. UV-Visible Spectrophotometer

The UV-Visible spectrophotometer was employed to confirm the production of Cu-NPs initially. Cu-NPs were characterized using UV-visible spectral analysis. These NPs were generated at a wavelength between 350 and 700 nm. The highest absorption is visible at a wavelength of around 560 nm, within 550 and 600 nm. It was demonstrated that the NPs were poly-dispersed. The metal NPs have free electrons (Ananthalekshmi et al., 2016). These electrons' reciprocal vibration results in a surface plasmon resonance (SPR), which in turn creates an absorption band. Additionally, it was discovered that as particle size increased, the peak value decrease substantially. This is a result of the fact that the particles grow opaque as the size increases (Goh et al., 2014). Studies have identified that spectrophotometric analysis is capable of detecting bigger molecules or clusters of microscopic molecules (Nakamur et al., 2013).

2. FTIR Spectroscopy

FTIR Spectroscopy is employed for the detection of the organic substances or functional groups present in the extract of *Ocimum sanctum* and is involved in the Cu-NPs synthesis process. The FTIR analysis was carried out within 4000–500 cm^{-1}. The peaks were obtained at 3799 cm^{-1} which determined the existence of the OH group, 3141 cm^{-1} determined the existence of the carboxylic acid group, 1581 cm^{-1} determined the existence of the amine group, 1405 cm^{-1} determined the existence of the alkanes, 1245 cm^{-1} determined the existence of the alkyl ketone group, 1044 cm^{-1} determined the existence of the C-O group, 854 cm^{-1} determined the existence of CH_2, and 805 cm^{-1} determined the existence of the aromatic compounds. As functional groups produce reducing factors that aid in the production of NPs, this suggests that functional groups play a significant part in the Cu-NPs synthesis.

3. *XRD Analysis*
The high degree of crystallinity displayed by the Cu-NPs complements the diffraction angles at 18°, 28°, and 32°. The Debye-Scherrer equation was utilized to estimate the average side of Cu-NPs.

$$D = k\lambda / \beta \, \mathrm{Cos}\theta$$

D, Crystalline size of NPs; k, Scherrer's constant ranges from 0.9; Cu-NP in the extract was measured to be 11 nm in size (Jayadev and Krishnan, 2021).

18.3 CONDITIONS AFFECTING THE GREEN SYNTHESIS OF PLANT EXTRACT-DERIVED CU-NPS

A. pH

The biosynthesis of plant extract-derived Cu-NPs must take the pH level of a solution into account (Gericke and Pinches, 2006). The dimensions, shape, and duration of synthesis of the NPs are impacted by the solution's pH (Vijayaraghavan and Ashokkumar, 2006). In the production of NPs, pH has a major effect on the occurrence of nucleation centers. The nucleation center may enlarge as the pH rises, which will speed up the conversion of metallic ions (Cu ions) to metal NPs (Cu-NPs). Due to the fact that pH affects the interaction of the ions of metals and the functional groups in the extract of the plant, the length of time that a metal salt takes to reduce has also been associated with the pH of the solution medium (Bali and Harris, 2010). Current studies had demonstrated that the development of small NPs takes place more frequently in the basic medium than in an acidic solution (Dubey et al., 2010). The impact of pH has been demonstrated on the SPR of Cu-NPs. The SPR peaks of Cu-NPs at 600, 590, and 584 nm have been observed at the synthesis pH of 3, 5, and 7 (Wu, 2007).

B. Reaction Time

The reaction and incubation period greatly impacts the yield, quality, and morphological characteristics of the Cu-NPs (Kuchibhatla et al., 2012). The properties of the synthesized NPs can also be affected by the changes in the incubation period that are associated with the storage conditions (Mudunkotuwa et al., 2012). In the generation of nanoparticles, an extended incubation time results in clustering and reduction potential during the synthesis (Baer, 2011). Few experiments have demonstrated that the effective development of NPs takes place at a high reaction time (Darroudi et al., 2011). It has been noted that the preparation of Cu-NPs employing UV analysis demonstrated the beginning of the reduction at 2 h and termination at 4 h. Additionally, the intensity of the absorption peaks was increased due to the increase in reaction time and it remained consistent at 3 h determining the stability of the Cu-NPs (Ngoc-Diep et al., 2019).

C. Temperature

Temperature also plays a key role during the Cu-NP synthesis, it impacts the morphological characteristics of the NPs as pH. It also influences the development of nucleation centers; as a result, low temperature leads to the development of a smaller number of nucleation centers which ultimately causes a decrease in the synthesis rate (Ngoc-Diep et al., 2019). Room temperature is preferred for the biosynthesis of NPs because of the existence of secondary metabolites in the plant extract. Room temperature does not cause the distortion and denaturation of the functional groups (Jemilugba et al., 2019). But, studies have demonstrated that the NPs of spherical shape are synthesized at elevated temperatures while NPs of triangular shape are synthesized at low temperatures (Raju et al., 2011). Iravani and Zolfaghari (2006) noted that at high temperatures, less amount of the plant extract is needed for the preparation of stable NPs; they also reported that at high temperatures, large NPs are synthesized.

D. Effect of Plant Extract Concentration

The amount of plant extract used greatly determines the Cu-NPs synthesis since it acts as the source of the production of electrons that are needed for the reduction of Cu ions. Less amount of plant extract results in less synthesis of Cu-NPs (Daniel et al., 2013). When generating Cu-NPs, a significant volume of plant extract produces more phytochemicals, which are essential for the quick reduction of Cu salt. But, the size of Cu-NPs decreased more rapidly as the metal salt was reduced (Din and Rehan, 2017).

E. Concentration and Nature of Copper Salt

The form, morphology, and dimensions of the Cu-NPs generated are influenced by the type of Cu salt utilized in the synthesis. Copper sulphate, copper nitrate, copper chloride, and copper acetate are the most frequently utilized Cu salts for the generation of NPs. When copper chloride salt is employed for the synthesis, tetrahedron and triangle-shaped Cu-NPs are produced and when copper acetate is utilized, rod-shaped Cu-NPs are produced (Shankar and Rhim, 2014). When copper sulphate salt is utilized, spherical-shaped NPs are produced (Shah et al., 2015). A study has demonstrated that the size of Cu-NPs increases with an elevation in the concentration of Cu salt (Muhammad et al., 2017).

F. Pressure

Some researchers have shown that the morphological characteristics of the Cu-NPs are influenced by pressure (Abhilash et al., 2012; Akintelu et al., 2021).

18.4 DIFFERENT ENVIRONMENTAL POLLUTANTS AND THEIR NEGATIVE IMPACT

Even while urbanization is closely tied to every nation's prosperity, it is frequently blamed as the main cause of toxic waste generation into land and water bodies. Groundwater, soil, and surface water are contaminated by hazardous contaminants that are released either intentionally or inadvertently into the untreated wastewater. Pesticides, heavy metals, pharmaceuticals and personal care products, volatile organic compounds, and hydrocarbons present in water bodies and land adversely affect the environmental quality and pose a major threat to both terrestrial and aquatic life (Fig. 18.2) (Nazir et al., 2019; Liu et al., 2019; Ali et al., 2020a, 2020b; Rasheed et al., 2020).

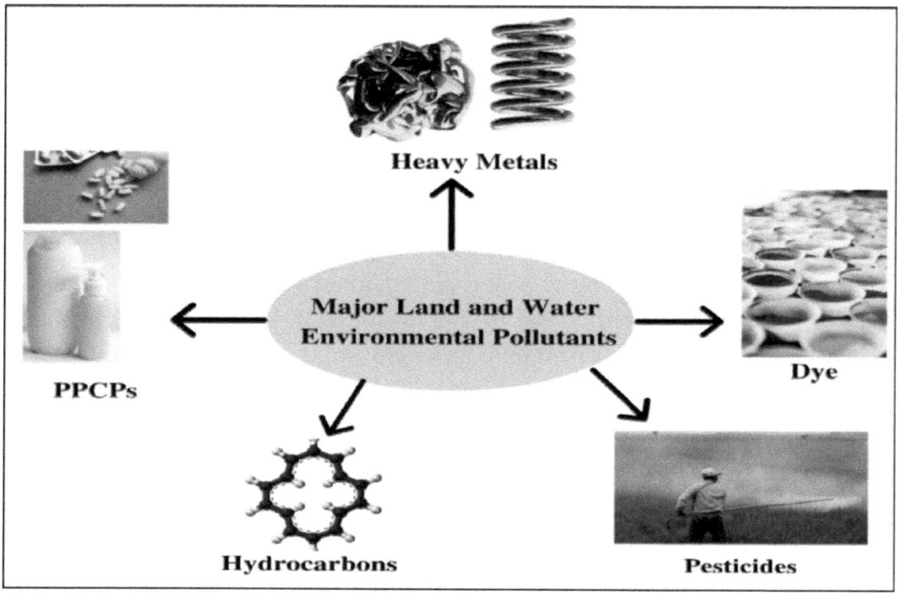

FIGURE 18.2 Different types of environmental pollutants.

Here are some of the major environmental pollutants of land and water bodies and their negative implications are discussed below.

A. **Pesticides**

Pesticides have significantly enhanced agriculture production by reducing pest infestations, as well as decreasing the incidence of insect-borne diseases that cause health risks (Rekha and Prasad, 2006). It is widely acknowledged that the globe needs to produce more food due to the population's rapid growth (Agoramoorthy et al., 2008). Moreover, the inconsistent utilization has a considerable negative influence on the quality of food as well as population health, putting stress on the environment as well as contributing to the emergence of tolerance to pesticides. There is a significant amount of waste as a result of overuse, which raises the cost and contributes to the negative effects on the environment and human health. When pesticides are applied improperly, they disrupt the habitat as a whole by contaminating the land, atmosphere, underground aquifers, and food web (Abhilash and Singh, 2009; UN/DESA, 2002).

B. Heavy Metals

Metals with a comparatively elevated density to water are referred to as heavy metals (Fergusson, 1990). Assuming that mass and hazard are connected, heavy metals also include metalloids like lead, mercury, and arsenic, which can cause harmful effects at low concentration levels (Duffus, 2002). Geological, commercial, pharmaceutical and personal care products, agricultural, and home effluents sources are some of the contributors of heavy metals in the ecosystem that have been noted (He ZL et al., 2005). Contamination of the pollution caused by these elements has recently been linked to rising environmental and worldwide health and safety concerns. Additionally, due to an enormous surge in their utilization in many commercial, agricultural, household, and technical applications, overexposure to humans has increased significantly (Bradl, 2002).

C. Pharmaceuticals and Personal Care Products (PPCPs)

Personal care products (PCPs) are utilized primarily to enhance everyday life, whereas pharmaceuticals are described as prescribed, over-the-counter, and veterinary treatment medications utilized to diagnose or manage human and animal disorders (Boxall et al., 2012). The unintended occurrence of PPCPs in the diversity of marine ecosystems (such as water, sand, and biosphere) with amounts potential of harming aquatic creatures has come to more people's attention in recent years. Due to PPCPs' expanding usage in human and veterinary pharmaceuticals, as well as their ongoing discharge into the habitat, this has grown to be a serious issue (Nikolaou et al., 2007).

PPCP chemicals are present in significant quantities in untreated wastewater as a result of ongoing medical and wellness product usage and discard. Some medications cannot be entirely metabolized by humans or animals. Most of the time, the hepatic or renal systems are used for the body's excretion of parent chemicals and their corresponding derivatives (Jjemba, 2006; Boxall et al., 2012). One method of PCPP chemicals entering wastewater lines is the direct discharge of expired medications (Petrie et al., 2016). The ingredients found in PCPs are often disposed of in wastewater after bathing or cleaning. Since these items, unlike medications, are often administered or utilized topically, normal cleaning or bathing can remove these substances from the body, allowing them to be readily released into the drainage system and contaminating water sources (Ternes et al., 2004).

D. Dye

Dyeing is the process of adding color to materials like clothes, sheets, and other items that can be colored. This is feasible because dyes can adhere to any materials that will comply with them (Yagub et al., 2014; Zeng et al., 2017). Due to the fact that dye-using organizations frequently keep dye effluents as industrial effluents after dyes have served their purpose in dyeing goods, dyes have grown to be a significant source of environmental toxins (Solís et al., 2012). Eventually, these pollutants are released into surrounding water sources, turning the clear water into polluted water. These enterprises should handle dye effluents as a proper step prior to discharging it into the surroundings (Rodríguez-Couto et al., 2009). These effluents cause harm to the ecosystem when they are discharged into the surroundings after use because they contaminate water bodies and make it difficult to use water. Dye effluents combined with water bodies cause a bad smell and eye soreness. Industrial effluents can impact both terrestrial and marine life, initially in marine habitats (Hirunpraditkoon et al., 2010; Zeng et al., 2017).

E. Hydrocarbons

The vapor pressure and overall liquid state of the aromatic hydrocarbons, which have molecular rings, make them visible at room temperature. Compounds like benzene, ethylbenzene, cumene, toluene, and xylene are known as aromatic hydrocarbons. Their discharge into the environment can cause severe hazards (Cerniglia et al., 2006). BTEX (benzene, toluene, ethylbenzene, and xylene) chemicals and other petrochemical hydrocarbons are recognized as indicators of contamination of volatile organic compounds and petroleum chemicals (Schneider et al., 1999). The consumption of BTEX products can occur via drinking polluted water, inhalation of polluted air particles or adsorption of BTEX compounds from the skin (Sairat et al., 2015). There are many different kinds of noxious chemicals for a variety of plant species and aquatic animals in crude oil-soluble compounds and processed compounds. The eggs, larvae, and juvenile organisms are more vulnerable than their mature and adult counterparts. Hydrocarbons produce behavioral and physiological anomalies in marine creatures at levels below the lethal limit, and at more late stages, they may result in the formation of deformities and/or the early mortality of those species (El-Din et al., 2001).

18.5 GREEN CU-NP APPLICATIONS IN NANOREMEDIATION

A. **Utilizing Green Cu-NPs in Agriculture**
It has been reported that Cu-NPs demonstrate anti-microbial properties against the organisms that harm the crops (Banik and Perez-de-Luque, 2017). In the agricultural sector, Cu-NPs are employed as pesticides and disinfectants (Mortazavi et al., 2010), fungicides (Garcia et al., 2003), algaecides (de Oliveira-Filho et al., 2004), and herbicides (Mastin and Rodgers, 2000). Cu-NPs-induced fruits demonstrate large amount of lycopene and vitamin-C content and better toughness. Cu-NPs-treated tomatoes exhibit better quality, increased amount of bioactive substances, and increased antioxidant activity (Lopez-Vargas et al., 2018).

B. **Filtration of Drinking Water by Utilizing Algae-Generated Green Synthesized Copper-Oxide Nanoparticles (CuO-NPs)**
The CuO-NPs demonstrate antimicrobial properties against harmful bacteria such as *E. coli*. A 50 mg/L of *E. coli* sample was evaluated on a sample of surface water. Fish blood was extracted for the evaluation of toxicity- development of mononuclei and evaluation of cell death was conducted on the Chinese hamster ovarian cell lines and human gastric epithelium cell lines. The domestic and industrial wastes discharged in the surface water is more susceptible to microbial growth. Bulking is the most frequent disadvantage of the activated sludge process, one of the most used disinfection techniques. The COD removal efficiency of a column type sequential batch reactor (SBR) is roughly 87%. Therefore, nowadays NPs and/or nanomembranes are used for the substantial eradication of harmful biological components and heavy metals (Shende et al., 2015).

C. **Coomassie Brilliant Blue Dye Photolytic Degradation Employing Green Synthesized CuO NPs**
For the photocatalytic degradation experiment, the Coomassie brilliant blue R-250 dye was utilized as a prototype (Arunachalam et al.,

2012). To accomplish equilibrium, 50 mL of the Coomassie brilliant blue dye (10 mg/L) was blended for 30 min. Moreover, in 50 mL of the Coomassie brilliant blue dye, 10 mg of the green CuO-NPs were put and blended thoroughly under sunlight. Sunlight is an important source for this study. A 2 mL of aliquots had undergone centrifugation for 10 min at 10,000 rpm at a particular duration of time (Arunachalam et al., 2012). Using a spectrophotometer, the absorbance spectrum of the supernatant was evaluated. Coomassie brilliant blue R-250 was determined to have an optimal absorbance peak at 559 nm. In the presence of CuO-NPs, the photocatalytic degradation of Coomassie brilliant blue R-250 was predicted by a drop in peak intensity within 2 h of incubation under direct sunlight. In the absence of treatment with CuO-NPs, the peak location of the Coomassie brilliant blue R-250 dye solution does not significantly change. The dimensions, shape, and charge density of the green-synthesized CuO-NPs are mostly responsible for color dye degradation (Arunachalam et al., 2012). Kumar et al., 2012, showed that the green metal NPs significantly caused dye degradation with an increase in time. From the aforementioned studies, it has been proven that the rod-shaped CuO-NPs can be utilized for the eradication of toxic and colored dyes (Meena et al., 2014).

D. Removal of Pharmaceutical Products Using Green Synthesized Cu Nano-Adsorbent From Wastewater

The Cu-NPs were biosynthesized from the extracts of Tilia leaves. Well dispersed NPs of diameter within the range of 4.7–17.4 nm of spherical and semi-spherical shape were synthesized. This method of adsorption is endothermic, physical in nature, and spontaneous. The release of non-steroidal anti-inflammatory drugs (NSAIDs) like diclofenac, naproxen, and ibuprofen in the water bodies causes severe environmental hazards. The green synthesized Cu-NPs act as non-adsorbents for the eradication of NSAIDs from water bodies. The efficient eradication of NSAIDs took place under the following conditions—pH ¼ 4.5, 298 K temperature, 10 mg Cu-NPs, and the percentage of eradication of diclofenac, naproxen, and ibuprofen was determined to be 91.4%, 86.9%, and 74.4%. The maximum monolayer adsorption capacities were evaluated for diclofenac, naproxen, and ibuprofen and were evaluated to be 36.0, 33.9, and 33.9 mg/g. From the aforementioned experiment, it can be deduced that the green synthesized Cu-NPs can be used for significant eradication of NSAIDs from wastewater (Husein et al., 2019).

E. **Eradication of Heavy Metal (Lead) Ions Through Adsorption Using Green Synthesized CuO-NPs**
In this experiment, three biosynthesized CuO-NPs taken from the extracts of leaves of rosemary (*Salvia rosmarinus*), astralgus (*Astragalus membranaceus*), and mallow (*Malva sylvestris*) were characterized. Three green synthesized NPs' efficiency at removing lead ions from wastewater was investigated. As per the outcomes, the eradication capacity of the CuO-NPs prepared from rosemary, astralgus, and mallow was evaluated to be 84.95%, 88.4%, and 69.6% at the initial maximum concentration of lead (1.5 mM). This may be because of the uniform shape and dimensions of the mallow synthesized CuO-NPs. As compared to the astralgus and rosemary synthesized CuO-NPs, the variations between equilibrium adsorption and equilibrium concentration were more constant. Thus, the Freundlich and Langmuir isotherms revealed the efficient adsorption of lead by mallow synthesized CuO-NPs. Hence, the biosynthesized CuO-NPs from various extracts of plants can be utilized for the adsorption of heavy metals from the contaminated water medium (Zarrabi and Ghasemi-Fasaei, 2022).

18.6 CONCLUSION AND FUTURE PERSPECTIVES

Plants are widely available, no harmful chemicals are used, and the process is environment-friendly, easy, safe, and inexpensive, making the green synthesis method of producing Cu-NPs the finest one so far. The conditions that influence the synthesis of green-synthesized Cu-NPs include pH, reaction time, temperature, the impact of concentration of the plant extract used, concentration and kind of copper salt, and pressure. The techniques of characterization employed for the evaluation of Cu-NPs have been discussed. The green Cu-NPs have various uses in nanoremediation like in agricultural applications, disinfection of drinking water, photolytic degradation of dyes, eradication of heavy metal ions, and pharmaceutical products from wastewater have been highlighted. The specific mechanism that governs this technique of synthesis is still not entirely known, even though there is an abundance of literature that emphasized the capabilities of many plant types in the preparation of Cu-NPs. Due to the diversity in the chemical components of plants, more research on the clarification of the process via experimental techniques should be supported. It is important to take into account the challenges of preserving plant extracts at extreme temperatures and for an extended duration of time. High oxidation of Cu-NPs upon introduction to the optimum atmosphere is a harmful consequence that requires attention, so it is necessary to design a conservation method that will be effective for the long-term sustainability of plant extracts even at elevated temperatures. Therefore, more research and literature are needed to advance the green synthesis of Cu-NPs so that it can operate more effectively than chemical and physical approaches.

KEYWORDS

- **copper**
- **nanoparticles**
- **green synthesis**
- **applications**

REFERENCES

Abd-Elkareem, J. I.; Bassuony, H. M.; Mohammed, S. M.; Fahmy, H. M.; AbdElkader, N. R. Eco-Friendly Methods of Copper Nanoparticles Synthesis. *J. Bionanosci.* **2016,** *10* (1), 15–37.

Abhilash, P. C.; Powell, J. R.; Singh, H. B.; Singh, B. K. Plant–Microbe Interactions: Novel Applications for Exploitation in Multipurpose Remediation Technologies. *Trends Biotechnol.* **2012,** *30* (8), 416–420.

Abhilash, P. C.; Singh, N. Pesticide Use and Application: An Indian Scenario. *J. Hazard. Mater.* **2009,** *165*, 1–12.

Agoramoorthy, G. Can India Meet the Increasing Food: A Demand by 2020? *Futures* **2008,** *40*, 503–506.

Ahmed, A.; Usman, M.; Q.-Liu, Y.; Y.-Shen, Q.; Yu, B.; H.-Cong, L. Plant Mediated Synthesis of Copper Nanoparticles by Using *Camelia sinensis* Leaves Extract and Their Applications in Dye Degradation. *Ferroelectrics* **2019,** *549* (1), 61–69.

Akintelu, S. A.; Folorunso, A. S.; Oyebamiji, A. K. Phytochemical and Antibacterial Investigation of *Moringa oleifera* Seed: Experimental and Computational Approaches. *Ecl. Quím. J.* **2021,** *46* (2), 17–25. https://doi.org/10.26850/1678-4618eqj.v46.2.2021.p17-25.

Ali, N.; Khan, A.; Bilal, M.; Malik, S.; Badshah, S.; Iqbal, H. Chitosan-Basedmposite Modified with Thiocarbamate Moiety for Decontamination of Cations from the Aqueous Media. *Molecules* **2020a,** *25* (1), 226.

Ali, N.; Khan, A.; Nawaz, S.; Bilal, M.; Malik, S.; Badshah, S.; Iqbal, H. M. Characterization and Deployment of Surface-Engineered Chitosan Triethylenetetramine Nanocomposite Hybrid Nano-Adsorbent for Divalent Cations Decontamination. *Int. J. Biol. Macromol.* **2020b,** *152*, 663–671.

Anandalakshmi, K.; Venugobal, J.; Ramasamy, V. Characterization of Silver Nanoparticles by Green Synthesis Method Using *Pedalium murex* Leaf Extract and Their Antibacterial Activity. *Appl. Nanosci.* **2016,** *6*, 399–408.

Arunachalam, R.; Dhanasingh, S.; Kalimuthu, B.; Uthirappan, M.; Rose, C.; A.Mandal, B. *Colloids Surf. B* **2012,** *94*, 226–230.

Baer, D. R. Surface Characterization of Nanoparticles. *J. Surf. Anal.* **2011,** *17* (3), 163–169.

Bali, R.; Harris, A. T. Biogenic Synthesis of Au Nanoparticles Using Vascular Plants. *Ind. Eng. Chem. Res.* **2010,** *49*, 12762–12772.

Banik, S.; Perez-de-Luque, A. In Vitro Effects of Copper Nanoparticles on Plant Pathogens, Beneficial Microbes and Crop Plants. *Spanish J. Agric. Res.* **2017,** *15* (2), e1005.

Bardos, P.; Merly, C.; Kvapil, P.; Koschitzky, H.-P. Status of Nanoremediation and Its Potential for Future Deployment: Risk-Benefit and Benchmarking Appraisals. *Remed. J.* **2018,** *28* (3), 43–56. https://doi.org/10.1002/rem.21559.

Batoool, M.; Masood, B. Green Synthesis of Copper Nanoparticles Using *Solanum Lycopersicum* (Tomato Aqueous Extract) and Study Characterization. *J. Nanosci Nanotechnol. Res.* **2017,** *1*, 1–5.

Boxall, A. B. A.; Rudd, M. A.; Brooks, B. W.; Caldwell, D. J.; Choi, K.; Hickmann, S.; Innes, E.; Ostapyk, K.; Staveley, J. P.; Verslycke, T.; Ankley, G. T.; Beazley, K. F.; Belanger, S. E.; Berninger, J. P.; Carriquiriborde, P.; Coors, A.; Deleo, P. C.; Dyer, S. D.; Ericson, J. F.; Gagne, F.; Giesy, J. P.; Gouin, T.; Hallstrom, L.; Karlsson, M. V.; Larsson, D. G.; Lazorchak, J. M.; Mastrocco, F.; Mclaughlin, A.; Mcmaster, M. E.; Meyerhoff, R. D.; Moore, R.; Parrott, J. L.; Snape, J. R.; R. Murray-Smith, Servos, M. R.; Sibley, P. K.; Straub, J. O.; Szabo, N. D.; Topp, E.; Tetreault, G. R.; Trudeau, V. L.; Van der Kraak, G. Pharmaceuticals and Personal Care Products in the Environment: What Are the Big Questions? *Environ. Health Perspect.* **2012,** *120*, 1221–1229.

Bradl, H., Ed. *Heavy Metals in the Environment: Origin, Interaction and Remediation*, Vol. 6; Academic Press: London, 2002.

Cai, C.; Zhao, M.; Yu, Z.; Rong, H.; Zhang, C. Utilization of Nanomaterials for In Situ Remediation of Heavy Metal(Loid) Contaminated Sediments: A Review. *Sci. Total Environ.* **2019,** *662*, 205–217. https://doi.org/10.1016/j.scitotenv.2019.01.180.

Cerniglia, C. E.; Sutherland, J. B. Relative Roles of Bacteria and Fungi in Polycyclic Aromatic Hydrocarbon Biodegradation and Bioremediation of Contaminated Soils. In *Fungi in Biogeochemical Cycles*; Gadd, G. M., Ed.; Cambridge University Press: Cambridge, 2006; pp 182–211.

Chauhan, R. P. S.; Gupta, C.; Prakash, D. Methodological Advancements in Green Nanotechnology and Their Applications in Biological Synthesis of Herbal Nanoparticles. *Int. J. Bioassays* **2012,** *1* (7), 6–10.

Chung, I.-M.; Rahuman, A. A.; Marimuthu, S.; Kirthi, A. V.; Anbarasan, K.; Padmini, P.; Rajakumar, G. Green Synthesis of Copper Nanoparticles Using *Eclipta prostrata* Leaves

Extract and Their Antioxidant and Cytotoxic Activities. *Exp. Ther. Med.* **2017,** *14* (1), 14–18.

Daniel, S. K.; Vinothini, G.; Subramanian, N.; Nehru, K.; Sivakumar, M. Biosynthesis of Cu, ZVI, and Ag Nanoparticles Using *Dodonaea viscosa* Extract for Antibacterial Activity Against Human Pathogens. *J. Nano Res.* **2013,** *15*, 1319.

Darroudi, M.; Ahmad, M. B.; Zamiri, R.; Zak, A.; Abdullah, A. H.; Ibrahim, N. A. Time Dependent Effect in Green Synthesis of Silver Nanoparticles. *Int. J. Nanomed.* **2011,** 6, 677–681.

Das, P. E.; Abu-Yousef, I. A.; Majdalawieh, A. F.; Narasimhan, S.; Poltronieri, P. Green Synthesis of Encapsulated Copper Nanoparticles Using a Hydroalcoholic Extract of *Moringa oleifera* Leaves and Assessment of Their Antioxidant and Antimicrobial Activities. *Molecules* **2020,** *25* (3), 555.

de Oliveira-Filho, E. C.; Lopes, R. M.; Paumgartten, F. J. R. Comparative Study on the Susceptibility of Freshwater Species to Copper-Based Pesticides. *Chemosphere* **2004,** *56*, 369–374.

de Souza, R. M.; Seibert, D.; Quesada, H. B.; Bassetti, F.; Fagundes-Klen, M. R.; Bergamasco, R. et al. Occurrence, Impacts and General Aspects of Pesticides in Surface Water: A Review. *Process Saf. Environ. Protect.* **2020,** *135*, 22–37. https://doi.org/10.1016/j.psep.2019.12.035.

Din, M. I.; Rehan, R. Synthesis, Characterization, and Applications of Copper Nanoparticles. *Anal. Lett.* **2017,** *50*, 50–62.

Długosz, O.; Chwastowski, J.; Banach, M. Hawthorn Berries Extract for the Green Synthesis of Copper and Silver Nanoparticles. *Chem. Pap.* **2020,** *74*, 239–252.

Dubey, S. P.; Lahtinen, M.; Sillanp€a€a, M. M. Fuit Mediated Greener Synthesis of Silver and Gold Nanoparticles. *Process Biochem.* **2010,** *45* (7), 1065–1071.

Duffus, J. H. Heavy Metals—A Meaningless Term? *Pure Appl Chem.* **2002,** *74* (5), 793807.

El-Din, N. M.; Abdel-Moati, M. A. R. Accumulation of Trace Metals, Petroleum Hydrocarbons, and Polycyclic Aromatic Hydrocarbons in Marine Copepods from the Arabian Gulf. *Bull. Environ. Contam. Toxicol.* **2001,** *66*, 110–117.

Elisma, N.; Labanni, A.; Rilda, Y.; Asrofi, M.; Arief, S. Green Synthesis of Copper Nanoparticles Using *Uncaria gambir* roxb. Leaf Extract and Its Characterization. *Rasayan J. Chem.* **2019,** *12* (4), 1752–1756.

Fajardo, C.; Sanchez-Fortun, S.; Costa, G.; Nande, M.; Botıas, P.; Garcıa-Cantalejo, J.; Mengs, G.; Martın, M. Evaluation of Nanoremediation Strategy in a Pb. *Sci. Total Environ.* **2020,** *706*, 136041. https://doi.org/10.1016/j.scitotenv.2019.136041.

Fergusson, J. E., Ed. *The Heavy Elements: Chemistry, Environmental Impact and Health Effects*; Pergamon Press: Oxford, 1990.

Garcia, P. C.; Rivero, R. M.; Ruiz, J. M.; Romero, L. The Role of Fungicides in the Physiology of Higher Plants: Implications for Defense Responses. *Bot. Rev.* **2003,** *69*, 162–172.

Gawande M. B.; Goswami, A.; Felpin, F. X.; Asefa, T.; Huang, X.; Silva, R.; Zou, X.; Zboril, R.; Varma, R. S. Cu and Cu-Based Nanoparticles: Synthesis and Applications in Catalysis. *Chem. Rev.* **2016,** *116*, 3722.

Gericke, M.; Pinches, A. Biological Synthesis of Metal Nanoparticles. *Hydrometallurgy* **2006,** *83* (1–4), 132–140.

Goh, E. G.; Xu, X.; Mccormick, P. The Effect of Particle Size on the UV Absorbance of Zinc Oxide Nanoparticles. *Scripta Materiala* **2014,** *78–79*, 49–52.

He, Z. L.; Yang, X. E.; Stoffella, P. J. Trace Elements in Agroecosystems and Impacts on the Environment. *J. Trace Element Med. Biol.* **2005,** *19* (2–3), 125–140.

Hirunpraditkoon, S.; Intiya, P.; Thonglert, W.; Nuithitikul, K. *Chemeca 2010*; Engineers Australia: Barton, 2010.

Husein, D. Z.; Hassanien, R.; Al-Hakkani, M. F. Green-Synthesized Copper Nano-Adsorbent for the Removal of Pharmaceutical Pollutants from Real Wastewater Samples. *Heliyon* **2019,** *5* (8), e02339. DOI: 10.1016/j.heliyon.2019.e02339.

Ingale A. G.; Chaudhari, A. N. Biogenic Synthesis of Nanoparticles and Potential Applications: An Eco-Friendly Approach. *J. Nanomed. Nanotechnol.* **2013,** *4* (2).

Iravani, I.; Zolfaghari, B. Green Synthesis of Silver Nanoparticles Using *Pinus eldarica* Bark Extract. *BioMed. Res. Int.* **2006,** 1–5. https://doi.org/10.1155/2013/639725.

Jamdade, D. A.; Rajpali, D.; Joshi, K. A.; Kitture, R.; Kulkarni, A. S.; Shinde, V. S.; Bellare, J.; Babiya, K. R.; Ghosh, S. *Gnidia glauca*—and *Plumbago zeylanica*—Mediated Synthesis of Novel Copper Nanoparticles as Promising Antidiabetic Agents. *Adv. Pharmacol. Sci.* **2019,** 2019, 1–11.

Jayadev, A.; Krishnan, N. Green Synthesis of Copper Nanoparticles and Its Characterization. *J. Sci. Res.* **2021,** *65* (1), 80–84.

Jemilugba, O. T.; Sakho, E. H. M.; Parani, S.; Mavumengwana, V.; Oluwafemi, O. S. Green Synthesis of Silver Nanoparticles Using Combretum Erythrophyllum Leaves and Its Antibacterial Activities. *Colloid Interface Sci. Commun.* **2019,** *31* (1), 100191.

Jjemba, P. K. Excretion and Ecotoxicity of Pharmaceutical and Personal Care Products in the Environment. *Ecotoxicol. Environ. Saf.* **2006,** *63*, 113–130.

Joerger R.; Klaus, T.; Granqvist, C. G. Biologically Produced Silver Carbon Composite Materials for Optically Functional Thin-Film Coatings. *Adv. Mater.* **2000,** *12* (6), 407–409.

Joseph, B.; Nair, V. M. Ethanopharmacological and Phytochemical Aspects of *Ocimum sanctum* Linn-the Elixir of Life. *Br. J. Pharm. Res.* **2013,** *3* (2), 273–292.

Karakoti, A. S.; Munusamy, P.; Hostetler, K.; Kodali, V.; Kuchibhatia, S.; Orr, G.; Pounds, J. G.; Teeguarden, J. G.; Thrall Baer, B. D. Preparation and Characterization Challenges to Understanding Environmental and Biological Impacts of Nanoparticles. *Surf Interface Anal.* **2013,** *44* (5), 882–889.

Kuchibhatla, S. V.; Karakoti, A. S.; Baer, D. R.; Samudrala, S.; Engelhard, M. H.; Amonette, J. E.; Thevuthasan, S.; Sea, S. Influence of Aging and Environment on Nanoparticle Chemistry: Implication to Confinement Effects in Nanoceria. *J. Phys. Chem. C* **2012,** *116* (26), 14108–14114.

Kumar, P.; Govindaraju, M.; Senthamilselvi, S.; Premkumar, K. *Colloids Surf. B* **2012,** *103*, 658–661.

Kuppusamy, P.; Yusoff, M. M.; Maniam, G. P.; Govindan, N. Biosynthesis of Metallic Nanoparticles Using Plant Derivatives and Their New Avenues in Pharmacological Applications—An Updated Report. *Saudi Pharm. J.* **2016,** *24* (4), 473–484.

Lellis, B.; Favaro-Polonio, C. Z.; Pamphile, J. A.; Polonio, J. C. Effects of Textile Dyes on Health and the Environment and Bioremediation Potential of Living Organisms. *Biotechnol. Res. Innov.* **2019,** *3* (2), 275–290. https://doi.org/10.1016/j.biori.2019.09.001.

Li, F.; Chen, J.; Hu, X.; He, F.; Bean, E.; Tsang, D. C. W.; Ok, Y. S.; Gao, B. Applications of Carbonaceous Adsorbents in the Remediation of Polycyclic Aromatic Hydrocarbon-Contaminated Sediments: A Review. *J. Clean. Product.* **2020,** *255*,120263. https://doi.org/10.1016/j.jclepro.2020.120263.

Liu, L.; Bilal, M.; Duan, X.; Iqbal, H. M. Mitigation of Environmental Pollution by Genetically Engineered Bacteria—Current Challenges and Future Perspectives. *Sci. Total Environ.* **2019,** *667*, 444–464.

Lopez-Vargas, E. R.; Ortega-Ortiz, H.; Cadenas-Pliego, G.; Romenus, K. A.; de la Fuente, M. C.; Benavides-Mendoza, A. Foliar Application of Copper Nanoparticles Increases the Fruit Quality and the Content of Bioactive Compounds in Tomatoes. *Appl. Sci.* **2018,** *8* (1020), 1–15.

Maleki, A.; Hamesadeghi, U.; Daraei, H.; Hayati, B.; Najafi, F.; McKay, G.; Rezaee, R. Amine Functionalized Multi-Walled Carbon Nanotubes: Single and Binary Systems for High Capacity Dye Removal. *Chem. Eng. J.* **2017,** *313*, 826–835. DOI:10.1016/j.cej.2016.10.058.

Mastin, B.; Rodgers, J. H. Toxicity and Bioavailability of Copper Herbicides (Clearigate, Cutrine-Plus, and Copper Sulfate) to Freshwater Animals. *Arch. Environ. Contam. Toxicol.* **2000,** *39*, 445–451.

Meena, K.; Muthu, K.; Meenatchi, V.; Rajasekar, M.; Bhagavannarayana, G.; Meenakshisundaram, S. P. Spectrochim Acta Part A: Mol. Biomol. Spectrosc. **2014,** *124*, 663–669.

Mehta, D.; Mazumdar, S.; Singh, S. K. Magnetic Adsorbents for the Treatment of Water/Wastewater—A Review. *J. Water Process Eng.* **2015,** *7*, 244–265. https://doi.org/10.1016/j.jwpe.2015.07.001.

Meyer, M. F.; Powers, S. M.; Hampton, S. E. An Evidence Synthesis of Pharmaceuticals and Personal Care Products (PPCPs) in the Environment: Imbalances Among Compounds, Sewage Treatment Techniques, and Ecosystem Types. *Environ. Sci. Technol.* **2019,** *53* (22), 12961–12973. https://doi.org/10.1021/acs.est.9b02966.

Mohamed, E. A. Green Synthesis of Copper & Copper Oxide Nanoparticles Using the Extract of Seedless Dates. *Heliyon* **2020,** *6* (1), e03123.

Mortazavi, M.; An, J.; Dubinett, S. p120-Catenin Is Transcriptionally Downregulated by FOXC2 in Nonsmall Cell Lung Cancer Cells. *Mol. Canc. Res.* **2010,** *8*, 762–774.

Mudunkotuwa, I. A.; Pettibone, J. M.; Grassian, V. H. Environmental Implications of Nanoparticle Aging in the Processing and Fate of Copper-Based Nanomaterials. *Environ. Sci. Technol.* **2012,** *46* (13), 7001–7010.

Muhammad, I. D.; Farhan, A.; Zaib, H.; Maria, M. Green Adeptness in the Synthesis and Stabilization of Copper Nanoparticles: Catalytic, Antibacterial, Cytotoxicity and Antioxidant Activities. *Nano. Res. Lett.* **2017,** *12*, 638. https://doi.org/10.1186/s11671-017-2399-8.

Nasrollahzadeh, M.; Sajadi, S. M.; Khalaj, M. Green Synthesis of Copper Nanoparticles Using Aqueous Extract of the Leaves of *Euphorbia esula* L. and Their Catalytic Activity for Ligand-Free Ullmann-Coupling Reaction and Reduction of 4-Nitrophenol, *RSC Adv.* **2014,** *4* (88), 47313–47318.

Nazir, M. S.; Mahdi, A. J.; Bilal, M.; Sohail, H. M.; Ali, N.; Iqbal, H. M. Environmental Impact and Pollution-Related Challenges of Renewable Wind Energy Paradigm—A Review. *Sci. Total Environ.* **2019,** *683*, 436–444.

Ngoc-Diep, P.; Minh-Man, D.; Minh-Vien, L.; Hoang, A. H.; Le-Kieu-Oanh, P. Preparation and Characterization of Antifungal Colloidal Copper Nanoparticles and Their Antifungal Activity Against *Fusarium oxysporum* and *Phytophthora capsici*. *Compt. Rendus Chem.* **2019,** 1–8. https://doi.org/10.1016/j.crci.2019.10.007.

Nikolaou, A.; Meric, S.; Fatta, D. Occurrence Patterns of Pharmaceuticals in Water and Wastewater Environments. *Anal. Bioanal. Chem.* **2007,** *387*, 1225e1234.

Noorjahan C. M.; Jasmine Shahina, S. K.; Deepika, T.; Rafiq, S. Green Synthesis and Characterization of Zinc Oxide Nanoparticles from Neem (Azadirachta indicia). *IJSETR* **2015,** 5751–5753.

Ocsoy, I.; Tasdemir, D.; Mazicioglu, S.; Celik, C.; Katı, A.; Ulgen, F. Biomolecules Incorporated Metallic Nanoparticles Synthesis and Their Biomedical Applications. *Mater. Lett.* **2018,** *212*, 45–50.

Padma, P.; Banu, S.; Kumari, S. Studies on Green Synthesis of Copper Nanoparticles Using *Punica granatum*. *Annu. Res. Rev. Biol.* **2018,** *23* (1), 1–10.

PEN. The Project on Emerging Nanotechnologies, 2013. http://www.nanotechproject.org/.

Petrie, B.; Youdan, J.; Barden, R.; Kasprzyk-Hordern, B. New Framework to Diagnose the Direct Disposal of Prescribed Drugs in Wastewater—A Case Study of the Antidepressant Fluoxetine. *Environ. Sci. Technol.* **2016,** *50*, 3781–3789.

Raju, D.; Mehta, U. J.; Hazra, S. Synthesis of Gold Nanoparticles by Various Leaf Fractions of *Semecarpus anacardium* Tree, L.. *Trees (Berl.)* **2011,** *25* (2), 145–151.

Rasheed, T.; Bilal, M.; Hassan, A. A.; Nabeel, F.; Bharagava, R. N.; Ferreira, L. F. R. et al. Environmental Threatening Concern and Efficient Removal of Pharmaceutically Active Compounds Using Metal-Organic Frameworks as Adsorbents. *Environ. Res.* **2020,** 109436.

Rekha, S. N.; Prasad, R. N. Pesticide Residue in Organic and Conventional Food-Risk Analysis. *Chem. Health Saf.* **2006,** *13*, 12–19.

Rodríguez-Couto, S.; Osma, J. F.; Toca-Herrera, J. L. Removal of Synthetic Dyes by an Eco-Friendly Strategy. *Eng. Life Sci.* **2009,** *9* (2), 116–123. DOI:10.1002/elsc.200800088.

Sairat, T.; Homwuttiwong, S.; Homwutthiwong, K.; Ongwandee, M. Investigation of Gasoline Distributions Within Petrol Stations: Spatial and Seasonal Concentrations, Sources, Mitigation Measures, and Occupationally Exposed Symptoms. *Environ. Sci. Pollut. Res.* **2015,** *22*, 13870–13880.

Schaefer, H. R.; Dennis, S.; Fitzpatrick, S. Cadmium: Mitigation Strategies to Reduce Dietary Exposure. *J. Food Sci.* **2020,** *85* (2), 260–267. https://doi.org/10.1111/1750-3841.14997.

Schneider, P.; Lorinci, G.; Gebefugi, I. L.; Heinrich, J.; Kettrup, A.; Wichmann, H. E. Vertical and Horizontal Variability of Volatile Organic Compounds in Homes in Eastern Germany. *J. Expo. Sci. Environ. Epidemiol.* **1999,** *9*, 282–292.

Shah, M.; Fawcett, D.; Sharma, S.; Tripathy, S.; Poinern, G. Green Synthesis of Metallic Nanoparticles via Biological Entities. Materials **2015,** *8* (11), 7278–7308.

Shankar, S.; Rhim, J. W. Effect of Copper Salts and Reducing Agents on Characteristics and Antimicrobial Activity of Copper Nanoparticles. *Mater. Lett.* **2014,** *132*, 307–311.

Shende, S.; Ingle, A. P.; Gade, A.; Rai, M. Green Synthesis of Copper Nanoparticles by *Citrus medica* Linn. (Idilimbu) Juice and Its Antimicrobial Activity. *World J. Microbiol. Biotechnol.* **2015,** *31* (6), 865–873. DOI: 10.1007/s11274-015-1840-3.

Solís, M.; Solís, A.; Inés Pérez, H.; Manjarrez, N.; Flores, M. Microbial Decolouration Of Azo Dyes: A Review. *Process Biochem.* **2012,** *47* (12), 1723–1748. DOI: 10.1016/j.procbio.2012.08.014.

Takahiro, N.; Herbani, Y.; Ursescu, D.; Banici, R.; Dabu R. V.; Sato, S. Spectroscopic Study of Gold Nanoparticle Formation Through High Intensity Laser Irradiation of Solution. *AIP Adv.* **2013,** *3*, 082101-1–082101-10.

Tamilvanan, A.; Kulendran, B.; Ponappa, K.; Kumar, B. Copper Nanoparticles: Synthetic Strategies, Properties, and Multifunctional Application. *Int. J. Nanosci.* **2014,** *13* (2), 1430001–1430022.

Ternes, T. A.; Joss, A.; Siegrist, H. Peer Reviewed: Scrutinizing Pharmaceuticals and Personal Care Products in Wastewater Treatment. *Environ. Sci. Technol.* **2004,** *38*, 392A–399A.

UN. The Sustainable Development Goals Report 2016; United Nations, 2016.

United Nations Department of Economic and Social Affairs (UN/DESA). Changing Unsustainable Patterns of Consumption and Production, Johannesburg Plan on Implementation of the World Summit on Sustainable Development, Johannesburg; Chapter III, Paragraph 14–23, 2002.

Vijayaraghavan, K.; Ashokkumar, T. Plant-Mediated Biosynthesis of Metallic Nanoparticles: A Review of Literature, Factors Affecting Synthesis, Characterization Techniques and Applications. *J. Environ. Chem. Eng.* **2006,** *21*, 1–78. https://doi.org/10.1016/j.jece.2017.09.026.

Wang, M.; Hossain, F.; Sulaiman, R.; Ren, X. Exposure to Inorganic Arsenic and Lead and Autism Spectrum Disorder in Children: A Systematic Review and Meta-Analysis. *Chem. Res. Toxicol.* **2019,** *32* (10), 1904–1919. https://doi.org/10.1021/acs.chemrestox.9b00134.

Wu, S. Preparation of Fine Copper Powder Using Ascorbic Acid as Reducing Agent and Its Application in MLLC. *Mater. Lett.* **2007,** *61*, 1125–1129.

Yagub, M. T.; Sen, T. K.; Afroze, S.; Ang, H. M. Dye and Its Removal from Aqueous Solution by Adsorption: A Review. *Adv. Colloid Interface Sci.* **2014,** *209*, 172–184. DOI: 10.1016/j.cis.2014.04.002.

Zarrabi, A.; Ghasemi-Fasaei, R. Preparation of Green Synthesized Copper Oxide Nanoparticles for Efficient Removal of Lead from Wastewaters. *Int. J. Phytoremediation.* **2022,** *24* (8), 855–866. DOI: 10.1080/15226514.2021.1984385.

Zeng, G.; Ye, Z.; He, Y.; Yang, X.; Ma, J.; Shi, H.; Feng, Z. Application of Dopamine-Modified Halloysite Nanotubes/PVDF Blend Membranes for Direct Dyes Removal From Wastewater. *Chem. Eng. J.* **2017,** *323*, 572–583. DOI:10.1016/j.cej.2017.04.13.

CHAPTER 19

Green Prussian Blue Nanoparticles for Nanoremediation

MEHAK HAFEEZ, IRFAN AHMAD, and SAIMA ANDLEEB

Division of Fish Genetics and Biotechnology, Faculty of Fisheries, Sher-e-Kashmir University of Agricultural Sciences and Technology of Kashmir, Rangil, Ganderbal, Jammu and Kashmir, India

ABSTRACT

Prussian blue nanoparticles (PBNPs) have distinctive attributes owing to their minute size, porosity, and huge surface area. PBNPs are nontoxic, decomposable, and biocompatible in nature, thus rendering them perfect players for in vivo and in vitro applications. Many studies regarding various biomedical applications of PBNPs were reported during the last twenty years and a relatively new significance of the utility of PBNPs for green nanoremediation is emerging. The other nanoparticles of metals such as copper (Cu), silver (Ag), gold (Au), and zero-valent iron (Fe^0) and metal oxides like titanium oxide (TiO_2), zinc oxide (ZnO) and copper oxide (CuO/Cu_2O) have already been comprehensively explored for remediating environmental pollution/contamination. The main application of PBNPs for green nanoremediation lies in their selective purging of hazardous ions from a contaminated ecosystem. They are excellent adsorbents to be used as antidotes for environmental contamination by radioactive as well as nonradioactive forms of cesium (Cs) and thallium (Tl) elements. Additional cations adsorbed by PBNPs are NH_4^+, Na^+, and K^+ among others. PBNPs are also able to trap small molecules of various gases, hydrocarbons, and even luminescent molecules such as 2-aminoanthracene. To add versatility to their function, PBNPs are encapsulated within a polymer or coated with magnetic particles.

Sustainable Nanoremediation: Modern Technologies for a Clean Environment.
Fernanda Maria Policarpo Tonelli, Arpita Roy, and Khalid Rehman Hakeem (Eds.)

However, broad investigative discernments need to be done to assess the possible perils concomitant with the extensive usage of these nanoparticles in ecological and biological systems.

19.1 INTRODUCTION

Abraham Werner, a German mineralogist of the 18th century fashioned a systematic method for distinguishing colors that was later modified and refined in the 1900's by a Scottish painter named Patrick Syme. At the 25th rank of this classification, is a typical blue color so-called Prussian blue (PB), which is a mishmash of velvet black with a mild concentration of indigo blue (Syme, 1821). Heinrich Diesbach, a Berlin colormaker/artist was the first to accidentally discover PB in 1704, who named it Berlin blue (Ware, 2008). This pigment is conceivably the first coordination compound synthesized artificially. Even after 300 years of its invention, PB is still a remarkable material for trailblazing studies.

Over the last 20 years, researchers have recognized PB as a promising nonorganic coordination polymer having profound industrial and scientific applications, such as hydrogen storage, photomagnets, ion- and biosensors, secondary batteries, electrocatalysts, and electrochromic displays (Gotoh et al., 2007). Furthermore, PB acts as an analog of peroxidase that is used as a catalyst for reducing hydrogen peroxide. Given the ability of PB to react in lower dosages, it has a great potential application in biosensors. Owing to its porosity, small ions and gaseous molecules can be adsorbed by PB and this property makes PB famous as a "chemical sponge" having wide applications in many fields, especially environmental remediation/cleanup and gas storage.

Nanosized structures of PB are architectured by assembling transition metal ions or lanthanides through cyano-bridged ligands and these are called as Prussian blue nanoparticles (PBNPs). This alluring family of inorganic nanoparticles is denoted by a common formula of $A_xM_y[M'(CN)_6]_z$ (where A is a monovalent cation, MII and M'III are di or trivalent transition metal ions or lanthanides). At present, the nanoscale assemblage of PB is of greater appeal to the scientific community due to its numerous peculiar properties including low density (~1.8 g cm^{-3}), porosity, flexible molecular structure, stability, solubility, and modifiable physicochemical characteristics. The development of PBNPs as a theranostic tool has found novel uses in the field of medicine such as photothermal healing, imaging, and delivery of drugs (Dacarro et al., 2018; Busquets and Estelrich, 2020). Due to their innate nanoenzyme bioactivity, PBNPs help in the treatment of ROS-related diseases in humans by the removal of superfluous reactive oxygen species (ROS) and also help in the release of oxygen from hydrogen peroxide to overcome the hypoxic conditions induced by tumors in the similar way as done by peroxidase, catalase, and superoxide dismutase enzymes. PBNPs are also endowed with the ability to transport active drug molecules due to their porosity and modifiable surface thus overcoming the drawbacks of usual medications such as toxic side-effects, less solubility, and inadequate target specificity. Furthermore, PBNPs have superb biocompatibility and biosafety due to which they were sanctioned in 2003 by the U.S. Food and Drug Administration (FDA) as safe for human use. The commercial nano-drug named Radiograde®, made of soluble PBNPs, is used for the treatment of patients suffering from thallium (Tl) or cesium (^{137}Cs) intoxication (Yang et al., 2007; Yang et al., 2008). PB is also present in the WHO Model List of Essential Medicines of the World Health Organization (WHO, 2013).

19.2 STRUCTURAL CHARACTERISTICS OF PRUSSIAN BLUE

Being a coordination compound, PB contains Fe^{3+} cation and the octahedral hexacyanoferrate complex ($[Fe(CN)_6]^{2+}$) with d6 Fe^{2+} cation in a low spin state. In this compound, the Fe^{2+} cations are coordinated by the carbon atom of the cyanide (CN^-) ligand, which acts as a bridge with Fe^{3+} cations, that are octahedrally coordinated by 6 nitrogen atoms (Kong et al., 2015). The Fe^{2+} and Fe^{3+} cations have unlike coordination spheres and therefore, can embrace many distinct magnetic configurations and oxidation states. This compound is concocted by combining any Fe^{3+} salt with potassium hexa-cyanoferrate ($K_4[Fe(CN)_6]$), giving rise to diverse formulas subject to the mixture ratio and the method of formulation. These formulas vary from $FeIII_4[FeII(CN)_6]_3 \cdot xH_2O$ (x = 14–16) to $KFeIII[FeII(CN)_6] \cdot xH_2O$ (x = 1–5), normally termed as "insoluble PB" and "soluble PB", respectively (Ware, 2008). On the contrary, if water molecules are neglected, PB can also be articulated as $KxFeIII[FeII(CN)_6]_{(3+x)/4}$, where x vacillates from 0 (insoluble PB) to 1 (soluble PB) (Catala and Mallah, 2017). Indeed, every different formula corresponds to a similar compound when observed by their crystal and molecular structures via the X-ray and neutron diffraction techniques (Buser et al., 1977).

In 1936, Keggin and Miles were the first to demonstrate the cubic structure of PB in a porous network using powder diffraction patterns, and the same was confirmed by Herren in 1980 using the single crystal diffraction technique. Each of the ~10-nm nanoparticles that aggregate to form a PB lattice contains a face-centered cubic unit cell consisting of alternate ferric (Fe^{3+}) and ferrous (Fe^{2+}) cations that are connected by cyanide (CN?) anion bridge. Each side of the unit cell is 10.2 Å and the average bond lengths of Fe(II)—, C—N, and Fe(III) —N—C are 1.90, 1.13, and 2.03 Å, respectively (Buser et al., 1977). Based on the structure, there are two forms of PB: insoluble PB ($FeIII_4[FeII(CN)_6]_3{\cdot}nH_2O$) and soluble PB ($KFeIII[FeII(CN)_6]$). In insoluble PB, with an ideal formulation of $FeIII_4[FeII(CN)_6]_3$ and high number of water molecules in the lattice, one-fourth of the $Fe(CN)_6]^{4-}$ complexes are absent and to balance the charge, water molecules either reside in the vacant defective sites and coordinate to Fe(II) sites (coordinative type) or move inside the voids without coordinating to Fe(II) sites (zeolitic/crystallized type). On the contrary, in the soluble PB, with an ideal formulation of $KFeIII[FeII(CN)_6]$ and less number of water molecules in the lattice, to balance the charge, K^+ replaces water molecules and occupies about 50% of the cubic cell centers. In reality, this categorization of PB based on solubility is not genuine as both soluble as well as insoluble PBs are insoluble in water (solubility product ~10^{-41}) and most common solvents. However, the soluble label is given to those PB that tend to form very small crystals that exist as nanoparticles and can be easily dispersed in water to form pure dark-blue colloidal suspensions. Large crystals are made by the insoluble form of PB that have the property to aggregate readily and result in the formation of a precipitate.

19.3 PHYSICOCHEMICAL CHARACTERISTICS OF PRUSSIAN BLUE

Due to its deep blue color, PB has been used by artists worldwide as a blue pigment. The blue color of PB is because of the transition of interval electrons from Fe(II) to Fe(III) which generates an absorption band typically around 700 nm in water although some subtle differences in the absorption maximum are reported; for example, 709 and 712 nm (Robin, 1962; Fu et al., 2012). The oxidation state of PB could change by modifying the external potentials, resulting in compounds with different colors like Prussian yellow $[FeIIIFeIII(CN)_6]$, Berlin green $[K_{1/3}Fe(FeIII(CN)_6)_{2/3}(FeII(CN)_6)_{1/3}]$ and Prussian white $[K_2FeIIFeII(CN)_6]$. While the Prussian white does not show any visible adsorption band, the absorption bands of both Prussian yellow as well as Berlin green lie in the visible spectrum, at 420 and 480 nm, respectively. Due to the presence of Fe ions, PB is ferromagnetic in behavior which can be revealed below 5.5 ± 0.5 K Curie temperature (T_C) (Ito et al., 1968).

19.4 SYNTHESIS OF PBNPS

S. P. Moulik and S. Mann were the first to publish an original document/report regarding the production of PBNPs and their analogous NPs in 1999 and 2000, respectively. Subsequently, many PBNPs were reportedly synthesized with varying shapes, sizes, molecular configurations, and adaptable surface features. The recent description of the various synthesizing procedures of PBNPs was given extensively in a publication by Dacarro et al. (Dacarro et al., 2018). The standard one-step technique (the coprecipitation method) to synthesize these nanoparticles requires combining an iron salt (Fe^{3+}/Fe^{2+}) and an alkali hexacyanoferrate ($[Fe(CN)_6]^{4-}$ or $[Fe(CN)_6]^{3-}$). If the same process is done at higher temperatures and pressure, then this modified technique is called as hydrothermal method. The third technique of synthesizing PBNPs is a single-precursor method where a single precursor in the form of $[Fe(CN)_6]^{4-}$ or $[Fe(CN)_6]^{3-}$ is used, that gradually releases Fe^{2+} or Fe^{3+} cations which undergo reduction or oxidation to form Fe^{3+} or Fe^{2+} ions, respectively, at pH below 7 (acidic) and that initiates a chain reaction with the unreacted iron ions. To render PBNPs soluble in water and organic solvents, some protective agents such as polymers, acids, or aliphatic amines are additionally used. Depending on the method of preparation, the structural formulas of PBNPs can differ but the shape remains more or less the same, that is, cubic or spheroidal.

19.4.1 COPRECIPITATION METHOD

When an iron (III) salt solution (i.e., $FeCl_3$) is mixed with a hexacyanoferrate (II) salt solution [i.e., $K_4Fe(CN)_6$], either a blue precipitate of PB is formed only if the former is in surplus or a colloidal suspension of PB is obtained if both solutions are used in the same molar ratio. There is an indirect method also which comprises of two steps only, involving a mixture of an iron (II) salt and a hexacyanoferrate (II) salt. This process forms "Berlin White," a white precipitate with the chemical name of ferrous ferrocyanide ($Fe_2[Fe(CN)_6]$). Then, this Berlin White undergoes oxidation by using an oxidizing agent, such as hydrogen peroxide, to finally generate PB. On the contrary, if the oxidation states of the precursors are interchanged, a blue pigment called

"Turnbulls Blue" is obtained, which is reported to be identical to PB as revealed by X-ray diffraction and Mössbauer spectroscopy (Ito et al., 1968). In general, the average temperature of the solution in the coprecipitation method is kept in the range of 27–60°C.

Sometimes, there is a need to synthesize PB-modified magnetite nanoparticles having the configuration of core–shell magnetite with PB outer coating. This type is yielded when magnetite NPs are suspended in $K_4Fe(CN)_6$ solution to which the ferric salt is added later on (Zhao et al., 2005). Although surface modification reduces the magnetization of magnetite, it still preserves its magnetism and can assemble the nanosystem. A similar type of nanosystem was prepared by Arun et al., who used iron (Fe^0) in place of the magnetite, which had a higher saturation magnetization (110 emu/g) than the same-sized magnetite (Arun et al., 2013).

PBNPs tend to aggregate with each other and are inherently unstable. On injecting these NPs into a host, its macrophage–phagocytic system will attack them, leading to their rapid removal from the body. Therefore, the surface of PBNPs is functionalized to impede their preclude coagulation and to provide a defense against physiological systems. The goal of functionalization is to make these NPs dispersible in water or organic media and this is achieved by using a plethora of oleylamines, organic acids, biomolecules, and polymers. Polymers such as polyethylene glycol (PEG), polyaniline, polystyrene sulfonate (PSS), polydiallyldimethylammonium (PDDA), polyallylamine (PAA), polyvinylalcohol (PVA), polyethyleneimine (PEI), and polyvinyl pyrrolidone (PVP) are mixed thoroughly with the solution of iron salt and then, hexacyanoferrate is mixed with this amalgam when these reagents are completely dissolved. Sometimes, additional molecules are conjugated with PB using PEI as a conjugate via a coupling reaction between the carboxyl group of the compound and the amino group of PEI. This reaction has led to the development of a multifunctional theranostic agent obtained by the conjugation of bovine serum albumin-coated quantum dots and hyaluronic acid on the surface of magnetic PBNPs (Yang et al., 2017). Reportedly, the size and the structure of functionalized PBNPs have been altered on varying the quantity of iron ion or polymer and temperature in the process. In addition to the previously mentioned polymers, various biomolecules like gelatin, poly(L-lysine), chitosan, avidin, and albumin can also be amassed on PBNP outer surfaces.

Another simple way to stabilize PBNPs is by capping them with citric acid as the small citrate anion is a frequently used stabilizer for many noble and transition metal NPs and is also used for the reduction reaction of metallic cations. This method is done at 27 or 60°C where Fe^{3+} aqueous solution is mixed with citric acid in the molar ratio of 1:2 and on thorough mixing, this mixture is mixed with an equal quantity of hexacyanoferrate (II) solution. The PBNPs thus obtained are cubic with a size of 25 nm [as determined by dynamic light scattering (DLS)] or 13 nm (as calculated via transmission electron microscopy (TEM)). Having a high negative z-potential of ~30 mV, citric acid-capped PBNPs are easily dispersed in water. The negative surface charge on these capped PBNPs, however, limits their cellular uptake in the living hosts due to repulsion by the negatively charged plasma membranes (Zhdanov et al., 2002). This can be avoided by derivatizing the carboxyl group of citrate with other compounds via a coupling reaction with 1-ethyl-3-(3-dimethylamonopropyl) carbodiimide (EDC). Using the same approach, Cheng et al. obtained a complex cubic nanosystem with a dimension of ~60 nm, consisting of polyacrylic acid (PAA) and PAH with a coating of PEG (Cheng et al., 2014). To label human mesenchymal stem cells (HMSCs), Kim et al. synthesized a complex of negatively charged and citrate-stabilized PBNPs (134 nm size) and coated with positively charged PLL (Kim et al., 2017).

Using aliphatic amines, the surface of PBNPs can be upgraded to enable them to get dispersed in nonaqueous solvents (Gotoh et al., 2007). This is only possible since the bonding of cyanide bridges in the 3D lattice is uneven and the surface of the cubic cell of PBNPs has only 15% active sites in the form of Fe(III)—OH_2. Hydrophobic NPs are yielded by the interaction of PBNPs with a long-chain alkylamine, namely, oleylamine. These modified PBNPs are incorporated into the lipid structures made from soybean phosphatidylcholine or disteroylphosphatydilethanolamine-polyethylene glycol (DSPE-PEG 2000), forming hydrophilic complexes (Chen et al., 2017; Xue et al., 2018; Busquets et al., 2019). It has been reported that isolated pH-responsive hydrogels containing encapsulated ferrous and hexacyanoferrate (III) ions can allow the in vivo synthesis of PB by injecting the two hydrogels into the tumor-bearing mice. Since the tumor site is acidic (pH < 7) in nature, these hydrogels degrade and release the encapsulated ions which result in the formation of PB (Cheng et al., 2017).

19.4.2 HYDROTHERMAL METHOD

In the hydrothermal method of synthesizing PBNPs, the iron salt and an alkali hexacyanoferrate are dissolved in water and the mixture is subjected to high pressure and temperature (>100°C) in an autoclave reactor. This technique was applied by Wu et al. to prepare PB mesocrystals while maintaining the mixture of PEG 4000 and potassium hexacyanoferrate (III) at 120°C for 20 h under low pH conditions. The PB mesocrystals were in the form of pseudotruncated cubes having 5–6 mm dimensions. Mesocrystals in the shape of truncated cubes (having both smooth and curved faces) were obtained when PEG 2000 was used in place of PEG 4000 (Hu et al., 2010). Qian et al. heated a mixture containing hydrochloric acid (HCl), PEG 2000, sodium dodecyl sulfonate (SDS), sodium acetate, ethylene glycol, ferric, and hexacyanoferrate (III) ions in an autoclave sealed with Teflon at 180°C for 6 h to synthesize PBNPs (Qian et al., 2017). Zhang et al. used an identical method, where the same blend was heated at 140°C for 24 h to formulate polyaniline-coated PBNPs (Zhang et al., 2007). PBNPs with hollow interiors were synthesized by Hu et al. using PVP (Hu et al., 2012). PBNPs with sizes between 110 and 190 nm were mixed with HCl and PVP and heated to 140°C for 4 h. On calcination by autoclaving, Fe_2O_3 particles with interior hollow cavities were obtained (Hu et al., 2012; Wu et al., 2015). Wang et al. used chloroplatinic acid in the mixture to synthesize PVP-capped PBNPs, which act as photo-Fenton reagents to destroy organic environmental contaminants (Wang et al., 2019).

19.4.3 SINGLE PRECURSOR METHOD

This process uses precursor ions ($[Fe(CN)_6]^{4-}$ or $[Fe(CN)_6]^{3-}$), that in acidic pH discharge ferric or ferrous ions which undergo reduction or oxidation to form ferrous or ferric ions, respectively. The newly released Fe cations, then undergo a reaction with unreacted cations of the medium to generate PBNPs. Since the process involves a relatively slow reaction using a polymer, the nanoparticles thus, synthesized have high monodispersity. One of the most common polymers used in this process is PVP (Cai et al., 2016). Vafakhah et al. used a similar approach to synthesize PBNPs embedded in graphene oxide aerogel which is efficient in removing the common salt (sodium chloride) from brackish water (Vafakhah et al., 2019). Magnetic PBNPs can also be prepared by this method where iron oxide NPs, hitherto formed on the hydrothermal process, are mixed briskly with potassium hexacyanoferrate (II) solution of pH = 2. Magnetic PBNPs are obtained through a shell-growing procedure on 1 h after the reaction (Yang et al., 2016). However, this process has some drawbacks as it takes a long time to complete and releases hydrogen cyanide into the medium. The free cyanide ions in the presence of pyridine, react with 1,3-diethyl barbituric acid and form a violet dye with a wavelength of 540 nm (Yang et al., 2007).

19.5 APPLICATION OF PBNPS FOR GREEN NANOREMEDIATION

The high porosity in the crystal lattice of PBNPs enables these particles to take up and trap metal ions and smaller molecules of water, various gases, and hydrocarbons in their vacant spaces. Since PBNPs have a great attraction for metallic ions, therefore these NPs function as an adsorbent for the exchange of ions. This zeolitic property renders them to be an excellent remedy for environmental contamination with heavy metals, like Tl and $^{137}Cs^{+}$ (harmful radioactive element) (Busquets et al., 2019).

The surface area assay of PBNPs by means of N_2 adsorption-desorption calculations at 77.350 K temperature provides substantial data on pore volume, pore size, Brunauer–Emmett–Teller (BET) specific surface area, and porosity. The adsorptive properties of PBNPs depend on their size and morphology. Studies show that the small- (~20 nm) and medium-sized PBNPs (~100 nm) exhibit considerably greater BET surface areas compared to larger ones (>200 nm). This could be probably because there is more hindrance in the removal of water molecules during degassing due to their large size and density. Moreover, the medium-sized PBNPs show the highest porosity (Ming et al., 2012). Because of more surface area, larger pore width, and even pore size, PBNPs act as zeolites. The Stokes radius of the hydrated cation influences its capability to be adsorbed by PBNPs. As opposed to the belief on hydrated ions, the stokes radii of alkali ions such as Na^{+} (0.184 nm), K^{+} (0.125 nm), Cs^{+} (0.119 nm), and Rb^{+} (0.118 nm) are smaller than their corresponding crystal radii (Pau et al., 1990). The term adsorption capacity [quantity of ions (in mg) adsorbed by volume of PBNPs (in g)] determines the adsorption ability of PBNPs and is calculated by the distribution coefficient (K_d). The adsorption capacity of PBNPs is high and varies according to the morphology of different PB aggregates, temperature, pH, time of equilibrium, and ratio of ion to adsorbent, among others.

19.5.1 ADSORPTION OF CESIUM IONS

The recent Fukushima nuclear accident triggered by a tsunami on 11 March 2011 in Japan and the Chernobyl nuclear reactor disaster on 26 April 1986 in Ukraine resulted in a significant discharge of radioactive elements into the environment. This environmental pollution by Cs isotopes, such as ^{134}Cs and ^{137}Cs with a half-life of 2 and 30.17 years, respectively, was one of the major concerns caused by these disasters. Because of its enormous fission output (6.09%), longer half-life, higher volatility, extraordinary activity, and higher solubility, Cs is one of the most significant fission materials (Delchet et al., 2012; Sheba, 2012). Cs emits strong gamma rays that are harmful to organisms. It also proffers a severe public health risk because it easily incorporates in the bodies of living beings and acts identically to Na^+ and K^+ and competes with these ions for passive and active membrane transport since these ions share similar Stokes radii of 0.119 nm in Cs^+, 0.125 nm in K^+ and 0.184 nm in Na^+. Thus, Cs^+ intoxication can lead to various disorders in living organisms. At present, different techniques such as ion exchange, adsorption, volatilization, extraction, and chemical precipitation are being recommended to remove Cs from the environment (Bok-Badura et al., 2020). Out of these, adsorption and ion exchange techniques are utmost easy, cost-effective, and effectual. The nanoparticles generated from PB are among the topmost significant sorbents formed to selectively remove the Cs^+ cations.

The decontamination of Cs by PBNPs ensues by the main mechanisms of ion exchange, surface adsorption, and mechanical trapping within the crystal structure. In the case of soluble PBNPs, ion exchange is reported to occur in which the Cs^+ ions get adsorbed by swapping with K^+ ions present in their cubic lattice (Hornok and Dékány, 2007). However, the adsorption mechanism of Cs^+ sorption via the insoluble PBNPs is still unknown. This mechanism is likely comprised of several self-regulating mechanisms such as physical sorption and Cs^+ entrapment and ion exchange. The former phenomenon indicates the physical sorption of hydrated Cs^+ into the regular lattice spaces surrounded by the Fe^{2+}–CN–Fe^{3+} bonds. The latter phenomenon necessitates defective spots in the crystal lattice, where Cs^+ is chemically adsorbed through proton elimination from the coordination water molecules. These huge defective sites play a vital role in PBNPs so as to effectively adsorb Cs^+ ions (Hayashi et al., 2019). The adsorption of Cs^+ ions by $KCu[Fe(CN_6)]$ is directed mainly via three ways: (a) ionic substitution of K^+ with Cs^+ (b) percolation of surface Cs^+ in vacant spots and (c) proton swap at less K^+ concentration with Cs^+ (Takahashi et al., 2018).

The pioneering work of Cs decontamination by employing PBNPs was anticipated about two decades back by Harjula et al. by the use of granulated potassium cobalt hexacyanoferrate ($K_2[CoFe(CN)_6]$ (Harjula et al., 2001). Takahashi et al. corroborated that insoluble PBNPs have more adsorption capacity than soluble PBNPs since the structure of insoluble PBNPs has more coordination with water molecules. They also found these adsorbed Cs^+ cations along with Fe^+ cations to be homogenously distributed all over these PBNPs and concluded that this process of adsorption is a highly selective, speed-limiting pseudosecond-order kinetic reaction step, with Cs^+ adsorbed by chemisorption instead of physical adsorption (Takahashi et al., 2018). Conversely, Fujita et al. suggested that the outer covering in PBNPs is the sole location for bonding of Cs^+ due to an exceedingly small intracrystalline diffusion coefficient ($<3.3 \times 10^{-22}$ m^2s^{-1}) and the rate of adsorption by PBNPs depends on these active spots and not on Cs^+ concentration of the medium (Fujita et al., 2014). Because of its small rate of adsorption, Cs^+ ions were found to infiltrate only 1–2 nm (1–2 crystalline lattice units) even after 14 days at 27°C. The extraordinary selectivity of PBNPs was verified by Fujita et al. as the equilibrium adsorption rate in the aqueous solution was abridged by a mere one-half to two-thirds than the pure system even if the molar concentration of K^+ and H_3O^+ was 50,000 and 200 times more than Cs^+ ions (Fujita et al., 2014). In most cases, the equilibrium adsorption isotherm process is based on the Langmuir model, with few studies showing isotherms based either on the Freundlich model or the Temkin model. Despite the high selectivity of PBNPs towards Cs^+, sometimes it is difficult to separate/filter the sorbent from the mixture if it is used as a fine powder because these tiny particles clump together into a compact cake. Moreover, the colloidal nature of the PBNP mixture prevents simple removal when used. These limitations are surmounted by synthesizing PBNPs encapsulated with a polymer matrix such as silica or glass (Delchet et al., 2012). The adsorbents can also be assimilated in the form of a crosslinked, bead-type copolymer resin made from polystyrene and divinylbenzene (Cho et al., 2020). PBNPs have also been immobilized by linking polydopamine to nanofiber surfaces that are engineered through polyacrylonitrile electrospinning (Gwon et al., 2020), on polyvinyl alcohol sponge (Wi et al., 2019; 2019), caged in sodium alginate gel (Vipin et al., 2013), in original colloidal solution for developing a delivery system for intestinal release (Sandal et al., 2017), in spongiform adsorbents (Hu et al., 2012), merged with hydrated magnesium silicate called sepiolite (Darder et al., 2014), embedded into a binder matrix (Chen et al., 2015), in porous cellulose aerogel (Lee et al., 2018) and in microporous nanofibril films made of carboxymethyl cellulose (Eun et al., 2020). PBNPs were discovered to precipitate spontaneously in agarose gel which in turn was capable of adsorbing Cs^+ cations more efficiently (Hayashi et al., 2019). A composite hydrogel of PBNPs with clay was prepared by Zhang et al., having an adsorption capacity of ~173 mg/g in water (Zhang et al., 2020).

Coating PB on superparamagnetic nanoparticles is an effective method for the removal of Cs from polluted waters. This paramagnetism of the core–shell nanocomposites makes them apt for green nanoremediation. Using the single-precursor method, Jang et al. synthesized magnetic nanoparticles using poly(diallyldimethylammonium chloride) (PDDA) in which the hexacyano-ferrate (II) ions react with the released Fe^{3+} cations. However, PDDA has no Cs affinity and lacks superparamagnetism, which diminishes the adsorption and magnetic capacity of these modified PBNPs (Jang et al., 2014). When the magnetic nanoclusters produced by the hydrothermal technique are coated with PB using a single-precursor process, PB-functionalized magnetic nanoclusters can be obtained (Yang et al., 2016). The same technique was used by Qian et al. for preparing PB-coated magnetic nanoparticles using PEG for the removal of Cs from human blood (Qian et al., 2017). The adsorption characteristics of some significant magnetic PBNPs have been summarized in Table 19.1.

TABLE 19.1 Comparison of Several Hybrid Structures of Magnetic PBNPs in Terms of Their Relevant Adsorption Properties.

Adsorbent	Synthesis method	Adsorption capacity (mg g−1)	Removal efficiency (%)	Equilibrium time	Kinetic model	Reference
Nanoparticles	Single-precursor	96.00		24 h	Langmuir	Thammawong et al. 2013
Nanoclusters	Single-precursor	45.87	>99.7	6 h	Langmuir	Yang et al. 2016
Nanocomposites with graphene oxide	Anchoring the magnetic PBNPs onto the surface of graphene	55.56	>90.0	12 h	Langmuir	Yang et al. 2014
Nanoparticles with PDDA as the interlayer	Single precursor	16.20	91.0	1 h	Freundlich	Sasaki et al. 2012
	Coprecipitation		94.0	3 h		Jang et al. 2014
	Coprecipitation		84.7–86.7			Namiki et al. 2012
Nanocomposites	Coprecipitation	280.82		24 h	Temkin	Jang et al. 2014
Nanoparticles with PEG	Hydrothermal	274.70	64.8	1 h		Qian et al. 2017

Kohiyama et al. have reported an alternate difficult mode for the adsorption of Cs^+ cations using PBNPs (Koshiyama et al., 2018). A hexacianoferrate (III) salt was hydrated with a lipid to form extruded liposomes, which were treated with Mohr's salt ($(NH_4)_2Fe(SO_4)_2 \cdot 6H_2O$) and antibiotic amphotericin B (AmB). AmB forms channels in the bilayer through which the salt enters the inner aqueous layer of the liposome, forming PBNPs inside it. The adsorption capacity of these liposomes for Cs^+ cations was found to be much greater than that of PBNPs in the medium. For augmenting the adsorption capacity of PBNP composites for Cs^+ cations, Bu et al. assembled a 3D hierarchical PB by interconnecting ultrathin nanosheets which had an adsorption capacity of 200 mg/g. In contrast with cubic PBNPs, the N_2 sorption assay of 3D hierarchical nanostructures indicated a sharp rise in the nitrogen adsorption-desorption isotherm, probably due to the more exposed surface of nanosheets (Bu et al., 2015). As reported by Lee and coworkers, the PBNP composite in the form of cellulose aerogel had the Cs^+ ions entirely dispersed into it, which is an essential criterion of an effective absorbent substance (Lee et al., 2018).

PBNPs have been used for green nanoremediation to rejuvenate nuclear waste-contaminated soil. In clay (2:1 type), potassium is usually bound with oxygen atoms of the inner layer. However, K gets replaced by Cs which in interaction with existent minerals results in nearly irrevocable contamination of soil (Evrard et al., 2015). For decontamination, the Cs^+ cations first need to be released free from the clay and then these liberated Cs^+ ions need to be adsorbed. The cationic polyelectrolytes such as ionized chitosan can cause Cs^+ ion desorption by increasing spaces in between the clay layers. Subsequent to this treatment, PB-functionalized magnetic microgels can be used for the adsorption of Cs^+ ions from the mixture (Qian et al., 2020). These microgels are recovered by desorbent, NaOH (0.1 M), and recycled magnetically for multiple uses on maintaining a constant adsorption capacity (149.70 mg/g).

The major drawback of PBNPs is their instability in basic solutions, which results in lethal cyanide toxicity (Faustino et al., 2008). According to the guidelines of WHO, the maximum quantity of total cyanide allowed in drinking water is 0.07 mg/L. To stabilize PBNPs at pH above 7, $CuSO_4$ salt can be added to the aqueous solution where Cu^{2+} cations shield them against excessive quantities of hydroxide anions (Manabe et al., 2020).

19.5.2 ADSORPTION OF THALLIUM IONS

The unique metallic element, Tl in univalent (thallous, Tl^{+1}) and trivalent (thallic, Tl^{3+}) forms is highly toxic to animals and their habitat. However, the neutral form of this element is principally not toxic. The human body is not able to distinguish between K^+ and Tl^{+1} since the ionic radii of both these ions are alike, resulting in terminal repercussions as Tl^{+1} disturbs the K^+ function pathway of various metabolic activities. As follows, Tl^{+1} results in severe immediate and long-term intoxication in humans by impacting renal, hepatic, and cardiac tissues. The bursting of a so-called "dirty bomb" made from radioactive isotopes of Cs and Tl might also cause Tl intoxication. Nonradioactive Tl finds its origin in our environs because of emissions from the combustion of coal, cement factories, and metallurgy of nonferrous metals. The predominant foundation of Tl contamination is wastewater from mining, beneficiation, and smelting of Tl-containing sulfide ores.

Several methods such as solvent extraction, electrochemical deposition, flotation, and precipitation are being established for the removal of Tl ions in wastewater from industries (Zhao et al., 2020). However, adsorption technology by the use of PBNPs exhibits great potential in terms of great refining efficacy, less power usage, and eco-friendliness (Yilmaz et al., 2019). The impact of pH and storage settings on the binding capacity of PBNPs was evaluated by Yang et al. by means of solutions of varying pH to determine the linkage of physicochemical features of insoluble PBNPs with their binding capability (Yang et al., 2008). The outcomes pointed out the influence of the hydration state of PBNPs on the uptake of Tl. PBNPs having around 17 mol of water were found to have a high binding rate constant of 0.52 while those having only 2.5 mol of water by undergoing dehydration had a low constant of only 0.32. It was then reported that soluble as well as insoluble PBNPs portraying identical features have almost similar binding capacities (Faustino et al., 2019). Immobilized on mesoporous silica, PBNPs, and a PB analog, Cu(II)ferrocyanide, were analyzed by Sangvanich et al. for removing Tl^{+1} ions from diverse aqueous solutions (Sangvanich et al., 2010). In comparison with PB, this analog exhibited a high adsorption capacity for Tl (5.82 mg/g for PB and 28.3 mg/g for PBA). Consistent with these fallouts, there was an increase in the value of the distribution coefficient of Tl for both sorbents when the pH of the medium was increased. Therefore, the cation ion exchange might be the major adsorption mechanism. Vincent et al. hypothesized an identical mechanism, where the exchange of Tl occurs with protons and alkali metal contaminants in the crystal lattice of PBNPs during the synthesis process (Vincent et al., 2014).

19.5.3 ADSORPTION OF CATIONS

PBNPs have an extra capability for adsorbing additional monovalent and divalent cations, besides Cs^+ and Tl^+ ions. These cations show less competition with Cs^+ and Tl^+ ions due to substitution by H^+ cations that are found in the PBNP lattice. In the case of magnetic PBNPs, there is a drastic reduction in the distribution coefficient of Cs due to the existence of other metallic cations, especially K^+ whose influence is quite prejudicial. This competition is probably due to the fact that the Stokes radii of both cations (Cs^+, 0.119 nm and K^+, 0.125 nm) or even their hydration radii (Cs^+, 0.325 nm and K^+, 0.330 nm) are almost similar. While Na^+ shows very little antagonism. A very effective hybrid deionization structure was established by Vafakhah et al. to remove common salt (NaCl) from brackish waters. In this system, for the removal of Na^+ cations an aerogel consisting of embedded redox-active PBNPs and very conductive 3D porous graphene oxide was employed as a binder-free intercalation anode. This mixture at 100 mA/g density generated an extraordinary salt confiscation capacity of 130 mg/g (Vafakhah et al., 2019).

In the absence of Cs^+ and Tl^+, other divalent cations such as Pb^{2+}, Ni^{2+}, Co^{2+}, and Cu^{2+} can also be adsorbed by PBNPs even though their dimensions are poles apart from Cs^+ and Tl^+. PBNPs having magnetic properties are reported to have prominent sorption capacities of 778 mg/g, 155 mg/g, 111 mg/g, and 138 mg/g, respectively, for Pb^{2+}, Ni^{2+}, Co^{2+} and Cu^{2+} (Uoginte et al., 2019). The adsorption of Cu^{2+} by PBNPs is noteworthy as Cu^{2+} cations show superior attraction for β-amyloid peptides forming complexes, as commonly seen in Alzheimer's syndrome. Moreover, it has been speculated that there exists a possible connection of Cu^{2+} ion incidence with the growth of these β-amyloid plaques (Atwood et al., 2000).

19.5.4 ADSORPTION OF GASES

PBNPs proffer a pronounced capacity to store various gases. This potential of adsorbing gases by PBNPs was reconnoitered for the first time by Long and coworkers by using them for storing the hydrogen gas at 77 K (Kaye and Long, 2005). PBNPs can also adsorb the CO_2 gas, however, it is quite challenging to trap and remove this gas from a gaseous mixture spurt due to its very low concentration (<15%). Several PBNP composites were employed by Thallapally et al. for capturing different gases like CO_2, H_2S, NO, N_2, and SO_2. These PBNPs were capable of adsorbing about 8–10% of CO_2 gas at standard conditions of atmospheric pressure and temperature and this adsorbed gas volume amplified with an increase in pressure (Thallapally et al., 2010). Using several techniques, Karadas et al. were able to increase the adsorption capacity of PBNPs for CO_2 at higher pressures of about 50 bar. On heating PBNP crystals, a large number of gaps are created for trapping the CO_2 gas molecules by the removal of the water molecules found in the open voids in their lattice, with no structural disruption. Nonetheless, there occurred unexpected fluctuations in its adsorption isotherm between pressures 35–40 bar (Karadas et al., 2012).

The chief antecedent of fine particulate matter is the gaseous form of ammonia, coming out of agricultural fields and animal ranches, causing air pollution in our environment and health problems in living organisms by affecting the lungs and heart. Being soluble in water, the incidence of water-dissolved ammonia is highly toxic for aquatic flora and fauna. PBNPs could be exploited to recuperate soluble forms of ammonia because the water from aquatic resources typically contains insignificant amounts of Cs and Tl elements (Takahashi et al., 2016). Selectivity is a crucial criterion while selecting an adsorbent for the removal of ammonia since the region/area of implementation commonly includes biomass matrices and synthetic fertilizers which causes discern trepidations. A PB analog amalgamated from copper hexacyanoferrate has the ability to capture water-soluble NH_4^+ and NH_3 concurrently when there is a large concentration of K present (Parajuli et al., 2016). Yet another PB analog synthesized by mixing a sodium salt with cobalt hexacyanoferrate has an advanced adsorption capacity for NH_4^+ than that for Cs^+ or K^+ (Zhang et al., 2019). The mentioned dissimilarities in PBNPs with their analogs are due to the presence of different types of metals in the ligand, the monovalent ion, and the coordination metal, which alter their crystal structures. Not long ago, cobalt hexacyanoferrate was found to show recycling via water flushing (Takahashi et al., 2020).

19.5.5 ADSORPTION OF MOLECULES

PBNPs can adsorb molecules other than cations (Cs^+, Tl^+, K^+, Na^+, and Cu^{2+}) or gases (CO_2, H_2, and NH_3). Several PB analogs were reported to be prepared by Boudjema et al., of which one particular analog having transition-metal ion of cobalt had the ability to segregate a hydrocarbon mixture into its constituents (cyclohexene, cyclohexane, *n*-hexane, and *n*-pentane) in muggy air condition (Boudjema et al., 2017). This accentuates the outstanding perspective of PBNPs for the separation of gases.

Other uses for PBNP adsorption also exist, besides purging harmful cations and molecules from our ecosystem. Larionova's group is used to render luminescence to magnetic as well as nonmagnetic PBNPs through an artificial functionalization inside the inner pores of the cyano-bridges by rhodamine B (RhB) or 2-aminoanthracene (AA) fluorophore (Mamontova et al., 2019). The presence of only one vacant site at the PBNP center allows the fluorophore to be accommodated within it. RhB binds to PBNP surfaces as it is larger than the pore size to enter inside. Both cases generate nanoprobes which can be utilized proficiently for monitoring cell internalization of the PBNPs via fluorescence imaging.

19.6 CONCLUSION

The book chapter furnishes an account of PBNPs employed for green nanoremediation. The nanotechnology-engineered PBNPs and PB analogs to remediate the polluted environmental resources is an inexpensive, time-saving, and environmentally safe option. The adsorptive properties of PBNPs enable them to capture cations and gaseous molecules. Since PBNPs are decomposable and nontoxic, these nanoparticles have been successfully used as an antidote for the remediation of our environment. The main characteristic feature of PBNPs is to confiscate Cs and Tl cations by adsorption. Nevertheless, PBNPs can also adsorb K^+ and Na^+ cations in natural environmental circumstances where harmful cations such as Cs^+ and Tl^+ are generally absent. The robustness in the adsorption capability of PB analogs for water, gaseous molecules, NH_3, and hydrocarbons is due to the swap of Fe cations with other cations. Interestingly, novel approaches are being developed that focus on reusing PBNPs for an environment-affable nanoremediation scheme.

The potential use of PBNPs for green nanoremediation has earned substantial scientific interest all over the world. However, more on-field/on-site studies need to be conducted in the future to determine the adeptness of PBNPs for environmental remediation as the majority of the research done so far has been carried out under controlled lab conditions. Moreover, attention must be focused on the interfering effects/risks of these nanoparticles on ecological and biological systems associated with their use for green nanoremediation.

KEYWORDS

- **Prussian blue**
- **nanoparticles**
- **nanoremediation**
- **pollution**
- **radioactive elements**

REFERENCES

Arun, T.; Prakash, K.; Joseyphus, R. J. Synthesis and Magnetic Properties of Prussian Blue Modified Fe Nanoparticles. *J. Magn. Magn. Mater.* **2013,** *345,* 100–105.

Atwood, C. C.; Scarpa, R. C.; Huang, X.; Moir, R. D.; Jones, W. D.; Fairli, D. P.; Tanzi, R. E.; Bush, A. I. Characterization of Copper Interactions with Alzheimer Amyloid ? Peptides. *J. Neurochem.* **2000,** *75,* 1219–1233.

Bok-Badura, J.; Jakóbik-Kolon, A.; Kazek-Kesik, A.; Karón, K. Hybrid-Pectin-Based Sorbents for Cesium Ion Removal. *Molecules* **2020,** *13,* 2160.

Boudjema, L.; Mamontova, E.; Long, J.; Larionova, J.; Guari, Y.; Trens, P. Prussian Blue Analogues for the Separation of Hydrocarbons in Humid Conditions. *Inorg. Chem.* **2017,** *56,* 7598–7601.

Bu, F. X.; Hu, M.; Zhang, W.; Meng, Q.; Xu, L.; Jiang, D. M.; Jiang, J. S. Three-Dimensional Hierarchical Prussian Blue Composed of Ultrathin Nanosheets: Enhanced Hetero-Catalytic and Adsorption Properties. *Chem. Commun.* **2015,** *51,* 17568–17571.

Buser, H. J.; Schwarzenbach, D.; Petter, W.; Ludi, A. The Crystal Structure of Prussian Blue: Fe4 [Fe(CN)6]3·Xh2o. *Inorg. Chem.* **1977,** *16,* 2704–2710.

Busquets, M. A.; Novella-Xicoy, A.; Guzman, V.; Estelrich, J. Facile Synthesis of Novel Prussian Blue–Lipid Nanocomplexes. *Molecules* **2019,** *24,* 4137.

Busquets, M. A.; Estelrich, J. Prussian Blue Nanoparticles: Synthesis, Surface Modification and Biomedical Applications. *Drug Discov. Today* **2020,** *25,* 1413–1443.

Cai, X.; Gao, W.; Zhang, L.; Ma, M.; Liu, T.; Du, W.; Zheng, Y.; Chen, H.; Shi, J. Enabling Prussian Blue with Tunable Localized Surface Plasmon Resonances: Simultaneously Enhanced Dual-Mode Imaging and Tumor Photothermal Therapy. *ACS Nano* **2016,** *10,* 11115–11126.

Catala, L.; Mallah, T. Nanoparticles of Prussian Blue Analogs and Related Coordination Polymers: From Information Storage to Biomedical Applications. *Coord. Chem. Rev.* **2017,** *346,* 32–61.

Chen, G. R.; Chang, Y. R.; Liu, X.; Kawamoto, T.; Tanaka, H.; Kitajima, A.; Parajuli, D.; Takasaki, M.; Yoshino, K.; Chen, M. L.; Lo, Y. K.; Lei, Z.; Lee D. J. Prussian Blue (PB) Granules for Cesium (Cs) Removal from Drinking Water. *Sep. Purif. Technol.* **2015,** *143,* 146–151.

Chen, H. Y.; Li, X. Y.; Liu, B. J.; Meng, X. H. Facile Synthesis of Prussian Blue Nanoparticles as pH-Responsive Drug Carriers for Combined Photothermal-Chemo Treatment of Cancer. *RSC Adv.* **2017,** *7,* 248–255.

Cheng, L.; Gong, H.; Zhu, W.; Liu, J.; Wang, X.; Liu, G.; Liu, Z. Pegylated Prussian Blue Nanocubes as a Theranostic Agent for Simultaneous Cancer Imaging and Photothermal Therapy. *Biomaterials* **2014,** *35,* 9844–9852.h

Cheng, M.; Peng, W.; Hua, P.; Chen, Z.; Sheng, J.; Yang, J.; Wu, Y. *In Situ* Formation of pH-Responsive Prussian Blue for Photoacoustic Imaging and Photothermal Therapy of Cancer. *RSD Adv.* **2017,** *7,* 18270–18276.

Cho, E.; Lee, J. J.; Lee, B. S.; Lee, K. W.; Yeom, B.; Lee, T. S. Cesium Ion-Exchange Resin Using Dodecylbenzenesulfonate for Binding to Prussian Blue. *Chemosphere* **2020,** *244,* 125589.

Dacarro, G.; Taglietti, A.; Pallavicini, P. Prussian Blue Nanoparticles as a Versatile Photothermal Tool. *Molecules* **2018,** *23,* 1414.

Darder, M.; González-Alfaro, Y.; Aranda, P.; Ruiz-Hitzky, E. Silicate-Based Multi-Functional Nanostructured Materials with Magnetite and Prussian Blue: Application to Cesium Uptake. *RSC Adv.* **2014,** *4,* 35415–35421.

Delchet, C.; Tokarev, A.; Dumail, X.; Toquer, C.; Barré, Y.; Guari, Y.; Guerin, Y.; Larionova, J.; Grandjean, A. Extraction of Radioactive Cesium Using Innovative Functionalized Porous Materials. *RSC Adv.* **2012,** *2,* 5707–5716.

Eun, S.; Hong, H. J.; Kim, H.; Jeong, H. S.; Kim, S.; Jung, J.; Ryu, J. Prussian Blue-Embedded Carboxymethyl Cellulose Nanofibril Membranes for Removing Radioactive Cesium from Aqueous Solution. *Carbohyd. Pol.* **2020,** *235,* 115984.

Evrard, O.; Laceby, J. P.; Lepage, H.; Onda, Y.; Cerdan, O.; Ayrault, S. Radiocesium Transfer From Hillslopes to the Pacific Ocean After the Fukushima Nuclear Power Plant Accident: A Review. *J. Environ. Radioact.* **2015,** *148,* 92–110.

Faustino, P. J.; Brown, A.; Lowry, B.; Yang, Y.; Wang, Y.; Khan, M. A.; Dumbar, K. R.; Mohammad, A. Quantitative Evaluation of the Thallium Binding of Soluble and Insoluble Prussian Blue Hexacyanoferrate Analogs: A Scientific Comparison Based on their Critical Quality Attributes. *Int. J. Pharm.* **2019,** *569,* 118600.

Faustino, P. J.; Yongsheng, Y.; Progar, J. J.; Brownell, C. R.; Sadrieh, N.; May, J. C.; Leutzinger, E.; Place, D. A.; Duffy, E. P.; Houn, F.; et al. Quantitative Determination of Cesium Binding to Ferric Hexacyanoferrate: Prussian Blue. *J. Pharm. Biomed. Anal.* **2008,** *47,* 114–125.

Fu, G.; Liu, W.; Feng, S.; Yue, X. Prussian Blue Nanoparticles Operate as a New Generation of Photothermal Ablation Agents for Cancer Therapy. *Chem. Commun.* **2012,** *48,* 11567–11569.

Fujita, H.; Sasano, H.; Miyajima, R.; Sazoka, A. Adsorption Equilibrium and Kinetics of Cesium Onto Insoluble Prussian Blue Synthesized by an Immediate Precipitation Reaction Between Fe^{3+} and $[Fe(CN)6]^{-4}$. *Adsorption* **2014,** *20,* 905–915.

Gotoh, A.; Uchida, H.; Ishizaki, M.; Satoh, T.; Kaga, S.; Okamoto, S.; Ohta, M.; Sakamoto, M.; Kawamoto, T.; Tanaka, H.; Tokumoto, M.; Hara, S.; Shiozaki, H.; Yamada, M.; Miyake, M.; Kurihara, M. Simple Synthesis of the Three Primary Colour Nanoparticle Inks of Prussian Blue and its Analogues. *Nanotechnology* **2007,** *18,* 345609.

Gwon, Y. J.; Lee, J. J.; Lee, K.-W.; Ogden, M. D.; Harwood, L. M.; Lee, T. S. Prussian Blue Decoration on Polyacrylonitrile Nanofibers Using Polydopamine for Effective Cs Ion Removal. *Ind. Eng. Chem. Res.* **2020,** *59,* 4872–4880.

Harjula, R.; Lehto, J.; Paajanen, A.; Brodkin, L.; Tusa, E. Removal of Radioactive Cesium from Nuclear Waste Solutions with the Transition Metal Hexacyanoferrate Ion Exchanger Cs Treat. *Nucl. Sci. Eng.* **2001,** *137,* 206–214.

Hayashi, H.; Aoki, S.; Takaishi, M.; Sato, Y.; Abe, H. An XAFS Study of Cs Adsorption by the Precipitation Bands of Mn-Fe-Based Prussian Blue Analogues Spontaneously Formed in Agarose Gel. *Phys. Che. Chem. Phys.* **2019,** *21,* 22553–22562.

Hayashi, H.; Sato, Y.; Aoki, S.; Takaishi, M. *In Situ* XRF Analysis of Cs Adsorption by the Precipitation Bands of Prussian Blue Analogues Formed in Agarose Gels. *J. Anal. At. Spectrom.* **2019,** *34,* 979–985.

Hornok, V.; Dékány, I. Synthesis and Stabilization of Prussian Blue Nanoparticles and Application for Sensors. *J. Coll. Interface Sci.* **2007,** *309,* 176–182.

Hu, B.; Fugetsu, B.; Yu, H.; Abe, Y. Prussian Blue Caged in Spongiform Adsorbents Using Diatomite and Carbon Nanotubes for Elimination of Cesium. *J. Hazard. Mater.* **2012,** 217–218.

Hu, M.; Jiang, J. S.; Lin, C. C.; Zeng, Y. Prussian Blue Mesocrystals: An Example of Self-Construction. *Cryst. Eng. Commun.* **2010,** *12,* 2679–2683.

Ito, A.; Suenaga, M.; Ono, K. Mössbauer Study of Soluble Prussian Blue, Insoluble Prussian Blue, and Turnbull's Blue. *J. Chem. Phys.* **1968,** *48,* 3597–3599.

Jang, S. C.; Hong, S. B.; Yang, H. M.; Lee, K. W.; Moon, J. K.; Seo, B. K.; Huh, Y. S.; Roh, C. Removal of Radioactive Cesium Using Prussian Blue Magnetic Nanoparticles. *Nanomaterials* **2014,** *4,* 894–901.

Karadas, F.; El-Faki, H.; Deniz, E.; Yavuz, C. T.; Aparicio, S.; Atilhan, M. CO2 Adsorption Studies On Prussian Blue Analogues. *Microporous Macroporous Mater.* **2012,** *162,* 91–97.

Kaye, S. S.; Long, J. R. Hydrogen Storage in the Dehydrated Prussian Blue Analogues M3 [Co(CN)6]2 (M) Mn, Fe, Co, Ni, Cu, Zn). *J. Am. Chem. Soc.* **2005,** *127,* 6506–6507.

Kim, T.; Lemaster, J. E.; Chen, F.; Li, J.; Jokerst, J. V. Photoacoustic Imaging of Human Mesenchymal Stem Cells Labeled with Prussian Blue-Poly(l-lysine) Nanocomplexes. *ACS Nano.* **2017,** *11,* 9022- 9032.

Kong, B.; Selomulya, C.; Zhenga, G.; Zhao, D. New Faces of Porous Prussian Blue: Interfacial Assembly of Integrated Hetero-Structures for Sensing Applications. *Chem. Soc. Rev.* **2015,** *44,* 7997–8018.

Koshiyama, T.; Tanaka, M.; Honjo, M.; Fukunaga, Y.; Okamura, T.; Ohba, M. Direct Synthesis of Prussian Blue Nanoparticles in Liposomes Incorporating Natural Ion Channels for Cs^+ Adsorption and Particle Size Control. *Langmuir* **2018,** *34,* 1666–1672.

Lee, I.; Kim, S. H.; Rethinasabapathy, M.; Haldorai, Y.; Lee, G. W.; Choe, S. R.; Jang, S. C.; Kang, S. M.; Han, Y. K.; Roh, C.; Cho, W. S.; Huh, Y. S. Porous·3D Prussian Blue/ Cellulose Aerogel as a Decorporation Agent for Removal of Ingested Cesium From the Gastrointestinal Tract. *Sci. Rep.* **2018,** *8,* 4540.

Mamontova, E.; Long, J.; Ferreira, R.; Botas, A. M. P.; Salles, F.; Guari, Y.; Carlos, L. A.; Larionova, J. Making Prussian Blue Analogues Nanoparticles Luminescent: Effect of the Luminophore Confinement Over the Properties. *Nanoscale* **2019,** *11,* 7097–7101.

Manabe, S.; Vipin, A. K.; Kumashiro, T.; Takiguchi, S.; Fugetsu, B.; Sakata, I. Stabilization of Prussian Blue Using Copper Sulfate for Eliminating Radioactive Cesium from a High pH Solution and Seawater. *J. Hazard. Mater.* **2020,** *386,* 121979.

Ming, H.; Torad, N. L. K.; Chiang, Y.-D.; Wu, K. C.-W.; Yamauchi, Y. Size- and Shape-Controlled Synthesis of Prussian Blue Nanoparticles by a Polyvinylpyrrolidone-Assisted Crystallization Process. *Cryst. Eng. Comm.* **2012,** *14,* 3387–3396.

Parajuli, D.; Noguchi, H.; Takahashi, A.; Tanaka, H.; Kawamoto, T. Prospective Application of Copper Hexacyanoferrate for Capturing Dissolved Ammonia. *Ind. Eng. Chem. Res.* **2016,** *55,* 6708–6715.

Pau, P. C. F.; Berg, J. O.; Mcmillan, W. G. Application of Stokes's Law to Ions in Aqueous Solution. *J. Phys. Chem.* **1990,** *94,* 2671–2679.

Qian, J.; Xu, J.; Kung, L.; Hua, D. Cesium Removal from Human Blood by poly(ethylene glycol)-Decorated Prussian Blue Magnetic Nanoparticles. *Chem. Plus Chem.* **2017,** *82,* 888–895.

Qian, J.; Zhou, L.; Yang, X.; Hua, D.; Wu, N. Prussian Blue Analogue Functionalized Magnetic Microgels with Ionized Chitosan for the Cleaning of Cesium-Contaminated Clay. *J. Hazard. Mater.* **2020,** *386,* 121965.

Robin, M. B. The Color and Electronic Configurations of Prussian Blue. *Inorg. Chem.* **1962,** *1,* 337–342.

Sandal, N.; Mittal, G.; Bhatnagar, A.; Pathak, D. P.; Singh, A. K. Preparation, Characterization, and In Vivo Pharmaco Scintigraphy Evaluation of an Intestinal Release Delivery System of Prussian Blue for Decorporation of Cesium and Thallium. *J. Drug Deliv.* **2017,** *4875784.*

Sangvanich, T.; Sukwarotwat, V.; Wiacek, R. J.; Grudzien, R. M.; Fryxell, G. E.; Addleman, R. S.; Timchalk, C.; Yantsee, W. Selective Capture of Cesium and Thallium From Natural Waters and Simulated Wastes with Copper Ferrocyanide Functionalized Mesoporous Silica. *J. Hazard. Mater.* **2010,** *182,* 225–231.

Sheba, R. R. Synthesis and Characterization of Magnetic Hexacyanoferrate (II) Polymeric Nanocomposite for Separation of Cesium From Radioactive Waste Solutions. *J. Colloid Interface Sci.* **2012,** *388,* 21–30.

Syme, P. Werners Nomenclature of Colours. Adapted to Zoology, Botany, Chemistry, Mineralogy, Anatomy, and the Arts (2nd edn), *Nat. Hist. Museum* **1821.**

Takahashi, A.; Minami, K.; Noda, K.; Sakurai, K.; Kawamoto, T. Trace Ammonia Removal from Air Selective Adsorbents Reusable with Water. *ACS Appl. Mater. Interfaces* **2020,** *12,* 15115–15119.

Takahashi, A.; Tanaka, H.; Minami, K.; Noda, K.; Ishizaki, M.; Kurihara, M.; Ogawa, H.; Kawamoto, T. Unveiling Cs-Adsorption Mechanism of Prussian Blue Analogs: Cs^+-Percolation via Vacancies to Complete Dehydrated State. *RSC Adv.* **2018,** *8,* 34808–34816.

Takahashi, A.; Tanaka, H.; Parajuli, D.; Nakamura, T.; Minami, K.; Sugiyama, Y.; Hakuta, Y.; Ohkoshi, S.; Kawamoto, T. Historical Pigment Exhibiting Ammonia Gas Capture Beyond Standard Adsorbents with Adsorption Sites of Two Kinds. *J. Am. Chem. Soc.* **2016,** *138,* 6376–6379.

Thallapally, P. K.; Motkuri, R. K.; Fernández, C. A.; Mcgrail, B. P.; Behrooz, G.S. Prussian Blue Analogues for CO2 and SO2 Capture and Separation Applications. *Inorg. Chem.* **2010**, *49,* 4909–4915.

Uoginte, I.; Lujaniene, G.; Mažeika, K. Study of Cu(II), Co(II), Ni(II) and Pb(II) Removal from Aqueous Solutions Using Magnetic Prussian Blue Nano-Sorbents. *J. Hazard. Mater.* **2019,** *369,* 226–235.

Vafakhah, S.; Guo, L.; Sriramulu, D.; Huang, S.; Saeedikhani, M.; Yang, H. Y. Efficient Sodium-Ion Intercalation Into the Freestanding Prussian Blue/Graphene Aerogel Anode in A Hybrid Capacitive Deionization System. *ACS Appl. Mater. Interfaces* **2019,** *11,* 5989–5998.

Vincent, T.; Taulemesse, J. M.; Dauvergne, A.; Chaunut, T.; Testa, F.; Guibal, E. Thallium (I) Sorption Using Prussian Blue Immobilized in Alginate Capsules. *Carbohyd. Polym.* **2014,** *99,* 517–526.

Vipin, A. K.; Hu, B.; Fugetsu, B. Prussian Blue Caged in Alginate/Calcium Beads as Adsorbents for Removal of Cesium Ions From Contaminated Water. *J. Hazard. Mater.* **2013,** 93–101.

Wang, N.; Ma, W.; Du, Y.; Ren, Z.; Han, B.; Zhang, L.; Sun, B.; Xu, P.; Han, X. Prussian Blue Microcrystals with Morphology Evolution as a High-Performance Photo-Fenton Catalyst for Degradation of Organic Pollutants. *ACS Appl. Mater. Interfaces* **2019,** *11,* 1174–1184.

Ware, M. Prussian Blue: Artists' Pigment and Chemists' Sponge. *J. Chem. Educ.* **2008,** *85,* 612–620.

WHO Model List of Essential Medicines, 3rd ed; WHO, 2013.

Wi, H.; Kang, S. W.; Hwang, Y. Immobilization of Prussian Blue Nanoparticles in Acrylic Acid-Surface Functionalized Poly(Vinyl Alcohol) Sponges for Cesium Adsorption. *Environ. Eng. Res.* **2019,** *24,* 173–179.

Wi, H.; Kim, H.; Oh, D.; Bae, S.; Hwang, Y. Surface Modification of Poly(Vinyl Alcohol) Sponge by Acrylic Acid to Immobilize Prussian Blue for Selective Adsorption of Aqueous Cesium. *Chemosphere* **2019,** *226,* 173–182.

Wu, Q.; Wu, G.; Wang, L.; Hu, W.; Wu, H. Facile Synthesis and Optical Properties of Prussian Blue Microcubes and Hollow Fe2O3 Microboxes. *Mater. Sci. Semicond. Process.* **2015,** *30,* 476–481.

Xue, P.; Sun, L.; Li, Q.; Zhang, L.; Xu, Z.; Li, C. M.; Kang, Y. Pegylated Magnetic Prussian Blue Nanoparticles as a Multifunctional Therapeutic Agent for Combined Targeted Photothermal Ablation and pH-Triggered Chemotherapy of Tumour Cells. *J. Colloid. Interface Sci.* **2018,** *509,* 384–394.

Yang, H. M.; Jang, S. C.; Hong, S. B.; Lee, K. W.; Roh, C.; Huh, Y. S.; Seo, B. K. Prussian Blue-Functionalized Magnetic Nanoclusters for the Removal of Radioactive Cesium from Water. *J. Alloys Compd.* **2016,** *657,* 387–393.

Yang, Y.; Jing, L.; Li, X.; Lin, L.; Yue, X.; Dai, Z. Hyaluronic Acid Conjugated Magnetic Prussian blue@quantum Dot Nanoparticles for Cancer Theranostics. *Theranostics* **2017,** *7,* 466–481.

Yang, Y.; Brownell, C.; Sadrieh, N.; May, J.; Del Grosso, A. Quantitative Measurement of Cyanide Released from Prussian Blue. *Clin. Toxicol.* **2007,** *45,* 776–781.

Yang, Y.; Faustino, P. J.; Progar, J. J.; Brownell, C. R.; Sadrieh, N.; May, J. C.; Leutzinger, E.; Place, D. A.; Duffy, E. P.; Yu, L. X.; Khan, M. A.; Lyon, R.C. Quantitative Determination of Thallium Binding to Ferric Hexacyanoferrate: Prussian Blue. *Int. J. Pharm.* **2008,** *353,* 187–194.

Yilmaz, S.; Zengin, A.; Akbulut, Y.; Sahan, T. Magnetic Nanoparticles Coated with Aminated Polymer Brush as a Novel Material for Effective Removal of Pb (II) Ions From Aqueous Environments. *Environ. Sci. Pollut. R.* **2019,** *26,* 1–15.

Zhang, H.; Hodges, C. S.; Kumar Misha, P.; Young Yoon, J.; Hunter, T. N.; Lee, J.; Harbottle, D. Bio-Inspired Preparation of Clay-Hexacyanoferrate Composite Hydrogels as Super Adsorbents for Cs^+. *ACS Appl. Mater. Interfaces* **2020,** *12,* 33173–33185.

Zhang, N.; Kawamoto, T.; Jiang, Y.; Takahashi, A.; Ishizaki, M.; Asai, M.; Kurihara, M.; Zhang, Z.; Lei, Z.; Parajuli, D. Interpretation of the Role of Composition on the Inclusion Efficiency of Monovalent Cations Into Cobalt Hexacyanoferrate. *Chem. Eur. J.* **2019,** *25,* 5950–5958.

Zhang, X.; Sui, C.; Gong, J.; Yang, R.; Luo, Y; Qu, L. Preparation, Characterization, and Property of Polyaniline/Prussian Blue Micro-Composites in a Low-Temperature Hydrothermal Process. *Appl. Surf. Sci.* **2007,** *253,* 9030–9034.

Zhao, G.; Feng, J. J.; Zhang, Q. L.; Li, C. P.; Chen, H. Synthesis and Characterization of Prussian Blue Modified Magnetite Nanoparticles and Its Application to the Electrocatalytic Reduction of H2O2. *Chem. Mater.* **2005,** *17,* 3154–3159.

Zhao, Z.; Xiong, Y.; Cheng, X.; Hou, X.; Yang, Y.; Tian, Y. M.; You, J.; Xu, L. Adsorptive Removal of Trace Thallium (I) from Wastewater: A Review and New Perspectives. *J. Hazard. Mater.* **2020,** *393,* 122378.

Zhdanov, R. I.; Podobed, O. V.; Vlassov, V.V. Cationic Lipid-DNA Complexes-Lipoplexes-for Gene Transfer and Therapy. *Bioelectrochemistry* **2002,** *58,* 53–64.

CHAPTER 20

Green Titanium Oxide Nanoparticles for Nanoremediation

ENEYEW TILAHUN BEKELE, YILKAL DESSIE SINTAYEHU, C. R. RAVIKUMAR and H. C. ANANDA MURTHY

Department of Applied Chemistry, School of Applied Natural Science, Adama Science and Technology University, Adama, Ethiopia

ABSTRACT

Several organic and inorganic pollutants have emerged as a result of various world factors including industrialization and populace growth, posing a series of risks to the natural environment and the health of human beings too. Through the last decades, numerous physicochemical wastewater treatment techniques that might be primary, secondary, and tertiary were adopted. However, those techniques have their own drawbacks and limitations. As a result, nanoremediation, which is defined as the use of designed materials to clean up the environment, is a potential technique for dealing with persistent chemicals originating from both organic and inorganic contaminants. Hence, green nanoparticles (NPs) synthesis has been a newly developed research trend in green nanoremediation technology because of its harmless nature, environmentally friendly, and efficient as compared to other counterparts. Plants, bacteria, algae, and fungi are among the biological agents that use the synthesis, although the green approach is garnering more interest across the world. Due to their great efficiency and biocompatibility, different types of green-based synthesized NPs are currently being employed in water-containing pollutants decontamination. As a result, the potential exposure of green synthesized titanium dioxide (TiO_2) NPs obtained using green alternative plant-based biological capping and reducing agents is discussed

Sustainable Nanoremediation: Modern Technologies for a Clean Environment.
Fernanda Maria Policarpo Tonelli, Arpita Roy, and Khalid Rehman Hakeem (Eds.)

in this book chapter. Furthermore, the advanced characterization techniques of green TiO_2 NPs such as thermal, structural crystallinity, optical, and functional group analysis have been well addressed. Applications of green TiO_2 NPs in environmental remediation such as green alternative catalysts for the decontamination of wastewater and as adsorbent for the remediation of water-contained pollutants are discussed. Finally, factors and parameters that affect the efficiency of green TiO_2 NPs in the decontamination of polluted water are also well addressed.

20.1 INTRODUCTION

Currently, scientists are on the way to creating a healthy environment to reduce the risk to the environment and human health (Del Prado-Audelo et al., 2021). Nowadays an environmental contamination sourced from extensive industrialization and intense agricultural activities is a series of issues worldwide (Roy et al., 2021). Hence, the fast improvement of industrial and economic development has led for the increase of energy consumption followed by emissions of toxic and poisoning gas species, and hazardous waste pollutants, into the atmosphere (Thangavelu et al., 2022). Contaminants are either microscopic or macroscopic-sized particles that result in the creation of a harsh ecosystem (Roy et al., 2021). In general, the primary sources of environmental pollutants are ranked as particulate matter, both ions of heavy and toxic metals, the compounds of pesticides and herbicides, fertilizers and toxic gases, industrial effluents, sewage, and organic compounds (Guerra et al., 2018). Therefore, the existence of these pollutants damages every living thing including humans (Roy et al., 2021). To treat those pollutants, environmental remediation and restoration to their original forms have now become a global focus for a healthy environment (Kalita and Baruah, 2020).

20.1.1 ENVIRONMENTAL POLLUTION

The fast expansion of the agroindustrial sector in both industrialized and emerging nations is one of the main causes of the worrying levels of environmental pollution in the globe today (Alayu and Yirgu, 2018). This causes a fast alteration of normal weather conditions that has a negative consequence for every living thing in the ecosystem. Environments such as air, soil, and water pollution have not only become the main serious risks and stress for human health but also for the entire ecosystem (Del Prado-Audelo et al., 2021; El-Ramady et al., 2017). The lack of cutting-edge and practical solutions that may be employed to lessen the current problem of environmental pollution and afterward to develop a healthy environmental system is the main rapidly growing concern in the modern society (Ganie et al., 2021). There is a widespread issue with contaminated soil and groundwater, in both industrialized and urban areas, which poses serious threats to both the environment and people (Alazaiza et al., 2021). So, for more understanding pollutants that affect air, soil, and water and their remediation solutions are discussed.

20.1.1.1 AIR POLLUTION

The pollution of atmospheric air is one of the most serious problems that has a negative impact on climate change and as well on human health. For this type of environmental pollution, the existence of particulate matter such as the oxides of nitrogen (NO_x), SO_2, CO, Pb, ground-level ozone, highly reactive and toxic organic compounds, dioxins, polycyclic aromatic hydrocarbons, and volatile organic compounds (VOCs) are all hazardous for mankind (Roy et al., 2021). Among, NO_x, SO_x, and VOCs as well as ammonia (NH_3) are considered as a secondary particulate matter precursors for air pollution (Del Prado-Audelo et al., 2021). Asthma, chest pain, congestion, throat inflammation, and cardiovascular disorders are only a few of the respiratory-related illnesses brought on by all these air pollution factors, which also affect other living things (Kalita and Baruah, 2020).

20.1.1.2 SOIL POLLUTION

Contaminated soil sourced from various industries and urban areas as well as agricultural sites has become a global issue that causes extreme harm to the environment and also human health (Alazaiza et al., 2021). Soil is highly affected by heavy metals from municipal solid waste (Vongdala et al., 2019), dentistry (Thangavelu et al., 2022), and surface sediment (Venkatramanan et al., 2018) sources. Such as Zhang et al. have shown the soil effect of hexavalent chromium, Cr(VI). Due to the high and severe toxicity and mobility nature in soils, the dangers of Cr(VI) to the environment and human health are significant (Zhang et al., 2019). Additionally, the cadmium and copper contamination in the soils of the dump and its vicinity reached the ecotoxicological concern level (Vongdala et al., 2019).

20.1.1.3 WATER POLLUTION

Water is a key natural resource on the planet, and its accessibility in its purest form is crucial for both human health and that of all living things. Without the intake of clean water, life as we know it would be unimaginable and impossible. Consequently, it is the main reason why there are not enough resources for safe drinking water (Das et al., 2020). Water (H_2O) is also known as the unique universal solvent due to its physicochemical properties such as solubility power, etc., which enables it to dissolve different types of solutes. A prevalent issue that affects many factories across the world is the release of untreated harmful chemical effluents into the soil and water bodies from businesses such as plastic, paper, medicines, leather, dyeing, and printing. Up to 20% of dyes are released during the dying and washing of clothing, which, if left untreated, causes the release of these persistent, carcinogenic, and poisonous compounds into the environment, especially in the textile sectors (Fulekar et al., 2018). Water pollution has a negative impact on the environment, including soil and air pollution, which has very serious effects on human health and has a negative effect on economic growth as well as the social outlook for the country in question (Roy et al., 2021). Due to this peculiar reason, the next section presents the causes of natural water contamination and its remediation/treatment mechanisms, specifically via the principle of green nanoremediation.

20.2 WASTEWATER AND ITS TREATMENT TECHNIQUES

20.2.1 SOURCE OF WASTEWATER AND ITS EFFECT

It is known that clean water consumption by humans and related living things has become one of the scarce and valuable resources of the 21st century. The common source for the contamination of water is the discharging of industrial effluents into the surface and groundwater without a strong management system. Most of these effluents contain inorganic and organic chemicals, biodegradable organic substances, toxic and heavy metal ions, and nutrients (Worku et al., 2018a, 2018b). Different types of heavy metal ions such as arsenic, lead, cadmium, mercury, nuclear waste, and pesticides contain organochlorines and organophosphates, fertilizers (nitrogenous fertilizers and phosphate fertilizers), and industrial dyes (azo dyes) are the most serious influents (Kalita and Baruah, 2020).

The existence of all pollutant sources has severe effects such as degrading the quality of water by decreasing the oxygen level and by altering the existing physicochemical composition, which might contribute to the damaging of aquatic ecosystems (Alemu et al., 2018), dental and skeletal fluorosis (Tor et al., 2009), a source of waterborne harmful bacteria (Das et al., 2020), a source of pathogens (Woldetsadik et al., 2017), carcinogenicity, methemoglobinemia, goiter, hypertension, birth malformations and anti-immune disorder in the human body; results in human health disorders (Kalita and Baruah, 2020).

20.2.2 TECHNIQUES OF WASTEWATER TREATMENT

Treatment of wastewater polluted by inorganic species might be performed via various conventional protocols such as chemical precipitation and ion exchange, adsorption and membrane filtration, coagulation–flocculation with flotation and also electrochemical approaches are some of them (Fu and Wang, 2011). However, most of them are inefficient in treating organic pollutants and their treatment performance even is too low. To solve these drawbacks, materials having nanosize ranges are found to be an effective and efficient way of strategy for the treatment of polluted water bodies. Due to the less costly and more effective uses of green nanomaterials, nanoremediation is a promising technology to clean up polluted media (Del Prado-Audelo et al., 2021). Hence, remediation of the various types of pollutants in the environment represents an advanced approach for the creation of safe and sustainable remediation of persistent nonbiodegradable organic compounds including pesticides, chlorinated solvents, brominated or halogenated chemicals, perfluoroalkyl and polyfluoroalkyl substances (PFAS) and finally the ions of both toxic and heavy metals are considered as the major once. So, nowadays, nanoremediation is one of the most promising emerging technologies to tackle environmental contamination, specifically dealing with recalcitrant contaminants (Ganie et al., 2021). Nanoremediation is the term used to describe the remediation process that uses nanomaterials and nanoparticles (NPs) to clean up various surroundings. This nanoremediation could be performed by numerous nanomaterials synthesized by using plants (phyto-nanoremediation), microbes (microbial nanoremediation), or animals (zoonanoremediation) (El-Ramady et al., 2017). In addition to its cost-effectiveness and environmentally friendly approach, enhanced sensitivity, high surface area and exceptional electronic properties and catalytic behavior of

NPs are the primary required mechanisms for wastewater remediation and pollutant degradation (Del Prado-Audelo et al., 2021).

20.3 NANOREMEDIATION OF WASTEWATER

Due to the innovative use of NPs in remediation techniques, scholars have been focusing on nanoremediation to block and solve formidable problems including pollution emergencies, contaminated land management and treatment, maintaining a saturated and balanced environment, which are suitable explanations for the quick and efficient treatment of the contaminated water sources. Therefore, using nanostructures in technology could greatly reduce the overall cost of large-scale contamination removal (Ganie et al., 2021).

20.3.1 GREEN NANOMATERIALS

Nanomaterials that have large surface area follow an eco-friendly green method of synthesis, and structural flexibility combined with novel chemical and physical properties in contrast to conventional approaches is termed as green nanomaterials (Khatami et al., 2018). They might be single (Nabi et al., 2020), hybrid (Khatami et al., 2018), and doped (Fazil et al., 2021) nanomaterials. Due to the use of green ecofriendly nature of reducing and stabilizing species, these nanomaterials are synthesized by using the biosynthesis method (Sila et al., 2019). So, due to the rapid, nontoxic with cost-effectiveness and environmentally eco-friendly alternative way of biosynthesis of NPs using various parts of green plant extract has expanded the novelty exploration by the researchers (Thakral et al., 2021). In addition, to plant extract, novel microbial mediated synthesis of nanomaterials is also being explored as a green alternative technique (Fulekar et al., 2018). Biomolecules present in different parts of plants and bacteria highly enhance the reduction and also the capping during synthesis (Verma et al., 2022). Since plant extract preparation is more time effective than microbe-based extracts, plant extract-based biosynthesis of nanomaterials has been given special attention for various applications. Maintaining organisms and optimum environmental conditions, such as temperature, pH, and other pertinent elements, is necessary for the safe mode of microbial development during the microbial templated synthesis of NPs. To develop a simple green synthesis approach, the plant extract mediated process is considered as a suitable method for the synthesis of green NPs (Krishnasamy et al., 2015).

20.3.2 GREEN NANOMATERIALS FOR WASTEWATER REMEDIATION

Remediation of wastewater which might be discharged from various industries and factories using green nanomaterials is cost-effective, safe for the environment, and it has enhanced treatment proficiency. Due to their tiny size and relatively high reactive surface area, green NPs among nanomaterials are of particular interest to scientists for the remediation of polluted water. Not only these, the presence of capping agents from plant extracts helps to prevent the formation of aggregation and is also used to attain the uniformity of the formed NPs beyond its environmentally friendly. Hence, green synthesis of nanomaterials is recommended as the superior and most attractive technique for the remediation of various contaminants from the wastewater environment (Pandey et al., 2020). Based on these, Singh et al. have reported on the green synthesis of TiO_2 NPs obtained in the presence of by using *Phyllanthus emblica* (Amla) leaves extract as a stable template. From their report, a new insight into the removal of Coralline Red dye with a maximum degradation efficiency of ~93% was recorded. They draw the conclusion that a platform for water filtration was created that was efficient, affordable, and sustainable (Singh et al., 2020).

Previously, Bhuyan et al. reported economical synthesis of green ZnO NPs in the presence of 25% (w/v) of *Azadirachta indica* leaf extract as a green templet for the treatment methylene blue dye present in polluted water. Because methylene blue dye is one of the main well-known water pollutants released by textile and related industries (Bhuyan et al., 2015). Later, Fulekar et al. have synthesized an effective green TiO_2 NPs by using Rhizosphere root extracts that are beneficial for methyl orange dye degradation for the polluted water remediation and then create a clean environment. Thus, the root synthesized TiO_2 NPs were reported to have degraded methyl orange dye with more than ~99% degradation in 120 min (Fulekar et al., 2018). In general, the use of green synthesized nano-based materials in waste-containing pollutant remediation are a very successful strategy for decomposing and eliminating resistant organic and inorganic contaminants (Verma et al., 2022). Hence, the next section clearly describes how metal oxide is going to be synthesized using green methods.

20.4 GREEN METAL OXIDE NANOPARTICLES

Over the previous decades, novel synthesis techniques of nanomaterials have become an interesting and very essential field in the area of nanoscience and technology. Of the various classes of nanomaterials, metal and metal oxide NPs have attracted the attention of researchers worldwide for diverse applications, specifically metal oxide nanoparticles (MONPs) fabricated via green protocols. Previous reports proved that the various classes of MONPs such as TiO_2, CuO, Co_3O_4, ZnO, MgO, Mn_2O, and Fe_3O_4 were synthesized using different types of green alternative templates as both capping and reducing agents for the targeted applications (Singh et al., 2018). A variety of live plants, algae, fungi, yeast, bacteria, and even viruses are among the biological resources that can be found in nature and used as efficient and environmentally beneficial templates for the synthesis of many MONPs. Since the availability of such various types of metabolites present within those green templates could be used during MONPs synthesis to prevent the formation of agglomeration and aggregation, control the size of NPs, and then obtain a uniformly homogenized surface NPs (Md Ishak et al., 2019). The chart shown in Fig. 20.1, shows the various synthesis techniques used for the preparation of MONPs.

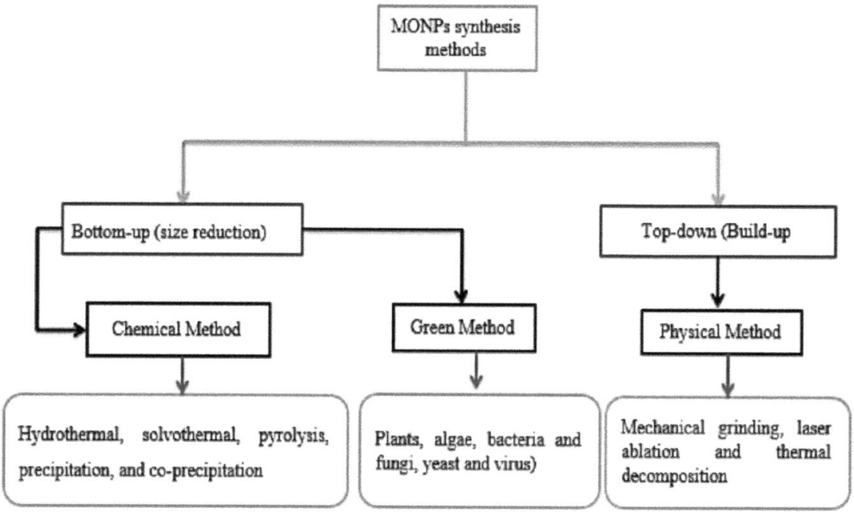

FIGURE 20.1 Synthesis techniques of MONPs (Tilahun Bekele et al., 2021).

Source: Reprinted with permission from Bekele et al., 2021. https://creativecommons.org/licenses/by/4.0/

20.4.1 SYNTHESIS OF TITANIUM OXIDE NANOPARTICLES VIA GREEN METHOD

TiO_2 NPs have been fabricated via various physicochemical approaches for the desired applications. However, those synthesis techniques involve the use of a large amount of pressure, temperature, too expensive, and toxic chemicals during the synthesis process as capping and reducing agents to control the overgrowth of the suspension. Instead, the green approach technique has been found to be the most approached and environmentally friendly for large-scale synthesis even with less or zero toxicity level. Due to the diversified nature of green plants, TiO_2 NPs synthesis in the presence of various parts of green plant extracts has been known as an interesting area worldwide as different plant species are being rapidly investigated (Hamed et al., 2021). Hence, Previously, TiO_2 NPs were synthesized using *Aegle marmelos* leaf extract and titanium isopropoxide precursor. The synthesized TiO_2 NPs were found to be spherical with an aggregated structure. The fabricated green TiO_2 NPs are used as an effective photo-catalyst for the degradation of nitroimidazole derivative drug ornidazole (ODZ) from wastewater (Ahmad et al., 2021). Moreover, *Syzygium cumini* leaf-mediated TiO_2 NPs were synthesized for the photocatalytic removal of Pb^{2+} from industrial polluted water. The synthesized TiO_2 NPs were characterized using numerous techniques and it has been reported that the synthesized MONPs were spherical with 10 nm average crystallite size. The high surface area green synthesized TiO_2 NPs were able to remove 82.53% of Pb^{2+} from the polluted water. This high percentage of removal was due to the enhanced physicochemical property of the green synthesized TiO_2 NPs (Sethy et al., 2020).

Furthermore, TiO_2 NPs were prepared within different composition using the root extract of *Kniphofia foliosa*. The green synthesized TiO_2 NPs of 1:1, 2:1, and 1:2 volume ratios were characterized via numerous characterization techniques. Green TiO_2 NPs were evaluated as an effective antibacterial agent. The green TiO_2 NPs were found to be thermally stable above 500°C with spherical structure having 10.2, 8.2, and 8.5 nm for the 1:2, 1:1, and 2:1, respectively. TiO_2 (1:1) NPs showed enhanced antibacterial activity against the targeted pathogens (Bekele et al., 2020). Moreover, the extract of *Calotropis gigantea* (L.) Dryand was investigated to keep and stabilize the bulk TiO_2 to TiO_2 NPs within 6 hours. The enhanced bioreduction potential was ascribed and possible due to the presence of phytochemicals in the leaf extract such as primary amines. The green obtained TiO_2 NPs provide an enhanced and improved acaricidal activity against the targeted species (Nadeem et al., 2018).

Current reports proved that an increased interest of the researchers toward the principle of green and environmental chemistry has led to the application of phytochemicals as green reducing agents for the synthesis of TiO_2 NPs for the desired applications. Extract of *Cicer arietinum* leaf was used to mediate the

TiO_2 NPs synthesis in the presence of titanium tetrachloride precursor salt. The report showed that the prepared green TiO_2 NPs is a highly crystal structure as confirmed via the characterization using Raman spectroscopy, XRD, TGA, TEM, and BET measurement system (Verma et al., 2022). In general, the green process of MONP synthesis such as TiO_2 NPs involves three important steps; reduction, nucleation, and capping (Aravind et al., 2021).

20.4.2 CHARACTERIZATION OF GREEN TITANIUM OXIDE NANOPARTICLES

After synthesis and before applying for the desired applications, confirming and gathering numerous physicochemical properties of metal oxide titanium oxide NPs could be done via thermogravimetric analysis–differential thermal analysis (TGA-DTA), X-ray diffraction analysis (XRD), scanning electron microscopy–energy dispersive X-ray spectroscopy (SEM–EDS), transmission electron microscopy (TEM) coupled with high-resolution transmission electron microscopy (HRTEM), selected area electron diffraction (SAED), ultraviolet–visible (UV–Vis) spectroscopy, diffuse reflectance spectroscopy (DRS), and Fourier transform infrared (FTIR) spectroscopy.

The TGA-DTA technique can be used to deal with the thermal stability of the synthesized TiO_2 NPs. This analysis technique is crucial for the identification of the various impurities and as well for the determination of calcination temperature before conducting other characterization techniques. In the previous report, TiO_2 NPs were synthesized using root extract of *Kniphofia schemperi* followed by TGA-DTA characterization (Bekele et al., 2021). The analysis showed that weight loss below 150°C was due to the removal of molecules of water from TiO_2 NPs, which might be introduced during synthesis process and washing steps. Moreover, the analysis depicts that within the 150–355°C range, the excess loss of weight within this range occurred which might result in due to the process of pyrolysis and carbonization from the surface of the green synthesized TiO_2 NPs, which could be attributed due to the use of root extract of *K. schemperi*. However, after 483.15°C, no significant weight loss was observed as supported in (Fig. 20.2a) ,and as a result, 500°C was chosen as calcination temperature.

Furthermore, TiO_2 NPs were green synthesized in the presence of *Aloe vera* leaf as a green template and characterized through the TGA-DTA technique (Venkatesh et al., 2015). The investigation revealed that the removal of the organic matrix, particularly bioactive molecules with O-H functional groups and little carbon, occurs around 235°C, approximately the same time that the physically and chemically bound water molecules are released. The observed exothermic region around 300°C is an indication of the crystallization of the TiO_2 NPs followed by the oxidation process. In the TGA curve (Fig. 20.2b), The additional weight loss is seen up to 420°C and is significantly correlated with the exothermic peak in the accompanying DTA curve. This association may be due to the sheath of biomass covering the TiO_2 NPs being completely removed.

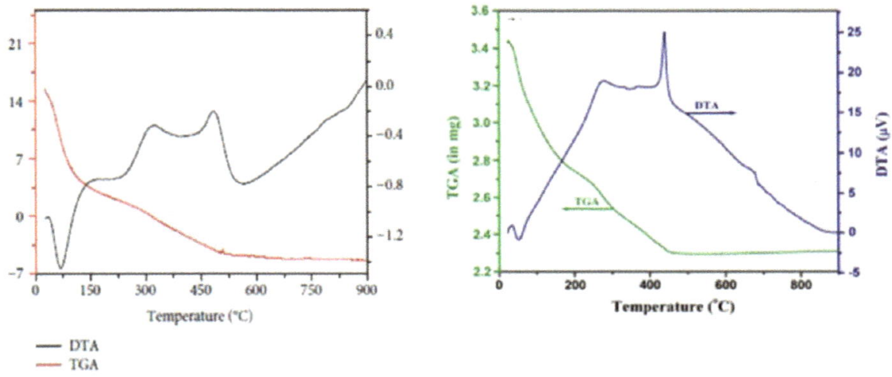

2a **2b**

FIGURE 20.2 TGA-DTA spectra of (a) root of *K. schemperi* and (b) *A. vera* leaf extract mediated synthesized TiO_2 NPs (Bekele et al., 2021) and (Venkatesh et al., 2015).

The degree of crystallinity nature and also the average crystalline size of green synthesized TiO_2 NPs could be obtained via XRD characterization technique. This technique of characterization also provides a degree of purity, since the absence of any secondary phase other than TiO_2 NPs is an indication of the purity of green synthesized TiO_2 NPs. TiO_2 NPs were synthesized by (Rajeshkumar et al., 2021) using suspension of *Andrographis paniculata* extract as an effective template as shown in Figure 20.3a. The XRD analysis depicts that the green TiO_2 NPs were found to be crystalline structures without any other additional distinctive peak, which might reveal the degree of purity. The crystalline size of these green TiO_2 NPs was estimated to be 22.97 nm. Moreover, studies showed that TiO_2 NPs were synthesized mediated by leaf extract of *Phyllanthus niruri* followed by XRD analysis (Shimi et al., 2022). The XRD result proved that the green fabricated TiO_2 NPs were found to be anatase phases having standard JCPDS card number 78-2486. As can be observed in (Fig. 20.3b), the absences of any foreign matter other than the targeted phases is a direct confirmation of the purity of the green MONPs. Furthermore, this analysis also depicts that the calculated average crystalline size was estimated to be 23 nm.

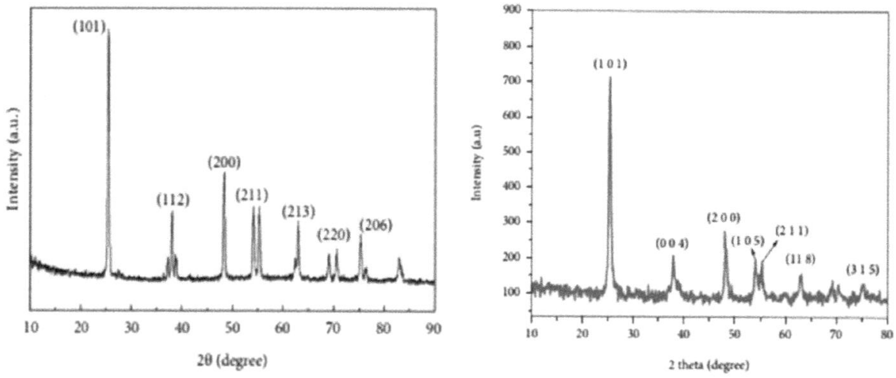

3a 3b

FIGURE 20.3 XRD spectra of (a) *A. paniculata* and (b) *P. niruri* leaf extract templated synthesized TiO_2 NPs (Rajeshkumar et al., 2021) and (Shimi et al., 2022).

Source: a) Reprinted with permission from Rajeshkumar et al., 2021. Copyright © 2021 S. Rajeshkumar et al. https://creativecommons.org/licenses/by/4.0/ b) Reprinted with permission from Shimi et al., 2022. Copyright © 2022 Annin K. Shimi et al. https://creativecommons.org/licenses/by/4.0/

For further investigations and analysis regarding the surface morphology and structure of green TiO_2 NPs, SEM–EDS characterization becomes the plausible technique. Previously, leaf extract of *P. emblica* mediated TiO_2 NPs were fabricated and characterized using FE-SEM at different magnification scales (Singh et al., 2020). The FE-SEM showed that green TiO_2 NPs have bunched spherical distorted structures. As can be confirmed from (Fig. 20.4a), the morphology of TiO_2 NPs possesses both agglomeration and aggregation with the presence of good dispersity. In addition, reports showed that TiO_2 NPs were synthesized in the presence of *Jatropha curcas* leaf extract. The prepared green TiO_2 NPs were then characterized via SEM technique (Goutam et al., 2018). The green TiO_2 NPs have spherical structures within a diameter range of 10–120 nm. The bioactive ingredients present within the extract such as polyphenols and flavonoids, alkaloids and terpenoids, free amino acids, and tannins present in the leaf extract of *J. curcas* played a key role as capping and reducing agents, which enhance the reaction of precursor salt of the NPs by reducing the formation of agglomeration and preventing the overgrowth of the suspension during synthesis process (Fig. 20.4b).

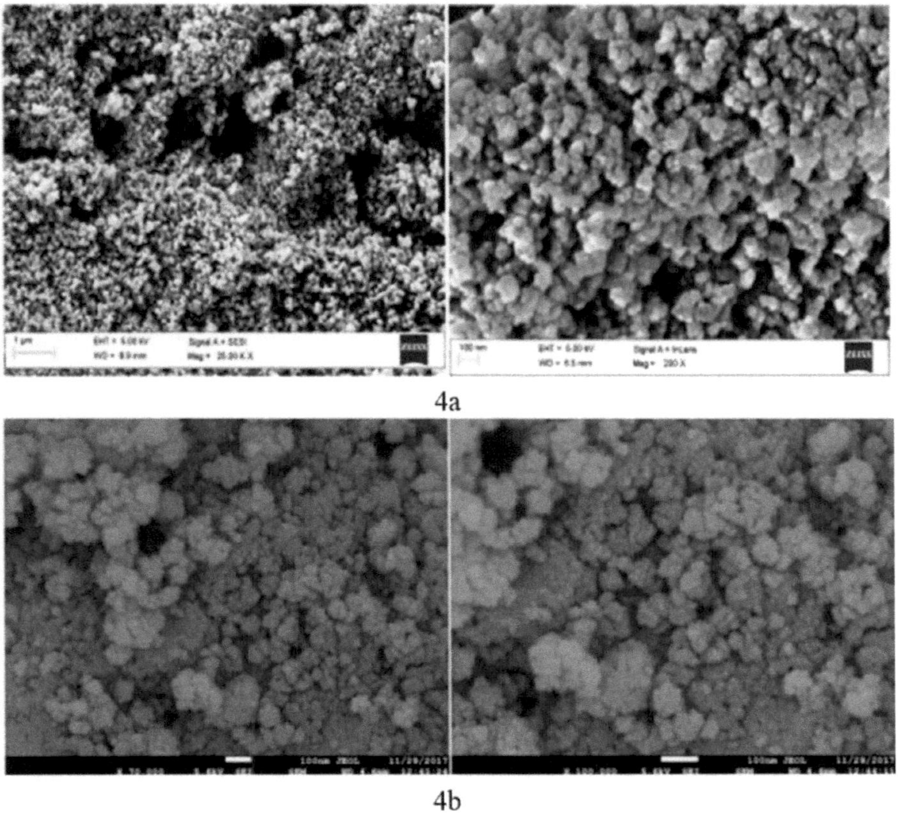

FIGURE 20.4 FE-SEM image of (a) *P. emblica* and (b) *J. curcas* leaf extract templated synthesized TiO_2 NPs.

Source: a) Reprinted with permission from Singh et al., 2020. Copyright © 2020 The Author(s). Published by Elsevier B.V. b) Reprinted with permission from Goutam et al., 2018. Copyright © 2017 Elsevier B.V.

The percent composition and as well as the degree of purity of green TiO_2 NPs might undergo EDS characterization technique. Previously, (Narayanan, Devi, et al., 2021) manufactured TiO_2 NPs using extract of *Pouteria campechiana* leaf. The TiO_2 NPs show up as the dominating signal in the EDS spectrum, and 58.44% of the reaction mixture was found to contain TiO_2 NPs, according to the signal strength. Additional minor peaks were observed, such as Ca, Mg, and Al, that might be contributed to the use of bioactive ingredients, and they served as a coating material for the synthesized TiO_2 NPs. Similar related works were also reported on the synthesis of TiO_2 NPs (Srinivasan et al., 2019) mediated by *Sesbania grandiflora* leaf extract. To confirm the composition, the green TiO_2 NPs were characterized by the EDS technique and the analysis proved that the strong signal of TiO_2 along with other weak signals existed, which might be found due to the biomolecules from the leaf extract that bound to the surface of TiO_2 NPs. The weaker signals observed (Fig. 20.5b) for the case of oxygen and chloride were contributed to the leaf extract.

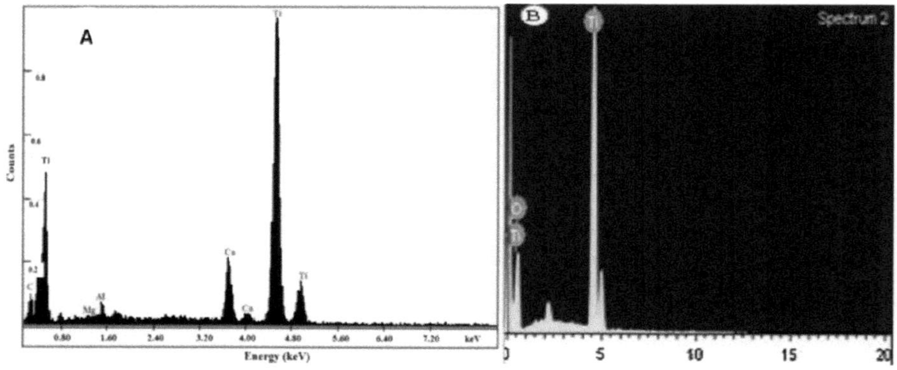

5a 5b

FIGURE 20.5 EDS spectra of TiO_2 NPs mediated by leaf extract of (a) *P. campechiana* and (b) *Sesbania grandiflora.*

Source: a) Reprinted with permission from Narayanan et al., 2021. Copyright © 2021 Elsevier Inc. b) Reprinted with permission from Srinivasan et al., 2019. Copyright © 2019 Elsevier Ltd.

Moreover, the internal structure and as well the crystallinity nature of metal oxide NPs are also confirmed by TEM, HRTEM, and SAED characterization. In addition, these techniques are employed to verify the particle size and particle distribution on the surface. In the previous report, TiO_2 NPs were synthesized in the presence of a leaf of *Cynodon dactylon*, as green alternative template. Then the green formed TiO_2 NPs were characterized via TEM, HRTEM, and SAED techniques as shown in Figure 20.6a. TEM and HRTEM analysis showed hexagonal and irregular shapes of TiO_2 NPs with moderate variation in size (13–34 nm). The SAED pattern of these green synthesized TiO_2 NPs was found to be anatase phase with good crystallinity. The white circular concentric rings on the center can be assigned to the corresponding nonspherical shape of anatase TiO_2 NPs (Hariharan et al., 2017). In the same manner, green TiO_2 NPs were obtained in the presence of *Cannabis sativa* (bhang) leaf extract and were characterized by TME-HRTEM and SAED analysis techniques. It has been found that the structural shape of green TiO_2 NPs was found to be spherical with anatase form (Fig. 20.6b). Moreover, from the HRTEM image, the calculated fringe width was estimated to be 0.35 nm. The SAED result supports that the presence of bright spherical rings clearly showed that the NPs are crystalline having an anatase phase (BinSabt et al., 2022).

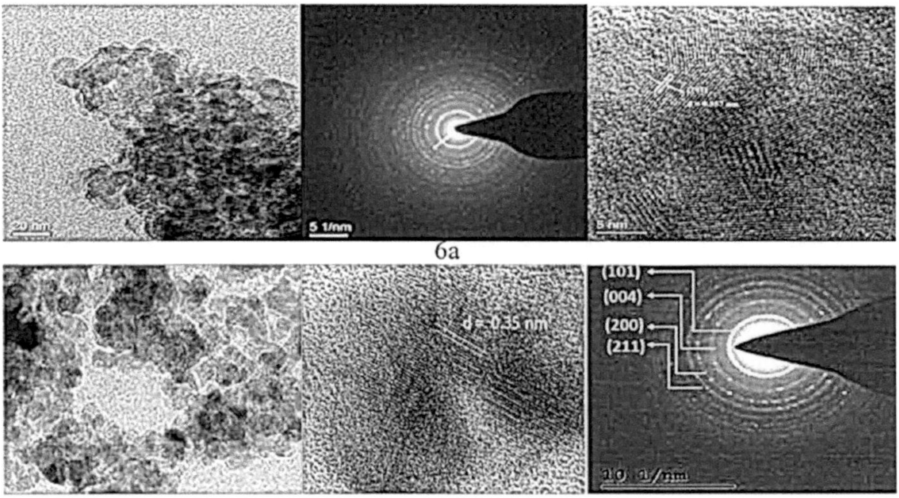

6b

FIGURE 20.6 TEM-HRTEM and SAED images of TiO_2 NPs mediated by leaf extract of (a) *C. dactylon* and (b) *C. sativa.*

Source: a) Reprinted with permission from Eneyew Tilahun Bekele et al., 2020. Copyright © 2020 by the authors. Licensee Hindawi Limited, United Kingdom. https://creativecommons.org/licenses/by/4.0/

b) Reprinted with permission from BinSabt et al., 2022. Copyright © 2022 by the authors. Licensee MDPI, Basel, Switzerland. https://creativecommons.org/licenses/by/4.0/

The optical behavior of metal oxide NPs are one of the crucial properties specifically for photocatalytic wastewater decontamination and as high surface area adsorbent. Reports showed that *Lagenaria siceraria* leaf extract mediated TiO_2 NPs were synthesized followed by optical property characterization in the range of 200–800 nm (Kaur et al., 2021). The spectrum of green TiO_2 NPs presents an intense absorption peak at the wavelength value of 258.5 nm, which suggests the formation of well-stable TiO_2 NPs. The calculated bandgap energy (E_g) was estimated to be 3.94 eV and as can be observed from the graph, it shows a blue shift; leading to the changes in the valence band and conduction band of the TiO_2 NPs utilizing *L. siceraria* aqueous leaf extract found in Figure 20.7a. (Nabi et al., 2020) have synthesized TiO_2 NPs using Cinnamon (*Cynoglossum zeylanicum*) bark extract. The absorbance was located within the wavelength range of 300 and 400 nm. The energy bandgap was determined from the Tauc plot and found to be 3.2 eV shown in Figure 20.7b.

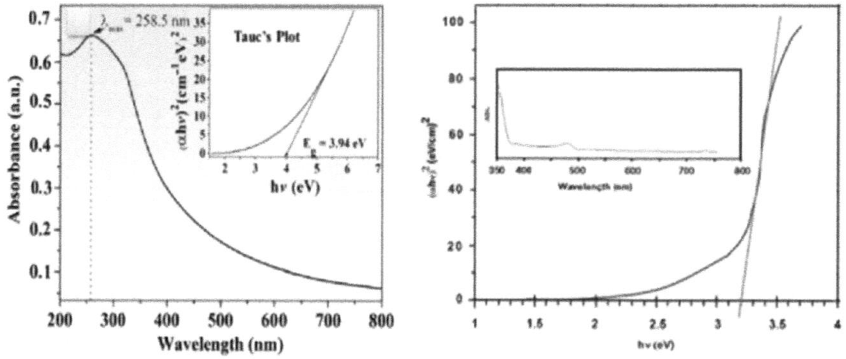

7a 7b

FIGURE 20.7 UV–Vis and the resulting Tauc plot of TiO_2 NPs mediated by (a) leaf of *L. siceraria* and (b) bark of *C. zeylanicum*.

Source: Reprinted with permission from Kaur et al., 2021. Copyright © 2020, Springer Science Business Media, LLC, part of Springer Nature; b) Reprinted with permission from Nabi et al., 2020. Copyright © 2019, Springer Science Business Media, LLC, part of Springer Nature

The influence of the plant extract during the synthesis process should be examined during the green synthesis of metal oxide NPs, such as TiO_2 NPs, using FTIR spectroscopy analysis. The TGA-DTA technique is also an option. This characterization technique could be used to identify the bioactive ingredients that were used to control the overgrowth of the TiO_2 NPs in the presence of various parts of green plant extracts. *P. niruri* leaf synthesized green TiO_2 NPs were reported by (Shanavas et al., 2019). The synthesized green TiO_2 NPs were then characterized by FTIR and this technique proved the presence of alkyl, amide, aldehyde, amino acid proteins, and terpenoids as found in Figure 20.8a. Moreover, the analysis also confirms that Ti–O–Ti is checked due to the presence of FTIR transmission peak around 400–800 cm?^{1} (Fig. 20.8b), which corresponds to the vibration of metal–oxygen bonding. Reports showed that the FTIR analysis technique not only showed the presence of phytochemicals, but also provide a hint in the formation of Ti–O–Ti. TiO_2 NPs were synthesized in the presence of leaf extract of *Coleus aromaticus* (Narayanan, Vigneshwari, et al., 2021). Furthermore, Ti–O–Ti stretching formation was observed at around 566.98 cm^{1}.

20.5 WASTEWATER REMEDIATION USING GREEN TiO_2 NANOPARTICLES

Currently, environmental pollution which is caused by numerous points and nonpoint sources might be attributed basically to the development of

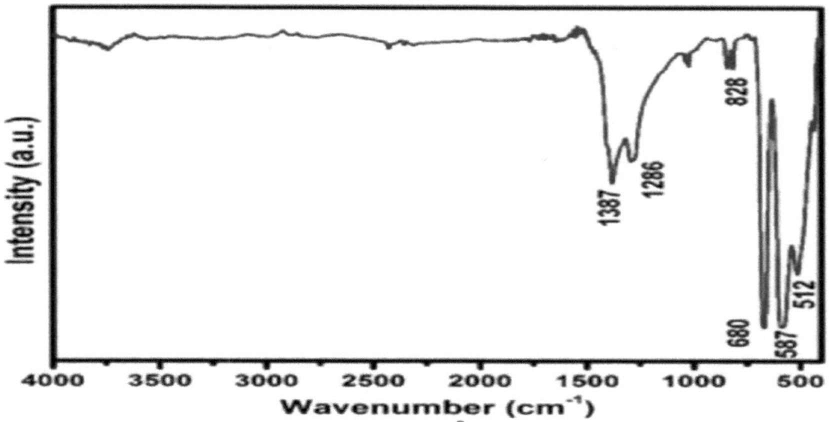

8a

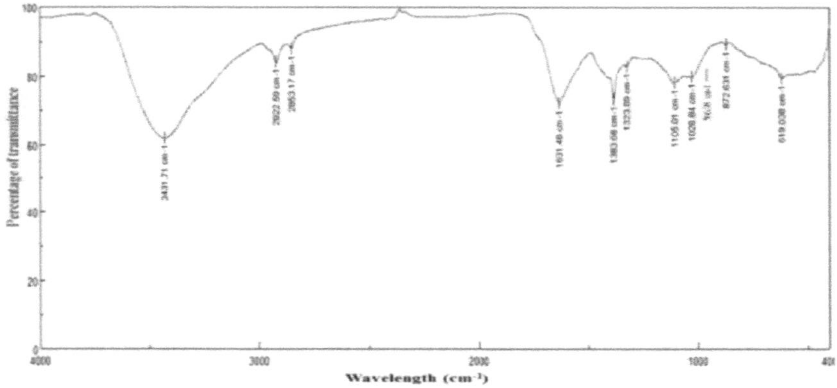

8b

FIGURE 20.8 FTIR spectra of green synthesized TiO_2 NPs in the presence of leaf of (a) *P. niruri* and (b) *C. aromaticus*.

Source: a) Reprinted with permission from Shanavas et al., 2019. Copyright © 2019 Elsevier Ltd; b) Reprinted with permission from Narayanan et al., 2021. Copyright © 2021 Elsevier Inc.

industrialization, urbanization, and the increment of the world's population size. This internship attracts researchers worldwide toward water pollution, air pollution, and soil pollution. The recent report proved that water contamination is the principal problem in accessing suitable drinking water resources, which needs the attention of researchers worldwide (Das et al., 2020). Previously, a number of miscellaneous techniques of wastewater treatment have been highly explored such as ion exchange, reverse osmosis, membrane filtration, adsorption processes, chlorination, chemical precipitation and electrochemical methods, sedimentation-flocculation and coagulation as well as molecular sieving (Angeler et al., 2019). However, those methods have a number of limitations, which limit their use for polluted water remediation. Instead, metal oxide nanomaterials, particularly MONPs obtained through green approach are found to be of prime importance in the science of wastewater remediation. They have increased capability for decontaminating and clearing away hazardous waste sites as well as for treating various contaminants that may be discharged from various chemical businesses.

Nowadays, bioremediation has been considered as one of the potential techniques for the treatment of water-containing contaminants. Since bioremediation involves the use of bacteria, fungi, algae, and enzymes, which needs advanced instrumentation and very well optimized conditions for the growth and preparation (Samuel et al., 2022). Instead, "nanoremediation" which is meant by the use of various classes of NPs for environmental remediations, such as to treat wastewater, soil, sediment, or other contaminated environments, has been considered as the most dominant technique (Sánchez et al., 2011).

20.5.1 WASTEWATER REMEDIATION THROUGH PHOTOCATALYSIS

Water that can be discharged from a number of chemical industries and factories that involves the use of toxic substances such as heavy and toxic metal ions, dyes, and organic contaminants at different stages of the production process should need to be treated before releasing into the environment (Pandey et al., 2020). So, now, the development of numerous wastewater remediation techniques specifically using green synthesized MONPs, and green nanoremediation, as cost-effective and efficient catalysts in the presence of light and as well as a high surface area adsorbent has been the currently most effective and green techniques.

Kaur et al. (2019) have synthesized TiO_2 nanocatalyst in the presence of leaf extract of *Carica papaya* as a template for the effective degradation of reactive orange-4 (RO-4) dye in the presence of ultraviolet (UV) light as a source. The degradation efficiency of green TiO_2 nanocatalyst was evaluated by taking 0.25-g catalyst at pH of 3.5 and 10.5. The adsorption-desorption equilibrium indicates that the optimum absorbance (1.964) was observed at around 489 nm and after the adsorption–desorption point, the absorbance decreased from 1.964 to 0.837 at pH 3.5 and 1.964 to 1.448 at pH 10.5 with 57.33% and 26.27% of RO-4 removal by adsorption in an acidic and basic environment, respectively. The report showed that as the dye was exposed to UV light, color fading was observed, which intensifies the degradation of the structure of the dye was takes place. After, 180 min of irradiation, 91.19% and 82.63% of degradation was achieved in acid and basic media, respectively.

Reports showed that TiO_2 NPs was *Acorus calamus* leaf extract as a green alternative catalyst for the photocatalytic degradation of rhodamine B (RhB) dye (Ansari et al., 2022). The maximum degradation efficiency of the green synthesized TiO_2 nanocatalyst showed 96.59% of RhB dye degradation after 120 min of UV light irradiation. It has been reported that the degradation efficiency of green TiO_2 nanocatalyst increased with increasing the irradiation time. Furthermore, studies depict that TiO_2 NPs were effectively synthesized using the aqueous extract of *P. emblica* leaf (Singh et al., 2020). Green TiO_2 NPs were investigated as green high surface area catalysts to degrade Corallene Red (CR) dye using UV light as a source. As irradiation time goes on, the absorbance of CR dye decreases, and maximum degradation efficiency (93%) was recorded after 140 min.

As many previous report supports, wastewater remediation in the presence of MONPs specifically, the TiO_2 NPs is found to be the most effective and too green alternative way even for the remediation of real wastewater that can be discharged from various factories and chemical industries. This intern confirms that various industries can manufacture different metal oxide NPs via the green method and scale up followed by treating the wastes before discharging them into the natural environment (Tilahun Bekele et al., 2021). Sethy et al. (2020) have synthesized TiO_2 NPs mediated by the extract of *S. cumini* leaf for the remediation of lead ion-containing water. The analysis showed that after 17 h treatment, a sharp minimization in Pb^{2+} concentration from 8.6 ppm to 1.5 ppm using TiO_2 nanocatalyst and in the presence of UV light as a source was achieved. The decrease in Pb^{2+} ion concentration during photolysis was contributed due to the degradation of Pb^{2+} ion concentration by using a strong UV light source.

Al-Hamoud et al. (2022) and his co-workers reported on the green synthesis of *Pulicaria undulate* leaf extracted templated synthesis of TiO_2 NPs for the photocatalytic decolorization of methylene blue (MB) and methyl orange (MO) dyes. The absorption peak of MB dye was observed at 664 nm followed by a gradual decrease and almost disappeared after 60 min irradiation. Moreover, the study proved that the degradation efficiency of green TiO_2 nanocatalyst toward MB was calculated to be 95%, which is also supported by a change in the color of the suspension from blue to colorless. The absorption spectrum of MO dye showed two characteristic bands, which were located at the wavelength range between 464 and 272 nm, attributed to the azo bond and phenyl ring of the MO dye, respectively. Similarly, the maximum Photodegradation of MO was achieved in the presence of green TiO_2 nanocatalyst after 120 min and was estimated to be 95%. From the report, it can be shown that 95% of MB and MO dye were degraded in the presence of green TiO_2 nanocatalyst at 60 and 120 min of irradiation time. The photodegradation of MO dye takes a longer irradiation time (120 min) as compared to MB dye photodegradation (60 min). This might be due to the existence of two bands in the case of MO dye, which takes much more time to break those bands relative to MB dye.

20.5.2 WASTEWATER REMEDIATION THROUGH ADSORPTION

In addition, water containing pollutants such as various organic dyes and ions of heavy and toxic metals can be also remediated in the presence of high surface green nanoadsorbents. The presence of such metal ions in excess and beyond the standard can cause carcinogenic, calcification of tissue, Wilson's disease, liver cirrhosis, hemochromatosis, kidney failure, corrosion of the gastrointestinal tract, and atrophy in the bone. So, developing environmentally being and cost-effective adsorbents to remediate natural water contaminates is one of the crucial issues worldwide (Irshad et al., 2021). Studies proved that the presence of lead (Pb) ions above 0.0 mg/L might contribute to the increase of blood pressure. Lead breaches the blood–brain barrier, causes kidney damage, and anemia, interferes with the calcium-regulated release of neurotransmitters, causes programmed cell death of the nervous system, causes digestive issues, neurological abnormalities, memory loss, and chronic pain in the muscles and joints in youngsters (Lu et al., 2022).

Reports support that TiO_2 NPs were fabricated in the presence of leaf of *Subabul* extract as a high surface area adsorbent for the decolonization of Solochrome Black (SCB), Titan Yellow (TY), Bromo Thymol Blue (BTB) and Bromo Cresol Purple (BCP) (Raju et al., 2007). The maximum % of removal was found to be 39% at pH 4 for SCB, 49% at pH 5 for TY, 45% at pH 6 for BTB, and 40% at pH 7 for BCP dyes, respectively. The analysis depicts that the optimum dosage was found to be selected as 5 mL and at this optimized dose, the maximum percent of removal was found to be 64, 76, 72, and 64% of SCB, TY, BTB, and BCP dye, respectively. The variation in the percent of removal for the same adsorbent but for different dyes might be attributed due to the variation in the structure of dye molecules, their bonding nature, and as well their charge variation. The enhanced remediations of those dyes are due to the larger surface area and availability of more sorption-active porous site of the green TiO_2 nanoadsorbent (Cao et al., 2014).

Moreover, Suriyaraj et al. have synthesized TiO_2 NPs using metal resistant Bacillus (NARW11) species as a template followed by calcination of the synthesized TiO_2 NPs at various temperatures and to investigate as an effective adsorbent for the improved removal of fluoride ion. All three phases of green synthesized TiO_2 nanoadsorbent showed enhanced adsorption of fluoride ions up to 80 min and beyond this time interval, the rate of adsorption decreases. The maximum fluoride ion removal (74.6%) was observed while using 0.5 mg/L adsorbate concentration. As the adsorbate concentration is increased from 0.5 to 4 mg/L, it has been found that the percentage of removal was constantly reduced up to 18.3%. This confirms that as the amount of the adsorbate is increased; the fluoride ion removal capacity gets decreased, which might be attributed due to the closing of pours sites of the adsorbent by the fluoride ion. Furthermore, as the calcination temperature is altered, the phase of green TiO_2 NPs also altered. So, the analysis proved that as the concentration of the adsorbate gets increasing from 0.5 to 4 mg/L, the percent of removal decreased from 78 to 20.8, 70.6 to 18.3, and 46.6 to 15% for the anatase, anatase mix rutile, and rutile phase adsorbents, respectively. This intern confirms that, as the phase is altered, the crystal structure of the adsorbent is also altered and this affects the adsorption capacity of the adsorbent (Suriyaraj et al., 2014).

The TiO_2 NPs were also synthesized via the simple sol-gel protocol using lemon leaf extract to modify the peel of orange (OP), lemon (LP), yam (YP), cassava (CP) for the remediation of nickel ion (Herrera-Barros et al., 2022). The adsorption capacities of the unmodified biomass of the peels varied in the range of 12-20 mg/g. While using with chemically modified green TiO_2 NPs orange, yam, cassava, and lemon peels biomass, the adsorption capacities were found to be 21.3, 18.01, 17.3, and 20 mg/g, respectively. The increase in the adsorption capacity of chemically modified green TiO_2 NPs biomass compared to the unmodified bass might be contributed due to the modification in the surface area and conductivity. Of the four biomass, OP modified with green TiO_2 NPs showed the highest adsorption capacity of nickel ion, which could be attributed due to the presence of highly cellulose-rich biomolecules in its structure. This intern showed that the available number of active sites required for the adsorption of nickel ions is increased (Chen et al., 2020).

20.6 CONCLUSION

Currently, environmental pollution has been found to be the most serious concern of the world society. Environmental pollution consists of air pollution, soil pollution, and water pollution. Of those classes of environmental pollution, water pollution has been taken as the most dominant series issue as compared to the relatives. Since pollution of water has a direct significant effect on the life and health of every living thing. Specifically, polluted water results in due to the development of industrialization, urbanization, and an increment in the population size of the world. Previously, researchers designed and developed numerous traditional wastewater treatment protocols. However, those techniques of wastewater decontamination protocols have a number of limitations and failures. Currently, the fabrication of nanomaterials, specifically green metal oxide nanoparticles such TiO_2 NPs has provided a promised approach for the decontamination of water-containing contaminants through catalysis, photocatalysis, and as an effective high surface area adsorbent. Due to their suitable physicochemical property, green TiO_2 NPs have been highly explored for the decolorization and decontamination of wastewater via the principle of photocatalytic degradation and then for the effective removal of carcinogenic organic dyes and heavy and toxic metals.

KEYWORDS

- **environmental pollution**
- **wastewater treatment**
- **nanoremediation**
- **green synthesis**
- **TiO2 NPs**

REFERENCES

Ahmad, W.; Singh, A.; Jaiswal, K. K.; Gupta, P. Green Synthesis of Photocatalytic TiO2 Nanoparticles for Potential Application in Photochemical Degradation of Ornidazole. *J. Inorg. Organomet. Polym. Mater.* **2021,** *31* (2), 614–623. https://doi.org/10.1007/s10904-020-01703-6.

Al-Hamoud, K.; Shaik, M. R.; Khan, M.; Alkhathlan, H. Z.; Adil, S. F.; Kuniyil, M.; Assal, M. E.; Al-Warthan, A.; Siddiqui, M. R. H.; Tahir, M. N.; Khan, S. T.; Mousa, A. A.; Khan, M. Pulicaria Undulata Extract-Mediated Eco-Friendly Preparation of TiO2 Nanoparticles for Photocatalytic Degradation of Methylene Blue and Methyl Orange. *ACS Omega* **2022,** *7* (6), 4812–4820. https://doi.org/10.1021/acsomega.1c05090.

Alayu, E.; Yirgu, Z. Advanced Technologies for the Treatment of Wastewaters from Agro-processing Industries and Cogeneration of by-products: A Case of Slaughterhouse, Dairy and Beverage Industries. *Int. J. Environ. Sci. Technol.* **2018,** *15* (7), 1581–1596. https://doi.org/10.1007/s13762-017-1522-9.

Alazaiza, M. Y. D.; Albahnasawi, A.; Ali, G. A. M.; Bashir, M. J. K.; Copty, N. K.; Amr, S. S. A.; Abushammala, M. F. M.; Al Maskari, T. Recent Advances of Nanoremediation Technologies for Soil and Groundwater Remediation: A Review. *Water (Switzerland)* **2021,** *13* (16), 2186. https://doi.org/10.3390/w13162186.

Alemu, K.; Assefa, B.; Kifle, D.; Kloos, H. Removal of Organic Pollutants from Municipal Wastewater by Applying High-Rate Algal Pond in Addis Ababa, Ethiopia. *Earth Syst. Environ.* **2018,** *2* (2), 377–386. https://doi.org/10.1007/s41748-018-0050-1.

Angeler, D. G.; Fried-petersen, H.; Allen, C. R.; Garmestani, A.; Chuang, W.; Donovan, V. M.; Eason, T.; Roberts, C. P.; Wonkka, C. L.; Survey, G.; Fish, N. C.; Management, R.; Agency, U. S. E. P.; Hall, K. EPA Public Access. *Adv.in Ecol. Res.* **2019,** *60* (218), 1–24. https://doi.org/10.1016/j.jhazmat.2020.123401.Green-synthesized.

Ansari, A.; Siddiqui, V. U.; Rehman, W. U.; Akram, M. K.; Siddiqi, W. A.; Alosaimi, A. M.; Hussein, M. A.; Rafatullah, M. Green Synthesis of TiO2 Nanoparticles Using *Acorus calamus* Leaf Extract and Evaluating its Photocatalytic and In Vitro Antimicrobial Activity. *Catalysts* **2022,** *12* (2). https://doi.org/10.3390/catal12020181.

Aravind, M.; Amalanathan, M.; Mary, M. S. M. Synthesis of TiO2 Nanoparticles by Chemical and Green Synthesis Methods and their Multifaceted Properties. *SN Appl. Sci.* **2021,** *3* (4), 409. https://doi.org/10.1007/s42452-021-04281-5.

Bekele, E. T.; Gonfa, B. A.; Zelekew, O. A.; Belay, H. H.; Sabir, F. K. Synthesis of Titanium Oxide Nanoparticles Using Root Extract of *Kniphofia foliosa* as a Template, Characterization, and Its Application on Drug Resistance Bacteria. *J. Nanomater.* **2020,** *2020,* 1–10. https://doi.org/10.1155/2020/2817037.

Bekele, E. T.; Zereffa, E. A.; Gultom, N. S.; Kuo, D. H.; Gonfa, B. A.; Sabir, F. K. Biotemplated Synthesis of Titanium Oxide Nanoparticles in the Presence of Root Extract of *Kniphofia schemperi* and Its Application for Dye Sensitized Solar Cells. *Int. J. Photoenergy* **2021,** *2021,* 1–12. https://doi.org/10.1155/2021/6648325.

Bhuyan, T.; Mishra, K.; Khanuja, M.; Prasad, R.; Varma, A. Biosynthesis of Zinc Oxide Nanoparticles from *Azadirachta indica* for Antibacterial and Photocatalytic Applications. *Mater. Sci. Semicond. Process.* **2015,** *32,* 55–61. https://doi.org/10.1016/j.mssp.2014.12.053.

BinSabt, M.; Sagar, V.; Singh, J.; Rawat, M.; Shaban, M. Green Synthesis of CS-TiO2 NPs for Efficient Photocatalytic Degradation of Methylene Blue Dye. *Polymers* **2022,** *14* (13), 2677. https://doi.org/10.3390/polym14132677.

Cao, C.; Xiao, L.; Chen, C.; Shi, X.; Cao, Q.; Gao, L. In Situ Preparation of Magnetic Fe3O4/Chitosan Nanoparticles via a Novel Reduction-Precipitation Method and their Application in Adsorption of Reactive Azo Dye. *Powder Technol* **2014,** *260,* 90–97. https://doi.org/10.1016/j.powtec.2014.03.025.

Chen, Y.; Shi, H.; Guo, H.; Ling, C.; Yuan, X.; Li, P. Hydrated Titanium Oxide Nanoparticles Supported on Natural Rice Straw for Cu (II) Removal from water. *Environ. Technol. Innov.* **2020,** *20,* 101143. https://doi.org/10.1016/j.eti.2020.101143.

Das, C.; Sen, S.; Singh, T.; Ghosh, T.; Paul, S. S.; Kim, T. W.; Jeon, S.; Maiti, D. K.; Im, J.; Biswas, G. Green Synthesis, Characterization and Application of Natural Product Coated

Magnetite Nanoparticles for Wastewater Treatment. *Nanomaterials* **2020,** *10* (8), 1–19. https://doi.org/10.3390/nano10081615.

Del Prado-Audelo, M. L.; García Kerdan, I.; Escutia-Guadarrama, L.; Reyna-González, J. M.; Magaña, J. J.; Leyva-Gómez, G. Nanoremediation: Nanomaterials and Nanotechnologies for Environmental Cleanup. *Front. Environ. Sci.* **2021,** *9* (December), 1–7. https://doi.org/10.3389/fenvs.2021.793765.

El-Ramady, H.; Alshaal, T.; El-Henawy, A.; Abdalla, N.; Taha, H.; Elmahrouk, M.; Shalaby, T.; Elsakhawy, T.; Omara, A. E.-D.; Elmarsafawy, S.; Elhawat, N.; Shehata, S. Environmental Nanoremediation under Changing Climate. *Env. Biodivers. Soil Secur.* **2017,** *1* (2017), 190–200. https://doi.org/10.21608/jenvbs.2017.1550.1009.

Fazil, A. A.; Narayanan, S.; Begum, M. S.; Manikandan, G.; Yuvashree, M. Green Synthesis Strategy for Producing Doped and Undoped ZnO Nanoparticles: Their Photocatalytic Studies for Industrial Dye Degradation. *Water Sci. Technol.* **2021,** *84* (10–11), 2958–2967. https://doi.org/10.2166/wst.2021.308.

Fu, F.; Wang, Q. Removal of Heavy Metal Ions from Wastewaters: A Review. *J. Environ. Manag.* **2011,** *92* (3), 407–418. https://doi.org/10.1016/j.jenvman.2010.11.011.

Fulekar, J.; Dutta, D. P.; Pathak, B.; Fulekar, M. H. Novel Microbial and Root Mediated Green Synthesis of TiO2 Nanoparticles and Its Application in Wastewater Remediation. *J. Chem. Technol. Biotechnol.* **2018,** *93* (3), 736–743. https://doi.org/10.1002/jctb.5423.

Ganie, A. S.; Bano, S.; Khan, N.; Sultana, S.; Rehman, Z.; Rahman, M. M.; Sabir, S.; Coulon, F.; Khan, M. Z. Nanoremediation Technologies for Sustainable Remediation of Contaminated Environments: Recent Advances and Challenges. *Chemosphere* **2021,** *275* (130065), 130065. https://doi.org/10.1016/j.chemosphere.2021.130065.

Goutam, S. P.; Saxena, G.; Singh, V.; Yadav, A. K.; Bharagava, R. N.; Thapa, K. B. Green Synthesis of TiO2 Nanoparticles Using Leaf Extract of *Jatropha curcas* L. for Photocatalytic Degradation of Tannery Wastewater. *Chem. Eng. J.* **2018,** *336,* 386–396. https://doi.org/10.1016/j.cej.2017.12.029.

Guerra, F. D.; Attia, M. F.; Whitehead, D. C.; Alexis, F. Nanotechnology for Environmental Remediation: Materials and Applications. *Molecules* **2018,** *23* (7), 1760. https://doi.org/10.3390/molecules23071760.

Hamed, M. T.; Bakr, B. A.; Shahin, Y. H.; Elwakil, B. H.; Abu-Serie, M. M.; Aljohani, F. S.; Bekhit, A. A. Novel Synthesis of Titanium Oxide Nanoparticles: Biological Activity and Acute Toxicity Study. *Bioinorg. Chem. Appl.* **2021,** *2021.* https://doi.org/10.1155/2021/8171786.

Hariharan, D.; Srinivasan, K.; Lc, N. Synthesis and Characterization of Tio2 Nanoparticles Using *Cynodon dactylon* Leaf Extract for Antibacterial and Anticancer (A549 Cell Lines) Activity. *J. Nanomed. Res.* **2017,** *5* (6), 4–8. https://doi.org/10.15406/jnmr.2017.05.00138.

Herrera-Barros, A.; Tejada-Tovar, C.; González-Delgado, Á. D. Green Nanoparticle-Aided Biosorption of Nickel Ions Using Four Dry Residual Biomasses: A Comparative Study. *Sustainability* **2022,** *14* (12), 7250. https://doi.org/10.3390/su14127250.

Irshad, M. A.; Nawaz, R.; Zia ur Rehman, M.; Adrees, M.; Rizwan, M.; Ali, S.; Ahmad, S.; Tasleem, S. Synthesis, Characterization and Advanced Sustainable Applications of Titanium Dioxide Nanoparticles: A Review. *Ecotoxicol. Environ. Saf.* **2021,** *212* (June 2022), 111978. https://doi.org/10.1016/j.ecoenv.2021.111978.

Kalita, E.; Baruah, J. Environmental Remediation. In *Colloidal Metal Oxide Nanoparticles*; Elsevier, 2020; pp 525–576). https://doi.org/10.1016/b978-0-12-813357-6.00014-0.

Kaur, H.; Kaur, S.; Kumar, S.; Singh, J.; Rawat, M. (2021). Eco-friendly Approach: Synthesis of Novel Green TiO2 Nanoparticles for Degradation of Reactive Green 19 Dye and Replacement of Chemical Synthesized TiO2. *J. Clust. Sci. 32* (5), 1191–1204. https://doi.org/10.1007/s10876-020-01881-w.

Kaur, H.; Kaur, S.; Singh, J.; Rawat, M.; Kumar, S. Expanding Horizon: Green Synthesis of TiO2 Nanoparticles Using Carica Papaya Leaves for Photocatalysis Application. *Mater. Res. Exp.* **2019,** *6* (9), 095034. https://doi.org/10.1088/2053-1591/ab2ec5.

Khatami, M.; Alijani, H. Q.; Nejad, M. S.; Varma, R. S. Core@shell Nanoparticles: Greener Synthesis Using Natural Plant Products. *Appl. Sci. (Switzerland)* **2018,** *8* (3), 411. https://doi.org/10.3390/app8030411.

Krishnasamy, A.; Sundaresan, M.; Velan, P. Rapid Phytosynthesis of Nano-sized Titanium Using Leaf Extract of *Azadirachta indica. Int. J. ChemTech Res.* **2015,** *8* (4), 2047–2052.

Lu, W.; Levin, R.; Schwartz, J. Lead Contamination of Public Drinking Water and Academic Achievements among Children in Massachusetts: A Panel Study. *BMC Public Health* **2022,** *22* (1), 107. https://doi.org/10.1186/s12889-021-12474-1.

Md Ishak, N. A. I.; Kamarudin, S. K.; Timmiati, S. N. Green Synthesis of Metal and Metal Oxide Nanoparticles via Plant Extracts: An Overview. *Mater. Res. Exp.* **2019,** *6* (11). https://doi.org/10.1088/2053-1591/ab4458.

Nabi, G.; Raza, W.; Tahir, M. B. Green Synthesis of TiO2 Nanoparticle Using Cinnamon Powder Extract and the Study of Optical Properties. *J. Inorg. Organomet. Polym. Mater.*, **2020,** *30* (4), 1425–1429. https://doi.org/10.1007/s10904-019-01248-3.

Nadeem, M.; Tungmunnithum, D.; Hano, C.; Abbasi, B. H.; Hashmi, S. S.; Ahmad, W.; Zahir, A. The Current Trends in the Green Syntheses of Titanium Oxide Nanoparticles and their Applications. *Green Chem. Lett. Rev.* **2018,** *11* (4), 492–502. https://doi.org/10.1080/17518253.2018.1538430.

Narayanan, M.; Devi, P. G.; Natarajan, D.; Kandasamy, S.; Devarayan, K.; Alsehli, M.; Elfasakhany, A.; Pugazhendhi, A. Green Synthesis and Characterization of Titanium Dioxide Nanoparticles Using Leaf Extract of *Pouteria campechiana* and Larvicidal and Pupicidal Activity on Aedes Aegypti. *Environ. Res.* **2021,** *200* (April), 111333. https://doi.org/10.1016/j.envres.2021.111333.

Narayanan, M.; Vigneshwari, P.; Natarajan, D.; Kandasamy, S.; Alsehli, M.; Elfasakhany, A.; Pugazhendhi, A. Synthesis and Characterization of TiO2 NPs by Aqueous Leaf Extract of *Coleus aromaticus* and Assess their Antibacterial, Larvicidal, and Anticancer Potential. *Environ. Res.* **2021,** *200* (April), 111335. https://doi.org/10.1016/j.envres.2021.111335.

Pandey, A.; Shukla, P.; Srivastava, P. K. Remediation of Dyes in Water using Green Synthesized Nanoparticles (NPs). *Int. J. Plant Environ.* **2020,** *6* (01), 68–84. https://doi.org/10.18811/ijpen.v6i01.08.

Rajeshkumar, S.; Santhoshkumar, J.; Jule, L. T.; Ramaswamy, K. Phytosynthesis of Titanium Dioxide Nanoparticles Using King of Bitter *Andrographis paniculata* and Its Embryonic Toxicology Evaluation and Biomedical Potential. *Bioinorg. Chem. Appl.* **2021,** *2021,* 1–11. https://doi.org/10.1155/2021/6267634.

Raju, C. A. I.; Prem, K.; Raju, P. R.; Kalyani, R. M.; Rao, J.; Student, P. G. Green Synthesis of Titanium Nano Particles and Its Applications for the Removal of Dyes from the Aqueous Solutions. *Int. J. Innov. Res. Sci. Eng. Technol.* **2007,** *3297,* 20543–20551. https://doi.org/10.15680/IJIRSET.2016.0512071.

Roy, A.; Sharma, A.; Yadav, S.; Jule, L. T.; Krishnaraj, R. Nanomaterials for Remediation of Environmental Pollutants. *Bioinorg. Chem. Appl.* **2021,** *2021,* 1–16. https://doi.org/10.1155/2021/1764647.

Samuel, M. S.; Ravikumar, M.; John, A.; Selvarajan, E.; Patel, H.; Chander, P. S.; Soundarya, J.; Vuppala, S.; Balaji, R.; Chandrasekar, N. A Review on Green Synthesis of Nanoparticles and Their Diverse Biomedical and Environmental Applications. *Catalysts* **2022,** *12* (5). https://doi.org/10.3390/catal12050459.

Sánchez, A.; Recillas, S.; Font, X.; Casals, E.; González, E.; Puntes, V. Ecotoxicity of, and Remediation with, Engineered Inorganic Nanoparticles in the Environment. *TrAC—Trends Anal. Chem.* **2011,** *30* (3), 507–516. https://doi.org/10.1016/j.trac.2010.11.011.

Sethy, N. K.; Arif, Z.; Mishra, P. K.; Kumar, P. Green synthesis of TiO2 Nanoparticles from *Syzygium cumini* Extract for Photo-Catalytic Removal of Lead (Pb) in Explosive Industrial Wastewater. *Green Proc. Synth.* **2020,** *9* (1), 171–181. https://doi.org/10.1515/gps-2020-0018.

Shanavas, S.; Priyadharsan, A.; Karthikeyan, S.; Dharmaboopathi, K.; Ragavan, I.; Vidya, C.; Acevedo, R.; Anbarasana, P. M. Green Synthesis of Titanium Dioxide Nanoparticles Using *Phyllanthus niruri* Leaf Extract and Study on Its Structural, Optical and Morphological Properties. *Mater. Today: Proc.* **2019,** *26* (July), 3531–3534. https://doi.org/10.1016/j.matpr.2019.06.715.

Shimi, A. K.; Wabaidur, S. M.; Siddiqui, M. R.; Islam, M. A.; Rane, K. P.; Jeevan, T. S. A. Photocatalytic Activity of Green Construction TiO2 Nanoparticles from *Phyllanthus niruri* Leaf Extract. *J. Nanomater.* **2022,** *2022,* 1–11. https://doi.org/10.1155/2022/7011539.

Sila, M. J.; Nyambura, M. I.; Abong'o, D. A.; Mwaura, F. B.; Iwuoha, E. Biosynthesis of Silver Nanoparticles from Eucalyptus corymbia Leaf Extract at Optimized Conditions. *Nano Hybrids Composites* **2019,** *25* (4), 32–45. https://doi.org/10.4028/www.scientific.net/nhc.25.32.

Singh, A.; Goyal, V.; Singh, J.; Rawat, M. Structural, Morphological, Optical and Photocatalytic Properties of Green Synthesized TiO2 NPs. *Curr. Res. Green Sustain. Chem.* **2020,** *3* (October), 100033. https://doi.org/10.1016/j.crgsc.2020.100033.

Singh, J.; Dutta, T.; Kim, K. H.; Rawat, M.; Samddar, P.; Kumar, P. "Green" Synthesis of Metals and Their Oxide Nanoparticles: Applications for Environmental Remediation. *J. Nanobiotechnol.* **2018,** *16* (1), 1–25. https://doi.org/10.1186/s12951-018-0408-4.

Srinivasan, M.; Venkatesan, M.; Arumugam, V.; Natesan, G.; Saravanan, N.; Murugesan, S.; Ramachandran, S.; Ayyasamy, R.; Pugazhendhi, A. Green Synthesis and Characterization of Titanium Dioxide Nanoparticles (TiO2 NPs) Using Sesbania Grandiflora and Evaluation of Toxicity in Zebrafish Embryos. *Proc. Biochem.* **2019,** *80* (January), 197–202. https://doi.org/10.1016/j.procbio.2019.02.010.

Suriyaraj, S. P.; Vijayaraghavan, T.; Biji, P.; Selvakumar, R. Adsorption of Fluoride from Aqueous Solution Using Different Phases of Microbially Synthesized TiO2 Nanoparticles. *J. Environ. Chem. Eng.* **2014,** *2* (1), 444–454. https://doi.org/10.1016/j.jece.2014.01.013.

Thakral, F.; Bhatia, G. K.; Tuli, H. S.; Sharma, A. K.; Sood, S. Zinc Oxide Nanoparticles: From Biosynthesis, Characterization, and Optimization to Synergistic Antibacterial Potential. *Curr. Pharmacol. Rep.* **2021,** *7* (1), 15–25. https://doi.org/10.1007/s40495-021-00248-7.

Thangavelu, L.; Veeraragavan, G. R.; Mallineni, S. K.; Devaraj, E.; Parameswari, R. P.; Syed, N. H.; Dua, K.; Chellappan, D. K.; Balusamy, S. R.; Bhawal, U. K. Role of Nanoparticles in Environmental Remediation: An Insight into Heavy Metal Pollution from Dentistry. Bioinorganic Chemistry and Applications, **2022,** *2022,* 1–13. https://doi.org/10.1155/2022/1946724.

Tilahun Bekele, E.; Gonfa, B. A.; Sabir, F. K. Use of Different Natural Products to Control Growth of Titanium Oxide Nanoparticles in Green Solvent Emulsion, Characterization, and Their Photocatalytic Application. *Bioinorg. Chem. Appl.* **2021,** 2021. https://doi.org/10.1155/2021/6626313.

Tor, A.; Danaoglu, N.; Arslan, G.; Cengeloglu, Y. Removal of Fluoride from Water by Using Granular Red Mud: Batch and Column Studies. *J. Hazard. Mater.* **2009,** *164* (1), 271–278. https://doi.org/10.1016/j.jhazmat.2008.08.011.

Venkatesh, K. S.; Krishnamoorthi, S. R.; Palani, N. S.; Thirumal, V.; Jose, S. P.; Wang, F. M.; Ilangovan, R. Facile One Step Synthesis of Novel TiO2 Nanocoral by Sol–Gel Method Using *Aloe vera* Plant Extract. *Indian J. Phys.* **2015,** *89* (5), 445–452. https://doi.org/10.1007/s12648-014-0601-8.

Venkatramanan, S.; Chung, S. Y.; Ramkumar, T.; Selvam, S. Ecological Risk Assessment of Selected Heavy Metals in the Surface Sediments of Three Estuaries in the Southeastern Coast of India. *Environ. Earth Sci.* **2018,** *77* (4). https://doi.org/10.1007/s12665-018-7294-9.

Verma, V.; Al-Dossari, M.; Singh, J.; Rawat, M.; Kordy, M. G. M.; Shaban, M. A Review on Green Synthesis of TiO2 NPs: Synthesis and Applications in Photocatalysis and Antimicrobial. *Polymers* **2022,** *14* (7), 1444. https://doi.org/10.3390/polym14071444.

Vongdala, N.; Tran, H. D.; Xuan, T. D.; Teschke, R.; Khanh, T. D. Heavy Metal Accumulation in Water, Soil, and Plants of Municipal Solid Waste Landfill in Vientiane, Laos. *Int. J. Environ. Res. Public Health* **2019,** *16* (1), 1–13. https://doi.org/10.3390/ijerph16010022.

Woldetsadik, D.; Drechsel, P.; Keraita, B.; Itanna, F.; Gebrekidan, H. Heavy Metal Accumulation and Health Risk Assessment in Wastewater-irrigated Urban Vegetable Farming Sites of Addis Ababa, Ethiopia. *Int. J. Food Contam.* **2017,** *4* (1). https://doi.org/10.1186/s40550-017-0053-y.

Worku, A.; Tefera, N.; Kloos, H.; Benor, S. Constructed Wetlands for Phytoremediation of Industrial Wastewater in Addis Ababa, Ethiopia. *Nanotechnol. Environ. Eng.* **2018a,** *3* (1), 9. https://doi.org/10.1007/s41204-018-0038-y.

Worku, A.; Tefera, N.; Kloos, H.; Benor, S. Bioremediation of brewery wastewater using hydroponics planted with vetiver grass in Addis Ababa, Ethiopia. *Bioresour. Bioprocess.* **2018b,** *5* (1), 39. https://doi.org/10.1186/s40643-018-0225-5.

Zhang, T.; Hu, L.; Zhang, M.; Jiang, M.; Fiedler, H.; Bai, W.; Wang, X.; Zhang, D.; Li, Z. Cr(VI) Removal from Soils and Groundwater Using an Integrated Adsorption and Microbial Fuel Cell (A-MFC) Technology. *Environ. Pollut.* **2019,** *252,* 1399–1405. https://doi.org/10.1016/j.envpol.2019.06.051.

CHAPTER 21

The Novel Technique and Capacity of Green-Gold Nanoparticles in Bioremediation

MEHREEN SHAH[1], SIRAJUDDIN AHMED[1], and ISHITA SHARAN SRIVASTAVA[2]

[1]*Department of environmental science and engineering, Jamia millia islamia, New Delhi, India*

[2]*Department of Environmental Science, Jawaharlal Nehru University, New Delhi, India*

ABSTRACT

"Green synthesis" is fast gaining popularity in the world of science as it offers a way to incorporate the ideas of eco-friendly and sustainability. In today's age of pollution as a primary issue it is important to devise new techniques that offer solutions to deal and mitigate this problem. Synthesis of gold nanoparticles (AuNPs) is achieved using reduction of sodium salts of gold with aqueous plant extracts. The extracts were obtained from various plants and bacterial, fungal, and yeast sources such as *Salvia officinalis, Escherichia coli, Pelargonium graveolens, Simarouba glauca*, and *Garcinia mangostana.* Since we are using gold nanoparticles for bioremediation purposes, we need to synthesize these particles in a way that they do not cause harm to the environment. They must be synthesized in a way that does not cause harm to the environment. Green-gold nanoparticles are synthesized by the use of plant extracts and reagents that do not cause pollution. Various compounds in the plant like saponins, proanthocyanidins are utilized for the reduction of gold nanoparticles, hence producing gold nanoparticles that do

Sustainable Nanoremediation: Modern Technologies for a Clean Environment.
Fernanda Maria Policarpo Tonelli, Arpita Roy, and Khalid Rehman Hakeem (Eds.)

not release toxins in the environment in the by-process of its synthesis. The characterization and analysis was then done by X-ray diffraction and UV analysis.

The results of the antibacterial lab analysis showed inhibition of *E. coli* indicating the bactericidal capacity of the green-gold nanoparticles. The degradation period was very short compared to other previous remediation tools affirming success of this method. Green-gold nanoparticles leave less polluting residues compared to conventional remediation tools like earlier used in wastewater management. Those conventional remediation techniques often require very high cost for post-processing and often cause pollution once released in water bodies. Hence, scientists and researchers around the globe have now worked on newer remediation tools that work with higher efficiency and also reduce post-processing costs and can later be disposed of without being deleterious to the environment in which they are being released. Green-gold nanoparticles also have an effect on degradation of synthetic dyes and detection of heavy metal ions. It also displayed a faster catalytic speed for degradation of pathogens.. Hence, green-gold nanoparticles show great potential for water remediation and are a unique solution for the future.

21.1 INTRODUCTION

From the last decade, new synthesis approaches and newer still methodologies for the creation of nanomaterials have been devised. This is a new avenue for environmental and chemistry research and offers new frontiers in technology. To obtain nanoparticles of requisite morphology, two main techniques are involved in synthesis namely the top-down and the bottom-up method. In the top-down method, nanoparticles are prepared through various methodologies and lithographic techniques which involve etching, sputtering, and ball-milling. In the bottom-up approach methods like sol–gel processes, laser pyrolysis, chemical and vapor deposition, spray pyrolysis, and molecular and atomic condensation are involved. They can be a victim of the following limitations such as sheer lack of stability in the medium of generation, lack of coherence in mechanisms, and intense levels of bioaccumulation in tissues of animals post release and toxic by-products, higher cost of processing and development, and issues in the recycling and disposal of the nanoparticles once consumed in remediation processes, recycle/reuse/regeneration.

These limitations are opening newer development in this emerging field of green environmental friendly research and development. Green synthesis helps us to prevent the detrimental loss to the environment that is caused due to the conventional methods that create pollution, hence defeating the original purpose of using nanoparticles for remediation. Green-gold nanoparticles offer ways to aid bioremediation of polluted water by its anti-oxidant activity, catalytic reduction, degradation of synthetic dyes in water, removal and detection of heavy metal and antibacterial activity. Some basic principles of "green synthesis" can be elucidated by several techniques like reduction of pollutants, use of safer chemicals for processing of ore into nanoparticles and remediation of local water bodies, and the use of safe reagents.

"Green synthesis" is a warranted way to reduce generation of toxic and harmful by-products through safe and sustainable, and eco-friendly generation of nanoparticles. Green synthesis of metallic nanoparticles has been used to include biologically originating materials such as fungi, plants, bacteria and endophytic associations found in plants. Using green nanotechnology we can also reduce economical costs that are involved in the synthesis of chemical reagents in industrial factories.

21.2 GREEN-GOLD NANOPARTICLE FOR BIOREMEDIATION

Compared to conventional methodologies of remediation, gold nanoparticles are having an edge over other methods due to their faster rate of degradation of pollutants, lysis of microbial cell walls, catalytic degradation of synthetic dyes among many various ways of remediation that are discussed in detail in this paper. As gold is an inert metal and resists corrosion despite a prolonged contact in water, it, once present in the supernatant solution, offers no adverse risk of toxicity or damage to animal health by bioaccumulation (uptake in ppm level in animal tissue). Owing to its numerous advantages in terms of economic feasibility, large-scale production, low-skilled labor, environmental-friendly, nontoxicity, easy operation, nonpolluting byproducts, and relative abundance of biologically compatible additives, it is highly preferred by scientists across the globe. Biologically synthesized Au-NPs have a greater antibacterial potency than synthetic AuNPs which are created using chemicals that are harmful to the environment. Several experiments conducted have confirmed that AuNPs synthesized from plant extracts are effective in disarming pathogens and causing death of harmful microbes or the test model organism compared to those gold nanoparticles synthesized by chemicals previously or conventionally.

Gold nanoparticles help in catalytic reduction of a major pollutant derivative of synthetic dye, 4-Nitrophenol into 4-Aminophenol. This helps in cleaning up polluted water and prevents toxic effects of such water when consumed by humans. Gold nanoparticles also help to detect heavy metals for example mercury, gallium, palladium, cadmium etc in water. Such detection is essential for remediation of water so as to improve the quality of water. Without proper detection of the presence of heavy metal ions in water no remediation technique is effective. The success of green-gold nanoparticles can be affirmed by using UV analysis by the spectroscopic method, X-ray diffraction, electron microscopy, and various other ways of analysis that can ensure the proper synthesis of gold nanoparticles and also affirm the presence of proper by-product formation in the waterbody.

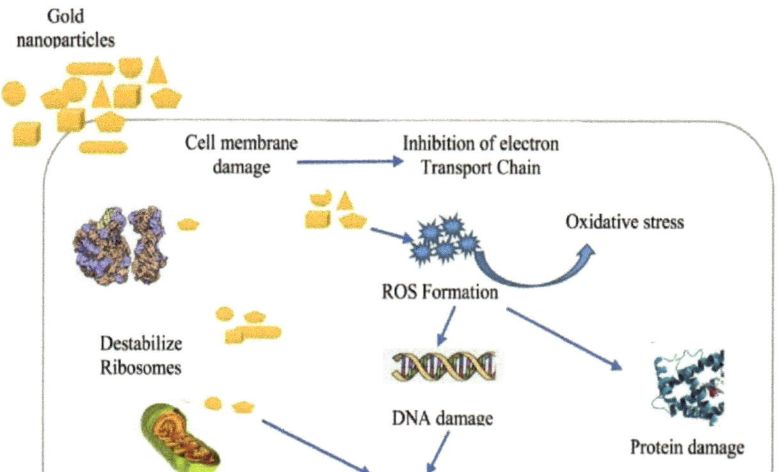

FIGURE 21.1 Depiction of antibacterial mechanisms adopted by gold nanoparticles in cell.

Source: Reprinted from Akintelu et al., 2021. https://creativecommons.org/licenses/by/4.0/

Gold nanoparticles exhibit antibacterial activity and hence are a solution to combat antibiotic-resistant strains of pathologically malignant bacteria that pose a rise of spread of contagious diseases. From lysis of cell membrane, to denaturation of cellular protein, damage to DNA, ribosomes, mitochondria etc various mechanisms are employed by the use of gold nanoparticles that lead to death of the bacteria cell and hence provide new techniques in remediation of water that is polluted by the presence of active bacteria which can cause spread of epidemic causing bacteria which pose a fatal risk to human health and environmental safety.

21.3 TECHNIQUES AND WAYS FOR BIOREMEDIATION

Bioremediation refers to clearing up of polluted and degraded water and soil by the use of nature and nature-derived biologically active/attenuated/extracted compounds or species. We are well aware of importance of aerobic and anaerobic bacteria in remediation of wastewater by municipalities.

Bioremediation can range from use of chitosan derived from fungi to various nanoparticles such as gold nanoparticles for improving the quality of water and polluted soil. Bioremediation can be achieved using organic acids, biopolymers derived from fungal sources, natural peptides derived from silk moth and of course as elucidated by this study, nanoparticles particularly gold nanoparticles.

- Microbial bioremediation uses microorganisms to digest organic waste matter and feed on the carbon substrate.
- Mycoremediation uses fungal origin enzymes to cause degradation of waste products such as hydrocarbons, pesticides, carbon matter, and both toxic and nontoxic heavy metals.
- Phytoremediation uses plants to break and remove such as metals, petroleum by-products, and harmful pollutants such as hydrocarbons, pesticides, or and chlorinated solvents.

Bioremediation can be both in-situ or ex-situ. This means in the former case, the contamination is cleared at the very site of generation and in the latter case, it is picked up and treated later at a facility.

In general, there are three types of bioremediation:

- Biostimulation—Microbes that are used for remediation are supplicated with chemicals and vital nutrients that activate them.
- Intrinsic Bioremediation—By using the application of the local microorganisms to the polluted area there is a breakdown and conversion of dangerous products into harmless ones.
- Bioaugmentation—Bacteria is allowed to replicate in the contaminated area where it then starts to break down the pollutants.

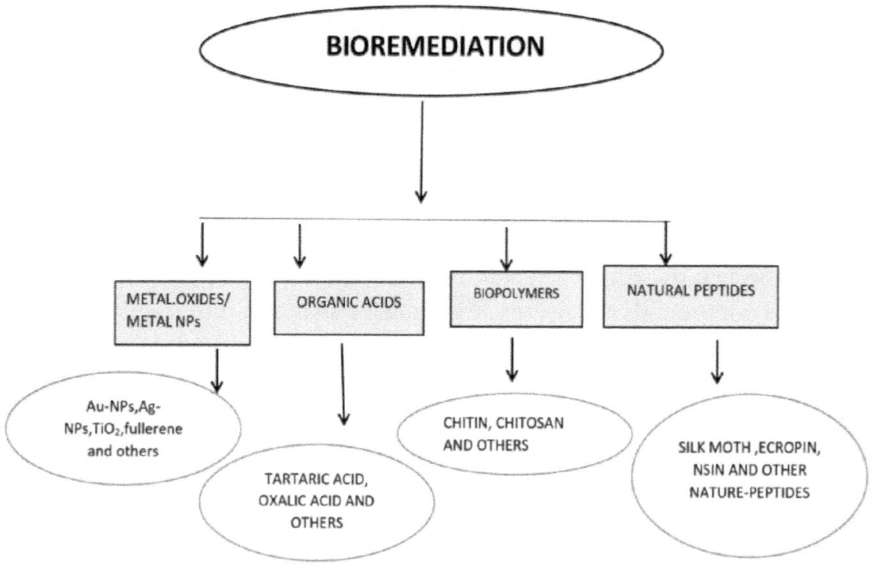

FIGURE 21.2 Depiction of various sources that enable the techniques of bioremediation.

21.4 APPLICATION IN ENVIRONMENTAL REMEDIATION

21.4.1 ANTIOXIDANT ACTIVITY

Gold nanoparticles offer a high speed of catalytic reduction of pollutants by the ability of their small size. Gold nanoparticles display good antioxidant ability which is referred to the ability to combat the generation and stop spread of free radicals. Free radicles such as nascent or free oxygen wreak havoc by subsequent chains of cascade reactions which lead to biologically degraded animal and cellular tissues. How effectively AuNPs demobilize the free radicals is governed by specific surface, size, and concentration as shown in an experiment by Singh et al. (2018). Furthermore, it is equally stated that AuNPs that have a high surface-to-volume ratio exhibit higher antioxidant activity. 2,2-diphenyl-1-picryl-hydrazyl-hydrate (DPPH) assay of AuNPs obtained from *Allium cepa* or common onion displayed an effective antioxidant activity of 14.44% at 100 Mg/mL as displayed in an experiment by Akintelu et al. (2021).

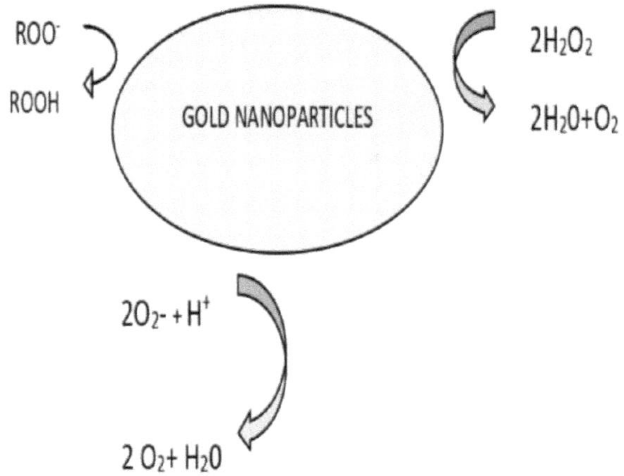

FIGURE 21.3 Depiction of antioxidant potency of green Au-NPs.

21.4.2 ANTIMICROBIAL ACTION

Green Au-NPs show a great affinity for pollutants such as microbes ranging from bacterial cells to protozoan sources. They exhibit a greater efficiency against gram-negative bacteria. The rate of bactericidal activity increases with the decrease in the size of the nanoparticles. Both size and dosage affect the activity. The smaller the size of the nanoparticles, the more is the rate of efficiency. Microbes are becoming a wide problem particularly the highly contagious ones that are precursors to infections and pandemics. Lot of bacterial strains are gaining resistance through natural selection to antibiotic strains and such strains can lead to occurrences of disease and epidemics. In-vitro studies are performed and metallic nanoparticles effectively cause death to microbes by various pathways. Nanoparticles cause cell lysis and apoptosis by making holes in the cell wall. They also cause denaturation of cellular proteins, uncoiling and damage to cellular DNA. They damage the mRNA. They stop cellular and genetic transcription. They cause disruption of enzymes by blocking the receptor sites.

There is damage to peptidoglycan and cellular biological cellular membranes. Many bacterial strains have genes that confer penicillin-resistance, sulfonamide-resistance, tryptophan-resistance etc. When these bacterial strains are treated with antibiotics, these do not die and only become more resistant over time and pose a great risk to human health and safety. Gold nanoparticles are nontoxic. Like most nanoparticles these generate oxygen-reactive species, which further enables the antibacterial action by immediate rupture of the peptidoglycan cell wall, disruption of bacterial replication mechanism, disruption of enzymes, and inhibition of biofilm formation.

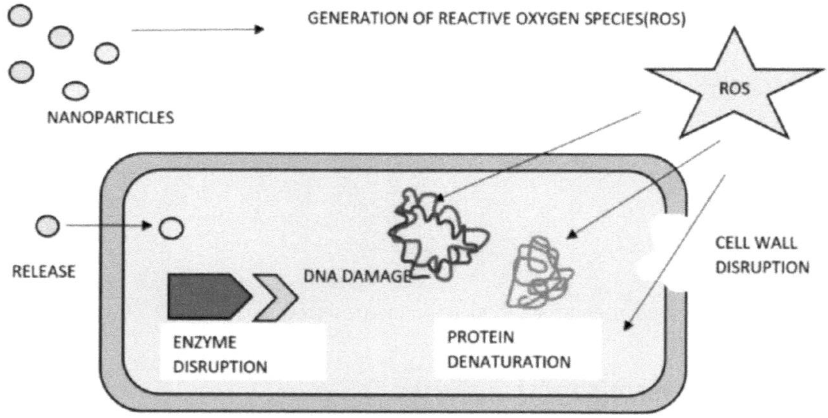

FIGURE 21.4 Depiction of antibacterial action and mechanism of gold nanoparticles on cellular contact.

21.4.3 REMOVAL OF POLLUTING DYES

Gold nanoparticles show absorbance for both visible light and ultraviolet light (UV) which leads to oxidation reaction for the synthetic dye present in the sample. AuNPs exhibit high catalytic action which results in the degradation of Congo red and methylene blue dye. Proanthocyanidin-chelated Au-NPs display a fast degradation for synthetic dyes such as methylene blue. This offers a remedy for the remediation of polluted wastewater that is degraded due to the presence of synthetic dyes present in it. Numerous industries use synthetic dyes such as paper, textile, pharmaceutical, food industries, hair coloring, toy production etc. Methylene blue (MB) is a cationic dye and Congo red (CR) is an anionic dye, both of which are notorious for polluting water.

If these dyes are not removed from the discharge of the industries then they can cause havoc in the delicate water ecology. Scientists have devised newer techniques for the degradation of synthetic dyes in water and Au-NPs have the perfect ability to cause catalytic reduction of these dyes in water, hence enabling the scientists to deal with polluted discharge from various industries. There are dipole–dipole interactions between the nitrogen group or N present in methylene blue and the phenol groups or --OH that are present via proanthocyanidins chelated to the Au-NPs.

Green synthesis of AuNPs has been achieved using a variety of plant secondary metabolites such as saponins including those extracted from plants such as *G. mangostana*, salvinia, etc. Removal of synthetic dyes is highly important to achieve remediation of polluted water as such water, if consumed, can cause damage to human health and under some conditions may even prove to be fatal.

21.5 HEAVY METAL ION DETECTION IN AQUEOUS MEDIUM

Heavy metals are a serious problem in wastewater management. Heavy metals wreak havoc on both human and animal health. They can cause problems such as methemoglobinemia, oxygen deficiency, cancer among many problems. Heavy metals such as Cr, Ni, Zn, Cu, Zn, Hg, and Cd can pollute water for several years and such water if consumed can be greatly toxic to health. Such water must be checked for the presence of heavy metal ions before any consumption can take place. Water that is fit for drinking is called potable water. Common methods used for the removal of heavy metal ions from aqueous solutions are numerous and include chemical precipitation, ultrafiltration, chemical oxidation/reduction, reverse osmosis, electrodialysis, ion exchangers, etc. But these conventional methodologies come with a set of challenges and limitations. It is imperative and highly important to develop new methods for the economical, high-efficiency, and environmentally friendly ways to minimize the chemical by-products and toxin compounds released due to these conventional methods and allow new techniques which lead to the protection of the environment and foster ecological balance. Proanthocyanidins-chelated gold nanoparticles display a high efficiency for the heavy metal ions and synthetic dyes removal, which indicate possible solution of advancement of bioremediation device for bioremediation.

Heavy metals can seep into groundwater and local water bodies due to mining activities, plastic, dye industries, vehicular emissions etc. Even at trace levels or ppm concentrations heavy metals such as mercury, cadmium, lead etc can be highly toxic. Detection is vital for proper remediation to take place.

Optical properties and refractive index of gold nanoparticles are helpful for heavy metal ion detection. Analysis using conventional techniques is time-consuming, expensive, skill-dependent, and highly complicated. Gold nanoparticles offer cheaper and more efficient methods for heavy metal ion detection. Colorimetric methods are usually employed where nanoparticles are helpful for heavy metal detection in water.

21.5.1 CATALYTIC REDUCTION

A synthetic compound that is widely indispensable for manufacture of pesticides, insecticides, drugs, fungicides, and leather tanning is 4-Nitrophenol. But if consumed or inhaled it can cause harmful effects on human health such as cyanosis or blueing of mouth, dizziness, headache, and nausea. Due to its toxic and damaging nature on health on humans and subsequently on ecology and yet its need in industrial manufacture it is a great issue for environmentalists. 4-nitrophenol is a cause of concern for scientists and environmentalists worldwide. It is important to cater to the reduction of 4-Nitrophenol before it can be properly disposed of without leading to pollution of natural water bodies or seepage into groundwater. When we reduce 4-nitrophenol the product so formed of the reduction is called 4-aminophenol, which is industrially important for manufacture of sulfur dyes, paracetamol (a therapeutic drug), rubber, natural antioxidants, preparation of black/white films, analgesic and antipyretic and stoppage of corrosion.

One way to reduce 4-nitrophenol is by using NaBH4 as a reductant and adding Au-NPs as a metal catalyst. Potential difference between donor (H3BO3/NaBH4) and nitrophenolate ions or the acceptor molecule may cause a requirement for a higher activation energy barrier which must be reduced so that the reaction may occur at a lower energy provided at substrate level and rate of reaction be at its utmost optimum speed.

Rate of reaction is significantly increased by increasing the adsorption of reactants on their surface as metallic NPs have a high surface/volume ratio, thereby lowering the activation energy significantly. 4-nitrophenol was characterized by a sharp band at 400 nm in its UV-visible spectra, hence affirming the success of usage of metallic nanoparticles for reduction of 4-Nitrophenol as performed in an experiment by Ganapuram et al. (2015). Green Ag-NPs (synthesized by plant stem extract) lead to a fast display in the absorption intensity at 400 nm, and a band at 313 nm indicates the formation of 4-aminophenol hence proving catalytic reduction potential of Au-NPs.

By subsequent lowering of the activation energy of the reaction we are able to initiate the enzyme-mediated reaction at a faster speed, which, however, can only be achieved if we use a catalyst.

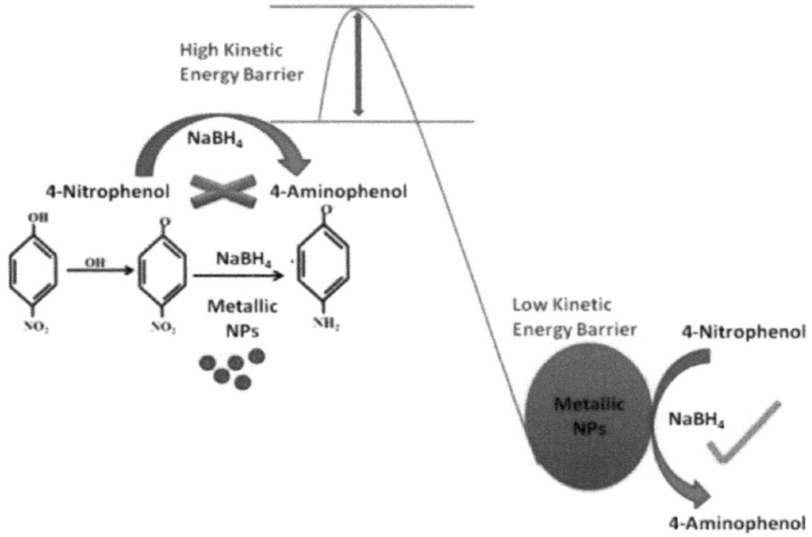

FIGURE 21.5 Depiction of catalytic reduction activity of Au-NPs.

21.6 GREEN SYNTHESIS OF GOLD NANOPARTICLES

Au-NPs nanoparticles can be synthesized using plant metabolites by various methods.

The green synthesis is preferred over chemical methods as it produces less toxic by-products. It is highly preferred for the same reasons as less toxic by-products mean easier ways to deal with pollution of the environment. Various green sources are used for the same such as fungi source mainly chitinase, bacterial sources, algal sources, and plant extracts containing compounds present in leaves such as amines, aldehydes, ketones, terpenoids, flavonoids, phenols, carboxylic acids, polyphenols, ascorbic acids, and saponins. Characterization of NPs is achieved by transmission electron microscopy or TEM, X-ray diffraction, UV-Vis spectroscopy which lead to the determination of gold nanoparticle formation as gold nanoparticles have a characteristic absorption spectrum.

21.7 METHODS

There are two main and different principles of synthesis namely bottom-up and top-down that are devised by research groups worldwide for various functions and modalities. Top-down method involves Nps that are prepared through a wide range of synthesis approaches and methodologies like lithographic techniques, sputtering, etching, and ball milling. Bottom-up approach also includes many methods like atomic/molecular condensation, laser pyrolysis, sol–gel, spray pyrolysis, and chemical-vapor deposition where Nps are made from simple molecules.

The morphology of the nanoparticles (the shape and size) hence formed can however be controlled and modified by various laboratorical conditions such as changes/alterations in temperature, pH, chemical reagents involved etc.

But conventional methods often lead to toxic by-products, hence green synthesis by fusion and reduction with plant extracts, bacterial strains, mycelium of fungi, yeast strains, and algal sources are being preferred nowadays. This is the green synthesis of nanomaterials that is eco-friendly compared to conventional methods of synthesis of nanomaterials that lead to generation of many toxic products.

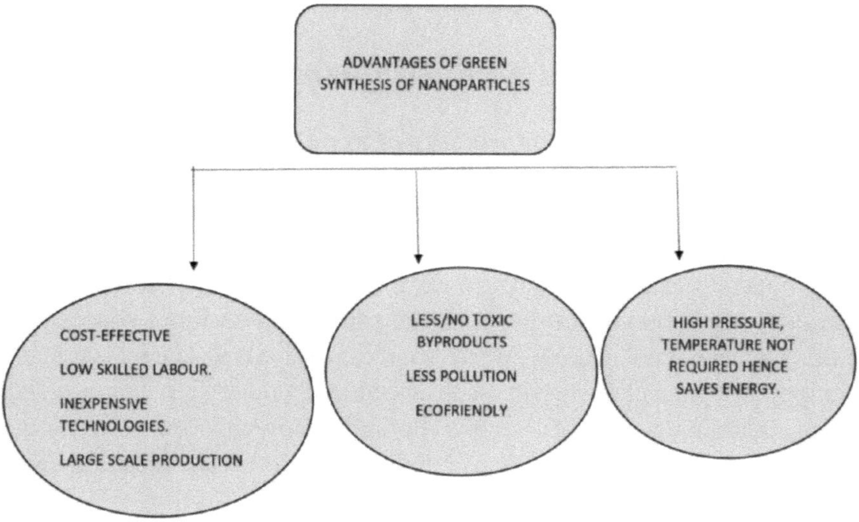

FIGURE 21.6 Several advantages of green synthesis over conventional methods.

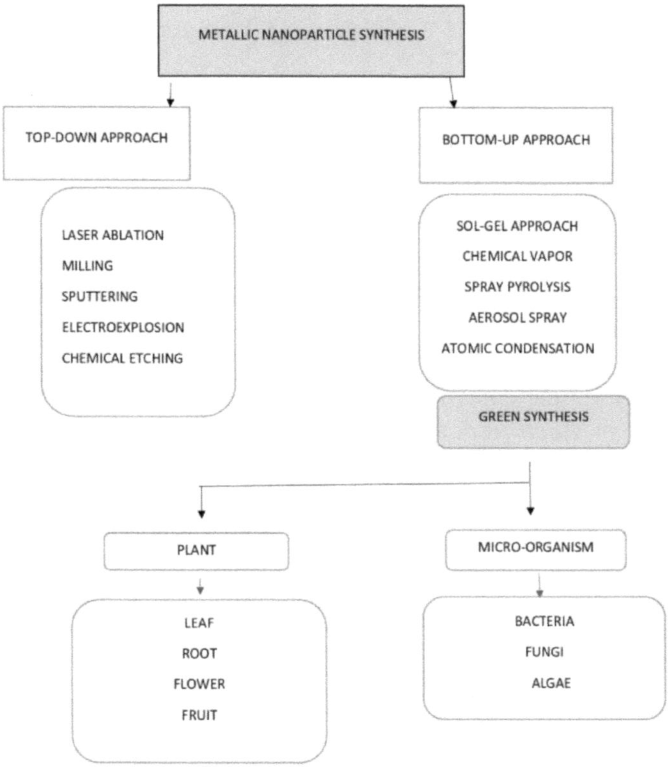

FIGURE 21.7 Depicting methods of synthesis of nanoparticles including green synthesis methods.

21.8 BIOLOGICAL COMPONENTS FOR GREEN SYNTHESIS

These methodologies are dependent on reactions of biologic specie and parameters such as pressure, solvent, pH conditions temperature. Plant metabolites such as aldehydes, ketones, amides, terpenoids flavonoids, phenols, carboxylic acids, polyphenols, saponins, ascorbic acids etc are widely employed. These help in reduction of metal ions into NPs.

Bioagent	Part used	Function
Simarouba glauca	Leaf	Antimicrobial
Streptomyces sp.	Mycelium	Antifungal
Garcinia magnostana	Peel extract	Antimicrobial
Salvinia officinalis	Leaf extract	Catalytic dye reduction

21.8.1 BACTERIA

Several bacterial species have been employed for commercial use and form the backbone of the biotechnology industry that includes bioremediation, bioleaching, genetic engineering etc. Reduction of NPs is achieved easily by bacterial species; hence, they are a better suited candidate for nanoparticle generation compared to chemical reduction methods. In a variety of metallic nanoparticle generation bacteria is widely used due to its less harmful effect on the environment. Actinomycetes and prokaryotic bacteria are used for synthesizing metal nanoparticles.

Bacteria can be genetically engineered, modified, and made to perform tailor-made, unique objectives in mind. *Escheichia coli, lactobacillus casei, aeromonas sp, Bacillus indicus* etc. are some bacterial strains that are widely employed for the generation of gold nanoparticles.

Bacillus subtilis, Shewanella alga, E. coli DH5a, Plectonema boryanum, Rhodopseudomonas capsulate etc. are widely used for metal NPs formation via reduction and form the conventional norm in the generation of gold nanoparticles using bacteria, hence forming a facet of green synthesis of nanoparticles.

21.8.2 FUNGI

Fungi offers a great way for fungi-mediated biosynthesis of nanoparticles as fungi have greater amounts of enzymes that are intracellular. They have reductase enzymes and intracellular spaces that offer ways for reduction of oxide particles into nanoparticles.

Inside the cell wall of fungi and cell spaces are enzymes that are utilized for nanoparticle synthesis. Chitanase forms a chelated agent along with the nanoparticles and hence helps in stabilizing the gold nanoparticles as performed by an experiment by Katas et al. (2019). The mycelium of fungi such as Streptomyces sp. are extracted for accomplishing this goal. The gold particles are present in trivalent form as Au^{+3}. These trivalent gold particles are then reduced into gold anionic nanoparticles. These are then stabilized by chitosan, a reductase enzyme present in fungi mycelium. These form a chelated complex around the gold anion nanoparticles. These are then stabilized. These gold nanoparticles can then be used for remediation purposes such as antifungal applications in remediation strategies. Since fungi is a natural source it helps us synthesize gold nanoparticles without using or primarily depending on chemicals and hence forms a way to synthesize gold nanoparticles in a "green way." That is without being polluting or harmful to the environment. This helps us achieve our goal to synthesize nanoparticles in a sustainable manner.

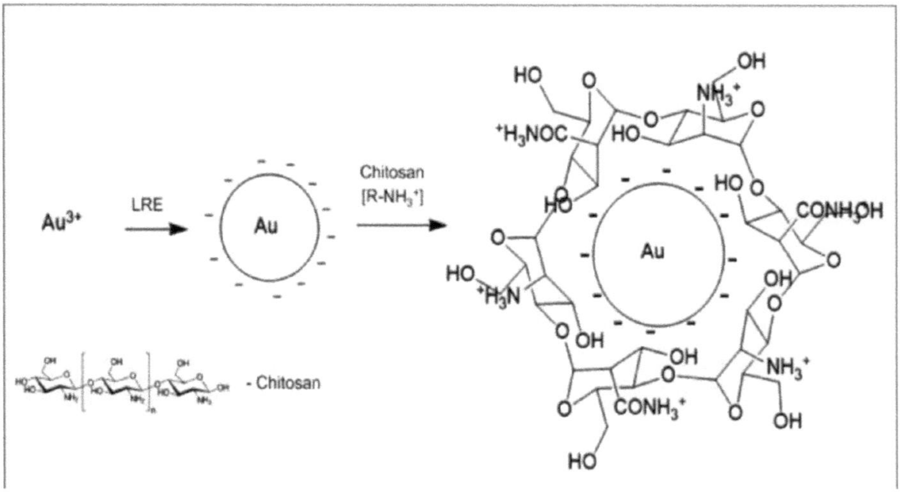

FIGURE 21.8 Depiction of using fungi chitosan source to create gold nanoparticles from gold ions.

21.8.3 YEAST

There are about 1500 yeast species that have been identified by scientists and are classified as single-celled eukaryotes. *Saccharomyces cerevisiae* or bakers yeast has been used by scientists for the reduction of metal ions into metal NPs successfully, hence affirming the entry of yeast as a suitable and alternative candidate for metal NPs formation. Via photo-extracellular synthesis bakers yeast has been used for capping and reduction of NPs via a pilot study by Attia et al. (2016) in which undecanoic acid, trimethylsilyl derivatives of butan-2,3-diol, indole-3-acetic acid, and glucose were isolated from yeast cells.

21.8.3.1 MICROORGANISM-BASED MECHANISM

Microbial enzymes help in reduction of the metallic NPs and allow in stabilization and capping of the isolate NPs. Sneha et al. have described the mechanism of Au/Ag NPs formation using Algae such as *Verticillium sp.*

as by priorly the gold and silver ions are captured on the surface of fungal cells via electrostatic interactions between negatively charged cell wall enzymes and ions. (b) Au/Ag nuclei is formed by the reduction of Au/Ag ion propagated without any external assistance. NADH (nicotinamide adenine dinucleotide) and NADH-dependent nitrate reductase forms the core of the reaction.

21.8.3.2 LEAF -EXTRACT (DERIVED FROM PLANT SOURCE) BASED MECHANISM

Parameters such as temperature, pH, solvent and metal ion concentration, and source of plant metabolite greatly affect the efficiency rates as compared to bacteria and fungal sources, which require a longer incubation time. Hence, plant extracts particularly leaf extracts are considered to be an indispensable source for the synthesis of metal nanoparticle synthesis. They are far more efficient and faster in rate of remediation compared to fungal or bacterial source; hence, plant sources remain a favorite with scientists. Using the ideals of green chemistry and ecologically favorable ways of generation that do not rely on harmful and potentially toxic chemicals that cause pollution on disposal to water bodies, plant metabolites open up newer avenues in the field of nanoparticle generation.

21.9 PREPARATION OF *SIMAROUBA GLAUCA* EXTRACT

Simarouba glouca leaves are dried, ground and the extract of leaves is isolated in an experiment by N Thangamani et al. (2019). Saponin extracts that were generated from the leaf of the plant were then fused with chloro auric acid. Reduction of Au ions in the solution is achieved and this leads to generation of gold nanoparticles.

21.9.1 PREPARATION USING GARCINIA MAGNOSTANA EXTRACT

Garcinia mangostana contains various metabolites and compounds that help in reduction of gold ions when in contact with its aqueous plant solution. There are several compounds in the leaf such as benzophenones, saponins, phenols, terpenes, flavonoids, and anthocyanins which are reducing agent for the gold ions in the aqueous solution. This method is nontoxic and eco-friendly

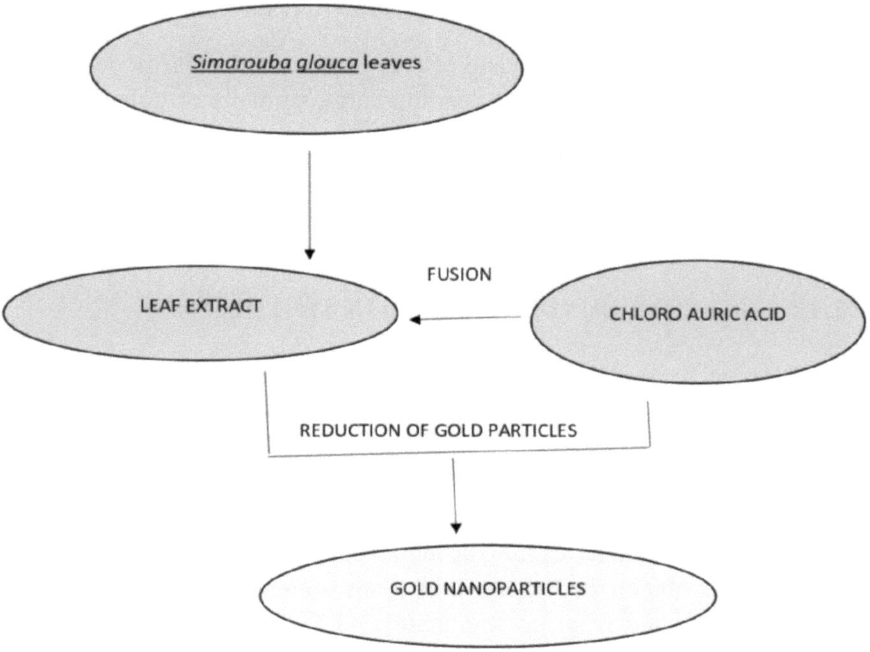

FIGURE 21.9 Gold ion reduction to produce gold nanoparticles by the use of leaf extract and chloroauric acid.

and is without the usage of synthetic materials which are harmful to the environment. Phenolic acid, alkaloids, flavonoids, terpenoids, alkaloids found in plant act as agents of reduction. Xanthone, a phenolic compound, is abundantly present in the fruit peel. α-mangostin and γ-mangostin are the most common xanthones isolated by researchers. There is a peak at 540 nm indicating formation of Au-NPs as seen in UV-visible spectroscopy.

21.10 PREPARATION USING SALVINIA OFFICINALIS

There is a safe environmentally friendly way of preparing Au-NPs using *salvia officinalis*. Polyphenols of alkaline pH around 11 and of acidic pH of around 5 were used which caused reduction of the gold ion salt. Colloidal stability of Au-NPs was achieved via polyphenol capping confirmed by IR or infrared analysis. Quantification of the result using UV-analysis also confirmed the same results by displaying a peak at 540 nm. The solvent used in the extraction of phenol from the dried powder of the plant was ethanol.

21.11 PREPARATION USING POMEGRANATE PEEL EXTRACT

Pomegranate is a popular fruit and often a large amount of fruit peel is found discarded. The peel of the fruit contains large amounts of polyphenols and proanthocyanidins. These form a chelated complex with the gold nanoparticles and offer it potential for degradation of synthetic dyes namely MB or methylene blue and help in removal of heavy metal ions such as mercury.

21.12 BIO-COMPOUNDS INVOLVED IN SYNTHESIS

Reduction of ions is achieved from a higher oxidation state to a lower oxidation state via plant secondary metabolites such as saponins, terpenes, ketones, aldehydes, phenols etc. These plant compounds are extracted from the plant source such as leaf and then added to the gold ion solution in apt experimental conditions such as temperature, pH, pressure etc. Using a solution of chloroauric acid this may be achieved.

These are obtained from several plant sources that help in Au-NPs synthesis that range from pomegranate peel to *salvinia* and several other plant sources and also bacterial, fungi, and other microorganisms. Fungal enzymes such as chitinase help in stabilizing Au-NPs by forming an aggregate around the nanoparticles.

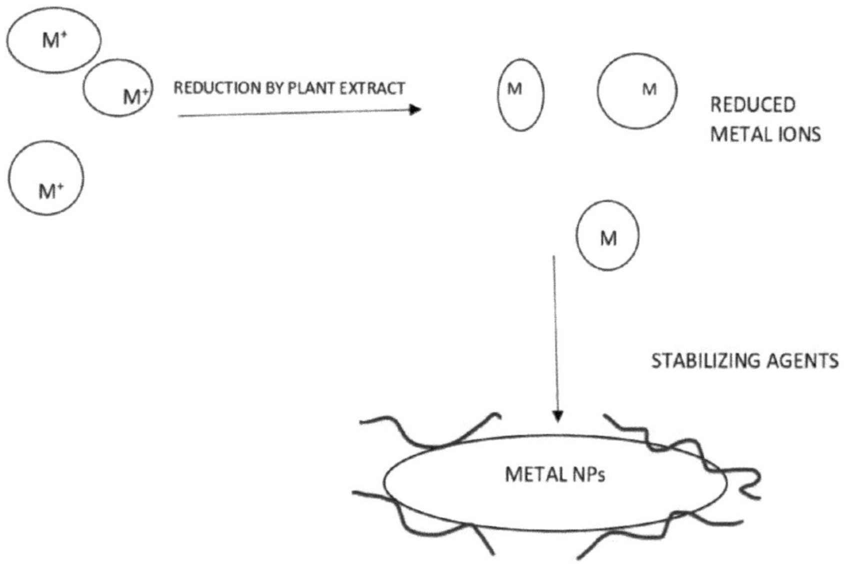

FIGURE 21.10 Depiction of gold ions reduction by plant extracts with stabilization.

21.12.1 CHARACTERIZATION OF GOLD NANOPARTICLES

21.12.1.1 X-RAY DIFFRACTION STUDIES

Crystalline nature of the gold nanoparticles formed can be observed using X-ray diffraction studies and is a popular characterization tool. Careful analysis of the XRD data can provide crucial data about the morphology of nanoparticles hence formed and its sizing in nm or nanometers and affirm the success of synthesis of Au-NPs from gold ions by detection using XRD. There is a reduction of Gold (III) to an atomic Au. The standard Bragg reflections (111), (200), (220), and (311) of the face-centered cubic lattice indicate its crystalline nature as obtained by Khalil et al. (2012) by reduction of gold ions with olive leaf extract. Preferential growth in the (111) direction is indicated by a sharp peak at 38.1 as represented. XRD is a great technique to characterize the synthesis of Au-Nps in the fusion with plant extract (a variety of fruit extracts) and chloro auric acid fusion, resulting in reduction of Au ions and formation of Au-NPs.

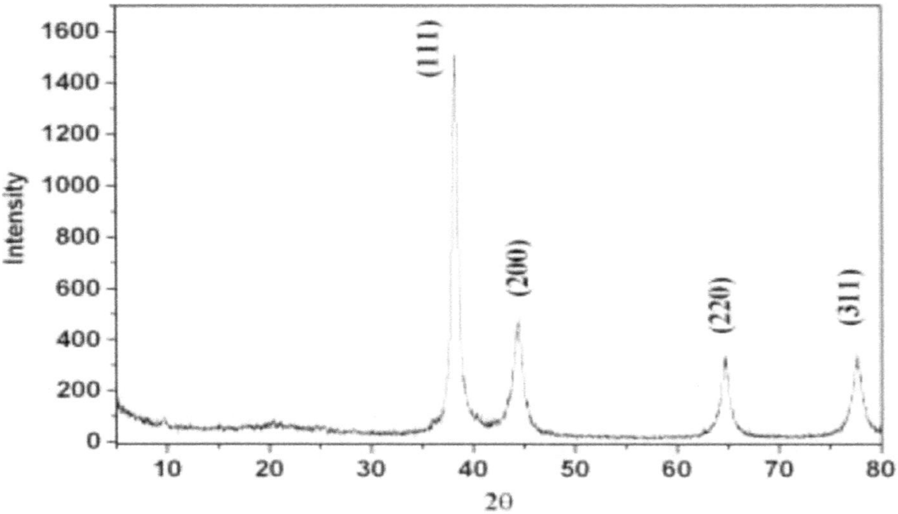

FIGURE 21.11 X-ray diffraction of Au-NPs displaying characteristic peaks.

21.12.1.2 UV-VISIBLE SPECTROSCOPY STUDY

NPs shape, size, concentration, agglomerate size, and its refractive index are some attributes of its optical properties that enable us to visualize and affirm NPs formation. This enables us to use successfully UV-Vis a scientific tool for identification, characterizing, and analysis of the nanomaterials. Localized surface plasmon resonance (LSPR), which refers to the collective oscillation of electrons in resonance with a unique wavelength of incident light, is seen in NPs. An absorbance band in the visible region (500 nm–600 nm) thanks to its LSPR, peak, or absorption maxima can be noted at 521 nm which is then confirmed by UV-Vis spectroscopy. The maximum absorption peaks are visible at 520–530 nm and there are red shifts that are gradually seen with the increase of diameters of the gold nanoparticles of range 12-41 nm as shown by Liu et al. (2010).

Light wave is trapped within nanoparticles NPs that are smaller than the incident light wavelength which causes LSPR. The LSPR spectrum is a feature of morphology, i.e., the shape and size of gold nanoparticles. As the diameter of the particle increases the absorbance wavelength also increases and the spectrum of absorbance shifts greatly into the spectrum's lower red region. There are increasing positive correlations of increasing NPs diameters and red shifts in the spectra.

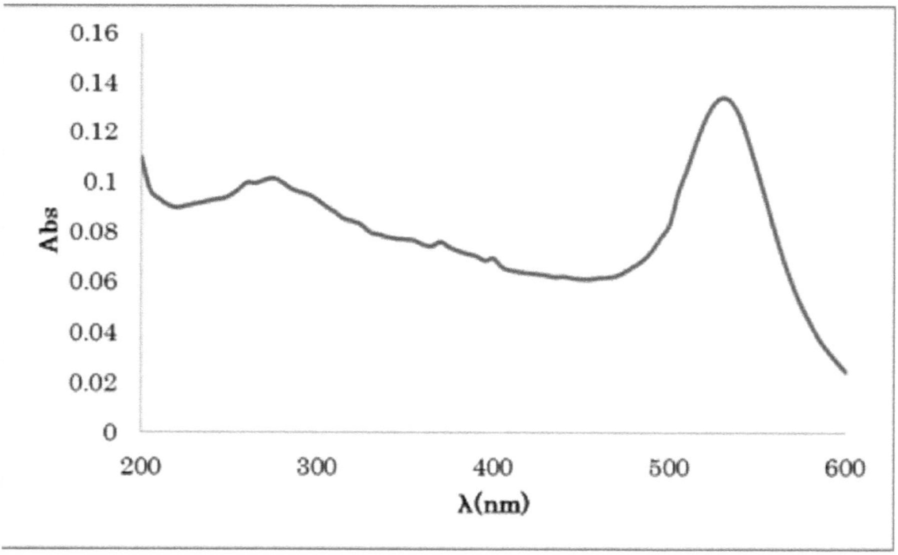

FIGURE 21.12 UV spectroscopy of gold nanoparticles showing absorption maxima at 520–530 nm.

21.12.2 TOXICITY AND STABILITY OF AU-NPS

Gold nanoparticles are nontoxic as they are inert metals. This makes them quite suitable for use by humans. From cancer treatments, drug delivery, tumor detection gold nanoparticles are widely used in various facets. For nanoparticles that are less than 12 nm in diameter, they may cross the blood–brain barrier (Semmler et al., 2014) and the gold nanoparticles of 30 nm or smaller can be endocytosed by cells. Modifying the surface of gold nanoparticles by incorporating immunogenic peptides caused great reduction in their toxicity. This reduction in the toxicity is linked with an increase in the ability to induce antibody response. The toxicity of Au-NPs at high doses in mice led to various results by Chen et al. (2009).

Keeping these factors in mind there is immense study of gold nanoparticles and it is greatly incorporated in human therapy. Biomedical applications of nanomaterials have been increasingly and widely studied. The environmental localization and transport of nanoparticles depend on their ability to make stable liquid suspension or aerosols in experimental media. The stability of the nanoparticles in the testing medium can therefore be evaluated by estimating their propensity to aggregate or interact with the surrounding media. Aggregation is a phenomenon that is associated with the rate of particles forming a mass with each other via collisions. Inter-particulate affinity and size of the particles is a key determinant of their conglomeration toward one another and their performance under experimental conditions such as ranges of various pH, temperature, type of suspension, extract and solvent used etc.

21.13 CONCLUSIONS

Green-gold nanoparticles/Au-NPs offer a novel breakthrough in the field of nanotechnology by recombining the facets of biotechnology and environmental sciences. It offers a great way to deal with the problem of bioremediation of contaminated and polluted wastewater by using nanoparticles. Compared to the previously and conventionally used nanoparticle synthesis by chemicals, green nanotechnology offers a much better alternative as it uses natural and nature-derived sources to reduce gold ions into gold nanoparticles. Green synthesis helps us to prevent the detrimental loss to the environment that is caused due to the conventional methods that create pollution, hence defeating the original purpose of using nanoparticles for remediation. Green-gold nanoparticles offer ways to aid bioremediation of polluted water by its antioxidant activity, catalytic reduction, degradation of synthetic dyes in water, removal and detection of heavy metal, and antibacterial activity. It uses a plethora of precursors such as plant extracts, bacteria, fungal enzymes, yeast, algae etc to help form Au-NPs. This results in the formation of NPs that are ecologically safe and help in improving the safety of environment by reducing the presence of toxic compounds in water by their remediation action. The small size and high catalytic, antioxidant activity of gold nanoparticles confer it a high advantage by degrading synthetic dyes such as methylene blue, destroying pathogenic microbes (even those that are antibiotic resistant) and detect heavy metal ions in water. The success of green-gold nanoparticles can be affirmed by using UV-analysis by spectroscopic method, X-ray diffraction, electron microscopy, and various other ways of analysis that can ensure the proper synthesis of gold nanoparticles and also affirm the presence of proper by-product formation in the waterbody.

This makes gold nanoparticles highly advantageous in the cause of bioremediation and makes it indispensable to the cause of green technology as shown in detail by this paper.

KEYWORDS

- **bioremediation**
- **gold nanoparticles**
- **saponins**
- **microorganism**
- **reduction**
- **plant**
- **dyes**
- **green**

REFERENCES

Akintelu, Sunday Adewale & Yao, Bo & Folorunso, Aderonke. (2021). Bioremediation and pharmacological applications of gold nanoparticles synthesized from plant materials. Heliyon. 7. e06591. 10.1016/j.heliyon.2021.e06591.

Akintelu, S. A.; Yao, B.; Folorunso, A. S. Bioremediation and Pharmacological Applications of Gold Nanoparticles Synthesized from Plant Materials. *Heliyon* **2021**.

Attia, Y.; Farag, Y.; Mohamed, Y.; Hussien, A.; Youssef, T. Photo-Extracellular Synthesis of Gold Nanoparticles Using Baker's Yeast and Their Anticancer Evaluation against Ehrlich Ascites Carcinoma Cells. *N. J. Chem.* **2016,** *40,* 9395–9402. doi: 10.1039/C6NJ01920J.

Chen, Y. S.; Hung, Y. C.; Liau, I.; Huang, G. S. Assessment of the In Vivo Toxicity of Gold Nanoparticles. *Nanosc. Res. Lett.* **2009**

Ganapuram, B.; Alle, M.; Dadigala, R.; et al. Catalytic Reduction of Methylene Blue and Congo Red Dyes Using Green Synthesized Gold Nanoparticles,. **2015**.

Katas, H.; Lim, C. S.; Nor Azlan, A. Y. H.; Buang, F.; Mh Busra, M. F., Antibacterial Activity of Biosynthesized Gold Nanoparticles using Biomolecules from Lignosus Rhinocerotis and Chitosan. *Saudi Pharm. J*, . **2019,** *27,* 2.

Khalil, M.; Ismail, E.; El-Magdoub, F. Biosynthesis of Au Nanoparticles Using Olive Leaf Extract **2012**.

Liu, Z.; Zu, Y.; Fu, Y.; Meng, R.; Guo, S.; Xing, Z.; Tan, Crystalline Gold Nanoparticles and their pH-dependent UV Absorption Characteristic. **2010**.

Menon, S.; Rajeshkumar, S.; Kumar, V. A Review on Biogenic Synthesis of Gold Nanoparticles, Characterization, and Its Applications. **2014**.

Semmler-Behnke, M.; Lipka, J.; Wenk, A.; et al. Size Dependent Translocation and Fetal Accumulation of Gold Nanoparticles from MATERNAL BLOOD in the Rat.

Singh, J.; Dutta, T.; Kim, K. H.; et al. 'Green' Synthesis of Metals and their Oxide Nanoparticles: Applications for Environmental Remediation. *J. Nanobiotechnol.* **2018,** *16,* 84.

Skerget, M.; Kotnik, P.; Hadolin, M.; Hras, A. R.; Simonic, M.; Knez, Z. Phenols, Proanthocyanidins, Flavones and Flavonols in Some Plant Materials and their Antioxidant Activities. *Food Chem.* **2005**.

Thangamani, N.; Bhuvaneshwari, N. Green Synthesis of Gold Nanoparticles using Simarouba Glauca Leaf Extract and their Biological Activity of Microorganism. *Chem. Phys. Lett.* **2019**.

CHAPTER 22

Green Nanomaterials: Biocompatible Materials in Nanoscale?

FERNANDA MARIA POLICARPO TONELLI

Federal University of São João del Rei, Divinópolis, Minas Gerais, Brazil

ABSTRACT

Nanomaterials have been synthesized since the 1950s to perform different desirable functions. Nanotechnology has shown promise in different fields of research such as delivery, diagnosis, energy storage, among others. However, many traditional protocols used to generate nanomaterials are expensive and environmentally harmful. Currently, the whole world aims at sustainable development to reduce pollution and preserve life on earth. Therefore, protocols that allow nanomaterials' green synthesis are preferable. Green nanotechnology is linked to principles of green engineering and also to principles of green chemistry, and its aim is synthesizing nanomaterials to contribute to environmental problems and that can be applied without adversely affecting the environment and the health of human beings. This chapter will be dedicated to discussing biocompatibility of green synthesized nanomaterials and future perspectives related to this relevant topic.

22.1 NANOTECHNOLOGY FIELD AND GREEN SYNTHESIS

In 1974, Norio Taniguchi started the concept of Nanotechnology. However, materials that show the size ranging from 1 to 100 nanometers are produced since the late 1950s to perform desirable properties such as delivery, diagnosis, energy storage, among others (Taniguchi, 1974; Tonelli et al., 2020). One interesting nanomaterials' characteristic is high surface area/volume ratio that enables the functionalizing to perform a desirable task and to

Sustainable Nanoremediation: Modern Technologies for a Clean Environment.
Fernanda Maria Policarpo Tonelli, Arpita Roy, and Khalid Rehman Hakeem (Eds.)

reduce off-target interactions (Subbiah et al., 2010; Albanese et al., 2012). In the field nanotechnology, the most commonly applied nanomaterials are silver (Ag) nanoparticles (NPs), zinc oxide (ZnO) nanoparticles, graphene oxide NPs, titanium dioxide (TiO_2) nanoparticles, and single- or multi-walled carbon nanotubes (CNTs) (Ahemed et al., 2022).

The traditional methods employed to allow the production of nanomaterials present high costs, require qualified operators, and lack consideration of fundamental mechanisms and modeling factors. Furthermore, the traditional methods are environmentally unsustainable because it may produce nanomaterials with bioaccumulative or toxic characteristic (Rafique et al., 2017; Singh et al., 2018; Ahemed et al., 2022). For sustainable development, green synthesized nanomaterials have emerged in the 1990s to reduce the cost and offer more efficient and eco-friendly materials (Anastas and Warner, 2020). Continuously, new protocols have been proposed as an interesting alternative to synthetize nanomaterials due to the current physical and/or chemical methods that use hazard reagents to the environment (Bolade et al., 2020; Nasrollahzadeh et al., 2021).

The green synthesis of nanomaterials aims to apply the principles of green chemistry (Figure 22.1) (Anastas and Eghbali, 2010) to nanomaterial synthesis. Green chemistry, also known as sustainable chemistry, aims to

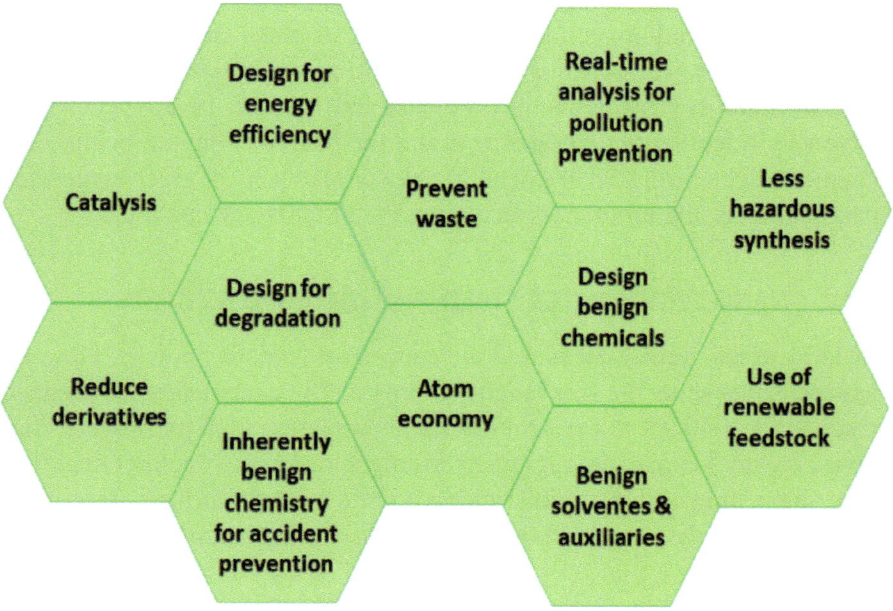

FIGURE 22.1 Green chemistry's principles.

establish protocols that use renewable energy sources to avoid the production of waste and offer optimized efficiency (Jahangirian et al., 2017). In summary, the green chemistry principles are to slow down hazardous chemical syntheses, employ safer reagents and solvents, and enhance energy efficiency by reducing negative effects (Anastas and Warner, 2020).

Several green systems have been employed to allow nanomaterial synthesis. Bioassist synthesis's proposed protocols may employ bacteria, cyanobacteria, fungi, seaweeds, living plants, plant extracts or biomass, and/or bio-derived products, to synthetize nanomaterial (Wadhwani et al., 2014; Babu-Maddinedi et al., 2015; Dhanasekar et al., 2015; Saratale et al., 2018; Jameel et al., 2020). For example, for the synthesis of gold nanoparticles fresh fruiting bodies from *Lentinula edodes* were used as a green reducer agent (Owaid et al., 2019).

Generally, green nanomaterials are classified according to its morphology and composition. Based on the composition, nanoparticles can be allocated into the following categories: liposome nanomaterials, organic–inorganic nanocomposites, organic nanomaterials, inorganic nanomaterials, carbon-rich nanomaterials and biological nanomaterials (Gour and Jain, 2019; Kaur et al., 2022).

In summary, green nanotechnology is linked to principles of green engineering and principles of green chemistry and its aims is to synthesize nanomaterials that contribute to environmental problems and that can be applied without adversely affecting the environment and the health of human beings (Krishnaswamy and Orsat, 2017). This chapter will be dedicated to discuss biocompatibility of green synthesized nanomaterials and future perspectives related to this relevant topic.

22.2 GREEN SYNTHESIZED NANOMATERIALS AND THEIR MAIN FUNCTIONS

Nanomaterials synthetized through green synthesis can be employed in several fields (Figure 22.2) such as catalyst for specific reactions, environmental pollution remediation, biosensor, antimicrobial, agriculture field, and drug delivery. Interestingly, green synthetized nanomaterials can also be used as drugs. Furthermore, green synthesized nanomaterials have been highlighted from the pharmaceutical industry one which can act in the aging and the diseases related to oxidative stress.

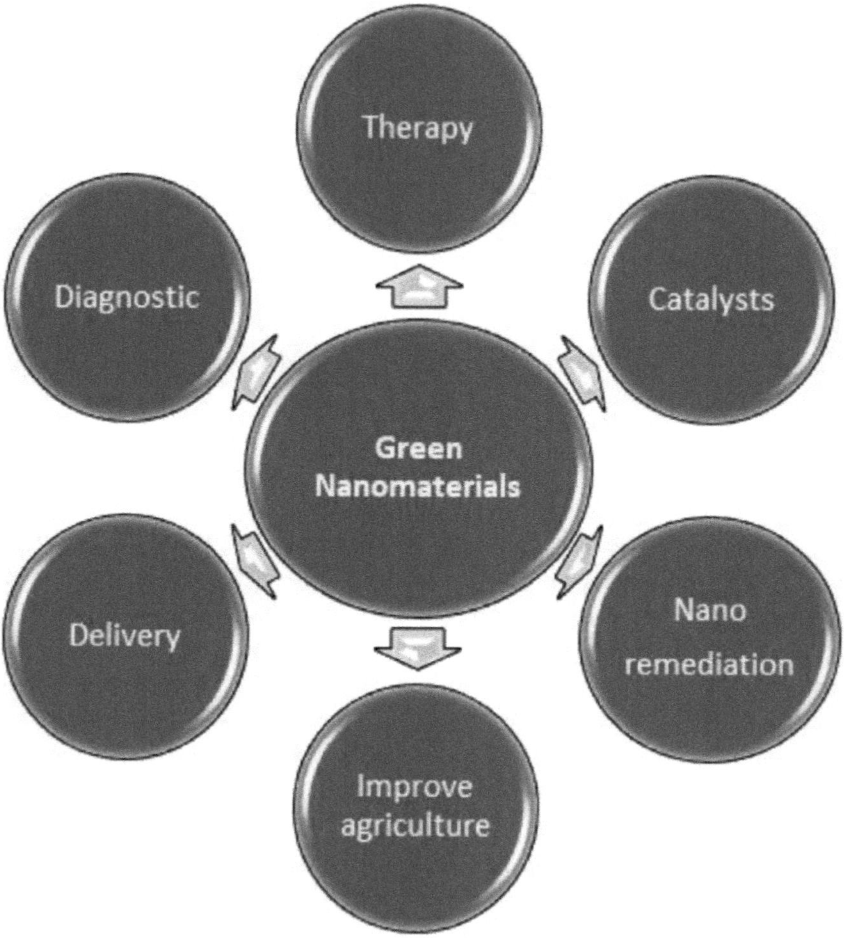

FIGURE 22.2 Examples of green nanomaterial's different types of use.

Nanomaterials produced for green synthesis can perform catalysts activity in specific reactions. Palladium nanoparticles synthetized through the dried fruit extract from *Gardenia jasminoides* showed catalytic activity on the hydrogenation reactions (Jia et al., 2009). Furthermore, Palladium nanoparticles were used as catalyst for reduction of nitroarenes and these nanoparticles could be synthetized from barberry fruit extract (Nasrollahzadeh et al., 2016).

Recently, many studies have pointed out the environmental pollution remediation potential from green synthesized nanomaterials. Silver nanoparticles (AgNPs) produced from leaf extracts of *Diospyros lotus* contributed to decolorization of industrial wastewater (Yasmin et al., 2020). Interestingly, nitrogen-doped quantum carbon dots synthetized through *Bombyx mori* silk fibroin enhance the photocatalytic activity of TiO_2 to promote remediation (Wang et al., 2020). A study conducted by Rabeea and collaborators (2020) demonstrated the degrading of organic pollutant methylene blue by gold nanoparticles originated from fresh fruiting bodies of *Flammulina velutipes*. Also, green synthesized nanomaterials can be employed to detect the contaminant to be degraded. For example, a study conducted by Alex and contributors (2020) reported a biosensor of Ag nanoparticles capable of detecting and degrading the Mancozed (Fungicide). These nanoparticles were originated from the leaves of Indian mint, aloe vera, guava, and neem.

Nanomaterials synthetized through green synthesis also could present antimicrobial activity. AgNPs synthesized by using the leaves of *Azadirachta indica* inhibited *Escherichia coli*'s growth (Verma and Mehata, 2016). Interestingly, these nanoparticles produced from the cellular extract of *Penicillium oxalicum* GRS-1 (that presents activity of the enzyme nitrate reductase) demonstrated strong antimicrobial activity against popular food-borne pathogens as well as against *Salmonella typhimurium, Staphylococcus aureus*, and *Escherichia coli* (Rose et al., 2019). Gold nanoparticles (AuNPs) produced by using *Agaricus bisporus'* extract through the hydrothermal process exhibited antimicrobial properties against fungal strains as well as *Aspergillus flavus* and *Aspergillus terreus* (Eskandari-Nojedehi, et al., 2018). Selenium nanoparticles synthetized through bacteria showed antimicrobial properties (Medina Cruz et al., 2018).

Surprisingly, green nanomaterials can simultaneously perform antimicrobial activity and be able to assist in the remediation of environmental pollution. For instance, non-cytotoxic coated magnetite nanoparticles synthetized from *Jatropha curcas'* crude latex as same as *Cinnamomum tamala*'s leaf extract were capable of remediating organic/inorganic pollution and acted as antibacterial agents (Das et al., 2020). Also, silver nanoparticles (AgNPs) synthetized from Zingiber officinale's rhizome extract showed antimicrobial properties and were capable of remediating pollution caused by methyl orange dye (Riaz et al., 2020).

Green synthesized nanomaterials can be employed in the agriculture field once its can improve seedling growth, yield, and quality of food. A study conducted by Acharya and collaborators (2019) demonstrated that gold and silver nanoparticles generated from onion extract enhanced the yield, growth, and seed germination (~23.9%) of *Allium cepa* L. Interestingly, chitosan nanoparticles which showed nontoxic and stable characteristics improved the rice germination (Divya et al., 2019). AgNPs generated from kaffir lime leaf extract was used as a nanopriming agent for rice aged seeds. Further analyses demonstrated the potential these nanomaterials to enhance seed germination rate (Mahakham et al., 2017). Surprisingly, silver nanoparticles generated from agro-industrial byproducts were used enhancing *Citrullus lanatus* yield, growth, and seed germination. It is noteworthy that quality of fruit was not altered by these nanoparticles (Acharya et al., 2020).

Nanostructures can be employed in the drug delivery once they present several advantages such as reduction of off-target delivery, enhancement in drug encapsulation quality, preventing premature drug degradation, and optimization of solubility (Hu et al., 2019; Kanwar et al., 2019). Quaternized chitosan–graphene oxide/bovine serum albumin/rectorite nanoparticles produced through a green and one-step synthesis method optimized drug delivery to intestines due to these nanoparticles improved resistance to gastric fluid and encapsulation efficiency (Tu et al., 2015). Polylactic-co-glycolic acid (PLGA)-based nanoparticles have shown efficiently to delivery drugs on humans and animals (Chariou et al., 2020). Moreover, chitosan-based nanomaterials present desirable properties to drug delivery such as nontoxic, biodegradable, and biocompatible (Tzeyung et al., 2019; Islam et al, 2019).

Nanomaterials to be employed in the diagnostic area also can be produced through green chemistry. For instance, *Olax scandens* leaf extract was applied to perform the generation of AgNPs which demonstrated fluorescence properties to diagnostic and were efficiently internalized by tumoral cells such as A549 and B16F10 (Mukherjee et al., 2014). Green carbon nanotags originated from cyanobacteria activated the fluorescence of G-tags to identify tumor cells. Moreover, these nanoparticles when conjugated to doxorubicin caused the death of tumor cells (Lee et al., 2014). Therefore, these green carbon nanotags showed teranostic property once they can act both in the treatment and in the diagnosis of disease.

Green synthetized nanomaterials can also be used as drugs. Zinc oxide nanoparticles generated through *Hibiscus subdariffa* leaf extracts demonstrated antidiabetic activity once it increased the insulin secretion from pancreatic islets (Bala et al., 2015). AuNPs produced through *Olea europeae*'s leaf extract reduced levels of glucose in blood and acted as a protector from oxidative stress to the liver and pancreas (Al-Azzawie and Yaaqoo, 2016). Biocompatible gold nanoparticles synthetized from leaves' extract of *Salix alba* showed a muscle relaxant effect (Islam et al., 2019). Silver nanoparticles originated from *Hypericum perforatum* L demonstrated antitumoral property (Alahmad et al., 2021).

Green synthesized nanomaterials are being target from the pharmaceutical industry to act in the diseases related to oxidative stress and aging. Bioavailability, half-life, stability, solubility, permeability, and tissue targeting of antioxidants are improved in the presence of nanostructures (Vaiserman et al., 2020). Supplements containing nanomaterials produced in the presence of bioactive compounds reduce the formation of reactive oxygen species. For instance, nanoparticles to deliver the antioxidant curcumin and the anticancer drug 5-fluorouracil were developed to treat hepatocarcinoma (Ni et al., 2019).

The applicability of the green synthetic nanomaterials presented earlier in this chapter are not the only possible ones. Recently, other applications of green nanomaterials to improve the quality of life have been reported (Abu Elella et al., 2022; Banerjee 2022; Klan et al., 2022).

22.3 NANOMATERIALS AND BIOCOMPATIBILITY

According to the European Society of Biomaterials' conference in 1986, biocompatibility is an adequate response of the host after contact with some substance. In other words, a nanomaterial is biocompatible when it promotes an appropriate response by the target cell and prevents an undesirable immunogenic response or other toxic effects. Certainly, when the nanomaterial is intended to be used in biomedical/clinical application or in any application in which human beings can be directly exposed to them, biocompatibility of nanomaterials is an issue that should receive attention to reduce the risks to human beings (Williams, 1989; Duncan and Izzo, 2005; Soufi and Iravani, 2020).

Several characteristics of nanomaterials can influence the response of cells and whole organisms such as the purity of nanomaterials, the presence of functional groups, size, shape, surface area, topography and charge, hydrophobicity (Shi et al., 2015; Adabi et al., 2016; Tonelli et al., 2020). Moreover, these nanomaterials' characteristics can influence whether the nanomaterials will be taken up by active or passive transport (Shang et al., 2014).

In the process of cellular internalization and intra/intercellular distribution, the nanomaterial size could interfere. A study conducted by Feng and collaborators (2018) pointed out that PEGylated-coated magnetic iron oxide nanoparticles with low diameter were better internalized by RAW264.7 macrophages and SKOV-3 cells (Feng et al., 2018). Also, smaller spherical gold nanoparticles were better internalized than larger nanoparticles by tumoral cells (Cho et al., 2010). Occasionally, factors such as functionalization and surface charge show little influence in the process of cellular uptake and intercellular distribution. However, the hydrodynamic diameter of gold nanoparticles attached to DNA was determinant to the cell internalization. In summary, size particles are more important than surface charge, DNA density, and conformation in the uptake and intercellular distribution (Wong and Wright, 2016).

In situations where there are polymorphs, they should also receive attention. For example, the anatase and rutile polymorphs of TiO2 nanoparticles showed different results in the nematode *Caenorhabditis elegans*. The anatases did not damage the worm's body length in the internalization process. Conversely, rutiles drastically diminished worm's population (Hu et al., 2018).

In order of optimize nanomaterial to desirable task, an important strategy is the functionalizing. For example, functionalization of carbon dots with phenylboronic acid contributed to interaction with sialic acid displayed/available on the target cell's surface. Therefore, this optimization of nanomaterial enables in the internalization process by J774.1 murine cells (Thoo et al., 2017). Also, in the work of ElZorkany and collaborators (2018), cadmiun/selenide quantum dots functionalized with zinc sulfide shell that was, on its turn, overcoated with silica were faster uptaken by HepG2 cells (ElZorkany et al., 2018). Furthermore, these functionalized nanoparticles showed smaller cytotoxicity on low concentration.

Target cell characteristics can influence in the internalization and action of nanomaterial on cell. For instance, human monocytes and granulocytes, graphene quantum dots showed better internalized by granulocytes and monocytes than by lymphocytes (Fasbender et al., 2017). Nanoparticles of polystyrene disrupted the zebrafish' development (Pitt et al., 2018). Conversely, these nanoparticles did not affect red blood cells (Pan et al., 2016).

There are protocols that use bio-based raw material to synthesize fewer toxic nanomaterials. To produce biocompatible polymeric nanoparticles it was employed a bio-based monomer derived from Castor oil in the process of thiol-ene polymerization. Fortunately, these nanoparticles did not show cytotoxicity to murine L929 cells and HeLa cells (Cardoso et al., 2016). Also, a great strategy is to employ substances such as chitosan once it is little toxic and naturally biocompatible. Glutaraldehyde-crosslinked chitosan nanoparticles decreased the toxicity of CuO NPs (Zhang et al., 2017). Interestingly, in the hydroxypropyl-β-cyclodextrin modified carboxymethyl chitosan that acted on the delivery of insulin no toxic effect was noted (Song et al., 2016).

In some cases, the cytotoxicity is a desirable feature in nanomaterials such as cancer treatment. However, methods are employed to provoke selective death on tumoral cells. Studies have reported mechanisms that contribute to high uptake rate by tumor cells (Cotin et al., 2018).

22.4 BIOCOMPATIBILITY OF GREEN SYNTHESIZED NANOMATERIALS

Green synthesis nanomaterials are a subgroup of nanoscale materials; then the aspects discussed previously about the biocompatibility of nanomaterials in the topic "Nanomaterials and biocompatibility" are also valid for this class.

Synthesis protocol can interfere in the nanomaterials' characteristics as well as in the internalization by cells. For instance, carbon dots originated from citric acid or bamboo leaves extract were not cytotoxic under a concentration of 0.5 mg/mL. Conversely, carbon dots originated from citric acid were internalized easily by J774.1 macrophages than that originated from citric acid using the plant extract (Thoo et al., 2017). Green synthesized Cr_2O_3 nanoparticles derived from the leaf extract of *Abutilon indicum* (L.) displayed promising antioxidant and anticancer activities when assayed on cancerous cells MCF-7 and the linoleic acid system. Surprisingly, green synthesized Cr_2O_3 nanoparticles showed great biocompatibility with Vero cell lines when compared to the one displayed by Cr_2O_3 nanoparticles chemically synthesized (Khan et al., 2021).

A characteristic from nanomaterials which can influence the cellular internalization is the morphology. A study conducted by Lee and collaborators (2019) pointed out that chitosan-capped gold nanoparticles originated from green tea extract showed different shapes as well as nanospheres, nanostars, or nanorods. Interestingly, these different shapes were uptaken by human hepatocyte carcinoma cells HepG2 in different rates of cellular uptake. Moreover, these nanomaterials showed different levels of cytotoxicity. Different sizes of gold nanoparticles originated from *Mimosa tenuiflora* bark extract showed moderate cytotoxicity and were internalized by human umbilical vein endothelial cells. However, the smallest nanoparticles were dispersed into the cytoplasm while bigger nanoparticles were accumulated in nuclear periphery (Rodríguez-León et al., 2019).

An important indicator of biocompatibility of the green synthesis nanomaterial is the chemical constitution. For instance, selenium nanoparticles are considered more stable and less toxic for human normal cells than other metallic nanoparticles. Moreover, selenium nanoparticles originated from *Ceropegia bulbosa* tuber's aqueous extract showed several biological properties such as antitumoral, antibacterial, and larvicidal. Besides, these nanomaterials demonstrated chemical property as photocatalytic activity (dealing with methylene blue) (Cittrarasu et al., 2021).

Recently, several studies have demonstrated the biocompatibility of the green nanomaterials. To evaluate the biocompatibility of these nanomaterials assays such as MTT on human normal cells and hemolysis of RBCs (human red blood cells) are made. In the RBCs assay, the hemolysis of the human red blood cells is analyzed against different concentrations of the NPs. If the sample presents the capacity to promote the rupture of RBCs, the hemolytic process will be noted. In accord with the guidelines for biocompatibility of the American Society for Testing substances that show more than 5% of hemolysis are considered noncompatible. In the MTT assay, the mitochondrial function of cells is evaluated after treatment with nanomaterials. If the mitochondrial function is reduced, probably the nanoparticles affected the cells' viability (Präbst et al., 2017). In general, normal cellular lineages such as Vero, fibroblasts, and keratinocytes are employed to evaluate the cytocompatibility. In Table 22.1, there are several green nanomaterials which were evaluated whether they are biocompatible.

It is noteworthy that to analyze nanomaterial's toxicity an attention should be dedicated to the principles of cytotoxicity assays once these assays could suffer interference from that type of nanomaterial (Lee et al., 2016).

22.5 CONCLUSIONS

We conclude that nanotechnology advances significantly in several areas to improve life quality for living beings including humans. In recent years, green nanotechnology's use has grown, aiming to achieve sustainable development. However, to achieve the safe use of these nanomaterials, it is important to, prior to promoting large-scale application, verify their safety for use in humans. Several cytotoxicity assays are used to evaluate biocompatibility and many green nanomaterials have shown biocompatibility. In fact, many green nanomaterials have shown better biocompatibility than chemically synthesized nanomaterials.

TABLE 22.1 Examples of Green Synthesized Nanomaterials with Great Biocompatibility.

Green synthesized nanomaterials	Biocompatibility	References
Silver nanoparticles originated from *Cistus incanus* leaves	Immune compatibility assays of nanoparticles pointed out no activation of the THP1-XBlue™ monocyte. Viability assays on L929 mice fibroblasts demonstrated which nanomaterials are secure up to a concentration of 16 ppm	Florkiewicz et al., 2021
Carbon dots originated from citric acid or bamboo leaves extract	Cytotoxicity assays demonstrated are secure under 0.5 mg/mL	Thoo et al., 2017
Selenium nanoparticles generated from the aqueous extract are obtained from *Ceropegia bulbosa* tuber	Cytotoxicity assays pointed out low toxicity for human normal cells.	(Cittrarasu et al., 2021).
Chromium oxide (Cr2O3) nanoparticles produced using *Abutilon indicum*'s leaf extract	Analysis demonstrated great cytocompatibility with Vero cell lines.	(Khan et al., 2021)
Nanoparticle of Zn oxide generated from solid coffee grounds and the peel of pomegranate fruit	Analysis demonstrated low toxicity on Vero cells	(Abdelmigid et al., 2022)
Au nanoparticles originated from different plant extracts (collected from Fertile Crescent area)	Cytocompatibility in human erythroleukemic cell lines and fibroblasts.	(Shnoudeh et al., 2022)
Iron (Fe) nanoparticles (NPs) originated of different plant extracts (collected from Fertile Crescent area)	Cytocompatibility in human erythroleukemic cell lines and fibroblasts.	Shnoudeh et al., 2022
Se nanoparticles originated of different plant extracts (collected from Fertile Crescent area)	Cytocompatibility in human erythroleukemic cell lines and fibroblasts.	Shnoudeh et al., 2022
Ag nanoparticles from *Barleria buxifolia*'s leaf extract (obtained through ultra-sonication-enhanced green method)	Low cytotoxicity against L929 (fibroblast-human normal cell line)	(Sekar et al., 2022)
Ag nanoparticles from the extract of *Conocarpus Lancifolius*	RBC lysis on human erythrocytes demonstrated low cytotoxicity up to a concentration of ~50 μg/ml	(Oves et al., 2022)
Gold nanoparticles from *Pseudobulbus Cremastrae* seu Pleiones	Analyses pointed out no apparent cytotoxic effect on (L02) normal liver cells.	(Zhu et al., 2022)

TABLE 22.1 *(Continued)*

Green synthesized nanomaterials	Biocompatibility	References
Carbon Dots originated from *Punica granatum* L. Peel	Good biocompatibility specially to be applied to perform fluorescence imaging in plants and to study mother–child barriers	(Zhang et al., 2022)
Hydroxyapatite nanoparticles originated from the coffee ground (CE) extracts	Analyses demonstrated a safer form of nano-hydroxyapatite on normal Vero E6 cells	(Abdelmigid et al., 2022)
AgVI and Ag/ZnOVI nanostructures obtained by applying *Vateria indica* (L.)	Studies performed on murine fibroblasts (normal NIH3T3 cells) and RBC (human cells) showed great biocompatibility.	(D'Souza et al., 2022)
Gold nanoparticles from *Gynostemma pentaphyllum*	In vitro analysis pointed out nontoxic effect on 3T3-L1 and RAW 264.7 cells up to a certain concentration	(Akter et al., 2022).
Silver nanoparticles from *Olea europaea* Leaf Extract	Biocompatibility was demonstrated on thMSC cells.	(Sellami et al., 2021)
Zinc oxide (ZnO) nanoparticles using aqueous fruit extracts of *Myristica fragrans*	Excellent biocompatible even in high dose	(Faisal et al., 2021).
MgO-NPs and Co3O4NPs from *Hibiscus rosa sinensis*	Both the nanoparticles showed excellent biocompatibility against RBCs from humans even on high concentrations.	(Kainat et al., 2021)
ZnO NPs originated from *L. acidophilus* and *L. plantarum*	These nanoparticles showed low cytotoxicity on the mouse embryonic fibroblasts cell line	(Harandi et al., 2021)

Source: Authors.

22.6 FUTURE PERSPECTIVES

The field of nanotechnology has advanced enormously and at a rapid pace; however, it has the potential to go even further, especially when it comes to offering sustainable and safe solutions. Cytotoxicity assays to assess the response of cells to the exposure to green synthesized nanomaterials are essential to allow the large-scale safe application of these nanostructures. Even for nanomaterials generated by green synthesis, these experiments are important to ensure safety in humans. It is also expected that new methods for generating green nanomaterials will continue to be proposed with the aim of optimizing the desirable effects of nanomaterials and these green nanomaterials will be biocompatible and eco-friendly to significantly improve significantly life on Earth.

KEYWORDS

- **green synthesized manomaterials**
- **nanotechnology**
- **biocompatibility**

REFERENCES

Abdelmigid, H. M.; Hussien, N. A.; Alyamani, A. A.; Morsi, M. M.; AlSufyani, N. M.; Kadi, H. A. Green Synthesis of Zinc Oxide Nanoparticles Using Pomegranate Fruit Peel and Solid Coffee Grounds vs. Chemical Method of Synthesis, with Their Biocompatibility and Antibacterial Properties Investigation. *Molecules (Basel, Switzerland).* **2022,** *27,* 1236.

Abdelmigid, H. M.; Morsi, M. M.; Hussien, N. A.; Alyamani, A. A.; Alhuthal, N. A.; Albukhaty, S. Green Synthesis of Phosphorous-Containing Hydroxyapatite Nanoparticles (nHAP) as a Novel Nano-Fertilizer: Preliminary Assessment on Pomegranate (*Punica granatum* L.). *Nanomaterials (Basel, Switzerland).* **2022,** *12,* 1527.

Abu Elella, M. H.; Shalan, A. E.; Sabaa, M. W.; Mohamed, R. R. One-pot green Synthesis of Antimicrobial Chitosan Derivative Nanocomposites to Control Foodborne Pathogens. *RSC Adv.* **2022,** *12,* 1095–1104.

Acharya, P.; Jayaprakasha, G.K.; Crosby, K.M.; Jifon, J.L.; Patil, B.S. Green-Synthesized Nanoparticles Enhanced Seedling Growth, Yield, and Quality of Onion (Allium cepa L.). *ACS Sustain. Chem. Eng.,* **2019,** *7,* 14580–14590.

Acharya, P.; Jayaprakasha, G. K.; Crosby, K. M.; Jifon, J. L.; Patil, B. S. Nanoparticle-Mediated Seed Priming Improves Germination, Growth, Yield, and Quality of Watermelons (Citrullus lanatus) at multi-locations in Texas. *Sci. Rep.* **2020,** *10,* 5037.

Adabi, M.; Naghibzadeh, M.; Adabi, M.; Zarrinfard, M. A.; Esnaashari, S. S.; Seifalian, A. M.; Faridi-Majidi, R.; Aiyelabegan, H. T.; Ghanbari, H. Biocompatibility and Nanostructured Materials: Applications in Nanomedicine. *Artif. Cells Nanomed. Biotechnol.* **2016,** *45,* 833–842.

Ahmed, S. F.; Mofijur, M.; Rafa, N.; Chowdhury, A. T.; Chowdhury, S.; Nahrin, M.; Islam, A.; Ong, H. C. Green Approaches in Synthesising Nanomaterials for Environmental Nanobioremediation: Technological Advancements, Applications, Benefits and Challenges. *Environ. Res.* **2022,** *204,* 111967.

Akter, R.; Ling, L.; Rupa, E. J.; KyuPark, J.; Mathiyalagan, R.; Nahar, J.; Won, L. J.; Hyun, K. D.; Murugesan, M.; Yang, D. C.; Kang, S. C.; Kwak, G. Y. Binary Effects of Gynostemma Gold Nanoparticles on Obesity and Inflammation via Downregulation of PPARγ/CEPBα and TNF-α Gene Expression. *Molecules* **2022,** *27,* 2795.

Al-Azzawie, H. F.; Yaaqoo, L. A. Hypoglycemic and Antioxidant Effects of Gold Nanoparticals in Alloxan-Induced Diabetes Rats. *Int. J. Res. Biotechnol. Biochem.* **2016,** *6,* 12–20.

Alahmad, A.; Feldhoff, A.; Bigall, N. C.; Rusch, P.; Scheper, T.; Walter, J. G. Hypericum perforatum L.-Mediated Green Synthesis of Silver Nanoparticles Exhibiting Antioxidant and Anticancer Activities. *Nanomaterials* **2021,** *11,* 487.

Albanese, A.; Tang, P. S.; Chan, W. C. W. The Effect of Nanoparticle Size, Shape, and Surface Chemistry on Biological Systems. *Annu. Rev. Biomed. Eng.* **2012,** *14,* 1–16.

Alex, K. V.; Pavai, P. T.; Rugmini, R.; Prasad, M. S.; Kamakshi, K.; Sekhar, K. C. Green Synthesized Ag Nanoparticles for Bio-Sensing and Photocatalytic Applications. *ACS Omega* **2020,** *5,* 13123–13129.

Anastas, P.; Eghbali, N. Green Chemistry: Principles and Practice. *Chem. Soc. Rev*. **2010,** *39,* 301–312.

Anastas, P. T.; Warner, J. C. *Green Chemistry: Theory and Practice*; Oxford University Press: New York, NY, 2020.

Babu-Maddinedi, S.; Mandal, B. K.; Ranjan, S.; Dasgupta, N. Diastase Assisted Green Synthesis of Size-controllable Gold Nanoparticles. *RSC Adv.* **2015,** *5,* 26727–26733.

Bala, N.; Saha, S.; Chakraborty, M.; Maiti, M.; Das, S.; Basu, R.; Nandy, P. Green Synthesis of Zinc Oxide Nanoparticles Using Hibiscus Subdariffa Leaf Extract: Effect of Temperature on Synthesis, Anti-bacterial Activity and Anti-diabetic Activity. *RSC Adv*. **2015,** *5,* 4993–5003.

Banerjee, A. N. Green Syntheses of Graphene and its Applications in Internet of Things (IoT)-a status Review. *Nanotechnology* **2022,** 33.

Bolade, O. P.; Williams, A. B.; Benson, N. U. Green Synthesis of Iron-based Nanomaterials for Environmental Remediation: A Review. *Environ. Nanotechnol. Monit. Manag.* **2020,** *13,* 100279.

Cardoso, P. B.; Machado, T. O.; Feuser, P. E.; Sayer, C.; Meier, M. A. R.; Araújo, P. H. H. Biocompatible Polymeric Nanoparticles From Castor Oil Derivatives via Thiol-Ene Miniemulsion Polymerization. *Eur. J. Lipid. Sci. Technol.* **2016,** *120,* 1700212.

Chariou, P. L.; Ortega-Rivera, O. A.; Steinmetz, N. F. Nanocarriers for the Delivery of Medical, Veterinary, and Agricultural Active Ingredients. *ACS Nano.* **2020,** *14,* 2678–2701.

Cho, E. C.; Au, L.; Zhang, Q.; Xia, Y. The Effects of Size, Shape, and Surface Functional Group of Gold Nanostructures on Their Adsorption and Internalization by Cells. *Small* **2010,** *6,* 517–522.

Cittrarasu, V.; Kaliannan, D.; Dharman, K.; Maluventhen, V.; Easwaran, M.; Liu, W. C.; Balasubramanian, B.; Arumugam, M. Green Synthesis of Selenium Nanoparticles Mediated from Ceropegia Bbulbosa Roxb Extract and its Cytotoxicity, Antimicrobial, Mosquitocidal and Photocatalytic Activities. *Sci. Rep.* **2021,** *11,* 1032.

Cotin, G.; Piant, S.; Mertz, D.; Felder-Flesch, D.; Begin-Colin, S. Iron Oxide Nanoparticles for Biomedical Applications: Synthesis, Functionalization, and Application. In: *Iron Oxide Nanoparticles for Biomedical Applications*; Elsevier: New York, 2018; pp 43–88.

Das, C.; Sen, S.; Singh, T.; Ghosh, T.; Paul, S. S.; Kim, T. W.; Jeon, S.; Maiti, D. K.; Im, J.; Biswas, G. Green Synthesis, Characterization and Application of Natural Product Coated Magnetite Nanoparticles for Wastewater Treatment. *Nanomaterials* **2020,** *10,* 1615.

Dhanasekar, N. N.; Rahul, G. R.; Narayanan, K. B.; Raman, G.; Sakthivel, N. Green Chemistry Approach for the Synthesis of Gold Nanoparticles Using the Fungus Alternaria sp. *J. Microbiol. Biotechnol.* **2015,** *25,* 1129–1135.

Divya, K.; Vijayan, S.; Nair, S. J.; Jisha, M. S. Optimization of Chitosan Nanoparticle Synthesis and its Potential Application as Germination Elicitor of Oryza Sativa L. *Int. J. Biol. Macromol.* **2019,** *124,* 1053–1059.

D'Souza, J. N.; Nagaraja, G. K.; Prabhu, A.; Navada, K. M.; Kouser, S.; Manasa, D. J. AgVI and Ag/ZnOVI Nanostructures from Vateria Indica (L.) Exert Antioxidant, Antidiabetic, Anti-inflammatory and Cytotoxic Efficacy on Triple Negative Breast Cancer Cells in Vitro. *Int. J. Pharm.* **2022,** *615,* 121450.

Duncan, R.; Izzo, L. Dendrimer Biocompatibility and Toxicity. *Adv. Drug. Deliv. Rev.* **2005,** *57,* 2215–2237.

ElZorkany, H. E.; El-Shorbagy, H. M.; Farroh, K. Y.; Youssef, T.; Sabet, S.; Salaheldin, T. A. Monitoring the Cellular Uptake of Silica-Coated CdSe/ZnS Quantum Dots by Time Lapse Confocal Laser Scanning Microscopy. *J. Appl. Pharm. Sci.* **2018,** *8,* 1–8.

Eskandari-Nojedehi, M.; Jafarizadeh-Malmiri, H.; Rahbar-Shahrouzi, J. Hydrothermal Green Synthesis of Gold Nanoparticles Using Mushroom (Agaricus bisporus) Extract: Physico-chemical Characteristics and Antifungal Activity Studies. *Green Proc. Synth.* **2018,** *7,* 38–47.

Faisal, S.; Jan, H.; Shah, S. A.; Shah, S.; Khan, A.; Akbar, M. T.; Rizwan, M.; Jan, F.; Wajidullah, Akhtar, N.; Khattak, A.; Syed, S. Green Synthesis of Zinc Oxide (ZnO) Nanoparticles Using Aqueous Fruit Extracts of *Myristica fragrans*: Their Characterizations and Biological and Environmental Applications. *ACS Omega* **2021,** *6,* 709–9722.

Fasbender, S.; Allani, S.; Wimmenauer, C.; Cadeddu, R. P.; Raba, K.; Fischer, J. C.; Bulat, B.; Luysberg, M.; Seidel, C. A. M.; Heinzel, T.; Haas, R. Uptake Dynamics of Graphene Quantum Dots into Primary Human Blood Cells Following in Vitro Exposure. *RSC Adv.* **2017,** *7,* 12208–12216.

Feng, Q.; Liu, Y.; Huang, J.; Chen, K.; Huang, J.; Xiao, K. Uptake, Distribution, Clearance, and Toxicity of Iron Oxide Nanoparticles with Different Sizes and Coatings. *Sci. Rep.* **2018,** *8,* 2082–2089.

Florkiewicz, W.; Pluta, K.; Malina, D.; Rudnicka, K.; Żywicka, A.; Guigou, M. D.; Tyliszczak, B.; Sobczak-Kupiec, A. Investigation on Green Synthesis, Biocompatibility, and Antibacterial Activity of Silver Nanoparticles Prepared Using *Cistus incanus*. *Materials (Basel).* **2021,** *14,* 5028.

Gour, A., Jain, N. K. Advances in Green Synthesis of Nanoparticles. *Artif. Cells Nanomed. Biotechnol.* **2019,** *47,* 844–851.

Harandi, F. N.; Khorasani, A. C.; Shojaosadati, S. A.; Hashemi-Najafabadi, S. Living Lactobacillus-ZnO Nanoparticles Hybrids as Antimicrobial and Antibiofilm Coatings for Wound Dressing Application. *Mater. Sci. Eng. C Mater. Biol. Appl.* **2021,** *130,* 112457.

Hu, Q.; Li, H.; Wang, L.; Gu, H.; Fan, C. DNA Nanotechnology-Enabled Drug Delivery Systems. *Chem. Rev.* **2019,** *119,* 6459–6506.

Hu, C. C.; Wu, G. H.; Hua, T. E.; Wagner, O. I.; Yen, T. J. Uptake of TiO2 Nanoparticles into C. elegans Neurons Negatively Affects Axonal Growth and Worm Locomotion Behavior. *ACS Appl. Mater. Inter.* **2018,** *10,* 8485–8495.

Islam, N. U.; Jalil, K.; Shahid, M.; Rauf, A.; Muhammad, N.; Khan, A.; Shah, M. R.; Khan, M. A. Green Synthesis and Biological Activities of Gold Nanoparticles Functionalized with Salix Alba. *Arab J. Chem.* **2019,** *12,* 2914–2955.

Jahangirian, H.; Lemraski, E. G.; Webster, T. J.; Rafiee-Moghaddam, R.; Abdollahi, Y. A Review of Drug Delivery Systems Based on Nanotechnology and Green Chemistry: Green Nanomedicine. *Int. J. Nanomed.* **2017,** *12,* 2957–2978.

Jameel, M. S.; Aziz, A. A.; Dheyab, M. A. Green Synthesis: Proposed Mechanism and Factors Influencing the Synthesis of Platinum Nanoparticles. *Green Proc. Synth.* **2020,** *9,* 41.

Jia, L.; Zhang, Q.; Li, Q.; Song, H. The Biosynthesis of Palladium Nanoparticles by Antioxidants in Gardenia Jasminoides Ellis: Long Lifetime Nanocatalysts Forp-Nitrotoluene Hydrogenation. *Nanotechnology* **2009,** *20,* 385601.

Kainat, Khan, M. A.; Ali, F.; Faisal, S.; Rizwan, M.; Hussain, Z.; Zaman, N.; Afsheen, Z.; Uddin, M. N.; Bibi, N. Exploring the Therapeutic Potential of *Hibiscus rosa sinensis* Synthesized Cobalt Oxide (Co3O4-NPs) and Magnesium Oxide Nanoparticles (MgO-NPs). *Saudi J. Biol. Sci.* **2021,** *28,* 5157–5167.

Kanwar, R.; Rathee, J.; Salunke, D.B.; Mehta, S. K. Green Nanotechnology-Driven Drug Delivery Assemblies. *ACS Omega* **2019,** *4,* 8804–8815.

Kaur, M.; Kumar, A.; Gupta, R. Updates on Health and Safety Aspects of Green Nanomaterials. In *Micro and Nano Technologies, Green Functionalized Nanomaterials for Environmental Applications*; Elsevier, 2022; pp 543–565.

Khan, A. H.; Khan, N. A.; Zubair, M.; Azfar Shaida, M.; Manzar, M. S.; Abutaleb, A.; Naushad, M.; Iqbal, J. Sustainable Green Nanoadsorbents for Remediation of Pharmaceuticals from Water and Wastewater: A Critical Review. *Environ Res*. **2022,** *204,* 112243.

Khan, S. A.; Shahid, S.; Hanif, S.; Almoallim, H. S.; Alharbi, S. A.; Sellami, H. Green Synthesis of Chromium Oxide Nanoparticles for Antibacterial, Antioxidant Anticancer, and Biocompatibility Activities. *Int. J. Mol. Sci.* **2021,** *22,* 502.

Krishnaswamy, K.; Orsat, V. Sustainable Delivery Systems Through Green Nanotechnology. In *Nano- and Microscale Drug Delivery Systems: Design and Fabrication*; Elsevier: New York, 2017; pp 17–32.

Lee, H. U.; Park, S. Y.; Park, E. S.; Son, B.; Lee, S. C.; Lee, J. W.; Lee, Y.; Kang, K. S.; Kim, M. I.; Park, H. G. Photoluminescent Carbon Nanotags from Harmful Cyanobacteria for Drug Delivery and Imaging in Cancer Cells. *Sci. Rep.* **2014,** *4,* 4665.

Lee, Y. H.; Chuang, S. M.; Huang, S. C.; Tan, X.; Liang, R. Y.; Yang, G. C. C.; Chueh, P. J. Biocompatibility Assessment of Nanomaterials for Environmental Safety Screening. *Environ. Toxicol.* **2016,** *32,* 1170–1182.

Lee, Y. J.; Ahn, E. Y.; Park, Y. Shape-dependent Cytotoxicity and Cellular Uptake of Gold Nanoparticles Synthesized using Green Tea Extract. *Nanoscale Res. Lett.* **2019,** *14,* 129.

Mahakham, W.; Sarmah, A. K.; Maensiri, S.; Theerakulpisut, P. Nanopriming technology for Enhancing Germination and Starch Metabolism of Aged Rice Seeds using Phytosynthesized Silver Nanoparticles. *Sci. Rep.* **2017,** *7,* 8263.

Medina Cruz, D.; Mi, G.; Webster, T. J. Synthesis and Characterization of Biogenic Selenium Nanoparticles with Antimicrobial Properties made by Staphylococcus Aureus, Methicillin-resistant Staphylococcus Aureus (MRSA), Escherichia Coli, and Pseudomonas Aeruginosa. *J. Biomed. Mater. Res. Part A.* **2018,** *106,* 1400–1412.

Mukherjee, S.; Chowdhury, D.; Kotcherlakota, R.; Patra, S.; Vinothkumar, B.; Bhadra, M. P.; Sreedhar, B.; Patra, C. R. Potential Theranostics Application of Bio-Synthesized Silver Nanoparticles (4-in-1 System). *Theranostics* **2014,** *4,* 316–335.

Nasrollahzadeh, M.; Sajadi, S. M.; Rostami-Vartooni, A.; Alizadeh, M.; Bagherzadeh, M. Green Synthesis of the Pd Nanoparticles Supported on Reduced Grapheme Oxide using Barberry Fruit Extract and its Application as a Recyclable and Heterogeneous Catalyst for the Reduction of Nitroarenes. *J. Colloid Interface Sci.* **2016,** *466,* 360–368.

Nasrollahzadeh, M.; Sajjadi, M.; Iravani, S.; Varma, R. S. Green-synthesized Nanocatalysts and Nanomaterials for Water Treatment: Current Challenges and Future Perspectives. *J. Hazard. Mater.* **2021,** *401,* 123401.

Ni, W.; Li, Z.; Liu, Z.; Ji, Y.; Wu, L.; Sun, S.; Jian, X.; Gao, X. Dual-targeting Nanoparticles: Codelivery of Curcumin and 5-fluorouracil for Synergistic Treatment of Hepatocarcinoma. *J. Pharm. Sci.* **2019,** *108,* 1284–1295.

Oves, M.; Ahmar Rauf, M.; Aslam, M.; Qari, H. A.; Sonbol, H.; Ahmad, I.; Sarwar Zaman, G.; Saeed, M. Green Synthesis of Silver Nanoparticles by *Conocarpus Lancifolius* Plant Extract and their Antimicrobial and Anticancer Activities. *Saudi J. Biol. Sci.* **2022,** *29,* 460–471.

Owaid, M. N.; Rabeea, M. A.; Aziz, A. A.; Jameel, M. S.; Dheyab, M. A. Mushroom-assisted Synthesis of Triangle Gold Nanoparticles Using the Aqueous Extract of Fresh Lentinula Edodes (shiitake), Omphalotaceae. *Env. Nanotechnol. Monit Manage.* **2019,** *12,* 100270.

Pan, D.; Vargas-Morales, O.; Zern, B.; Anselmo, A. C.; Gupta, V.; Zakrewsky, M.; Mitragotri, S.; Muzykantov, V. The Effect of Polymeric Nanoparticles on Biocompatibility of Carrier Red Blood Cells. *PLoS One* **2016,** *11,* e0152074.

Pitt, J. A.; Kozal, J. S.; Jayasundara, N.; Massarsky, A.; Trevisan, R.; Geitner, N.; Wiesner, M.; Levin, E. D.; Di Giulio, R. T. Uptake, Tissue Distribution, and Toxicity of Polystyrene Nanoparticles in Developing Zebrafish (Danio rerio). *Aquat Toxicol.* **2018,** *194,* 185–194.

Präbst, K.; Engelhardt, H.; Ringgeler, S.; Hübner, H. Basic Colorimetric Proliferation Assays: MTT, WST, and Resazurin. *Methods in Mole. Biol. (Clifton, N.J.).* **2017,** *1601,* 1–17.

Rabeea, M. A.; Owaid, M. N.; Aziz, A. A.; Jameel, M. S.; Dheyab, M. A. Mycosynthesis of Gold Nanoparticles using the Extract of Flammulina Velutipes, Physalacriaceae, and their Efficacy for Decolorization of Methylene Blue. *J. Environ. Chem. Eng.* **2020,** *8,* 103841.

Rafique, M., Sadaf, I., Rafique, M. S., Tahir, M. B. A Review on Green Synthesis of Silver Nanoparticles and their Applications. *Artif Cells Nanomed. Biotechnol.* **2017,** *45,* 1272–1291.

Riaz, M.; Ismail, M.; Ahmad, B.; Zahid, N.; Jabbour, G.; Khan, M.S.; Mutreja, V.; Sareen, S.; Rafiq, A.; Faheem, M.; Shah, M. M.; Khan, M. I.; Bukhari, S. A. I.; Park, J. Characterizations and Analysis of the Antioxidant, Antimicrobial, and Dye Reduction Ability of Green Synthesized Silver Nanoparticles. *Green Process. Synth.* **2020,** *9,* 64.

Rodríguez-León, E. Rodríguez-Vázquez, B. E.; Martínez-Higuera, A.; Rodríguez-Beas, C.; Larios-Rodríguez, E.; Navarro, R. E.; López-Esparza, R.; Iñiguez-Palomares, R.

A. Synthesis of Gold Nanoparticles Using Mimosa tenuiflora Extract, Assessments of Cytotoxicity, Cellular Uptake, and Catalysis. *Nanosc. Res. Lett.* **2019,** *14,* 334.

Rose, G. K.; Soni, R.; Rishi, P.; Soni, S. K. Optimization of the Biological Synthesis of Silver Nanoparticles using Penicillium Oxalicum GRS-1 and their Antimicrobial Effects Against Common Food-borne Pathogens. *Green Process. Synth.* **2019,** *8,* 42.

Saratale, R. G.; Karuppusamy, I.; Saratale, G. D., Pugazhendhi, A.; Kumar, G.; Park, Y.; Ghodake, G. S.; Bharagava, R. N.; Banu, J. R.; Shin, H. S. A Comprehensive Review on Green Nanomaterials Using Biological Systems: Recent Perception and their Future Applications. *Colloids Surf. B Biointer.* **2018,** *170,* 20–35.

Sekar, V.; Balakrishnan, C.; Kathirvel, P.; Swamiappan, S.; Alshehri, M. A.; Sayed, S.; Panneerselvam, C. Ultra-sonication-enhanced Green Synthesis of Silver Nanoparticles using *Barleria buxifolia* Leaf Extract and their Possible Application. *Artif. Cells Nanomed. Biotechnol.* **2022,** *50,* 177–187.

Sellami, H.; Khan, S. A.; Ahmad, I.; Alarfaj, A. A.; Hirad, A. H.; Al-Sabri, A. E. Green Synthesis of Silver Nanoparticles Using *Olea europaea* Leaf Extract for Their Enhanced Antibacterial, Antioxidant, Cytotoxic and Biocompatibility Applications. *Int. J. Mol. Sci.* **2021,** *22,* 12562.

Shang, L.; Nienhaus, K.; Nienhaus, G. U. Engineered Nanoparticles Interacting with Cells: Size Matters. *J. Nanobiotechnol.* **2014,** *12,* 5–9.

Shi, Y.; Miller, M. L.; Di Pasqua, A. J. Biocompatibility of Mesoporous Silica Nanoparticles? *Comment Inorg. Chem.* **2015,** *36,* 61–80.

Shnoudeh, A. J.; Qadumii, L.; Zihlif, M.; Al-Ameer, H. J.; Salou, R. A.; Jaber, A. Y.; Hamad, I. Green Synthesis of Gold, Iron and Selenium Nanoparticles Using Phytoconstituents: Preliminary Evaluation of Antioxidant and Biocompatibility Potential. *Molecules* **2022,** *27,* 1334.

Singh, J.; Dutta, T.; Kim, K. H.; Rawat, M.; Samddar, P.; Kumar, P. 'Green' Synthesis of Metals and their Oxide Nanoparticles: Applications for Environmental Remediation. *J. Nanobiotechnol.* **2018,** *16,* 84.

Song, H.; Ma, X.; Xiong, F.; Hong, H.; Li, C.; Li, L.; Wu, S.; Zhan, X.; Zhang, J.; Hu, J. Preparation and Evaluation of Insulin-loaded Nanoparticles based on Hydroxypropyl-β-Cyclodextrin Modified Carboxymethyl Chitosan for Oral Delivery. *J. Wuhan Univ. Technol. Mater Sci. Ed.* **2016,** *31,* 1394–1400.

Soufi, G. J.; Iravani, S. Eco-friendly and Sustainable Synthesis of Biocompatible Nanomaterials for Diagnostic Imaging: Current Challenges and Future Perspectives. *Green Chem.* **2020,** *22,* 2662–2687.

Subbiah, R.; Veerapandian, M.; Yun, K. S. Nanoparticles: Functionalization and Multifunctional Applications in Biomedical Sciences. *Curr. Med. Chem.* **2010,** *17,* 4559–4577.

Taniguchi, N. In *On the Basic Concept of Nanotechnology*; Proceedings of the International Conference on Production Engineering. **1974,** 18–23.

Thoo, L.; Fahmi, M. Z.; Zulkipli, I. N.; Keasberry, N.; Idris, A. Interaction and Cellular Uptake of Surface-modified Carbon Dot Nanoparticles by J774.1 Macrophages. *Cent. Eur. J. Biol.* **2017,** *42,* 324–330.

Tonelli, F. M. P.; Tonelli, F. C. P.; Ferreira, D. R. C.; da Silva, K. E.; Cordeiro, H. G.; Ouchida, A. T.; Nunes, N. A. M. Biocompatibility and Functionalization of Nanomaterials. In *Intelligent Nanomaterials for Drug Delivery Applications*; Elsevier: New York, EUA, 2020; p 204.

Tu, H.; Lu, Y.; Wu, Y.; Tian, J.; Zhan, Y.; Zeng, Z.; Deng, H.; Jiang, L. Fabrication of Rectorite-contained Nanoparticles for Drug Delivery with a Green and One-step Synthesis method. *Int. J. Pharm.* **2015,** *493,* 426–433.

Tzeyung, A. S.; Md, S.; Bhattamisra, S. K.; Madheswaran, T.; Alhakamy, N. A.; Aldawsari, H. M.; Radhakrishnan, A. K. Fabrication, Optimization, and Evaluation of Rotigotine-Loaded Chitosan Nanoparticles for Nose-To-Brain Delivery. *Pharmaceutics* **2019,** *11,* 26.

Vaiserman, A.; Koliada, A.; Zayachkivska, A.; Lushchak, O. Nanodelivery of Natural Antioxidants: An Anti-aging Perspective. *Front. Bioeng. Biotechnol.* **2020,** *7,* 447.

Verma, A.; Mehata, M. S. Controllable Synthesis of Silver Nanoparticles using Neem Leaves and their Antimicrobial Activity. *J. Radiat. Res. Appl. Sci.* **2016,** *9,* 109–115.

Wadhwani, S. A.; Shedbalkar, U. U.; Singh, R.; Karve, M. S.; Chopade, B. A. Novel Polyhedral Gold Nanoparticles: Green Synthesis, Optimization and Characterization by Environmental Isolate of Acinetobacter sp. SW30. *World J. Microbiol. Biotechnol.* **2014,** *30,* 2723–2731.

Wang, Q.; Cai, J.; Biesold-McGee, G. V.; Huang, J.; Ng, Y. H.; Sun, H.; Wang, J.; Lai, Y.; Lin, Z. Silk Fibroin-derived Nitrogen-doped Carbon Quantum Dots Anchored on TiO2 Nanotube Arrays for Heterogeneous Photocatalytic Degradation and Water Splitting. *Nano Energy* **2020,** *78,* 105313.

Williams, D. A Model for Biocompatibility and its Evaluation. *J. Biomed. Eng.* **1989,** *11,* 185–191.

Wong, A. C.; Wright, D. W. Size-dependent Cellular Uptake of DNA Functionalized Gold Nanoparticles. *Small.* **2016,** *12,* 5592–5600.

Yasmin, S.; Nouren, S.; Bhatti, H. N.; Iqbal, D. N.; Iftikhar, S.; Majeed, J.; Mustafa, R.; Nisar, N.; Nisar, J.; Nazir, A.; Iqbal, M.; Rizvi, H. Green Synthesis, Characterization and Photocatalytic Applications of Silver Nanoparticles using Diospyros Lotus. *Green Process. Synth.* **2020,** *9,* 87–96.

Zhang, Y.; Lu, Y.; Zeng, W.; Cheng, J.; Zhang, M.; Kong, H.; Qu, H.; Lu, T.; Zhao, Y. Fluorescence Imaging, Metabolism, and Biodistribution of Biocompatible Carbon Dots Synthesized Using *Punica granatum* L. Peel. *J. Biomed. Nanotechnol.* **2022,** *18,* 381–393.

Zhang, Y.; Xu, Y.; Xi, X.; Shrestha, S.; Jiang, P.; Zhang, W.; Gao, C. Amino Acid-modified Chitosan Nanoparticles for Cu^{2+} Chelation to Suppress CuO Nanoparticle Cytotoxicity. *J. Mater Chem. B.* **2017,** *5,* 3521–3530.

Zhu, J.; Liu, Z.; Pu, Y.; Xu, J.; Zhang, S.; Bao, Y. Green Synthesized gold Nanoparticles from *Pseudobulbus Cremastrae seu Pleiones* show Efficacy Against Hepatic Carcinoma Potentially through Immunoregulation. *Drug Delivery.* **2022,** *29,* 1983–1993.

CHAPTER 23

Practical Difficulties in the Application of Green Nanoparticles to Environmental Pollution Remediation

FUMEENA HASHIM[#], SHAFIYA[#], JAHANGEER, NEHA SAMI, and SIRAJUDDIN AHMED

Department of Environmental Science, Jamia Millia Islamia, New Delhi, India

[#]*Both authors have contributed equally.*

ABSTRACT

The past decennium has witnessed an outstanding thrust in nanotechnology research because of its extensive scope of applications in various fields, one of which includes environmental pollution remediation, particularly termed nanoremediation. It is known that metal nanoparticles have a huge surface area, sturdy adsorption ability, and excessive reducibility; thus, they can be utilized in the degradation of pollutants. Recently, the nanoparticles are being synthesized utilizing the plant and microbes which are called green nanoparticles (GNPs). They have been employed as a pollution abatement tool due to the existence of several functional groups that can conveniently act on pollutants and help ineffective bioremediation and also promote an immaculate environment. However, not all GNPs can eliminate environmental pollutants like heavy metals and display low pollutant elimination efficiency with limited application conditions. Moreover, the green synthesis is constrained by production time and location, as well as exhibits low purity and low yield. The size and shape of nanoparticles synthesized with the aid of using different extracts are enormously inconstant and the properties determined are inadequate. Modern reports display massive variations in particle size that makes controlling particle size throughout manufacturing

Sustainable Nanoremediation: Modern Technologies for a Clean Environment.
Fernanda Maria Policarpo Tonelli, Arpita Roy, and Khalid Rehman Hakeem (Eds.)

a prime challenge. Further, nanoremediation is a very expensive technology that limits its efficiency. Apart from these difficulties, the most challenging one is the massive release of nanoparticles into the environment that results in the production of nano-waste, referred to as nanopollution that is affecting living organisms and also inflicting a threat to the ecosystem.

23.1 INTRODUCTION

Nanotechnology has gained lots of attention in the past few decades due to the unique physical properties of nanoparticles (NPs). They are small particles that range between 1 and 100 nanometers (nm) in size and exist in an unbound state or in the form of aggregates or agglomerates that are undetectable by the human eye. They are taken into consideration to be the basic element of nanotechnology as they can give rise to diverse nanostructured devices or nanomaterials. On the basis of the type of materials used for their designing and construction, the nanomaterials can be principally categorized into three types viz., inorganic nanomaterials (metal oxides and metals), polymer nanomaterials, and carbon nanomaterials [graphene, fullerene (C60, C540); single-walled and multi-walled nanotubes] (Fig. 23.1) (Guerra et al., 2018). Out of them, polymer-based nanomaterials have been considered the best type that can be used in detecting and removing pollutants such as heavy metals and separating gases like CO, NOx, and SO_2 (Luo et al., 2021a, 2021b). The chemical and physical properties of nanomaterials along with their shape and morphology determine their effectiveness as a good adsorbent (Oghyanous, 2022; Ningthoujam et al., 2022; Zhang et al., 2021). Since they possess a higher surface-to-volume ratio, elasticity, many sorption sites, adjustable pore size, and unique surface chemistry (functionalization with other functional groups so that they may target specific pollutants), they can be considered a potential candidate for dirty groundwater and wastewater purification (Jain et al., 2021; Yogalakshmi et al., 2020). Furthermore, they can also be employed in the filtration of sediments polluted with hydrocarbons and heavy metals, for isolating organic or inorganic compounds from soil (Bharagava et al., 2020; De Gisi et al., 2017). Selecting the best nanomaterial to mitigate a particular pollutant in a specific environmental context calls for complete evaluation and assessment of many factors viz., the type of contaminant to be removed, the accessibility to the remediation site, the amount of material needed to implement efficient remediation, and whether it is advantageous to recover the remediation nanomaterial or not (Guerra et al., 2018; Roy et al., 2021). Moreover, the materials used for the production of nanomaterials should also be nontoxic, clean, and a low-cost alternative. The use of green source substances is more advantageous as they are safer and can be accepted under any situations as they produce no material waste that needs to be disposed of after the treatment. It has been found that green polymeric nanoparticles apart from inorganic, carbonaceous nanomaterials have been effectively employed in environmental remediation (Ali, 2020; Pandey and Jain, 2020; Rizwan et al., 2014).

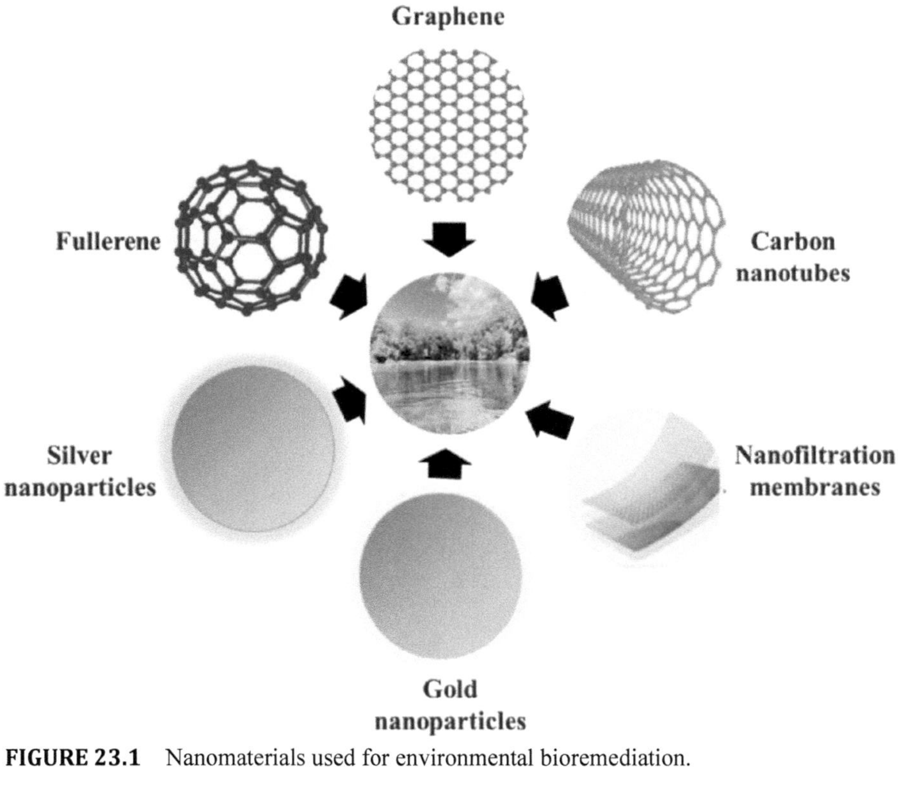

FIGURE 23.1 Nanomaterials used for environmental bioremediation.

23.2 SYNTHESIS OF NANOPARTICLES

The two fundamental strategies utilized for the synthesis of nanoparticles include the bottom-up and the top-down technique. The bottom-up technique includes the miniaturization of material components (up to the atomic level) with a further self-assembly process ensuing in the formation of nanostructures. During self-assembly, the physical forces running at the nanoscale are re-accustomed to amalgamate basic units into larger stable structures. Typical examples are quantum dot formation in the course of epitaxial growth and the production of nanoparticles from colloidal dispersion. But the top-down technique uses larger (macroscopic) initial structures, which might be externally controlled within the processing of nanostructures, typical examples are etching via the mask, ball milling, and application of intense plastic deformation (Agarwal, et al., 2017) (Fig. 23.2).

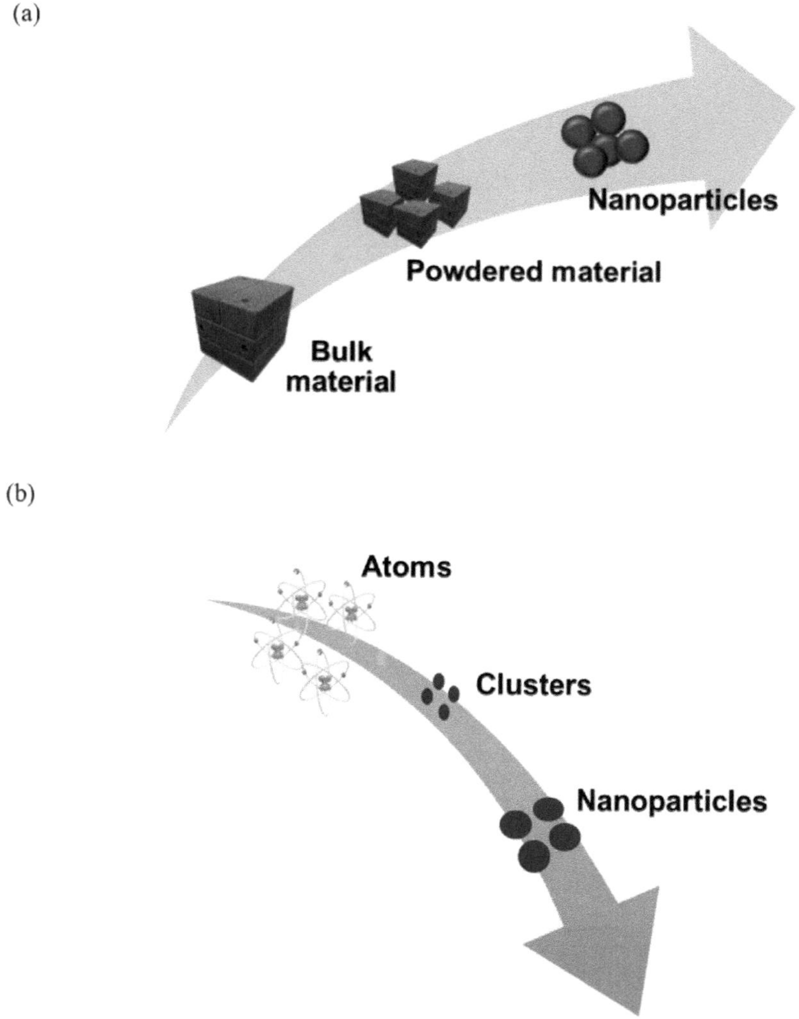

FIGURE 23.2 Strategies utilized for the synthesis of nanoparticles: (a) Top-down technique, (b) bottom-up technique.

The nanoparticles are often synthesized fundamentally by using physical and chemical methods. However, the biosynthesis of nanoparticles, also referred to as green synthesis, has captivated the attention of the scientific community owing to its reduced toxicity and cost efficiency (Fig. 23.3). Green synthesis of nanoparticles is an alternative and growing area of research. Green synthesis involves a bottom-up technique and presently microorganisms including algae, bacteria, and fungi (Arsiya et al., 2017; Afzal et al., 2019; Kumar et al., 2007; Subramaniyam et al., 2015); extracts from leaves (Agarwal et al., 2020; Chahardoli et al., 2018; Leili et al., 2018), flowers (Sone et al., 2020; Thovhogi et al., 2015, 2016), roots, peelings (Ehrampoush et al., 2015), fruits (Kumar et al., 2013), and seeds (Dhand et al., 2016; Gao et al., 2016) of various plants have been utilized to synthesize green nanoparticles (GNPs) (Tables 23.1–23.4).

No doubt green synthesis has proved to be economical and clean; however, there are some drawbacks that have been reported that need to be addressed to make the process more efficient. Some of them are discussed below:

23.2.1 REACTION CONDITIONS

Detailed and intricate optimization studies on reactants (extract concentration, microorganism inoculum, fermentation medium composition, etc.) and method parameters (temperature, pH, rational speed, time, pressure, etc.) are required to govern the dimensions and shape of the GNPs, which makes this process tedious and very difficult (Ahmed et al., 2022).

23.2.2 SIZE AND SHAPE OF NANOPARTICLES

The size and shape of GNPs synthesized by different extracts are enormously variable, and the properties determined are inadequate. Current reviews display huge variations in particle size, which makes green technology unfit for wide-scale production or controlling particle size at some point of production a major challenge. The synthesized NZVIs with the use of Nettle and Thyme leaf extracts appear to form abnormal clusters (Leili et al., 2018), and the silver NPs (Ag NPs) synthesized with the aid of using *Colocasia esculenta* leaf extract also have abnormal shapes (Agarwal et al., 2020). The smaller Pd NPs synthesized with the aid of using black tea (*Camellia sinensis)* extract generally tend to agglomerate right into a nonspherical shape. Defects

TABLE 23.1 Plant Sources for Nanoparticle Synthesis.

Nanoparticle	Shape	Size (nm)	Source	Application	Reference
Au	Spherical Triangular	10–15	Leaves of *P. benghalensis*	Degradation of methylene blue	Paul et al. (2015)
Ag	NA	20–30	*Acacia nilotica*	Electrocatalytic reduction of benzyl chloride	Jebakumar Immanuel Edison and Sethuraman (2013)
CuO	NA	27.6–8.2	*Centella asiatica* leaves	Photocatalytic degradation of methyl orange	Markus et al. (2016)
FeNPs	NA	NA	*Spinacia oleracea* leaves	Reduction of BOD and COD	Turakhia et al. (2018)
NZVI (nano zero valent iron)	Collar-type	61.1–100.6	Blueberry (*Vaccinium SPP*)	Removal of arsenate	Manquian-Cerda et al. (2017)
NZVI	NA	NA	*Urtica dioica*	Removal of Cr (VI)	Fazlzadeh et al. (2017)
Fe NPs	Polydispersed	20–80	Eucalyptus leaf	Removal of Cr (VI) and Cu (II)	Weng et al. (2016)
Fe NPs	Irregular	40–60	Green tea leaves	Removal of MB MO	Shahwan et al. (2011)
Fe_3O_4	Spherical	33	Carica papaya leaves	Eutrophic wastewater	Hussain et al. (2016)
NZVI	NA	15–45	Grape (*Vitis vinifera L).* marc and black tea (*Camellia sinensis)* vine leaves	Degradation of ibuprofen	Machado et al. (2014)

TABLE 23.2 Bacterial Source for Nanoparticle Synthesis

Bacteria	Nanoparticle	Location	Size (nm)	Shape	References
Bacillus cereus	Ag	NA	5	Spherical	Ganesh Babu et al. (2009)
E. coli	CdS	NA	NA	NA	Sweeney et al. (2004)
Lactobacillus strains	Ag, Au	NA	15/40	NA	Sintubin et al. (2009)
Actinobacter spp.	Magnetite	Extracellular	10–40	Spherical	Bharde et al. (2005)
Thermomonospora sp.	Au	Extracellular	8	Spherical	Ahmad et al. (2003b)
P. stutzeri	Cu	Extracellular	50–150	NA	Mohseniazar et al. (2011)
Bacillus subtilis	Fe3O4	Extracellular	NA	NA	Sundaram et al. (2012)
Anabaena	akageneite β-FeOOH	NA	NA	Nanorods	Vijayakumar et al. (2009)
Shewanella oneidensis	Uranium IV	Extracellular	NA	NA	Marshall et al. (2007)
T. ethanolicus TOR-39	Zinc ferrite	Extracellular	0.02, 0.23, 0.30	NA	Yeary et al. (2011)

TABLE 23.3 Fungal Source of Nanoparticle Synthesis.

Fungi	Nanoparticle	Location	Size (nm)	Shape	References
Aspergillus niger	Ag	NA	20	Spherical	Gade et al. (2008)
Klebsiella pneumoniae	Se	NA	100/400	Spherical	Fesharakil et al. (2010)
F. oxysporum	Au	Extracellular	20–40	NA	Ahmad et al. (2003a)
A. oryzae	Fe	Extracellular	10–24.6	Spherical	Tarafdar et al. (2013)
Neurospora crassa	Au, Au/Ag	Intracellular, extracellular	32,20–50	Spherical	Castro-Longoria et al. (2011)
Coriolus versicolor	CdS	Extracellular	100–200	Spherical	Sanghi et al. (2009)
F. oxysporum	$PbCO_3$, $CdCO_3$	Extracellular	120–200	Spherical	Sanyal et al. (2005)
F. oxysporum	$SrCO_3$	Extracellular	10–50	Needlelike	Rautaray et al. (2004)
F. oxysporum	CdSe	Extracellular	9–15	Spherical	Kumar et al. (2007)
Schizosaccharomyces pombe	CdS	Intracellular	1–1.5	Hexagonal lattice	Kowshik et al. (2002)

TABLE 23.4 Algal Source for Nanoparticle Synthesis.

Algae	Nanoparticles	Location	Size (nm)	Shape	References
Sargassum wightii	Ag	Extracellular	8–12	Spherical	Thirumalairaj et al. (2014)
Klebsormidium	Akageneite β-FeOOH nanorods	Intracellular	NA	NA	Luna et al. (2014)
Nannochloropsis oculate	Ag	Extracellular	15	NA	Mohseniazar et al. (2011)
Chlorella vulgaris	Ag	Extracellular	15	NA	Mohseniazar et al. (2011)
Dunaliella salina	Ag	Extracellular	15	NA	Mohseniazar et al. (2011)
Sargassum plagiophyllum	AgCl	Intracellular	NA	Spherical	Dhas et al. (2014)
Sargassum wightii	Au	Extracellular	8–12	Planar	Singaravelu et al. (2007)
Shewanella algae	Pt	Intracellular	5	NA	Konishi et al. (2007)
Shewanella oneidensis	Fe_3O_4	Extracellular	40–50	Rectangular, rhombic, hexagonal	Perez-Gonzalez et al. (2010)
HSMV-1	Fe_3O_4	Intracellular	50–70	Bullet-shaped	Lefèvre et al. (2010)

in products formed limit practical benefits, restrict their applications, and make it difficult to perform large-scale production (Lebaschi et al., 2017).

23.2.3 YIELD OF PRODUCE

The conversion rate and yield of nanoparticles produced from green materials are low. In all situations investigated, there was constrained reduction of iron using *Methaspicata* (spearmint, SM), *Syzygium aromaticum* (clove, CL), and *Camellia sinensis* (green tea, GT) extracts. The end result confirmed that the conversion rate of iron-to-iron NPs did not exceed 50% (Wang et al., 2017). However, the nanoparticles synthesized through a physical approach can exhibit high purity and uniform particle size distribution. This phenomenon displays the low conversion rate and low utilization rate of metal ions; this means that only a small number of nanoparticles can be synthesized from a large concentration of metal ions, and hence the economic benefit is low (Xu et al., 2020).

23.3 APPLICATION OF GNPS IN ENVIRONMENTAL REMEDIATION

Globally, the use of GNPs for everything from medical implants to cleaning up environmental pollution has increased dramatically (Fig. 23.3). They have transformed the way that various techniques operate. Since the surface area of particles is enormous compared to the volume of nanoscale materials, their reactivity in chemical or biological surface-mediated reactions can be greatly increased compared to the same materials of very large size and can be used to remove contaminants (U.S. EPA 2007). They can be modified for specific purposes to provide unique functionality not found in microscale or macroscale particles of the same material. The high surface area to volume ratio and the abundance of reactive sites in nanoscale materials can make them extremely reactive (Afzal and Fatma, 2021; Afzal et al., 2022).

23.3.1 GREEN NANOCATALYSTS

It is known that a catalyst delivers sustainable, cost-effective, and effective means to get important fuels and chemicals from the raw/source materials. Catalysts also help in speeding up the chemical reactions that change toxic exhausts from vehicles and factories into harmless gases (Ningthoujam et

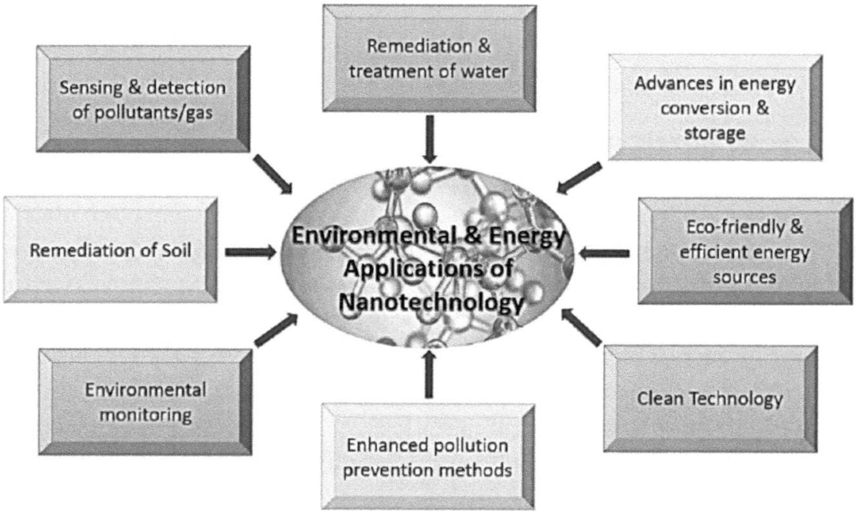

FIGURE 23.3 Green nanoparticles targeting environmental bioremediation.

al., 2022). But, the usage of toxic chemicals, industrial solvents, harmful and/or harsh reaction conditions besides the time-consuming and challenging wasteful separations reduce the effectiveness of a catalyst and have resulted in the popularity of greener and environmentally friendly catalytic procedures (Chaturvedi and Dave, 2021). Nanocatalyst is generally made up of semiconducting materials and metal nanoparticles that possess the remarkable potential for activity against radionuclides, microbes, and highly concentrated organic contaminants (Chiang and Mohan Sankaran, 2009; Kuna et al., 2021). However, recently, green nanocatalysts that are made up of nanomaterials have been used as catalysts in homogeneous and heterogeneous catalyses (Astruc, 2020; Miceli et al., 2021; Rao, 2010). Most nanocatalysts include antimicrobial catalysts, Fenton-based catalysts, and photocatalysts that have been used in solving the worldwide water pollution problem (Guerra et al., 2018).

23.3.2 NANOMEMBRANES

Nanomembranes have been used in different environmental sectors to separate and purify polluting gases, vapors, and prevent their release into the environment. They have a width of less than 100 nm and have a larger surface. The volume ratio is large, and it can be used for various shapes such as conical and cylinders (Pavani, 2015). The surface of these films can be changed to a porous surface that can be used for detection, separation, and purification of fiber, polymer, and bio molecules and specific unique functions (Kumar et al., 2014). Various kinds of nanomembranes such as nanocomposite membranes, nano biomembranes, carbon nanotube membranes, and polymer nanomembranes have been used for the remediation of pollutants present in the environment from pollutants (Corry, 2008; Gobato et al., 2018; Lindemann et al., 2014; Yin and Deng, 2015). They have been also used to remove salt from seawater after which the filtered water has been used for drinking purposes (Tarabara, 2009).

23.3.3 NANOFILTERS

The pores of the nanofilters range from 1 to 10 nm, which makes them more effective than the conventional filters for removing a variety of microorganisms and organic pollutants (Mondal and Wickramasinghe, 2008; Worou et al., 2021). Nanofilters work on the basis of the following three mechanisms to remove toxins, i.e., surface contact with the filter structure, application of inertial force while the gas is moving, and production of electric charge during particle and filter construction.

Recently, new ultra-low pressure reverse osmosis (ULPRO) surface layers and nanofilter membranes have been used to desalinate even brackish water under high pressure and shield the surface from contaminants. It works on oils, hydrophobic colloids, proteins, and other organics (Diallo et al., 2005; Hassan et al., 2000).

23.3.4 NANOSENSORS

A state-of-the-art tool that can recognize and respond to a physical change that can be felt on a nanometer is called a nanosensor. GNPs, which are tiny, circular, 3D substances, are used as sensors that reveal the presence of toxins in an area. These biosensors can capture harmful gas molecules and can be made of single or multiple membrane nanotubes. They have been used to detect pesticide pollution, heavy metal pollution, and even plant pathogens. Immuno chromatographic strip (ICTS) nanosensors have been used to deal with pesticide pollution (Adam and Gopinath, 2022; Sharma et al., 2021).

23.4 PRACTICAL DIFFICULTIES IN THE APPLICATION OF GNPS IN ENVIRONMENTAL REMEDIATION

Green nanoparticles are now an important component of developing industries in both the environmental and technological fields. However, the synthesis of nanoparticles can affect how well they perform in practical applications (Khan et al., 2019), which can have unfavorable effects. Due to their synthetical, intellectual, optical, and organic properties, NPs have garnered a lot of attention. In light of this, safety evaluation emerges as a crucial concern for the successful application of these novel materials. The complex interactions taking place at various biological levels, as well as the chemical composition, size, and shape of GNPs, may have an effect on the environment (Shinde et al., 2012). These drawbacks include the inability to withstand harsh conditions, bioaccumulation, toxicity, and the difficulty of recycling, reusing, and regeneration, among others.

23.4.1 THE EUTECTIC ENVIRONMENT AND THE FEED-THROUGH PROBLEM

GNP operations are designed to take place in areas where certain parameters, especially pressure, temperature, and isolation, need to be controlled and isolated from the outside world. This is the so-called "eutectic" environment (Pan et al., 2019). There are good reasons for this: the presence of uncontrolled, foreign chemical species will almost certainly result in molecular adsorption on any exposed surfaces which will be followed by uncontrolled mechanochemistry, resulting in irreversible chemical damage to the mechanisms (Harrison et al., 2018). GNPs will require an extremely high vacuum to function (Jin et al., 2012).

23.4.2 STABILITY ISSUE

GNPs are quite unstable and can quickly alter their own structure or interact with other compounds to sometimes take on a rather stable form (Liu et al., 2021). This alters GNPs in desired or undesirable ways, causing them to be highly reactive and poorly stable. Due to the physical instability of GNPs, which eventually causes them to lose their unique nanoscale features and form aggregate particles with a high polydispersity index (PDI), the challenge of large-scale production of GNPs remains challenging (Phan and Haes, 2019).

23.4.3 THERMAL NOISE AND BROWNIAN MOTION

Close dimensional tolerances are required by the mechanical engineering paradigm that underpins GNPs. However, at room temperature, Brownian motion (Gruenert et al., 2010), and effective "thermal tolerance" is substantially worse than the mechanical tolerances employed in macroscopic engineering because thermal noise causes components to bend and vary in size (Stuart et al., 2000).

23.4.4 FRICTION

The relative amount of interfacial area increases as mechanisms become smaller and surface forces become stronger (Sasaki et al., 1996). As people strive to reduce micro-electromechanical systems (MEMS) to the nanoscale, many devices fail because of the combining of friction and irreversible sticking known in the field as "stiction" (So et al., 1996).

23.4.5 BIOACCUMULATION

Environmental fate and transport estimates show that GNPs may infiltrate soil and streams either directly or indirectly as nanowaste. Thus, these GNPs may be released into the environment through washing, rain, and other processes, particularly into the aquatic system, in which they could potentially be hazardous to biota and have negative ecological consequences in addition to negative economical repercussions on humans. One of the most common nanomaterials found in consumer goods is silver nanoparticles (AgNPs) (Siddiqi et al., 2018). AgNPs have the potential to be toxic to aquatic biota when released into the environment. Because of their extensive use in commercial products and strong antibacterial properties, AgNPs are present in the environment in steadily rising amounts (Piccinno et al., 2012). The majority of environmental worries center on what will happen to AgNPs in applications like washing machines, the textile industry, and other similar ones. AgNPs were submitted to a variety of treatments, including interactions with surfactants, oxidizing agents, and various pH levels, in order to assess the release of AgNPs during washing procedures and the transit of these NPs in human perspiration. Ag^{+} release was greatly boosted by these treatments, and AgNPs were a key pathway for NPs to enter the environment (Egorova et al., 2019).

23.4.6 POOR REMOVAL EFFICIENCY OF GNPS

Even after 15 days, the COD and BOD removal efficiencies of the zero valent iron nanoparticles (NZVI) generated by *Spinacia oleracea* (spinach) are only 73.82 and 60.31%, respectively (Turakhia et al., 2018). The chromium gets degraded by the NZVIs produced by mandarin, lemon, lime, and orange with degradation efficiencies ranging from 12 to 37% (Machado et al., 2014). Using NZVI made from eucalyptus leaf extract, 30.4% of the total phosphorus in wastewater has been removed (Wang et al., 2017). These poor NZVI clearance rates raise issues, particularly about the wasteful use or waste of iron. Separate treatments for Cu and Cr(VI) had removal efficiencies of 45.2% and 74.2%, respectively. The removal efficiency of combined metal ions clearly diminishes. Heavy metals and other contaminants in sewage are combined for practical use and seldom does a waste contain only one heavy metal. As a result, in order to employ green-produced nanoscale materials for this sort of practical application, efficiency must be enhanced (Weng et al., 2016). Although green synthesis of nanoparticles has great potential, there are still difficulties with selection of materials, synthesis conditions, quality control, and application. These issues make it difficult to apply green synthesis of nanoparticles for large-scale environmental remediation.

23.4.7 QUALITY OF GNPS

The characteristics mentioned are not sufficient since different extracts can yield nanoparticles of different shapes and sizes. According to recent research, there are significant disparities in particle size, which makes using green technology to clean up environmental pollutants or regulate particle size throughout production difficult. NZVI nanoparticles made from grape seeds ranged in size from 63 to 381 nm, whereas Ag nanoparticles made from *Nigella arvensis* leaves ranged in size from 5 to 100 nm (Gao et al., 2016; Chahardoli et al., 2018). Particle sizes of gold nanoparticles produced from *Pistacia integerrimagall*, aqueous *Elaise guineensis* (oil palm), and *Galaxauraelongata* varied widely and ranged from 20 to 200 nm, 13 to 97 nm, and 2 to 8 nm, respectively. Some items have flaws in the way they are made. The palladium NPs produced by *Colocasia esculenta* leaf extract and the NZVIs created utilizing nettle and oregano leaf extracts seem to cluster in an atypical way (Leili et al., 2018). Defects in the product will severely restrict its usefulness, limit its applications, and make it challenging to clean up environmental damage.

23.4.8 TOXICITY OF METAL NANOPARTICLES

Due to their distinct physicochemical features, NPs are widely used in a variety of industries, including electrical, horticulture, chemicals, pharmaceutical, nutrition, and environmental pollution remediation. Silicon dioxide (SiO_2), titanium dioxide (TiO_2), zinc oxide (ZnO), aluminum hydroxide [$Al(OH)_3$], cerium oxide (CeO), copper oxide (CuO), silver (Ag), nano clays, carbon nanotubes, and nanocellulose are among the most widely utilized NPs (Giorgetti et al., 2019; Pulit-Prociak and Banach., 2016). However, widespread industrial NPs releases into the environment (air, water, and soil) are producing nano pollution, which is harmful for living things and endangers the stability of ecosystems. Size, reactivity, mobility, form, nature stability, aggregation, storage duration, surface chemistry, surface charge, catalytic properties, and numerous other characteristics of NPs influence their toxicity. NPs have bad impacts on both human and animal health. The risk of several illnesses in humans, including diabetes, cancer, respiratory disease, infections, swelling, fibrosis in multicellular organisms, cytotoxicity, etc., has increased due to the usage of NPs. Humans who are exposed to carbon black particles (NPs), the majority of which are carcinogenic, like as polycyclic aromatic hydrocarbons, are at danger of developing a number of cardiovascular and respiratory diseases (PAHs). They are extremely small soot particles that can enter the lungs deeply and have harmful consequences. Workers in the gas, coal, asphalt, and carbon electrode manufacturing sectors, as well as those employed in those facilities, may pass away too soon from illnesses linked to air pollution brought on by nanoparticles. Through ingestion and inhalation, NPs enter an animal's body and are then taken up by cells through the processes of phagocytosis and endocytosis. This causes the production of reactive oxygen species (ROS), which causes lipid peroxidation, mitochondrial damage, and other problems.

Numerous bacteria's enzymatic activity has also been reduced by several NPs including Ag, Cu, ZnO, and Ni, among others. The ecosystem's food chain is also being impacted by the overproduction of NPs (Das et al., 2020). The toxic effects of NPs on microbes, plants, and mammals have been shown in Figure 23.4.

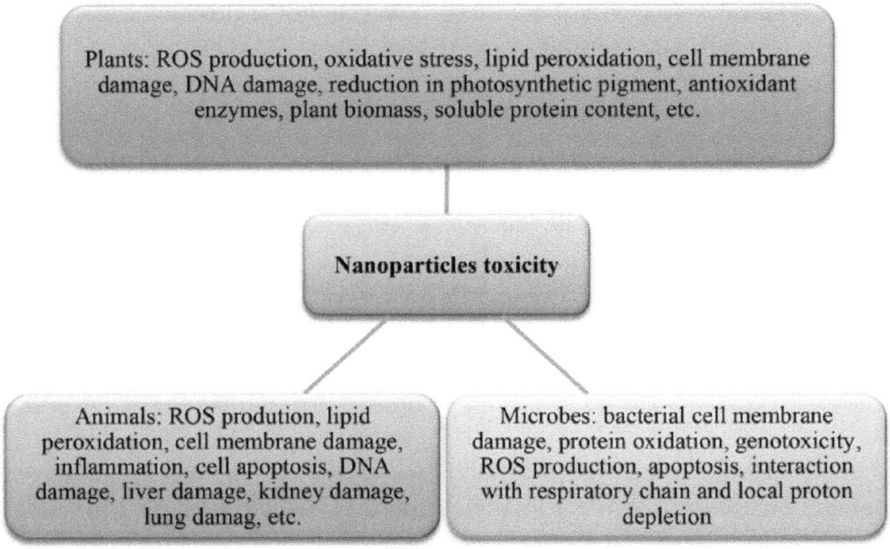

FIGURE 23.4 Toxicity effects of NPs on microbes, plants, and animals.

23.4.9 TOXIC EFFECTS OF NANOPARTICLES ON HUMAN BEINGS

The nanoparticles residue left in the environment poses toxicity to human beings and other organisms, which makes it the most challenging barriers to its practical application.

23.4.9.1 CARDIOVASCULAR MORBIDITY AND MORTALITY

Adults and vulnerable groups have experienced a consistent rise in morbidity and death over the past few decades, which has been linked to air pollution in both industrialized and developing countries. Numerous epidemiological studies have discovered a link between NP exposure and the development of respiratory and cardiovascular conditions (Wichmann et al., 2000; Peters et al., 1997; Penntinen et al., 2001, 2002). According to the Cancer Prevention Study II, an increase in PM2.5 of 10 g/m^3 was linked to an increase in mortality from heart failure, cardiac arrest, dysrhythmias, and ischemic heart disease of 8–18%. Three theories have been put out to explain why certain circumstances can result in cardio-pulmonary morbidity and death in a population exposed to NPs. The NPs may be able to excite lung neurons, which would have an impact on the central nervous system and cardiac autonomic function, according to the first theory. The second theory holds that inhaled NPs directly enter the systemic flow and reach their intended organs, where they cause infection, the release of cytokines, reactive oxygen species (ROS), C-reactive proteins, and cardiac problems. The third theory holds that breathing inhaled NPs causes a lung-wide acute inflammatory response, which in turn stimulates the release of cytokinesis, chemokines, ROS, and transcription factors (Dockery et al., 2005). Nuclear factor kappa B (NF-KB), mitogen-activated protein kinase (MAPK), redox-sensitive transcription factors, and Activating Protein-1 (AP-1) are all crucially activated by the surge of subsequent events and infection, encouraging pulmonary infection that leads to cardiac events. Due to the direct relationship between infection and atherosclerosis, there is strong evidence linking infections to coronary heart disease (Sun et al., 2005).

23.4.9.2 PULMONARY MORBIDITY AND MORTALITY

The lungs are the primary location of NPs, and there is strong evidence linking elevated NP levels to adverse health outcomes in children, persons with asthma, and susceptible adults (Nel, 2005). The main detrimental health impacts of ambient air pollution caused by NPs are increased respiratory symptoms, reduced lung function, increased respiratory problems, chronic obstructive pulmonary disease (COPD), and increased mortality (Gong et al., 2005; Koenig et al., 2005; Pietropaoli et al., 2004; Silkoff et al., 2005). As a portion of the inhaled ultrafine particles gets translocated to different extra-pulmonary locations such the blood, liver, coronary heart, spleen, and brain, experimental studies have consistently shown that exposure to ultra-fine particles and NPs increases lung inflammation (Nemmar et al., 2003; Renwick et al., 2004).

23.4.9.3 TRANSLOCATION AND TOXICITY TO OTHER ORGANS

Recent studies on both humans and animals have demonstrated the movement of ultrafine particles to extrapulmonary regions such as liver, brain, coronary heart, and systemic flow (Kreyling et al., 2002; Nemmar et al., 2002b; Oberdörster et al., 2002). Infection of the lungs with histological extrade and transport of debris to extrapulmonary regions have been shown in NPs investigations on animals. Inhaled NPs that translocate to the circulation affect endothelial function, encourage thrombosis, and cause other blood system issues, including accelerated blood coagulation (Elder et al.,

2000, 2002, 2004; Kreyling et al., 2002; Nemmar et al., 2002a; Zhou et al., 2003). A recent study by Geiser et al. (2005) has shown that inhaled TiO_2 NPs cross cellular membranes by non-phagocytic mechanisms in the lungs and were present even in capillaries.

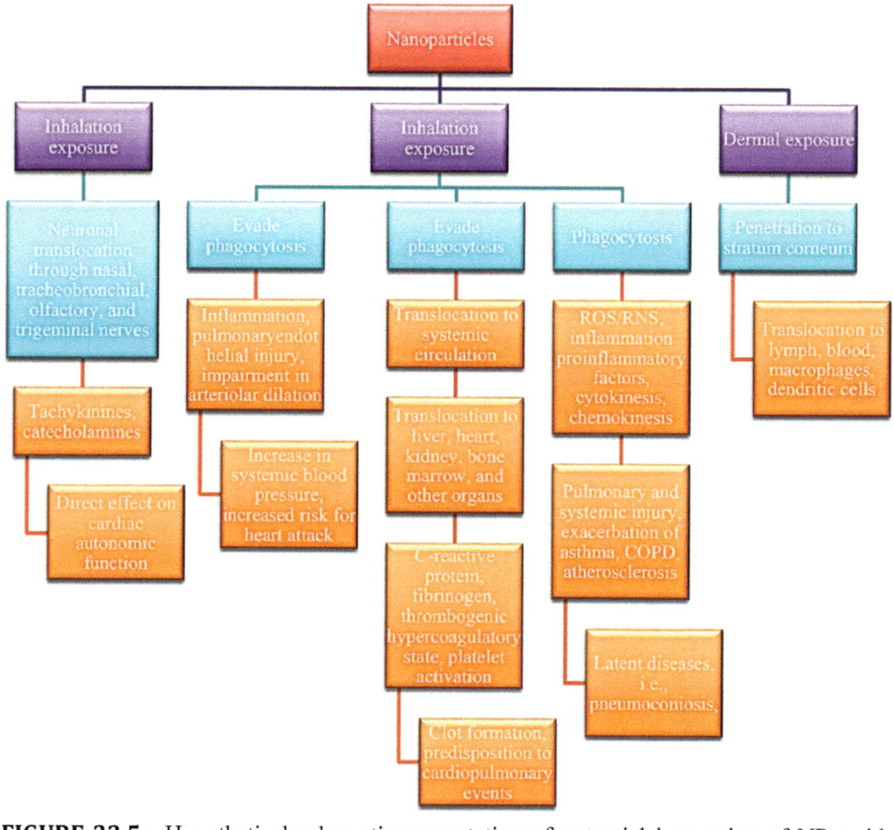

FIGURE 23.5 Hypothetical schematic presentation of potential interaction of NPs with human beings. Abbreviations: RNS, reactive nitrogen species; COPD, chronic obstructive pulmonary diseases.

23.4.10 *HIGH MANUFACTURING COST*

Nanotechnology-based systems, regardless of their advantages, face several limitations before they can be safely brought to the market. Scalability and manufacturing costs are the most common issues, as are intellectual and regulatory properties. Some of its techniques are still objects of academic laboratory studies, simplicity of fabrication, lowering of cost, and commercial availability of applied nano materials and their precursors are still a major target of current investigation. Nowadays, nanotechnology is quite costly, and it may be extremely expensive to create. Nanotechnology also has some manufacturing challenges, which contributes to the higher cost of nanotechnology-based goods. The technology requires huge funds to operate and the raw material required is also very costly. Hence, the overall acquisition of nanotechnology for an ordinary man becomes expensive. Advancement in nanotechnology for various fields including environmental pollution remediation may also result in unemployment. Due to innovation in science and technology human labor work has been reduced a lot and oftentimes people lose their jobs, especially in the areas of engineering nanotechnology and environmental engineering nanotechnology where there has been a decrease in labor-intensive work.

23.4.11 COMMON DIFFICULTIES IN PRACTICAL APPLICATION OF GNPs

GNPs have huge surface area, sturdy adsorption, and high reducibility that can be utilized in eliminating pollutants. However, not all metal nanoparticles synthesized via green synthesis can remove environmental pollutants. Some of the studies highlighting the same have been mentioned below:

- The COD and BOD elimination efficiencies of NZVI synthesized by *spinacia oleracea* (spinach) were not high and found as 73.82% and 60.31%, respectively, after 15 days (Türakhia et al., 2018).
- NZVI synthesized from eucalyptus leaf extract could remove only 30.4% of total phosphorus in wastewater (Wang et al., 2017). These low elimination rates for NZVIs caused issues especially on inefficient use or waste of iron.
- GNPs showed low efficiency in eliminating mixed toxic metals. Weng et al. (2016) used eucalyptus leaf extracts to synthesize FeNPs and implemented them to cast off mixed Cu and Cr (VI) at efficiencies of 33.0% and 58.9%, respectively. However, when Cu and Cr (VI) were treated separately, the removal efficiency increased to 45.2% and 74.2% respectively. Clearly, removal efficiency was found to be less for mixed metal ions. For practical applications, heavy metals and other pollutants in sewage are mixed waste that consists of different heavy metals. Therefore, efficiency needs to

be advanced with a purpose to use GNPs for this type of practical application.

- Lack of monodispersity and prolonged duration of synthesis is the practical limitation of biosynthesis of GNPs.
- The iron content in Fe-P NPs using *Eucalyptus tereticornis, Melaleuca nesophila,* and *Rosemarinus officinalis* leaf extract were 0.24%, 8.58%, and 0.53%, respectively (Arsiya et al., 2017). However, the nanoparticles synthesized by the physical method exhibited high purity and uniform particle size distribution (Xu et al., 2020). This phenomenon reflects the low conversion rate and low utilization rate of metal ions, which means that only a small number of nanoparticles can be synthesized from a large concentration of metal ions, and thus the economic benefit is low.

23.5 CONCLUSION

The study of GNPs and potential applications has advanced significantly in recent years. In the present chapter, the environmentally friendly synthesis of metallic NPs using a variety of biological sources, including plants, fungi, yeast, and bacteria, has been investigated. However, a number of issues still exist that prevent its widespread production and ensuing use. The use depends on a number of experimental characteristics, including friction, Brownian motion, the instability of GNPs, high manufacturing costs, and other challenges that must be solved for the practical manufacture and use of green nanoparticles. To fully harness the potential of nanoparticles for environmental applications, these obstacles must be overcome.

KEYWORDS

- **green nanoparticles**
- **environmental pollution**
- **green technology**
- **nanotechnology**
- **nanoremediation**
- **problems**

REFERENCES

Adam, T.; Gopinath, S. C., Nanosensors: Recent Perspectives on Attainments and Future Promise of Downstream Applications. *Process Biochem.* **2022**.

Afzal, B.; Fatma, T. Selenium Nanoparticles: Green Synthesis and Exploitation. In *Emerging Technologies for Nanoparticle Manufacturing*; Springer: Cham, 2021; pp 473–484.

Afzal, B.; Naaz, H.; Sami, N.; Yasin, D.; Khan, N. J.; Fatma, T. Mitigative Effect of Biosynthesized SeNPs on Cyanobacteria under Paraquat Toxicity. *Chemosphere* **2022,** *293,* 133562.

Afzal, B.; Yasin, D.; Husain, S.; Zaki, A.; Srivastava, P.; Kumar, R.; Fatma, T. Screening of Cyanobacterial Strains for the Selenium Nanoparticles Synthesis and their Anti-oxidant Activity. *Biocat. Agri. Biotechnol.* **2019**, *21,* 101307.

Agarwal, H.; Kumar, S. V.; Rajeshkumar, S. A Review on Green Synthesis of Zinc Oxide Nanoparticles—An Eco-friendly Approach. *Resour. Efficient Technol.* **2017**, *3* (4), 406–413.

Agarwal, S.; Gogoi, M.; Talukdar, S.; Bora, P.; Basumatary, T. K.; Devi, N. N. Green Synthesis of Silver Nanoplates using the Special Category of Plant Leaves Showing the Lotus Effect. *RSC Adv*. **2020**, *10* (60), 36686–36694.

Ahmad, A.; Mukherjee, P.; Senapati, S.; Mandal, D.; Khan, M. I.; Kumar, R.; Sastry, M. Extracellular Biosynthesis of Silver Nanoparticles using the Fungus *Fusarium oxysporum*. *Colloid. Surface B*. **2003a,** *28,* 313–318.

Ahmad, A.; Senapati, S.; Khan, M. I.; Kumar, R.; Sastry, M. Extracellular Biosynthesis of Monodisperse Gold Nanoparticles by a Novel Extremophilic Actinomycete, *Thermomonospora* sp. *Langmuir* **2003b,** *19,* 3550–3553.

Ahmed, S. F.; Mofijur, M.; Rafa, N.; Chowdhury, A. T.; Chowdhury, S.; Nahrin, M.; Islam, A. S.; Ong, H. C. Green Approaches in Synthesising Nanomaterials for Environmental Nanobioremediation: Technological Advancements, Applications, Benefits and Challenges. *Env. Res.* **2022,** *204,* 111967.

Ali, A. S. Application of Nanomaterials in Environmental Improvement. *Nanotechnol. Environ.* **2020**.

Arsiya, F.; Sayadi, M. H.; Sobhani, S. Green Synthesis of Palladium Nanoparticles using Chlorella Vulgaris. *Mater. Lett.* **2017,** *186,* 113–115.

Astruc, D. Introduction: Nanoparticles in Catalysis. *Chem. Revi.* **2020,** *120* (2), 461–463.

Benn, T. M.; Westerhoff, P. Nanoparticle Silver Released into Water from Commercially Available Sock Fabrics. *Environ. Sci. Technol*. **2008,** *42,* 4133–4139. doi: 10.1021/es7032718.

Bharagava, R. N.; Saxena, G.; Mulla, S. I. Introduction to Industrial Wastes Containing Organic and Inorganic Pollutants and Bioremediation Approaches for Environmental Management. In *Bioremediation of Industrial Waste for Environmental Safety*; Springer: Singapore, 2020; pp 1–18.

Bharde, A.; Wani, A.; Shouche, Y.; Pattayil, A.; Bhagavatula, L.; Sastry, M. Bacterial Aerobic Synthesis of Nanocrystalline Magnetite. *JACS* **2005,** *127,* 9326–9327.

Castro-Longoria, E.; Vilchis-Nestor, A. R.; Avalos Borja, M. Biosynthesis of Silver, Gold and Bimetallic Nanoparticles using the Filamentous Fungus *Neurospora Crassa*. *Colloid. Surf. B* **2011,** *83,* 42–48.

Chahardoli, A.; Karimi, N.; Sadeghi, F.; Fattahi, A.; Green Approach for Synthesis of Gold Nanoparticles from *Nigella Arvensis* Leaf Extract and Evaluation of their Antibacterial,

Antioxidant, Cytotoxicity and Catalytic Activities. *Artif. Cells Nanomed. Biotechnol.* **2018,** *46* (3), 579–588.

Chaturvedi, S.; Dave, P. N. Chapter 13—Nanocatalyst: As Green Catalyst. In *Handbook of Greener Synthesis of Nanomaterials and Compounds*; Kharisov, B., Kharissova, O., Eds.; Elsevier, 2021; pp 445–458. ISBN 9780128219386, https://doi.org/10.1016/B978-0-12-821938-6.00013-X.

Chiang, W. H.; Mohan Sankaran, R. Linking Catalyst Composition to Chirality Distributions of As-grown Single-walled Carbon Nanotubes by Tuning NixFe1 − x Nanoparticles. *Nat. Mater.* **2009,** *8* (11), 882–886.

Corry, B. Designing Carbon Nanotube Membranes for Efficient Water Desalination. *J. Phys. Chem. B* **2008,** *112* (5), 1427–1434.

Dash, S. R.; Kundu, C. N. Promising Opportunities and Potential Risk of Nanoparticle on the Society. *IET Nanobiotechnol.* **2020,** *14,* 253–260.

De Gisi, S.; Minettoa, D.; Lofranoab, G.; Libralatoac, G.; Contea, B.; Todaroa, F.; Notarnicolaa, M. Nano-scale Zero Valent Iron (nZVI) Treatment of Marine Sediments Slightly Polluted by Heavy Metals. *Chem. Eng. Trans.* **2017,** *60.*

Diallo, M. S.; Christie, S.; Swaminathan, P.; Johnson, J. H.; Goddard, W. A. Dendrimer Enhanced Ultrafiltration. 1. Recovery of Cu (II) from Aqueous Solutions using PAMAM Dendrimers with Ethylene Diamine Core and Terminal NH2 Groups. *Environ. Sci. Technol.* **2005,** *39* (5), 1366–1377.

Dhas, T. S.; Kumar, V. G.; Karthick, V. A. K. J.; Angel, K. J.; Govindaraju, K. Facile Synthesis of Silver Chloride Nanoparticles using Marine Alga and Its Antibacterial Efficacy. *Spectrochim. Acta A Mol. Biomol. Spectrosc.* **2014,** *120,* 416–420.

Dockery, D. W.; Luttmann-Gibson, H.; Rich, D. Q.; Link, M. S.; Mittleman, M. A.; Gold, D. R.; Koutrakis, P.; Schwartz, J. D.; Verrier, R. L. Association of Air Pollution with Increased Incidence of Ventricular Tachyarrhythmias Recorded by Implanted Cardioverter Defibrillators. *Environ. Health Perspect.* **2005,** *113* (6), 670–674.

Egorova, E. M.; Kaba, S. I. The Effect of Surfactant Micellization on the Cytotoxicity of Silver Nanoparticles Stabilized with Aerosol-OT. *Toxicol. Vitr.* **2019,** *57,* 244–254. doi: 10.1016/j.tiv.2019.03.006.

Elder, A. C. P.; Gelein, R.; Finkelstein, J. N.; Cox, C.; Oberdörster, G. Pulmonary Inflammatory Response to Inhaled Ultrafine Particles is Modified by Age, Ozone Exposure, and Bacterial Toxin. *Inhal. Toxicol.* **2000**, *12* (Suppl. 4), 227–246.

Elder, A. C. P.; Gelein, R.; Azadniv, M.; Frampton, M.; Finkelstein, J.; Oberdorster, G. Systemic Interactions between Inhaled Ultrafine Particles and Endotoxin. *Ann. Occup. Hyg.* **2002,** *46* (Suppl 1), 231–234.

Elder, A. C. P.; Gelein, R.; Azadniv, M.; Frampton, M.; Finkelstein, J.; Oberdorster, G. Systemic Effects of Inhaled Ultrafine Particles in Two Compromised, Aged Rat Strains. *Inhal. Toxicol.* **2004,** *16,* 461–471.

Fazlzadeh, M.; et al. A Novel Green Synthesis of Zero Valent Iron Nanoparticles (NZVI) using Three Plant Extracts and their Efficient Application for Removal of Cr(VI) from Aqueous Solutions. *Adv. Powder Technol.* **2017,** *28* (1), 122–130.

Fesharaki, P. J.; Nazari, P.; Shakibaie, M.; Rezaie, S.; Banoee, M.; Abdollahi, M.; Shahverdi, A. R. Biosynthesis of Selenium Nanoparticles using *Klebsiella pneumoniae* and their Recovery by a Simple Sterilization Process. *Braz. J. Microbiol.* **2010,** *41,* 461–466.

Gade, A. K.; Bonde, P. P.; Ingle, A. P.; Marcato, P.; Duran, N.; Rai, M. K. Exploitation of *Aspergillus niger* for Synthesis of Silver Nanoparticles. *J. Biobased Mater. Bio.* **2008,** *2,* 1–5.

Ganesh Babu, M. M.; Gunasekaran, P. Production and Structural Characterization of Crystalline Silver Nanoparticles from *Bacillus cereus* Isolate. *Colloid. Surf. B* **2009,** *74,* 191–19.

Geiser, M.; Rothen-Rutishauser, B.; Kapp, N.; Schurch, S.; Kreyling, W.; Schulz, H.; et al. Ultrafine Particles Cross Cellular Membranes by Nonphagocytic Mechanisms in Lungs and in Cultured Cells. *Environ. Health Perspect.* **2005,** *113,* 1555–1560.

Giorgetti, L. Effects of Nanoparticles in Plants: Phytotoxicity and Genotoxicity Assessment. *Nanomater. Plants Algae Microorg.* **2019**, *2*, 65–87.

Gobato, R.; Heidari, A.; Mitra, A. Using the Quantum Chemistry for Genesis of a Nano Biomembrane with a Combination of the Elements Be, Li, Se, Si, C and H. *J. Nanomed. Res.* **2018,** *7* (4), 241–252.

Gong, H.; Linn, W. S.; Clark, K. W.; Anderson, K. R.; Geller, M. D.; Sioutas, C. Respiratory Responses to Exposures with Fine Particulates and Nitrogen Dioxide in the Elderly with and without COPD. *Inhal. Toxicol.* **2005,** *17* (3), 123–132.

Gruenert, G.; Ibrahim, B.; Lenser, T.; Lohel, M.; Hinze, T.; Dittrich, P. Rule-based Spatial Modeling with Diffusing, Geometrically Constrained Molecules. *BMC Bioinformat.* **2010,** *11* (1), 1–14.

Guerra, F. D.; Attia, M. F.; Whitehead, D. C.; Alexis, F. Nanotechnology for Environmental Remediation: Materials and Applications. *Molecules* **2018,** *23* (7), 1760.

Harrison J. A.; Schall, J. D.; Maskey, S.; Mikulski, P. T.; Knippenberg, M. T.; Morrow, B. H. Review of Force Fields and Intermolecular Potentials used in Atomistic Computational Materials Research. *Appl. Phys. Revi.* **2018,** *5* (3), 031104.

Hassan A. M.; Farooque, A. M.; Jamaluddin, A. T. M.; Al-Amoudi, A. S.; Al-Sofi, M. A. K.; Al-Rubaian, A. F.; Kither, N. M.; Al-Tisan, I. A. R.; Rowaili, A. A Demonstration Plant Based on the New NF—SWRO Process. *Desalination.* **2000,** *131* (1–3), 157–171.

Hussain, I.; Singh, N. B.; Singh, A.; Singh, H.; Singh, S. C. Green Synthesis of Nanoparticles and its Potential Application. *Biotechnol. Let.* **2016,** *38* (4), 545–560.

Jain, K.; Patel, A. S.; Pardhi, V. P.; Flora, S. J. S. Nanotechnology in Wastewater Management: A New Paradigm Towards Wastewater Treatment. *Molecules* **2021,** *26* (6), 1797.

Jebakumar Immanuel Edison, T. N.; Sethuraman, M. G. Electrocatalytic Reduction of Benzyl Chloride by Green Synthesized Silver Nanoparticles using Pod Extract of *Acacia nilotica*. *ACS Sustain. Chem. Engineer.* **2013,** *1* (10), 1326–1332.

Khan, I.; Saeed, K.; Khan, I. Nanoparticles: Properties, Applications and Toxicities. *Arab. J. Chem.* **2019,** *12,* 908–931.

Koenig, J. Q.; Mar, T. F.; Allen, R. W.; Jansen, K.; Lumley, T.; Sullivan, J. H.; Trenga, C. A.; Larson, T. V.; Liu, L. J. S. Pulmonary Effects of Indoor-and Outdoor-generated Particles in Children with Asthma. *Environ. Health Perspect.* **2005,** *113* (4), 499–503.

Konishi, Y.; Ohno, K.; Saitoh, N.; Nomura, T.; Nagamine, S.; Hishida, H.; Takahashi, Y.; Uruga, T. Bioreductive Deposition of Platinum Nanoparticles on the Bacterium *Shewanella* Algae. *J. Biotechnol.* **2007,** *128,* 648–653.

Kowshik, M.; Deshmukh, N.; Vogel, W.; Urban, J.; Kulkarni, S. K.; Paknikar, K. M. Microbial Synthesis of Semiconductor CdS Nanoparticles, their Characterization, and their use in the Fabrication of an Ideal Diode. *Biotechnol. Bioeng.* **2002,** *78* (5), 583–588.

Kreyling, W. G.; Semmler, M.; Erbe, F.; Mayer, P.; Takenaka, S.; Schulz, H.; Oberdörster, G.; Ziesenis, A. Translocation of Ultrafine Insoluble Iridium Particles from Lung Epithelium to Extrapulmonary Organs is Size Dependent but Very Low. *J. Toxicol. Environ. Health, A.* **2002,** *65* (20), 1513–1530.

Kumar, A.; Gayakwad, A.; Nagale, B. D. A review: Nanomembrane and Application. *Int. J. Innov. Res. Sci. Eng. Technol.* **2014,** *3,* 8373–8381.

Kumar, K. M.; Mandal, B. K.; Kumar, K. S.; Reddy, P. S.; Sreedhar, B. Biobased Green Method to Synthesise Palladium and Iron Nanoparticles using *Terminalia chebula* Aqueous Extract. *Spectrochim. Acta A Mol. Biomol. Spectrosc.* **2013,** *102,* 128–133.

Kumar, S. A.; Ansary, A. A.; Ahmad, A.; Khan, M. I. Extracellular Biosynthesis of CdSe Quantum Dots by the Fungus, *Fusarium oxysporum. J. Biomed. Nanotechnol.* **2007,** *3* (2), 190–194.

Kuna, E.; Mrdenovic, D.; Jönsson-Niedziółka, M.; Pieta, P.; Pieta, I. S. Bimetallic Nanocatalysts Supported on Graphitic Carbon Nitride for Sustainable Energy Development: The Shape-structure–activity Relation. *Nanosc. Adv.* **2021,** *3* (5), 1342–1351.

Lefèvre, C. T.; Abreu, F.; Schmidt, M. L.; Lins, U.; Frankel, R. B.; Hedlund, B. P.; Bazylinski, D., A Moderately Thermophilic Magnetotactic Bacteria from Hot Springs in Nevada. *Appl. Environ. Microbiol.* **2010,** *76,* 3740–3743.

Leili, M.; Fazlzadeh, M.; Bhatnagar, A. Green Synthesis of Nano-zero-valent Iron from Nettle and Thyme Leaf Extracts and their Application for the Removal of Cephalexin Antibiotic from Aqueous Solutions. *Environ. Technol.* **2018,** *39* (9), 1158–1172.

Lindemann, P.; Tsotsalas, M.; Shishatskiy, S.; Abetz, V.; Krolla-Sidenstein, P.; Azucena, C.; Monnereau, L.; Beyer, A.; Gölzhäuser, A.; Mugnaini, V.; Gliemann, H. Preparation of Freestanding Conjugated Microporous Polymer Nanomembranes for Gas Separation. *Chem. Mater.* **2014,** *26* (24), 7189–7193.

Liu, X.; Yao, G.; Cheng, X.; Xu, J.; Cai, X.; Hu, W.; Xu, W. W.; Zhang, C.; Zhu, Y. Cd-driven Surface Reconstruction and Photodynamics in Gold Nanoclusters. *Chem. Sci.* **2021,** *12* (9), 3290–3294.

Luna, C.; Ilyn, M.; Vega, V.; Prida, V. M.; González, J.; MendozaReséndez R. Size Distribution and Frustrated Antiferromagnetic Coupling Effects on the Magnetic Behavior of Ultrafine Akaganéite (β-FeOOH) Nanoparticles. *J. Phys. Chem. C.* **2014,** *118,* 21128–21139.

Luo, J.; Fu, K.; Yu, D.; Hristovski, K. D.; Westerhoff, P.; Crittenden, J. C. Review of Advances in Engineering Nanomaterial Adsorbents for Metal Removal and Recovery from Water: Synthesis and Microstructure Impacts. *ACS ES&T Eng.* **2021a**, *1* (4), 623–661.

Luo, J.; Yu, D.; Hristovski, K. D.; Fu, K.; Shen, Y.; Westerhoff, P.; Crittenden, J. C. Critical Review of Advances in Engineering Nanomaterial Adsorbents for Metal Removal and Recovery from Water: Mechanism Identification and Engineering Design. *Environ. Sci. Technol.* **2021b,** *55* (8), 4287–4304.

Machado, S.; Grosso, J. P.; Nouws, H. P. A.; Albergaria, J. T.; Delerue-Matos, C. Utilization of Food Industry Wastes for the Production of Zero-valent Iron Nanoparticles. *Sci. Total Environ.* **2014,** *496,* 233–240.

Manquián-Cerda, K.; Cruces, E.; Rubio, M. A.; Reyes, C.; Arancibia-Miranda, N. Preparation of Nanoscale Iron (oxide, oxyhydroxides and zero-valent) Particles Derived from Blueberries: Reactivity, Characterization and Removal Mechanism of Arsenate. *Ecotoxicol. Environ. Safety.* **2017,** *145,* 69–77.

Markus, J.; Mathiyalagan, R.; Kim, Y. J.; Abbai, R.; Singh, P.; Ahn, S.; Perez, Z. E. J.; Hurh, J.; Yang, D. C. Intracellular Synthesis of Gold Nanoparticles with Antioxidant Activity by

Probiotic *Lactobacillus kimchicus* DCY51T Isolated from Korean kimchi. *Enzyme Microb. Technol.* **2016,** *95,* 85–93.
Marshall, M.; Beliaev, A.; Dohnalkova, A.; David, W.; Shi, L.; Wang, Z.; Boyanov, M. I.; Lai, B.; Kemner, K. M.; McLean, J. S.; Reed, S. B.; Culley, D. E.; Bailey, V. L.; Simonson, C. J.; Saffarini, D. A.; Romine, M. F.; Zachara, J. M.; Fredrickson, J. K. c-Type Cytochrome-dependent Formation of UIV Nanoparticles by Shewanella Oneidensis. *PLoS Biol.* **2007,** *8,* 1324–1333.
Miceli, M.; Frontera, P.; Macario, A.; Malara, A. Recovery/reuse of Heterogeneous Supported Spent Catalysts. *Catalysts* **2021,** *11* (5), 591.
Mohseniazar, M.; Barin, M.; Zarredar, H.; Alizadeh, S.; Shanehbandim, D. Potential of Microalgae and Lactobacilli in Biosynthesis of Silver Nanoparticles. *BioImpacts* **2011,** *1,* 149–152.
Mondal, S.; Wickramasinghe, S. R. Produced Water Treatment by Nanofiltration and Reverse Osmosis Membranes. *J. Membr. Sci.* **2008,** *322* (1), 162–170.
Nel, A. Atmosphere. Air Pollution-related Illness: Effects of Particles. *Science* **2005,** *308* (5723), 804–806.
Nemmar, A.; Hoet, P. M.; Vanquickenborne, B.; Dinsdale, D.; Thomeer, M.; Hoylaerts, M. F.; Vanbilloen, H.; Mortelmans, L.; Nemery, B. Passage of Inhaled Particles into the Blood Circulation in Humans. *Circulation* **2002**a, *105* (4), 411–414.
Nemmar, A.; Hoylaerts, M. F.; Hoet, P. H.; Dinsdale, D.; Smith, T.; Xu, H.; Vermylen, J.; Nemery, B. Ultrafine Particles Affect Experimental Thrombosis in an In Vivo Hamster Model. *Am. J. Respirat. Critic. Care Med.* **2002**b, *166* (7), 998–1004.
Nemmar, A.; Hoylaerts, M. F.; Hoet, P. H.; Vermylen, J.; Nemery, B. Size Effect of Intratracheally Instilled Particles on Pulmonary Inflammation and Vascular Thrombosis. *Toxicol. Appl. Pharmacol.* **2003,** *186* (1), 38–45.
Ningthoujam, R.; Singh, Y. D.; Babu, P. J.; Tirkey, A.; Pardhan, S.; Sarma, M. Nanocatalyst in Remediating Environmental Pollutants. *Chem. Phys. Impact.* **2022,** *100064.*
Oberdörster, G.; Utell, M. J. Ultrafine Particles in the Urban Air: To the Respiratory Tract--and Beyond? *Environ. Health Perspec.* **2002,** *110* (8), A440–A441.
Oghyanous, F. A. Nanoparticles in Wastewater Treatment. In *Water Conservation: Inevitable Strategy;* 2022, p 107.
Pan, Q.; Shang, X.; Li, J.; Ma, S.; Li, L.; Sun, L. Energy-efficient Separation Process and Control Scheme for Extractive Distillation of Ethanol-water using Deep Eutectic solvent. *Separat. Purificat. Technol.* **2019,** *219,* 113–126.
Pandey, G.; Jain, P. Assessing the Nanotechnology on the Grounds of Costs, Benefits, and Risks. *Beni-Suef Univ. J. Basic Appl. Sci.* **2020,** *9* (1), 1–10.
Paul, B.; Bhuyan, B.; Purkayastha, D. D.; Dhar, S. S. Photocatalytic and Antibacterial Activities of Gold and Silver Nanoparticles Synthesized using Biomass of *Parkia roxburghii* leaf. *J. Photochem. Photobiol. B: Biol.* **2016,** *154,* 1–7.
Perez-Gonzalez, T.; Jimenez-Lopez, C.; Neal, A. L.; Rull-Perez, F.; Rodriguez-Navarro, A.; Fernandez-Vivas, A.; Iañez-Pareja, E. Magnetite Biomineralization Induced by *Shewanella oneidensis*. *Geochim. Cosmochimica Acta* **2010,** *74* (3), 967–979.
Peters, A.; Wichmann, H. E.; Tuch, T.; Heinrich, J.; Heyder, J. Respiratory Effects are Associated with the Number of Ultrafine Particles. *Am. J. Respirat. Critic. Care Medi.* **1997,** *155* (4), 1376–1383.
Phan, H. T.; Haes, A. J. What does Nanoparticle Stability Mean? *J. Phys. Chem. C* **2019,** *123* (27), 16495–16507.

Piccinno, F.; Gottschalk, F.; Seeger, S.; Nowack, B. Industrial Production Quantities and Uses of Ten Engineered Nanomaterials in Europe and the World. *J. Nanopart. Res.* **2012,** *14,* 1109. doi: 10.1007/s11051-012-1109-9.

Pietropaoli, A. P.; Frampton, M. W.; Hyde, R. W.; Morrow, P. E.; Oberdörster, G.; Cox, C.; Speers, D. M.; Frasier, L. M.; Chalupa, D. C.; Huang, L. S.; Utell, M. J. Pulmonary Function, Diffusing Capacity, and Inflammation in Healthy and Asthmatic Subjects Exposed to Ultrafine Particles. *Inhal. Toxicol.* **2004**, *16* (Suppl 1), 59–72.

Pulit-Prociak, J.; Banach, M. Silver nanoparticles—A material of the Future? *Open Chem.* **2016,** *14,* 76–91.

Rao, P. N. Nanocatalysis: Applications in the Chemical Industry. 2010.

Rautaray, D.; Sanyal, A.; Adyanthaya, S. D.; Ahmad, A.; Sastry, M. Biological Synthesis of Strontium Carbonate Crystals Using the Fungus *Fusarium Oxysporum. Langmuir* **2004,** *20* (16), 6827–6833.

Renwick, L. C.; Brown, D.; Clouter, A.; Donaldson, K. Increased Inflammation and Altered Macrophage Chemotactic Responses Caused by Two Ultrafine Particle Types. *Occupat. Environ. Med.* **2004,** *61* (5), 442–447.

Rizwan, M.; Singh, M.; Mitra, C. K.; Morve, R. K. Ecofriendly Application of Nanomaterials: Nanobioremediation. *J. Nanopart.* **2014,** *2014.*

Roy, A.; Sharma, A.; Yadav, S.; Jule, L. T.; Krishnaraj, R. Nanomaterials for Remediation of Environmental Pollutants. *Bioinorgan. Chem. Appl.* **2021,** *2021.*

Sanghi, R.; Verma, P. A Facile Green Extracellular Biosynthesis of CdS Nanoparticles by Immobilized Fungus. *Chem. Eng. J.* **2009,** *155,* 886–891.

Sanyal, A.; Rautaray, D.; Bansal, V.; Ahmad, A.; Sastry, M. Heavy Metal Remediation by a Dungus as a Means of Production of Lead and Cadmium Carbonate Crystals. *Langmuir* **2005,** *21,* 7220–7224.

Sasaki, N.; Kobayashi, K.; Tsukada, M. Atomic-scale Friction Image of Graphite in Atomic-force Microscopy. *Phys. Rev. B* **1996,** *54* (3), 2138.

Shahwan, T.; Sirriah, S. A.; Nairat, M.; Boyacı, E.; Eroğlu, A. E.; Scott, T. B.; Hallam, K. R. Green Synthesis of Iron Nanoparticles and their Application as a Fenton-like Catalyst for the Degradation of Aqueous Cationic and Anionic Dyes. *Chem. Engineer. J.* **2011,** *172* (1), 258–266.

Sharma, P.; Pandey, V.; Sharma, M. M. M.; Patra, A.; Singh, B.; Mehta, S.; Husen, A. A Review on Biosensors and Nanosensors Application in Agro Ecosystems. *Nanoscale Res. Let.* **2021,** *16* (1), 1–24.

Shinde, N. C.; Keskar, N. J.; Argade, P. D. Nanoparticles: Advances in Drug Delivery Systems. *Res. J. Pharm. Biol. Chem. Sci.* **2012,** *3* (1), 922–929.

Siddiqi, K. S.; Husen, A.; Rao, R. A. A Review on Biosynthesis of Silver Nanoparticles and their Biocidal Properties. *J. Nanobiotechnol.* **2018,** *16,* 14. doi: 10.1186/s12951-018-0334-5.

Silkoff, P. E.; Zhang, L.; Dutton, S.; Langmack, E. L.; Vedal, S.; Murphy, J.; Make, B. Winter Air Pollution and Disease Parameters in Advanced Chronic Obstructive Pulmonary Disease Panels Residing in Denver, Colorado. *J. Allergy Clin. Immunol.* **2005,** *115* (2), 337–344.

Singaravelu, G.; Arockiamary, J. S.; Kumar, V. G.; Govindaraju, K. A Novel Extracellular Synthesis of Monodisperse Gold Nanoparticles using Marine Alga, *Sargassum wightii* Greville. *Colloids Surface B: Biointerf.* **2007,** *57* (1), 97–101.

Sintubin, L.; De Windt, W.; Dick, J.; Mast, J.; van der Ha, D.; Verstraete, W.; Boon, N. Lactic Acid Bacteria as Reducing and Capping Agent for the Fast and Efficient Production of Silver Nanoparticles. *Appl. Microbiol. Biotechnol.* **2009,** *84,* 741–749.

So, M. R.; Jacobsen, K. W.; Stoltze, P. Simulations of Atomic-scale Sliding Friction. *Phys. Rev. B* **1996,** *53* (4), 2101.

Stuart, S. J.; Tutein, A. B.; Harrison, J. A. A reactive Potential for Hydrocarbons with Intermolecular Interactions. *J. Chem. Phys.* **2000,** *112* (14), 6472–6486.

Sun, Q.; Wang, A.; Jin, X.; Natanzon, A.; Duquaine, D.; Brook, R. D.; Aguinaldo, J. G. S.; Fayad, Z. A.; Fuster, V.; Lippmann, M.; Chen, L. C. Long-term Air Pollution Exposure and Acceleration of Atherosclerosis and Vascular Inflammation in an Animal Model. *JAMA* **2005,** *294* (23), 3003–3010.

Sundaram, P. A.; Augustine, R.; Kannan, M. Extracellular Biosynthesis of Iron Oxide Nanoparticles by *Bacillus subtilis* Strains Isolated from Rhizosphere Soil. *Biotechnol. Bioproc. Engineer.* **2012,** *17* (4), 835–840.

Sweeney, R. Y.; Mao, C.; Gao, X.; Burt, J. L.; Belcher, A. M.; Georgiou, G.; Iverson, B. L. Bacterial Biosynthesis of Cadmium Sulfide Nanocrystals. *Chem. Biol.* **2004,** *11* (11), 1553–1559.

Tarabara, V. V. Multifunctional Nanomaterial-enabled Membranes for Water Treatment. In *Nanotechnology Applications for Clean Water*; William Andrew Publishing, 2009; pp 59–75.

Tarafdar, J. C.; Raliya, R.; Rapid, Low-cost, and Ecofriendly Approach for Iron Nanoparticle Synthesis Using *Aspergillus oryzae* TFR9. *J. Nanopart.* **2013,** *2013.*

Thirumalairaj, V. K.; Vijayan, M. P.; Durairaj, G.; Shanmugaasokan, L.; Yesudas, R.; Gunasekaran, S. Potential Antibacterial Activity of Crude Extracts and Silver Nanoparticles Synthesized from *Sargassum wightii. Int. Curr. Pharm. J.* **2014,** *3* (10), 322–325.

Turakhia, B.; Turakhia, P.; Shah, S. Green Synthesis of Zero Valent Iron Nanoparticles from *Spinacia oleracea* (spinach) and its Application in Waste Water Treatment. *J. Adv. Res. Appl. Sci.* **2018,** *5* (1), 46–51.

Vijayakumar, P. S.; Prasad, B. L. V. Intracellular Biogenic Silver Nanoparticles for the Generation of Carbon Supported Antiviral and Sustained Bactericidal Agents. *Langmuir* **2009,** 25 (19), 11741–11747.

Wang, X.; Wang, A.; Ma, J.; Fu, M.,Facile Green Synthesis of Functional Nanoscale Zero-valent Iron and Studies of Its Activity Toward Ultrasound-enhanced Decolorization of Cationic Dyes. *Chemosphere* **2017,** *166*, 80–88.

Weng, X.; Jin, X.; Lin, J.; Naidu, R.; Chen, Z. Removal of Mixed Contaminants Cr (VI) and Cu (II) by Green Synthesized Iron Based Nanoparticles. *Ecol. Engineer.* **2016,** *97,* 32–39.

Wichmann, H. E.; Peters, A. Epidemiological Evidence of the Effects of Ultrafine Particle Exposure. Philosophical Transactions of the Royal Society of London. *Series A: Math. Phys. Engineer. Sci.* **2000,** *358* (1775), 2751–2769.

Worou, C. N.; Chen, Z. L.; Bacharou, T. Arsenic Removal from Water by Nanofiltration Membrane: Potentials and Limitations. *Water Pract. Technol.* **2021,** *16* (2), 291–319.

Yeary, L. W.; Moon, J. W.; Rawn, C. J.; Love, L. J.; Rondinone, A. J.; Thompson, J. R.; Chakoumakos, B. C.; Phelps, T. J. Magnetic Properties of Bio-synthesized Zinc Ferrite Nanoparticles. *J. Magnet. Magnet. Mater.* **2011,** *323* (23), 3043–3048.

Yin, J.; Deng, B. Polymer-matrix Nanocomposite Membranes for Water Treatment. *J. Membrane Sci.* **2015,** *479,* 256–275.

Yogalakshmi, K. N.; Das, A.; Rani, G.; Jaswal, V.; Randhawa, J. S. Nano-bioremediation: A New Age Technology for the Treatment of Dyes in Textile Effluents. In *Bioremediation of Industrial Waste for Environmental Safety*; Springer: Singapore, 2020; pp 313–347.

Zhang, X.; Ma, G.; Wei, W. Simulation of Nanoparticles Interacting with a Cell Membrane: Probing the Structural basis and Potential Biomedical Application. *NPG Asia Mater.* **2021,** *13* (1), 1–18.

Zhou, Y. M.; Zhong, C. Y.; Kennedy, I. M.; Leppert, V. J.; Pinkerton, K. E. Oxidative Stress and NFκB Activation in the Lungs of Rats: A Synergistic Interaction between Soot and Iron Particles. *Toxicol. Appl. Pharmacol.* **2003,** *190* (2), 157–169.

Index